# List of abbreviations

**Abbreviations, commonly used in Astronomy,** and not always explained explicitly in this book:

| | |
|---|---|
| AU | Astronomical Unit (= Distance Earth–Sun) |
| B.C. | Bolometric Correction |
| BD | Bonner Durchmusterung |
| CDS | Centre de Données Stellaires, Observatoire de Strasbourg, 11, rue de l'Université, F-67000 Strasbourg |
| CLV | Center-limb variation |
| CMD | Colour-magnitude-diagram |
| CNO | Carbon, Nitrogen, and Oxygen (*not* as molecule) e.g. CNO cycle, CNO anomalies |
| ESA | European Space Agency |
| ESO | European Southern Observatory |
| ET | or E.T. Ephemeris Time |
| EUV | Extreme ultraviolet |
| FWHM | Full Width of Half Maximum |
| HD | Henry Draper Catalogue |
| HR | Harvard Revised Catalogue |
| HRD | Hertzsprung-Russell Diagram |
| IAU | International Astronomical Union |
| IR | Infrared |
| ISM | Interstellar Matter |
| JD | Julian Date |
| LB, NS or LB | Landolt-Börnstein, Numerical Data and Functional Relationships in Science and Technology, New Series or: Landolt-Börnstein, NS |
| LC | Luminosity Class |
| LF | Luminosity Function |
| LMC | Large Magellanic Cloud |
| LTE | Local thermodynamic equilibrium |
| M | Messier Catalogue |
| MHD | Magneto-hydrodynamics |
| MMT | Multi-Mirror-Telescope |
| MPI | Max-Planck-Institut |
| NASA | National Aeronautics and Space Administration |
| NEP | Noise Equivalent Power |
| NGC | New General Catalogue |
| NLTE | Non-local thermodynamic equilibrium |
| NRAO | National Radio Astronomy Observatory, Green Bank, W. Va., USA |
| POSS | Palomar Observatory Sky Survey |
| RV | Radial velocity |
| SMC | Small Magellanic Cloud |
| Sp | Spectral type |
| URSI | International Union of Radio Science |
| UT | Universal time |
| UV | Ultraviolet |
| VLBI | Very Long Baseline Interferometry |
| XUV | X-ray and ultraviolet region |
| ZAMS | Zero Age Main Sequence |

Abbreviations of further Star Catalogues: see 8.1.1

For abbreviations of special star types (e.g. WR stars), see "Spectralclassification" (4.1.1), "Variable stars" (5.1), "Peculiar stars" (5.2) and subject index.

**Some important Astronomical Artificial Satellites,** mentioned in this book

| | |
|---|---|
| ANS | Astronomical Netherlands Satellite (The Netherlands NASA) |
| ATS | Applications Technology Satellite |
| COS | Cosmic Ray Satellite (ESA) |
| ESRO-TD | European Space Research Organisation, Thor Delta |
| GIRL | German Infrared Laboratory |
| HEAO | High Energy Astrophysical Observatory (NASA) |
| HEOS | High Eccentricity Earth-Orbiting Satellite (ESA) |
| IMP | Interplanetary Monitoring Platform |
| IRAS | Infrared Astronomical Satellite |
| IUE | International Ultraviolet Explorer (NASA–UK–ESA) |
| OAO | Orbiting Astronomical Observatory (NASA) |
| OGO | Orbiting Geophysical Observatory |
| OSO | Orbiting Solar Observatory |
| MTS | Meteoroid Technology Satellite (NASA) |
| RAE | Radio Astronomy Explorer |
| SAS | Small Astronomy Satellite (NASA) |

# LANDOLT-BÖRNSTEIN

Numerical Data and Functional Relationships in Science and Technology

*New Series*

Editor in Chief: K.-H. Hellwege

Group VI: Astronomy · Astrophysics and Space Research

## Volume 2

## Astronomy and Astrophysics

Extension and Supplement to Volume 1

Subvolume c

## Interstellar Matter
## Galaxy · Universe

P. Biermann · H. H. Fink · K. J. Fricke · W. Gliese
M. Grewing · W. K. Huchtmeier · B. F. Madore · H. Netzer
J. Rahe · H. Scheffler · L. D. Schmadel · J. Schmid-Burgk
G. A. Tammann · J. Trümper · R. Wielen · A. Witzel · G. Zech

Editors: K. Schaifers and H. H. Voigt

Springer-Verlag Berlin · Heidelberg · New York 1982

# LANDOLT-BÖRNSTEIN

## Zahlenwerte und Funktionen aus Naturwissenschaften und Technik

*Neue Serie*

Gesamtherausgabe: K.-H. Hellwege



Gruppe VI: Astronomie · Astrophysik
und Weltraumforschung

## Band 2

## Astronomie und Astrophysik

Weiterführung und Ergänzung von Band 1

Teilband c

## Interstellare Materie
## Die Galaxis · Universum

P. Biermann · H. H. Fink · K. J. Fricke · W. Gliese
M. Grewing · W. K. Huchtmeier · B. F. Madore · H. Netzer
J. Rahe · H. Scheffler · L. D. Schmadel · J. Schmid-Burgk
G. A. Tammann · J. Trümper · R. Wielen · A. Witzel · G. Zech

Herausgeber: K. Schaifers und H. H. Voigt



Springer-Verlag Berlin · Heidelberg · New York 1982

CIP-Kurztitelaufnahme der Deutschen Bibliothek
*Zahlenwerte und Funktionen aus Naturwissenschaften und Technik*/Landolt-Börnstein. – Berlin; Heidelberg; New York: Springer
Parallelt.: Numerical data and functional relationships in science and technology

NE: Landolt, Hans [Begr.]; PT N.S./Gesamthrsg.: K.-H. Hellwege. Gruppe 6, Astronomie, Astrophysik und Weltraumforschung. Bd. 2. Astronomie und Astrophysik: Weiterführung u. Erg. von Bd. 1. Teilbd. c. Interstellare Materie, die Galaxis, Universum/ P. Biermann ... Hrsg.: K. Schaifers u. H. H. Voigt. – 1982.

ISBN 3-540-10977-3 (Berlin, Heidelberg, New York)
ISBN 0-387-10977-3 (New York, Heidelberg, Berlin)

NE: Hellwege, Karl-Heinz [Hrsg.]; Biermann, Peter [Mitverf.]; Schaifers, Karl [Hrsg.]



Printed in Germany



Typesetting, printing and bookbinding: Brühlsche Universitätsdruckerei, 6300 Giessen
2163/3020—543210

## Editors

**K. Schaifers,** Landessternwarte, Königstuhl, 6900 Heidelberg, FRG

**H. H. Voigt,** Universitätssternwarte, Geismarlandstraße 11, 3400 Göttingen, FRG

## Contributors

**P. Biermann,** Max-Planck-Institut für Radioastronomie, Auf dem Hügel 69, 5300 Bonn, FRG

**H. H. Fink,** Max-Planck-Institut für Physik und Astrophysik, Institut für Extraterrestrische Physik, 8046 Garching b. München, FRG

**K. J. Fricke,** Universitätssternwarte, Geismarlandstraße 11, 3400 Göttingen, FRG

**W. Gliese,** Astronomisches Recheninstitut, Mönchhofstraße 12–14, 6900 Heidelberg, FRG

**M. Grewing,** Astronomisches Institut der Universität, Waldhäuser Straße 64, 7400 Tübingen, FRG

**W. K. Huchtmeier,** Max-Planck-Institut für Radioastronomie, Auf dem Hügel 69, 5300 Bonn, FRG

**B. F. Madore,** David Dunlap Observatory, University of Toronto, Richmond Hill, ON L4C 4Y6, Canada

**H. Netzer,** Department of Physics and Astronomy and The Wise Observatory, Tel Aviv University, Ramat Aviv, 69978 Tel Aviv, Israel

**J. Rahe,** Dr. Remeis-Sternwarte, Sternwartstraße 7, 8600 Bamberg, FRG

**H. Scheffler,** Landessternwarte, Königstuhl, 6900 Heidelberg, FRG

**L. D. Schmadel,** Astronomisches Recheninstitut, Mönchhofstraße 12–14, 6900 Heidelberg, FRG

**J. Schmid-Burgk,** Max-Planck-Institut für Radioastronomie, Auf dem Hügel 69, 5300 Bonn, FRG

**G. A. Tammann,** Astronomisches Institut der Universität Basel, Venusstraße 7, 4102 Binningen, Switzerland

**J. Trümper,** Max-Planck-Institut für Physik und Astrophysik, Institut für Extraterrestrische Physik, 8046 Garching b. München, FRG

**R. Wielen,** Institut für Astronomie und Astrophysik, Technische Universität Berlin, Ernst-Reuter-Platz 7, 1000 Berlin

**A. Witzel,** Max-Planck-Institut für Radioastronomie, Auf dem Hügel 69, 5300 Bonn, FRG

**G. Zech,** Astronomisches Recheninstitut, Mönchhofstraße 12–14, 6900 Heidelberg, FRG

# Vorwort

Der hiermit vorgelegte Teilband VI/2c vervollständigt den Band VI/2 „Astronomie und Astrophysik". Auch für ihn gilt das im Vorwort von Teilband VI/2a Gesagte.

In dem Kapitel 9 „Galaxien und Universum" werden dem aufmerksamen Leser gewisse Unterschiede zwischen den einzelnen Abschnitten hinsichtlich ihrer Vollständigkeit und Aktualität nicht entgehen. Bei diesem wieder in Bewegung geratenen großen Forschungsgebiet, das starke und schnelle Änderungen und Fortschritte erlebt, ist eine absolute Homogenität heute wohl auch noch nicht zu erreichen. Außerdem hat ein Autor sehr kurzfristig mitgeteilt, daß er von seinem ursprünglich umfangreichen Beitrag nur einen kleinen Teil liefern könne. Wir danken daher insbesondere den Herren Huchtmeier, Netzer und Witzel, die daraufhin einsprangen und in sehr kurzer Zeit Beiträge oder Bibliographien zur Verfügung stellten. Die restlichen Lücken mußten, um die Herausgabe des Bandes nicht noch mehr zu verzögern, schließlich von einem der Herausgeber gefüllt werden. Speziell diese Beiträge sind nicht erschöpfend und beschränken sich weitgehend auf eine Bibliographie. Sie sollten später, nach Konsolidierung des Gebietes, ausgebaut werden.

Unser Dank gilt ferner den Herren Schmadel und Zech und ihren Mitarbeitern am Astronomischen Recheninstitut in Heidelberg, die in Zusammenarbeit mit einem der Herausgeber das Gesamtregister bearbeitet und mit den Hilfsmitteln des Astronomischen Recheninstituts erstellt haben. Der Aufbau und die Benutzung des Registers sind im einzelnen auf S. 417 erläutert.

Heidelberg und Göttingen, September 1982 **Die Herausgeber**

# Preface

The publication of part VI/2c completes volume VI/2 "Astronomy and Astrophysics". The statements in the preface of volume VI/2a also apply in this case.

In chapter 9 "Galaxies and Universe", the reader will note differences in the individual sections with regard to their comprehensiveness and actuality. An absolute homogeneity is perhaps not to be obtained at the present time in this large field of research which has become very active recently and undergoes large and quick developments. Furthermore, one author announced at short notice that he was unable to submit, except a small part, what was originally conceived to be an extensive contribution. Therefore we are particularly indebted to Messrs. Huchtmeier, Netzer, and Witzel for stepping into the breach and supplying contributions and bibliographies within a remarkably short time. In order to avoid further delay, one of the editors was obliged to fill the remaining gaps. His contributions in particular are consequently not exhaustive and concentrate primarily upon presenting a bibliography. After the consolidation of the field these topics will later have to be treated in detail.

Furthermore, we thank Messrs. Schmadel and Zech and their collaborators at the Astronomisches Recheninstitut in Heidelberg, who edited together with one of the volume editors the comprehensive subject index and compiled it with the aid of the Astronomisches Recheninstitut. The organization and the use of this subject index are explained in detail on p. 417.

Heidelberg and Göttingen, September 1982 **The Editors**

## Survey

Subvolume a: Methods. Constants. Solar system
1 Astronomical instruments
2 Positions and time determination, astronomical constants
3 The solar system

Subvolume b: Stars and star clusters
4 The stars
5 Special types of stars
6 Double stars and star clusters

**Subvolume c: Interstellar matter. Galaxy. Universe**
**7 Interstellar matter**
**8 Our Galaxy**
**9 Galaxies and universe**
**Comprehensive index**

## Übersicht

Teilband a: Methoden. Konstanten. Sonnensystem
1 Astronomische Instrumente
2 Orts- und Zeitbestimmung, astronomische Konstanten
3 Das Sonnensystem

Teilband b: Sterne und Sternhaufen
4 Die Sterne
5 Spezielle Sterntypen
6 Doppelsterne und Sternhaufen

**Teilband c: Interstellare Materie. Die Galaxis. Universum**
**7 Interstellare Materie**
**8 Unsere Galaxis**
**9 Galaxien und Universum**
**Gesamtregister**

# Contents

## 5 Special types of stars

## 6 Double stars and star clusters

see Subvolume b, p. 381ff.

## 7 Interstellar matter

## 8 Our Galaxy

## 9 Galaxies and universe

# List of abbreviations

**Abbreviations, commonly used in Astronomy,** and not always explained explicitly in this book.

| | |
|---|---|
| AU | Astronomical Unit (= Distance Earth–Sun) |
| B.C. | Bolometric Correction |
| BD | Bonner Durchmusterung |
| CDS | Centre de Données Stellaires, Observatoire de Strasbourg, 11, rue de l'Université, F-67000 Strasbourg |
| CLV | Center-limb variation |
| CMD | Colour-magnitude-diagram |
| CNO | Carbon, Nitrogen, and Oxygen (*not* as molecule) e.g. CNO cycle, CNO anomalies |
| ESA | European Space Agency |
| ESO | European Southern Observatory |
| ET | or E.T. Ephemeris Time |
| EUV | Extreme ultraviolet |
| FWHM | Full Width of Half Maximum |
| HD | Henry Draper Catalogue |
| HR | Harvard Revised Catalogue |
| HRD | Hertzsprung-Russell Diagram |
| IAU | International Astronomical Union |
| IR | Infrared |
| ISM | Interstellar Matter |
| JD | Julian Date |
| LB, NS or LB | Landolt-Börnstein, Numerical Data and Functional Relationships in Science and Technology, New Series or: Landolt-Börnstein, NS |
| LC | Luminosity Class |
| LF | Luminosity Function |
| LMC | Large Magellanic Cloud |
| LTE | Local thermodynamic equilibrium |
| M | Messier Catalogue |
| MHD | Magneto-hydrodynamics |
| MMT | Multi-Mirror-Telescope |
| MPI | Max-Planck-Institut |
| NASA | National Aeronautics and Space Administration |
| NEP | Noise Equivalent Power |
| NGC | New General Catalogue |
| NLTE | Non-local thermodynamic equilibrium |
| NRAO | National Radio Astronomy Observatory, Green Bank, W. Va., USA |
| POSS | Palomar Observatory Sky Survey |
| RV | Radial velocity |
| SMC | Small Magellanic Cloud |
| Sp | Spectral type |
| URSI | International Union of Radio Science |
| UT | Universal time |
| UV | Ultraviolet |
| VLBI | Very Long Baseline Interferometry |
| XUV | X-ray and ultraviolet region |
| ZAMS | Zero Age Main Sequence |

Abbreviations of further Star Catalogues: see 8.1.1
For abbreviations of special star types (e.g. WR stars), see "Spectralclassification" (4.1.1), "Variable stars" (5.1), "Peculiar stars" (5.2) and subject index.

**Some important Astronomical Artificial Satellites,** mentioned in this book

| | |
|---|---|
| ANS | Astronomical Netherlands Satellite (The Netherlands NASA) |
| ATS | Applications Technology Satellite |
| COS | Cosmic Ray Satellite (ESA) |
| ESRO-TD | European Space Research Organisation, Thor Delta |
| GIRL | German Infrared Laboratory |
| HEAO | High Energy Astrophysical Observatory (NASA) |
| HEOS | High Eccentricity Earth-Orbiting Satellite (ESA) |
| IMP | Interplanetary Monitoring Platform |
| IRAS | Infrared Astronomical Satellite |
| IUE | International Ultraviolet Explorer (NASA–UK–ESA) |
| OAO | Orbiting Astronomical Observatory (NASA) |
| OGO | Orbiting Geophysical Observatory |
| OSO | Orbiting Solar Observatory |
| MTS | Meteoroid Technology Satellite (NASA) |
| RAE | Radio Astronomy Explorer |
| SAS | Small Astronomy Satellite (NASA) |

## Additions

In the meantime published:

p. 28 ref. 4: Mon. Not. R. Astron. Soc. **197** (1981) 247.
ref. 14: Astrophys. J. **251** (1981) 630.

p. 29 ref. 21: Galactic X-Ray Sources (Sanford, P.W., Laskarides, P., Salton, J., eds.), John Wiley & Sons, Chichester (1982) p. 159.
ref. 25: Astron. Astrophys. **89** (1980) 249.
ref. 34: Galactic X-Ray Sources (Sanford, P.W., Laskarides, P., Salton, J., eds.), John Wiley & Sons, Chichester (1982) p. 205.

p. 32 ref. 5: Space Sci. Rev. **29** (1981) 221.

p. 43 ref. 31: Nature **275** (1978) 626.

p. 44 ref. 46: Bull. American Astron. Soc. **12** (1980) 463; Astrophys. J. Suppl. (in press).
ref. 70: Astron. Astrophys. **105** (1982) 164.

p. 48, 51, 54, 60, 64, 66
In the ref. list *add*: For general references, see 7.1.1.

p. 102 In the ref. list *add*: For general references, see 7.3.1.

p. 104 ref. 118: Astron. Astrophys. **97** (1981) 334.

p. 127 ref. 0: Astrophys. J. **248** (1981) 373.

p. 146 ref. 67: Astrophys. J. **252** (1982) 179.

p. 146 ref. 68: Astrophys. J. **249** (1981) 134.
ref. 69: Astrophys. J. **247** (1981) L77.

p. 384 line 2 *read*: Halley's comet was re-discovered on October 16, 1982 by a group of astronomers from Cal. Tech. (G. E. Danielson, D. C. Jewitt et al.) with the 200″ telescope of the Palomar Observatory, with magnitude $m_v = 24.2$, 8″ west and 0″ north of the predicted position.

## Corrections

p. 49 Table 3, column "Region", last line: *instead of* catalogues 7 and 9
*read* catalogues 7 and 10

p. 85 Table 7, column "Publications", line 8: *delete the page number* 786

p. 219 Table 9, right-hand side, line two from the bottom: *read:* 41.722 $R^2$

# 5 Special types of stars

5.1···5.5 in Subvolume b; see table of contents.

## 5.6 Compact objects

### 5.6.1 Neutron stars

#### 5.6.1.1 General properties

The concept of neutron stars as theoretically possible stable structures was introduced by Landau [1]. Baade and Zwicky [2] suggested that such stars might be formed in supernova explosions, and Oppenheimer and Volkoff [3] calculated the first neutron star model using a Fermi gas of non-interacting neutrons.

Neutron star models are computed by integrating the general-relativistic equation of hydrostatic equilibrium which was derived by Tolbert, Oppenheimer, and Volkoff (T.O.V. equation)

$$-\frac{\mathrm{d}P}{\mathrm{d}r}=G\frac{(\varrho+P/c^2)(\mathfrak{M}+4\pi r^3P/c^2)}{r^2(1-2G\mathfrak{M}/rc^2)}. \tag{1}$$

Here $\varrho(r)$ is the rest mass plus potential and kinetic energy density ($\varrho=mn+\varepsilon/c^2$ with $m$ = rest mass of neutron, $n$ = number density of neutrons), and $G$ the constant of gravity. $\mathfrak{M}(r)$ is the mass within radius $r$, and $P(r)$ denotes the pressure. The radius $R$ of the star is defined as usual by $P(R)=0$.

A schematic illustration of the structure of neutron stars computed from (1) is shown in Fig. 1. Typical radii are 10···20 km, masses $\approx 1\,\mathfrak{M}_\odot$, and central densities exceed the density of nuclear matter, $\varrho_0=2.8\cdot 10^{14}$ g cm$^{-3}$.

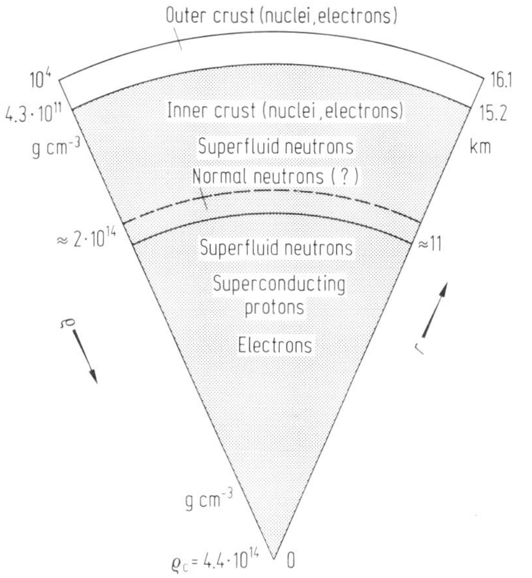


Fig. 1. Cross-section through a representative neutron star with a mass of 1.3 $\mathfrak{M}_\odot$.
$r$ = radius, $\varrho$ = density, $\varrho_c$ = central density.

In the outermost regions, the structure of a neutron star is expected to be the same as in the interior of white dwarf stars: $^{56}$Fe nuclei surrounded by a sea of degenerate electrons and a density of about $10^4$ g cm$^{-3}$ at the surface. The iron nuclei, because of their mutual electrostatic repulsion, are expected to form a crystalline lattice (as soon as the temperature has dropped below about $10^{10}$ K), thereby giving the neutron star a solid outer crust.

Beneath this crust the density increases, causing an increased electron Fermi energy which in turn results in electron capture by nuclei, leading to neutron-rich nuclei. These form the inner crust, another crystalline lattice, and are embedded in a sea of degenerate electrons. Still further inward, the further density increase causes the number of free neutrons to rise rapidly until, at about $2 \cdot 10^{14}$ g cm$^{-3}$, the nuclei completely dissolve into a neutron sea. The neutrons are likely to be superfluid. At densities higher than $2 \cdot 10^{14}$ g cm$^{-3}$ the properties of matter are still not completely clear but it is expected that a sea of electrons, superfluid neutrons and superconducting protons exists. If the core density is even higher, as some models suggest, new particles such as muons and hyperons will be created.

Whereas this coarse description of the structure of neutron stars applies to most model calculations, there are important differences in detail between the various computations. These result from the differences in the equations of state used which relate the pressure $P(r)$ to the density $\varrho(r)$ in the different density regimes. In the low density regime the equation of state is well understood, and the principal progress has been a better description of the properties of matter near the surface in the strong magnetic fields that prevail (see below). At densities near $\varrho_0$ recent work has brought out the sensitivity of the stellar radius and crust thickness to microscopic details of the equation of state in this region. At the highest densities, the recent discussion has concentrated e.g. on the pion condensation and quark matter (see e.g. [4]).

The properties of matter at high densities and their implication for the structure of neutron stars have recently been reviewed, e.g. by Canuto [5], and Baym and Pethick [6, 4]. These authors have also compiled extensive lists of references to the original literature.

## 5.6.1.2 Results from model calculations

### 5.6.1.2.1 Masses, radii and moments of inertia

Theoretically determined masses, radii and moments of inertia depend on the choice of the equation of state. As some uncertainty still remains about the correct form of the equation of state at the very high densities that occur in neutron stars, we give here the results of Arnett and Bowers [7] who calculated the structure of slowly rotating neutron stars for a variety of equations of state listed in Table 1, p. 3.

Eleven of these models (models A···M) employ a non-relativistic many-body formalism in constructing the equation of state. Two models (N and O) are based on a relativistic description of the interactions and a consistent (but simplified) relativistic many-body theory. With two exceptions (models G and M), all of the equations of state represent normal systems, i.e. Fermi fluids. Models H and L represent a very soft and a very stiff equation of state, respectively. Some of the models can hardly be considered as realistic, e.g. model H; they have, however, been included as a reference. At densities below about $10^{14}$ g cm$^{-3}$ the equation of state is believed to be much better known, and in general for these densities the equations of state as given by Baym, Pethick, and Sutherland [18] and by Baym, Bethe, and Pethick [19] have been used. Model F represents a reasonable alternative to this, and it has been included to test the sensitivity of models near mass peak on the low-density region. For further details, see [7].

Table 1. Equations of state, considered by Arnett and Bowers [7].

| | Model | Ref. | Interactions | Many-body theory | Density range $\varrho$ in [g cm$^{-3}$] | Composition*) |
|---|---|---|---|---|---|---|
| A | Pandharipande | 8 | Reid soft core – adapted to nuclear matter | Variational principle applied to correlation function | $\varrho > 6.97 \cdot 10^{14}$ | n |
| B | Pandharipande | 9 | Same as A; arbitrary reduction for hyperon-hyperon attraction | Same as A | $\varrho > 7.000 \cdot 10^{14}$ | n, p, $\Lambda$, $\Sigma^{\pm,0}$, $\Delta^{-,0}$ |
| C | Bethe-Johnson | 10 | Modified Reid soft core | Constrained variational principle | $1.71 \cdot 10^{14} \cdots 3.23 \cdot 10^{16}$ | n, p, $\Lambda$, $\Sigma^{\pm,0}$, $\Delta^{\pm,0}$, $\Delta^{++}$ |
| D | Bethe-Johnson | 10 | Same as C; more realistic adaptation to hyperon matter | Same | $1.7 \cdot 10^{14} \cdots 2.26 \cdot 10^{16}$ | Same |
| E | Moszkowski | 11 | Reid soft core; modified hyperon interactions based on quark theory | Reaction matrix | $2.2 \cdot 10^{14} \cdots 3.08 \cdot 10^{15}$ | n, p, $\Sigma^-$, $\Lambda^0$, $\Delta^-$ |
| F | Arponen | 12 | Thomas Fermi model | Brueckner *G*-matrix | $3.1 \cdot 10^{11} \cdots 2.0 \cdot 10^{15}$ | n, p, $e^-$, $\mu^{\pm}$ |
| G | Canuto-Chitre | 13 | Modified Reid soft core. Localization via non-relativistic harmonic oscillators | *T*-matrix; includes spin dependence | $2.37 \cdot 10^{15} \cdots 7.23 \cdot 10^{15}$ | n (solid) |
| H | Ideal neutron gas | 3 | None | Fermi statistics | Entire range | n |
| I | Cameron-Cohen-Langer-Rosen | 14 | Levinger-Simmons velocity-dependent | Hartree-Fock approximation with two-body potential | $1.0 \cdot 10^{14} \cdots 5.35 \cdot 10^{15}$ | n, p, $e^-$, $\mu^{\pm}$ |
| L | Pandharipande and Smith | 15 | Nuclear attraction due to scalar exchange | Mean-field approximation for scalar; variational method | $\varrho > 4.386 \cdot 10^{11}$ | n |
| M | Pandharipande and Smith | 15 | Nuclear attraction due to pion-exchange tensor interactions | Constrained variational method | $\varrho > 8.428 \cdot 10^{13}$ | n |
| N | Walecka | 16 | Relativistic mean-field scalar plus vector exchange fitted to nuclear matter | Mean-field approximation (relativistic) | $\varrho > 1.723 \cdot 10^{14}$ | n |
| O | Bowers, Gleeson, and Pedigo | 17 | Non-perturbative, phenomenological approximation to relativistic meson exchange | Relativistic finite-density Green's functions | $\varrho > 1.732 \cdot 10^{14}$ | n, p, $\Lambda$, $\Sigma^{\pm,0}$, $\Xi^{-,0}$ |

*) n, p: nucleons; $e^-$, $\mu^+$: leptons; $\Lambda$, $\Sigma$, $\Xi$, $\Delta$: hyperons.

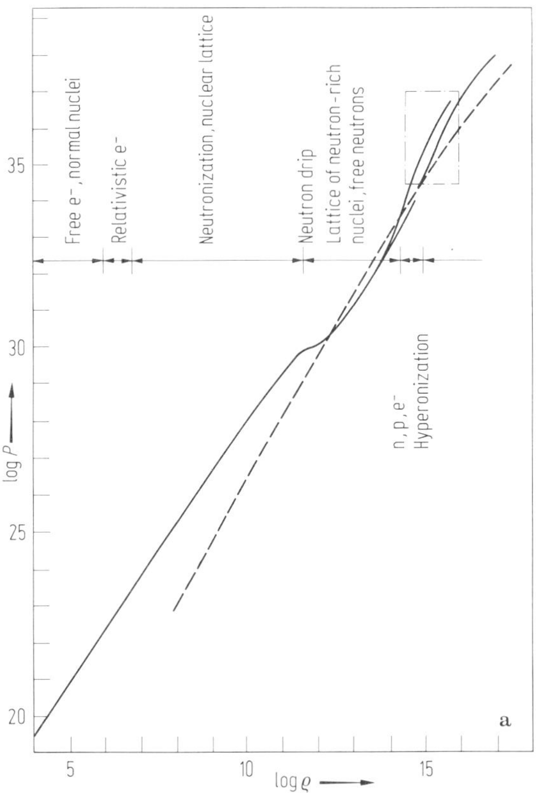


Figs. 2a···c. Representative equations of state for cold neutron stars. In 2a, the equation of state for a free gas of neutrons is shown by the dashed curve for comparison. The region contained in the rectangular box in 2a is shown enlarged in 2b and 2c. In 2b, results are shown for models A···I, in 2c results for models L···O (from [7], see also Table 1). $P$ in [dyn cm$^{-2}$], $\varrho$ in [g cm$^{-3}$].

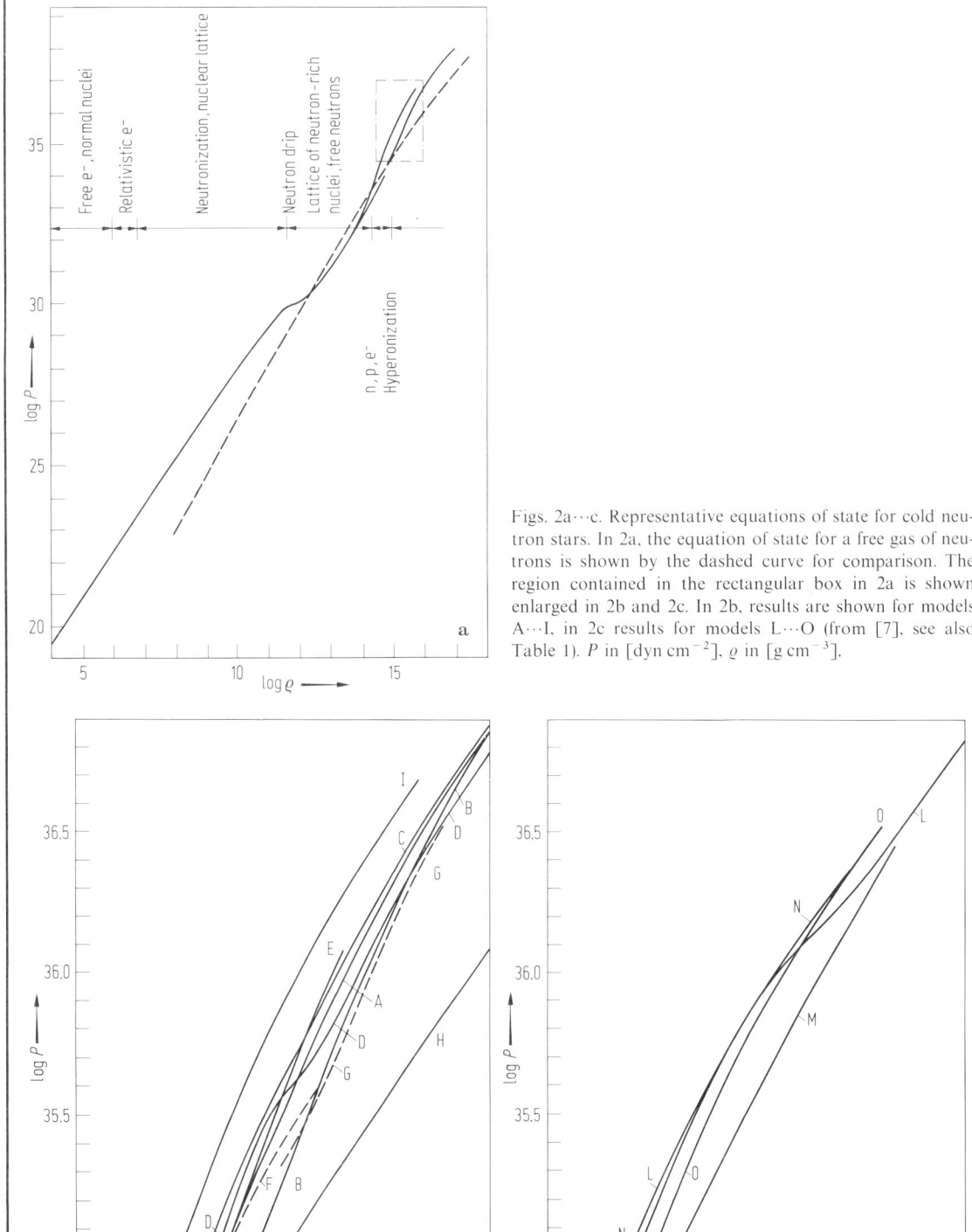

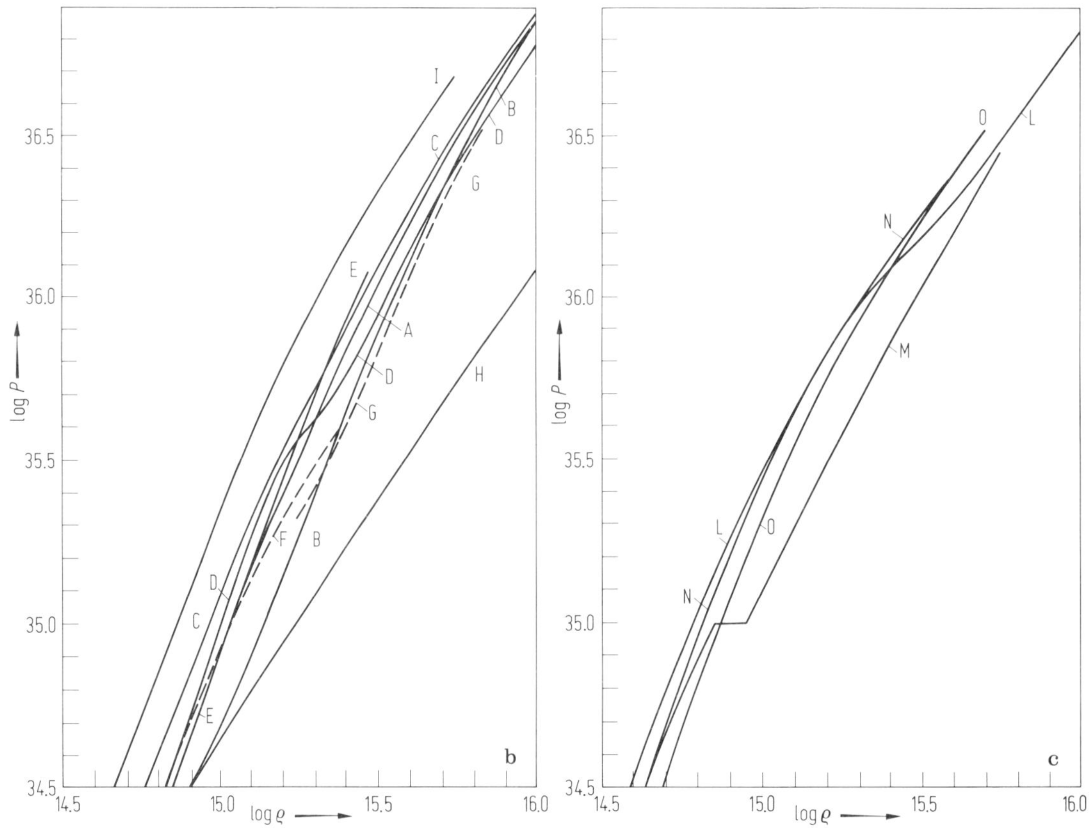

Figs. 2a···c illustrate the equations of state used in the calculations by Arnett and Bowers [7]. Figs. 3a, b give the resulting neutron star masses as a function of the stars' central densities $\varrho_c$. The equations of state based on non-relativistic interactions (A···G) produce maximum masses $1.36 < \mathfrak{M} < 1.85\,\mathfrak{M}_\odot$, and have central densities in the range $3.0 \cdot 10^{15} \cdots 6.3 \cdot 10^{15}$ g cm$^{-3}$. Models L, N, O, the last two of which are relativistic, lead to neutron stars with $2.39 \leqq \mathfrak{M}_{max} \leqq 2.70\,\mathfrak{M}_\odot$, having central densities in the range $1.4 \cdot 10^{15} \cdots 2 \cdot 10^{15}$ g cm$^{-3}$. With the exception of models A, G, and M, there is a marked tendency for the value of $\mathfrak{M}_{max}$ to increase and move to lower central densities as the equation of state becomes stiffer.

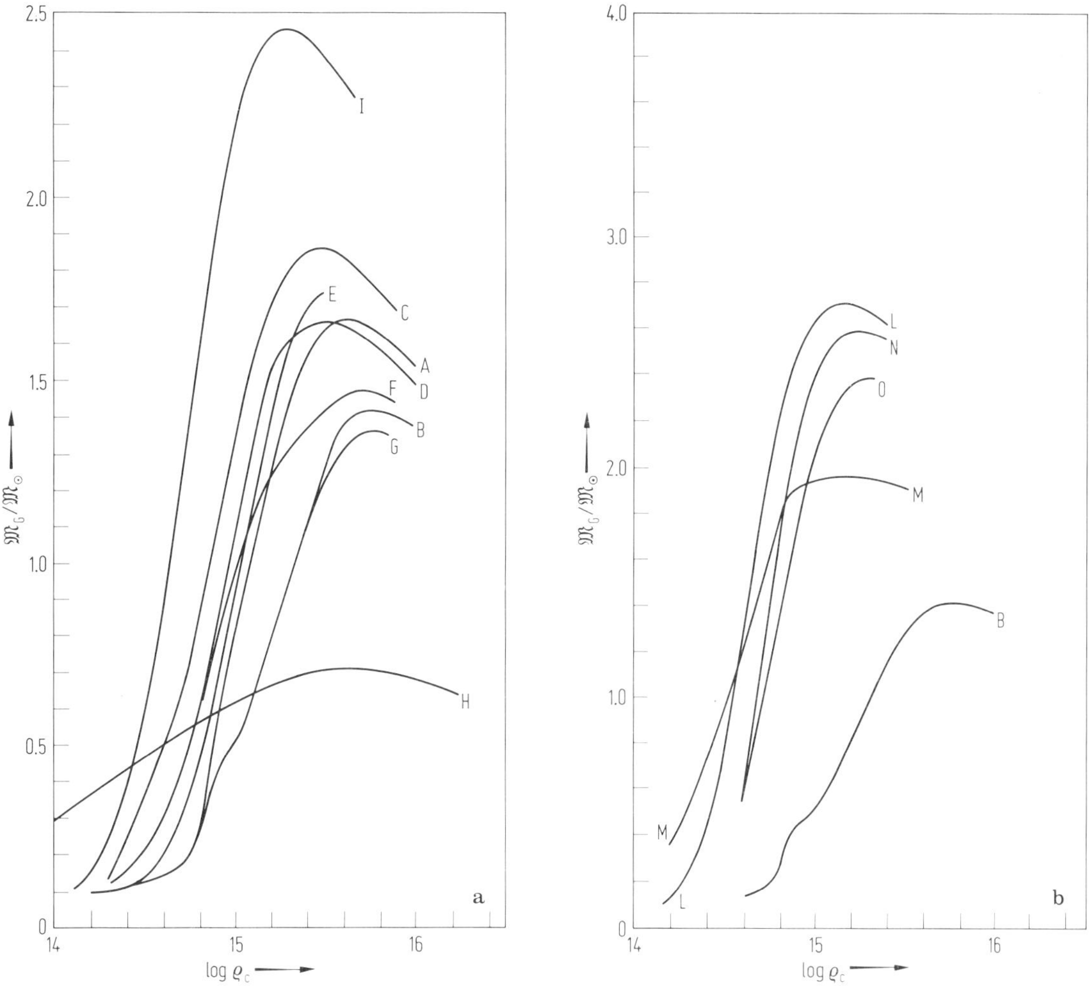


Figs. 3a, b. The gravitational mass of neutron stars $\mathfrak{M}_G$ versus their central density $\varrho_c$ for models A through I (3a), and models L through O (3b). For comparison model B is repeated in both figures (from [7], see also Table 1).
$\varrho_c$ in [g cm$^{-3}$]

In Fig. 4 the gravitational mass-radius relation is shown for a total of nine different equations of state. From these results it is evident that the range of radii is fairly limited and lies for most models in the range 8···20 km.

Fig. 5 shows the resulting moment of inertia as a function of the central density for all the models that have been considered by Arnett and Bowers [7]. For central densities below about $10^{15}$ g cm$^{-3}$ there is a strong dependence of the calculated moment of inertia on the central density which in turn depends on the stiffness of the equation of state. As there is some hope of determining the moment of inertia of neutron stars from observations (see below) this should reveal the behaviour of matter at very high densities.

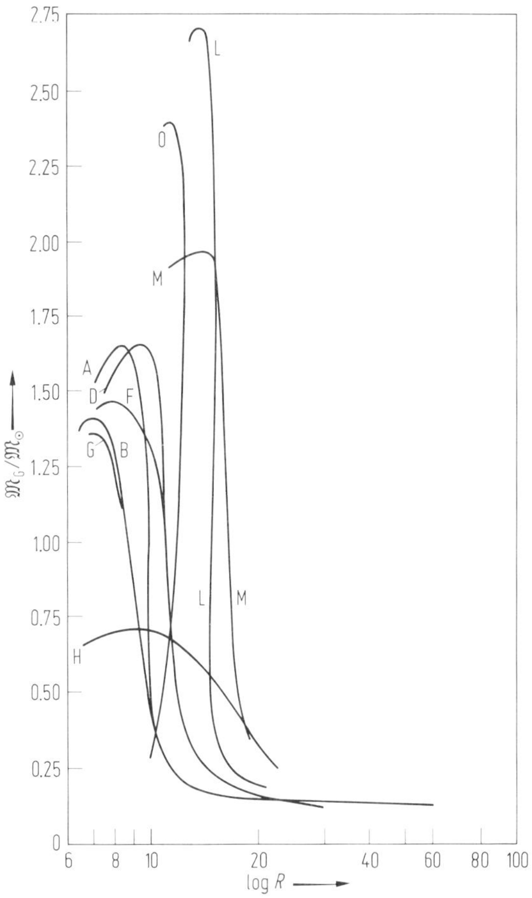


Fig. 4. The gravitational mass-radius relation for selected equations of state (from [7], see also Table 1). $R$ in [km]

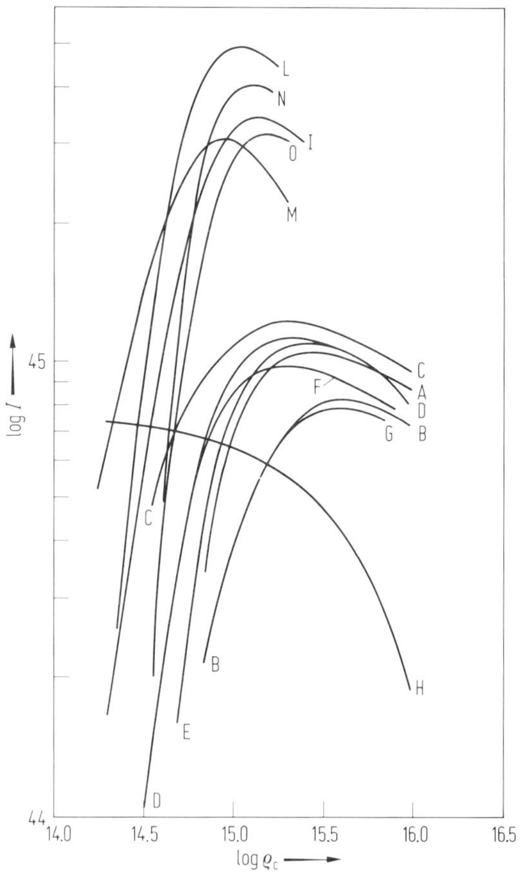


Fig. 5. The moment of inertia $I$ versus central density $\varrho_c$ for all models calculated by Arnett and Bowers ([7], see also Table 1). $I$ in [g cm$^2$], $\varrho_c$ in [g cm$^{-3}$]

### 5.6.1.2.2 Maximum-mass considerations

Due to the uncertainties that still remain in the equation of state that applies at the very high densities that occur in neutron stars, it is not yet clear what maximum mass stable neutron stars can have. A number of attempts have, however, been made to derive such a limit. Rhoades and Ruffini [20] applied an extremal principle in selecting an equation of state that produces a maximum critical mass in the framework of Einstein's theory of general relativity, le Chatelier's principle and the principle of causality. They find a value $\mathfrak{M}_{max} = 3.2\,\mathfrak{M}_\odot$. Very similar limits were found in two more recent calculations [21, 22]. However, under extreme conditions, these limits can be pushed to roughly $8\,\mathfrak{M}_\odot$ as shown in [22], or even $11\,\mathfrak{M}_\odot$ as shown in [23].

### 5.6.1.2.3 Magnetic fields

Magnetic fields are observed on main-sequence stars. The interior of such stars can be safely assumed to remain in a highly conducting state until the ultimate collapse of the core. Ruderman and Sutherland [24] suggest that turbulent convection in the core of presupernova stars results in the field building up to an equipartition value of $\approx 3 \cdot 10^9$ Gauss. Conservation of flux during the collapse to a neutron star then results in a surface field strength of $\approx 4 \cdot 10^{12}$ Gauss. The field probably has significant multipole structure, but in the outer magnetosphere it is usually assumed to be bipolar. Probably, there is also a huge torodial field interior to the neutron star. The magnitude of this is unknown but is probably at least as strong as the exterior polodial field [25].

Several scientists have raised the question of magnetic field decay in neutron stars. The only region within such stars where the field could decay due to ohmic dissipation on a reasonably short time scale is the crust (e.g. [26, 27]).

#### 5.6.1.2.4 Cooling calculations for neutron stars

The significantly increased sensitivity of recent X-ray experiments and the anticipated sensitivities of future X-ray experiments have greatly stimulated new calculations of the cooling of young neutron stars. Recent computations are given in [28···31]. References to earlier papers are given by these authors.

The most recent calculation by Nomoto and Tsuruta [31] is based on an exact stellar evolution code, incorporating the full general relativistic version of the stellar structure equations with the best physical input currently available. This approach takes into account the effect of the finite time scale of thermal conduction which has been neglected in most of the earlier computations.

In Fig. 6 both the surface temperature and the photon luminosity are given as a function of age for a neutron star with $\mathfrak{M}_G = 1.3\,\mathfrak{M}_\odot$ according to calculations by Nomoto and Tsuruta [31].

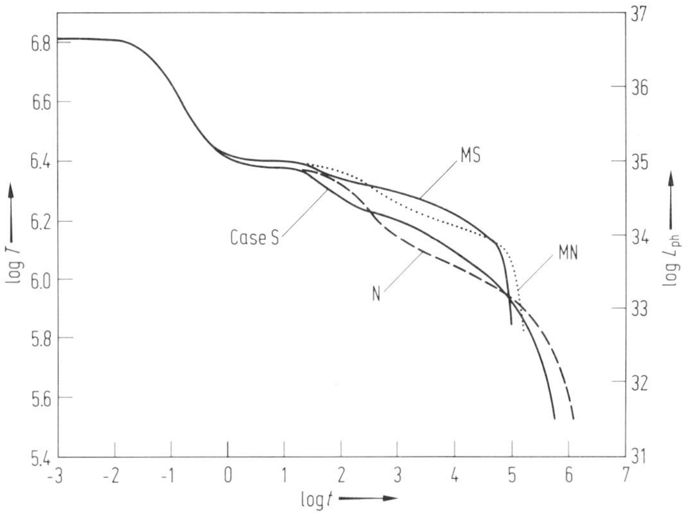


Fig. 6. The surface temperature $T$ and the photon luminosity $L_{ph}$ from a representative neutron star model calculation ($\mathfrak{M}_G = 1.3\,\mathfrak{M}_\odot$) after Nomoto and Tsuruta [31].
Standard cooling curves are shown for four cases: Case S with superfluid nucleons and case N with non-superfluid (normal) nucleons both for non-magnetized neutron stars (zero magnetic field), and cases MS (superfluid) and MN (non-superfluid) for the magnetic field $H = 4.4 \cdot 10^{12}$ Gauss. For details of the equation of state used, see [31]. $t$ in [a], $T$ in [K], $L_{ph}$ in [erg s$^{-1}$]

## 5.6.1.3 Observational results

### 5.6.1.3.1 Masses, radii and moments of inertia

Optical and X-ray observations of binary X-ray sources provide the possibility of determining the masses of neutron stars (see also 6.1.4). For a recent review, see Bahcall [33] who refers extensively to the original literature.

The mass determination involves the derivation of either an optical or an X-ray mass function, or both. This function is defined by

$$F(\mathfrak{M}_1, \mathfrak{M}_2, i) = \mathfrak{M}_1^3 \sin^3 i / (\mathfrak{M}_1 + \mathfrak{M}_2)^2\,, \qquad (1)$$

where $\mathfrak{M}_1$, $\mathfrak{M}_2$ are the masses of the two stars in binary orbit and $i$ is the inclination of the plane of the orbit to the plane of the sky (see also 6.1.3.1). The mass function for either star is determined directly from observations by measuring either the size of the star's orbit or its radial velocity. From Newton's laws for two mass points moving in elliptical orbits under the influence of their mutual gravitational attraction one finds that

$$F(\mathfrak{M}_1, \mathfrak{M}_2, i) = (4\pi/GP^2)\,(a_2 \sin i)^3 \qquad (2)$$

and

$$F(\mathfrak{M}_1, \mathfrak{M}_2, i) = (P/2\pi G)\,(1-e^2)^{3/2}\,(v_2 \sin i)^3 \qquad (3)$$

with $a_2$ = semi-major axis of the orbit of star 2, $P$ = orbital period, $e$ = orbit eccentricity, $v_2 \sin i$ = amplitude of the orbital velocity curve, and $G$ = constant of gravity.

For pulsating X-ray sources one can determine the X-ray mass function $F_X$ from the amplitude of the observed delays of the pulse arrival time, $a_2 \sin i/c$, and the measured orbital period. From Eq. (1) we then have

$$\mathfrak{M}_X = F_X q(1+q)^2/\sin^2 i\,, \tag{4}$$

where $q$ is the ratio of the mass of the X-ray star, $\mathfrak{M}_X$, to the mass of the optical star, $\mathfrak{M}_{opt}$. In order to obtain the individual masses of the components, also the optical mass function must be known. For a careful discussion of the possible sources of error that enter into such an analysis, see Bahcall [33]:

Table 2. Masses of neutron stars in X-ray binaries [33].

| Source | $\mathfrak{M}_X$ [$\mathfrak{M}_\odot$] | Comments |
|---|---|---|
| 3U 0900-40 | $1.0 \leqq \mathfrak{M}_X \leqq 3.4$ | Successful test of method of analysis. |
| SMC X-1 | $0.5 \leqq \mathfrak{M}_X \leqq 1.8$ | X-ray heating and possible extra light source complicate system. |
| Cen X-3 | $0.7 \leqq \mathfrak{M}_X \leqq 4.4$ | Optical velocity amplitude predicted to be $15 \cdots 80$ km s$^{-1}$. |
| Her X-1 | $0.4 \leqq \mathfrak{M}_X \leqq 2.2$ | Measurements required during X-ray "off" period of optical velocity amplitude and ellipsoidal light variations. The optical velocity amplitude predicted to be $50 \cdots 130$ km s$^{-1}$. |
| 3U 1700-37 | $0.6 \lesssim \mathfrak{M}_X$ | Assumed mass of the optical primary exceeds 10 $\mathfrak{M}_\odot$. |

These results are basically in agreement with the model predictions discussed in 5.6.1.2.1. However, the range of uncertainties is still too large to favour any of the particular neutron star models discussed in 5.6.1.2.1.

Observations of X-ray burst sources can in principle be used to independently constrain the mass-radius relation for neutron stars as was pointed out by van Paradijs [34].

Information on moments of inertia of neutron stars may be obtained from observations of the secular rates of change of their spin periods. A comparison of the slow-down rate of the Crab pulsar with the luminosity of the Crab nebula yields a lower bound on its moment of inertia of $> 1.5 \cdot 10^{44}$ g cm$^2$ [25]. Observed speed-ups of pulsating X-ray sources, that indicate a time scale of $\approx 10^2 \cdots 10^5$ years, combined with model descriptions of accretion torques, indicate moments of inertia which are consistent with this result as well as with the model predictions [35···37].

### 5.6.1.3.2 Magnetic fields

Surface magnetic fields of neutron stars in active radiopulsars and binary X-ray sources are inferred from the rates of energy loss (see 5.6.2) and from the structure of the radiation and the spin-up rates (see 5.6.3), respectively. In either case the results are consistent with fields in the order of $10^{11} \cdots 10^{13}$ Gauss; the values depend, however, on a number of model assumptions which enter into the analysis.

A totally independent determination can be derived from the observation (Trümper et al. [38]) of an emission (or absorption) feature in the X-ray spectrum of Her X-1 at $\approx$ 58 keV ($\approx$ 42 keV) if this is interpreted as cyclotron emission (absorption) by electrons in the strong magnetic field of the neutron star. The field strength turns out to be $\approx 6 \cdot 10^{12}$ Gauss ($\approx 4 \cdot 10^{12}$ Gauss), respectively for Her X-1.

### 5.6.1.3.3 Surface temperatures

In Fig. 7, we have compiled the presently available results for a total of 11 objects.

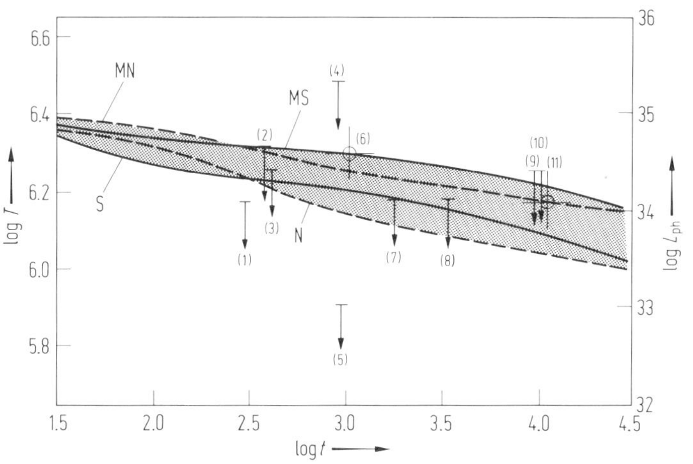


Fig. 7. Nine upper limits and two possible measurements of surface temperatures $T$ of neutron stars (and corresponding photon luminosities $L_{ph}$) as function of their age. For comparison the theoretical cooling curves [31] discussed in 5.6.1.2.4 have been repeated (see Fig. 6).
Identification of the sources:

| | | |
|---|---|---|
| (1) Cas A [39] | (5) SN 1006 [42] | (9) G 350.0-18 [40] |
| (2) Kepler [40] | (6) RCW 103 [43] | (10) G 22.7-0.2 [40] |
| (3) Tycho [40] | (7) RCW 86 [40] | (11) Vela [44] |
| (4) Crab [41] | (8) W 28 [40] | |

$t$ in [a], $T$ in [K], $L_{ph}$ in [erg s$^{-1}$]

## 5.6.1.4 References for 5.6.1


1 Landau, L.: Phys. Z. Sowjetunion **1** (1932) 285.
2 Baade, W., Zwicky, F.: Proc. Nat. Acad. Sci. **20** (1934) 254.
3 Oppenheimer, J.R., Volkoff, G.M.: Phys. Rev. **55** (1939) 374.
4 Baym, G., Pethick, C.J.: Annu. Rev. Astron. Astrophys. **17** (1979) 415.
5 Canuto, V.: Annu. Rev. Astron. Astrophys. **12** (1974) 167.
6 Baym, G., Pethick, C.J.: Annu. Rev. Nucl. Sci. **25** (1975) 27.
7 Arnett, W.D., Bowers, R.L.: Astrophys. J. Suppl. **33** (1977) 411.
8 Pandharipande, V.: Nucl. Phys. **A174** (1971) 641.
9 Pandharipande, V.: Nucl. Phys. **A178** (1971) 123.
10 Bethe, H.A., Johnson, M.: Nucl. Phys. **A230** (1974) 1.
11 Moszkowski, S.: Phys. Rev. **D9** (1974) 1613.
12 Arponen, J.: Nucl. Phys. **A191** (1972) 257.
13 Canuto, V., Chitre, S.M.: Phys. Rev. **D9** (1974) 1587.
14 Cohen, J.M., Langer, W.D., Rosen, L.C., Cameron, A.G.W.: Astrophys. Space Sci. **6** (1970) 228.
15 Pandharipande, V., Smith, R.A.: Nucl. Phys. **A175** (1975) 225.
16 Walecka, J.D.: Ann. Phys. **83** (1974) 491.
17 Bowers, R.L., Gleeson, A.M., Pedigo, R.D.: Phys. Rev. **D12** (1975) 3043.
18 Baym, G., Pethick, C., Sutherland, P.: Astrophys. J. **170** (1972) 299.
19 Baym, G., Bethe, H.A., Pethick, C.: Nucl. Phys. **A175** (1971) 225.
20 Rhoades, C.E., Ruffini, R.: Phys. Rev. Lett. **32** (1974) 324.
21 Chitre, D.M., Hartle, J.B.: Astrophys. J. **207** (1976) 592.
22 Durgapal, M.C., Rawat, P.S.: Mon. Not. R. Astron. Soc. **192** (1980) 659.
23 Hegyi, D.J.: Astrophys. J. **217** (1977) 244.
24 Ruderman, M.A., Sutherland, P.G.: Nature Phys. Sci. **246** (1973) 93.
25 Ruderman, M.A.: Annu. Rev. Astron. Astrophys. **10** (1972) 427.
26 Heintzmann, H., Grewing, M.: Z. Physik **250** (1971) 254.
27 Ewart, G.M., Guyer, R.A., Greenstein, G.: Astrophys. J. **202** (1975) 238.

28 Tsuruta, S.: Phys. Rep. **56** (1979) 273.
29 Glen, G., Sutherland, P.G.: Astrophys. J. **239** (1980) 671.
30 Riper, K.A. van, Lamb, D.Q.: Astrophys. J. (Lett.) **244** (1981) L13.
31 Nomoto, K., Tsuruta, S.: NASA TM 82145 (1981).
32 Pandharipande, V.R., Pines, D., Smith, R.A.: Astrophys. J. **208** (1976) 550.
33 Bahcall, J.N.: Annu. Rev. Astron. Astrophys. **16** (1978) 241.
34 Paradijs, J. van: Astrophys. J. **234** (1979) 609.
35 Elsner, R.F., Lamb, F.K.: Nature **262** (1976) 356.
36 Rappaport, S., Joss, P.: Nature **266** (1977) 123.
37 Ghosh, P., Lamb, F.K.: Astrophys. J. **232** (1979) 259.
38 Trümper, J., Pietsch, W., Reppin, C., Voges, W., Staubert, R., Kendziorra, E.: Astrophys. J.(Lett.) **219**(1978) L105.
39 Murray, S.S., Fabbiano, G., Fabian, A.C., Epstein, A., Giacconi, R.: Astrophys. J. (Lett.) **234** (1979) L69.
40 Helfand, D.J., Chanan, G.A., Novick, R.: Nature **283** (1980) 337.
41 Toor, A., Seward, F.J.: Astrophys. J. **216** (1977) 560.
42 Pye, J.P., Pounds, K.A., Rolf, D.P., Seward, F.D., Smith, A., Willingate, R.: Mon. Not. R. Astron. Soc. **194** (1981) 569.
43 Tuohy, I., Garmire, G.: Astrophys. J. (Lett.) **239** (1980) L107.
44 Harnden, F.R., jr., Hertz, P., Gorenstein, P., Grindlay, J., Schreier, E., Seward, F.: Bull. American Astron. Soc. **11** (1979) 424.

## 5.6.2 Radiopulsars

### 5.6.2.1 General properties

Discovered by Hewish et al. [1], these objects show the following basic characteristics:

broad-band radio noise in the form of a periodic sequence of pulses
highly variable pulse intensities
a pulse duration that is typically 3% of the pulse period $P$
individual pulses that often consist of two or more subpulses with a duration of typically $1\cdots2\%$ of the period
within the subpulses some pulsars show microstructure
the radio frequency spectrum of pulsars is fairly steep with a power law index of typically $-1.5$
only very few pulsars have been detected outside the radio band (e.g. the Crab and the Vela pulsar)
in all cases where accurate observations have been made, the pulsar periods are found to increase in a regular way with time. The typical rate of change is $10^{-15}$ s s$^{-1}$
with only two exceptions all known radiopulsars are single neutron stars without a companion.

For a detailed description of these and other characteristics and a compilation of further references, see [2, 3].

### 5.6.2.2 Surveys

Following the detection of the first pulsar in 1967 [1], and successful pulsar searches by many different groups, the vast majority of all known pulsars was first detected in one of five major surveys: the first and second Molonglo survey (Large and Vaughan [4] and references therein: 36 pulsars in total, 31 new; and Manchester et al. [5]: 224 total, 155 new); the Jodrell Bank survey (Davies et al. [6, 7]: 51 total, 31 new); the University of Massachusetts Arecibo survey (Hulse and Taylor [8, 9]: 50 total, 40 new); and the University of Massachusetts NRAO survey (Damashek et al. [10]: 43 total, 17 new). Except for the Molonglo surveys which cover most of the sky south of $\delta = +20°$, all the other searches concentrated on regions in the vicinity of the galactic plane. The original Molonglo survey and the Jodrell Bank survey had reduced sensitivity for pulsars with short periods ($\leqq 0.3$ s) and/or high dispersion ($DM \geqq 150$ cm$^{-3}$ pc, see below). These limits have been pushed considerably further in the more recent surveys, the Arecibo survey e.g. maintaining its full sensitivity up to $DM = 1280$ cm$^{-3}$ pc, and it was less sensitive by only about 3 dB for periods as short as 0.06 s.

### 5.6.2.3 Radiopulsar positions, periods, period changes, and dispersion measures

In Table 1 data for 321 radiopulsars have been compiled from the literature. Whereas the bulk of the data has been taken from the references [1···10], data of individual pulsars have also been taken from the references [11···46].

In Table 1 the position of the objects is given by their right ascension and declination for epoch 1950. Also given is the pulse period $P$ and the rate of change of the period $\Delta P$ and an epoch of observation to which these data refer if this information was available. Finally the dispersion measure is given which is defined as

$$DM = \int_0^d N_e \, dl \,, \qquad (1)$$

where $N_e$ is the number density of free electrons in interstellar space and $d$ the distance of the pulsar from the sun. Due to these electrons, the group velocity of radio waves is slightly less than the velocity of light and is a function of frequency of radiation. As a consequence, the pulse arrival time for two different observing frequencies ($\omega_i = 2\pi \cdot \nu_i$, i = 1, 2) differs by

$$t_2 - t_1 = (2\pi e^2/m_e c) \cdot (\omega_2^{-2} - \omega_1^{-2}) \cdot DM \,. \qquad (2)$$

Table 1. Data for 321 radiopulsars.

$\alpha, \delta$ = position
$P$ = pulse period
$\Delta P$ = rate of change of the period
JD = epoch of observation (Julian Date); see 2.2.2.5)
$DM$ = dispersion measure; for definition, see Eq. (1)

| Pulsar PSR | α (1950.0) | δ (1950.0) | P s | ΔP $10^{-9}$ s/d | JD − 2440000 | DM pc cm$^{-3}$ |
|---|---|---|---|---|---|---|
| 0012+47 | 0$^h$12$^m$30$^s$.00 | 47°30′ 0″.0 | 1.2407 | – | 3513 | 25 |
| 0031−07 | 0 31 36.48 | − 7 38 32.7 | 0.94295078485 | 0.0346 | 690 | 10.890 |
| 0105+65 | 1 5 0.00 | 65 50 0.0 | 1.2836506982 | 1.074 | 2006 | 30.150 |
| 0136+57 | 1 36 40.00 | 57 50 0.0 | 0.2725 | – | 3513 | 60 |
| 0138+59 | 1 38 20.00 | 59 52 0.0 | 1.2229480053 | 0.015 | 2006 | 34.800 |
| 0148−06 | 1 48 54.00 | − 6 55 0.0 | 1.464651 | – | 3415 | 27 |
| 0149−16 | 1 49 46.43 | −16 52 39.8 | 0.83274108859 | 0.11197 | 3556 | 12.400 |
| 0153+61 | 1 53 0.00 | 61 55 0.0 | 2.3516185238 | 16.358 | 2007 | 60.000 |
| 0203−40 | 2 3 57.20 | −40 42 21.8 | 0.63054980456 | 0.10351 | 3556 | 12.900 |
| 0254−53 | 2 54 24.17 | −53 16 25.4 | 0.44770843528 | 0.00272 | 3556 | 15.900 |
| 0301+19 | 3 1 42.42 | 19 21 12.5 | 1.3875836669693 | 0.11199 | 2326 | 15.690 |
| 0320+39 | 3 20 20.00 | 39 40 0.0 | 3.0321 | – | 3513 | 25 |
| 0329+54 | 3 29 11.00 | 54 24 36.7 | 0.714518663928 | 0.1772 | 622 | 26.776 |
| 0340+53 | 3 40 10.00 | 53 10 0.0 | 1.9355 | – | 3513 | 85 |
| 0355+54 | 3 55 0.45 | 54 4 42.6 | 0.15638005592 | 0.379 | 1536 | 57.030 |
| 0403+61 | 4 3 10.00 | 61 40 0.0 | 0.5946 | – | 3513 | 60 |
| 0403−76 | 4 3 15.17 | −76 16 25.8 | 0.54525231097 | 0.13306 | 3556 | 21.700 |
| 0447−12 | 4 47 49.00 | −12 55 0.0 | 0.438014 | – | 3325 | 40 |
| 0449+55 | 4 49 50.00 | 55 50 0.0 | 0.3408 | – | 3513 | 15 |
| 0450−18 | 4 50 21.43 | −18 4 20.8 | 0.5489353684 | 0.36 | 1536 | 39.930 |
| 0459+47 | 4 59 30.00 | 47 0 0.0 | 0.6386 | – | 3513 | 40 |
| 0523+11 | 5 23 9.00 | 11 15 0.0 | 0.354437 | – | 3260 | 80 |
| 0525+21 | 5 25 51.75 | 21 58 18.0 | 3.74549702902671 | 3.46088 | 958 | 50.955 |
| 0531+21 | 5 31 31.43 | 21 58 0.7 | 0.0332003858 | 36.46 | 1994 | 56.791 |
| 0538−75 | 5 38 18.90 | −75 45 38.0 | 1.2458554349 | 0.04925 | 3556 | 18.900 |

continued

Table 1, continued

| Pulsar | α (1950.0) | δ (1950.0) | *P* s | Δ*P* $10^{-9}$ s/d | JD − 2440000 | *DM* pc cm$^{-3}$ |
|---|---|---|---|---|---|---|
| 0540+23 | 5 40 6.94 | 23 27 46.7 | 0.24596610516121 | 1.33304 | 2392 | 72.000 |
| 0559−05 | 5 59 33.00 | − 5 30 0.0 | 0.395968 | – | 3325 | 75 |
| 0559−57 | 5 59 59.40 | −57 56 52.0 | 2.261364513 | 0.24019 | 3558 | 29.500 |
| 0611+22 | 6 11 14.95 | 22 26 6.2 | 0.33492505400698 | 5.15206 | 2881 | 96.700 |
| 0621−04 | 6 21 52.00 | − 4 15 0.0 | 1.039086 | – | 3325 | 60 |
| 0626+24 | 6 26 0.00 | 24 40 0.0 | 0.4767 | – | 3513 | 70 |
| 0628−28 | 6 28 50.80 | −28 34 8.1 | 1.244414895981 | 0.217 | 242 | 34.360 |
| 0656+14 | 6 56 57.00 | 14 25 0.0 | 0.38486 | – | 3260 | 9 |
| 0656+64 | 6 56 30.00 | 64 20 0.0 | 0.1955 | – | 3513 | 5 |
| 0727−18 | 7 27 19.00 | −18 20 0.0 | 0.51015 | – | 3325 | 61 |
| 0736−40 | 7 36 50.87 | −40 35 46.8 | 0.37491871069 | 0.1399 | 2555 | 161.000 |
| 0740−28 | 7 40 47.87 | −28 15 32.6 | 0.166753902088 | 1.45479 | 3556 | 73.770 |
| 0743−53 | 7 43 50.50 | −53 44 1.0 | 0.2148363514 | 0.23587 | 3780 | 122.300 |
| 0752+32 | 7 52 10.00 | 32 40 0.0 | 1.4425 | – | 3513 | 50 |
| 0756−15 | 7 56 11.00 | −15 20 0.0 | 0.682265 | – | 3335 | 62 |
| 0808−47 | 8 8 12.58 | −47 45 0.7 | 0.54719837196 | 0.26654 | 3557 | 228.300 |
| 0809+74 | 8 9 2.83 | 74 38 12.5 | 1.29224132381 | 0.0138 | 689 | 5.757 |
| 0818−13 | 8 18 6.00 | −13 41 22.8 | 1.23812810723 | 0.182 | 1006 | 40.900 |
| 0818−41 | 8 18 29.74 | −41 5 4.0 | 0.5454455279 | 0.00233 | 3558 | 111.300 |
| 0820+02 | 8 20 35.00 | 2 10 0.0 | 0.864878 | – | 3430 | 24 |
| 0823+26 | 8 23 50.46 | 26 47 19.0 | 0.53065995905517 | 0.14892 | 2717 | 19.463 |
| 0826−34 | 8 26 19.10 | −34 7 8.0 | 1.848918399 | 0.0864 | 3436 | 51.400 |
| 0833−45 | 8 33 39.35 | −45 0 8.6 | 0.08923471388 | 10.8026 | 2701 | 69.080 |
| 0834+06 | 8 34 26.17 | 6 20 43.7 | 1.27376417151777 | 0.58745 | 1708 | 12.855 |
| 0835−41 | 8 35 33.27 | −41 24 42.0 | 0.75162112843 | 0.30637 | 3558 | 147.600 |
| 0839−53 | 8 39 9.20 | −53 21 51.9 | 0.7206117681 | 0.14265 | 3558 | 155.600 |
| 0840−48 | 8 40 29.72 | −48 40 32.3 | 0.6443540558 | 0.82054 | 3558 | 197.000 |
| 0844−35 | 8 44 7.87 | −35 22 38.8 | 1.1160964496 | 0.13565 | 3558 | 96.400 |
| 0853−33 | 8 53 38.00 | −33 25 0.0 | 1.267539 | – | 3335 | 86 |
| 0855−61 | 8 55 54.02 | −61 26 15.8 | 0.9625085573 | 0.14489 | 3558 | 95.100 |
| 0901−63 | 9 1 31.80 | −63 13 17.1 | 0.66031337153 | 0.00924 | 3558 | 75.700 |
| 0903−42 | 9 3 8.19 | −42 34 13.0 | 0.9651709987 | 0.16304 | 3558 | 146.300 |
| 0904+77 | 9 4 0.00 | 77 40 0.0 | 1.57905 | – | 222 | – |
| 0904−74 | 9 4 28.76 | −74 47 40.9 | 0.54955311909 | 0.04 | 3556 | 51.100 |
| 0905−51 | 9 5 40.59 | −51 45 50.8 | 0.25355550115 | 0.15841 | 3556 | 104.000 |
| 0906−17 | 9 6 19.00 | −17 25 0.0 | 0.401625 | – | 3335 | 16 |
| 0909−71 | 9 9 22.32 | −71 59 51.1 | 1.3628899542 | 0.02877 | 3558 | 54.300 |
| 0919+06 | 9 19 34.98 | 6 51 4.1 | 0.43061391485 | 1.18576 | 3556 | 28.100 |
| 0922−52 | 9 22 30.79 | −52 49 47.2 | 0.74629520954 | 3.06521 | 3556 | 152.900 |
| 0923−58 | 9 23 4.90 | −58 1 8.2 | 0.7394995609 | 0.41731 | 3556 | 60.300 |
| 0932−52 | 9 32 46.85 | −52 36 2.5 | 1.4447714964 | 0.40202 | 3558 | 99.400 |
| 0940−55 | 9 40 37.76 | −55 39 10.7 | 0.66436112446 | 1.96465 | 3556 | 180.200 |
| 0940+16 | 9 40 46.00 | 16 55 0.0 | 1.08741 | – | 3335 | 28 |
| 0941−56 | 9 41 18.65 | −56 43 57.4 | 0.80811632004 | 3.42292 | 3558 | 160.800 |
| 0942−13 | 9 42 3.00 | −13 35 0.0 | 0.570265 | – | 3430 | 13 |
| 0943+10 | 9 43 27.12 | 10 5 45.9 | 1.09770363952461 | 0.30492 | 1665 | 15.350 |
| 0950−38 | 9 50 11.80 | −38 25 2.1 | 1.3738151214 | 0.05011 | 3558 | 166.900 |
| 0950+08 | 9 50 30.53 | 8 9 45.2 | 0.25306506818905 | 0.0198 | 1501 | 2.969 |
| 0935−52 | 9 53 41.78 | −52 50 0.9 | 0.86211773933 | 0.30387 | 3556 | 156.900 |
| 0957−47 | 9 57 30.08 | −47 55 23.0 | 0.6700857604 | 0.00708 | 3558 | 92.700 |

continued

Table 1, continued

| Pulsar | α (1950.0) | δ (1950.0) | $P$ s | $\Delta P$ $10^{-9}$ s/d | JD − 2440000 | $DM$ pc cm$^{-3}$ |
|---|---|---|---|---|---|---|
| 0959−54 | 9 59 51.62 | −54 52 39.1 | 1.4365681929 | 4.46386 | 3558 | 130.600 |
| 1001−47 | 10 1 23.82 | −47 32 29.0 | 0.30707169975 | 1.90717 | 3556 | 98.100 |
| 1010−23 | 10 10 10.00 | −23 30 0.0 | 2.51793 | – | 3440 | 20 |
| 1014−53 | 10 14 37.40 | −53 30 13.9 | 0.7695835272 | 0.16641 | 3559 | 66.800 |
| 1015−56 | 10 15 22.78 | −56 6 29.4 | 0.50345881546 | 0.27043 | 3558 | 439.100 |
| 1030−58 | 10 30 15.00 | −58 55 0.0 | 0.46420801 | 0.2592 | 4096 | 418.900 |
| 1039−19 | 10 39 12.00 | −19 20 0.0 | 1.386414 | – | 3415 | 60 |
| 1039−55 | 10 39 59.35 | −55 5 22.3 | 1.170859077 | 0.58173 | 3558 | 306.400 |
| 1044−57 | 10 44 19.46 | −57 58 2.4 | 0.3694266962 | 0.09884 | 3556 | 240.200 |
| 1054−62 | 10 54 28.06 | −62 42 44.7 | 0.42244618669 | 0.30845 | 3556 | 323.400 |
| 1055−52 | 10 55 48.59 | −52 10 51.9 | 0.197107608187 | 0.50401 | 3556 | 30.100 |
| 1056−78 | 10 56 27.90 | −78 58 19.7 | 1.3474021612 | 0.11465 | 3559 | 51.700 |
| 1056−57 | 10 56 55.25 | −57 26 8.5 | 1.18499740291 | 0.37048 | 3558 | 107.900 |
| 1105−59 | 11 5 51.10 | −59 30 49.0 | 1.5165309654 | 0.02938 | 3559 | 105.200 |
| 1110−65 | 11 10 36.33 | −65 56 44.9 | 0.33421283978 | 0.07119 | 3556 | 249.300 |
| 1110−69 | 11 10 54.07 | −69 10 10.9 | 0.82048420411 | 0.24546 | 3558 | 148.400 |
| 1112+50 | 11 12 48.40 | 50 46 36.0 | 1.6564380807 | 0.224 | 1536 | 9.160 |
| 1114−41 | 11 14 20.63 | −41 6 20.9 | 0.94315457524 | 0.68602 | 3565 | 41.400 |
| 1118−79 | 11 18 11.10 | −79 20 4.2 | 2.2805971963 | 0.31709 | 3558 | 27.400 |
| 1119−54 | 11 19 1.67 | −54 27 38.5 | 0.53578313115 | 0.23906 | 3557 | 204.700 |
| 1133+16 | 11 33 27.44 | 16 7 36.1 | 1.18791153608267 | 0.32251 | 1665 | 4.848 |
| 1133−55 | 11 33 38.92 | −55 8 32.2 | 0.36470327369 | 0.71093 | 3556 | 85.500 |
| 1143−60 | 11 43 41.70 | −60 14 19.4 | 0.27337242988 | 0.15501 | 3556 | 112.800 |
| 1154−62 | 11 54 43.72 | −62 8 8.3 | 0.40052094651 | 0.33955 | 3556 | 325.200 |
| 1159−58 | 11 59 54.08 | −58 3 51.3 | 0.452800508195 | 0.18373 | 3556 | 145.800 |
| 1221−63 | 12 21 34.71 | −63 51 16.3 | 0.21647484297 | 0.42809 | 3556 | 96.900 |
| 1222−63 | 12 22 54.53 | −63 52 6.7 | 0.4196177915 | 0.08173 | 3559 | 416.200 |
| 1232−55 | 12 32 31.00 | −55 0 0.0 | 0.63823381 | – | 3559 | 100 |
| 1236−68 | 12 36 57.00 | −68 15 59.6 | 1.3019081139 | 1.02747 | 3559 | 96.200 |
| 1237+25 | 12 37 12.02 | 25 10 16.9 | 1.38244861210284 | 0.0829 | 611 | 9.179 |
| 1237−41 | 12 37 33.93 | −41 8 24.0 | 0.51224209318 | 0.15025 | 3565 | 44.100 |
| 1240−64 | 12 40 20.00 | −64 6 51.0 | 0.388479 | – | 1638 | 297.4 |
| 1256−67 | 12 56 8.61 | −67 25 29.4 | 0.6633290494 | 0.1042 | 3565 | 95.000 |
| 1302−64 | 13 2 10.90 | −64 39 22.7 | 0.5716469444 | 0.34836 | 3556 | 5.800 |
| 1309−53 | 13 9 3.10 | −53 46 47.3 | 0.7281542173 | 0.01279 | 3558 | 133.200 |
| 1309−55 | 13 9 50.67 | −55 0 53.2 | 0.8492356848 | 0.493 | 3557 | 134.100 |
| 1317−53 | 13 17 49.07 | −53 43 24.0 | 0.279726553496 | 0.79997 | 3557 | 97.600 |
| 1322−66 | 13 22 33.20 | −66 45 16.0 | 0.5430085792 | 0.4587 | 3556 | 211.000 |
| 1323−63 | 13 23 9.20 | −63 53 9.3 | 0.7926701073 | 0.2681 | 3559 | 505.300 |
| 1323−58 | 13 23 44.30 | −58 43 56.4 | 0.4779896901 | 0.27743 | 3556 | 283.200 |
| 1323−62 | 13 23 57.10 | −62 7 10.0 | 0.5299062943 | 1.6321 | 3556 | 310.000 |
| 1325−43 | 13 25 9.05 | −43 42 1.8 | 0.53269880337 | 0.26041 | 3614 | 42.000 |
| 1325−49 | 13 25 31.09 | −49 6 2.7 | 1.4787213559 | 0.0527 | 3558 | 117.800 |
| 1336−64 | 13 36 27.50 | −64 41 31.4 | 0.37862206628 | 0.43642 | 3565 | 77.300 |
| 1352−51 | 13 52 44.32 | −51 39 13.9 | 0.64430109622 | 0.24304 | 3557 | 112.100 |
| 1354−62 | 13 54 10.00 | −62 13 0.0 | 0.455761 | – | 2017 | 434 |
| 1356−60 | 13 56 26.18 | −60 23 36.2 | 0.12750077685 | 0.54765 | 3556 | 295.000 |
| 1358−63 | 13 58 10.47 | −63 43 17.2 | 0.84278490076 | 1.46059 | 3556 | 98.000 |
| 1359−51 | 13 59 41.00 | −51 10 0.0 | 1.380177 | – | 3420 | 39 |
| 1417−54 | 14 17 2.74 | −54 2 38.8 | 0.9357719221 | 0.02048 | 3558 | 129.600 |

continued

Table 1, continued

| Pulsar | α (1950.0) | δ (1950.0) | P<br>s | ΔP<br>$10^{-9}$ s/d | JD − 2440000 | DM<br>pc cm$^{-3}$ |
|---|---|---|---|---|---|---|
| 1424−55 | 14 24 54.68 | −55 17 26.3 | 0.5702897883 | 0.18032 | 3556 | 82.400 |
| 1426−66 | 14 26 35.30 | −66 9 45.6 | 0.78543998083 | 0.23941 | 3556 | 65.300 |
| 1436−63 | 14 36 31.54 | −63 31 55.8 | 0.45960536385 | 0.09678 | 3556 | 124.200 |
| 1449−64 | 14 49 25.30 | −64 1 0.5 | 0.179483983915 | 0.23742 | 3556 | 70.500 |
| 1451−68 | 14 51 29.08 | −68 31 31.2 | 0.26337677868 | 0.0084 | 2554 | 8.600 |
| 1454−51 | 14 54 8.56 | −51 10 5.2 | 1.7483010801 | 0.45706 | 3564 | 37.100 |
| 1503−51 | 15 3 4.30 | −51 46 37.0 | 0.840738545 | 0.55037 | 3614 | 61.000 |
| 1503−66 | 15 3 23.16 | −66 29 26.0 | 0.35565490117 | 0.09996 | 3565 | 129.800 |
| 1504−43 | 15 4 14.01 | −43 40 33.1 | 0.28675704522 | 0.13863 | 3556 | 48.700 |
| 1507−44 | 15 7 27.30 | −44 10 47.6 | 0.9438712975 | 0.0527 | 3558 | 83.500 |
| 1508+55 | 15 8 3.74 | 55 42 55.8 | 0.73967789849 | 0.4352 | 624 | 19.599 |
| 1510−48 | 15 10 44.61 | −48 23 10.5 | 0.45483920949 | 0.07992 | 3565 | 51.500 |
| 1523−55 | 15 23 50.31 | −55 41 42.0 | 1.0487048866 | 0.97459 | 3559 | 357.600 |
| 1524−39 | 15 24 42.11 | −39 21 12.0 | 2.4175822701 | 1.64765 | 3559 | 48.500 |
| 1529+28 | 15 29 30.00 | 28 0 0.0 | 1.1248 | – | 3513 | 15 |
| 1530−53 | 15 30 22.46 | −53 24 16.7 | 1.368880509 | 0.12338 | 3559 | 24.800 |
| 1540−06 | 15 40 50.00 | − 6 15 0.0 | 0.709064 | – | 3430 | 18 |
| 1541−52 | 15 41 12.59 | −52 59 22.9 | 0.178553799688 | 0.00524 | 3556 | 35.200 |
| 1541+09 | 15 41 14.35 | 9 38 42.8 | 0.74844817748045 | 0.03718 | 2304 | 34.990 |
| 1550−54 | 15 50 5.30 | −54 47 15.0 | 1.0813282469 | 1.35216 | 3559 | 210.000 |
| 1552−31 | 15 52 10.00 | −31 25 0.0 | 0.518111 | – | 3330 | 72 |
| 1552−23 | 15 52 32.00 | −23 35 0.0 | 0.532576 | – | 3415 | 55 |
| 1555−55 | 15 55 23.35 | −55 37 9.0 | 0.9572424424 | 1.76939 | 3559 | 211.300 |
| 1556−44 | 15 56 11.05 | −44 30 16.0 | 0.2570557234 | 0.089 | 2554 | 58.800 |
| 1556−57 | 15 56 14.66 | −57 42 47.4 | 0.19445388693 | 0.18358 | 3557 | 176.900 |
| 1557−50 | 15 57 8.83 | −50 35 55.7 | 0.19259831885 | 0.4375 | 2554 | 270.000 |
| 1558−50 | 15 58 33.89 | −50 51 44.8 | 0.8642020784 | 6.01102 | 3559 | 165.000 |
| 1600−27 | 16 0 5.00 | −27 0 0.0 | 0.778311 | – | 3325 | 48 |
| 1600−49 | 16 0 42.03 | −49 1 45.7 | 0.327417287977 | 0.08741 | 3557 | 140.800 |
| 1601−52 | 16 1 25.38 | −52 49 26.0 | 0.6580131007 | 0.02212 | 3557 | 32.000 |
| 1604−00 | 16 4 37.86 | − 0 24 41.7 | 0.42181611019652 | 0.02644 | 2307 | 10.720 |
| 1609−47 | 16 9 51.08 | −47 6 49.5 | 0.38237595278 | 0.0546 | 3557 | 161.300 |
| 1612+07 | 16 12 15.00 | 7 45 0.0 | 1.206807 | – | 3415 | 21 |
| 1612−29 | 16 12 45.00 | −29 45 0.0 | 2.4776 | – | 3430 | 40 |
| 1620−42 | 16 20 18.10 | −42 49 57.0 | 0.3645903553 | 0.0883 | 3557 | 295.200 |
| 1620−08 | 16 20 34.00 | − 8 45 0.0 | 1.276438 | – | 3415 | 70 |
| 1630−59 | 16 30 48.44 | −59 48 29.1 | 0.52912125685 | 0.1182 | 3565 | 134.900 |
| 1641−45 | 16 41 10.31 | −45 53 38.7 | 0.45505432741 | 1.7384 | 2563 | 475.000 |
| 1641−68 | 16 41 40.67 | −68 26 25.6 | 1.7856112396 | 0.14688 | 3558 | 42.600 |
| 1642−03 | 16 42 24.65 | − 3 12 31.1 | 0.38768879135 | 0.15382 | 622 | 35.665 |
| 1647−52 | 16 47 43.29 | −52 50 45.0 | 0.8905339578 | 0.17885 | 3559 | 163.900 |
| 1647−52 | 16 47 46.67 | −52 17 55.8 | 0.63505597647 | 0.15656 | 3557 | 179.100 |
| 1648−42 | 16 48 17.00 | −42 40 0.0 | 0.8440794 | – | 4096 | 524.6 |
| 1648−17 | 16 48 39.00 | −17 5 0.0 | 0.973388 | – | 3330 | 31 |
| 1649−23 | 16 49 57.00 | −23 50 0.0 | 1.70374 | – | 3330 | 70 |
| 1659−60 | 16 59 47.90 | −60 12 43.0 | 0.3063230254 | 0.07828 | 3558 | 53.800 |
| 1700−32 | 17 0 7.00 | −32 40 0.0 | 1.2117846864 | 0.059 | 2005 | 103.000 |
| 1700−18 | 17 0 56.00 | −18 40 0.0 | 0.80434 | – | 3440 | 45 |
| 1701−75 | 17 1 15.70 | −75 35 23.0 | 1.1910239966 | 0.16243 | 3566 | 36.700 |
| 1702−18 | 17 2 41.00 | −18 55 0.0 | 0.298986 | – | 3325 | 23 |

continued

Table 1, continued

| Pulsar | α (1950.0) | δ (1950.0) | $P$ s | $\Delta P$ $10^{-9}$ s/d | JD − 2440000 | $DM$ pc cm$^{-3}$ |
|---|---|---|---|---|---|---|
| 1706−16 | 17 6 33.16 | −16 37 11.7 | 0.65305047326 | 0.5502 | 622 | 24.880 |
| 1707−53 | 17 7 50.22 | −53 46 38.1 | 0.8992177993 | 1.33868 | 3566 | 106.100 |
| 1717−29 | 17 17 23.00 | −29 32 0.0 | 0.6204479069 | 0.065 | 2005 | 40.000 |
| 1718−02 | 17 18 22.00 | − 2 5 0.0 | 0.477717 | – | 3325 | 70 |
| 1718−32 | 17 18 48.00 | −32 5 0.0 | 0.4771570987 | 0.059 | 2005 | 120.000 |
| 1719−37 | 17 19 35.47 | −37 9 6.0 | 0.23616867232 | 0.9345 | 3558 | 99.700 |
| 1727−47 | 17 27 56.00 | −47 42 22.0 | 0.829699077 | 15.54 | 1638 | 121.900 |
| 1729−41 | 17 29 17.69 | −41 26 42.0 | 0.6279806844 | 1.1092 | 3566 | 195.300 |
| 1730−22 | 17 30 25.00 | −22 29 0.0 | 0.8716828585 | 0.002 | 2005 | 45.000 |
| 1732−07 | 17 32 22.00 | − 7 15 0.0 | 0.419335 | – | 3325 | 72 |
| 1737+13 | 17 37 48.00 | 13 15 0.0 | 0.803049 | – | 3415 | 50 |
| 1737−39 | 17 37 49.52 | −39 26 8.6 | 0.51221006003 | 0.1563 | 3558 | 158.500 |
| 1738−08 | 17 38 39.00 | − 8 35 0.0 | 2.04309 | – | 3330 | 75 |
| 1740−03 | 17 40 30.00 | − 3 45 0.0 | 0.444643 | – | 3330 | 35 |
| 1742−30 | 17 42 42.00 | −30 43 0.0 | 0.3674214886 | 0.921 | 2005 | 86.000 |
| 1745−13 | 17 45 29.00 | −13 5 0.0 | 0.394135 | – | 3335 | 100 |
| 1745−56 | 17 45 31.51 | −56 4 25.0 | 1.3323097198 | 0.18317 | 3559 | 57.500 |
| 1747−46 | 17 47 57.01 | −46 56 39.5 | 0.74235203333 | 0.11189 | 3558 | 21.700 |
| 1749−28 | 17 49 49.23 | −28 5 50.0 | 0.5625555824 | 0.7007 | 3559 | 50.880 |
| 1754−24 | 17 54 37.00 | −24 21 0.0 | 0.234091 | – | 2200 | 188 |
| 1756−22 | 17 56 23.00 | −22 5 0.0 | 0.460969 | – | 3330 | 175 |
| 1804−27 | 18 4 2.00 | −27 15 0.0 | 0.827766 | – | 3325 | 305 |
| 1804−08 | 18 4 53.91 | − 8 48 10.5 | 0.163727360007 | 0.00248 | 3557 | 112.800 |
| 1806−53 | 18 6 39.98 | −53 38 45.7 | 0.26104931796 | 0.03309 | 3559 | 44.500 |
| 1811+40 | 18 11 50.00 | 40 20 0.0 | 0.9311 | – | 3513 | 40 |
| 1813−26 | 18 13 28.00 | −26 50 0.0 | 0.5928851483 | | 2005 | 130 |
| 1813−36 | 18 13 43.02 | −36 19 11.3 | 0.38701692095 | 0.17682 | 3559 | 94.300 |
| 1818−04 | 18 18 13.61 | − 4 29 3.3 | 0.59807263977 | 0.5461 | 622 | 84.480 |
| 1819−22 | 18 19 57.00 | −22 53 0.0 | 1.8742677952 | 0.05 | 2005 | 125.000 |
| 1820−31 | 18 20 31.00 | −31 0 0.0 | 0.284053 | – | 3325 | 52 |
| 1821−19 | 18 21 2.78 | −19 47 29.0 | 0.18933213477 | 0.45255 | 3557 | 225.700 |
| 1821+05 | 18 21 5.00 | 5 55 0.0 | 0.752906 | – | 3335 | 66 |
| 1822−09 | 18 22 46.20 | − 9 37 31.0 | 0.7689585865 | 4.52045 | 3781 | 21.900 |
| 1826−17 | 18 26 48.00 | −17 53 0.0 | 0.3071292501 | 0.483 | 2005 | 215.000 |
| 1828−60 | 18 28 44.50 | −60 25 19.0 | 1.88944355004 | 0.02333 | 3559 | 34.900 |
| 1831−03 | 18 31 4.00 | − 3 40 0.0 | 0.6866768343 | 3.582 | 2004 | 235.000 |
| 1831−04 | 18 31 47.00 | − 4 31 0.0 | 0.2901081509 | 0.009 | 2004 | 79.000 |
| 1834−10 | 18 34 8.00 | −10 0 0.0 | 0.56272 | – | 3330 | 290 |
| 1839+09 | 18 39 34.00 | 9 10 0.0 | 0.381319 | – | 3415 | 48 |
| 1842+14 | 18 42 38.00 | 14 55 0.0 | 0.375461 | – | 3335 | 44 |
| 1844−04 | 18 44 45.00 | − 4 5 32.0 | 0.5977389937 | 4.48 | 2004 | 141.900 |
| 1845−19 | 18 45 20.55 | −19 55 49.0 | 4.308179262 | 2.01398 | 3557 | 20.300 |
| 1845−01 | 18 45 49.00 | − 1 27 0.0 | 0.65942849 | 0.453 | 2005 | 163.000 |
| 1846−06 | 18 46 26.00 | − 6 43 0.0 | 1.451292705 | 3.949 | 2005 | 152.000 |
| 1851−79 | 18 51 46.80 | −79 55 51.1 | 1.2791931935 | 0.1607 | 3566 | 39.300 |
| 1851−14 | 18 51 53.00 | −14 35 0.0 | 1.14659 | – | 3420 | 130 |
| 1857−26 | 18 57 44.00 | −26 4 49.0 | 0.6122090765 | 0.014 | 2005 | 36.000 |
| 1859+03 | 18 59 1.95 | 3 26 45.6 | 0.65544511516946 | 0.64691 | 2100 | 402.000 |
| 1900+05 | 19 0 15.54 | 5 52 0.9 | 0.74656970998652 | 1.11422 | 2855 | 170.000 |
| 1900+01 | 19 0 57.96 | 1 31 9.4 | 0.72930163273662 | 0.34839 | 2346 | 228.000 |

continued

Table 1, continued

| Pulsar | α (1950.0) | δ (1950.0) | *P* s | Δ*P* $10^{-9}$ s/d | JD − 2440000 | *DM* pc cm$^{-3}$ |
|---|---|---|---|---|---|---|
| 1900−06 | 19 0 59.00 | − 6 35 0.0 | 0.4318847828 | 0.298 | 2005 | 190.000 |
| 1901+10 | 19 1 40.00 | 10 0 0.0 | 1.856568 | – | 2322 | 140 |
| 1904+12 | 19 4 57.00 | 12 40 0.0 | 0.827096 | – | 2322 | 260 |
| 1905+40 | 19 5 50.00 | 40 10 0.0 | 1.2358 | – | 3513 | 30 |
| 1906+09 | 19 6 35.38 | 9 11 21.5 | 0.83026988470746 | 0.00853 | 2833 | 250.000 |
| 1907+00 | 19 7 1.59 | 0 3 3.6 | 1.01694545765411 | 0.47651 | 2647 | 111.000 |
| 1907+02 | 19 7 7.75 | 2 49 56.3 | 0.49491386400505 | 0.23884 | 2416 | 190.000 |
| 1907+10 | 19 7 27.33 | 10 57 7.7 | 0.2836386708361 | 0.22778 | 2540 | 144.000 |
| 1907+03 | 19 7 39.00 | 3 45 0.0 | 2.33028 | – | 3335 | 95 |
| 1907−03 | 19 7 52.00 | − 3 10 0.0 | 0.504603 | – | 3325 | 205 |
| 1907+12 | 19 7 53.95 | 12 26 43.0 | 1.44173740006922 | 0.71234 | 2846 | 260.000 |
| 1910+10 | 19 10 30.00 | 10 30 0.0 | 0.409338 | – | 2322 | 140 |
| 1910+20 | 19 10 34.05 | 20 59 26.1 | 2.23296353405424 | 0.87972 | 2492 | 84.000 |
| 1911+13 | 19 11 6.41 | 13 55 42.4 | 0.52147222249774 | 0.06957 | 2827 | 145.000 |
| 1911−04 | 19 11 15.17 | − 4 45 59.7 | 0.82593368968 | 0.3509 | 624 | 89.430 |
| 1911+09 | 19 11 30.00 | 9 30 0.0 | 1.241964 | – | 2322 | 155 |
| 1911+03 | 19 11 48.00 | 3 54 0.0 | 2.3303 | – | 2340 | 35 |
| 1911+11 | 19 11 49.06 | 11 16 49.9 | 0.60099749633819 | 0.05664 | 2827 | 80.000 |
| 1913+16 | 19 13 4.44 | 16 41 50.4 | 1.61623128388884 | 0.03522 | 2840 | 55.000 |
| 1913+10 | 19 13 7.61 | 10 4 25.4 | 0.40453042033029 | 1.31709 | 2847 | 240.000 |
| 1913+16 | 19 13 12.48 | 16 1 8.4 | 0.059029995272 | 0.00076 | 2322 | 167.000 |
| 1914+09 | 19 14 9.51 | 9 46 2.7 | 0.27025289788764 | 0.21758 | 2825 | 60.000 |
| 1914+13 | 19 14 39.72 | 13 7 25.3 | 0.281840278652 | 0.31253 | 2847 | 230.000 |
| 1915+13 | 19 15 21.57 | 13 48 28.7 | 0.194626341491 | 0.62233 | 2302 | 94.000 |
| 1916+14 | 19 16 7.12 | 14 39 21.3 | 1.18088366474985 | 18.26065 | 2956 | 30.000 |
| 1917+00 | 19 17 17.22 | 0 16 3.3 | 1.27225573196516 | 0.66324 | 2426 | 85.000 |
| 1918+19 | 19 18 52.60 | 19 43 2.5 | 0.82103460382547 | 0.07735 | 2577 | 140.000 |
| 1919+14 | 19 19 6.38 | 14 13 32.2 | 0.61817973001943 | 0.48488 | 2834 | 95.000 |
| 1919+21 | 19 19 36.16 | 21 47 16.2 | 1.33730131688477 | 0.1165 | 1760 | 12.431 |
| 1919+20 | 19 19 40.00 | 20 0 0.0 | 0.760682 | – | 2322 | 70 |
| 1920+20 | 19 20 8.00 | 20 10 0.0 | 1.172761 | – | 2322 | 215 |
| 1920+21 | 19 20 43.98 | 21 4 52.1 | 1.07791915514292 | 0.7076 | 2547 | 220.000 |
| 1921+17 | 19 21 6.00 | 17 0 0.0 | 0.547209 | – | 2322 | 135 |
| 1922+20 | 19 22 30.00 | 20 30 0.0 | 0.23779 | – | 2322 | 215 |
| 1923+04 | 19 23 56.00 | 4 20 0.0 | 1.07408 | – | 3420 | 100 |
| 1924+19 | 19 24 16.00 | 19 20 0.0 | 1.34601 | – | 2314 | 420 |
| 1924+16 | 19 24 30.30 | 16 42 27.2 | 0.57981189556372 | 1.5555 | 2835 | 170.000 |
| 1924+14 | 19 24 39.58 | 14 28 48.5 | 1.32492185956581 | 0.01929 | 2824 | 205.000 |
| 1925+18 | 19 25 0.00 | 18 50 0.0 | 0.482765 | – | 2322 | 250 |
| 1925+22 | 19 25 0.10 | 22 28 49.7 | 1.43106641423616 | 0.06658 | 2831 | 180.000 |
| 1925+18 | 19 25 27.00 | 18 50 0.0 | 0.298312 | – | 2322 | 90 |
| 1927+18 | 19 27 18.00 | 18 40 0.0 | 1.220466 | – | 2322 | 110 |
| 1927+13 | 19 27 41.71 | 13 9 52.2 | 0.76003196831149 | 0.31631 | 2829 | 200.000 |
| 1929+15 | 19 29 30.00 | 15 30 0.0 | 0.314351 | – | 2314 | 120 |
| 1929+10 | 19 29 51.91 | 10 53 3.5 | 0.22651715301438 | 0.09994 | 1704 | 3.176 |
| 1929+20 | 19 29 57.16 | 20 14 17.6 | 0.26821490433979 | 0.36103 | 3029 | 200.000 |
| 1930+22 | 19 30 12.49 | 22 15 19.0 | 0.14442788985753 | 4.99217 | 2676 | 219.000 |
| 1930+13 | 19 30 58.00 | 13 0 0.0 | 0.928325 | – | 2322 | 165 |
| 1933+17 | 19 33 15.00 | 17 40 0.0 | 0.654408 | – | 2322 | 210 |
| 1933+16 | 19 33 31.86 | 16 9 58.3 | 0.35873624827023 | 0.51871 | 2265 | 158.530 |

continued

Table 1, continued

| Pulsar | α (1950,0) | δ (1950,0) | P<br>s | ΔP<br>$10^{-9}$ s/d | JD − 2440000 | DM<br>pc cm$^{-3}$ |
|---|---|---|---|---|---|---|
| 1933+15 | 19 33 44.78 | 15 29 52.8 | 0.96733836953229 | 0.34894 | 2831 | 165.000 |
| 1937−26 | 19 37 58.00 | −26 10 0.0 | 0.402857 | – | 3420 | 50 |
| 1939+17 | 19 39 47.00 | 17 40 0.0 | 0.696261 | – | 2322 | 175 |
| 1940−12 | 19 40 37.00 | −12 45 0.0 | 0.972427 | – | 3325 | 29 |
| 1941−17 | 19 41 13.00 | −17 45 0.0 | 0.84116 | – | 3330 | 56 |
| 1942+17 | 19 42 15.00 | 17 50 0.0 | 1.996898 | – | 2322 | 160 |
| 1942−00 | 19 42 54.00 | − 0 55 0.0 | 1.045621 | – | 3330 | 55 |
| 1943+18 | 19 43 18.00 | 18 30 0.0 | 1.068707 | – | 2322 | 240 |
| 1943−29 | 19 43 44.00 | −29 5 0.0 | 0.95942 | – | 3420 | 30 |
| 1944+22 | 19 44 16.16 | 22 37 34.8 | 1.33444985909 | 0.07678 | 2855 | 140.000 |
| 1944+17 | 19 44 38.75 | 17 58 15.4 | 0.44061846173304 | 0.00208 | 1501 | 16.300 |
| 1946−25 | 19 46 24.00 | −25 45 0.0 | 0.957615 | – | 3420 | 20 |
| 1946+35 | 19 46 33.95 | 35 32 38.3 | 0.71730676525717 | 0.6093 | 2221 | 129.100 |
| 1952+29 | 19 52 21.85 | 29 15 22.2 | 0.42667678559459 | 0.00017 | 2434 | 7.914 |
| 1954+51 | 19 54 10.00 | 51 10 0.0 | 0.5189 | – | 3513 | 25 |
| 2002+31 | 20 2 53.70 | 31 28 34.5 | 2.11121695230296 | 6.4429 | 2304 | 233.000 |
| 2003−08 | 20 3 34.00 | − 8 0 0.0 | 0.580866 | – | 3325 | 25 |
| 2016+28 | 20 16 0.17 | 28 30 30.2 | 0.55795342110241 | 0.01292 | 1761 | 14.176 |
| 2020+28 | 20 20 33.29 | 28 44 43.2 | 0.34340081651279 | 0.16385 | 1501 | 24.620 |
| 2021+51 | 20 21 25.30 | 51 45 7.9 | 0.52919532808 | 0.2635 | 625 | 22.580 |
| 2024+21 | 20 24 55.00 | 21 40 0.0 | 0.3981726 | – | 2322 | 90 |
| 2028+22 | 20 28 28.54 | 22 18 13.2 | 0.63051210272684 | 0.07636 | 2841 | 60.000 |
| 2043−04 | 20 43 22.00 | − 4 30 0.0 | 1.546937 | – | 3330 | 38 |
| 2044+15 | 20 44 20.00 | 15 45 0.0 | 1.138286 | – | 3330 | 35 |
| 2045−16 | 20 45 45.00 | −16 25 46.0 | 1.961569594 | 0.96768 | 3565 | 11.510 |
| 2048−72 | 20 48 41.43 | −72 12 2.6 | 0.34133616161 | 0.01693 | 3558 | 18.100 |
| 2106+44 | 21 6 30.00 | 44 30 0.0 | 0.4148704991 | 0.005 | 2694 | 129.000 |
| 2111+46 | 21 11 37.77 | 46 31 42.4 | 1.01468444504 | 0.0619 | 1006 | 141.500 |
| 2113+14 | 21 13 50.00 | 14 5 0.0 | 0.440152 | – | 3415 | 54 |
| 2123−67 | 21 23 19.62 | −67 1 30.8 | 0.32577128742 | 0.01953 | 3558 | 34.700 |
| 2148+63 | 21 48 40.00 | 63 15 0.0 | 0.38014003094 | 0.014 | 2006 | 125.000 |
| 2151−56 | 21 51 34.00 | −56 56 10.0 | 1.373654387 | 0.36547 | 3558 | 14.200 |
| 2152−31 | 21 52 18.43 | −31 33 8.6 | 1.0300015591 | 0.1067 | 3558 | 16.200 |
| 2154+40 | 21 54 57.21 | 40 3 26.1 | 1.5252634548 | 0.283 | 2006 | 71.000 |
| 2217+47 | 22 17 45.89 | 47 39 48.1 | 0.538467394565 | 0.2388 | 624 | 43.540 |
| 2223+65 | 22 23 30.00 | 65 22 0.0 | 0.6825333245 | 0.825 | 2006 | 88.000 |
| 2255+58 | 22 55 46.00 | 58 54 30.0 | 0.3682433246 | 0.497 | 2006 | 148.000 |
| 2303+30 | 23 3 34.10 | 30 43 48.6 | 1.57588474427198 | 0.25019 | 2341 | 49.900 |
| 2305+55 | 23 5 0.00 | 55 26 0.0 | 0.4750674553 | 0.006 | 3513 | 45.000 |
| 2310+42 | 23 10 30.00 | 42 40 0.0 | 0.3494 | – | 3513 | 15 |
| 2315+21 | 23 15 30.00 | 21 20 0.0 | 1.4447 | – | 3513 | 20 |
| 2319+60 | 23 19 41.40 | 60 8 2.3 | 2.2564837049 | 0.588 | 1536 | 96.000 |
| 2321−61 | 23 21 33.70 | −61 10 33.0 | 2.347485196 | 0.22464 | 3558 | 15.500 |
| 2323+63 | 23 23 10.00 | 63 0 0.0 | 1.4368 | – | 3513 | 120 |
| 2324+60 | 23 24 25.00 | 60 56 30.0 | 0.2336517104 | 0.031 | 2006 | 120.000 |
| 2327−20 | 23 27 49.72 | −20 22 4.1 | 1.643619656 | 0.40038 | 3557 | 8.300 |

## 5.6.2.4 Distance estimates

In one case, PSR 1929+10, it has been possible to determine a parallax. Salter et al. [47] have given the value of 0.″0215±0.″0080 for this pulsar which places it at a distance of 47 pc. The same authors report statistically less significant results for several other pulsars.

In a few cases distance estimates can be derived from pulsar – supernova remnant associations. The two best studied cases are the Crab pulsar (PSR 0531+21) for which Wyckoff and Murray [48] have given a distance of 2000 pc, and the Vela pulsar (PSR 0833−45) for which Milne [49] has given a distance of 500 pc.

A third method to derive pulsar distances relies on the observation of neutral hydrogen 21-cm absorption line profiles against pulsars and their interpretation via a rotation model of our Galaxy. This method has been applied by several groups and yielded the results listed in Table 2.

Table 2. Pulsar distance *d* from 21-cm absorption.

| PSR | *d* [kpc] | Ref. | PSR | *d* [kpc] | Ref. | PSR | *d* [kpc] | Ref. | PSR | *d* [kpc] | Ref. |
|---|---|---|---|---|---|---|---|---|---|---|---|
| 0138+59 | ≈3 | 52 | 0833−45 | 0.5 | 51 | 1818−04 | <1.5 | 50, 51 | 2020+28 | >2 | 51, 52 |
| 0329+54 | 1.3···2.0 | 50 | 0835−41 | 2.4 ··· 5.0 | 53 | 1822−09 | <1.5 | 51 | 2021+51 | <1 | 50, 51 |
| | 1 ···2 | 51 | 1154−62 | 10.5 ···12.5 | 55 | 1826−17 | >1.5 | 52 | 2111+45 | 4 ···6 | 52 |
| | 2.3···2.9 | 54 | 1240−64 | 12 ···16 | 55 | 1859+03 | 19···21 | 55 | 2319+60 | >2.5 | 51, 52 |
| 0355+54 | 1.5···2.5 | 51 | 1323−62 | 6.5 ··· 9.5 | 55 | | 6···18 | 56 | | 2.8···3.8 | 54 |
| | 1 ···2 | 52 | 1557−50 | 8 ···10 | 55 | 1900+01 | 3···(5) | 56 | | | |
| 0525+21 | ≈2 | 50, 51 | 1641−45 | 4.5 ··· 5.3 | 55 | 1929+10 | <1 | 50,51 | | | |
| 0531+21 | ≈2 | 51 | 1642−03 | 0.15··· 0.17 | 52 | 1933+16 | >6 | 51,54 | | | |
| 0611+22 | ≈2 | 51 | 1718−32 | >1 | 51 | 1946+35 | (>8.5) | 52 | | | |
| 0736−40 | 1.5 | 50 | | ≈0.2 | 53 | 2002+31 | 8···13 | 56 | | | |
| | 1.7···2.3 | 55 | 1749−28 | <1.5 | 50 | 2016+28 | <0.8 | 50 | | | |
| 0740−28 | 1.5···2.5 | 50, 51 | | <1 | 51 | | <1 | 51 | | | |
| 0809+74 | ≈0.2 | 53 | | | | | | | | | |

A fourth method to estimate pulsar distances uses the observed dispersion measures. If the density of electrons in interstellar space were constant, one could immediately derive a distance from

$$d = DM/\langle N_e \rangle \,.$$

This method has indeed been applied many times using, e.g. $\langle N_e \rangle = 0.03\ \mathrm{cm}^{-3}$. Obviously, this is an oversimplification which will in general tend to overestimate the actual distances for low-latitude pulsars due to intervening H II-regions along the line of sight with higher than average electron density (e.g. [57, 58]), whereas for distant high-latitude pulsars the true distances will be underestimated due to scale height effects. For a recent discussion of this problem, see [59, 60, 61, 62] and further references quoted therein.

## 5.6.2.5 Proper motions and space velocities

In Table 3 we have compiled some of the proper motion results on radiopulsars from the literature. Only the more significant results have been listed here whereas many more measurements can be found in the references quoted. Also listed is a distance estimate which is based on the dispersion measure except for two cases (PSR 0833−45 and PSR 1929+10) where the distance is taken from the pulsar-supernova remnant association and from the parallax measurement, respectively. With the distances *d* [pc] being known, the proper motions $\mu$ [″/a] can be converted into tangential velocities by $v_t = 4.74 \cdot \mu \cdot d$ [km s$^{-1}$]. These values are also given. They prove that at least some of the radiopulsars have large space velocities which may indicate their escape from binary systems.

Table 3. Radiopulsar proper motion $\mu$, distance $d$ and tangential velocity $v_t$.

| PSR | $\mu$ 0″001 $a^{-1}$ | Ref. | $d$ pc | $v_t$ km $s^{-1}$ | PSR | $\mu$ 0″001 $a^{-1}$ | Ref. | $d$ pc | $v_t$ km $s^{-1}$ |
|---|---|---|---|---|---|---|---|---|---|
| 0329+54 | 14 ± 4 | 63 | 893 | 59 | 1133+16 | 365 ± 36 | 63 | 160 | 277 |
| 0450−18 | 1400 700 | 68 | 1317 | 8740 | 1237+25 | 102 18 | 63 | 310 | 150 |
| 0531+21 | 11 2 | 64 | 1893 | 99 | 1508+55 | 129 32 | 65 | 653 | 400 |
| 0809+74 | 260 130 | 68 | 191 | 236 | 1604−00 | 55 26 | 67 | 357 | 93 |
| 0823+26 | 135 10 | 63 | 650 | 416 | 1929+10 | 164 16 | 67 | 47 | 37 |
| 0833−45 | 15 5 | 66 | 500 | 40 | 1952+29 | 180 98 | 67 | 333 | 221 |
| 0834+06 | 52 14 | 63 | 430 | 106 | 2016+28 | 20 8 | 63 | 473 | 45 |
| 0950+08 | 448 135 | 68 | 100 | 210 | | | | | |

## 5.6.2.6 Statistical properties

### 5.6.2.6.1 Galactic distribution

In Fig. 1 the pulsar positions are shown in galactic coordinates for 224 objects, detected in the second Molonglo survey [5]. This is the largest homogeneous set of data which at the same time is not restricted to the immediate vicinity of the galactic plane but covers essentially the sky south of $\delta = +20°$. Fig. 1 illustrates a concentration of the pulsars towards the galactic plane, showing that these sources are located within the Galaxy.

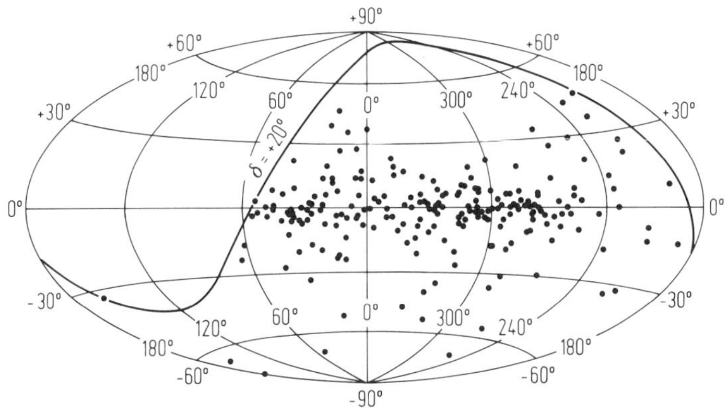


Fig. 1. Equal-area plot in galactic coordinates of the 224 pulsars with $\delta < +20°$ detected in the second Molonglo pulsar survey [5].

In Fig. 2 the distribution of pulsars as a function of their distance from the galactic plane, $z$, is shown. This result is based on the crude assumption that the distances of individual pulsars, $d$, can be calculated from the dispersion measure by assuming a constant value for the interstellar electron density (see above). Whereas any particular distance derived in this way can be wrong by a factor of two or more, the statistical result given in Fig. 2 should not be seriously affected. It clearly shows that the $z$-distribution of pulsars has a half-width of about 300 pc. This value is large compared to that of the progenitor stars which must be massive early-type stars. The wider $z$-distribution can at least partly be accounted for by the large space velocities observed for some radiopulsars (see above) which, combined with typical pulsar ages of $10^6 \cdots 10^7$ years (see below) can produce $z$-values in the order of several hundred pc.

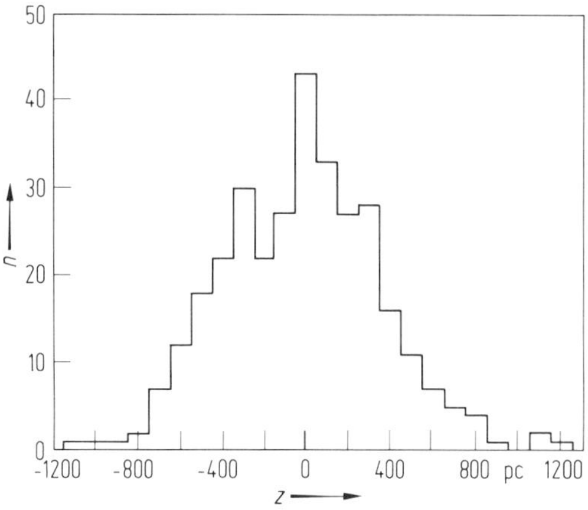


Fig. 2. The distribution of pulsars as a function of their distance from the galactic plane.
$z$ = distance from the galactic plane, $n$ = number of pulsars

### 5.6.2.6.2 Period distribution

Fig. 3 shows the distribution of pulsar periods. The shortest period is that of the Crab pulsar (PSR 0531 + 21, $P = 0.033$ s), the longest period that of PSR 1845 − 19 ($P = 4.308$ s). The overall distribution is thought to be intrinsic, i.e. it should almost be free from selection effects.

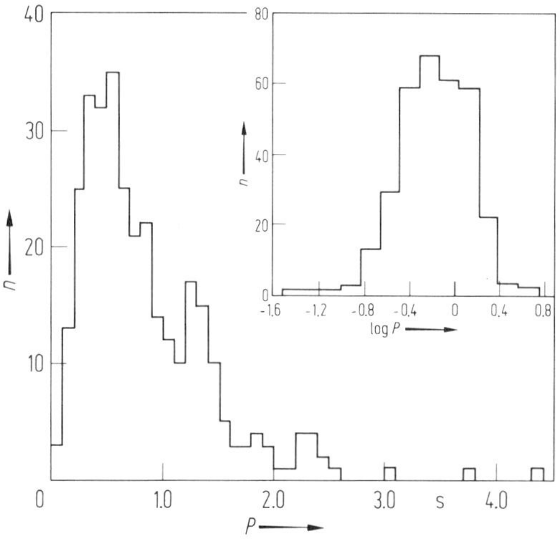


Fig. 3. The period distribution for 321 radiopulsars.
$n$ = number of pulsars

### 5.6.2.6.3 Characteristic pulsar ages

The fact that, in all cases where accurate observations have been made, the pulsar periods have been found to increase with time in a regular way offers the possibility of calculating characteristic pulsar ages. The observed increase of the periods $P$ is most readily attributed to the loss of rotational energy and angular momentum via ejected particles and/or electromagnetic radiation at the rotation frequency. In most models the braking is proportional to the power n of the rotational frequency

$$\dot{\Omega} = -K\Omega^{\mathrm{n}} \quad \text{or} \quad \dot{P} = (2\pi)^{\mathrm{n}-1} K P^{2-\mathrm{n}} . \tag{3}$$

Integration of (3) leads to

$$t = -\frac{\Omega}{(\mathrm{n}-1)\dot{\Omega}} (1 - (\Omega/\Omega_{\mathrm{i}})^{\mathrm{n}-1}) \qquad (\mathrm{n} \neq 1), \tag{4}$$

where $\Omega_i$ is the pulsation frequency at time $t = 0$, and $\Omega$ and $\dot{\Omega}$ are the current frequency and its derivative, respectively. For $\Omega_i \gg \Omega$ and for n = 3, i.e. its value for magnetic dipole braking, Eq. (4) gives the so called characteristic pulsar age

$$\tau = -0.5\Omega/\dot{\Omega} = 0.5P/\dot{P} . \tag{5}$$

The distribution of these values is shown in Fig. 4, illustrating the fact that the age of most of the known radio pulsars is in the range of $10^6 \cdots 10^7$ years. In a few cases, the pulsar age can be determined independently, as in the case of the Crab pulsar which has in 1982 an age of 928 years if one assumes that it was formed during the supernova explosion that was observed in the year 1054.

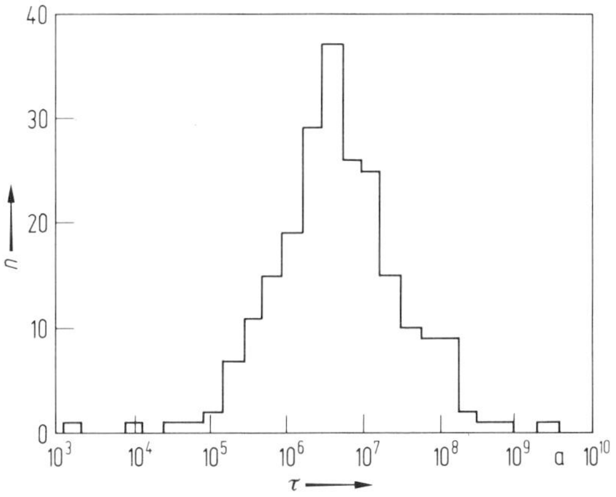


Fig. 4. The distribution of pulsar ages estimated from the rate of change of the pulse periods.
$n$ = number of pulsars

### 5.6.2.6.4 Characteristic surface magnetic fields

If one makes the simplifying assumption that the observed loss of rotational energy of pulsars is caused by magnetic dipole radiation at the rotational frequency, one can use the relation

$$L = -I\Omega\dot{\Omega} = (2/3c^3)\cdot(D\sin\alpha)^2\Omega^4, \tag{6}$$

where $I$ is the moment of inertia of the neutron star and $D$ its magnetic moment with the angle $\alpha$ between the rotational and magnetic axes. Replacing $D$ by $D = B_0R^3$, where $B_0$ is the surface magnetic field and $R$ the radius of the neutron star, one can solve the above equation for

$$\frac{(B_0\sin\alpha)\cdot R^3}{I^{1/2}} = -(3c^3/2)^{1/2}\cdot(\dot{\Omega}/\Omega^3)^{1/2} = (3c^3/8\pi^2)^{1/2}\cdot(P\dot{P})^{1/2}. \tag{7}$$

From neutron star model calculations (see 5.6.1) it is known that the moment of inertia for neutron stars is of the order of $10^{45}$ g cm$^2$ and that the radius is a few $10^6$ cm. Normalizing to such values we obtain the simple relation

$$\frac{(B_{12}\sin\alpha)\cdot R_6^3}{(I_{45})^{1/2}} = 1.01\ (P\dot{P}_{-15})^{1/2} \tag{8}$$

with $B_{12} = B_0/10^{12}$ Gauss, $I_{45} = I/10^{45}$ g cm$^2$, $\dot{P}_{-15} = \dot{P}/10^{-15}$ s s$^{-1}$, and $R_6 = R/10^6$ cm. The values calculated from this relation are plotted in Fig. 5. This figure illustrates the fact that the surface fields of the magnetized neutron stars that give rise to the radiopulsar phenomenon must be of the order of $10^{11}\cdots10^{13}$ Gauss.

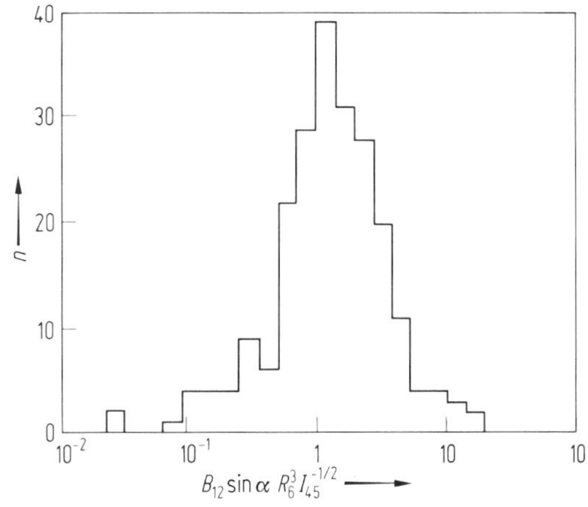


Fig. 5. The distribution of the estimated surface magnetic field strength of radiopulsars. For explanation of symbols, see text.

### 5.6.2.7 Radiopulsars in binary systems

During a systematic search for new pulsars [8, 9], Taylor et al. [69] discovered the first radiopulsar that is a member in a binary system: PSR 1913+16. More recently, a second pulsar was shown to have a companion star: PSR 0820+02. However, as pointed out by Manchester et al. [70], the characteristics of this second system are completely different from those of PSR 1913+16. In Table 4 we have compiled some of the system parameters. For a more detailed discussion of PSR 1913+16 and various implications of this discovery, see [3] and further references given therein.

Table 4. Radiopulsars in binary systems.
$a_1$ = semi-axis of primary component

| PSR | Pulsar period [s] | Orbital period [d] | $a_1 \sin i$ [km] | Mass function |
|---|---|---|---|---|
| 0820+02 | $0.864879 \pm 3 \cdot 10^{-6}$ | $1710 \pm 160$ | $(1.2 \pm 0.3) \cdot 10^8$ | $0.023\,\mathfrak{M}_\odot$ |
| 1913+16 | $0.059029995272 \pm 5 \cdot 10^{-12}$ | $0.322997537 \pm 2 \cdot 10^{-8}$ | $(7.0043 \pm 0.0004) \cdot 10^5$ | $0.13126\,\mathfrak{M}_\odot$ |

### 5.6.2.8 Theoretical models of the magnetosphere of radiopulsars

The possibility that neutron stars might possess very strong magnetic fields, which – as a consequence of the expected rapid rotation of these stars – could lead to the emission of intense low-frequency radiation, was first pointed out by Pacini [71] in 1967. Immediately after the discovery of the Crab pulsar and the observation that it was slowing down, Gold [72] drew attention to the model of a spinning magnetized neutron star and its potential to explain the pulsed nature of the emission from radio pulsars as being caused by a lighthouse effect. This lighthouse effect could result e.g. from a dipole magnetic field. Ostriker and Gunn [73] considered the oblique rotator model quantitatively and demonstrated that a strong magnetic dipole rotating in vacuum could be a very efficient particle accelerator (see also [74]). Goldreich and Julian [75] were the first to consider in more detail the consequences of such strong magnetic and, because of the rapid rotation, also electric fields. They showed that the surrounding region of the star cannot be a vacuum but must contain a substantial space charge. Their calculation is based on the simplifying assumption of an axisymmetric rotator which cannot account for the pulsed nature of the emission. Since then many attempts have been made to calculate more realistic models. For a recent discussion and also for further references, see [3].

### 5.6.2.9 References for 5.6.2


1 Hewish, A., Bell, S.J., Pilkington, J.D.H., Scott, P.F., Collins, R.A.: Nature **217** (1968) 709.
2 Smith, F.G.: Pulsars, Cambridge Univ. Press (1977).
3 Manchester, R.N., Taylor, J.H.: Pulsars, W. H. Freeman and Comp., San Francisco (1977).
4 Large, M.I., Vaughan, A.E.: Mon. Not. R. Astron. Soc. **151** (1971) 277.
5 Manchester, R.N., Lyne, A.G., Taylor, J.H., Durdin, J.M., Large, M.I., Little, A.G.: Mon. Not. R. Astron. Soc. **185** (1978) 409.
6 Davies, J.G., Lyne, A.G., Seiradakis, J.H.: Nature **240** (1972) 229.
7 Davies, J.G., Lyne, A.G., Seiradakis, J.H.: Nature Phys. Sci. **244** (1973) 84.
8 Hulse, R.A., Taylor, J.H.: Astrophys. J. (Lett.) **191** (1974) L59.
9 Hulse, R.A., Taylor, J.H.: Astrophys. J. (Lett.) **201** (1975) L55.
10 Damashek, M., Taylor, J.H., Hulse, R.A.: Astrophys. J. (Lett.) **225** (1978) L31.
11 Lyne, A.G., Rickett, B.J.: Nature **219** (1968) 1339.
12 Goldstein, S.J., jr., James, J.T.: Astrophys. J. (Lett.) **158** (1969) L179.
13 Huguenin, G.R., Taylor, J.H.: Int. Astron. Union Circ. No. 2128 (1969).
14 Davies, J.G., Large, M.I.: Mon. Not. R. Astron. Soc. **149** (1970) 301.
15 Davies, J.G., Large, M.I., Pickwick, A.C.: Nature **227** (1970) 1123.
16 Komesaroff, M.M., Ables, J.G., Morris, D., Cooke, D.J., Schwartz, U.J., Hamilton, P.A.: Int. Astron. Union Circ. No. 2201 (1970).
17 Vaughan, A.E., Large, M.I.: Nature **225** (1970) 167.
18 Vitkevich, V.V., Shitov, Y.P.: Nature **225** (1970) 248.
19 Vitkevich, V.V., Shitov, Y.P.: Nature **226** (1970) 1235.
20 Hunt, G.C.: Mon. Not. R. Astron. Soc. **153** (1971) 119.
21 Krishnan Mohan, S., Balasubramanian, V., Swarup, G.: Nature Phys. Sci. **234** (1971) 151.

22 Manchester, R.N.: Astrophys. J. Suppl. **23** (1971) 283.
23 McNamara, B.J.: Publ. Astron. Soc. Pacific **83** (1971) 491.
24 Salter, C.F., Facondi, S.R.: Mon. Not. R. Astron. Soc. **152** (1971) 5P.
25 Swarup, G., Mohandy, D.K., Balasubramanian,V.: Int. Astron. Union Circ. No. 2356 (1971).
26 Komesaroff, M.M., Hamilton, P.A., Ables, J.G.: Australian J. Phys. **25** (1972) 759.
27 Manchester, R.N.: Astrophys. J. **172** (1972) 43.
28 Manchester, R.N., Peters, W.L.: Astrophys. J. **173** (1972) 224.
29 Manchester, R.N., Taylor, J.H.: Astrophys. Lett. **10** (1972) 67.
30 Manchester, R.N., Taylor, J.H., Huguenin, G.R.: Nature Phys. Sci. **240** (1972) 74.
31 Graham, D.A., Hunt, G.C.: Nature Phys. Sci. **242** (1973) 86.
32 Komesaroff, M.M., Ables, J.G., Cooke, D.J.: Astrophys. Lett. **15** (1973) 169.
33 Komesaroff, M.M., Hamilton, P.A., McCulloch, P.M., Ables, J.G., Cooke, D.J.: Int. Astron. Union Circ. No. 2505 (1973).
34 Komesaroff, M.M., Hamilton, P.A., McCulloch, P.M., Ables, J.G., Cooke, D.J.: Int. Astron. Union Circ. No. 2563 (1973).
35 McCulloch, P.M., Komesaroff, M.M., Ables, J.G., Hamilton, P.A., Rankin, J.M.: Astrophys. Lett. **14** (1973) 169.
36 Vaughan, A.E., Disney, M.J., Nicholson, P.: Mon. Not. R. Astron. Soc. **168** (1974) 361.
37 Lyne, A.G., Ritchings, R.T., Smith, F.G.: Mon. Not. R. Astron. Soc. **171** (1975) 579.
38 Mohandy, D.K., Balasubramanian, V.: Mon. Not. R. Astron. Soc. **171** (1975) 17P.
39 Taylor, J.H., Manchester, R.N.: Astron. J. **80** (1975) 794.
40 Manchester, R.N., Goss, W.M., Hamilton, P.A.: Nature **259** (1976) 291.
41 Manchester, R.N., Goss, W.M., Newton, L.M., Hamilton, P.A.: Proc. Astron. Soc. Australia **3** (1976) 81.
42 Helfand, D.J.: Ph. D. Thesis, FCRAO, Massachusetts, USA (1977).
43 Manchester, R.N., Newton, L.M., Goss, W.M., Hamilton, R.A.: Mon. Not. R. Astron. Soc. **184** (1978) 35P.
44 Smith, F.G., Disney, M.J., Hartley, K.F., Jones, D.H.P., King, D.J., Wellgate, G.B., Manchester, R.N., Lyne, A.G., Goss, W.M., Wallace, P.T., Peterson, B.A., Murdin, P.G., Danziger, I.J.: Mon. Not. R. Astron. Soc. **184** (1978) 39P.
45 Gullahorn, G.E., Rankin, J.M.: Astron. J. **83** (1978) 1219.
46 Newton, L.M., Manchester, R.N., Cooke, D.J.: Mon. Not. R. Astron. Soc. **194** (1981) 841.
47 Salter, M.J., Lyne, A.G., Anderson, B.: Nature **280** (1979) 477.
48 Wyckoff, S., Murray, C.A.: Mon. Not. R. Astron. Soc. **180** (1977) 717.
49 Milne, D.K.: Australian J. Phys. **23** (1970) 425.
50 Gordon, K.J., Gordon, C.P.: Astron. Astrophys. **27** (1973) 119.
51 Gómez-Gonzàlez, J., Guélin, M.: Astron. Astrophys. **32** (1974) 441.
52 Graham, D.A., Mebold, U., Hesse, K.H., Hills, D.L., Wielebinski, R.: Astron. Astrophys. **37** (1974) 405.
53 Gordon, K.J., Gordon, C.P.: Astron. Astrophys. **40** (1975) 27.
54 Booth, R.S., Lyne, A.G.: Mon. Not. R. Astron. Soc. **174** (1976) 53P.
55 Ables, J.G., Manchester, R.N.: Astron. Astrophys. **50** (1976) 177.
56 Weisberg, J.M., Boriakoff, V., Rankin, J.: Astron. Astrophys. **77** (1979) 204.
57 Prentice, A.J.R., ter Haar, D.: Mon. Not. R. Astron. Soc. **146** (1969) 423.
58 Grewing, M., Walmsley, C.M.: Astron. Astrophys. **11** (1971) 65.
59 Roberts, D.H.: Astrophys. J. (Lett.) **205** (1976) L29.
60 Taylor, J.H., Manchester, R.N.: Astrophys. J. **215** (1977) 885.
61 Grewing, M., Schrüfer, E.: Mitt. Astron. Ges. **45** (1979) 27.
62 Hall, A.N.: Mon. Not. R. Astron. Soc. **191** (1980) 751.
63 Anderson, B., Lyne, A.G., Peckham, R.J.: Nature **258** (1975) 215.
64 Wyckoff, S., Murray, C.A.: Mon. Not. R. Astron. Soc. **180** (1977) 717.
65 Helfand, D.J., Taylor, J.H., Manchester, R.N.: Astrophys. J. (Lett.) **213** (1977) L1.
66 Helfand, D.J., Tademaru, E.: Astrophys. J. **216** (1977) 842.
67 Gullahorn, G.E., Rankin, J.M.: Astrophys. J. **225** (1978) 963.
68 Helfand, D.J., Taylor, J.H., Backus, P.R., Cordes, J.M.: Astrophys. J. **237** (1980) 206.
69 Taylor, J.H., Hulse, R.A., Fowler, I.A., Gullahorn, G.E., Rankin, J.M.: Astrophys. J. (Lett.) **206** (1976) L53.
70 Manchester, R.N., Newton, L.M., Cooke, D.J.: Astrophys. J. (Lett.) **236** (1980) L25.
71 Pacini, F.: Nature **216** (1967) 567.
72 Gold, T.: Nature **218** (1968) 731.
73 Ostriker, J.P., Gunn, J.E.: Astrophys. J. **157** (1969) 1395.
74 Grewing, M., Schrüfer, E., Heintzmann, H.: Z. Physik **260** (1973) 375.
75 Goldreich, P., Julian, W.H.: Astrophys. J. **157** (1969) 869.

# 5.6.3 Pulsating X-ray sources

## 5.6.3.1 General properties

Up to the end of 1980, more than 500 galactic X-ray sources had been detected (see 5.7.1.2). The majority of them are probably compact stars (white dwarfs, neutron stars, black holes) in close binary systems accreting matter from their normal companion (see also 6.1.4). Some of these binary X-ray sources show periodic pulsations. Since the first discovery (Cen X-1 [16]), about 17 of these sources have been detected. Table 1 summarizes their basic properties and those of the corresponding binary systems. In general, the pulsating X-ray sources show the following observational characteristics:

Pulsational periods ranging from 0.7 s to 835 s.
Most sources show a long-term decrease of pulse periods (spin-up) (see Fig. 1), superimposed by short-term period changes in both directions.
A rather large duty cycle (20 %···50 %).
A fractional pulsed flux in the range from 25 %···100 %.
A variety of pulse profiles from simple sinusoidal shapes to complex patterns (see Fig. 2). In general, the pulse profiles become simpler with increasing photon energy. There is no systematic dependence of pulse shape on pulse period.
There is evidence for pulse-to-pulse variations of the pulse shape (Her X-1 [64]; Vela X-1 [65]).
Generally the sources have rather hard (flat) power law spectra with spectral breaks at 15 keV···30 keV, above which the intensity falls off steeply.

Table 1, see p. 25.

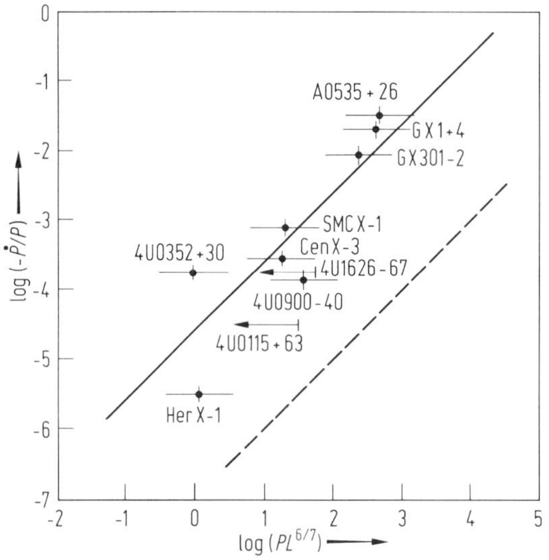


Fig. 2: see p. 26.

Fig. 1. Empirical relation between the fractional rate of change of pulse period in $[a^{-1}]$ and the product of pulse period $P$ in [s] and the (6/7)th power of the luminosity $L$ in $[10^{37}\,\mathrm{erg\,s^{-1}}]$ for ten pulsating X-ray sources [21]. The solid line is the best power law fit with a logarithmic slope of 1.0; the intercept at $\log(\dot{P}/P) = -4.6$ agrees well with the expected value if the X-ray source is an accreting neutron star [7]. The dashed line is that expected for a 1 $\mathfrak{M}_\odot$ white dwarf.

Detailed investigations, in particular of the four best measured systems (SMC X-1, Her X-1, Cen X-3, Vela X-1), have led to the following qualitative understanding of the basic properties [66···76]: The X-ray luminosity is derived from matter accretion onto a compact, highly magnetized neutron star. The matter is transferred from the companion star either in the form of a stellar wind or by Roche lobe overflow. In the latter case the transferred matter, when entering the potential lobe of the compact object, forms an accretion disk which extends down to the magnetospheric boundary of the neutron star. By processes which are still unknown in detail the plasma gets into the magnetosphere and is channeled along the field lines down to the stellar surface, thus forming accretion columns above the magnetic poles of the neutron star. The kinetic energy of the incident material is converted into thermal energy at or near the surface of the neutron star; this leads to the formation of a hot plasma region which is cooled by X-ray emission. The X-ray beam pattern is determined by the radiation transfer in the strongly magnetized plasma. Depending on the detailed conditions fan beams, pencil beams or more complicated patterns may be expected. The observed pulses are a consequence of stellar rotation and an inclination between magnetic dipole and rotation axis.

Table 1. Basic properties of the pulsating X-ray sources.

| Pulsating X-ray source | | Pulse period | | Period change | | Orbital period | | X-ray luminosity *) | | Optical counterpart | | | | Remarks |
|---|---|---|---|---|---|---|---|---|---|---|---|---|---|---|
| Catalogue name **) | Synonym | $P$ [s] | Ref. | $-\frac{\dot{P}}{P}$ [a$^{-1}$] | Ref. | $P$ [d] | Ref. | $L_X$ [erg s$^{-1}$] | Ref. | Name | Ref. | Spectral class | Ref. | |
| 4U0115−73 | SMC X-1 | 0.71 | 1 | $5\cdot10^{-4}$ | 2 | 3.89 | 3 | $1\cdot10^{39}$ | 4 | Sk 160 | 5 | B0 I | 5 | |
| 4U1656+35 | Her X-1 | 1.24 | 6 | $3\cdot10^{-6}$ | 7 | 1.70 | 8 | $1\cdot10^{37}$ | 9 | HZ Her | 10 | A8···F0 | 11 | |
| 4U0115+63 | | 3.6 | 12 | $3\cdot10^{-5}$ | 13 | 24.3 | 14 | $3\cdot10^{37}$ | 13 | | | B3 ? | 15 | Transient |
| 4U1118−60 | Cen X-3 | 4.84 | 16 | $1\cdot10^{-4}$ | 2 | 2.09 | 17 | $3\cdot10^{37}$ | 18 | Krzeminski's star | 19 | O6.5 III | 18 | |
| 4U1626−67 | | 7.68 | 20 | $2\cdot10^{-4}$ | 21 | 0.03 | 22 | $3\cdot10^{37}\cdot d_{10}^2$ | 23 | Late-type dwarf | 22 | | | $\mathfrak{M}_X < 0.1\mathfrak{M}_\odot$ [22] |
| 2S1417−624 | | 17.64 (July 1978) | 24 | $2\cdot10^{-2}$ ? | 24 | > 10 | 24 | $1\cdot10^{37}\cdot d_{10}^2$ | 25 | | | | | |
| OAO1653−40 | | 38.2 | 26 | $5\cdot10^{-3}$ | 27 | | | $1\cdot10^{37}\cdot d_{10}^2$ | 26 | | | | | |
| A0535−26 | | 104 | 28 | $\approx10^{-2}$ | 29 | 17···77 | 30 | $6\cdot10^{37}$ | 7 | HDE 245770 | 31 | Bpe | 32 | Transient |
| 4U1728−24 | GX 1+4 | 135 (Oct. '70)<br>114 (Nov. '78) | 33<br>34 | $2\cdot10^{-2}$ | 62 | | | $5\cdot10^{37}\cdot d_{10}^2$ | 36 | | | M6 III | 37 | True pulse period possibly twice the given value [34, 35] |
| 4U1258−61 | GX 304−1 | 272 | 38 | | | > 15 | 38 | | | MMV star | 39 | B2···A0 | 39 | |
| 4U0900−40 | Vela X-1 | 283 | 40 | $1\cdot10^{-4}$ | 7 | 8.97 | 41 | $2\cdot10^{36}$ | 42 | HD 77581 | 43 | B0.5Ib | 44 | GX 263+3 |
| 4U1145−61 | | 291 | 45 | $< 10^{-4}$ | 45 | 100···200 ? | 45 | $3\cdot10^{35}$ | 46 | HEN 715 | 45 | B1 Vne | 47 | 2S1145−619 |
| 1E1145.1−6141 | | 297 | 45 | | | | | $10^{35}$ ? | 47 | | | ≈ B3 | 63 | |
| A1118−61 | | 405 | 48 | | | | | | | { RSCen ?<br>Be star? | 49<br>50 | | | Transient |
| 4U1538−52 | Nor XR-2 | 529 | 51 | $< 10^{-3}$ | 51 | 3.75 | 52 | $4\cdot10^{36}$ | 53 | | | B0.2Ia | 53 | A1540−53; 2S1538−522 |
| 4U1223−62 | GX 301−2 | 700 (Jan. '75) | 54 | $3\cdot10^{-3}$ | 55 | 35 | 21 | $3\cdot10^{36}$ | 56 | WRA977 | 57 | B1.5Ia | 57 | |
| 4U0352+30 | | 835 | 58 | $2\cdot10^{-4}$ | 59 | ≈ 580 ? | 59 | $4\cdot10^{33}$ | 58 | X Per | 60 | B0 9.5 (III···V)e | 61 | 2S0352+308 |

*) X-ray luminosity is given for a distance of 10 kpc in those cases where the optical companion star has not yet been identified.
**) For list of catalogues, see 5.7.1.2.

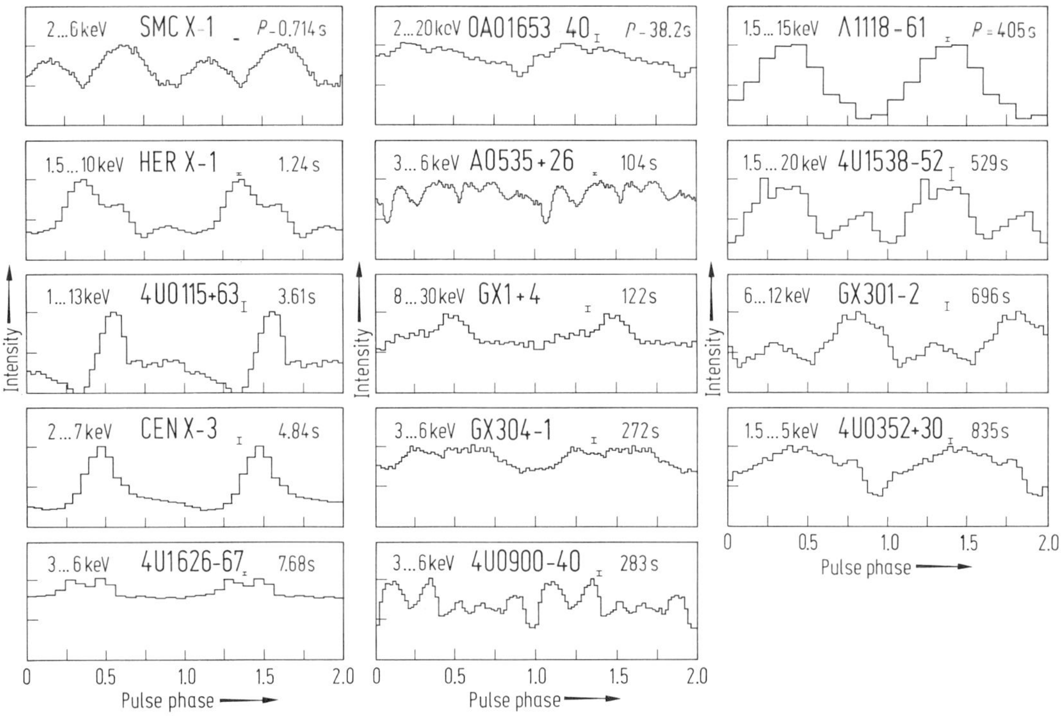


Fig. 2. Sample pulse profiles for 14 pulsating X-ray sources [21].

The observed secular spin-up of the neutron stars is qualitatively well understood in terms of the effects of torques exerted on the star's magnetosphere by the accreting matter at the inner edge of the accretion disk. A quantitative analysis shows that the observed spin-up rates are consistent with those expected for neutron stars, but much too high for white dwarfs. The observed luminosities are also larger than expected for white dwarfs.

### 5.6.3.2 Determination of binary-system parameters

In order to determine the basic binary-system parameters the following key measurements have to be made:
Orbital period.
Projected semi-major axis of the neutron star orbit.
Semi-amplitude of the Doppler velocity curve for the optical companion star.
Duration of the X-ray eclipse.

The basic measurement for all further evaluations is the tracking of the heliocentric corrected arrival times of a large number of consecutive X-ray pulses. Due to secular changes in the intrinsic pulse period and due to the orbital motion these arrival times vary during the orbital period. The individual pulse arrival time can be described by the function [13]:

$$t_{\mathrm{n}} = a_0 + \mathrm{n}P + \frac{1}{2} P\dot{P}\mathrm{n}^2 + \frac{a_{\mathrm{X}} \sin i}{c} F(e, \omega, \theta), \tag{1}$$

where

$t_n$ = arrival time of an arbitrary, but fixed, fiducial point on the $n^{th}$ pulse profile
$a_0$ = constant
$P$ = intrinsic pulse period for a specified epoch
$\dot{P}$ = rate of change in the intrinsic pulse period
$a_X \sin i$ = projected semi-major axis of the neutron star orbit
$F(e, \omega, \theta)$ = function representing an elliptic orbit with
$e$ = eccentricity
$\omega$ = longitude of periastron
$\theta = \frac{2\pi}{P_{orb}}(t-\tau)$
$P_{orb}$ = orbital period
$\tau$ = time of orbit phase 0.0.

By fitting this function to the data set of actually measured pulse arrival times, the pulse period $P$ and its secular change $\dot{P}/P$ are determined for the specified epoch together with the crucial quantity $(a_X \sin i)$ and the parameters which describe the ellipticity of the orbit. The orbital period $P_{orb}$ is often determined separately by analyzing the time sequence of consecutive X-ray eclipse centers. The measured orbital period and the semi-major axis of the neutron star orbit derived from the fit procedure mentioned above yield the mass function:

$$f(\mathfrak{M}) = \frac{4\pi(a_X \sin i)^3}{G \cdot P_{orb}^2} = \frac{\mathfrak{M}_C \sin^3 i}{(1+q)^2}. \qquad (2)$$

Here $G$ denotes the universal gravitational constant and $q$ the ratio of the mass $\mathfrak{M}_X$ of the compact X-ray star to the mass $\mathfrak{M}_C$ of the optical companion. This ratio can be written [21] as:

$$q \equiv \frac{\mathfrak{M}_X}{\mathfrak{M}_C} = \frac{K_C P_{orb} \sqrt{1-e^2}}{2\pi\, a_X \sin i}, \qquad (3)$$

where $K_C$ is the semi-amplitude of the Doppler velocity curve of the optical companion and e the orbit's eccentricity. The final critical system parameter, the orbital inclination angle $i$, is given in first approximation by the expression, [77, 78]:

$$\sin i = \sqrt{1-\beta^2 \cdot (a+b \log q)^2} / \cos\theta_{ecl}. \qquad (4)$$

$\theta_{ecl}$ is the measured X-ray eclipse half-angle. The constants $a$ and $b$ describe the effective radius of the companion's Roche lobe. For large inclination angles and given mass ratio $q$ they depend only on the ratio $\Omega$ of the rotation frequency of the optical star to the orbital frequency:

$$(a; b) = \begin{cases} (0.384; 0.200) & \text{for} \quad \Omega = 0 \\ (0.405; 0.230) & \text{for} \quad \Omega = 1. \end{cases}$$

This is correct for $0.02 < q < 0.2$ and $50° < i \leqq 90°$ [21].
Furthermore, in Eq. (4) $\beta$ stands for the ratio of the radius of the optical star to the radius of its Roche lobe and thus represents a measure of the extent to which the companion star fills its critical lobe. From the observation of the ellipsoidal optical light curves in several of these systems it is found that $\beta$ is probably greater than 0.9 [78].
By using the additional approximate expression for the radius of the optical companion star ($D$ denotes the stellar separation)

$$R_C \approx D \sqrt{\cos^2 i + \sin^2 i \cdot \sin^2 \theta_{ecl}} \qquad (5)$$

the nominal values of all system parameters can easily be found by inserting the measured quantities of $K_C$, $a_X \sin i$, and $\theta_{ecl}$ together with a best guess of $\beta$ and $\Omega$ into equations (2)···(5).

## 5.6.3.3 Neutron star properties as derived from observations of pulsating X-ray sources

Masses

The derivation of neutron star masses from observations has already been discussed in 5.6.1.3.1.

Magnetic fields

Magnetic dipole moments of accreting neutron stars may be estimated from the observed spin-up rates. Using simplifying assumptions for the interaction between the accretion disk and the magnetosphere one finds [7, 68] (see Fig. 1)

$$\frac{\dot{P}}{P} = -3\cdot 10^{-5}\left(\frac{\xi v_r}{v_{ff}}\right)^{1/7} \cdot \mathfrak{M}_X^{-3/7} \cdot I^{-1} \cdot R^{6/7} \cdot \mu^{2/7} \cdot L^{6/7} \cdot P\,[\mathrm{a}^{-1}],$$

where

$\xi$ = solid angle subtended at the compact star by the incident matter at the magnetopause

$\frac{v_r}{v_{ff}}$ = ratio of the average radial velocity of incidence of a particle to its free-fall velocity just outside the magnetopause

$R$ = radius of the neutron star in $[10^6\,\mathrm{cm}]$

$I$ = moment of inertia of the neutron star in $[10^{12}\,\mathfrak{M}_\odot\,\mathrm{cm}^2]$

$\mu$ = magnetic dipole moment of the neutron star in $[10^{30}\,\mathrm{Gauss\,cm}^3]$

$P$ = rotation period of the neutron star (= period of pulses) in [s]

$\mathfrak{M}_X$ = mass of the neutron star in $[\mathfrak{M}_\odot]$

$L$ = accretion-driven luminosity in $[10^{37}\,\mathrm{erg\,s}^{-1}]$.

For a standard neutron star ($\mathfrak{M}_X = 1.4\,\mathfrak{M}_\odot$; $R = 10^6$ cm) the resulting magnetic dipole moment is of the order of $10^{30}$ Gauss cm$^3$. (For an alternative model, see also [70].)

Surface magnetic fields of neutron stars may be measured by cyclotron resonance spectroscopy [79] (see 5.6.1.3.2 and 5.7.1.3.3).

## 5.6.3.4 References for 5.6.3


1 Lucke, R., Yentis, D., Friedman, H., Fritz, G., Shulman, S.: Astrophys. J. **206** (1976) L25.
2 Mason, K.O.: Mon. Not. R. Astron. Soc. **178** (1977) 81p.
3 Tuohy, I.R., Rapley, C.G.: Astrophys. J. **198** (1975) L69.
4 Coe, M.J., Bell-Burnell, S.J., Engel, A.R., Evans, A.J., Quenby, J.J.: Mon. Not. R. Astron. Soc. (1981), in press.
5 Webster, B.L., Martin, W.L., Feast, M.W., Andrews, P.J.: Nature Phys. Sci. **240** (1972) 183.
6 Tananbaum, H., Gursky, H., Kellogg, E.M., Levinson, R., Schreier, E., Giacconi, R.: Astrophys. J. **174** (1972) L143.
7 Rappaport, S., Joss, P.C.: Nature **266** (1977) 683.
8 Giacconi, R., Gursky, H., Kellogg, E., Levinson, R., Schreier, E., Tananbaum, H.: Astrophys. J. **184** (1973) 227.
9 Bahcall, J.N., Joss, P.C., Avni, Y.: Astrophys. J. **191** (1974) 211.
10 Bahcall, J.N., Bahcall, N.A.: Astrophys. J. **178** (1972) L1.
11 Crampton, D.: Astrophys. J. **187** (1974) 345.
12 Cominsky, L., Clark, G.W., Li, F., Mayer, W., Rappaport, S.: Nature **273** (1978) 367.
13 Rappaport, S., Clark, G.W., Cominsky, L., Joss, P.C., Li, F.: Astrophys. J. **224** (1978) L1.
14 Kelley, R.L., Rappaport, S., Brodheim, M.J., Cominsky, L., Stothers, R.: Astrophys. J. (1981), in press.
15 Hutchings, J.B., Crampton, D.: Astrophys. J. **247** (1981) 222.
16 Giacconi, R., Gursky, H., Kellogg, E., Schreier, E., Tananbaum, H.: Astrophys. J. **167** (1971) L67.
17 Schreier, E.J., Fabbiano, G.: Recent Uhuru Results on Centaurus X-3. In: X-Ray Binaries, NASA SP-389 (1975) 197.
18 Osmer, P.S., Hiltner, W.A., Whelan, J.A.J.: Astrophys. J. **195** (1975) 705.
19 Krzeminski, W.: Astrophys. J. **192** (1974) L135.
20 Rappaport, S., Markert, T., Li, F.K., Clark, G.W., Jernigan, J.G., McClintock, J.E.: Astrophys. J. **217** (1977) L29.

21 Rappaport, S.: Binary X-Ray Pulsars. In: NATO Advanced Study Institute on Galactic X-Ray Sources, Cape Sounion (1979), in press.
22 Middleditch, J., Mason, K.O., Nelson, J.E., White, N.E.: Astrophys. J. **244** (1981) 1001.
23 Pravdo, S.H., White, N.E., Boldt, E.A., Holt, S.S., Serlemitsos, P.J., Swank, J.H., Szymkoviak, E.A., Tuohy, I., Garmire, A.: Astrophys. J. **231** (1979) 912.
24 Kelley, R., Apparao, K., Bradt, H., Jernigan, J.G., Naranan, S., Rappaport, S.: Bull. Am. Astron. Soc. **12** (1980) 512.
25 Apparao, K.M.V., Naranan, S., Kelley, R.L.: Nature (1980), in press.
26 White, N.E., Pravdo, S.H.: Astrophys. J. **233** (1979) L121.
27 Parmar, A.N., Branduardi-Raymont, G., Pollard, G.S.G., Sanford, P.W., Fabian, A.C., Stewart, G.C., Schreier, E.J., Polidan, R.S., Oegerle, W.R., Locke, M.: Mon. Not. R. Astron. Soc. **193** (1980) 49p.
28 Rosenberg, F.D., Eyles, C.J., Skinner, G.K., Willmore, A.P.: Nature **256** (1975) 628.
29 Li, F., Rappaport, S., Clark, G.W., Jernigan, J.G.: Astrophys. J. **228** (1979) 893.
30 Rappaport, S., Joss, P.C., Bradt, H., Clark, G.W., Jernigan, J.G.: Astrophys. J. **208** (1976) L119.
31 Liller, W.: Int. Astron. Union Circ. No. 2780 (1975).
32 Stier, M., Liller, W.: Astrophys. J. **206** (1976) 257.
33 Lewin, W.H.G., Ricker, G.R., McClintock, J.E.: Astrophys. J. **169** (1971) L17.
34 Kendziorra, E., Staubert, R., Reppin, C., Pietsch, W., Voges, W., Trümper, J.: The Pulsating X-Ray Source GX 1+4 (4U1728–24). In: NATO Advanced Study Institute on Galactic X-Ray Sources, Cape Sounion (1979), in press.
35 Koo, J.W.C., Haymes, R.C.: Astrophys. J. **239** (1980) L60.
36 Doty, J.P., Hoffman, J.A., Lewin, W.H.G.: Astrophys. J. **243** (1981) 257.
37 Davidsen, A., Malina, R., Bowyer, S.: Astrophys. J. **211** (1977) 866.
38 McClintock, J.E., Rappaport, S., Nugent, J.N., Li, F.K.: Astrophys. J. **216** (1977) L15.
39 Mason, K.O., Murdin, P.G., Visvanathan, N.: Int. Astron. Union Circ. No. 3054 (1977).
40 Rappaport, S., McClintock, J.E.: Int. Astron. Union Circ. No. 2794 (1975).
41 Hutchings, J.B.: Astrophys. J. **192** (1974) 685.
42 McClintock, J.E., Rappaport, S., Joss, P.C., Bradt, H., Buff, J., Clark, G.W., Hearn, D., Lewin, W.H.G., Matilsky, T., Mayer, W., Primini, F.: Astrophys. J. **206** (1976) L99.
43 Hiltner, W.A.: Int. Astron. Union Circ. No. 2502 (1973).
44 Morgan, W., Code, A., Whitford, A.: Astrophys. J. Suppl. **2** (1955) 41.
45 White, N.E., Pravdo, S.H., Becker, R.H., Boldt, E.A., Holt, S.S., Serlemitsos, P.J.: Astrophys. J. **239** (1980) 655.
46 Moffat, A.F.J., Haupt, W., Schmidt-Kaler, T.: Astron. Astrophys. **23** (1973) 433.
47 Lamb, R.C., Markert, T.H., Hartman, R.C., Thompson, D.J., Bignami, G.F.: Astrophys. J. **239** (1980) 651.
48 Ives, I.C., Sanford, P.W., Bell-Burnell, S.J.: Nature **254** (1975) 578.
49 Fabian, A.C., Pringle, J.E., Webbink, R.F.: Nature **255** (1975) 208.
50 Chevalier, C., Ilovaisky, S.A.: Int. Astron. Union Circ. No. 2778 (1975).
51 Davison, P.J.N., Pye, J.P., Watson, M.G.: Mon. Not. R. Astron. Soc. **181** (1977) 73p.
52 Becker, R.H., Swank, J.H., Boldt, E.A., Holt, S.S., Pravdo, S.H., Saba, J.R., Serlemitsos, P.J.: Astrophys. J. **216** (1977) L11.
53 Parkes, G.E., Murdin, P.G., Mason, K.O.: Mon. Not. R. Astron. Soc. **184** (1978) 73p.
54 White, N.E., Huckle, H.E., Mason, K.O., Charles, P.A., Pollard, G., Sanford, P.W.: Int. Astron. Union Circ. No. 2870 (1975).
55 Kelley, R., Rappaport, S., Petre, P.: Astrophys. J. **238** (1980) 699.
56 Ghosh, P., Lamb, F.K.: Astrophys. J. **234** (1979) 296.
57 Vidal, N.V.: Astrophys. J. **186** (1973) L81.
58 White, N.E., Mason, K.O., Sanford, P.W., Murdin, P.: Mon. Not. R. Astron. Soc. **176** (1976) 201.
59 Mason, K.O.: Mon. Not. R. Astron. Soc. **178** (1977) 81p.
60 Hutchings, J.B., Cowley, A.P., Crampton, D., Redman, R.O.: Astrophys. J. **191** (1974) L101.
61 Chevalier, C., Ilovaisky, S.A.: Int. Astron. Union Circ. No. 2778 (1975).
62 Becker, R.H., Boldt, E.A., Holt, S.S., Pravdo, S.H., Rothschild, R.E., Serlemitsos, R.J., Swank, J.H.: Astrophys. J. **207** (1976) L167.
63 Hutchings, J.B., Cowley, A.P., Crampton, D.: Int. Astron. Union Circ. No. 3544 (1980).
64 Friedman, H., Wood, K.S.: Sky and Telescope **56** (1978) 490.
65 Staubert, R., Kendziorra, E., Pietsch, W., Reppin, C., Trümper, J., Voges, W.: Astrophys. J. **239** (1980) 1010.
66 Pringle, J.E., Rees, M.J.: Astron. Astrophys. **21** (1972) 1.
67 Davidson, K., Ostriker, J.P.: Astrophys. J. **179** (1973) 585.

68 Lamb, F.K., Pethick, C.J., Pines, D.: Astrophys. J. **184** (1973) 271.
69 Baan, W.A., Treves, A.: Astron. Astrophys. **22** (1973) 421.
70 Anzer, U., Börner, G.: Astron. Astrophys. **83** (1980) 133.
71 Alme, M., Wilson, J.: Astrophys. J. **186** (1973) 1015.
72 Elsner, R., Lamb, F.K.: Nature **262** (1976) 356; Astrophys. J. **215** (1977) 897.
73 Lamb, F.K., Pines, D., Shaham, J.: Astrophys. J. **224** (1978) 969; **225** (1978) 582.
74 Ghosh, P., Lamb, F.K.: Astrophys. J. **232** (1979) 239; **234** (1979) 296.
75 Arons, J., Lea, S.M.: Astrophys. J. **207** (1976) 914.
76 Pines, D.: Science **207** (1980) 597.
77 Paczyński, B.: Annu. Rev. Astron. Astrophys. **9** (1971) 183.
78 Avni, Y., Bahcall, J.N.: Astrophys. J. **197** (1975) 675; **202** (1975) L131.
79 Trümper, J., Pietsch, W., Reppin, C., Voges, W., Staubert, R., Kendziorra, E.: Astrophys. J. **219** (1978) L105.

## 5.6.4 X-ray bursters

X-ray bursters were discovered by Grindlay et al. [1] and studied in detail with the instruments aboard the SAS-3 and Hakucho satellites [2···5]. The bursters are characterized by impulsive emission of X-rays (rise time less than or about a few seconds; decay time, seconds to minutes) (see Fig. 3). Most of the 50 known X-ray burst sources are found within 30° around the galactic center; at least seven of them are located in globular clusters. This indicates that bursters are members of an old stellar population.

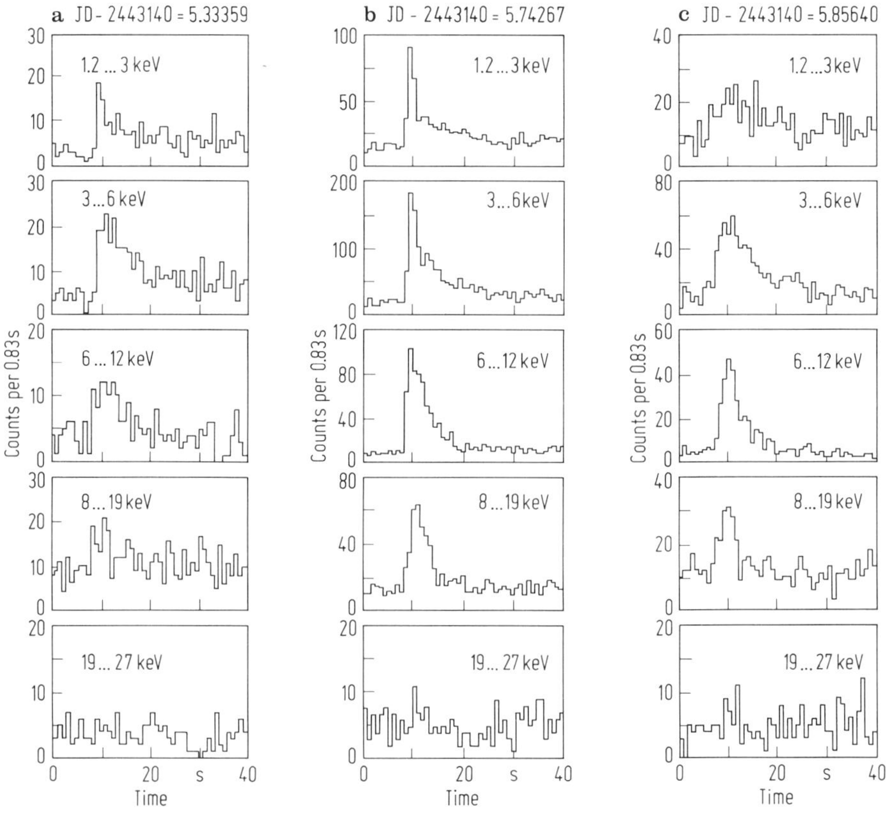


Fig. 3. Raw counting rate plots of five energy channels (1.2···27 keV) from the SAS-3 X-ray detector system for six X-ray bursts from MXB 1636-53, January '77. In each column a···f of time bins, the time 0 corresponds to the reduced Julian day given at the head of the respective column [2].

Further characteristics are [6]:

Rather soft spectra ($kT \approx$ several keV); softening during the burst's decay.

Black-body type spectra.

Peak luminosities of about $10^{38}$ erg $s^{-1}$ (at 10 kpc distance).

Radii of about 10 km of the black-body radiation emitting region (neutron stars).

No periodic pulsations (weak magnetic fields?).

Optical bursts delayed by a few seconds with respect to the X-ray bursts (re-emission from an accretion disk; binary system).

No eclipses (low-mass binary systems, see [7]).

Steady X-ray luminosities of about $10^{36}$ erg $s^{-1}$ (at 10 kpc distance).

Ratio of time averaged burst luminosity to steady luminosity of about $10^{-2}$.

Interpretation:

X-ray bursters are low-mass binary systems with neutron stars accreting matter via an accretion disk. The X-ray bursts may be produced either by thermonuclear explosions (helium flashes) of accreted matter [8, 9] or by instabilities in the accretion process [10].

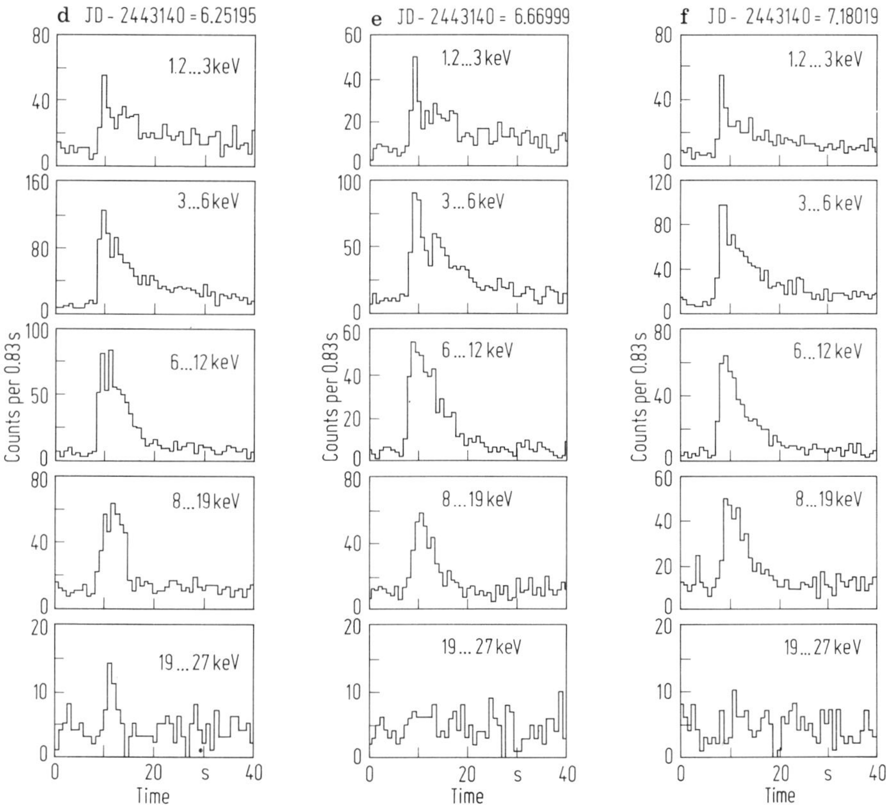


Fig. 3.

## References for 5.6.4

1 Grindlay, J., Gursky, H., Schnopper, H., Parsignault, D.R., Heise, J., Brinkman, A.C., Schrijver, J.: Astrophys. J. **205** (1976) L127.

2 Hoffman, A.J., Lewin, W.H.G., Doty, J.: Astrophys. J. **217** (1977) L23.

3 Hoffman, A.J., Marshall, H.L., Lewin, W.H.G.: Nature **271** (1978) 630.

4 Lewin, W.H.G., Joss, P.C.: Nature **270** (1977) 211.
5 Hayakawa, S.: Space Sci. Rev. (1981), in press.
6 Lewin, W.H.G., Clark, G.W.: Ann. New York Acad. Sci. **336** (1980) 451.
7 Milgrom, M.: Astron. Astrophys. **67** (1978) L25.
8 Joss, P.C.: Astrophys. J. **225** (1978) L123.
9 Maraschi, L., Cavaliere, A.: Highlights of Astronomy **4** (1977).
10 Lamb, F.K., Fabian, A.C., Pringle, J.E., Lamb, D.Q.: Astrophys. J. **217** (1977) 197.


## 5.6.5 Black holes

A black hole is a collapsed object whose total mass is confined inside an "event horizon", i.e. a surface from which no signals can escape to infinity. In the spherically symmetric case the metric is given by the Schwarzschild line element

$$ds^2 = -(1 - R_{\mathrm{Sch}}/r)\,dt^2 + \frac{dr^2}{1 - R_{\mathrm{Sch}}/r} + r^2(d\theta^2 + \sin^2\theta\, d\varphi^2),$$

where $r$, $\theta$, $\varphi$ are polar coordinates, and $R_{\mathrm{Sch}} = \frac{2G\mathfrak{M}}{c^2}$ ($G$ = gravitational constant, $\mathfrak{M}$ = mass, $c$ = velocity of light) is the Schwarzschild radius [1].

Outside a rotating collapsed object the metric is given by the Kerr line element which can be written in Boyer-Lindquist form [2, 3] as

$$\begin{aligned} ds^2 = &-\left(\frac{\Delta}{\varrho^2}\right)[dt - a\sin^2\theta\, d\varphi]^2 \\ &+\left(\frac{\sin^2\theta}{\varrho^2}\right)\cdot[(r^2+a^2)\,d\varphi - a\,dt]^2 \\ &+\frac{\varrho^2}{\Delta}dr^2 + \varrho^2\,d\theta, \end{aligned}$$

$$\Delta = r^2 - R_{\mathrm{Sch}} + a^2,$$

$$\varrho^2 = r^2 + a^2\cos^2\theta,$$

where $a = \frac{I}{\mathfrak{M}c}\left(a \leqq \frac{G\mathfrak{M}}{c^2}\right)$ is the angular momentum per unit mass of the black hole.

Taking into account the quantum statistical fluctuations of the vacuum, a black hole emits thermal radiation at a black-body temperature of [4]

$$T_{\mathrm{BH}} = \frac{\hbar\cdot\varkappa}{2\pi\cdot ck},$$

where $\varkappa$ is the surface gravity of the black hole and $k$ is Boltzmann's constant. For a Schwarzschild black hole with

$$\varkappa = \frac{c^4}{4G\mathfrak{M}}$$

this temperature is $T_{\mathrm{Sch}} \approx 10^{-6}\left(\frac{\mathfrak{M}_\odot}{\mathfrak{M}}\right)$ K.

Because of the emission of thermal radiation the black hole will loose mass and therefore have a finite lifetime of about $\tau \approx 10^{71}\left(\frac{\mathfrak{M}}{\mathfrak{M}_\odot}\right)^3$ s.

Black holes can grow by accretion of matter and radiation. Accreting black holes may be visible through the radiation emitted by the inflowing gas which can be heated by compression, shock waves, turbulence etc. to temperatures sufficiently high to allow the emission of X- and $\gamma$-rays.

Black holes of stellar mass have probably been detected as compact X-ray sources in binary systems (see 6.1.4). The best candidate is Cyg X-1 [5, 6].

Accretion onto supermassive black holes may be responsible for the enormous power radiated from small regions of space in nuclei of active galaxies and quasars [7, 8], (see 9.5).

## References for 5.6.5


1 Schwarzschild, K.: Sitz. Ber. Preuss. Akad. Wiss. **189** (1916).
2 Kerr, R.P.: Phys. Rev. Lett. **11** (1963) 237.
3 Boyer, R.H., Lindquist, R.W.: J. Math. Phys. **8** (1967) 265.
4 Hawking, S.W.: Nature **248** (1974) 30.
5 Oda, M.: Space Sci. Rev. **20** (1977) 757.
6 Eardley, D.M., Lightman, A.P., Shakura, N.L., Shapiro, S.L., Sunyaev, R.A.: Comments on Astrophysics (1978).
7 Lynden-Bell, D.: Nature **233** (1969) 690.
8 Rees, M.J.: Ann. New York Acad. Sci. **320** (1978) 613.


# 5.7 X-ray and γ-ray sources

## 5.7.1 X-ray sources

### 5.7.1.1 Overview

Since the discovery of the first cosmic X-ray source in 1962 [1, 2], numerous observations by means of rocket and balloon-borne instruments have been performed, thus gradually exploring the X-ray sky. After the launch of the Uhuru satellite in 1970 [3] the knowledge of the X-ray sky has rapidly grown due to various satellite missions (see also: X-ray and γ-ray satellites, 1.7.3). The increasing sensitivity and angular resolution of X-ray detectors has led to the discovery of a large number of X-ray sources and their identification with objects of optical and radio astronomy. At the same time the spectral band explored in X-ray astronomy has been broadened and now covers the range from 0.1 keV to 500 keV. Improved spectral resolution has led to the discovery of line features in the spectra of various sources. Studies of time variability have proven to be of prime diagnostic value in X-ray astronomy. By 1980, the total number of known X-ray sources was more than a thousand, comprising nearly all kinds of astronomical objects. A list of different classes of observed X-ray sources is given in Table 1. (For X-ray binaries, see also 6.1.4 .)

Table 1. Classes of observed X-ray sources, and their typical members.
$L_X$ = X-ray luminosity for the spectral band given in the preceding column.
X-ray luminosity range = general luminosity range of the class as a whole.
In some cases the proposed radiation processes are given in footnotes.

Table 1a. Galactic objects.

| Object class | Typical object | X-ray luminosity | | | X-ray luminosity range | |
|---|---|---|---|---|---|---|
| | | Spectral band keV | $L_X$ erg s$^{-1}$ | Ref. | erg s$^{-1}$ | Ref. |
| Sun a) quiet corona (see 3.1.1.5) | | 0.5···10 | $10^{24}$ | 4 | | [1]) |
| b) solar flares (see 3.1.2.7) | | 0.5···10 | $10^{26}\cdots10^{28}$ | 4 | | [2]) |
| Main-sequence stars | O3 V HD93205 | 0.5···4.5 | $1.0\cdot10^{33}$ | 5 | $10^{26}\cdots10^{33}$ | 5 |
| | B9.5 V $\alpha^2$ CVn | | $1.0\cdot10^{29}$ | 5 | | |
| | A1 V α C Ma | | $8.0\cdot10^{26}$ | 5 | | |
| | F8 V β Vir | | $2.5\cdot10^{28}$ | 5 | | |
| | G0 V β Com | | $2.0\cdot10^{28}$ | 5 | | |
| | K2 V ε Eri | | $2.0\cdot10^{28}$ | 6 | | |
| | M1 Ve YY Gem | | $4.0\cdot10^{29}$ | 5 | | |

[1]) Thermal radiation from a hot plasma; line spectra.
[2]) Bremsstrahlung.

continued

Table 1a, continued

| Object class | Typical object | X-ray luminosity: Spectral band keV | $L_X$ erg s$^{-1}$ | Ref. | X-ray luminosity range: erg s$^{-1}$ | Ref. |
|---|---|---|---|---|---|---|
| Giants | O5 IIIf HD93403 | 0.5···4.5 | $1.0\cdot10^{33}$ | 5 | $10^{28}\cdots10^{33}$ | 5 |
| | B5 III τ Ori | | $1.6\cdot10^{30}$ | | | |
| | A5 III α Oph | | $4.0\cdot10^{28}$ | | | |
| | G5 III β Lep | | $1.6\cdot10^{29}$ | | | |
| | K2 III α Ser | | $6.3\cdot10^{27}$ | | | |
| Supergiants | O3 If HD93129A | 0.5···4.5 | $4.0\cdot10^{33}$ | 5 | $10^{30}\cdots10^{34}$ | 5 |
| | B5 Ib 67 Oph | | $2.0\cdot10^{31}$ | | | |
| | F0 Ib α Car | | $1.0\cdot10^{30}$ | | | |
| RS CVn stars | RS CVn | 0.4···4 | $\leqq1.3\cdot10^{31}$ | 7 | $2\cdot10^{30}\cdots4\cdot10^{31}$ | 8 |
| T Tauri stars (see 5.1.3.9) | T Tau | 0.5···4.5 | $3\cdot10^{30}$ | 9 | $<3\cdot10^{29}\cdots5\cdot10^{31}$ | 9, 10 |
| Cataclysmic variables (see 5.1.3.3; 6.1.3.6; 6.1.4) | | | | | | |
| a) AM Her type | AM Her | 0.1···4.5 | $3.6\cdot10^{32}$ | 11 | $10^{31}\cdots5\cdot10^{32}$ | 11 |
| | | 2···6 | $7.5\cdot10^{31}$ | 11 | | |
| b) SS Cyg type | SS Cyg | 0.4···2 | $5.7\cdot10^{31}$ | 11 | $10^{30}\cdots10^{32}$ | 11 |
| | | 2···6 | $8.3\cdot10^{31}$ | 11 | | |
| c) U Gem type | U Gem | 0.1···4.5 | $1.8\cdot10^{30}$ | 12 | $10^{29}\cdots10^{32}$ | 12 |
| Single hot white dwarfs (see 5.5) | Hz 43 | 0.1···1 | $1.6\cdot10^{32}$ | 13 | | |
| | Sirius B | 0.2···0.284 | $9\cdot10^{27}$ | 14 | | |
| Neutron stars (see 5.6.1) | | | | | | |
| Cooling neutron stars | 1E161348−5055.1 (SNR RCW 103) | 0.3···2 | $2.5\cdot10^{34}$ | 15 | | [3] |
| Accreting neutron stars | | | | | | |
| a) pulsating X-ray sources | Her X-1 | 0.2···80 | $10^{37}$ | 16 | $10^{36}\cdots10^{38}$ | 17 [4] |
| b) bursters (see 5.6.4) | MXB 1728−34 | 2···11 | steady: $10^{37}$<br>bursting: $5\cdot10^{38}$ | 18 | $10^{38}\cdots10^{39}$ | 19 [5] |
| Globular clusters (see 6.2.1) | Terzan 2 | 1.1···10 | $5\cdot10^{38}$ | 20 | $10^{36}\cdots10^{39}$ | 21 |
| Novae (see 5.1.3.2) | | | | | | |
| a) X-ray novae | Nova Mon 1975 | 0.3···10 | outburst: $3\cdot10^{38}$ | 22 | | |
| b) classical novae | CP Pup | 0.15···10 | $10^{31}$ | 23 | $3\cdot10^{30}\cdots10^{32}$ | 23 |
| Supernova remnants (see 5.1.3.1) | Puppis A | 0.2···1.5 | $3.6\cdot10^{36}$ | 24 | $10^{34}\cdots10^{37}$ | [6] 25 [7] |
| | | 1.5···6 | $1.9\cdot10^{36}$ | 24 | | |
| | Crab nebula | 1···200 | $10^{37}$ | 25 | | |

| | Energy band keV | Surface brightness range photons cm$^{-2}$ s$^{-1}$ sr$^{-1}$ | Ref. |
|---|---|---|---|
| Diffuse galactic emission | 0.15···0.28 | 40···280 | 26 [8] |
| | 1.15···2 | 60···120 | |

Footnotes [3]...[8]: p. 35.

Footnotes to Table 1a:

[3]) The identification of 1E161348–5055.1 with a neutron star is tentative.

[4]) Thermal radiation from a hot, strongly magnetized plasma.

[5]) Thermal radiation. Thermonuclear explosions on the surface of neutron stars.

[6]) Thermal radiation from a hot plasma; line spectra.

[7]) Synchrotron radiation from relativistic electrons.

[8]) Strongly non-uniform background. Thermal radiation from the hot component of interstellar matter.

Table 1b. Extragalactic objects.

| Object class | Typical object | X-ray luminosity | | | X-ray luminosity range | |
|---|---|---|---|---|---|---|
| | | Spectral band keV | $L_X$ erg s$^{-1}$ | Ref. | erg s$^{-1}$ | Ref. |
| Normal galaxies (see 9.1) | M 31 | 0.5···4.5<br>2 ···6 | $2.7\cdot10^{39}$<br>$2.3\cdot10^{39}$ | 27 | $10^{38}\cdots5\cdot10^{39}$ | 28 |
| Narrow emission-line galaxies | NGC 7582 | 0.2···3.5<br>2 ···10 | $4\cdot10^{41}$<br>$5\cdot10^{42}$ | 29<br>30 | $<5\cdot10^{42}\cdots10^{44}$ | 30 |
| N-type galaxies (see 9.1.3.2) | 3C 111 | 2 ···10 | $5.6\cdot10^{44}$ | 31 | $10^{44}\cdots10^{45}$ | 31 |
| Radio galaxies (see 9.6) | Cen A | 2··· 6<br>12··· 140<br>140···2300 | $5\cdot10^{41}$<br>$6\cdot10^{42}$<br>$1\cdot10^{43}$ | 32 | | |
| BL Lacertae objects (see 9.5) | Mkn 421 | 2 ···10 | $2\cdot10^{44}$ | 30 | | |
| Seyfert galaxies (see 9.5) a) type 1 | Mkn 376 | 0.5···4.5<br>2 ···10 | $3\cdot10^{44}$<br>$3\cdot10^{44}$ | 33<br>30 | $<10^{42}\cdots7\cdot10^{44}$ | 33 |
| b) type 2 | NGC 2992 | 2 ···10 | $2\cdot10^{43}$ | 30 | | |
| QSO (see 9.5) a) radio quiet | TON 155 | 0.5···4.5 | $1.4\cdot10^{46}$ | 34 | $5\cdot10^{42}\cdots5\cdot10^{47}$ | 34 |
| b) radio loud | 3C 273 | 0.5···4.5<br>2 ···10 | $1.7\cdot10^{46}$<br>$1\ \cdot10^{46}$ | 35<br>30 | | |
| Clusters of galaxies (see 9.3) a) rich | Coma | 0.2···3<br>2 ···10 | $6\ \cdot10^{44}$<br>$9\ \cdot10^{44}$ | 36<br>37 | $<5\cdot10^{43}\cdots2\cdot10^{45}$ | 38 |
| b) poor | AWM 7 | 2 ··· 6 | $1\ \cdot10^{44}$ | 39 | $<10^{42}\cdots10^{44}$ | 39<br>40 |

| | Energy keV | Surface brightness photons cm$^{-2}$ s$^{-1}$ sr$^{-1}$ | Ref. |
|---|---|---|---|
| Diffuse extragalactic emission | 3<br>10<br>50<br>200<br>1000<br>10000 | 5<br>3<br>0.8<br>0.09<br>0.01<br>0.001 | 41 |

### 5.7.1.2 Sky surveys

The hosts of known celestial X-ray sources have been discovered in the course of several more or less complete sky surveys carried out since 1970 using satellite-borne detectors. Table 2 lists the main surveys and catalogues of X-ray sources.

Table 2. Main surveys and catalogues of X-ray sources.

| Satellite / experiment | Catalogue designation | Energy band keV | Number of detected sources | Sensitivity *) erg cm$^{-2}$ s$^{-1}$ keV$^{-1}$ | Remarks | Ref. |
|---|---|---|---|---|---|---|
| Uhuru | 4U | 2···6 | 339 | $5\cdot10^{-12}$ | All-sky survey | 42 |
| OSO-7 | 1M | 3···10 | 184 | $4\cdot10^{-11}$ | Galactic plane | 43 |
| Ariel-V/SSI | 2A | 2···18 | 105 | $7\cdot10^{-12}$ | High galactic latitude $\|b\|>10°$ | 44 |
| /SSI | A | 1···20 | 16 | $4\cdot10^{-12}$ | Galactic plane $\|b\|<10°$ | 45 |
| HEAO-1/A2 | H | 2···10 | 85 | $4\cdot10^{-12}$ | High galactic latitude $\|b\|>20°$ | 71 |
| HEAO-1/A4 | | 13···180 | 87 | | All-sky survey | 46 |

*) For a Crab-like spectrum.

### 5.7.1.3 X-ray emission processes

#### 5.7.1.3.1 Bremsstrahlung from a hot plasma

Thermal emission of X-rays requires temperatures in the range of $T \approx 10^6 \cdots 10^9$ K, i.e. $kT \approx 0.1 \cdots 100$ keV. For an optically thin plasma consisting of nuclei with atomic number $Z$ and electrons having a Maxwellian velocity distribution of temperature $T$, the spectrum of the bremsstrahlung emitted per unit volume is given by [47]

$$\frac{\mathrm{d}P_\mathrm{B}(E,T)}{\mathrm{d}V\,\mathrm{d}E} = \sum_{N_z} 1.0\cdot10^{-11}\cdot N_\mathrm{e} N_z Z^2 \cdot \bar{g}_\mathrm{B}(E,T)\, T^{-1/2}\cdot \mathrm{e}^{-E/kT}\left[\frac{\mathrm{erg}}{\mathrm{cm^3\,s\,erg}}\right].$$

Here $N_\mathrm{e}$ and $N_z$ are the electron and ion number densities [$\mathrm{cm}^{-3}$], respectively, and $\bar{g}_\mathrm{B}$ is the temperature-averaged Gaunt factor. The latter is approximately given by

$$\bar{g}_\mathrm{B}(E,T) = \begin{cases} \dfrac{\sqrt{3}}{\pi}\ln\left(2.25\dfrac{kT}{E}\right) & \text{for} \quad E \ll kT \;[48] \\[2ex] \left(\dfrac{kT}{E}\right)^{0.4} & \text{for} \quad E \gtrsim kT \;[49]. \end{cases}$$

For temperatures above $10^6$ K the Gaunt factor becomes constant within a 10% accuracy

$$\bar{g}_\mathrm{B}(T>10^6\,\mathrm{K}) \approx 1.2 .$$

In the X-ray emitting regime the total bremsstrahlung emission of a fully ionized plasma with cosmic abundances of the elements $\left(\sum_{N_z} N_\mathrm{e} N_z Z^2 = 1.4\, N_\mathrm{e}^2\right)$ can therefore be written as

$$\frac{\mathrm{d}P_\mathrm{B}(T)}{\mathrm{d}V} = 2.4\cdot10^{-27}\cdot T^{1/2}\cdot N_\mathrm{e}^2\left[\frac{\mathrm{erg}}{\mathrm{cm^3\,s}}\right].$$

The corresponding cooling time of the plasma by bremsstrahlung is

$$t_\mathrm{c} = \frac{3N_\mathrm{e}kT}{\left(\dfrac{\mathrm{d}P_\mathrm{B}}{\mathrm{d}V}\right)} \approx 2\cdot10^{11}\cdot\frac{T^{1/2}}{N_\mathrm{e}}\,[\mathrm{s}].$$

### 5.7.1.3.2 Recombination radiation from a hot plasma

For a hot plasma with cosmic abundances and for a Maxwellian distribution of electron velocities, the continuous recombination radiation spectrum is given by [50, 51, 52, 53].

$$\frac{\mathrm{d}P_{\mathrm{RR}}(E,T)}{\mathrm{d}V\,\mathrm{d}E}=2.8\cdot 10^{-6}\cdot N_{\mathrm{e}}N_{\mathrm{H}}T^{-3/2}\cdot \mathrm{e}^{-E/kT}\cdot X\left[\frac{\mathrm{erg}}{\mathrm{cm}^3\,\mathrm{s\,erg}}\right],$$

where

$$X=\sum_{Z,l,n} nf(n)\cdot\left(\frac{n(Z,i)}{N_{\mathrm{Z}}}\right)\left(\frac{N_{\mathrm{Z}}}{N_{\mathrm{H}}}\right)\left(\frac{I(Z,i-1,n)}{I_{\mathrm{H}}}\right)^2\cdot \mathrm{e}^{\frac{I(Z,i-1,n)}{kT}}.$$

The number $n$ is the principal quantum number of the bound state of the ion into which the electron is captured, whereas $f(n)$ represents the incompleted fraction of the shell $n$. $I(Z,i,n)$ is the corresponding ionization energy of level $n$ of the ion with nuclear charge $Z$ and net ionic charge $i$, the number density of which is $n(Z,i)$. The sum has to be carried out over all ions and all levels for which $I(Z,i-1,n)>E$ holds.

The ratio of recombination radiation to bremsstrahlung radiation at a given photon energy can then be written as

$$r_{\mathrm{RR,B}}=\frac{\left(\dfrac{\mathrm{d}P_{\mathrm{RR}}}{\mathrm{d}V\,\mathrm{d}E}\right)}{\left(\dfrac{\mathrm{d}P_{\mathrm{B}}}{\mathrm{d}V\,\mathrm{d}E}\right)}\approx\frac{10^5}{T}\cdot X.$$

For a given temperature $T$, recombination dominates bremsstrahlung radiation at shorter wavelengths $\left(\lambda\lessapprox\frac{3\cdot 10^7}{T}[\text{Å}]\right)$.

### 5.7.1.3.3 Line emission from a hot plasma

Atomic lines

At temperatures necessary for the production of X-rays, matter is partly ($T\lessapprox 10^8$ K) or completely ionized ($T>10^8$ K). The intensity of line radiation depends on the plasma density and temperature, abundance of elements, ionization equilibrium, oscillator strength and energy of the lines. In the case of collisional excitation of level $n'$ of ion $Z$ in the ground state $n$ by electrons having a Maxwellian distribution of velocities, the power emitted in the resonance line per unit volume is given by [54]

$$\frac{\mathrm{d}P_{\mathrm{L}}}{\mathrm{d}V}=2.7\cdot 10^{-15}N_{\mathrm{e}}N_{\mathrm{Z}}T^{-1/2}\cdot f_{\mathrm{nn'}}\overline{g_{\mathrm{nn'}}}\cdot \mathrm{e}^{-\frac{E_{\mathrm{nn'}}}{kT}}\left[\frac{\mathrm{erg}}{\mathrm{cm}^3\,\mathrm{s}}\right].$$

Here $E_{\mathrm{nn'}}$ is the energy of excitation and $f_{\mathrm{nn'}}$ the oscillator strength for the transition. $\overline{g_{\mathrm{nn'}}}$ represents a mean Gaunt factor, which amounts to about 0.2 for $kT<E_{\mathrm{nn'}}$.

For a line which is produced by a transition to a state other than the ground state the branching ratio of the state of interest has to be taken into account.

Lines of Fe, O, S, and Si ions are of particular importance in X-ray astronomy. As an example, Fig. 1 gives the intensities of the strongest iron lines as a function of temperature for a plasma with solar coronal abundances. X-ray line spectroscopy may be used to determine the elemental abundance in X-ray sources.

Fig. 2 shows the total emission from a hot plasma of solar coronal abundances in lines (L), recombination radiation (RR), and bremsstrahlung radiation (B).

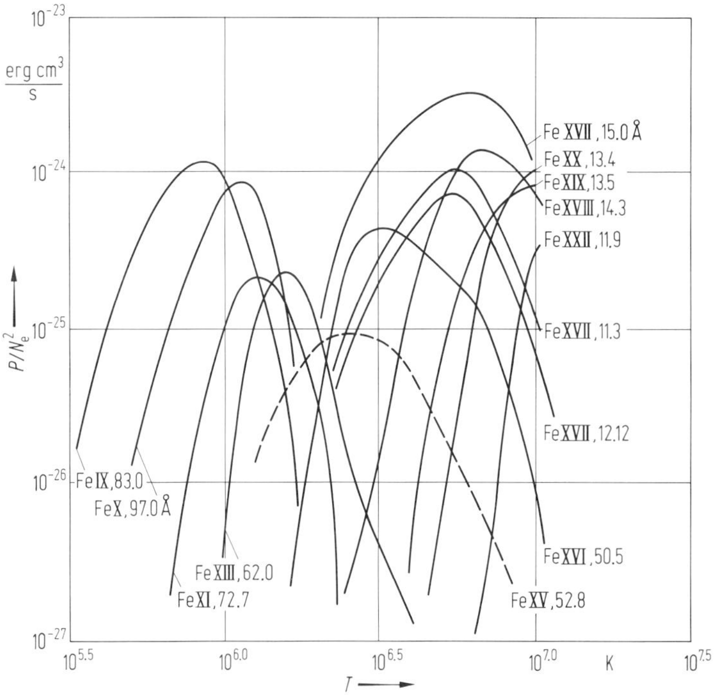


Fig. 1. Line intensity for some strong lines of Fe ions as a function of temperature in the wavelength range of 1···100 Å [50]. Numbers behind the ions give the wavelengths in [Å].

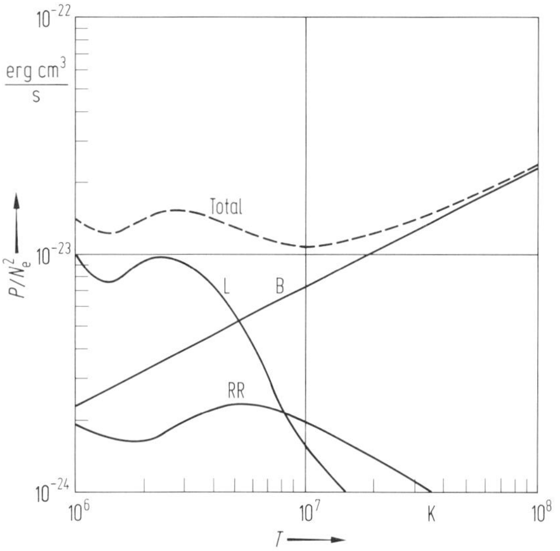


Fig. 2. Total emission from a hot plasma [55].
L: lines; RR: recombination radiation; B: bremsstrahlung

Cyclotron lines

In a homogeneous magnetic field the energy eigenvalues of an electron are given by [56]

$$E_{j,s}=mc^2\left[\sqrt{1+\left(\frac{p_z}{mc}\right)^2+(2j+s+1)\frac{B}{B_{cr}}}-1\right],$$

where:

| | |
|---|---|
| $mc^2$ | electron rest energy |
| $p_z$ | continuous momentum along the magnetic field lines |
| $B$ | magnetic field strength |
| $B_{cr}=(m^2c^3)/eh$ | critical magnetic field strength |
| $j=0,1,2,\dots$ | angular momentum quantum number |
| $s=\pm 1$ | spin quantum number. |

In the non-relativistic limit ($E_{j,s} \ll mc^2$) the transverse energy is quantized in multiples of the cyclotron frequency

$$E^{\perp}_{j,s} = \hbar \cdot \frac{eB}{mc}\left(j + \frac{s}{2} + \frac{1}{2}\right) \qquad \text{(Landau levels)}$$

with a level spacing of

$$\Delta E^{\perp} = 11.6 \cdot \left(\frac{B}{10^{12}}\right)[\text{keV}]$$

with $B$ in [G].

In the superstrong magnetic field at the polar cap of a neutron star ($B \approx 10^{12}$ G) the corresponding Landau transitions lie in the X-ray domain. Therefore, the observation of cyclotron lines in emission or absorption can make possible the determination of the magnetic field of a matter accreting neutron star in a very direct way [57].

#### 5.7.1.3.4 Synchrotron radiation

A highly relativistic electron ($E = \gamma mc^2$; $\gamma \gg 1$) moving in a magnetic field $B$ along an helical path of curvature $r_c \propto E/eB \sin\alpha$ with a pitch angle $\alpha$ emits a synchrotron radiation spectrum which peaks at a frequency [58]

$$\nu_m \propto \frac{c\gamma^3}{2\pi r_c} = \frac{eB\gamma^2 \sin\alpha}{2\pi mc} \approx 2.8 \cdot 10^6 \gamma^2 B \sin\alpha\,[\text{Hz}].$$

In order to produce X-rays the product ($E^2B$) has to lie in the range of $10^{11} \cdots 10^{14}$ (MeV)$^2$ G.

The total energy loss of an ultrarelativistic electron due to synchrotron radiation is given by [58]

$$P = 1.6 \cdot 10^{-15} \gamma^2 B^2 \sin^2\alpha \left[\frac{\text{erg}}{\text{s}}\right].$$

The lifetime of such an energetic electron with respect to synchrotron losses can then be defined as

$$\tau \propto \frac{E}{P} \approx \frac{3 \cdot 10^8}{\gamma B^2 \sin^2\alpha}[\text{s}] \propto (\nu_m \cdot B^3)^{-1/2}.$$

In general, synchrotron radiation is elliptically polarized.

An incoherently radiating ensemble of relativistic electrons with an isotropic distribution of pitch angles having a power law spectrum of energy

$$N(E)\,dE = N_0 \cdot E^{-\Gamma}\,dE$$

gives rise to a synchrotron radiation spectrum which is again given by a power law

$$P(\nu) \propto \nu^{-q}$$

whereby the spectral indices are related by [47] $q = \dfrac{\Gamma - 1}{2}$.

#### 5.7.1.3.5 Inverse Compton effect

X-rays can be produced by Compton scattering of lower energy photons with energetic electrons. This process, which converts a low-frequency radiation field into a higher frequency radiation (EUV and X-rays), can take place under a variety of conditions. Here we discuss only the two extreme cases which are of particular importance in X-ray astronomy.

Inverse Compton scattering by relativistic electrons

If an isotropic low-frequency radiation field of mean frequency $\bar{\nu}_0$ and energy density $u_{ph}$ interacts with a beam of ultrarelativistic electrons ($E = \gamma mc^2$; $\gamma \gg 1$), the mean frequency of the radiation after the Compton scattering is shifted to [59]

$$\bar{\nu} = \tfrac{4}{3}\gamma^2 \bar{\nu}_0.$$

In order to obtain X-rays ($\nu = 10^{16} \cdots 10^{19}$ Hz) a typical energy of the scattering electrons is required, for example,

$$E = \gamma mc^2 = \begin{cases} 0.2 \cdots 6\,\text{GeV} & \text{for} \quad \bar{\nu}_0 = 6 \cdot 10^{10}\,\text{Hz} \quad \text{(3 K background radiation)} \\ 1 \cdots 50\,\text{MeV} & \text{for} \quad \bar{\nu}_0 = 7 \cdot 10^{14}\,\text{Hz} \quad \text{(3 eV, starlight)}. \end{cases}$$

In the Thomson limit, $\gamma h\nu_0 \ll mc^2$, the scattered Compton power generated per fast electron, i.e. the total energy loss rate of the relativistic electron, is given by [5]

$$P_{\mathrm{iC}} = -\frac{\mathrm{d}E}{\mathrm{d}t} = \frac{4}{3}\sigma_{\mathrm{T}} c \gamma^2 u_{\mathrm{ph}} = 2.6\cdot 10^{-14}\gamma u_{\mathrm{ph}} \left[\frac{\mathrm{erg}}{\mathrm{s}}\right],$$

where $\sigma_{\mathrm{T}} = 6.65\cdot 10^{-25}\ \mathrm{cm}^2$ is the Thomson cross-section.
The lifetime of the electrons due to inverse Compton losses is then

$$\tau_{\mathrm{iC}} = \frac{E}{P_{\mathrm{iC}}} = \frac{3mc}{4\sigma_{\mathrm{T}}\gamma u_{\mathrm{ph}}} \approx \frac{2.1\cdot 10^7}{\gamma\cdot u_{\mathrm{ph}}}[\mathrm{s}].$$

As in the case of synchrotron radiation an ensemble of relativistic electrons having a power law spectrum $N(E)\,\mathrm{d}E = N_0 E^{-\Gamma}\,\mathrm{d}E$ gives rise to an inverse Compton spectrum of $P(\nu)\propto\nu^{-q}$ whereby the spectral indices are again related by $q=\frac{1}{2}(\Gamma-1)$.

Comptonization of soft photons by a hot plasma

Consider a source of low-energy photons of frequency $\nu_0$ which is embedded in a hot spherical cloud of non-relativistic, Maxwellian-distributed electrons ($kT_{\mathrm{e}} \gg h\nu_0$; $kT_{\mathrm{e}} \ll mc^2$). Photons diffusing through the electron cloud will be Thomson-scattered gaining an average energy amount of $\Delta\nu/\nu_0 \approx kT_{\mathrm{e}}/mc^2$ in each scattering event. Under a wide range of conditions the spectrum of the photons escaping the electron cloud is given by [60···63]

$$P(\nu)\propto\begin{cases}\nu^{-\alpha} & \text{for } \dfrac{h\nu}{kT}<1\\[2ex] \left(\dfrac{h\nu}{kT_{\mathrm{e}}}\right)^3\cdot \mathrm{e}^{-h\nu/kT_{\mathrm{e}}} & \text{for } \dfrac{h\nu}{kT}\gtrsim 1\end{cases}\qquad \text{with}\qquad \begin{aligned}\alpha &= -\frac{3}{2}+\sqrt{\frac{9}{4}-\gamma}\\ \gamma &= \frac{\pi^2}{3}\,\frac{mc^2}{kT_{\mathrm{e}}(\tau_0+\frac{2}{3})^2},\end{aligned}$$

where $\tau_0$ represents the optical thickness of the cloud with respect to Thomson scattering.

In order to produce X-rays the Comptonizing electron cloud should have a temperature of the order of $kT_{\mathrm{e}} \approx 1\cdots 100$ keV. Such conditions may be found in the inner parts of accretion disks around compact objects.

## 5.7.2 γ-ray sources

### 5.7.2.1 Overview

The sky surveys of the OSO-III, SAS-2, and COS-B satellites as well as a number of balloon experiments have revealed the existence of different components of cosmic γ-rays:

Extragalactic γ-ray sources:

Isotropic γ-ray background
Quasar 3C 273 (see 9.5.1.2)

Galactic γ-ray sources:

| | |
|---|---|
| Diffuse emission from the Galactic disk (see Fig. 3) | Cosmic ray interactions with interstellar matter |
| Discrete Galactic sources: | |
| Crab pulsar } (see 5.6.2) | Period: $P=33$ ms |
| Vela pulsar } (see 5.6.2) | $P=89$ ms |
| ϱ Oph molecular cloud, Orion cloud complex (see 7.2) | Cosmic ray interactions |

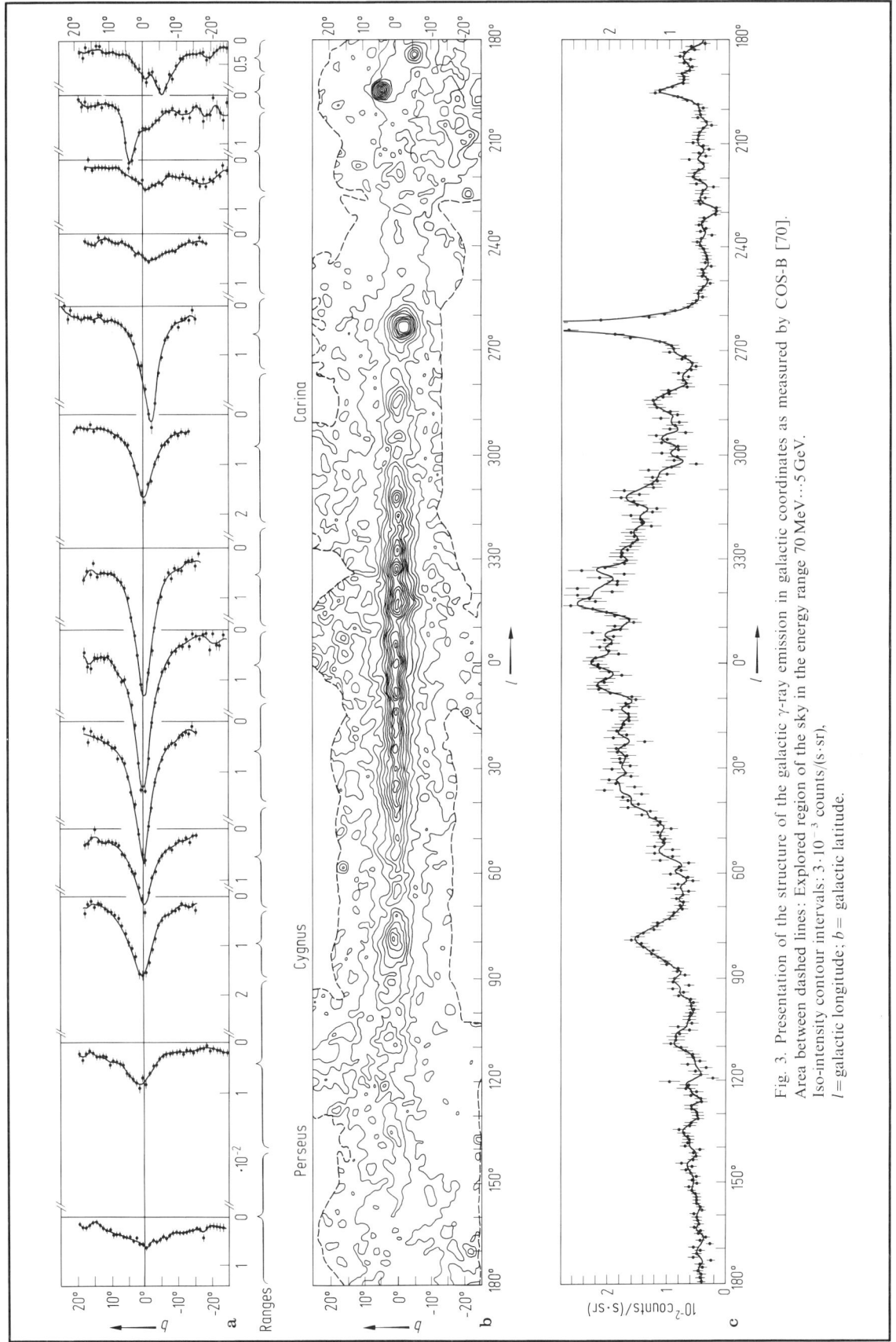

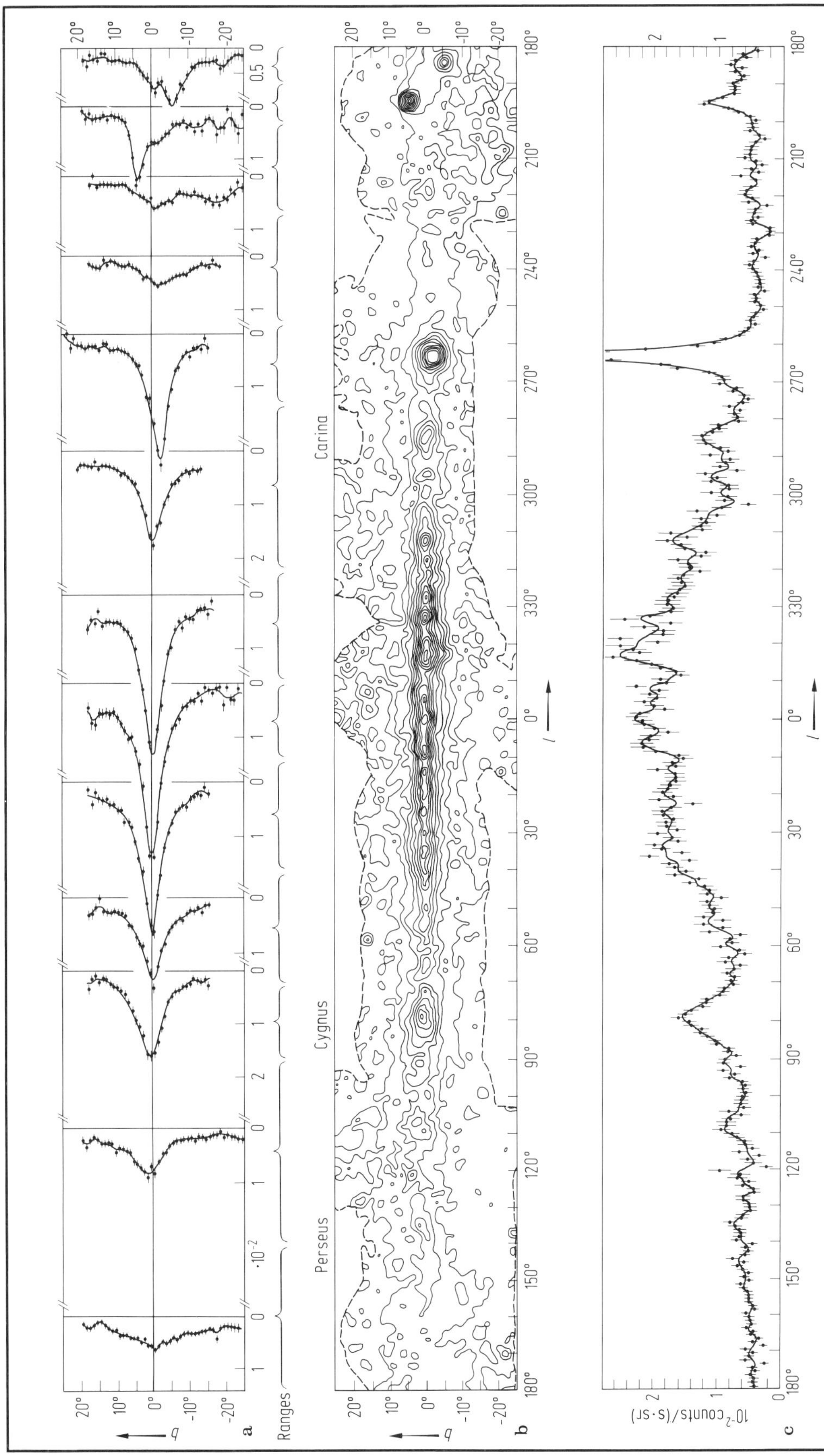


Fig. 3. Presentation of the structure of the galactic γ-ray emission in galactic coordinates as measured by COS-B [70]. Area between dashed lines: Explored region of the sky in the energy range 70 MeV···5 GeV. Iso-intensity contour intervals: $3\cdot10^{-3}$ counts/(s·sr), $l$ = galactic longitude; $b$ = galactic latitude.

γ-ray bursters:

The Vela satellites have discovered cosmic γ-ray bursts [64], which have been further studied by a number of satellites.

The events are characterized by time scales of seconds and a quasi-isotropic distribution in the sky. Recent observations with the Venus 11 and 12 space probes suggest that at least part of the γ-ray bursts are produced on magnetic neutron stars, probably by thermonuclear explosions of accreted matter [65].

### 5.7.2.2 γ-radiation processes

Both continuous and line spectra are expected from cosmic γ-ray sources. The main continuum emission processes are:

1. Neutral pion decay
Collisions between the relativistic nuclei of cosmic radiation with interstellar matter produce $\pi^0$-mesons amongst other secondary particles. The neutral pion decay, which has a lifetime of $10^{-16}$ s, yields two photons of $E_\gamma = 67.5$ MeV, in the rest frame of the decaying particle. The γ-ray spectrum produced by cosmic-ray collisions is proportional to the cosmic-ray spectrum at high energies ($E_\gamma \gg 70$ MeV), and shows a broad peak at $E_\gamma = 67.5$ MeV, [66].

2. Bremsstrahlung
Bremsstrahlung in the γ-ray range arises whenever a relativistic cosmic-ray electron is decelerated in the Coulomb field of a target nucleus of the interstellar matter. The radiation cross section for ultra-relativistic electrons is given by [67]

$$\sigma_B(E_e, E_\gamma)\mathrm{d}E_\gamma \approx \frac{\langle M\rangle}{XE_\gamma}\mathrm{d}E_\gamma \quad \text{for} \quad E_e \gg \frac{mc^2}{\alpha Z^{1/3}}$$

where $\alpha$ is the fine structure constant, $\langle M\rangle$ the average mass of the target nucleus, $Z$ its ionic charge, and $X$ the average radiation length of the stopping gas. Hereby the radiation length of each constituent of the interstellar matter is given by

$$\frac{1}{X} \approx 6 \cdot 10^{-3} Z^2 \ln(191\, Z^{-1/3}) \cdot \frac{1}{A},$$

where $A$ denotes the mass number of the target nucleus. The average radiation length for interstellar matter amounts to

$$X_{\mathrm{ISM}} \approx 65\ \mathrm{g\,cm^{-2}}$$

based on values calculated for different cosmic abundant elements by Nishimura [68].

3. Inverse Compton scattering
A continuous γ-radiation is produced if ambient low-energy photons, e.g. from stellar or universal black-body radiation fields, are scattered by relativistic cosmic-ray electrons (see also 5.7.1.3.5).

The first two processes are probably responsible for a large fraction of the observed galactic-disk emission, whereas the Compton scattering is expected to be a powerful γ-ray producing mechanism in compact extragalactic sources.

4. De-excitation of excited nuclei

5. Electron-positron annihilation
The annihilation of an electron-positron pair gives rise to a γ-ray spectrum the shape of which is determined by the detailed kinematics of the pair collision. The total cross section for the annihilation of a free electron-positron pair into two γ-photons was first calculated by Dirac [69] and is given by

$$\sigma_{e^-e^+} \approx \frac{3}{8}\sigma_T \frac{1}{\beta} \quad \text{at non-relativistic energies } (\Gamma \approx 1), \text{ where } \Gamma = \frac{E_{e^+}}{mc^2}$$

and

$$\sigma_{e^-e^+} \approx \frac{3}{8}\sigma_T \frac{[\ln(2\Gamma) - 1]}{\Gamma} \quad \text{at ultrarelativistic energies } (\Gamma \gg 1).$$

In the first case, the two-photon annihilation of the electron-positron pair almost at rest gives simply a rather sharp line feature at 0.511 MeV, isotropically emitted and occasionally Doppler broadened.

## 5.7.3 References for 5.7


1 Giacconi, R., Gursky, H., Paolini, F.R., Rossi, B.B.: Phys. Rev. Lett. **9** (1962) 439.
2 Gursky, H., Giacconi, R., Paolini, F.R., Rossi, B.B.: Phys. Rev. Lett. **11** (1963) 524.
3 Giacconi, R., Kellogg, E., Gorenstein, P., Gursky, H., Tananbaum, H.: Astrophys. J. **165** (1971) L27.
4 Vaiana, G.: Solar X-Ray Emission. In: X-Ray Astronomy (Giacconi, R., Gursky, H., eds.), Reidel, Dordrecht (1974) p. 169.
5 Vaiana, G.S., Cassinelli, J.P., Fabbiano, G., Giacconi, R., Golub, L., Gorenstein, P., Haisch, B.M., Harnden, F.R., Johnson, H.M., Linsky, J.L., Maxson, C.W., Mewe, R., Rosner, R., Seward, F., Topka, K., Zwaan, C.: Astrophys. J. **245** (1981) 163.
6 Johnson, J.M.: Astrophys. J. **243** (1981) 234.
7 Swank, J.H., White, N.E., Holt, S.S., Becker, R.H.: Astrophys. J. **246** (1981) 208.
8 Walter, F., Charles, P., Bowyer, P., Cash, W.: Astrophys. J. **236** (1980) 212.
9 Feigelson, E.D., DeCampli, W.M.: Astrophys. J. **243** (1981) L89.
10 Gahm, G.F.: Astrophys. J. **242** (1980), L163.
11 Fabbiano, G., Hartmann, L., Raymond, J., Steiner, J., Branduardi-Raymont, G., Matilsky, T.: Astrophys. J. **243** (1981) 911.
12 Cordova, F.A., Mason, K.O., Nelson, J.E.: Astrophys. J. **245** (1981) 609.
13 Hearn, D.R., Richardson, J.A., Bradt, H.V.D., Clark, G.W., Lewin, W.H.G., Mayer, W.F., McClintock, J.E., Primini, F.A., Rappaport, S.A.: Astrophys. J. **203** (1976) L21.
14 Mewe, R., Heise, J., Gronenschild, E.H.B.M., Brinkman, A.C., Schrijver, J., den Boggende, A.F.J.: Astrophys. J. **202** (1975) L67.
15 Tuohy, I., Garmire, G.: Astrophys. J. **239** (1980) L107.
16 Ghosh, P., Lamb, F.K.: Astrophys. J. (in press).
17 Pines, D.: Science **207** (1980) 597.
18 Hoffman, J.A., Lewin, W.H.G., Doty, J., Hearn, D.R., Clark, G.W., Jernigan, G., Li, F.K.: Astrophys. J. **210** (1976) L13.
19 Lewin, W.H.G., Joss, P.C.: Nature **270** (1977) 211.
20 Grindlay, J.E., Marshall, H.L., Hertz, P., Soltan, A., Weisskopf, M.C., Elsner, R.F., Ghosh, P., Darbro, W., Sutherland, P.G.: Astrophys. J. **240** (1980) L121.
21 Clark, G.W., Markert, T.H., Li, F.K.: Astrophys. J. **199** (1975) L93.
22 Doxsen, R., Jernigan, G., Hearn, D., Bradt, H., Buff, J., Clark, G.W., Delvaille, J., Epstein, A., Joss, P.C., Matilsky, T., Mayer, W., McClintock, J., Rappaport, S., Richardson, J., Schnopper, H.: Astrophys. J. **203** (1976) L9.
23 Becker, R.H., Marchall, F.E.: Astrophys. J. **244** (1981) L93.
24 Burginyon, G.A., Hill, R.W., Seward, F.: Astrophys. J. **199** (1975) 680.
25 Friedman, H.: Science **181** (1973) 395.
26 Fried, P.M., Nousek, J.A., Sanders, W.P., Kraushaar, W.L.: Astrophys. J. **242** (1980) 987.
27 Van Speybroeck, L., Epstein, A., Forman, W., Giacconi, R., Jones, C., Liller, W., Smarr, L.: Astrophys. J. **234** (1979) L45.
28 Silk, J.: Annu. Rev. Astron. Astrophys. **11** (1973) 269.
29 Maccacaro, T., Perola, G.C.: Astrophys. J. **246** (1981) L11.
30 Marshall, N., Warwick, W., Pounds, K.A.: Mon. Not. R. Astron. Soc. **194** (1981) 987.
31 Marshall, F.E., Mushotzky, R.F., Boldt, E.A., Holt, S.S., Rothschild, R.E., Serlemitsos, P.J.: Astrophys. J. (in press).
32 Baity, W.A., Rothschild, R.E., Lingenfelter, R.E., Stein, W.A., Nolan, P.L., Gruber, D.E., Knight, K.F., Matteson, J.L., Peterson, L.E., Primini, F.A., Levine, A.M., Lewin, W.H.G., Mushotzky, R.F., Tennant, A.F.: Astrophys. J. **244** (1981) 429.
33 Kriss, G.A., Canizares, C.R., Ricker, G.R.: Astrophys. J. **242** (1980) 492.
34 Zamorani, G., Henry, J.P., Maccacaro, T., Tananbaum, H., Soltan, A., Avni, Y., Liebert, J., Stocke, J., Strittmatter, P.A., Weymann, R.J., Smith, M.G., Condon, J.J.: Astrophys. J. **245** (1981) 357.
35 Tananbaum, H., Avni, Y., Branduardi, G., Elvis, M., Fabbiano, G., Feigelson, E., Giacconi, R., Henry, J.P., Pye, J.P., Soltan, A., Zamorani, G.: Astrophys. J. **234** (1979) L9.
36 Ku, W.: NASA CP-2113 (1979).
37 Mitchell, R.J., Dickens, R.J., Bell-Burnell, S.J., Culhane, J.L.: Mon. Not. R. Astron. Soc. **189** (1979) 329.
38 McKee, J.D., Mushotzky, R.F., Boldt, E.A., Holt, S.S., Marshall, F.E., Pravdo, S.H., Serlemitsos, P.J.: Astrophys. J. **242** (1980) 843.

39 Schwartz, D.A., Davis, M., Doxsey, R.E., Griffiths, R.E., Huchra, J., Johnston, M.D., Mushotzky, R.F., Swank, J., Tonry, J.: Astrophys. J. **238** (1980) L53.
40 Kriss, G.A., Canizares, C.R., McClintock, J.E., Feigelson, E.D.: Astrophys. J. **235** (1980) L61.
41 Boldt, E.A.: NASA TM-80659 (1980).
42 Forman, W., Jones, C., Cominsky, L., Julien, P., Murray, S., Peters, G., Tananbaum, H., Giacconi, R.: Astrophys. J. Suppl. **38** (1978) 357.
43 Markert, T.H., Winkler, P.F., Laird, F.N., Clark, G.W., Hearn, D.H., Sprott, G.F., Li, F.K., Bradt, H.V., Lewin, W.H.G., Schnopper, W.H.: Astrophys. J. Suppl. **39** (1979) 573.
44 Cooke, B.A., Ricketts, M.J., Maccacaro, T., Pye, J.P., Elvis, M., Watson, M.G., Griffiths, R.E., Pounds, K.A., McHardy, I., Maccagni, D., Seward, F.D., Page, C.G., Turner, M.J.L.: Mon. Not. R. Astron. Soc. **182** (1978) 489.
45 Villa, G., Page, C.G., Turner, M.J.L., Cooke, B.A., Ricketts, M.J., Pounds, K.A.: Mon. Not. R. Astron. Soc. **176** (1976) 609.
46 Levine, A., Bautz, M., Howe, S., Lang, F., Primini, F., Lewin, W.H.G., Baity, W., Gruber, D., Knight, F., Peterson, L., Rothschild, R., Matteson, J., Nolan, P.: Astrophys. J. Suppl. (in press).
47 Tucker, W.H.: Radiation Processes in Astrophysics, MIT Press, Cambridge, Mass. (1975).
48 Karzas, W., Latter, R.: Astrophys. J. Suppl. **6** (1961) 167.
49 Gorenstein, P., Gursky, H., Garmire, G.: Astrophys. J. **153** (1968) 885.
50 Kato, K.: Astrophys. J. Suppl. **30** (1976) 397.
51 Elwert, G.: Z. Naturforschung **53** (1954) 637.
52 Tucker, W.H., Gould, R.J.: Astrophys. J. **144** (1966) 244.
53 Culhane, J.: Mon. Not. R. Astron. Soc. **144** (1969) 375.
54 Tucker, W.H., Koren, M.: Astrophys. J. **168** (1971) 283.
55 Blumenthal, G.R., Tucker, W.H.: Mechanisms for the Production of X-Rays in Cosmic Setting. In: X-Ray Astronomy (Giacconi, R., Gursky, H., eds.), Reidel, Dordrecht (1974) p. 144.
56 Canuto, V., Ventura, J.: Quantizing Magnetic Fields in Astrophysics, New York, GSFC/NASA (1976).
57 Trümper, J., Pietsch, W., Reppin, C., Voges, W., Staubert, R., Kendziorra, E.: Astrophys. J. **219** (1978) L105.
58 Jackson, J.D.: Classical Electrodynamics, John Wiley & Sons, New York, London (1962).
59 Felten, J.E., Morrison, P.: Astrophys. J. **146** (1966) 686.
60 Katz, J.I.: Astrophys. J. **206** (1976) 910.
61 Shapiro, S.L., Lightman, A.P., Eardley, D.M.: Astrophys. J. **204** (1976) 187.
62 Sunyaev, R.A., Titarchuk, L.G.: Academia Nauk., UdSSR, IKI Paper No. 441 (1979).
63 Sunyaev, R.A., Trümper, J.: Nature **279** (1979) 506.
64 Klebesadel, R.W., Strong, I.B., Olson, R.A.: Astrophys. J. **182** (1973) L85.
65 Mazets, E.P., Golenetskii, S.V., Aptekar, R.L., Guryan, Yu.A., Ilyinskii, V.N.: Nature **290** (1981) 378.
66 Stecker, F.W.: Cosmic Gamma Rays, NASA SP-249 (1971).
67 Heitler, W.: The Quantum Theory of Radiation, Oxford Press, London (1966).
68 Nishimura, J.: Theory of Cascade Showers. In: Hdb. d. Physik, Vol. XLVI/2, Springer, Berlin, New York (1967).
69 Dirac, P.A.M.: Proc. Cambridge Phil. Soc. **26** (1930) 361.
70 Mayer-Hasselwander, H.A., Bennett, K., Bignami, G.F., Buccheri, R., Caraveo, P.A., Hermsen, W., Kanbach, G., Lebrun, F., Lichti, G.G., Masnou, J.L., Paul, J.A., Pinkau, K., Sacco, B., Scarsi, L., Swanenburg, B.N., Wills, R.D.: Astron. Astrophys. (1982) in press.
71 Piccinotti, G., Mushotzky, R.F., Boldt, E.A., Holt, S.S., Marshall, F.E., Serlemitsos, P.J., Shafer, R.A.: NASA TN-82168 (1981).

# 7 Interstellar matter

## 7.1 Phenomena of the generally distributed medium

For definition of the different colour systems (*UBV*, *RGU* ...), see 4.2.5.1 ... 4.2.5.17.

### 7.1.1 General references


Aannestad, P.A., Purcell, E.M.: Interstellar Grains, Annu. Rev. Astron. Astrophys. **11** (1973) 309.

Gehrels, T. (ed.): Planets, Stars and Nebulae studied with Photopolarimetry, Int. Astron. Union Coll. **23**, Univ. Arizona Press, Tucson (1974).

Greenberg, J.M., van de Hulst, H.C. (eds.): Interstellar Dust and Related Topics, Int. Astron. Union Symp. **52** (1974).

Greenberg, J.M.: Interstellar Dust, in: Cosmic Dust (McDonnell, J.A.M., ed.), Wiley, New York (1978).

Kerr, F.J., Simonson, S.C. (eds.): Galactic Radio Astronomy, Int. Astron. Union Symp. **60** (1974).

Lynds, B.T., Wickramasinghe, N.C.: Interstellar Dust, Annu. Rev. Astron. Astrophys. **6** (1968) 215.

Lynds, B.T. (ed.): Dark Nebulae, Globules and Protostars, Univ. Arizona Press, Tucson (1971).

Middlehurst, B.M., Aller, L.H. (eds.): Nebulae and Interstellar Matter, Stars and Stellar Systems Vol. VII, Univ. Chicago Press (1968).

Pinkau, K. (ed.): The Interstellar Medium (NATO Advanced Study Inst. Schliersee, April 1973), Reidel, Dordrecht (1974).

Savage, B.D., Mathis, J.S.: Observed Properties of Interstellar Dust, Annu. Rev. Astron. Astrophys. **17** (1979) 73.

Van Woerden, H. (ed.): Topics in Interstellar Matter (Astrophys. and Space Science Library Vol. **70**), Reidel, Dordrecht (1977).

Verschuur, G.L., Kellermann, K.I. (eds.): Galactic and Extragalactic Radio Astronomy, Springer, Berlin (1974).

Wickramasinghe, N.C.: Interstellar Grains, Chapman and Hall, London (1967).

Wickramasinghe, N.C., Kahn, F.D., Mezger, P.G.: Interstellar Matter (Swiss Soc. Astron. Astrophys. Course, Saas-Fee, March 1972) Geneva Obs. Sauverny (1972).

Wickramasinghe, N.C., Nandy, K.: Recent Work on Interstellar Grains, Rep. Prog. Phys. **35** (1972) 157.


### 7.1.2 Interstellar extinction

#### 7.1.2.1 Mean values and fluctuations

Table 1. Mean colour excess $\bar{E}_{B-V}(r)$, mean visual extinction $\bar{A}_V(r)$ and rms fluctuations $\sigma_{A_V}(r)=\langle(A_V-\bar{A}_V)^2\rangle^{1/2}$ for stars at low galactic latitudes (distance from the galactic plane $|z|\lessapprox 75$ pc) at different distances $r$ [9, 16].

| $r$ [pc] | $\bar{E}_{B-V}$ | $\bar{A}_V$ | $\sigma_{A_V}$ |
|---|---|---|---|
| 200 | 0ᵐ050 | 0ᵐ15 | 0ᵐ3 |
| 300 | 0.14 | 0.4 | 0.4 |
| 500 | 0.30 | 0.9 | 0.9 |
| 800 | 0.52 | 1.6 | 1.2 |
| 1000 | 0.65 | 2.0 | 1.2 |
| 1300 | 0.72 | 2.2 | 1.2 |
| 2000 | 0.72 | 2.2 | 1.0 |

The mean colour excess $\bar{E}_{B-V}(b)$ at galactic latitude $b$ for objects outside the absorbing layer can be roughly estimated by [3, 8, 26]; see also LB, NS, Vol. VI/1 (1965) p. 665,

$$\bar{E}_{B-V}(b)=0.05 \text{ cosec } |b| \text{ for middle latitudes } b$$

or, for all $b$, since $E_{B-V}\approx 0$ at $|b|\approx 90°$, by [23]

$$\bar{E}_{B-V}(b)=0.06 \text{ cosec } |b|-0.06.$$

Mean half-width of the absorbing layer at 1/e of the maximum extinction for $r\lessapprox 0.5$ kpc [13]: $z_0\approx 40$ pc.

Visual extinction to the galactic center [1]: $A_V\approx 30^m$.

Relation of reddening $E_{B-V}$ [mag] to hydrogen column density $\mathfrak{N}(H)$ (including molecular and ionized hydrogen) [2]:

$$\mathfrak{N}(H)=5.8\cdot 10^{21}\, E_{B-V} \text{ [atoms cm}^{-2}\text{]}.$$

### 7.1.2.2 Interstellar reddening law

Symbols

$A_\lambda$ interstellar extinction in [mag] at wavelength $\lambda$

$A_V$ interstellar extinction in [mag] for the visual ($V$, $\lambda \approx 0.55$ µm) wavelength of the $UBV$ system

$E_{\lambda-V}$ colour excess in [mag], defined as the difference between the extinction $A_\lambda$ at an arbitrary wavelength $\lambda$ and $A_V$.

The ratio of total extinctions $A_\lambda/A_V$ can be expressed by the normalized colour excess $E_{\lambda-V}/E_{B-V}$ as derived from observations and the ratio $R = A_V/E_{B-V}$ (cf. 7.1.2.3):

$$\frac{A_\lambda}{A_V} = 1 + \frac{E_{\lambda-V}}{E_{B-V}} \cdot \frac{1}{R}.$$

The observed reddening "law" shows variations in the sky from region to region. Large deviations occur due to very localized conditions (H II regions, dense regions of star formation). For UV extinction, see [27, 28].

Table 2. Mean wavelength-dependence of the extinction ratio $A_\lambda/A_V$ derived from observations in the UV, optical and infrared regions [1, 7, 12, 17].

| $\lambda$ [µm] | $1/\lambda$ [µm$^{-1}$] | $A_\lambda/A_V$ | $\lambda$ [µm] | $1/\lambda$ [µm$^{-1}$] | $A_\lambda/A_V$ |
|---|---|---|---|---|---|
| 0.1375 | 7.28 | 2.73 | 0.5480 | 1.82 | 1.00 |
| 0.1590 | 6.30 | 2.57 | 0.5840 | 1.71 | 0.91 |
| 0.1830 | 5.46 | 2.53 | 0.6050 | 1.65 | 0.88 |
| 0.2000 | 5.00 | 2.80 | 0.6436 | 1.55 | 0.83 |
| 0.2080 | 4.80 | 2.98 | 0.7100 | 1.41 | 0.71 |
| 0.2140 | 4.67 | 3.09 | 0.7550 | 1.32 | 0.68 |
| 0.2190 | 4.57 | 3.10 | 0.8090 | 1.24 | 0.62 |
| 0.2230 | 4.49 | 3.07 | 0.8446 | 1.18 | 0.58 |
| 0.2360 | 4.23 | 2.68 | 0.871 | 1.15 | 0.54 |
| 0.2500 | 4.00 | 2.39 | 0.970 | 1.03 | 0.47 |
| 0.2740 | 3.65 | 2.03 | 1.061 | 0.94 | 0.40 |
| 0.3200 | 3.13 | 1.74 | 1.087 | 0.92 | 0.38 |
| 0.3400 | 2.94 | 1.64 | 1.25 | 0.80 | 0.30 |
| 0.3636 | 2.75 | 1.56 | 2.2 | 0.455 | 0.15 |
| 0.4036 | 2.48 | 1.43 | 3.4 | 0.294 | 0.10 |
| 0.4255 | 2.35 | 1.35 | 4.9 | 0.204 | 0.05 |
| 0.4566 | 2.19 | 1.26 | 8.7 | 0.115 | 0.01 |
| 0.5000 | 2.00 | 1.16 | 10.0 | 0.100 | 0.01 |

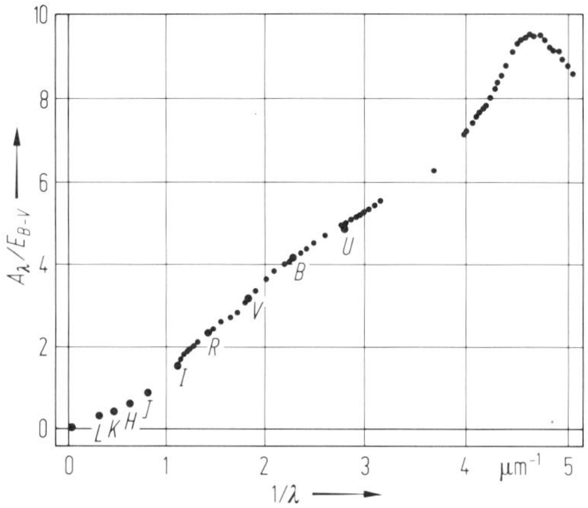


Fig. 1. The interstellar reddening law expressed by $A_\lambda/E_{B-V}$, the extinction in [mag] for a colour excess $E_{B-V} = 1.0$, as a function of $1/\lambda$ [µm$^{-1}$] [17].

### 7.1.2.3 Ratio of total to selective extinction

The ratio $R = A_V/E_{B-V}$ seems to be fairly constant, $R \approx 3$, in the solar neighbourhood. Deviations up to $R \approx 6$ are well documented for some regions of star formation. Recent determinations of $R$ for normal regions have been summarized by Crawford and Mandwewala (1976) [4] indicating

$$R = 3.2 \pm 0.1.$$

Many authors use the value $R = 3.1$, e.g. [17, 18, 20].
Reddening ratio and relations between $UBV$, $RGU$ and $uvby$ systems [4, 21]:

$$\frac{E_{U-B}}{E_{B-V}} = 0.72 + 0.05\, E_{B-V}$$

$$E_{G-R} = 1.39\, E_{B-V} + 0.015\, E_{B-V}^2$$

$$E_{U-G} = 0.70\, E_{G-R}, \quad A_G = 2.69\, E_{G-R}$$

$$E_{b-y} = 0.74\, E_{B-V}, \quad A_V = 4.3\, E_{b-y}$$

$$E_{u-b} = 1.5\, E_{b-y}, \quad E_{C_1} = 0.19\, E_{b-y}, \quad E_{m_1} = -0.33\, E_{b-y}.$$

Relation to interstellar polarization: The wavelength $\lambda_{max}$ at which the maximum interstellar linear polarization occurs (cf. 7.1.3.3) is well correlated with the ratios $E_{\lambda-V}/E_{B-V}$. In particular, the ratio $R = A_V/E_{B-V}$ can be estimated for any individual star from the relationship [19, 22]

$$R = 5.5\, \lambda_{max}$$

with the mean observed value $\lambda_{max} \approx 0.55$ µm corresponding to $R \approx 3.0$.
For regions of very high extinction, $A_V$ can be roughly estimated from the optical depth, $\tau$, of the absorption feature at $\lambda \approx 9.7$ µm using the relation [6]

$$A_V/\tau_{9.7\,\mu m} \approx 10 \cdots 20.$$

### 7.1.2.4 Visual extinction or colour excess as a function of position ($l$, $b$) and distance $r$

| Author | Number of stars | Presentation of data |
|---|---|---|
| Neckel, T. (1966, 1967) [13, 14] | 4697 | Distribution over the sky ($r \lessapprox 1$ kpc). Extinction-distance diagrams for 207 fields, 160 within $\lvert b \rvert < 7°$. Distribution in the galactic plane ($r \lessapprox 3$ kpc). |
| FitzGerald, M.P. (1968) [5] | 7835 | Distribution over the sky for fixed distances $r = 0.5$, 1.0, 2.0 kpc. Colour excess-distance diagrams for 74 fields near the galactic equator. Distribution in the galactic plane ($r \lessapprox 3$ kpc). |
| Deutschman, W.A., Davis, R.J., Schild, R.E. (1976) [24] | 2846 | Extinction-distance diagrams from $UBV$ and $H\beta$ observations for stars of the Celescope Catalogue. |
| Lucke, P.B. (1978) [11] | ≈ 4000 | Distribution of mean $E_{B-V}$ per kpc for $r \lessapprox 2$ kpc over the sky. Distribution in the galactic plane and in 6 vertical sections. |
| Krautter, J. (1980) [10] | 3748 | Distribution of extinction in the galactic plane ($r < 3$ kpc) as derived from polarization data. |
| Neckel, T., Klare, G. (1980) [15] | 11072 | Extinction-distance diagrams. Distribution over the sky and in the galactic plane. |

Catalogues of $A_V$ values:

Individual extinction values $A_V$ for 4697 stars and clusters of the whole sky have been published by T. Neckel (1967) [14]. An extension of this catalogue including interstellar extinction data for about 12000 stars is prepared by G. Klare, T. Neckel and M. Sarcander and can be ordered from the Centre de Données Stellaires, in Strasbourg (Bull. Inform. CDS No. 19, 1980, p. 61).

Extinction for galactic clusters: see [25].

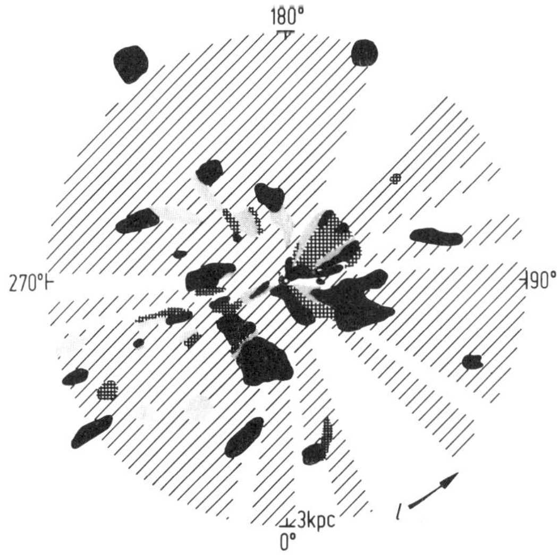


Fig. 2. Distribution of reddening material in the galactic plane derived by Neckel and Klare (1980) [15].
Hachured areas: $a_v = A_v/\text{kpc} < 1^m.0/\text{kpc}$
dotted areas: $1^m.0/\text{kpc} \leqq a_v < 2^m.0/\text{kpc}$
criss-crossed areas: $2^m.0/\text{kpc} \leqq a_v < 3^m.0/\text{kpc}$
black areas: $a_v \geqq 3^m.0/\text{kpc}$

## 7.1.2.5 References for 7.1.2


1 Becklin, E.E., Neugebauer, G.: Astrophys. J. **151** (1968) 145.
2 Bohlin, R.C., Savage, B.D., Drake, J.F.: Astrophys. J. **224** (1978) 132.
3 Burstein, D., Heiles, C.: Astrophys. Lett. **19** (1978) 69.
4 Crawford, D.L., Mandwewala, N.: Publ. Astron. Soc. Pac. **88** (1976) 917.
5 FitzGerald, M.P.: Astron. J. **73** (1968) 983.
6 Gillett, F.C., Forrest, W.J., Merrill, K.M., Capps, R.W., Soifer, B.T.: Astrophys. J. **200** (1975) 609.
7 Hackwell, J.A., Gehrz, R.G.: Astrophys. J. **194** (1974) 49.
8 Heiles, C.: Astrophys. J. **204** (1975) 379.
9 Knude, J.: Astron. Astrophys. **71** (1979) 344.
10 Krautter, J.: Astron. Astrophys. **89** (1980) 74.
11 Lucke, P.B.: Astron. Astrophys. **64** (1978) 367.
12 Nandy, K., Thompson, G.I., Jamar, C., Monfils, A., Wilson, R.: Astron. Astrophys. **44** (1975) 195.
13 Neckel, T.: Z. Astrophys. **63** (1966) 221.
14 Neckel, T.: Veröff. Landessternwarte Heidelberg-Königstuhl Band **19** (1967).
15 Neckel, T., Klare, G.: Astron. Astrophys. Suppl. **42** (1980) 251.
16 Scheffler, H.: Z. Astrophys. **65** (1967) 60.
17 Schild, R.E.: Astron. J. **82** (1977) 337.
18 Schultz, G.V., Wiemer, W.: Astron. Astrophys. **43** (1975) 133.
19 Serkowski, K., Mathewson, D.S., Ford, V.L.: Astrophys. J. **196** (1975) 261.
20 Sneden, C., Gehrz, R.D., Hackwell, J.A.: Astrophys. J. **223** (1978) 168.
21 Steinlin, U.W.: Z. Astrophys. **69** (1968) 276.
22 Whittet, D.C.B., van Breda, I.G.: Astron. Astrophys. **66** (1978) 57.
23 Woltjer, L.: Astron. Astrophys. **42** (1975) 109.
24 Deutschman, W.A., Davis, R.J., Schild, R.E.: Astrophys. J. Suppl. **30** (1976) 97.
25 Spaenhauer, A.M.: Astron. Astrophys. **83** (1980) 234.
26 Milne, D.K., Aller, L.H.: Astron. J. **85** (1980) 17.
27 Seab, C.G., Snow, T.P., Joseph, C.L.: Astrophys. J. **246** (1981) 788.
28 Mezger, P.G., Mathis, J.S., Panagia, N.: Astron. Astrophys. **105** (1982), Table A1, p. 383.

# 7.1.3 Interstellar polarization of starlight

## 7.1.3.1 Symbols and definitions

$I_{max}$ intensity of the partially linearly polarized starlight in the plane of polarization (=plane of vibration of the electric vector)

$I_{min}$ intensity in the orthogonal plane

$I = I_{max} + I_{min}$ = total intensity

$P = (I_{max} - I_{min})/(I_{max} + I_{min})$, degree of linear polarization

$\Delta m_P = -2.5 \log \frac{I_{min}}{I_{max}} = 2.172\,P \quad (P \ll 1)$, polarization expressed by a magnitude difference

$\theta_E$ position angle of polarization in the equatorial coordinate system

$\theta_G$ position angle of polarization in the galactic coordinate system; $\theta_G$ can be computed from $\theta_E$ by means of the formula [2]

$$\cot(\theta_E - \theta_G) = \frac{\cos b \tan b_N - \cos(l - l_N) \sin b}{\sin(l - l_N)}$$

where $l$, $b$ are the galactic coordinates of the star, and $l_N$, $b_N$ are the galactic coordinates of the equatorial north pole for the equinox of the observations.

$V/I$ degree of circular polarization (V = Stokes parameter)

## 7.1.3.2 Degree and angle of linear polarization

Table 3. Catalogues.
$n$ = number of stars

| | Author | Publication | $n$ | Region |
|---|---|---|---|---|
| 1 | Hiltner, W.A. | Astrophys. J. Suppl. **2** (1956) 389 | 1259 | $l = 350 \cdots 0 \cdots 240°$ |
| 2 | Smith, Elske van P. | Astrophys. J. **124** (1956) 43 | ≈ 200 | $l = 230 \cdots 0 \cdots 20°$ |
| 3 | Hall, J.S. | Publ. U.S. Naval Obs. Washington 2nd Ser. Vol. **17**, No. 6 (1958) (includes 1 and 2) | 2592 | $l = 0 \cdots 360°$ |
| 4 | Behr, A. | Veröffentl. d. Univ. Sternwarte Göttingen Nr. 126 (1959) | 550 | $l = 0 \cdots 360°$ ($r < 250$ pc) |
| 5 | Lodén, L.O. | Stockholm Obs. Ann. **21**, No. 7 (1961) | 1300 | mainly Selected Areas $l = 50 \cdots 125°$ |
| 6 | Appenzeller, I. | Z. Astrophys. **64** (1966) 269 | 320 | Cygnus and Orion |
| | | Astrophys. J. **151** (1968) 907 | 308 | $130 \cdots 180°$ ($r < 200$ pc) |
| 7 | Mathewson, D.S., Ford, V.L. | Mem. R. Astron. Soc. **74** (1970) 139 | 1800 | southern sky: $\delta \lessapprox +30°$ |
| 8 | Schmidt, Th. | Z. Astrophys. **68** (1968) 380 | 143 | $l = 127 \cdots 133°$, $\|b\| \lessapprox 25°$ |
| 9 | Schöder, R. | Astron. Astrophys. Suppl. **23** (1976) 499 | 511 | $b < -45°$ |
| 10 | Klare, G., Neckel, T. | Astron. Astrophys. Suppl. **27** (1977) 215 | 1660 | southern sky |
| 11 | Markkannen, T. | Obs. and Astrophys. Laboratory Univ. of Helsinki, Report 1, 1977 | 81 | $l = 130 \cdots 170°$ |
| 12 | Krautter, J. | Astron. Astrophys. Suppl. **39** (1980) 167 | 132 | $l = 35 \cdots 113°$, $178 \cdots 203°$ |
| | | | 181 | $l = 231 \cdots 299°$ |
| 13 | Mathewson, D.S., Ford, V.L., Klare, G., Neckel, T., Krautter, J. | Polarization Catalogue Bull. d'Inf. Centre de Données Stellaires, Strasbourg **15** (1978) 125 on microfiche (C 2034.1) | 7503 | whole sky (includes catalogues 7 and 9) |

The largest polarization recorded in these surveys is about $P=0.07$. Recent observations have shown, however, that stars in dusty H II regions may exhibit linear polarizations up to about 25% [9, 12].

The distribution of polarization position angles $\theta_G$ over the sky for the stars included in catalogues 1 to 7 of Table 3 is shown in Fig. 3. Corresponding plots for the stars of catalogue 9 have been given by Klare and Neckel (1971) [6].

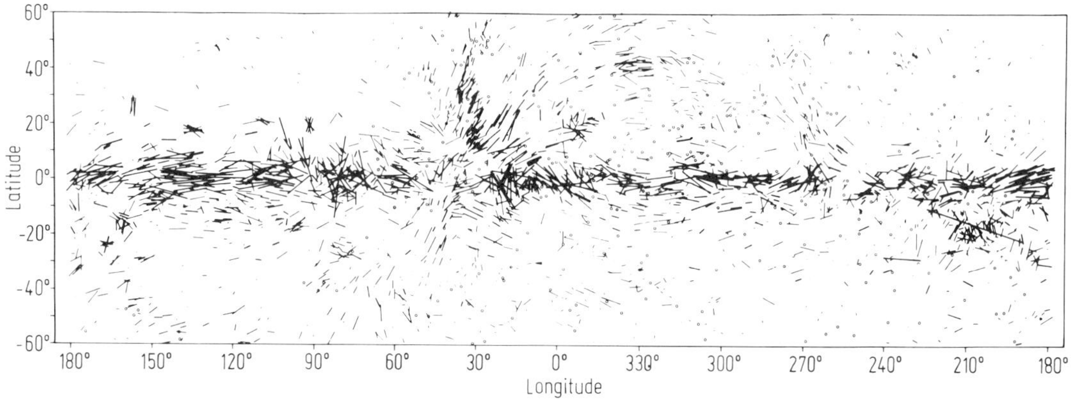


Fig. 3. The results for $P$ and $\theta_G$ from Catalogues 1 to 7 plotted in galactic coordinates. The length of each line is proportional to $P$, and the angle at which each line is drawn relative to the galactic north pole in the direction of increasing longitude is $\theta_G$. Stars with $P<0.08\%$ are represented by small circles [8].

### 7.1.3.3 Wavelength-dependence of linear polarization

Generally the polarization degree $P(\lambda)$ shows a broad maximum somewhere between $\lambda=0.5\cdots0.7\ \mu m$, but there are significant variations from star to star. A striking similarity in behavior emerges if $P(\lambda)/P_{max}$ is plotted against $\lambda_{max}/\lambda$, where $\lambda_{max}$ is the wavelength of maximum polarization of the star $P_{max}=P(\lambda_{max})$ (Fig. 4).

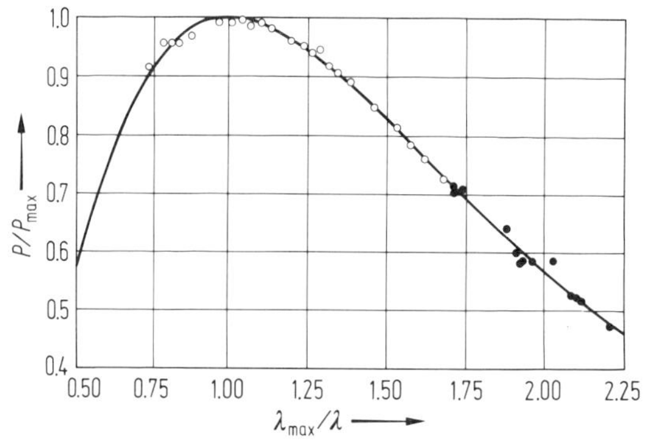


Fig. 4. Normalized wavelength-dependence of interstellar linear polarization: $P(\lambda)/P_{max}$ versus $\lambda_{max}/\lambda$. Every open circle is based on 20 stars; full circles represent the observations of an individual star.

The observed polarizations normalized in this way follow closely the curve [10, 3]

$$\frac{P(\lambda)}{P_{max}}=\exp\{-1.15(\ln(\lambda_{max}/\lambda))^2\}.$$

$0.45\ \mu m \leqq \lambda_{max} \leqq 0.8\ \mu m$; $\langle\lambda_{max}\rangle=0.55\ \mu m$.

More recent measurements and discussion: see [12, 14].

### 7.1.3.4 Interdependence of linear polarization and extinction

There is never interstellar polarization without extinction, but there may be extinction with little or no polarization (Fig. 5). Ratio of polarization in [mag] to visual extinction [1, 4]

$$\frac{\Delta m_P(V)}{A_V} \lessapprox 0.065\,.$$

Higher values are exceedingly few and unreliable. On the average one may take

$$\frac{\Delta m_P(V)}{A_V} \approx 0.03\,.$$

For the ratio of maximum polarization $P_{max} = P(\lambda_{max})$ in [%] to colour excess $E_{B-V}$ in [mag] Serkowski, Mathewson and Ford (1975) [10] find (see also Fig. 5)

$$P_{max}\,[\%] \lessapprox 9.0\,E_{B-V}\,.$$

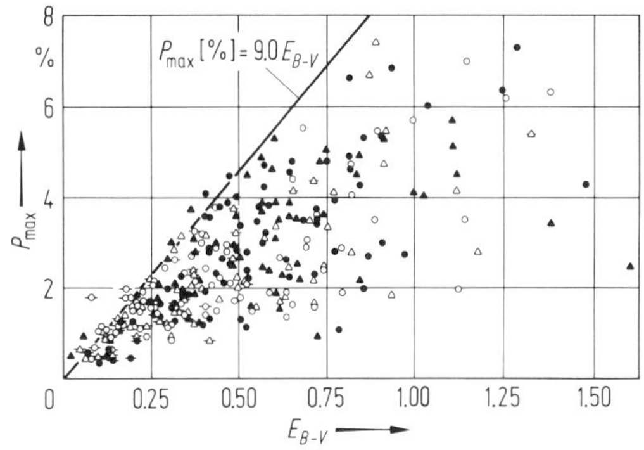


Fig. 5. Correlation between maximum percentage polarization $P_{max}$ [%] and colour excess $E_{B-V}$ [10].

| Symbol for star | $\lambda_{max}$ |
|---|---|
| full triangle | $\leqq 0.51$ µm |
| full circle | 0.52···0.54 µm |
| open circle | 0.55···0.57 µm |
| open triangle | $\geqq 0.58$ µm |

Symbols with a horizontal bar denote stars nearer than 400 pc.

### 7.1.3.5 Circular polarization

Weak circular polarizations $V/I \approx 1 \cdots 3 \cdot 10^{-4}$ with errors of about $0.3 \cdot 10^{-4}$ have been detected for several stars by Kemp and Wolstencroft (1972) [5] and by Martin (1974) [7]; see also [13] and contributions of these authors in [11]. The observed circular polarization changes with wavelength showing a reversal of sign at about 6000 Å $\approx \lambda_{max}$.

### 7.1.3.6 References for 7.1.3


1 Aannestad, P.A., Purcell, E.M.: Annu. Rev. Astron. Astrophys. **11** (1973) 309.
2 Appenzeller, I.: Astrophys. J. **151** (1968) 907.
3 Coyne, G.V., Tapia, S., Vrba, F.J.: Astron. J. **84** (1979) 356.
4 Greenberg, J.M. in: Cosmic Dust (McDonnell, J.A.M., ed.), Wiley, New York (1978) p. 195.
5 Kemp, J.C., Wolstencroft, R.D.: Astrophys. J. Lett. **176** (1972) L115.
6 Klare, G., Neckel, T.: Astron. Astrophys. **11** (1971) 155.
7 Martin, P.G.: Astrophys. J. **187** (1974) 461.
8 Mathewson, D.S., Ford, V.L.: Mem. R. Astron. Soc. **74** (1970) 139.
9 Schulz, A., Proetel, K., Schmidt, Th.: Astron. Astrophys. Lett. **64** (1978) L13.
10 Serkowski, K., Mathewson, D.S., Ford, V.L.: Astrophys. J. **196** (1975) 261.
11 Gehrels, T. (ed.): Planets, Stars and Nebulae studied with Photopolarimetry, Int. Astron. Union Coll. **23**, Univ. Arizona Press, Tucson (1974).
12 Schulz, A., Lenzen, R., Schmidt, Th., Proetel, K.: Astron. Astrophys. **95** (1981) 94.
13 Martin, P.G., Angel, J.R.P.: Astrophys. J. **207** (1976) 126.
14 Wilking, B.A., Lebofsky, M.J., Martin, P.G., Rieke, G.H., Kemp, J.C.: Astrophys. J. **235** (1980) 905.

# 7.1.4 Scattering of starlight by interstellar dust

## 7.1.4.1 Reflection nebulae

Table 4. Catalogues of reflection nebulae.
$n$ = number of objects

| Author | Publication | $n$ | Region | Data |
|---|---|---|---|---|
| Dorschner, J., Gürtler, J. | Astron. Nachr. **287** (1964) 257; **289** (1965) 57 | 192 | $\delta \gtrapprox -33°$ | $\alpha, \delta, l, b$; size, brightness of illuminating star |
| Lynds, B.T. [1]) | Astrophys. J. Suppl. **12**, No. 105 (1965) 163 | 1163 | $\delta \gtrapprox -33°$ | $\alpha, \delta, l, b$; size, area, colour class, brightness class |
| Van den Bergh, S. | Astron. J. **71** (1966) 990; Data of illuminating stars see: Racine, R.: Astron. J. **73** (1968) 233 | 158 | $\delta \gtrapprox -33°$ | $l, b$; size; colour class, brightness class |
| Rozhkovski, D.A., Kurchakov, A.V. | Trudy Astrophyiz. Inst. Akad. Nauk Kazakh. S.S.R. Alma Ata **11** (1963) 3 | 118 | $\delta \gtrapprox -25°$ | $\alpha, \delta; l, b$; size, surface brightness, data of illuminating stars |
| Van den Bergh, S., Herbst, W. | Astron. J. **80** (1975) 208; Data of illuminating stars see: Herbst, W.: Astron. J. **80** (1975) 212 | 93 | $\delta < -33°$ $l = 250° \cdots 360°$ | $l, b$; size; colour class, brightness class |
| Bernes, C. [2]) | Astron. Astrophys. Suppl. **29** (1977) | 160 | nearly whole Milky Way | $\alpha, \delta; l, b$; size, brightness class |

[1]) Catalogue contains all types of bright diffuse nebulae.
[2]) Catalogue of bright nebulosities (of all types) in opaque dust clouds.

Apparent distribution over the sky: Rozhkovski and Kurchakov (1968) (see Table 4); van den Bergh and Racine (1973) [15].

Hubble relation: Dorschner and Gürtler (1965) (see Table 4) find

$$m_B = -(4.95 \pm 0.17) \log a + 12.01 \pm 0.16$$

where $m_B$ is the apparent $B$ magnitude of the illuminating star and $a$ denotes the angular distance between the star and the most distant visible point of the nebula in arc minutes seen on the prints of the Palomar Observatory Sky Survey.

Distribution of diameters: see LB, NS, Vol. VI/I (1965) p. 656.

Table 5: see next page.

Detailed quantitative photometric and polarimetric measurements of typical reflection nebulae:

Merope nebula: [2, 10, 18]; NGC 7023, NGC 2068: [2, 20]. Results prior to 1965: [4].

Table 5. Selected bright reflection nebulae and illuminating stars.

Angular size — dimensions in [arc min]; mostly largest and smallest values estimated from the blue prints of the Palomar Observatory Sky Survey
$V/\square''$ — visual surface brightness in [mag per square second of arc]
$\Delta(B-V)$ — color difference between nebulae and illuminating star $(B-V)_n-(B-V)_s$, mean values
HD/BD — number in HD or BD catalogue
Sp — spectral type

| Name | $\alpha_{1950}$ | $\delta_{1950}$ | Angular size [1, 3, 7] | $V/\square''$ [3, 5] | $\Delta(B-V)$ [3, 5] | Illuminating stars [12, 14] HD/BD | Sp | $V$ | $B-V$ |
|---|---|---|---|---|---|---|---|---|---|
| IC 348 | $3^h41$ | $+31°59'$ | $10'\times 10'$ | $23^m.3$ | $-0^m.71$ | 281 159 | B5 V | $8^m.53$ | $+0^m.69$ |
| Electra nebula | 3 42 | +23 57 | 20×16 | 21.2 | −0.19 | 23 302 | B6 III | 3.71 | −0.11 |
| Maia nebula | 3 42 | +24 24 | 30×30 | 21.4 | −0.45 | 23 408 | B7 III | 3.88 | −0.07 |
| Merope nebula | 3 43 | +23 43 | 30×30 | 21.0 | −0.11 | 23 480 | B6 IV nn | 4.18 | −0.06 |
| NGC 1788 | 5 04 | − 3 25 | 7× 4 | | | 293 815 | B9 V | 10.12 | +0.24 |
| Cederblad 44 | 5 19 | + 8 20 | 22×16 | 23.6 | −0.59 | 34 989 | B1 V | 5.77 | −0.13 |
| IC 435 | 5 40 | − 2 23 | 5× 5 | 22.6 | | 38 087 | B3 n | 8.30 | +0.13 |
| NGC 2068 = M 78 | 5 44 | 0 00 | 8× 6 | 20.6 | −0.37 | 38 563 | B5 | 10.49 | +0.62 |
| NGC 2071 | 5 45 | + 0.12 | 7× 5 | | | 290 861 | F8 | 9.90 | +1.05 |
| IC 2169 | 6 28 | +10 03 | 25×20 | | | 258 686 | B7 III p | 9.12 | −0.06 |
| NGC 2247 | 6 30 | +10 23 | 4× 4 | 21.8 | | 259 431 | B3 pe | 8.74 | +0.28 |
| Antares nebula | 16 26 | −26 20 | 126×78 | 23.7 | −1.19 | 148 478/9 | M1 Ib (+B2.5 V) | 0.92 | +1.84 |
| IC 1287 | 18 29 | −10 50 | 20×10 | 23.2 | −0.49 | 170 740 | B2 V | 5.72 | +0.24 |
| NGC 7023 | 21 01 | +68 00 | 10× 8 | 22.0 | −0.24 | 200 775 | B5 e | 7.39 | +0.38 |
| NGC 7129 | 21 41 | +65 50 | 8× 6 | 20.8 | | + 65°1637/8 | B2 nne + B3 | {10.15<br>10.18 | +0.41<br>+0.45 |

### 7.1.4.2 Diffuse galactic light

Table 6. Typical values of the components of the surface brightness of the sky for $|b| \approx 1°$ between $l=65°$ and $l=145°$ in units of $[S_{10}(\lambda)/\square°]$, the equivalent numbers of A0 V stars of $m(\lambda)=10.0$ mag per square degree [6].

$I_*$ = surface brightness of the line-of-sight integrated starlight
$I_{zod}$ = surface brightness of the zodiacal light
$I_{DGL}$ = surface brightness of the diffuse galactic light

| $\lambda$ [Å] | 4250 | 3320 | 2980 | 2940 | 2460 | 2390 | 2040 | 1910 | 1550 |
|---|---|---|---|---|---|---|---|---|---|
| | $I\ [S_{10}(\lambda)/\square°]$ | | | | | | | | |
| $I_{zod}$ | 41 | 31 | 17 | 14 | 3 | 4 | 13 | 4 | 2 |
| $I_{DGL}$ | 47 | 55 | 39 | 39 | 22 | 21 | 19 | 22 | 51 |
| $I_*$ | 113 | 108 | 93 | 93 | 88 | 84 | 80 | 86 | 140 |
| $I_{DGL}/I_*$ | 0.42 | 0.51 | 0.42 | 0.42 | 0.25 | 0.25 | 0.24 | 0.26 | 0.36 |

Table 7. Diffuse galactic light $I_{DGL}$ in the 1550···4250 Å region and line-of-sight starlight $I_*$ at 4250 Å averaged over galactic longitudes $l=65°\cdots145°$ for seven galactic latitudes [6].

| $\lambda$ [Å] | 4250 | 3320 | 2980 | 2940 | 2460 | 2390 | 2040 | 1910 | 1550 | (4250) |
|---|---|---|---|---|---|---|---|---|---|---|
| $\|b\|$ | $I_{DGL}\ [S_{10}(\lambda)/\square°]$ | | | | | | | | | $I_*\ [S_{10}/\square°]$ |
| 1° | 47 | 55 | 39 | 39 | 22 | 21 | 19 | 22 | 51 | 113 |
| 4 | 37 | 45 | 37 | 30 | 26 | 22 | 20 | 27 | 53 | 94 |
| 8 | 38 | 40 | 33 | 31 | 20 | 19 | 18 | 23 | 47 | 76 |
| 13 | 31 | 22 | 15 | 15 | 7 | 11 | 10 | 13 | 31 | 57 |
| 20 | 17 | 16 | 12 | 11 | 9 | 10 | 6 | 9 | 25 | 55 |
| 31 | 11 | 12 | 9 | 9 | 4 | 5 | 0 | 0 | 11 | 32 |
| 58 | 2 | 7 | 4 | 7 | 4 | 3 | 0 | 0 | 8 | 23 |

Results of a recent rocket photometry at $\lambda=1800$, 2200, and 2600 Å are given in [11]; results from ESRO TD-1 satellite observations for $\lambda=1565$, 1965, 2365, and 2740 Å: see [19].

Table 8. Conversion of $[S_{10}(\lambda)/\square°]$ into units of $[1\cdot10^{-9}\ \text{erg cm}^{-2}\ \text{s}^{-1}\ \text{sterad}^{-1}\ \text{Å}^{-1}]$ for some wavelength bands used in surface photometry of the sky. The conversion coefficients listed correspond to 1 $S_{10}(\lambda)/\square°$ (A0V) [9, 17].

| $\lambda$ [Å] | 5480 | 4700 | 4250 | 3320 | 2960 | 2400 | 2040 | 1910 | 1550 |
|---|---|---|---|---|---|---|---|---|---|
| Coefficient | 1.20 | 1.87 | 2.66 | 1.18 | 1.38 | 1.50 | 1.77 | 1.85 | 2.90 |

For albedo of interstellar grains, see 7.4.1.3.

### 7.1.4.3 References for 7.1.4


1 Dorschner, J., Gürtler, J.: Astron. Nachr. **287** (1964) 257; **289** (1965) 57.
2 Elvius, A., Hall, J.S.: Lowell Bull. **6**, No. 135 (1966) 257; **7**, Nr. 137 (1967) 17.
3 Johnson, H.M.: Publ. Astron. Soc. Pacific **72** (1960) 10.
4 Johnson, H.M.: Stars and Stellar Systems V11 (1968) 75.
5 Landolt-Börnstein, NS, Vol. V1/1 (1965) p. 656.
6 Lillie, C.F., Witt, A.N.: Astrophys. J. **208** (1976) 64.
7 Lynds, B.T.: Astrophys. J. Suppl. **12** (1965) 163.
8 Mattila, K.: Astron. Astrophys. **9** (1970) 53.
9 Mattila, K.: Astron. Astrophys. **47** (1976) 77.
10 O'Dell, C.R.: Astrophys. J. **142** (1965) 604.
11 Pitz, E., Leinert, C., Schulz, A., Link, H.: Astron. Astrophys. **72** (1979) 92.
12 Racine, R.: Astron. J. **73** (1968) 233.
13 Roach, F.E., Smith, L.L., Pfleiderer, J., Batishko, C., Batishko, K.: Astrophys. J. **173** (1972) 343.

14 Van den Bergh, S.: Astron. J. **71** (1966) 990.
15 Van den Bergh, S., Racine, R.: Int. Astron. Union Symp. **52** (1973) 123.
16 Witt, A.N.: Astrophys. J. **152** (1968) 59.
17 Witt, A.N., Johnson, M.W.: Astrophys. J. **181** (1973) 366.
18 Witt, A.N.: Publ. Astron. Soc. Pacific **89** (1977) 750.
19 Gondhalekar, P.M., Phillips, A.P., Wilson, R.: Astron. Astrophys. **85** (1980) 272.
20 Witt, A.N., Cottrell, M.J.: Astron. J. **85** (1980) 22.

## 7.1.5 Interstellar absorption lines and bands in stellar spectra

### 7.1.5.1 Lines of the spectral region 3000···8000 Å (ground-based observations)

Table 9. Visible and near-ultraviolet atomic and molecular interstellar lines and their equivalent widths $W_\lambda$ (sum of all components) in the spectrum of the bright O9.5 V star ζ Ophiuchi [6, 8].

| Atomic lines | | | Molecular lines | | | |
|---|---|---|---|---|---|---|
| Ion | $\lambda_{lab}$ [Å] (air) | $W_\lambda$ [mÅ] | Molecule | $\lambda_{lab}$ [Å] (air) | Transition | $W_\lambda$ [mÅ] |
| Li I | 6707.91<br>6707.96 | 0.68 | CH | 4300.31 | $A^2\Delta - X^2\Pi$ (0,0) $J''$ = 1/2 | 20.8 |
| | | | | 3890.23 | $B^2\Sigma^- - X^2\Pi$ (0,0) $J''$ = 1/2 | 5.6 |
| | | | | 3886.39 | 1/2 | 5.9 |
| Na I D1 | 5895.92 | 210.5 | | 3878.77 | 1/2 | 3: |
| D2 | 5889.95 | 266.2 | | 3146.01 | $C^2\Sigma^+ - X^2\Pi$ (0,0) $J''$ 1/2 | 5.0 |
| | 3302.98 | 22.3 | | 3143.15 | 1/2 | 7.4 |
| | 3302.37 | 28.7 | | 3137.53 | 1/2 | 4.0 |
| K I | 7698.96 | 84 | $CH^+$ | 4232.54 | $A^1\Pi - X^1\Sigma^+$ (0,0) R (0) | 22.3 |
| | 7664.91 | | | 3957.70 | (1,0) R (0) | 14.0 |
| | 4047.21 | 0.3 | | 3745.31 | (2,0) R (0) | 7.2 |
| | 4044.14 | 0.8 | | 3579.02 | (3,0) R (0) | 3.7 |
| | | | | (3447.08) | (4,0) R (0) | 1.2: |
| Ca I g | 4226.73 | 1.3 | | (4229.34) | (0,0) R (1) | < 0.6 |
| Ca II H | 3968.47 | 26.0 | $^{13}CH^+$ | 4232.28 | $A^1\Pi - X^1\Sigma^+$ (0,0) R (0) | 0.35 |
| K | 3933.66 | 44.5 | | | | |
| Ti II | 3383.76 | 6.9 | CN | 3874.61 | $B^2\Sigma^+ - X^2\Sigma^+$ (0,0) $N''$ = 0 | 8.40 |
| | 3241.98 | 5.9 | | 3875.77 | 1 | 1.49 |
| | 3229.19 | < 3 | | 3874.00 | 1 | 2.74 |
| | 3072.97 | < 3 | | 3873.37 | 2 | < 0.43 |
| Fe I | 3859.91 | 1.0 | | | | |
| | 3719.94 | 1.9 | | | | |

Table 10. Surveys: see p. 56.

Relation of line strength to distance

The total equivalent width of an interstellar line observed with low resolution (sum of all components) is correlated with the distance of the star. For stars at low galactic latitudes the mean relations for the K and D lines are approximately [LB, NS, Vol. VI/1 (1965) p. 280]

$$W_\lambda(\mathrm{K}) \approx 0.3\, r$$
$$W_\lambda(\mathrm{D}) \approx 0.4\, r$$

where $r$ = distance in [pc], $W_\lambda$ (K) = equivalent width of the Ca II K line in [mÅ], $W_\lambda(\mathrm{D}) = \frac{1}{2}[W_\lambda(\mathrm{D1}) + W_\lambda(\mathrm{D2})]$ = mean equivalent width of the D lines in [mÅ]. For $r > 1$ kpc the relations tend to a slower increase of $W_\lambda$. The scatter about these mean relations is very large. The mean error in a distance determination based on the equivalent width of the K line is about 35% at 1 kpc, but increases to nearly 100% at 200 pc. For further discussion, see [8].

Table 10. Surveys of interstellar lines of the region 3000···8000 Å.
The work prior to 1965 has been summarized by Münch [8].
$n$ = number of stars

The notes indicate: h = extremely high spectral resolving power
RV = radial velocities only
s = southern stars

| Lines | Author | Ref. | $n$ | Notes |
|---|---|---|---|---|
| Na I D $\lambda$ 5890/96 | Hobbs, L.M. | Astrophys. J. **157** (1969) 135 | 77 | h |
| | | Astrophys. J. **203** (1976) 143; Astrophys. J. Lett. **206** (1976) L117 | 14 | h |
| | | Astrophys. J. **222** (1978) 491 | 18 | h |
| | | Astrophys. J. Suppl. **38** (1978) 129 | 44 | h |
| | Vaughan, A.H., Münch, G. | Astron. J. **71** (1966) 184 | 20 | h |
| | Cohen, J.G. | Astrophys. J. **186** (1973) 149 | 30 | |
| | | Astrophys. J. **194** (1974) 37 | 14 | |
| | | Astrophys. J. **197** (1975) 117 | 36 | |
| Na I DU $\lambda$ 3302/03 | de Boer, K.S., Pottasch, S.R. | Astron. Astrophys. **32** (1974) 1 | 13 | |
| | Crutcher, R.M. | Astrophys. J. **202** (1975) 634 | 21 | |
| K I $\lambda$ 7699 | Hobbs, L.M. | Astrophys. J. Lett. **188** (1974) L67 | 26 | h |
| | | Astrophys. J. **203** (1975) 143; Astrophys. J. Lett. **206** (1976) L117 | 14 | h |
| | Lutz, B.L. | Astrophys. J. Lett. **191** (1974) L131 | 26 | |
| K I $\lambda$ 4044 | Crutcher, R.M. | Astrophys. J. **219** (1978) 72 | 3 | |
| Ca I $\lambda$ 4227 | White, R.E. | Astrophys. J. **183** (1973) 81 | 10 | h |
| Ca II K $\lambda$ 3934 | Greenstein, J.L. | Astrophys. J. **152** (1968) 431 | 38 | |
| | Marschall, L.A., Hobbs, L.M. | Astrophys. J. **173** (1972) 43 | 65 | h |
| | Chu-Kit, M. | Astron. Astrophys. **22** (1973) 69 | 68 | RV, s |
| | Rickard, J.J. | Astron. Astrophys. **31** (1974) 47 | 168 | RV, s |
| | Balona, L.A. | Mem. R. Astron. Soc. **78** (1975) 51 | 585 | s |
| | Cohen, J.G. | Astrophys. J. **186** (1973) 149 | 30 | |
| | | Astrophys. J. **194** (1974) 37 | 14 | |
| | | Astrophys. J. **197** (1975) 117 | 36 | |
| | Hobbs. L.M. | Astrophys. J. Suppl. **38** (1978) 129 | 7 | h |
| Ti II $\lambda$ 3384 | Wallerstein, G., Goldsmith, D. | Astrophys. J. **187** (1974) 237 | 15 | |
| | Stokes, G.M. | Astrophys. J. Suppl. **36** (1978) 115 | 68 | |
| Fe I $\lambda$ 3720/3860 | Chaffee, F.H. | Astrophys. J. **189** (1974) 427 | 3 | |
| CH, $CH^+$ | Frisch, P. | Astrophys. J. **173** (1972) 301 | 30 | |
| | Hobbs, L.M. | Astrophys. J. **181** (1973) 79 | 28 | h |
| | Cohen, J.G. | Astrophys. J. **186** (1973) 149 | 30 | |
| | | Astrophys. J. **197** (1975) 117 | 36 | |
| | Chaffee, F.H. | Astrophys. J. **199** (1975) 379 | 3 | |

Structure of the lines

Observations with extremely high spectral resolving power show that nearly all interstellar K and D lines are multiple [5, 8]. Detection of hyperfine structure of interstellar NaI: see [16].

Half-width of a single component $\approx 1\ \mathrm{km\,s^{-1}}$ [5].

Mean number of clouds (components) in the line of sight $\nu = 8 \cdots 10\ \mathrm{kpc^{-1}}$ (for $b \approx 0°$) [11].

Dispersion of radial velocities between different clouds $\sigma \approx 6\ \mathrm{km\,s^{-1}}$ (for $|b| < 25°$) [3].

### 7.1.5.2 Lines of the UV region $\lambda < 3000$ Å

Most useful instruments have been the UV spectrometers on the NASA Orbiting Astronomical Observatory Copernicus (OAO—3) for 912···3200 Å (resolution 0.05 Å from 950 to 1500 Å) and on the ESRO TD—1 satellite for 2000···3000 Å [2, 10].

Table 11. Far-ultraviolet interstellar atomic lines in the spectrum of ζ Ophiuchi with well-measured equivalent widths $W_\lambda$. For $\lambda_{lab}$, vacuum wavelengths are quoted for $\lambda < 2000$ Å and air values for $\lambda > 2000$ Å [6].

| Ion | $\lambda_{lab}$ [Å] | $W_\lambda$ [mÅ] |
|---|---|---|
| H I | 1215.670 | 16700 |
| | 1025.722 | |
| C I | 1328.833 | 52 |
| | 1280.135 | 34 |
| | 1277.245 | 74 |
| | 1276.482 | 24.9 |
| | 1270.143 | 10.6 |
| | 1260.736 | 38 |
| | 1193.996 | 18.6 |
| | 1193.031 | 54 |
| | 1192.218 | 15.3 |
| | 1190.021 | 11.1 |
| | 1188.833 | 25.3 |
| | 1158.324 | 16.1 |
| | 1157.910 | 30 |
| | 1157.186 | 7.9 |
| | 1155.809 | 21.5 |
| | 1140.010 | 14.3 |
| | 1139.789 | 42 |
| | 1138.383 | 16.8 |
| | 1129.318 | 14.4 |
| | 1129.196 | 31 |
| | 1128.477 | 9.2 |
| | 1128.075 | 7.9 |
| | 1128.171 | 9.1 |
| | 1122.447 | 16.2 |
| | 1121.713 | 10.0 |
| | 1121.520 | 11.4 |
| C I *) | 1560.70 | 50: |
| | 1329.101 | 60 |
| | 1280.597 | 20.4 |
| | 1280.404 | 15.6 |
| | 1279.890 | 18.5 |
| | 1279.056 | 10.8 |
| | 1277.513 | 39 |
| | 1276.750 | 10.6 |
| | 1261.122 | 15.2 |
| | 1260.996 | 18.6 |
| | 1260.927 | 22.4 |
| | 1194.406 | 11.9 |
| | 1194.229 | 7.9 |
| | 1193.679 | 15.8 |
| | 1194.301 | 10.4 |
| | 1192.451 | 12.5 |
| C I *) | 1189.249 | 12.2 |
| | 1189.065 | 16.3 |
| | 1188.992 | 12.7 |
| | 1158.544 | 5.7 |
| | 1158.035 | 23.4 |
| | 1158.130 | 11.0 |
| | 1157.770 | 12.6 |
| | 1157.405 | 7.3 |
| | 1156.00 | 37 |
| | 1139.514 | 6.7 |
| | 1139.300 | 5.4 |
| | 1138.595 | 19.4 |
| | 1129.749 | 3.4 |
| | 1129.078 | 16.0 |
| | 1128.686 | 9.9 |
| | 1128.27 | 10.7 |
| | 1122.75 | 10.1 |
| C I **) | 1561.42 | 50: |
| | 1329.584 | 33 |
| | 1280.847 | 9.9 |
| | 1280.333 | 12.2 |
| | 1279.498 | 3.7 |
| | 1279.229 | 9.3 |
| | 1277.954 | 2.4 |
| | 1277.723 | 13.8 |
| | 1261.552 | 15.5 |
| | 1261.426 | 11.6 |
| | 1194.614 | 6.4 |
| | 1193.393 | 9.1 |
| | 1191.838 | 5.9 |
| | 1189.631 | 8.7 |
| | 1189.447 | 7.8 |
| | 1158.967 | 4.1 |
| | 1158.397 | 2.5 |
| | 1156.765 | 5.0 |
| | 1156.560 | 10.9 |
| | 1156.389 | 7.0 |
| | 1123.065 | 5.2 |
| | 1122.33 | 7.5 |
| | 1122.260 | 9.5 |
| C II | 1334.532 | 189 |
| | 1036.337 | |
| C II *) | 1335.703 | 140 |
| | 1037.018 | |

continued

Table 11, continued

| Ion | $\lambda_{lab}$ [Å] | $W_\lambda$ [mÅ] |
|---|---|---|
| N I | 1200.711 | 133 |
| | 1200.224 | 131 |
| | 1199.549 | 146 |
| | 1134.980 | 126 |
| | 1134.415 | 122 |
| | 1134.165 | 119 |
| N II | 1083.990 | 125 |
| N II *) | 1084.575 | 97···25 |
| N II **) | 1085.701 | 35 |
| | 1085.545 | 14.9 |
| O I | 1302.169 | 201 |
| | 1039.230 | 108 |
| Mg I | 2852.127 | 218 |
| | 2025.824 | 31: |
| Mg II | 2802.704 | 290 |
| | 2795.528 | 312 |
| | 1240.395 | 13.2 |
| | 1239.925 | 18.7 |
| Al II | 1670.786 | 55: |
| Al III | 1862.790 | 34: |
| | 1854.716 | 57: |
| Si II | 1808.012 | 87 |
| | 1526.708 | 190: |
| | 1304.372 | 132 |
| | 1260.421 | 170 |
| | 1193.289 | 147 |
| | 1190.416 | 132 |
| | 1020.699 | 74 |
| Si II *) | 1264.737 | 15.4 |
| | 1197.394 | 4.8 |
| | 1194.500 | 14.3 |
| Si III | 1206.510 | 123 |
| Si IV | 1402.770 | 17.9 |
| | 1393.755 | 24.7 |
| P II | 1301.87 | 15.9 |
| | 1152.81 | 61 |
| S I | 1807.311 | 27.8 |
| | 1316.553 | 17.8 |
| | 1296.174 | 14.2 |
| | 1295.653 | 15.3 |
| | 1270.782 | 8.7 |
| S II | 1259.520 | 112 |
| | 1253.812 | 106 |
| | 1250.586 | 100 |
| S III | 1190.206 | 67 |
| | 1012.504 | |
| S IV | 1062.672 | |
| Cl I | 1347.240 | 20.3 |
| | 1188.772 | 22.7 |
| | 1094.769 | 13···15 |
| | 1097.369 | 10.3 |
| Cl II | 1071.036 | 4.1 |
| Ar I | 1066.660 | 85 |
| | 1048.218 | 97 |
| Mn II | 2605.697 | 111 |
| | 2593.731 | 116 |
| | 2576.107 | 138 |
| | 1201.124 | 12.1 |
| | 1199.388 | 20.4 |
| | 1197.172 | 18.7 |
| Fe II | 2599.395 | 226 |
| | 2585.876 | 214 |
| | 2382.034 | 238 |
| | 2373.733 | 120 |
| | 2343.495 | 186 |
| | 1608.456 | 85: |
| | 1260.542 | 38: |
| | 1144.946 | 91 |
| | 1143.235 | 38 |
| | 1142.334 | 18.1 |
| | 1133.678 | 15.7 |
| | 1121.987 | 42 |
| | 1096.886 | 52 |
| | 1055.269 | 21.1 |
| Fe III | 1122.526 | 21.3 |
| Ni II | 1370.136 | 19.7 |
| Cu II | 1358.773 | 7.4 |
| Zn II | 2062.016 | 95: |
| | 2025.512 | 100: |

*) First excited fine-structure state.
**) Second excited fine-structure state.

Highly ionized species

Broad shallow features of the lines O VI $\lambda$ 1031.9/1037.6 Å were seen in many stars. Corresponding velocity dispersion: 20···50 km s$^{-1}$ [10]. More recently evidence has been given for interstellar absorption in the lines C IV $\lambda$ 1550. 77/1548.20 Å and Si IV $\lambda$ 1402.77/1393.75 Å [14, 15].

Table 12. Vacuum wavelengths and equivalent widths $W_\lambda$ of far-ultraviolet interstellar molecular lines without blending in the spectrum of ζ Ophiuchi [6].

| Molecule | Transition | $\lambda_{lab}$ [Å] | $W_\lambda$ [mÅ] |
|---|---|---|---|
| $H_2$ | $B\,^1\Sigma_u^+ - X\,^1\Sigma_g^+$ | | |
| | (0,0)P(3) | 1115.896 | 50 |
| | (0,0)R(4) | 1116.013 | 27.3 |
| | (0,0)P(4) | 1120.247 | 29.7 |
| | (0,0)R(5) | 1120.399 | 5.6 |
| | (0,0)P(5) | 1125.539 | 4.8 |
| | (1,0)P(2) | 1096.439 | 166 |
| | (1,0)R(3) | 1096.725 | 75 |
| | (1,0)P(3) | 1094.788 | 78 |
| | (1,0)R(4) | 1100.165 | 35 |
| | (1,0)P(4) | 1104.084 | 45 |
| | (1,0)R(5) | 1104.547 | 14.2 |
| | (1,0)P(5) | 1109.313 | 22.5 |
| | (2,0)R(2) | 1079.226 | 131 |
| | (2,0)P(2) | 1081.265 | 202 |
| | (2,0)R(3) | 1081.710 | 73 |
| | (2,0)R(4) | 1085.144 | 43 |
| | (2,0)P(4) | 1088.794 | 43 |
| | (2,0)R(5) | 1089.512 | 27.3 |
| | (3,0)R(2) | 1064.995 | 200 |
| | (3,0)P(2) | 1066.901 | 312 |
| | (3,0)R(3) | 1067.480 | 80 |
| | (3,0)P(3) | 1070.142 | 80 |
| | (3,0)R(4) | 1070.898 | 48 |
| | (3,0)P(4) | 1074.313 | 43 |
| | (3,0)R(5) | 1075.244 | 35 |
| | (3,0)P(5) | 1079.399 | 24.5 |
| | (4,0)R(2) | 1051.499 | 228 |
| | (4,0)P(2) | 1053.281 | 267 |
| | (4,0)R(3) | 1053.976 | 105 |
| | (4,0)P(3) | 1056.469 | 79 |
| | (4,0)R(4) | 1057.379 | 50 |
| | (4,0)P(4) | 1060.580 | 44 |
| | (4,0)R(5) | 1061.697 | 40 |
| | (4,0)P(5) | 1065.598 | 29.3 |
| | (5,0)R(2) | 1038.690 | 222 |
| | (5,0)P(2) | 1040.367 | 290 |
| | (5,0)R(3) | 1041.156 | 78 |
| | (5,0)P(3) | 1043.498 | 82 |
| | (5,0)R(4) | 1044.546 | 48 |
| | (5,0)P(4) | 1047.554 | 44 |
| | (5,0)P(5) | 1052.499 | 28.1 |
| $H_2$ | $B\,^1\Sigma_u^+ - X\,^1\Sigma_g^+$ | | |
| | (6,0)P(2) | 1028.106 | 415 |
| | (6,0)R(3) | 1028.989 | 86 |
| | (6,0)P(3) | 1031.191 | 82 |
| | (6,0)R(4) | 1032.356 | 57 |
| | (6,0)P(4) | 1035.184 | 50 |
| | (6,0)P(5) | 1040.062 | 32 |
| | (6,0)R(6) | 1041.728 | 7.4 |
| | (7,0)P(2) | 1016.472 | 227 |
| | (7,0)P(3) | 1019.506 | 80 |
| | (7,0)P(4) | 1023.437 | 58 |
| | (7,0)P(5) | 1028.250 | 32 |
| | (8,0)R(2) | 1003.989 | 232 |
| | (8,0)P(2) | 1005.397 | 233 |
| | (9,0)R(3) | 995.975 | 60 |
| | (9,0)P(3) | 997.829 | 52 |
| | (9,0)R(4) | 999.273 | 42 |
| $H_2$ | $C\,^1\Pi_u^+ - X\,^1\Sigma_g^+$ | | |
| | (0,0)Q(2) | 1010.941 | 210 |
| | (0,0)R(3) | 1010.132 | 192 |
| | (0,0)R(4) | 1011.817 | 38 |
| | (0,0)Q(5) | 1017.836 | 40 |
| HD | $B\,^1\Sigma_u^+ - X\,^1\Sigma_g^+$ | | |
| | (3,0)R(0) | 1066.271 | 14.3 |
| | (4,0)R(0) | 1054.286 | 19.5 |
| | (5,0)R(0) | 1042.847 | 24.5 |
| | (6,0)R(0) | 1031.909 | 21.2 |
| | (7,0)R(0) | 1021.453 | 14.4 |
| | (8,0)R(0) | 1011.457 | 13.9 |
| CO | $A\,^1\Pi - X\,^1\Sigma^+$ | | |
| | (4,0) | 1419.044 | 66 |
| | (5,0) | 1392.529 | 68 |
| | (6,0) | 1367.617 | 28.3 |
| | (7,0) | 1344.184 | 40 |
| | (8,0) | 1322.147 | 17.2 |
| | (9,0) | 1301.401 | 17.9 |
| | (10,0) | 1281.866 | 9.8 |
| | $B\,^1\Sigma^+ - X\,^1\Sigma^+$ | | |
| | (0,0) | 1150.48 | 35 |
| | $C\,^1\Sigma^+ - X\,^1\Sigma^+$ | | |
| | (0,0) | 1087.867 | 44 |
| | $E\,^1\Pi - X\,^1\Sigma^+$ | | |
| | (0,0) | 1076.033 | 51 |

## 7.1.5.3 Diffuse interstellar absorption bands

Symbols

$W_\lambda$ = equivalent width  $\Delta\lambda$ = total width at half-depth  $A_C$ = central depth relative to the continuum

Table 13. Certain or very probably diffuse interstellar absorption bands in the region 4400···6850 Å, observed in the star HD 183143 with $E_{B-V} = 1^m.28$ [4].

| $\lambda$ Å | $W_\lambda$ Å | $A_C$ | $\Delta\lambda$ Å | $\lambda$ Å | $W_\lambda$ Å | $A_C$ | $\Delta\lambda$ Å |
|---|---|---|---|---|---|---|---|
| 4428 | 3.4 | 0.16 | 20 | 5844.1 | 0.14 | 0.032 | 4.5 |
| 4501 | 0.25 | 0.065 | 3.0 | 5849.79 | 0.10 | 0.095 | 1.0 |
| 4726 | 0.20 | 0.05 | 5.0 | 6010.9 | 0.19 | 0.04 | 4.2 |
| 4754.9 | 0.17 | 0.03 | 5.6 | 6042 | 0.31: | 0.02 | 14.0: |
| 4763.0 | 0.29 | 0.05 | 5.3 | 6113.0 | 0.072 | 0.05 | 0.85 |
| 4779.7 | 0.065 | 0.035 | 1.8 | 6177.1 | 1.85 | 0.07 | 30.0 |
| 4882 | 1.27 | 0.06 | 17.0 | 6195.95 | 0.10 | 0.14 | 0.70 |
| 5362 | 0.15: | 0.025: | 4.4: | 6203.06 } 6206.49 } | 0.43 | 0.16 | 2.3 |
| 5404.3 | 0.07 | 0.055 | 1.0 | | | | |
| 5420 | 0.18 | 0.02 | 11.0: | 6269.77 | 0.39 | 0.17 | 1.4 |
| 5449 | 0.56: | 0.045: | 14.0 | 6283.91 | 2.0 | 0.38 | 3.8 |
| 5487.31 | 0.30 | 0.06 | 4.4 | 6314 | 0.8 | 0.04 | 19.0: |
| 5493.8 | 0.038 | 0.05 | 0.8 | 6353.5 | 0.06 | 0.03 | 3.1 |
| 5535 | 0.53 | 0.025: | 23.0: | 6376.08 | 0.091 | 0.057 | 1.5 |
| 5544.6 | 0.035 | 0.035 | 0.8 | 6379.30 | 0.16 | 0.16 | 0.86 |
| 5705.12 | 0.29 | 0.075 | 3.5 | 6425.7 | 0.035 | 0.04 | 1.1 |
| 5778.3 | 0.95 | 0.06 | 17.0 | 6597.4 | 0.02 | 0.03 | 0.60 |
| 5780.41 | 0.88 | 0.37 | 2.6 | 6613.63 | 0.40 | 0.34 | 1.1 |
| 5794.96 } 5797.03 } | 0.39 | 0.22 | 1.3 | 6660.71 | 0.1 | – | – |

Catalogues

Diffuse band data have been collected up to 1976 and reduced to a common measurement system by Snow et al. [9]. They have tabulated central depths for $\lambda$ 4430 Å and equivalent widths for $\lambda\lambda$ 5780, 5797 and 6284 Å for a large number of stars. Observed infrared bands: see [17].

Relations between equivalent width $W_\lambda$ in [Å] or central depth $A_C$ and colour excess $E_{B-V}$ in [mag] [1, 4, 9, 12, 13].

$W_\lambda$ (4430 Å) $\approx 2.5\, E_{B-V}$
$A_C$ (4430 Å) $\approx 7.2\, E_{B-V} + 2.8$
$W_\lambda$ (5780 Å) $\approx 0.3\, E_{B-V} + 0.2$
$W_\lambda$ (6284 Å) $\approx 1.4\, E_{B-V}$

## 7.1.5.4 References for 7.1.5


1 Aannestad, P.A., Purcell, E.M.: Interstellar Grains, Annu. Rev. Astron. Astrophys. **11** (1973) 328.
2 De Jager, C., Hoekstra, R., van der Hucht, K.A., Kamperman, T.M., Lamers, H.J.G.L.M., Hammerschlag, A., Werner, W., Emming, J.G.: Astrophys. Space Sci. **26** (1974) 207.
3 Falgarone, E., Lequeux, J.: Astron. Astrophys. **25** (1973) 253.
4 Herbig, G.H.: Astrophys. J. **196** (1975) 129.
5 Hobbs, L.M.: Astrophys. J. **157** (1969) 165.
6 Morton, D.C.: Astrophys. J. **197** (1975) 85.
7 Münch, G.: Galactic Structure and Interstellar Absorption Lines, in: Stars and Stellar-Systems V (1965) 203.
8 Münch, G.: Interstellar Absorption Lines, in: Stars and Stellar Systems VII (1968) 365.
9 Snow, Th.P., York, D.G., Welty, D.E.: Astron. J. **82** (1977) 113.
10 Spitzer, L., Jenkins, E.B.: Ultraviolet Studies of the Interstellar Gas, Annu. Rev. Astron. Astrophys. **13** (1975) 133.
11 Van Woerden, H.: Int. Astron. Union Symp. **31** (1967) 3.
12 Schmidt-Kaler, Th., Tüg, H., Buchholz, M., Schlosser, W.: Astron. Astrophys. Suppl. **39** (1980) 305.
13 Tüg, H., Schmidt-Kaler, Th.: Astron. Astrophys. **94** (1981) 16.
14 Savage, B.D., de Boer, K.S.: Astrophys. J. Lett. **230** (1979) L77.

15 Bruhweiler, F.C., Kondo, Y., McCluskey, G.E.: Astrophys. J. **237** (1980) 19.
16 Wayte, R.C., Wynne-Jones, I., Blades, J.C.: Mon. Not. R. Astron. Soc. **182** (1978) 5 P.
17 Merrill, K.M. in: The interaction of variable stars with their environment, Proc. Int. Astron. Union Coll. No. **42**, (Kippenhahn, R., Rahe, J., Strohmeier, W., eds.), Bamberg (1977) p. 446.

## 7.1.6 Radio line emission and absorption

### 7.1.6.1 21-cm line of neutral hydrogen

Frequency from laboratory measurements [1]:
$\nu_0 = 1420.405752$ MHz

Table 14. Large surveys of 21-cm hydrogen line emission from the Milky Way. Earlier surveys: see Kerr (1968) [1]. Presentation of the data:
1) maps: brightness temperature as a function of radial velocity, $v_r$, and one coordinate, the other position coordinate being kept constant;
2) line profiles at a given position;
3) drift scans for fixed velocity with variable position.
$\Delta\nu$ = band width; $\theta$ = beam width

| Author | Publication | $\theta$ | $\Delta\nu$ kHz | Region | Presentation of the data |
|---|---|---|---|---|---|
| Henderson, A.P. | Doctoral thesis 1967, University of Maryland | 10′ | 8 | $l=16\cdots230°$<br>$b=-10\cdots10°$ | $(b, v_r)\|_l$ maps |
| Westerhout, G. | Maryland-Green Bank Galactic 21-cm Line Survey, 2nd ed. Univ. of Maryland College Park (1969) | 13′ | 8 | $l=11\cdots235°$<br>$b=-1\cdots1°$ | $(\alpha, v_r)\|_\delta$ maps |
| Kerr, F.J. | Parkes Hydrogen-Line Survey of the Milky Way I, Austral. J. Phys. Astrophys. Suppl. **9** (1969) 1 | 14ʹ.5 | 38 | $l=296\cdots63°.5, b=0°$<br>$l=300\cdots60°, b=-2\cdots2°$ | $(l, v_r)\|_b$ maps |
| Burton, W.B. | Astron. Astrophys. Suppl. **2** (1970) 261, 291 | 0°.6 | 8 | $l=354\cdots120°, b=0°$;<br><br>$l=43\cdots56°$,<br>$b=-4.5\cdots4°.5$ | profiles;<br>$(l, v_r)\|_b$ maps<br>$(b, v_r)\|_l$ maps |
| Hindman, J.V., Kerr, F.J. | Austral. J. Phys. Astrophys. Suppl. **18** (1970) 1, 43 | 14ʹ.5 | 38 | $l=190\cdots299$,<br>$b\approx-5\cdots5°$;<br>$l=185\cdots63°$,<br>$b=0°$ | $(b, v_r)\|_l$ maps<br><br>$(l, v_r)\|_b$ maps |
| Velden, L. | Beiträge zur Radioastronomie, Max-Planck-Institut für Radioastronomie, Bonn, Vol. **1**, Part 7 (1970) | 0°.6 | 12 | $l=120\cdots240°$<br>$b=-30\cdots30°$ | profiles |
| Dieter, N.H. | Berkeley Survey of High Velocity Interstellar Neutral Hydrogen, Astron. Astrophys. Suppl. **5** (1972) 21, 313 | 0°.6 | 10 | $l=10\cdots250°$<br>$b=-15\cdots15°$;<br>$\|b\| \gtrapprox 15°$ | $(b, v_r)\|_l$ maps |
| Venugopal, V.R., Shuter, W.L.H. | Mem. R. Astron. Soc. **74** (1970) 1 | 0°.6 | 10 | $\delta > -29°$ | profiles |
| Burton, W.B., Verschuur, G.L. | Astron. Astrophys. Suppl. **12** (1973) 145 | 30′ | 10 | $l=20\cdots230°$<br>$b=-20\cdots20°$ | $(l, v_r)\|_b$ maps |
| | | 10′ | 10 | $l=15\cdots140°$<br>$b=-30\cdots30°$ | $(b, v_r)\|_l$ maps |
| Lindblad, P.O. | Astron. Astrophys. Suppl. **16** (1973) 207 | 12ʹ.5 | 10 | $l=12\cdots72°$<br>$b=1\cdots10°$ | $(b, v_r)\|_l$ maps |
| | | 26′ | 5 | $l=339\cdots12°$<br>$b=-15\cdots15°$ | |

continued

Table 14, continued

| Author | Publication | $\theta$ | $\Delta\nu$ kHz | Region | Presentation of the data |
|---|---|---|---|---|---|
| Weaver, H.F., Williams, D.R.W. | Berkeley Low-Latitude Survey of Neutral Hydrogen. Astron. Astrophys. Suppl. **8** (1973) 1; **17** (1974) 1, 251 | 36′ | 10 | $l = 10 \cdots 250°$ <br> $b = -30 \cdots 30°$ | profiles <br> $(l, v_r)\|_b$ maps |
| Simonson, S.C., Sancisi, R. | Astron. Astrophys. Suppl. **10** (1973) 283 | 0.°6 | 16 | $l = 354 \cdots 24°$ <br> $b = -5 \cdots 5°$ | profiles <br> $(l, v_r)\|_b$, $(b, v_r)\|_l$ maps |
| Tuve, M.A., Lundsager, S. | Astron. J. **77** (1972) 652; Monograph No. 630, Department of Terrestrial Magnetism, Carnegie Inst. of Washington (1973) | 0.°9 | 10 | $l = 336 \cdots 270°$ <br> $b = -16 \cdots 16°$ | profiles |
| Heiles, C., Habing, H.J. | An almost complete survey of 21-cm line radiation; I. Atlas of contour maps. Astron. Astrophys. Suppl. **14** (1974) 1–555; Photographic Presentation: Heiles, C., Jenkins, E.B.: Astron. Astrophys. **46** (1976) 333 | 0.°6 | 10 | $\delta > -30°$ <br> $\|b\| \geqq 10°$ | $(l, v_r)\|_b$ maps; column densities: see Astron. Astrophys. Suppl. **20** (1975) 37 |
| Westerhout, G | Maryland-Green Bank Galactic 21-cm Line Survey, 3rd. ed. Univ. of Maryland, College Park (1973) | 13′ | 10 | $l = 13 \cdots 235°$ <br> $b = -2 \cdots 2°$ | $(l, v_r)\|_b$ maps |
| | Maryland-Bonn 21-cm Line Survey of the Galactic Plane, University of Maryland, College Park (1976) | 9′ | | $l = 355 \cdots 249°$ <br> $b = 0°$ | |
| Bystrova, N.V., Rakhimov, I.A. | Pulkovo Sky Survey in the Interstellar Neutral Hydrogen Line III. Akademiya Nauk SSSR, Leningrad 1977, 62 pp. | $7' \times 5°$ | 20 | $\delta = -29 \cdots +40°$ | drift curves for fixed radial velocities |
| Kerr, F.J., Harten, R.H., Ball, D.M. | Parkes Hydrogen-Line Survey of the Milky Way II. Astron. Astrophys. Suppl. **25** (1976) 391 | 14.′5 | 10 | $l = 236 \cdots 345°$ <br> $b = -2 \cdots 2°$ | $(b, v_r)\|_l$ maps |
| Bajaja, E., Colomb, F.R., Gil, M., Morras, R. | Carnegie Instn. Wash. Publ. No. 632 (1973); Astron. Astrophys. Suppl. **26** (1976) 195; **41** (1980) 67, 121 | 0.°5 | 10 | $l = 220 \cdots 325°$ <br> $b = -32 \cdots +2°$ | $(b, v_r)\|_l$ maps <br> $(l, v_r)\|_b$ maps |
| Cleary, M., Heiles, C., Haslam, C.G.T. | Astron. Astrophys. Suppl. **36** (1979) 95 | 48′ | 33 | $\delta < -30°$ <br> $\|b\| > 10°$ | photographic presentation; for column densities, see Austral. J. of Phys., Astrophys. Suppl. No. **47** (1979) |
| Colomb, F.R., Pöppel, W.G.L., Heiles, C. | Astron. Astrophys. Suppl. **40** (1980) 47 | 0.°5 | 10 | $\delta < -30°$ <br> $\|b\| > 10°$ | photographic presentation |

The Maryland-Greenbank and Maryland-Bonn surveys and some other 21-cm line surveys are available in machine-readable form as tape-records. See Westerhout, G., Jaschek, C.: Bull. d'Inf. Centre de Données Stellaires, Strasbourg, No. 13 (1977) 28; No. 14 (1978) 20; No. 15 (1978) 99.

Table 15. Some recent surveys of 21-cm line emission from the galactic center (see also Table 14). $\theta$ = beam width; $\Delta v$ = velocity resolution

| Author | Publication | $\theta$ | $\Delta v$ $km\,s^{-1}$ | Sensitivity K |
|---|---|---|---|---|
| Garzoli, S.L. | Carnegie Instn. Wash. Publ. No. 629 (1972) | 30′ | | |
| Sanders, R.H., Wrixon, G.T., Penzias, A.A. | Astron. Astrophys. **16** (1972) 322 | 2° | 15.8 | 0.1 |
| Wrixon, G.T., Sanders, R.H. | Astron. Astrophys. Suppl. **11** (1973) 339 | 21′ | 5.5 | 0.2 |
| Simonson, S.C., Sancisi, R. | Astron. Astrophys. Suppl. **10** (1973) 283 | 36′ | 3.4 | 2.0 |
| Weaver, H.F., Williams, D.R.W. | Astron. Astrophys. Suppl. **17** (1974) 1 | 36′ | 2.1 | 0.4 |
| Lindblad, P.O. | Astron. Astrophys. Suppl. **16** (1974) 207 | 13′, 21′ | 1 | (2.0) |
| Cohen, R.J., Davis, R.D. | Mon. Not. R. Astron. Soc. **171** (1975) 659, **186** (1979) 453 | 31′ × 35′; 13′ | 7.3 | 0.3 |
| Sanders, R.H., Wrixon, G.T., Mebold, U. | Astron. Astrophys. **61** (1977) 329 | 9′ | 2.75 | 0.1 |
| Mirabel, I.F. | Astron. Astrophys. Suppl. **28** (1977) 327 | 28′ | | |
| Burton, W.B., Gallagher, J.S., McGrath, M.A. | Astron. Astrophys. Suppl. **29** (1977) 123 | 21′ | 5.5 | 0.2 |
| Burton, W.B., Liszt, H.S. | Astrophys. J. **225** (1978) 815 | 24′ | 11 | 0.03 |
| Sinha, R.P. | Astron. Astrophys. Suppl. **37** (1979) 403 | 24′ | 5 | 0.4 |
| Braunsfurth, E., et al. | Astron. Astrophys. Suppl. **44** (1981) 437 | 9′ | 3.0/6.0 | 2.5 |

21-cm absorption line in the spectra of discrete continuum sources:

New high-resolution surveys have been provided by Crovisier et al. (1978) [2] and by Dickey et al. (1978) [3]. Earlier work is summarized in these papers. Crovisier et al. give a catalogue including information on 819 sources.

### 7.1.6.2 Diffuse ("weak") recombination line emission

Optical and radio recombination line emission from the interstellar medium outside the higher electron density H II regions has been observed recently.

Diffuse optical lines

Contour map and distribution of radial velocities of Hα emission [4, 5].

Diffuse radio lines

Summary of results up to 1974 (e.g. for H 166 α): [6].

More recent surveys of H 166 α emission ($\nu = 1425$ MHz, $\lambda \approx 21$ cm) for $l = 5 \cdots 70°$ and $358 \cdots 51°$, $b = 0°$: [7, 8].

Velocity-longitude diagram for the galactic plane: [9].

Surveys of radio recombination line emission from the well-defined H II regions of higher density, see 7.3.2, Table 2.

### 7.1.6.3 References for 7.1.6


1 Kerr, F.J.: Radio-Line Emission and Absorption by the Interstellar Gas, in: Stars and Stellar Systems VII (1968) 575.
2 Crovisier, J., Kazés, I., Aubry, D.: Astron. Astrophys. Suppl. **32** (1978) 205.
3 Dickey, J.M., Salpeter, E.E., Terzian, Y.: Astrophys. J. Suppl. **36** (1978) 77.
4 Reynolds, R.J., Scherb, F., Roesler, F.L.: Astrophys. J. **185** (1973) 869.
5 Reynolds, R.J., Roesler, F.L., Scherb, F.: Astrophys. J. **192** (1974) L53.
6 Guélin, M.: Int. Astron. Union Symp. No. **60** (1974) 51.
7 Hart, L., Pedlar, A.: Mon. Not. R. Astron. Soc. **176** (1976) 547.
8 Lockman, F.J.: Astrophys. J. **209** (1976) 427.
9 Hart, L., in: Topics in Interstellar Matter (van Woerden, H., ed.), Reidel Publ. Comp., Dordrecht, Holland (1977) 187.


## 7.1.7 Continuous radio emission of interstellar origin: nonthermal background

Table 16. Large radio continuum surveys. For surveys prior to ≈1970, see Price (1974) [7]; surveys of high angular resolution using aperture synthesis techniques are mainly concerned with discrete sources (see 7.3.2). $\nu$ = frequency; $\theta$ = beam width

| Author | Publication | $\nu$ [MHz] | $\theta$ | Region |
|---|---|---|---|---|
| Altenhoff, W.J., Downes, D., Pauls, T., Schraml, J. | Astron. Astrophys. Suppl. **35** (1979) 23 | 5000 | 2′.6 | $l = 357.5\cdots0\cdots60°$<br>$b = -1\cdots+1°$ |
| Haynes, R.F., Caswell, J.L., Simons, L.W.J. | Australian J. Phys., Astrophys. Suppl. **45** (1978) 1 | 5000 | 4′ | $l = 190\cdots0\cdots40°$<br>$b = -2\cdots2°$ |
| Hughes, V.A., Routledge, D. | Astron. J. **74** (1969) 604 | 3200 | 9′ | $l = 32\cdots49°$, $b \approx 0°$ |
| Altenhoff, W.J., Downes, D., Goad, L., Maxwell, A., Rinehart, R. | Astron. Astrophys. Suppl. **1** (1970) 319 | 2700 | 11′ | $l = 345\cdots0\cdots75°$<br>$b = -2\cdots2°$ |
| Day, G.A., Caswell, J.L., Cooke, D.J. | Australian J. Phys., Astrophys. Suppl. **25** (1972) 1 | 2700 | 8′ | $l = 46\cdots61°$, $190\cdots290°$<br>$b = -2\cdots2°$ |
| Berkhuijsen, E.M. | Astron. Astrophys. Suppl. **5** (1972) 263 | 820 | 1°.2 | $\delta = -7\cdots+85°$ |
| Haslam, C.G.T., Wilson, W.E., Salter, C.J., Stoffel, H. | Astron. Astrophys. Suppl. **47** (1982) 1 | 408 | 0°.85 | whole-sky |
| Landecker, T.L., Wielebinski, R. | Australian J. Phys., Astrophys. Suppl. **16** (1970) | 85<br>150 | 3°.7<br>2°.2 | $\delta = -25\cdots+25°$ |
| Milogradov-Turin, J., Smith, F.G. | Mon. Not. R. Astron. Soc. **161** (1973) 269 | 38 | 7°.5 | $\delta = -25\cdots+70°$ |
| Jones, B.B., Finley, E.A. | Australian J. Phys. **27** (1974) 687 | 30 | 0°.8 | $l = 225\cdots0\cdots30°$<br>$b \approx 0°$ |
| Caswell, J.L. | Mon. Not. R. Astron. Soc. **177** (1976) 601 | 10 | 2° | $\delta > -5°$ |

The observed radio continuum radiation can be described in terms of a superposition of two components with different spectral distributions: The so-called thermal (free-free) disk component dominating at high frequencies $\nu$, and a nonthermal component being well represented by a power law spectrum

$$I_\nu \propto \nu^{-\alpha}$$

with the spectral index $\alpha > 0$ and, therefore, most intense at low frequencies. Most of the thermal disk radiation has been resolved into discrete sources (dense H II regions), but it also includes a weak diffuse thermal component which appears to be emitted from an extended low-density ionized gas (cf. 7.1.6.2). The nonthermal continuum radiation also consists of discrete sources (galactic supernovae remnants, extragalactic sources) and of a diffuse background continuum. The values of the spectral index of this nonthermal background are $\alpha = 0.4 \cdots 0.9$ for frequencies $\nu = 200$ MHz$\cdots$3 GHz; the mean value is $\alpha \approx 0.6$. It can be interpreted as synchrotron radiation from relativistic cosmic ray electrons radiating in galactic magnetic fields (see also 7.6).

The distribution of this nonthermal background in latitude is wider than that of the thermal sources; its longitude distribution near $b = 0°$ exhibits some "step" maxima which are presumed to coincide with spiral arms seen tangentially. In many parts of the sky, loops or spur features are present. They are probably nearby supernova remnants (Table 17).

At a frequency near 3 MHz the spectrum shows a peak brightness, and for lower frequencies it exhibits a positive slope ($\alpha < 0$), which is caused by attenuation of the radio emission. Observations at these very low frequencies ($\nu \lessapprox 10$ MHz) have been made by the aid of earth satellites.

Early galactic continuum surveys of the northern part of the Milky Way with relatively low angular resolution: for 1.4 GHz [9], for 2.7 GHz [1].

Whole-sky maps of radio continuum radiation: for 408 MHz (Table 16), 150 MHz (Fig. 6), 85 MHz [8], 30 MHz [4], and 10 MHz (northern sky) [5], see also [3]. For continuum maps of the galactic central region, see 7.3.2.

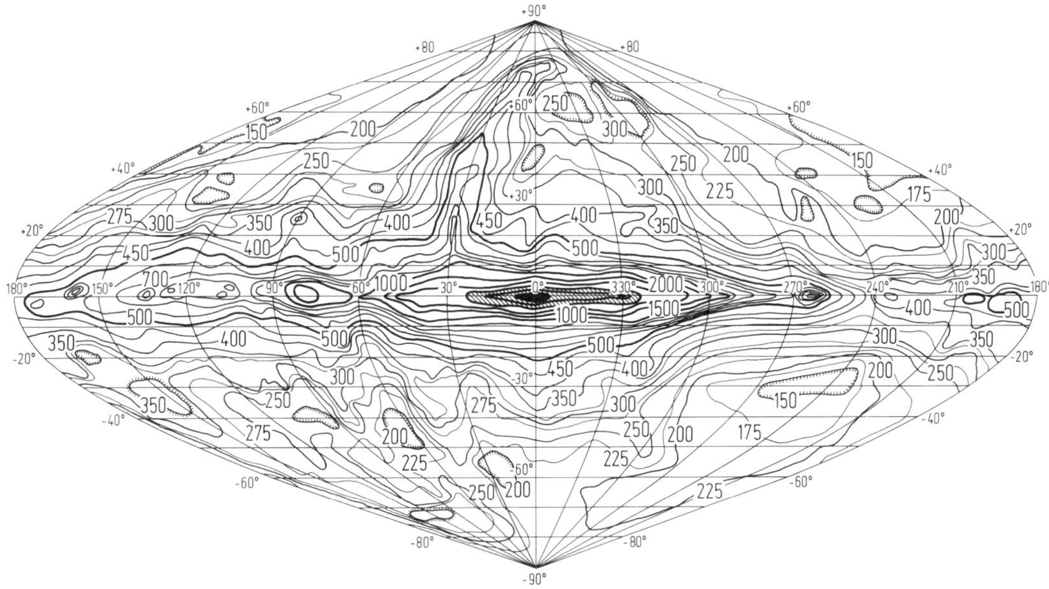


Fig. 6. Sky map of 150 MHz brightness temperatures in galactic coordinates prepared by Landecker and Wielebinski (1970) [6].

Table 17. Galactic spurs [2, 7].

| Feature | Center | | Diameter | Estimated distance [pc] |
|---|---|---|---|---|
| | $l$ | $b$ | | |
| Loop I (north polar spur) | 329° | +17°.5 | 116° | 130 |
| Loop II (Cetus Arc) | 100 | −32.5 | 91 | 110 |
| Loop III | 124 | +15.5 | 65 | 150 |
| Loop IV | 315 | +48.5 | 40 | 250 |
| Origem loop | 194.5 | + 0.5 | 5 | 1000 |
| Lupis loop | 330 | +15 | 4.5 | 400 |
| Monoceros loop | 206 | 0 | 3.5 | 600 |

### References for 7.1.7

1 Altenhoff, W., Mezger, P.G., Wendker, H., Westerhout, G.: Veröffentl. Univ.-Sternwarte Bonn Nr. 59 (1960).
2 Berkhuijsen, E.M.: Astron. Astrophys. **24** (1973) 143.
3 Cane, H.V.: Proc. Astron. Soc. Australia **2** (1975) 330.
4 Cane, H.V.: Australian J. Phys. **31** (1978) 561.
5 Caswell, J.L.: Mon. Not. R. Astron. Soc. **177** (1976) 601.
6 Landecker, T.L., Wielebinski, R.: Australian J. Phys. Astrophys. Suppl. **16** (1970).
7 Price, R.M., in: Int. Astron. Union Symp. **60** (1974) 637.
8 Yates, K.W.: Australian J. Phys. **21** (1968) 167.
9 Westerhout, G.: Bull. Astron. Inst. Netherland **14** (1958) 215.

## 7.1.8 UV and visual interstellar radiation field

Table 18. Spectral energy density, $U_\lambda$, of the average interstellar radiation field in the solar neighborhood for the wavelength region 950···8000 Å [2, 4, 6, 7].

| $\lambda$ Å | $U_\lambda$ $10^{-17}$ erg cm$^{-3}$ Å$^{-1}$ | $\lambda$ Å | $U_\lambda$ $10^{-17}$ erg cm$^{-3}$ Å$^{-1}$ |
|---|---|---|---|
| 950 | 5 | 2400 | 2.9 |
| 1000 | 7 | 2500 | 3.0 |
| 1100 | 9 | 3000 | 3.0 |
| 1200 | 9.4 | 3300 | 3.1 |
| 1300 | 9.6 | 4250 | 6.8 |
| 1400 | 9.7 | 4500 | 7.2 |
| 1550 | 9.2 | 5000 | 6.6 |
| 1700 | 7.1 | 6000 | 5.8 |
| 1900 | 3.9 | 7000 | 5.2 |
| 2000 | 3.1 | 8000 | 4.8 |

The values given in Table 18 for the far-ultraviolet field, $\lambda < 2000$ Å, represent probably a kind of upper limit. They are based on [6] and [2], and on the calculations in [4]. For more recent results, see [9, 10].

For approximate representation by a sum of dilute black-body spectra, see, e.g. [1] or [8].

Recent measurements of the diffuse EUV background radiation field by Sandel et al. (1979) [5] give lower values by a factor of about 2.5. These observations confirm the predicted lack of radiation in the region below 875 Å; near the Lyman limit of 912 Å the spectral intensity rises steeply to 970 Å, where it takes on a lower slope.

The far-ultraviolet interstellar radiation field without the contribution of the diffusely scattered starlight has been derived by integration of the light of stars included in the Smithsonian Astrophysical Observatory (SAO) Star Catalogue by Henry (1977) [3]. In the region 1000···1700 Å the resulting energy density is approximately constant $U_\lambda \approx 6 \cdot 10^{-17}$ erg cm$^{-3}$ Å$^{-1}$. At $\lambda = 975$ Å only spectral classes earlier than ≈ B5 contribute significantly, and at $\lambda \approx 2000$ Å early B stars still dominate.

### References for 7.1.8

1 Drapatz, S.: Astron. Astrophys. **75** (1979) 26.
2 Henry, R.C., Swandic, J.R., Shulman, S.D., Fritz, D.: Astrophys. J. **212** (1977) 707.
3 Henry, R.C.: Astrophys. J. Suppl. **33** (1977) 451.
4 Jura, M.: Astrophys. J. **191** (1974) 375.
5 Sandel, B.R., Shemansky, D.E., Broadfoot, A.L.: Astrophys. J. **227** (1979) 808.
6 Witt, A.N., Johnson, M.W.: Astrophys. J. **181** (1973) 363.
7 Zimmermann, H.: Astron. Nachr. **288** (1965) 95, 99.
8 Greenberg, J.M.: Interstellar dust, in: Cosmic Dust (McDonnell, J.A.M., ed.), Wiley, New York (1978) p. 208.
9 Gondhalekar, P.M., Phillips, A.P., Wilson, R.: Astron. Astrophys. **85** (1980) 272.
10 Henry, R.C.: Astrophys. J. Lett. **244** (1981) L69.

# 7.2 Cool interstellar clouds

## 7.2.1 General references

Andrew, B.H. (ed.): Interstellar Molecules, Int. Astron. Union Symp. **87** (1980).
Gehrels, T. (ed.): Protostars and Planets, Univ. of Arizona Press, Tucson (1978) Part I–III.
Heiles, C.: Physical Conditions and Chemical Constitution of Dark Clouds, Annu. Rev. Astron. Astrophys. **9** (1971) 293.
Heiles, C.: A modern look at "interstellar clouds", in: Int. Astron. Union Symp. **60** (1974) 13.
Lynds, B.T.: Dark Nebulae, in: Stars and Stellar Systems VII (1968) 119.
Lynds, B.T. (ed.): Dark Nebulae, Globules, and Protostars, Univ. of Arizona Press, Tucson (1970).
Robinson, B.J.: Molecular Astronomy, Proc. Astron. Soc. Austral. **3** (1976) 12.
Solomon, P.M., Edmunds, M.G. (eds.): Giant Molecular Clouds in the Galaxy; Third Gregynog Astrophysics Workshop, Pergamon Press, Oxford (1980).
Thaddeus, P.: Molecular Clouds, Int. Astron. Union Symp. **75** (1977) 37.
Turner, B.E.: General Physical Characteristics of the Interstellar Molecular Gas, Int. Astron. Union Symp. **84** (1979) 257.
Van de Hulst, H.C., Greenberg, J.M. (eds.): Interstellar Dust and Related Topics, Int. Astron. Union Symp. **52** (1972).
Van Woerden, H. (ed.): Radio Astronomy and the Galactic System, Part 1: Interstellar Clouds, Int. Astron. Union Symp. **31** (1967) 3.
Winnewisser, G.: Interstellar Molecules in the Galaxy, Naturwissenschaften **62** (1975) 200.
Zuckerman, B.: Observations of Molecular Clouds, in: Topics in Interstellar Matter (van Woerden, H., ed.), Reidel, Dordrecht, Holland (1977) p. 107.

## 7.2.2 Dark nebulae and globules

Table 1. Catalogues of dark nebulae.
$n$ = number of objects

| Author | Publication | $n$ | Data |
|---|---|---|---|
| Barnard, E.E. | Atlas of Selected Regions of the Milky Way (Frost, E.B., Calvert, M.R., eds.), Publ. Carnegie Inst. Washington No. 247 (1927) | 349 | position, size, description |
| Lundmark, K. | Medd. Obs. Uppsala No. 12 (1926) | 1550 | position, size, description |
| Khavtassi, J. Sh. | Bull. Abastumani Obs. No. 18 (1958); Atlas of Galactic Dark Nebulae, Abastumani Astrophys. Obs. Tbilisi (1960) | 797 | position, area, opacity |
| Lynds, B.T. | Astrophys. J. Suppl. **7** (1962) 1 | 1802 | position, area, opacity class |
| Schoenberg, E.E. | Veröff. Sternwarte München **5**, No. 21 (1964) | 1456 | position, area, shape, extinction |
| Sandqvist, Aa., Lindroos, K.P. | Astron. Astrophys. **53** (1976) 179 | 42 | position, area, opacity class |
| Sandqvist, Aa. | Astron. Astrophys. **57** (1977) 467 (southern Milky Way) | 95 | position, area, opacity class |
| Schneider, S., Elmegreen, B.G. | A Catalogue of Dark Globular Filaments; Astrophys. J. Suppl. **41** (1979) 87 | 23 | position, size, structure, opacity |

Distribution of dark nebulae over the sky: Maps in galactic coordinates have been given by Lynds (1962, 1968) [32, 33] for the northern and by Sandqvist (1977) [49] for the southern Milky Way. Distribution with galactic latitude: see Landolt-Börnstein, NS, Vol. VI/1 (1965) p. 653.

The frequency distribution of apparent areas of dark nebulae is also given in Landolt-Börnstein, NS, Vol. VI/1 (1965) p. 653.

Table 2. Selected dark nebulae or complexes of individual objects that have been studied in detail using star counts and, partly, colour excesses [33].
$l, b$ = galactic coordinates
$r$ = distance estimate
$a_{pg}$ = total photographic extinction of the dark nebula

| Constellation or designation | $l$ | $b$ | $r$ [pc] | $a_{pg}$ [mag] | Ref. |
|---|---|---|---|---|---|
| Scutum | 28° | −2° | 180···265 | 3.3 | 66 |
| Aquila | 42 | +0.5 | 110···300 | 4 | 60 |
| | 52 | 0 | 500 | 2 | 4 |
| Cygnus | 71 | −3 | 200, 500···700 | 2 | 7 |
| 52 Cygni | 72 | −7 | 400···560 | 1 | 64 |
| Cygnus | 75 | −2 | 600 | 1.5 | 50 |
| Great Rift | | | | | 36, 65 |
| E, S and W of NGC 7000 | | | 550 | 4.0 | 33 |
| Rift, $l$=73°···79° | | | 1200 | 3.0 | 33 |
| Rift, $l$=65°···73° | | | 800···1600 | 2.0 | 33 |
| Rift, $l$=56°···65° | | | 600 | 2···3 | 33 |
| Cygnus | 92 | +3 | 250, 630 | 1.0, 1.0 | 58 |
| Cepheus | 99 | +3 | 150···250 | 0.5 | 56 |
| | 100 | +2 | 200···500 | 0.9 | 50 |
| | 107 | +2 | 200···600; 300 | 0.8, 1.7 | 61 |
| Cassiopeia | 120 | +5 | 500 | 2 | 40 |
| | 130 | 0 | 500, 800 | 1.5, 3 | 5 |
| | 132 | +5 | 300 | 1.8 | 40 |
| Taurus | 165 | −16 | 50, 175; 600 | 4, 1, 1.5 | 1 |
| | 174 | −13 | ≈ 120 | 2···8 | 35, 9 |
| Auriga | 165 | +4 | 130 | 0.6 | 15 |
| | 166 | −6 | 300 | 2.5 | 23 |
| | 167 | −6 | 115, 340 | 1.0, 1.1 | 15 |
| S Monocerotis | 202 | +1 | 410···575 | 2.1 | 67 |
| | | | 500···1000 | 1.5···2 | 2 |
| Orion | 207 | −20 | 300···400 | 1 | 3 |
| Vela | 256 | 0 | 500···750 | 2 | 21 |
| η Carinae | 286 | 0 | 800 | 0.8 | 8 |
| "Coalsack" | 302 | 0 | 170 | 0.7···2.4 | 46 |
| ϱ Ophiuchi | 353 | +17 | 200 | 2···8 | 9, 17 |
| θ Ophiuchi | 1 | +5 | 250 | 2.0 | 59 |
| Corona Australis complex | ≈ 2 | ≈ −21 | ≈ 150 | 0.4···3.5 (isophotes) | 47 |
| Scorpius, Ophiuchus complex | 350···20 | +15···25 | ≈ 200 | 0.4···4.0 (isophotes) | 48 |

H I observations of dark nebulae: see [69].

Table 3. Selected globules [10, 11, 14, 34, 72].

$d$ = approximate angular diameter  $R$ = radius
$a_{pg}$ = photographic extinction  $\mathfrak{M}$ = most probable mass (= 100·mass of the dust)
$r$ = assumed distance

| Object | $\alpha_{1975}$ | $\delta_{1975}$ | $d$ | $a_{pg}$ mag | $r$ pc | $R$ pc | $\mathfrak{M}/\mathfrak{M}_\odot$ | Comments |
|---|---|---|---|---|---|---|---|---|
| IC 1848 | $3^h01^m.0$ | +60°26′ | 1′ | 10 *) | 1700 | 0.3 | 20 | large globule associated with nebulosity |
| Orion 1 | 5 37.0 | − 1 48 | 0′.7 | 10 *) | 400 | 0.1 | 3 | globule with bright emission rim south and east |
| Barnard 34 | 5 41.8 | +32 39 | 20′ | 2.5 | 600 | 1.1 | 70 | larger isolated globule |
| Barnard 227 | 6 05.9 | +19.29 | 4′ | 4.0 | 400 | 0.3 | 20 | isolated globule |
| NGC 2237–44 (Rosette nebula) | 6 30.8 | + 4 53 | 0′.1···0′.2 | 10 *) | 1660 | 0.02···0.06 | 0.1···0.8 | bunches of small globules in the emission nebula |
| NGC 2264 | 6 39.9 | + 9 31 | 1′ | 10 *) | 900 | 0.14 | 4 | square globule within an elephant trunk structure |
| Barnard 255 | 17 19.2 | −23 27 | 4′ | 1.3 | 200 | 0.14 | ≳ 1 | larger globules |
| Barnard 68 | 17 21.1 | −23 48 | 3′ | 13.0 | 200 | 0.09 | ~ 1 | larger globules |
| Barnard 72 | 17 22.1 | −23 39 | 3′.5 | 4.4 | 200 | 0.08 | ≳ 1 | larger globules |
| M 8 | 18 01.8 | −24 23 | 0′.1···0′.4 | 10 *) | 1260 | 0.02···0.08 | 0.1···1.4 | emission nebula contains some small globules |
| Barnard 92 | 18 14.2 | −18 16 | 8′ | 2···13 | 200 | 0.26 | ≳ 20 | dense core + envelope |
| Barnard 133 | 19 04.8 | − 6 57 | 4′ | 13.0 | 400 | 0.26 | ≳ 50 | larger globules |
| Barnard 134 | 19 05.6 | − 6 17 | 4′ | 1.3 | 400 | 0.22 | 20 | larger globules |
| Barnard 335 | 19 35.7 | + 7 31 | 3′.8 | 1.3···13 | 400 | 0.24 | 20 | larger globules |
| Barnard 361 | 21 11.9 | +47 18 | 20′ | 2.5 | 600 | 0.9 | 50 | isolated globule |

*) Assumed value for "opaque" globule.

A table for the classification of Barnard objects in the catalogue of Lynds is given by Lynds [32].

Mapping of several molecular lines of CO, CS, $H_2CO$, and OH in globules: see [34].

Far-infrared observations of globules: see [73].

## 7.2.3 Statistical description of interstellar cloud structure

Each of the different methods of observation gives information only on a limited section of the whole interstellar cloud spectrum. For example, by using star counts relatively dense clouds with visual extinction above ≈0.3 mag (and with distances below ≈500 pc) can be detected, while the more frequent very tenuous diffuse clouds produce still traceable narrow absorption lines in stellar spectra. These selection effects must be kept in mind when the numbers of Table 4 are compared with each other – for example, the different values for the number of small clouds in the line-of-sight per kpc. Table 4: see next page.

Table 4. Statistical properties of cool interstellar clouds. (Explanation of symbols and footnote: see next page.)

| Type of cloud | Observational data used | $a_V$ mag | $\mathfrak{N}_H$ atoms $cm^{-2}$ | $2R$ pc | $N_H$ atoms $cm^{-3}$ | $\mathfrak{M}$ $\mathfrak{M}_\odot$ | $T$ K | $\nu$ $kpc^{-1}$ | Comments | Ref. |
|---|---|---|---|---|---|---|---|---|---|---|
| Small clouds or diffuse clouds | colour excesses | 0.19 | $3.6\cdot10^{20}$ *) | — | — | — | — | 6 | distance $\lessapprox 1$ kpc | 37 |
| | | 0.26 | $4.9\cdot10^{20}$ *) | 3 | 50 | 20 | — | 5 | distance $\lessapprox 1$ kpc | 51 |
| | | 0.16 | $3.0\cdot10^{20}$ *) | 4 | 30 | 30 | — | 7 | distance $\lessapprox 0.3$ kpc | 27, 28 |
| | interstellar absorption lines in stellar spectra | — | $5\cdot10^{20}$ | — | 30 | — | 80 | 8···10 | NaD, CaK lines; | 54, 55 |
| | | | | | | | | 5 (for $\mathfrak{N}_H > 1\cdot10^{20}$) | rotational excitation of $H_2$ | 26 |
| | 21-cm line emission | — | $2\cdot10^{19}$ | 5 | 2 | 4 | — | — | smallest detectable irregularities ("cloudlets") | 25 |
| | | | $3\cdot10^{20}$ | 5 | 20 | 40 | 60 | — | values compatible with observation | 6, 22 |
| | 21-cm line absorption | | $3\cdot10^{20}$ | — | — | — | 80 | 4 | | 55 |
| Large clouds or cloud complexes | colour excesses | 0.9 | $1.7\cdot10^{21}$ | — | — | — | — | 0.8 | | 37 |
| | | 1.6 | $3.0\cdot10^{21}$ | 70 | 20 | $1\cdot10^5$ | — | 0.5 | | 51 |
| | 21-cm line emission | — | $5\cdot10^{20}$ | 30 | 5 | $3\cdot10^3$ | — | — | | 24 |
| Typical dark nebula | star counts and colour excesses | 4 | $8\cdot10^{21}$ | 8 | $5\cdot10^2$ | $4\cdot10^3$ | — | — | | 10, 55 |
| | line emission of simple molecules CO, OH, $H_2CO$, CS etc. | — | $5\cdot10^{21}\cdots5\cdot10^{22}$ | 1 core | $10^3\cdots10^4$ | $10^2$ | 20···40 | — | probably continuous transition to "massive" (giant) molecular clouds | 55, 17, |
| | | | | 10 (envelope) | $10^2\cdots10^3$ | $10^3\cdots10^4$ | 10···20 | — | | 38, 70 |
| Typical large globule | star counts, CO line emission | 4 | $8\cdot10^{21}$ | 0.6 | $7\cdot10^3$ | 30 | 10 | — | | 55, 34 |
| Molecular cloud | CO, CS line emission; higher excited molecular lines | $\approx10^2$ | $\approx10^{23}$ | 5 | $2\cdot10^4$ | $4\cdot10^4$ | $\approx50$ (core) | $\approx0.5$ | central condensations with peak density $> 10^4$ $cm^{-3}$ | 53, 13, 70 |
| | | | | 1 (core) 10 (overall) | | | 10···20 (envelope) | | | |
| Giant molecular cloud | CO line emission | — | $\approx10^{23}$ | 40 | $3\cdot10^2$ | $5\cdot10^5$ | 10 | | | 53a |

Symbols and footnote to Table 4:
$a_V$ = mean visual extinction
$\mathfrak{N}_H$ = mean column density of H atoms including hydrogen in $H_2$
$2R$ = mean diameter
$N_H$ = mean density of H atoms including hydrogen in $H_2$
$\mathfrak{M}$ = mean mass
$T$ = kinetic temperature
$\nu$ = mean number of clouds in the line-of-sight per kpc

*) Calculated from $a_V$ using $\mathfrak{N}_H = 5.8 \cdot 10^{21}\ E_{B-V}$ and $A_V/E_{B-V} = 3.1$ (see 7.1.2.1).

Volume fraction, $f$, occupied by the clouds (filling factor):

| | | |
|---|---|---|
| "small clouds" | $f \approx 0.02$ | [51, 28] |
| "large clouds" | $f \approx 0.1$ | [51] |

(see also 8.3.4.1, Table 16)

Mass spectrum of interstellar clouds represented by

$$F(\mathfrak{M}) \propto \mathfrak{M}^{-\beta}.$$

Estimates of $\beta$ from observations are compiled in Table 5. Excluding the total masses of the "large clouds" which are composed of several single clouds, a value near $\beta = 1.5$ seems to be a reasonable approximation.

Table 5. Determination of the exponent $\beta$ of the cloud mass spectrum $F(\mathfrak{M}) \propto \mathfrak{M}^{-\beta}$.

| $\beta$ | Mass range $\mathfrak{M}_\odot$ | Observational data | Ref. |
|---|---|---|---|
| 2 | $<10^3$ | extinction by dust | 52 |
| 1.0 ···1.5 | $>10^3$ | extinction by dust | |
| 1.65···2 | 1···24 | 21 cm emission | 24 |
| | | | 20 |
| 1.5 | $20 \cdots 2 \cdot 10^3$ | $Ca^+$ absorption lines | 41 |
| 2.2 | $6 \cdots 2 \cdot 10^4$ | extinction by dust | 45 |
| 1.75 | $\geqq 10^3$ | 21-cm emission | 42 |
| 1.3 ···1.7 | $<10^5$ | 21-cm emission | 12 |
| 1.3 | $\lessapprox 10^2$ | extinction by dust | 28 |

Velocities of the clouds

Internal dispersion of radial velocities within clouds of central optical thickness in the 21-cm line $\tau_0 \gtrapprox 0.2$ [43, 44, 26] — 1.8 km s$^{-1}$

External dispersion of radial velocities between different clouds [19] — 6 km s$^{-1}$

Three-dimensional velocity dispersion on the assumption of isotropy — 10 km s$^{-1}$

Intercloud medium [55, 39, 71] (see also 7.5.4):

Outside the cool clouds one recognizes at least three constituents of an "intercloud medium":

(1) a "warm" low-density neutral gas with fluctuating density ($\lessapprox 0.3$ cm$^{-3}$) and temperature (1000···10000 K). Mean density and temperature: $N_H \approx 0.1$ cm$^{-3}$, $T \approx 6000$ K.

(2) Extended low-density H II regions surrounding OB stars etc., which are believed to produce the observed diffuse H$\alpha$ and H 166$\alpha$ emission. Mean electron density $N_e \approx N_H \approx 0.1$ cm$^{-3}$.

(3) A very hot and tenuous plasma ("coronal gas") with $T \approx 6 \cdot 10^5$ K, $N_H \approx 3 \cdot 10^{-3}$ cm$^{-3}$.

## 7.2.4 Molecular clouds

Some of the simpler molecules like OH, CO, HCN, $HCO^+$, $H_2CO$ have been observed in diffuse interstellar clouds of low extinction and low density. The term "molecular cloud" is generally used for clouds with densities greater than $100\,cm^{-3}$ which are primarily composed of molecular hydrogen. These include at least two categories:

(1) dark nebulae with high extinction ("dark dust clouds") found by optical means

(2) dense clouds which were found by molecular spectroscopy, often named "massive" or "giant" molecular clouds, showing line emission from more than about ten molecular species.

Since some clouds can be put into both classes – for example, the ϱ Ophiuchi cloud – another categorization has been proposed [18, 68]:

(A) dense clouds with temperatures less than 20 K throughout

(B) dense clouds with temperatures rising above 20 K somewhere in the cloud, including condensations of higher temperature such as far-infrared sources, compact H II regions and masers ("hot-centered clouds").

Molecular clouds have a wide range of sizes and masses. Large clouds appear generally to consist of one or more dense cores and more extended envelopes; in this sense they should be regarded as complexes. The largest and most massive clouds, referred to as giant molecular clouds, have masses of about $10^5 \cdots 10^6\,\mathfrak{M}_\odot$. These are the most massive objects in the galaxy [53a].

Table 6. Molecular microwave lines observed in interstellar space up to April 1979 and their distribution for a selected number of interstellar sources. A few transitions whose detection is anticipated have been included. Unless otherwise indicated the electronic ground state is $^1\Sigma$, and the vibrational state is $v=0$. The numbers in parentheses following the nuclear hyperfine quantum number, $F$, give the relative intensities of the transitions [63a, 31a].
$J$ = total rotational angular momentum quantum number, excluding spin
$F$ = total angular momentum quantum number including spin
$\nu$ = rest frequency; estimated uncertainties in parentheses
E, A, ME: column indicates whether the interstellar line occurs in emission (E), absorption (A) or maser emission (ME)

| Molecule | Transition | | | ν [MHz] | E, A, ME | Sgr A | Sgr B2 | Ori A | W 51 | W 3 | DR 21 | NGC 2264 | NGC 2024 | IRC+10216 | L 134 | Cloud 2 |
|---|---|---|---|---|---|---|---|---|---|---|---|---|---|---|---|---|
| OH | $^2\Pi_{3/2}$; | $J=3/2$; | $F=1$–2 | 1612.2310 (2) | | × | × | × | × | × | × | | × | | | |
| | | | 1–1 | 1665.4018 (2) | E, A, ME | × | × | × | × | × | × | × | × | | × | × |
| | | | 2–2 | 1667.3590 (2) | | × | × | × | × | × | × | × | × | | × | × |
| | | | 2–1 | 1720.5300 (2) | | × | × | × | × | × | × | | × | | × | |
| | | $J=5/2$; | $F=2$–3 | 6016.746 (5) | | | | | | | | | | | | |
| | | | 2–2 | 6030.747 (5) | E | | | | × | × | | | | | | |
| | | | 3–3 | 6035.092 (5) | E | | × | | × | × | × | | | | | |
| | | | 3–2 | 6049.084 (8) | | | | | | | | | | | | |
| | | $J=7/2$; | $F=3$–4 | 13433.930 (25) | | | | | | | | | | | | |
| | | | 3–3 | 13434.596 (10) | | | | | | | | | | | | |
| | | | 4–4 | 13441.365 (10) | | | | | | × | | | | | | |
| | | | 4–3 | 13442.030 (25) | | | | | | | | | | | | |
| | | $J=9/2$; | $F=5$–4 | 23805.2077 (10) | | | | | | | | | | | | |
| | | | 4–4 | 23817.6153 (2) | A | | | | | × | | | | | | |
| | | | 5–5 | 23826.6211 (2) | A | | | | | × | | | | | | |
| | | | 4–5 | 23838.933 (10) | | | | | | | | | | | | |

continued

Table 6, continued

| Molecule | Transition | | | ν [MHz] | E, A, ME | Sgr A | Sgr B2 | Ori A | W 51 | W 3 | DR 21 | NGC 2264 | NGC 2024 | IRC+10216 | L 134 | Cloud 2 |
|---|---|---|---|---|---|---|---|---|---|---|---|---|---|---|---|---|
| OH | $^2\Pi_{1/2}$; | $J=1/2$; | $F=0$–1 | 4660.242 (3) | | | × | | | | | | | | | |
| | | | 1–1 | 4750.656 (3) | | | × | | | | | | | | | |
| | | | 1–0 | 4765.562 (3) | | | × | | | | | | | | | |
| | | $J=3/2$; | $F=1$–2 | 7749.909 | | | | | | | | | | | | |
| | | | 1–1 | 7761.747 | | | | | | | | | | | | |
| | | | 2–2 | 7820.125 (5) | | | | | | | | | | | | |
| | | | 2–1 | 7831.962 (5) | | | | | | | | | | | | |
| | | $J=5/2$; | $F=2$–3 | 8118.052 (5) | | | | | | | | | | | | |
| | | | 2–2 | 8135.868 (5) | | | | | | | | | | | | |
| | | | 3–3 | 8189.586 (5) | | | | | | | | | | | | |
| | | | 3–2 | 8207.401 (5) | | | | | | | | | | | | |
| $^{18}$OH | $^2\Pi_{3/2}$; | $J=3/2$; | $F=1$–1 | 1637.564 (2) | A | × | × | | | | | | | | | |
| | | | 2–2 | 1639.503 (2) | A | × | × | | | | | | | | | |
| $^{17}$OH | $^2\Pi_{3/2}$; | $J=3/2$; | $F=9/2$–9/2 | 1626.161 (10) | | × | | | | | | | | | | |
| | | | 7/2–7/2 | 1624.518 (10) | | × | | | | | | | | | | |
| $^{28}$SiO | $v=0$, | $J=2$–1 | | 86846.891 (28) | E | | × | × | | | | | | × | | |
| | | 3–2 | | 130268.574 (37) | E | | | × | | | | | | × | | |
| | $v=1$, | $J=1$–0 | | 43122.027 (15) | ME | | | × | | | | | | | | |
| | | 2–1 | | 86243.350 (20) | ME | | | × | | | | | | | | |
| | | 3–2 | | 129363.262 (27) | E | | | × | | | | | | | | |
| | $v=2$, | $J=1$–0 | | 42820.539 (15) | ME | | | × | | | | | | | | |
| | $v=3$, | $J=1$–0 | | 42519.332 (15) | E | | | | | | | | | | | |
| $^{29}$SiO | $v=0$, | $J=2$–1 | | 85758.953 (50) | E | | | × | | | | | | | | |
| $^{30}$SiO | $v=0$, | $J=2$–1 | | 84746.009 (50) | E | | × | × | | | | | | | | |
| SiS | | $J=5$–4 | | 90771.546 (26) | E | | | | | | | | | × | | |
| | | $J=6$–5 | | 108924.267 (48) | E | | × | | | | | | | × | | |
| NO | $^2\Pi^+_{1/2}$; | $J=3/2$–1/2; | $F=5/2$–3/2 | 150176.54 (25) | E | | × | | | | | | | | | |
| | $^2\Pi^-_{1/2}$; | $J=3/2$–1/2; | $F=5/2$–3/2 | 150546.50 (25) | E | | × | | | | | | | | | |
| NS | $^2\Pi_{1/2}$; | $J=5/2$–3/2; | $F=7/2$–5/2 | 115153.835 (80) | E | | × | | | | | | | | | |
| | | | 5/2–3/2 | 115156.799 (80) | E | | × | | | | | | | | | |
| | | | 3/2–1/2 | 115163.07 | | | | | | | | | | | | |
| | | | 3/2–3/2 | 115185.55 | E | | × | | | | | | | | | |
| | | | 5/2–5/2 | 115191.07 | | | | | | | | | | | | |
| NS | $^2\Pi_{1/2}$; | $J=5/2$–3/2; | $F=5/2$–5/2 | 115489.61 | E | | | | | | | | | | | |
| | | | 3/2–3/2 | 115524.82 | E | | × | | | | | | | | | |
| | | | 7/2–5/2 | 115556.312 (60) | E | | × | | | | | | | | | |
| | | | 5/2–3/2 | 115570.762 | E | | × | | | | | | | | | |
| | | | 3/2–1/2 | 115571.93 (6) | E | | × | | | | | | | | | |

continued

Table 6, continued

| Molecule | Transition | | $\nu$ [MHz] | E, A, ME | Sgr A | Sgr B2 | Ori A | W 51 | W 3 | DR 21 | NGC 2264 | NGC 2024 | IRC+10216 | L 134 | Cloud 2 |
|---|---|---|---|---|---|---|---|---|---|---|---|---|---|---|---|
| SO | $^3\Sigma^-$; $J, N=1,2–1,1$ | | 13043.70 (10) | E | | × | | | | | | | | | |
| | 2,2–1,1 | | 86093.95 (10) | E | | × | × | | | | | | × | | |
| | 3,2–2,1 | | 99299.87 (10) | E | | × | × | × | | × | × | × | | | |
| | 2,3–1,2 | | 109252.10 (10) | E | | | × | | | | | | | | |
| | 4,3–3,2 | | 138178.60 (50) | E | | × | × | × | | × | | × | | | |
| | 4,5–4,4 | | 100029.644 (34) | E | | | × | | | | | | | | |
| | 1,0–0,1 | | 30001.523 (10) | E | | × | | | | | | | | | |
| $^{34}SO$ | $^3\Sigma^-$; $J, N=2,3–1,2$ | | 106742.92 (15) | E | | | × | | | | | | | | |
| | 3,2–2,1 | | 97715.39 (15) | E | | | × | | | | | | | | |
| CO | $J=1–0$ | | 115271.204 (5) | E | × | × | × | × | × | × | × | × | × | × | × |
| | 2–1 | | 230537.974 (30) | E | | | × | × | | × | | × | | | |
| | 3–2 | | 345795.900 (13) | E | | | × | | | | | | | | |
| $^{12}C^{17}O$ | $J=1–0$ | | 112359.276 (80) | E | | × | × | | | × | × | | × | | |
| $^{13}C^{16}O$ | $J=1–0$ | | 110201.370 (8) | E | × | × | × | × | × | × | × | × | × | × | × |
| | 2–1 | | 220398.714 (50) | E | | | × | × | | × | | × | | | |
| $^{12}C^{18}O$ | $J=1–0$ | | 109782.182 (8) | E | × | × | × | × | | × | × | × | × | | |
| | 2–1 | | 219560.369 (50) | E | | | | | | × | | | | | |
| CN | $^2\Sigma^+$; $N=1–0$; $J=3/2–1/2$; | $F=3/2–1/2$ (10) | 113488.140 (5) | E | | | × | | | | | | | | |
| | | 5/2–3/2 (27) | 113490.982 (3) | E | | | × | × | × | × | × | | × | | |
| | | 1/2–1/2 (8) | 113499.639 (5) | E | | | × | | | | | | | | |
| | | $F=3/2–3/2$ (8) | 113508.944 (13) | E | | | × | | | | | | | | |
| | | 1/2–3/2 (1) | 113520.34 | E | | | | | | | | | | | |
| | $J=1/2–1/2$; | $F=1/2–1/2$ (1) | 113123.83 | E | | | | | | | | | | | |
| | | 1/2–3/2 (8) | 113144.192 (9) | E | | | × | | | | | | | | |
| | | 3/2–1/2 (10) | 113170.528 (20) | E | | | × | | | | | | | | |
| | | 3/2–3/2 (10) | 113191.317 (40) | E | | | × | | | | | | | | |
| $^{12}CH$ | $^2\Pi_{1/2}$; $J=1/2$; | $F=1–0$ | 3349.193 (3) | A, E | | × | | | | | | | | | |
| | | 1–1 | 3335.481 (2) | A, E | | × | | | | | | × | | | |
| | | 0–1 | 3263.794 (3) | A, E | × | × | × | × | × | × | | | | × | |
| $^{12}C^{32}S$ | $J=1–0$ | | 48991.000 (12) | E | × | × | × | × | × | × | × | × | × | × | |
| | 2–1 | | 97981.007 (16) | E | × | × | × | × | × | × | × | × | × | × | |
| | 3–2 | | 146969.039 (60) | E | | × | × | × | × | × | | × | × | | |
| $^{12}C^{34}S$ | $J=1–0$ | | 48206.948 (12) | E | | × | | × | | × | | | | | |
| | 2–1 | | 96412.953 (30) | E | | × | × | × | × | | | | | | |
| | 3–2 | | 144617.117 (16) | E | | × | × | | | | | | × | | |
| $^{13}C^{32}S$ | $J=1–0$ | | 46247.472 (40) | E | | × | | | | | | | | | |
| | 2–1 | | 92494.084 (81) | E | | | × | | | | | | × | | |
| | 3–2 | | 138738.97 (13) | E | | × | × | × | | | × | | × | | |

continued

Table 6, continued

| Molecule | Transition | | ν [MHz] | E, A, ME | Sgr A | Sgr B2 | Ori A | W 51 | W 3 | DR 21 | NGC 2264 | NGC 2024 | IRC+10216 | L 134 | Cloud 2 |
|---|---|---|---|---|---|---|---|---|---|---|---|---|---|---|---|
| $^{12}C^{33}S$ | $J=1–0$ | | 48586.520 (25) | | | | | | | | | | | | |
| | | $F=1/2–3/2$ (17) | 48583.264 (10) | E | | | | | | | | | | | |
| | | 5/2–3/2 (50) | 48585.906 (10) | E | | × | | | | | | | | | |
| | | 3/2–3/2 (33) | 48589.068 (10) | E | | | | | | | | | | | |
| | $J=2–1$ | | 97172.090 | E | | | | | | | | | | | |
| | 3–2 | | 145755.760 | E | | × | × | | | | | | | | |
| $H_2O$ | $6_{16}–5_{23}$ | | 22235.07985 | ME | | × | × | × | × | × | | | | | |
| | $3_{12}–2_{20}$ | | 183310.0906 (15) | E | | | × | | | | | | | | |
| $H_2{}^{18}O$ | $3_{13}–2_{20}$ | | 203407.52 (2) | E | | | × | | | | | | | | |
| HDO | $1_{10}–1_{11}$ | | 80578.283 (53) | E | | | × | | | | | | | | |
| $H_2S$ | $1_{11}–1_{10}$ | | 168762.76237 (2) | E | | × | × | × | × | × | × | | | | |
| | $1_{10}–1_{10}$ | | 195558.96 (12) | E | | | × | | | | | | | | |
| $SO_2$ | $8_{17}–8_{08}$ | | 83688.074 (30) | E | | × | × | | | | | | | | |
| | $8_{35}–9_{28}$ | | 86639.100 (50) | E | | | × | | | | | | | | |
| | $7_{35}–8_{26}$ | | 97702.352 (50) | E | | | × | | | | | | | | |
| | $3_{13}–2_{02}$ | | 104029.417 (50) | E | | | × | | | | | | | | |
| | $12_{2,12}–14_{1,13}$ | | 132744.800 | E | | | × | | | | | | | | |
| | $5_{15}–4_{04}$ | | 135695.985 | E | | | × | | | | | | | | |
| | $6_{24}–6_{15}$ | | 140306.160 | E | | | × | | | | | | | | |
| | $16_{2,14}–16_{1,15}$ | | 143057.045 | E | | | × | | | | | | | | |
| | $8_{26}–8_{17}$ | | 134004.802 | E | | | × | | | | | | | | |
| $N_2D^+$ | $J=1–0$; | $F=1–1$ | 77107.86 (9) | E | | | | | | | | | | × | |
| | | 2–1 | 77109.61 (8) | E | | | | | | | | | | × | |
| | | 0–1 | 77112.2 (1) | E | | | | | | | | | | × | |
| HNO | $1_{01}–0_{00}$ | | 81477.49 (10) | | | × | | | | | × | | | | |
| $N_2H^+$ | $J=1–0$; | $F_1=1–1$ (33) | 93171.88 (4) | E | × | | × | × | × | × | × | | | | |
| | | 2–1 (55) | 93173.75 (4) | E | × | × | × | × | × | × | × | | | | |
| | | 0–1 (11) | 93176.13 (22) | E | × | | × | × | × | × | × | | | | |
| OCS | $J=6–5$ | | 72976.784 (2) | E | | × | | | | | | | | | |
| | 7–6 | | 85139.107 (2) | E | | × | | | | | | | | | |
| | 8–7 | | 97301.212 (2) | E | | × | | | | | | | | | |
| | 9–8 | | 109463.067 (2) | E | | × | | | | | | | | | |
| | 12–11 | | 145946.819 (3) | E | | × | | | | | | | | | |
| HCN | $J=1–0$ | | 88631.6624 (10) | E | | | | | | | | | | | |
| | | $F=1–1$ (33) | 88630.4157 | E | | | × | | | | | | | | |
| | | 2–1 (55) | 88631.8473 | E | × | × | × | × | × | × | × | × | × | × | × |
| | | 0–1 (11) | 88633.9360 | E | | | × | | | | | | | | |
| | $J=3–2$ | | 265886.432 (10) | E | | | × | | | | | | | | |

continued

Table 6, continued

| Molecule | Transition | | ν [MHz] | E, A, ME | Sgr A | Sgr B2 | Ori A | W 51 | W 3 | DR 21 | NGC 2264 | NGC 2024 | IRC+10216 | L 134 | Cloud 2 |
|---|---|---|---|---|---|---|---|---|---|---|---|---|---|---|---|
| $H^{13}CN$ | $J=1–0$ | | 86339.944 (25) | E | | | | | | | | | | | |
| | | $F=1–1$ (33) | 86338.767 | E | | | | | | | | | | | |
| | | 2–1 (55) | 86340.184 | E | × | × | × | × | × | × | × | | × | | |
| | | 0–1 (11) | 86342.274 | E | | | | | | | | | | | |
| | $J=2–1$; | $F=1–0$, 2–2 | 172676.573 (50) | E | | | × | | | | | | | | |
| | | 2–1, 3–2 | 172677.959 (50) | E | | | | | | | | | | | |
| | | 1–1 | 172680.209 (50) | E | | | | | | | | | | | |
| $HC^{15}N$ | $J=1–0$ | | 86054.961 (25) | | | × | × | × | | | | | × | | |
| | 2–1 | | 172107.956 (45) | | × | | × | | | | | | | | |
| DCN | $J=1–0$ | | 72414.6941 (7) | | | | | | | | | | | | |
| | $F_N=1–1$ (33); | $F_D=1–0, 1, 2$ | 72413.4843 | | × | | × | | × | | × | | | | |
| | | 2–1, 2 | 72413.5143 | | × | | × | | × | | × | | | | |
| | | 0–0, 1 | 72413.5584 | | | | | | | | | | | | |
| | $F_N=2–1$ (55); | $F_D=1–0, 1, 2$ | 72414.9054 | | | | × | × | | × | | | | | |
| | | 2–1, 2 | 72414.9270 | | | | × | × | | × | | | | | |
| | | 3–2 | 72414.9732 | | | | × | × | | × | | | | | |
| | $F_N=0–1$ (11); | $F_D=1–0, 1, 2$ | 72417.0297 | | | | × | | | | | | | | |
| | $J=2–1$ | | 144828.0003 (15) | | | | | | | | | | | | |
| | | $F_N=2–2$ (8) | 144826.5727 | E | | × | × | × | | × | × | | × | | |
| | | 1–0 (11) | 144826.8161 | E | | × | × | × | | × | × | | × | | |
| | | 2–1 (25) | 144828.0003 | E | | × | × | × | | × | × | | × | | |
| | | 3–2 (47) | 144828.1093 | E | | × | × | × | | × | × | | × | | |
| | | 1–1 (8) | 144830.3358 | E | | × | × | × | | × | × | | × | | |
| | $J=3–2$ | | 217238.531 (10) | | | | | | | | | | | | |
| | | $F_N=4–3$ (43) | 217238.5951 | E | | | × | | | | | | | | |
| | | 3–2 (30) | 217238.531 | E | | | × | | | | | | | | |
| | | 2–1 (21) | 217238.288 | E | | | × | | | | | | | | |
| HNC | $J=1–0$ | | 90663.592 (40) | E | × | × | × | × | × | × | × | × | × | × | × |
| | 3–2 | | 271981.1 (6) | E | | | × | | | | | | | | |
| $HN^{13}C$ | $J=1–0$ | | 87090.851 (40) | E | | | | | | | | | | | |
| | | $F=1–1$ | 87090.950 (46) | E | | | | | | | | | | × | × |
| | | 2–1 | 87090.864 (46) | E | | | | | | | | | | × | × |
| | | 0–1 | 87090.734 (46) | E | | | | | | | | | | × | × |
| DNC | $J=1–0$ | | 76305.727 (46) | E | | × | | | × | | × | | | | |
| | 2–1 | | 152609.774 (50) | E | | × | | | × | | × | | | | |
| $HCO^+$ | $J=1–0$ | | 89188.545 (20) | E | × | × | × | × | × | × | × | × | × | × | × |
| | 3–2 | | 267557.2 (5) | E | | | × | | | | | | | | |
| $H^{13}CO^+$ | $J=1–0$ | | 86754.341 (5) | E | | × | × | × | × | × | × | | | | |
| $HC^{18}O^+$ | $J=1–0$ | | 85162.157 (50) | E | | | × | | | | × | | | × | |

continued

Table 6, continued

| Molecule | Transition | | | ν [MHz] | E, A, ME | Sgr A | Sgr B2 | Ori A | W 51 | W 3 | DR 21 | NGC 2264 | NGC 2024 | IRC+10216 | L 134 | Cloud 2 |
|---|---|---|---|---|---|---|---|---|---|---|---|---|---|---|---|---|
| $DCO^+$ | | $J$=1–0 | | 72039.35 | E | | × | × | | | × | × | | | × | × |
| | | 2–1 | | 144077.3 | E | | | × | | | × | × | | | × | × |
| HCO | $1_{01}$–$0_{00}$; | $J$=3/2–1/2; | $F$=2–1 | 86670.65 (10) | | | | | × | × | | | × | | | |
| $C_2H$ | $^2\Sigma$; $N$=1–0; | $J$=3/2–1/2; | $F$=1–1 (4) | 87284.38 | | | | | | | | | | | | |
| | | | 2–1 (42) | 87317.05 (4) | E | × | × | × | × | × | × | × | × | × | | |
| | | | 1–0 (21) | 87328.70 | E | | × | × | × | × | × | × | × | × | | |
| | | $J$=1/2–1/2; | $F$=1–1 (21) | 87402.10 | E | | | × | × | | × | × | × | × | | |
| | | | 0–1 (8) | 87407.23 | E | | | × | × | | × | × | × | | | |
| | | | 1–0 (4) | 87446.42 | | | | | | | | | | | | |
| $NH_3$ | | $J$, $K$=1, 1; | $F$=0–1 (11) | 23692.969 (1) | E, A | | × | × | | × | | | | | × | × |
| | | | 2–1 (14) | 23693.905 (1) | E, A | | × | × | | × | × | × | | | × | × |
| | | | 1–1 (8) | 23694.486 (1) | E, A | | × | × | | × | × | × | | | × | × |
| | | | 2–2 (42) | 23694.506 (1) | E, A | × | × | × | | × | | | | | × | × |
| | | | 1–2 (14) | 23695.113 (1) | E, A | | × | × | | × | | | | | × | × |
| | | | 1–0 (11) | 23696.041 (1) | E, A | | × | × | | × | | | | | × | × |
| | | 2, 2; | $F$=1–2 (5) | 23720.575 | E, A | | × | × | | × | | | | | | |
| | | | 3–2 (5.2) | 23721.337 | E, A | | × | × | | × | | | | | | |
| | | | 2–2 (23) | 23722.632 | E, A | | × | × | | × | × | × | | | | × |
| | | | 3–3 (42) | 23722.634 | E, A | × | × | × | | × | × | × | | | | × |
| | | | 1–1 (15) | 23722.634 | E, A | | × | × | | × | × | × | | | | × |
| | | | 2–3 (5.2) | 23723.929 | E, A | | × | × | | × | | | | | | |
| | | | 2–1 (5.0) | 23724.692 | E, A | | × | × | | × | | | | | | |
| | | 3, 3; | $F$=2–3 (2.65) | 23867.805 (5) | E | | | | | × | | | | | | |
| | | | 4–3 (2.68) | 23868.450 (5) | E | | | | | × | | | | | | |
| | | | 2–2 (21.2) | 23870.130 (5) | E, A | | | | | × | × | | | | | |
| | | | 3–3 (28.0) | 23870.128 (5) | E, A | × | × | × | | × | × | | | | | |
| | | | 4–4 (40.2) | 23870.130 (5) | E, A | | | | | × | × | | | | | |
| | | | 3–4 (2.68) | 23871.807 (5) | E | | | | | × | | | | | | |
| | | | 3–2 (2.65) | 23872.453 (5) | E | | | | | × | | | | | | |
| | | 4, 4 | | 24139.416 (5) | E, A | | × | × | | | | | | | | |
| | | 5, 5 | | 24532.989 (5) | E, A | | × | × | | | | | | | | |
| | | 6, 6 | | 25056.025 (5) | E | | × | × | | | | | | | | |
| | | 2, 1 | | 23098.819 (5) | E | | × | × | | | | | | | | |
| | | 3, 2 | | 22834.182 (5) | E | | × | × | | | | | | | | |
| | | 4, 2 | | 21703.358 (5) | E | | | × | | | | | | | | |
| | | 4, 3 | | 22688.312 (5) | E | | | × | | | | | | | | |
| | | 5, 4 | | 22653.022 (5) | E | | | × | | | | | | | | |
| | | 6, 5 | | 22732.429 (5) | E | | | × | | | | | | | | |
| | | 7, 6 | | 22924.940 (5) | E | | | × | | | | | | | | |
| | | 8, 7 | | 23232.238 (5) | E | | | × | | | | | | | | |
| | | 9, 8 | | 23657.471 (5) | E | | | × | | | | | | | | |
| | | 10, 9 | | 24205.287 (5) | E | | | × | | | | | | | | |
| $^{15}NH_3$ | | 1, 1 | | 22624.91 | E | | | × | | | | | | | | |
| | | 2, 2 | | 22649.843 | E | | | × | | | | | | | | |
| | | 3, 3 | | 22789.442 | E | | | × | | | | | | | | |

continued

Table 6, continued

| Molecule | Transition | ν [MHz] | E, A, ME | Sgr A | Sgr B2 | Ori A | W 51 | W 3 | DR 21 | NGC 2264 | NGC 2024 | IRC+10216 | L 134 | Cloud 2 |
|---|---|---|---|---|---|---|---|---|---|---|---|---|---|---|
| $NH_2D$ | $1_{11}^{(+)}–1_{01}^{(-)}$ | 85926.2 (2) | E | | | × | | | | | | | | |
| | $1_{11}^{(-)}–1_{01}^{(+)}$ | 110153.4 (2) | E | | × | × | | | | | | | | |
| $H_2CO$ | $1_{10}–1_{11}$ | 4829.660 | E, A | × | × | × | × | × | × | × | × | | × | × |
| | $2_{11}–2_{12}$ | 14488.474 | A | × | × | × | × | × | × | × | × | | × | × |
| | $3_{12}–3_{13}$ | 28974.805 | A | | × | | | | | | | | | |
| | $5_{14}–5_{15}$ | 72409.090 | E | | | × | | | | | | | | |
| | $1_{01}–0_{00}$ | 72837.948 (10) | E | | × | × | | | | | | | | |
| | $2_{02}–1_{01}$ | 145602.949 (10) | E | | | × | | | | | | | | |
| | $4_{13}–4_{14}$ | 48284.519 (7) | E | | | × | | | | | | | | |
| | $6_{15}–6_{16}$ | 101332.964 | E | | | × | | | | | | | | |
| | $3_{12}–2_{11}$ | 225647.795 (20) | E | | | × | | | | | | | | |
| | $2_{12}–1_{11}$ | 140839.502 (10) | E | × | | × | × | × | | | | | | |
| | $2_{11}–1_{10}$ | 150498.334 (10) | E | | | × | | | | | | | | |
| $H_2{}^{13}CO$ | $1_{10}–1_{11}$ | 4593.089 (1) | A | × | × | | × | × | × | × | × | | × | |
| | $2_{12}–1_{11}$ | 137449.963 (10) | E | | | × | | | | | | | | |
| $H_2C^{18}O$ | $1_{10}–1_{11}$ | 4388.797 (1) | A | × | × | | × | | | | | | | |
| $C_3N$ | $^2\Sigma$; $J=9–8$ | 89045.7 (3) | E | | | | | | | | | × | | |
| | 9–8 | 89064.4 (3) | E | | | | | | | | | × | | |
| | 10–9 | 98939.9 (5) | E | | | | | | | | | × | | |
| | 10–9 | 98958.6 (5) | E | | | | | | | | | × | | |
| HNCO | $1_{01}–0_{00}$ | 21981.574 | E | | | | | | | | | | | |
| | $F=2–1$ (55) | 21981.47055 | E | | × | | | | | | | | | |
| | 1–1 (33) | 21982.08535 | E | | × | | | | | | | | | |
| | 0–1 (11) | 21980.54533 | E | | × | | | | | | | | | |
| | $2_{02}–1_{01}$ | 43963.042 (2) | E | | × | | | | | | | | | |
| | $4_{04}–3_{03}$ | 87925.252 (10) | E | | × | | | | | | | | | |
| | $4_{14}–3_{13}$ | 87597.342 (10) | | | × | | | | | | | | | |
| | $5_{05}–4_{04}$ | 109905.758 (10) | E | | × | | | | | | | | | |
| $H_2CS$ | $1_{10}–1_{11}$ | 1046.488 (1) | A | | | | | | | | | | | |
| | $2_{11}–2_{12}$ | 3139.380 (30) | A | | × | | | | | | | | | |
| | $3_{12}–3_{13}$ | 6278.650 (8) | | | × | | | | | | | | | |
| | $4_{13}–4_{14}$ | 10463.970 (5) | A | | × | | | | | | | | | |
| | $3_{12}–2_{11}$ | 104616.977 (51) | E | | × | × | × | | × | | | × | | |
| $C_4H$ | $N=9–8$ | 85633.9 (1) | E | | | | | | | | | × | | |
| | 9–8 | 85672.4 (1) | E | | | | | | | | | × | | |
| | 10–9 | 95149.5 (1) | E | | | | | | | | | × | | |
| | 10–9 | 95189.0 (1) | E | | | | | | | | | × | | |
| | 11–10 | 104667.3 | E | | | | | | | | | × | | |
| | 11–10 | 104706.0 | E | | | | | | | | | × | | |
| | 12–11 | 114182.0 | E | | | | | | | | | × | | |
| | 12–11 | 114221.0 | E | | | | | | | | | × | | |

continued

Table 6, continued

| Molecule | Transition | | ν [MHz] | E, A, ME | Sgr A | Sgr B2 | Ori A | W 51 | W 3 | DR 21 | NGC 2264 | NGC 2024 | IRC+10216 | L 134 | Cloud 2 |
|---|---|---|---|---|---|---|---|---|---|---|---|---|---|---|---|
| $H_2CNH$ | $1_{10}$–$1_{11}$; | $F$=0–1 (11.1) | 5288.980 (40) | E | | × | | | | | | | | | |
| | | 1–0 (11.1) | 5289.786 (40) | E | | × | | | | | | | | | |
| | | 2–2 (41.7) | 5289.786 (40) | E | | × | | | | | | | | | |
| | | 2–1 (13.9) | 5290.726 (40) | E | | × | | | | | | | | | |
| | | 1–2 (13.9) | 5290.726 (40) | E | | × | | | | | | | | | |
| | | 1–1 (8.3) | 5291.646 (40) | E | | × | | | | | | | | | |
| $NH_2CN$ | $5_{14}$–$4_{13}$ | | 100629.500 (120) | E | | × | | | | | | | | | |
| | $4_{13}$–$3_{12}$ | | 80504.600 (120) | E | | × | | | | | | | | | |
| | $4_{14}$–$3_{13}$ | | 79449.73 (9) | E | | × | | | | | | | | | |
| | $4_{04}$–$3_{03}$ | | 79979.596 (90) | E | | × | | | | | | | | | |
| | $v$=1; $4_{04}$–$3_{03}$ | | 79915.101 (50) | E | | × | | | | | | | | | |
| HCOOH | $1_{10}$–$1_{11}$ | | 1638.805 (3) | E | | × | | | | | | | | | |
| | $2_{11}$–$2_{12}$ | | 4916.312 (8) | E | | × | | | | | | | | | |
| HCCCN | $J$=1–0 | | 9098.1152 (2) | | | | | | | | | | | | |
| | | $F$=1–1 (33) | 9097.0346 (3) | | × | × | | | | | | | | × | |
| | | 2–1 (55) | 9098.3321 (3) | | × | × | | | | | | | | × | |
| | | 0–1 (11) | 9100.2727 (5) | | × | × | | | | | | | | × | |
| | $J$=2–1 | | 18196.226 (10) | | | | | | | | | | | | |
| | | $F$=2–2 (8.3) | 18194.936 (20) | | | | | | | | | | | | |
| | | 1–0 (11.1) | 18195.190 (50) | | | | | | | | | | | | |
| | | 2–1 (25.0) | 18196.279 (50) | | | × | | | | | | | | | |
| | | 3–2 (46.7) | 18196.279 (50) | | | × | | | | | | | | | |
| | | 1–1 (8.3) | 18198.366 (20) | | | | | | | | | | | | |
| | $J$=5–4 | | 45490.307 (4) | | | × | × | × | | | | | | | × |
| | 6–5 | | 54588.247 (5) | | | × | | | | | | | × | | × |
| | 8–7 | | 72783.822 (15) | | × | × | × | × | | | | | × | | |
| | 9–8 | | 81881.458 (7) | | | × | × | × | | | | | | | |
| | 10–9 | | 90979.023 (20) | | | × | × | × | × | | | | × | | |
| | 11–10 | | 100076.392 (15) | | × | × | × | × | | | | | | | |
| | 12–11 | | 109173.634 (4) | | | × | | | | | | | × | | |
| | 14–13 | | 127367.666 (30) | | | × | × | | | | | | | | |
| | 15–14 | | 136464.404 (13) | E | | × | | × | | | | | | | |
| | 16–15 | | 145560.946 (30) | E | | × | | × | | | | | | | |
| $H^{13}CCCN$ | $J$=1–0 | | 8816.901 (20) | | | | | | | | | | | | |
| | | $F$=1–1 (33) | 8815.824 (40) | | | × | | | | | | | | | |
| | | 2–1 (55) | 8817.115 (20) | | | × | | | | | | | | | |
| | | 1–0 (11) | 8819.065 (50) | | | × | | | | | | | | | |
| $H^{13}CCCN$ | $J$=9–8 | | 79350.477 (12) | E | | × | × | | | | | | | | |
| $HC^{13}CCN$ | 9–8 | | 81534.126 (18) | E | | × | × | | | | | | | | |
| $HCC^{13}CN$ | 9–8 | | 81541.983 (5) | E | | × | × | | | | | | | | |

continued

Table 6, continued

| Molecule | Transition | | ν [MHz] | E, A, ME | Sgr A | Sgr B2 | Ori A | W 51 | W 3 | DR 21 | NGC 2264 | NGC 2024 | IRC+10216 | L 134 | Cloud 2 |
|---|---|---|---|---|---|---|---|---|---|---|---|---|---|---|---|
| $HC^{13}CCN$ | $J=1$–$0$ | | 9059.543 (20) | | | | | | | | | | | | |
| | | $F=1$–$1$ (33) | 9058.498 (30) | E | | × | | | | | | | | | |
| | | 2–1 (55) | 9059.739 (50) | E | | × | | | | | | | | | |
| | | 1–0 (11) | 9061.700 (50) | E | | × | | | | | | | | | |
| $HCC^{13}CN$ | $J=1$–$0$ | | 9060.403 (10) | | | | | | | | | | | | |
| | | $F=1$–$1$ (33) | 9059.412 (10) | E | | × | | | | | | | | | |
| | | 2–1 (55) | 9060.624 (10) | E | | × | | | | | | | | | |
| | | 1–0 (11) | 9062.555 (10) | E | | × | | | | | | | | | |
| $H_2C_2O$ | $4_{14}$–$3_{13}$ | | 80076.611 (26) | E | | × | | | | | | | | | |
| | $4_{04}$–$3_{03}$ | | 80832.076 (28) | E | | × | | | | | | | | | |
| | $4_{13}$–$3_{12}$ | | 81586.193 (26) | E | | × | | | | | | | | | |
| | $5_{15}$–$4_{14}$ | | 100094.466 (28) | E | | × | | | | | | | | | |
| | $5_{05}$–$4_{04}$ | | 101036.558 (30) | E | | × | | | | | | | | | |
| | $5_{14}$–$4_{13}$ | | 101981.387 (28) | E | | × | × | | | | | | | | |
| $CH_3OH$ | $1_1$ –$1_1$ | | 834.267 (2) | E | × | × | | | | | | | | | |
| | $3_1$ –$3_1$ | | 5005.3208 (2) | E | | × | | | | | | | | | |
| | $1_0$ –$0_0$ | | 48372.60 | E | | × | | | | | | | | | |
| | $1_0$ –$0_0$ | | 48377.09 | E | | × | | | | | | | | | |
| | $3_0$ –$2_0$ | | 145093.75 | E | | | × | | | × | | | | | |
| | $3_{-1}$–$2_{-1}$ | | 145097.47 | E | × | × | × | | | × | | | | | |
| | $3_0$ –$2_0$ | | 145103.23 | E | | | × | | | × | | | | | |
| | $3_2$ –$2_2$ | | 145124.41 | E | | | | | | | | | | | |
| | $3_2$ –$2_2$ | | 145126.37 | E | | | × | | | × | | | | | |
| | $3_{-2}$–$2_{-2}$ | | 145126.37 | E | | | | | | | | | | | |
| | $3_1$ –$2_1$ | | 145131.88 | E | | | × | | | × | | | | | |
| | $3_{+2}$–$2_{+2}$ | | 145133.46 | E | | | × | | | × | | | | | |
| | $3_2$ –$3_1$ | | 24928.70 (10) | | | | × | | | | | | | | |
| | $4_2$ –$4_1$ | | 24933.468 (2) | | | | × | | | | | | | | |
| | $2_2$ –$2_1$ | | 24934.382 (5) | | | | × | | | | | | | | |
| | $5_2$ –$5_1$ | | 24959.080 (2) | | | | × | | | | | | | | |
| | $6_2$ –$6_1$ | | 25018.123 (2) | | | | × | | | | | | | | |
| | $7_2$ –$7_1$ | | 25124.873 (2) | | | | × | | | | | | | | |
| | $8_2$ –$8_1$ | | 25294.411 (3) | | | | × | | | | | | | | |
| | $4_{-1}$–$3_0$ | | 36169.24 | | | × | | | | | | | | | |
| | $5_{-1}$–$4_0$ | | 84521.21 (8) | | | × | | | | | | | | | |
| | $6_{-2}7_{-1}$ | | 85568.2 (1.0) | | | | × | | | | | | | | |
| | $7_2$ –$6_3$ | | 86615.4 (5) | | | | × | | | | | | | | |
| | $7_2$ –$6_3$ | | 86903.5 (5) | | | | × | | | | | | | | |
| | $8_{-4}$–$9_{-3}$ | | 89505.86 (4) | | | | × | | | | | | | | |
| | $8_3$ –$9_2$ | | 94540.7 (10) | | | | × | | | | | | | | |
| | $8_0$ –$7_1$ | | 95169.44 (10) | | | | × | | | | | | | | |
| | $10_1$ –$10_0$ | | 169335.34 (10) | | | | × | | | | | | | | |
| | $9_2$ –$9_1$ | | 25541.43 | ME | | | × | | | | | | | | |
| | $10_2$ –$10_1$ | | 25878.18 | ME | | | × | | | | | | | | |
| | $11_2$ –$11_1$ | | 26313.11 | ME | | | × | | | | | | | | |
| | $12_2$ –$12_1$ | | 26847.27 | ME | | | × | | | | | | | | |
| | $13_2$ –$13_1$ | | 27472.54 | ME | | | × | | | | | | | | |

continued

Table 6, continued

| Molecule | Transition | | ν [MHz] | E, A, ME | Sgr A | Sgr B2 | Ori A | W 51 | W 3 | DR 21 | NGC 2264 | NGC 2024 | IRC+10216 | L 134 | Cloud 2 |
|---|---|---|---|---|---|---|---|---|---|---|---|---|---|---|---|
| $CH_3OH$ | $14_2-14_1$ | | 28169.31 | ME | | | × | | | | | | | | |
| | $15_2-15_1$ | | 28905.70 | ME | | | × | | | | | | | | |
| | $11_1-10_2$ | | 76247.4 | E | | | × | | | | | | | | |
| | $5_0-4_1$ | | 76509.9 | E | | | × | | | | | | | | |
| | $2_{-1}-1_{-1}$ | | 96739.39 (10) | E | × | × | × | × | × | × | × | × | | | |
| | $2_0-1_0$ | | 96741.42 (10) | E | × | × | × | × | × | × | × | × | | | |
| | $2_0-1_0$ | | 96744.58 (10) | E | × | × | × | × | × | × | × | × | | | |
| | $2_0-1_0$ | | 96755.51 (10) | E | | × | × | × | | | | | | | |
| $^{13}CH_3OH$ | $2_{-1}-1_{-1}$ | | 94405.17 (17) | E | | × | | | | | | | | | |
| | $2_0-1_0$ | | 94407.02 (10) | E | | × | | | | | | | | | |
| | $2_0-1_0$ | | 94410.76 (10) | E | | × | | | | | | | | | |
| | $2_1-1_1$ | | 94420.36 (5) | | | | | | | | | | | | |
| $CH_3OD$ | $2_{-1}-1_{-1}$ | | 90703.78 (5) | E | | × | | | | | | | | | |
| | | | 90705.77 (5) | E | | × | | | | | | | | | |
| $NH_2{}^{13}CHO$ | $1_{11}-1_{10}$; | $F=2$–2 | 1570.825 (30) | E | | × | | | | | | | | | |
| $CH_3CN$ | $J, K=5,0$–4,0 | | 91987.054 (60) | E | | | × | | | | | | | | |
| | 5,1–4,1 | | 91985.284 (60) | E | | | × | | | | | | | | |
| | 5,2–4,2 | | 91979.998 (60) | E | | | × | | | | | | | | |
| | 5,3–4,3 | | 91971.374 (60) | E | | | × | | | | | | | | |
| | 5,4–4,4 | | 91959.206 (60) | E | | | × | | | | | | | | |
| | $J, K=6,0$–5,0 | | 110383.494 (60) | | | × | | | | | | | | | |
| | 6,1–5,1 | | 110381.362 (60) | | | × | | | | | | | | | |
| | 6,2–5,2 | | 110374.086 (60) | | | × | | | | | | | | | |
| | 6,3–5,3 | | 110364.490 (60) | | | × | | | | | | | | | |
| | 6,4–5,4 | | 110349.706 (60) | | | × | | | | | | | | | |
| | 6,5–5,5 | | 110330.728 (60) | | | × | | | | | | | | | |
| $NH_2CHO$ | $1_{10}-1_{11}$ | | 1539.543 (3) | | | | | | | | | | | | |
| | | $F=1$–1 (8) | 1538.135 (20) | | × | × | | | | | | | | | |
| | | 1–2 (14) | 1538.693 (20) | | × | × | | | | | | | | | |
| | | 2–1 (14) | 1539.295 (20) | | | × | | | | | | | | | |
| | | 1–0 (11) | 1539.570 (20) | | | × | | | | | | | | | |
| | | 2–2 (42) | 1539.851 (20) | | × | × | | | | | | | | | |
| | | 0–1 (11) | 1541.018 (20) | | × | × | | | | | | | | | |
| | $2_{11}-2_{12}$ | | 4618.553 (9) | | | | | | | | | | | | |
| | | $F=2$–2 (23) | 4617.14 (5) | E | | × | | | | | | | | | |
| | | 3–3 (42) | 4619.00 (5) | E | | × | | | | | | | | | |
| | | 1–1 (15) | 4620.01 (5) | E | | × | | | | | | | | | |
| | $3_{12}-3_{12}$ | | 9236.548 (17) | | | | | | | | | | | | |
| | | $F=3$–3 (28) | 9235.130 (30) | E | | × | | | | | | | | | |
| | | 4–4 (40) | 9237.045 (30) | E | | × | | | | | | | | | |
| | | 2–2 (21) | 9237.705 (30) | E | | × | | | | | | | | | |

continued

Table 6, continued

| Molecule | Transition | | ν [MHz] | E, A, ME | Sgr A | Sgr B2 | Ori A | W 51 | W 3 | DR 21 | NGC 2264 | NGC 2024 | IRC+10216 | L 134 | Cloud 2 |
|---|---|---|---|---|---|---|---|---|---|---|---|---|---|---|---|
| $CH_3NH_2$ | $2_{02}$–$1_{01}$ | | | | | | | | | | | | | | |
| | | $F$=1–0 (11) | 8775.06 (3) | E | | × | | | | | | | | | |
| | | 3–2 (47) | 8777.38 (3) | E | | × | × | | | | | | | | |
| | | 1–1 (8)<br>2–2 (8) | 8778.18 (3) | E | | × | | | | | | | | | |
| | | 2–1 (25) | 8779.47 (3) | E | | × | | | | | | | | | |
| | $5_{15}$–$5_{05}$ | | | | | | | | | | | | | | |
| | | $F$=4–4 (26) | 73044.01 (10) | E | | × | × | | | | | | | | |
| | | 5–5 (31) | 73045.15 (10) | E | | × | × | | | | | | | | |
| | | 6–6 (38) | 73044.20 (10) | E | | × | × | | | | | | | | |
| | $4_{14}$–$4_{04}$ | | | | | | | | | | | | | | |
| | | $F$=3–3 (24) | 86074.20 | E | | × | × | | | | | | | | |
| | | 4–4 (30) | 86075.43 | E | | × | × | | | | | | | | |
| | | 5–5 (39) | 86074.44 | E | | × | × | | | | | | | | |
| | $2_{02}$–$1_{01}$ | | | | | | | | | | | | | | |
| | | $F$=1–1 (8) | 88666.82 (50) | E | | × | | | | | | | | | |
| | | 3–2 (47) | 88668.03 (50) | E | | × | | | | | | | | | |
| | | 2–1 (25) | 88668.03 (50) | E | | × | | | | | | | | | |
| | | 1–0 (11) | 88668.59 (50) | E | | × | | | | | | | | | |
| | | 2–2 (8) | 88668.59 (50) | E | | × | | | | | | | | | |
| $CH_3CHO$ | $1_{10}$–$1_{11}$ | | 1065.075 (5) | E | × | × | | | | | | | | | |
| | | | 1853.399 (4) | | | | | | | | | | | | |
| | $2_{11}$–$2_{12}$ | | 3195.167 (10) | E | | × | | | | | | | | | |
| | | | 3527.106 (698) | | | | | | | | | | | | |
| | $3_{12}$–$3_{13}$ | | 6389.167 (481) | E | | × | | | | | | | | | |
| | | | 6546.454 (519) | | | × | | | | | | | | | |
| | $4_{13}$–$4_{14}$ | | 10648.428 (20) | E | | × | | | | | | | | | |
| | $2_{12}$–$1_{01}$ | | 83584.26 (18) | E | | | × | | | | | | | | |
| | $6_{16}$–$5_{15}$ | | 112248.720 (160) | E | | × | | | | | | | | | |
| | | | 112254.480 (160) | E | | × | | | | | | | | | |
| | $6_{06}$–$5_{05}$ | | 114939.900 (100) | | | × | | | | | | | | | |
| | | | 114959.650 (100) | E | | × | | | | | | | | | |
| $CH_3CCH$ | $J, K$=5,0–4,0 | | 85457.29 | E | | × | × | | | | | | | | |
| | 5,1–4,1 | | 85455.67 | E | | × | | | | | | | | | |
| | 5,2–4,2 | | 85450.78 | | | × | | | | | | | | | |
| | 5,3–4,3 | | 85442.61 | E | | × | | | | | | | | | |
| $CH_2CHCN$ | $2_{11}$–$2_{12}$ | | | | | | | | | | | | | | |
| | | $F$=1–1 (15) | 1371.709 (2) | E | | × | | | | | | | | | |
| | | 3–3 (42) | 1371.794 (2) | E | | × | | | | | | | | | |
| | | 2–2 (23) | 1371.947 (2) | E | | × | | | | | | | | | |
| | $10_{0,10}$–$9_{0,9}$ | | 94276.638 (18) | E | | × | | | | | | | | | |

continued

Table 6, continued

| Molecule | Transition | | $\nu$ [MHz] | E, A, ME | Sgr A | Sgr B2 | Ori A | W 51 | W 3 | DR 21 | NGC 2264 | NGC 2024 | IRC+10216 | L 134 | Cloud 2 |
|---|---|---|---|---|---|---|---|---|---|---|---|---|---|---|---|
| $HC_5N$ | $J=1–0$ | | 2662.662 | E | | × | | | | | | | | | × |
| | 4–3 | | 10650.657 | E | | × | | | | | | | | | × |
| | 8–7 | | 21301.257 | E | | × | | | | | | | | | × |
| | 9–8 | | 23963.897 | E | | | | | | | | | | | |
| | 31–30 | | 82539.32 | E | | × | | | | | | | | | |
| | 33–32 | | 87863.97 | E | | × | | | | | | | | | |
| | 37–36 | | 98513.01 | E | | × | | | | | | | | | |
| | $J=2–1$ | | 5325.330 (10) | | | | | | | | | | | | |
| | | $F=3–2$ | 5325.421 | E | | × | | | | | | | | | × |
| | | 2–1 | 5325.330 | E | | × | | | | | | | | | × |
| | | 1–0 | 5324.270 | E | | × | | | | | | | | | × |
| | | 2–2 | 5324.058 | E | | × | | | | | | | | | × |
| | | 1–1 | 5327.451 | | | | | | | | | | | | |
| $HC_7N$ | $J=8–7$ | | 9024.014 | E | | | | | | | | | | | × |
| | 9–8 | | 10152.022 | E | | | | | | | | | | | × |
| | 13–12 | | 14663.986 | E | | | | | | | | | | | × |
| | 20–19 | | 22559.907 | E | | | | | | | | | | | × |
| | 21–20 | | 23687.890 | E | | | | | | | | | | | × |
| | 22–21 | | 24815.82 | E | | | | | | | | | | | × |
| $C_2H_5CN$ | $10_{0,10}–9_{0,9}$ | | 88323.72 (8) | E | | | × | | | | | | | | |
| | $10_6\ –9_6$ | | 89562.24 (7) | E | | | × | | | | | | | | |
| | $10_7\ –9_7$ | | 89564.95 (3) | E | | | × | | | | | | | | |
| | $10_5\ –9_5$ | | 89568.03 (6) | E | | | × | | | | | | | | |
| | $10_8\ –9_8$ | | 89572.98 (11) | E | | | × | | | | | | | | |
| | $10_{47}\ –9_{46}$ | | 89589.98 (6) | E | | | × | | | | | | | | |
| | $10_{46}\ –9_{45}$ | | 89590.96 (6) | E | | | × | | | | | | | | |
| | $10_{38}\ –9_{37}$ | | 89628.41 (7) | E | | | × | | | | | | | | |
| | $11_{1,10}–10_{1,10}$ | | 95442.44 (8) | E | | | × | | | | | | | | |
| | $11_{0,11}–10_{0,10}$ | | 96919.71 (8) | E | | | × | | | | | | | | |
| | $11_{2,10}–10_{2,9}$ | | 98177.55 (7) | E | | | × | | | | | | | | |
| | $11_6\ –10_6$ | | 98523.80 (7) | E | | | × | | | | | | | | |
| | $11_7\ –10_7$ | | 98524.58 (9) | E | | | × | | | | | | | | |
| | $11_8\ –10_8$ | | 98532.01 (12) | E | | | × | | | | | | | | |
| | $11_5\ –10_5$ | | 98533.91 (6) | E | | | × | | | | | | | | |
| | $11_{4,8}\ –10_{4,7}$ | | 98564.78 (6) | E | | | × | | | | | | | | |
| | $11_{47}\ –10_{48}$ | | 98566.74 (6) | E | | | × | | | | | | | | |
| | $11_{39}\ –10_{38}$ | | 98610.07 (6) | E | | | × | | | | | | | | |
| | $11_{38}\ –10_{37}$ | | 98701.07 (6) | E | | | × | | | | | | | | |
| | $11_{1,10}–10_{1,9}$ | | 100614.28 (7) | E | | | × | | | | | | | | |
| | $12_{1,12}–11_{1,11}$ | | 104051.24 (8) | E | | | × | | | | | | | | |
| | $12_{1,11}–11_{1,10}$ | | 109650.29 (6) | E | | | × | | | | | | | | |
| | $13_{0,13}–12_{0,12}$ | | 113978.21 (10) | E | | | × | | | | | | | | |
| | $13_{2,12}–12_{2,11}$ | | 115894.35 (9) | E | | | × | | | | | | | | |
| $HCOOCH_3$ | $1_{10}–1_{11}$ | | 1610.249 (3) | E | | × | | | | | | | | | |
| | | | 1610.906 (3) | E | | × | | | | | | | | | |

continued

Table 6, continued

| Molecule | Transition | ν [MHz] | E, A, ME | Sgr A | Sgr B2 | Ori A | W 51 | W 3 | DR 21 | NGC 2264 | NGC 2024 | IRC+10216 | L 134 | Cloud 2 |
|---|---|---|---|---|---|---|---|---|---|---|---|---|---|---|
| $(CH_3)_2O$ | $2_{02}$–$1_{11}$ | 9118.818 (15) | E | | × | | | | | | | | | |
| | | 9196.670 (15) | E | | × | | | | | | | | | |
| | | 9120.517 (15) | E | | × | | | | | | | | | |
| | $2_{11}$–$2_{02}$ | 31105.26 (10) | E | | × | × | | | | | | | | |
| | | 31106.20 (5) | E | | × | × | | | | | | | | |
| | | 31107.12 | E | | × | × | | | | | | | | |
| | $2_{20}$–$2_{11}$ | 86223.76 (10) | E | | | × | | | | | | | | |
| | | 86225.67 (12) | E | | | × | | | | | | | | |
| | | 86226.728 (96) | E | | | × | | | | | | | | |
| | | 86228.72 (2) | E | | | × | | | | | | | | |
| | $3_{13}$–$2_{02}$ (6) | 82651.08 (10) | E | | × | × | | | | | | | | |
| | (16) | 82650.18 (10) | E | | × | × | | | | | | | | |
| | (6) | 82649.30 (10) | E | | × | × | | | | | | | | |
| | $4_{22}$–$4_{13}$ (6) | 82691.14 (10) | E | | | × | | | | | | | | |
| | (16) | 82688.77 (10) | E | | | × | | | | | | | | |
| | (6) | 82686.50 (10) | E | | | × | | | | | | | | |
| | $3_{21}$–$3_{12}$ (16) | 84634.40 (10) | E | | | × | | | | | | | | |
| | (10) | 84632.02 (10) | E | | | × | | | | | | | | |
| | $6_{06}$–$5_{15}$ (6) | 90937.539 (40) | E | | × | × | | | | | | | | |
| | (16) | 90938.099 (30) | E | | × | × | | | | | | | | |
| | (6) | 90938.674 (50) | E | | × | × | | | | | | | | |
| | $4_{23}$–$4_{14}$ (16) | 93857.11 (10) | E | | | × | | | | | | | | |
| | (10) | 93854.44 (10) | E | | | × | | | | | | | | |
| | $5_{23}$–$5_{14}$ | 80538.54 (10) | E | | × | | | | | | | | | |
| | $5_{24}$–$5_{15}$ (6) | 96852.46 (10) | E | | | × | | | | | | | | |
| | (16) | 96849.85 (10) | E | | | × | | | | | | | | |
| | (6) | 96847.25 (10) | E | | | × | | | | | | | | |
| | $4_{14}$–$3_{03}$ | 99326.00 (20) | E | | | × | | | | | | | | |
| | | 99325.25 (20) | E | | | × | | | | | | | | |
| | | 99324.43 (20) | E | | | × | | | | | | | | |
| $C_2H_5OH$ | $6_{06}$–$5_{15}$ | 85265.46 (15) | E | | × | × | | | | | | | | |
| | $4_{14}$–$3_{03}$ | 90117.51 (20) | E | | × | × | | | | | | | | |
| | $5_{15}$–$4_{04}$ | 104808.58 (28) | E | | × | × | | | | | | | | |
| $HC_9N$ | $J = 18$–$17$ | 10458.634 | E | | | | | | | | | | | × |
| | 25–24 | 14525.862 | E | | | | | | | | | | | × |

Interstellar molecular transitions in the microwave spectral region listed according to increasing frequency: see [31a].

Table 7. Surveys of 2.6 mm CO line emission ($J=1\rightarrow0$) with observations spaced uniformly along the galactic equator.
$\theta$ = beam width; $\Delta v$ = spectral resolution

| Author | Publication | $\theta$ | $\Delta v$ [km s$^{-1}$] | Region |
|---|---|---|---|---|
| Scoville, N.Z., Solomon, P.M. | Astrophys. J. Lett. **199** (1975) L105 | 1'.2 | 0.65 | $l=348\cdots0\cdots90°$ sampling every degree |
| Burton, W.B., Gordon, M.A., Bania, T.M., Lockman, F.J. | Astrophys. J. **202** (1975) 30<br>Astrophys. J. Lett. **207** (1976) L189 | 1'.2 | 0.65; 1.3; 2.6 | $l=10\cdots200°$, partly $\|b\|\leqq2°$, in 5° longitude intervals away from known thermal sources |
| Burton, W.B., Gordon, M.A. | Astrophys. J. **208** (1976) 346, 786<br>Astron. Astrophys. **63** (1978) 7 | 1'.2 | 1.3 | $l=352\cdots0\cdots82°$, sampling interval $\Delta l=0°.2$ |
| Bania, T.M. | Astrophys. J. **216** (1977) 381 | 1'.2 | 2.6 | central region: $352\cdots10°$, $\Delta l=0°.2$ |
| Cohen, R.S., Thaddeus, P. | Astrophys. J. Lett. **217** (1977) L155 | 8' | 2.6 | $l=15°\cdots60°$, $b=-1°.5\cdots+1°.5$ |
| Solomon, P.M., Sanders, D.B., Scoville, N.Z. | Astrophys. J. Lett. **232** (1979) L89 | 1'.1 | 0.65 | $l=10°\ldots44°$ |

Relation between column densities $\mathfrak{N}(CO)$, $\mathfrak{N}(H_2)$ and visual extinction $A_V$ of dense clouds:

Above a threshold at $\mathfrak{N}(CO)\approx5\cdot10^{20}$ cm$^{-2}$, corresponding to $A_V\approx0.5$ mag, an appreciable amount of CO is present. The relations to visual extinction are approximately given by [17, 14a]

$$\mathfrak{N}(^{13}CO)\approx2.5\cdot10^{15}\,A_V$$
$$\mathfrak{N}(CO)\approx1\cdot10^{17}\,A_V$$
$$\mathfrak{N}(H_2)\approx1.25\cdot10^{21}\,A_V$$

with $\mathfrak{N}$ in [cm$^{-2}$], $A_V$ in [mag].

For cool clouds with $1.5\lessapprox A_V\lessapprox10$ mag the average relationship between molecular hydrogen and $^{13}CO$ (LTE) column densities is [14a]

$$\mathfrak{N}(H_2)=5.0\cdot10^5\,\mathfrak{N}(^{13}CO)\ [\mathrm{cm}^{-2}]$$

For hot-centered clouds with $A_V\approx10^2$ mag the conversion factor seems to be about $2\cdot10^6$ [68].

Taking the isotope ratio $^{12}C/^{13}C\approx40$ [59a], instead of the terrestrial value $^{12}C/^{13}C\approx89$, one obtains for cool clouds

$$\frac{\mathfrak{N}(CO)}{\mathfrak{N}(H_2)}\approx8\cdot10^{-5}$$

Table 8. Properties of selected star-forming molecular clouds.

$N(H_2)$ = mean number density of $H_2$ molecules
$T_{kin}$ = kinetic temperature of the gas taken equal to the brightness temperature of the $^{12}CO$ emission line
$T_d$ = temperature of the dust
$\mathfrak{M}$ = total mass

| Molecular cloud and/or fragment | Size pc | $N(H_2)$ $cm^{-3}$ | $T_{kin}$ K | $T_d$ K | $\mathfrak{M}$ $\mathfrak{M}_\odot$ | Ref. |
|---|---|---|---|---|---|---|
| Orion A complex | ≈ 10 × 50 | $5 \cdot 10^2$ | – | – | $10^5$ | 29 |
| OMC 1 | ≈ 5 (envelope) | $10^3$ | 20 | 85 | $10^4$ | 62, 63, 70 |
| | ≲ 0.5 (core) | $2 \cdot 10^5$ | 80 | | 500 | |
| OMC 2 | 0.15 | $10^4$ | 60 | 40 | 5 | 62, 70 |
| M 17 complex | ≈ 22 × 85 | $10^3$ | – | – | ≈ $10^6$ | |
| M 17 SW | 5 | $5 \cdot 10^3$ | 50 | 75 | $5 \cdot 10^4$ | 30, 16, 75 |
| M 17 N | 5 | $4 \cdot 10^3$ | 35 | – | $5 \cdot 10^4$ | |
| W 3 fragment | ≈ 5 | $2 \cdot 10^3$ | 30 | 80 | $10^4$ | 31, 62, 68 |
| W 3 (OH) fragment | ≈ 3 | $2 \cdot 10^3$ | 25 | 60 | $5 \cdot 10^3$ | |
| W 3 core | 1 | $5 \cdot 10^4$ | 35 | – | $10^4$ | 74 |
| Sgr B 2 | 40 (envelope) | $10^3$ | 20 | 35 | $10^6$ | 62, 63, 70 |
| | 6 (core) | ≳ $10^5$ | ≈ 100 | | $5 \cdot 10^5$ | |

Recent review on physical conditions in cool clouds: Evans [76].

Far-infrared observations of hot-centered clouds: see [68].

## 7.2.5 References for 7.2

For general references, see 7.2.1 .


1 Adolfsson, T.: Ark. Astron. **1** (1955) 495.
2 Andrews, L.B.: Publ. Am. Astron. Soc. **7** (1933) 211.
3 Asklöf, S.: Medd. Obs. Uppsala No. 51 (1930).
4 Baker, R.H.: Astrophys. J. **94** (1941) 493.
5 Baker, R.H., Nantkes, E.: Astrophys. J. **99** (1944) 125.
6 Baker, P.L., Burton, W.B.: Astrophys. J. **198** (1975) 281.
7 Balanovsky, I., Hase, V.: Pulkovo Bull. **14** (1935) 2.
8 Bok, B.J.: Harvard Reprints No. 77 (1932).
9 Bok, B.J.: Astron. J. **61** (1956) 309.
10 Bok, B.J., Cordwell, C.S., Cromwell, R.H. in: Dark Nebulae, Globules and Protostars (Lynds, B.T., ed.), Univ. of Arizona Press, Tucson (1970) p. 33.
11 Bok, B.J., McCarthy, C.C.: Astron. J. **79** (1974) 42.
12 Braunsfurth, E.: Zur Wolkenstruktur des neutralen Wasserstoffs, Dissertation Bonn (1975).
13 Clayton, D.D. in: Protostars and Planets (Gehrels, T., ed.), Univ. of Arizona Press, Tucson (1978) p. 13.
14 Dickman, R.L.: Astron. J. **83** (1978) 363.
14a Dickman, R.L.: Astrophys. J. Suppl. **37** (1978) 407.
15 Eklöf, O.E.: Medd. Obs. Uppsala No. 119 (1958).
16 Elmegreen, B.G., Lada, C.J.: Astron. J. **81** (1976) 1089.
17 Encrenaz, P.J., Falgarone, E., Lucas, R.: Astron. Astrophys. **44** (1975) 73.
18 Evans, N.J.: Star Formation in Molecular Clouds, in: Protostars and Planets (Gehrels, T., ed.), Univ. of Arizona Press, Tucson (1978) p. 153.
19 Falgarone, E., Lequeux, J.: Astron. Astrophys. **25** (1973) 253.
20 Field, G.B., Hutchins, J.: Astron. Astrophys. **153** (1968) 737.
21 Greenstein, J.L.: Ann. Harvard **105** (1936) 359.
22 Hachenberg, O., Mebold, U.: Die Struktur und der physikalische Zustand des interstellaren Gases aus Beobachtungen der 21 cm H I Linie, Max-Planck-Institut f. Radioastronomie, Bonn, Sonderdruck Nr. 119, Ser. A (1976).
23 Heeschen, D.S.: Astrophys. J. **114** (1951) 132.
24 Heiles, C.: Astrophys. J. Suppl. **15** (1967) 97.

25 Heiles, C.: Int. Astron. Union Symp. **60** (1974) p. 31.
26 Hobbs, L.M.: Astrophys. J. **191** (1974) 395.
27 Knude, J.: Astron. Astrophys. **71** (1979) 344.
28 Knude, J.: Astron. Astrophys. Suppl. **38** (1979) 407.
29 Kutner, M.L., Tucker, K.D., Chin, G., Thaddeus, P.: Astrophys. J. **215** (1977) 521.
30 Lada, C.J.: Astrophys. J. Suppl. **32** (1976) 603.
31 Lada, C.J., Elmegreen, B.G., Cong, H.-I., Thaddeus, P.: Astrophys. J. Lett. **226** (1978) L39.
31a Lovas, F.J., Snyder, L.E., Johnson, D.R.: Astrophys. J. Suppl. **41** (1979) 451.
32 Lynds, B.T.: Astrophys. J. Suppl. **7** (1962) 1.
33 Lynds, B.T. in: Nebulae and Interstellar Matter, Stars and Stellar Systems **VII** (1968) p. 130, 131.
34 Martin, R.N., Barrett, A.H.: Astrophys. J. Suppl. **36** (1978) 1.
35 McCuskey, S.W.: Astrophys. J. **94** (1941) 468.
36 Miller, F.D.: Proc. Nat. Acad. Sci. **23** (1937) 405.
37 Münch, G.: Astrophys. J. **116** (1952) 575.
38 Myers, P.C., Ho, P.T.P., Schneps, M.H.: Astrophys. J. **220** (1978) 864.
39 Myers, P.C.: Astrophys. J. **225** (1978) 380.
40 Nantkes, E., Baker, R.H.: Astrophys. J. **107** (1948) 113.
41 Penston, M.V., Munday, V.A., Stickland, D.J., Penston, M.J.: Mon. Not. R. Astron. Soc. **142** (1969) 355.
42 Perry, J.F.W., Helfer, H.L.: Astrophys. J. **174** (1972) 341.
43 Radhakrishnan, V., Goss, W.M.: Astrophys. J. Suppl. **24** (1972) 161.
44 Radhakrishnan, V., Murray, J.D., Lockhart, P., Whittle, R.P.J.: Astrophys. J. Suppl. **24** (1972) 15.
45 Reddish, V.C., Sloan, C.: Observatory **91** (1971) 70.
46 Rodgers, A.W.: Mon. Not. R. Astron. Soc. **120** (1960) 163.
47 Rossano, G.S.: Astron. J. **83** (1978) 234.
48 Rossano, G.S.: Astron. J. **83** (1978) 241.
49 Sandqvist, Aa.: Astron. Astrophys. **57** (1977) 467.
50 Schalén, C.: Medd. Obs. Uppsala **58** (1934).
51 Scheffler, H.: Z. Astrophys. **65** (1967) 60.
52 Scheffler, H.: Z. Astrophys. **66** (1967) 33.
53 Scoville, N., Solomon, P.: Astrophys. J. Lett. **199** (1976) L105.
53a Solomon, P.M., Sanders, D.B., Scoville, N.Z.: Int. Astron. Union Symp. **84** (1979) p. 35.
54 Spitzer, L., Jenkins, E.B.: Annu. Rev. Astron. Astrophys. **13** (1975) 133.
55 Spitzer, L.: Physical Processes in the Interstellar Medium, Wiley, New York (1978).
56 Sticker, B.: Veröff. Sternwarte Bonn **30** (1937).
57 Thaddeus, P.: Int. Astron. Union Symp. **75** (1977) p. 40.
58 Vanäs, E.: Ann. Obs. Uppsala **1** (1939) No. 1.
59 Wallenquist, A.: Ann. Bosscha Sterrew Lembang (Java) **5** (1939) p. E1.
59a Wannier, P.G., Penzias, A.A., Linke, R.A., Wilson, R.W.: Astrophys. J. **204** (1976) 26.
60 Weaver, H.F.: Astrophys. J. **110** (1949) 190.
61 Wernberg, G.: Ann. Obs. Uppsala **1** (1939) No.4.
62 Werner, M.W., Becklin, E.E., Neugebauer, G.: Science **197** (1977) 723.
63 Winnewisser, G.: Naturwissenschaften **62** (1975) 200.
63a Winnewisser, G., Churchwell, E., Walmsley, C.M. in: Modern aspects of Microwave Spectroscopy (Chantry, G.W., ed.), Academic Press Ltd., London (1979) p. 317.
64 Wolf, M.: Astron. Nachr. **219** (1923) 109.
65 Wolf, M.: Astron. Nachr. **223** (1924) 89.
66 Wolf, M.: Astron. Nachr. **229** (1926) 1.
67 Wolf, M. in: Probleme der Astronomie, Springer, Berlin (1926) p. 312.
68 Rowan-Robinson, M.: Astrophys. J. **234** (1979) 111.
69 Knapp. G.R.: Astron. J. **79** (1974) 527, 541.
70 Linke, R.A., Goldsmith, P.F.: Astrophys. J. **235** (1980) 437.
71 McCray, R., Snow, T.P.: Annu. Rev. Astron. Astrophys. **17** (1979) 213.
72 Bok, B.J., Cordwell, C.S. in: Molecules in the Galactic Environment (Gordon, M.A., Snyder, L.E., eds.), Wiley Interscience, New York (1973) p. 53.
73 Keene, J.: Astrophys. J. **245** (1981) 115.
74 Dickel, H.R.: Astrophys. J. **238** (1980) 829.
75 Elmegreen, B.G., Lada, Ch.J., Dickinson, D.F.: Astrophys. J. **230** (1979) 415.
76 Evans, N.J.: Int. Astron. Union Symp. **87** (1980) 1; see Andrew, B.H. in 7.2.1.

# 7.3 H II regions

## 7.3.1 General references

Brown, R.L., Lockman, F.J., Knapp, G.R.: Radio Recombination Lines, Annu. Rev. Astron. Astrophys. **16** (1978) 445.
Dupree, A.K., Goldberg, L.: Radiofrequency Recombination Lines, Annu. Rev. Astron. Astrophys. **8** (1970) 231.
Fazio, G.G. (ed.): Infrared and Submillimeter Astronomy, Reidel, Dordrecht, Holland (1977).
Felli, M.: Properties of H II regions, in: Stars and Starsystems (Westerlund, B.E., ed.), Reidel, Dordrecht, Holland (1979).
Gordon, M.A.: The radio characteristics of H II regions and the diffuse thermal background, in: Galactic and Extra-Galactic Radio Astronomy (Verschuur, G.L., Kellermann, K.I., eds.), Springer, Berlin (1974) p. 51.
Habing, H.J.: H II Regions, in: The Interstellar Medium (Pinkau, K., ed.), Reidel, Dordrecht, Holland (1974) p. 91.
Habing, H.J., Israel, F.P.: Compact H II Regions and OB Star Formation, Annu. Rev. Astron. Astrophys. **17** (1979) 345.
Johnson, H.M.: Diffuse Nebulae, Stars and Stellar Systems VII (1968) 65.
Lemke, D., Harris, A.W.: Infrared Observations of H II Regions, Naturwiss. **66** (1979) 73.
Mezger, P.G. in: Interstellar Matter (Wickramasinghe, N.C., Kahn, F.D., Mezger, P.G., eds.), Swiss Soc. Astron. Astrophys., Second Advanced Course, Saas-Fee (1972) p. 152.
Pottasch, S.R.: The Diffuse Emission Nebulae, in: Vistas in Astronomy (Beer, A., ed.), **6** (1965) 149.
Setti, G., Fazio, G.G. (eds.): Infrared Astronomy, Reidel, Dordrecht, Holland (1978).
Shaver, P.A. (ed.): Radio Recombination Lines, Proceedings of a Workshop held in Ottawa, Canada; Reidel, Dordrecht, Holland (1980).
Terzian, Y. (ed.): Interstellar Hydrogen, W. A. Benjamin, New York (1968).
Terzian, Y.: Emission Nebulae at Radio Wavelengths, in: Vistas in Astronomy (Beer, A., ed.), **16** (1974) 279.
Werner, M.W., Becklin, E.E., Neugebauer, G.: Infrared Studies of Star Formation, Science **197** (1977) 723.
Wilson, T.L., Downes, D. (eds.): H II Regions and Related Topics (Lecture Notes in Physics No. 42), Springer, Berlin (1975).

## 7.3.2 Catalogues, surveys, statistical data

Table 1: see p. 89.

Table 2. Surveys of radio continuum and H 109α recombination line observations of H II regions with single dishes (angular resolution $\gtrsim 1'$).

| Author | Ref. | Region of the Milky Way | λ cm | Continuum, line | Remarks |
|---|---|---|---|---|---|
| Kuzmin et al. | 68a | $l=330\cdots0\cdots260°$ | 9.6 | cont. | |
| Westerhout | 98 | $l=353\cdots0\cdots89°$ | 21 | cont. | |
| Schraml and Mezger | 86 | $l=351\cdots0\cdots207°$ | 2 | cont. | with contour maps |
| Altenhoff et al. | 3 | $l=335\cdots0\cdots75°$ | 21/11/6 | cont. | with contour maps |
| Reifenstein et al. | 82 | $l=348\cdots0\cdots209°$ | 6 | H 109α | |
| Goss and Shaver | 38 | $l=206\cdots0\cdots49.°5$ | 6 | cont. | with contour maps |
| Wilson et al. | 99 | $l=189\cdots0\cdots49.°5$ | 6 | H 109α | |
| Felli and Churchwell | 26 | $l=4\cdots235°$ | 21 | cont. | with contour maps |
| Caswell | 12 | 13 southern sources | 6 | H 109α, cont. | |
| Altenhoff et al. | 4 | $l=359\cdots0\cdots55°$ | 6 | cont. | contour maps |
| Haynes et al. | 61 | $l=190\cdots0\cdots40°$ | 6 | cont. | contour maps |
| Large et al. | 69a | $l=350\cdots0\cdots56°$ | 74 | cont. | with contour maps |

Recent recombination line observations of individual H II regions with single dishes: see e.g. [137, 138, 139].

Table 1. Catalogues of optical observations of H II regions.
$n$ = number of objects   $r$ = distance
$l$ = galactic longitude   $v_r$ = radial velocity
$\delta$ = declination   exc. = exciting

| Author | Ref. | $n$ | Spectral region | Region of the Milky Way | Data | Notes |
|---|---|---|---|---|---|---|
| Cederblad | 13 | 215 | photogr. | | $r$, diameter, exc. stars (includes emission and reflection nebulae) | |
| Gum | 50 | 85 | Hα | $l = 223 \cdots 0 \cdots 23°$ | $r$, diameter, exc. stars | |
| Hase and Shain | 53 | 286 | Hα | $l = 0 \cdots 360°$ | angular diameter, brightness, exc. stars | furthermore: Atlas of Diffuse Gaseous Nebulae, Moscow: Acad. Sci. USSR 1952 |
| Sharpless | 87 | 313 | red (Hα) | $\delta > -27°$ | angular diameter, structure, brightness, exc. stars | based on Palomar Schmidt plates or prints |
| Rodgers, Campbell, and Whiteoack | 84 | 181 | Hα | $l = 223 \cdots 0 \cdots 39°$ | angular diameter, brightness class | furthermore: Atlas of Hα Emission in the Southern Milky Way: [120] |
| Lynds | 73 | 1100 | red, blue | $l = 337 \cdots 0 \cdots 265°$ | angular dimensions, areas, colour and brightness classes (includes emission and reflection nebulae) | based on the red and blue prints of the Palomar Obs. Sky Atlas |
| Georgelin and Georgelin | 33 | 174 | Hα | $l = 6 \cdots 360°$ | $v_r$, $r$, exc. stars | Fabry-Perot method |
| Georgelin, Georgelin, and Roux | 36 | 60 | Hα | $l = 4 \cdots 268°$ | $v_r$, $r$, exc. stars | Fabry-Perot method |
| Georgelin and Georgelin | 34 | 45 | Hα | $l = 250 \cdots 360°$ | exc. stars | including photographs of the nebulae, atlas of this region: [119] |
| Crampton, Georgelin, and Georgelin | 18 | 23 | Hα | $l = 20 \cdots 115°$ | $v_r$, $r$, exc. stars | Fabry-Perot method |
| Marsalkova (1974) (comparison catalogue compiled from 13 catalogues and lists) | 74 | 698 | Hα etc. | $l = 0 \cdots 360°$ | angular dimensions, brightness class | with charts of the apparent distribution |
| Dubout-Crillon | 109 | 527 | Hα | $l = 10 \cdots 215°$ | position, angular diameter | including 19 Hα-photographs |

Very-wide field photographic Hα survey of the whole Milky Way: [117, 118]. Faint galactic Hα background: [118, 125].

Imagery of H II regions through narrow passband interference filters centered on several emission lines and continuum at visual wavelengths: [135].

Ultraviolet imagery of H II regions: see [134].

High-resolution observations with synthesis radio telescopes:

A useful compilation of synthesis maps of galactic H II regions published up to ≈1976 has been given by Mezger and Wink [76]. More recent high-resolution studies for compact H II regions at several wavelengths: [74a, 26a, 140, 146].

Observations at mm-wavelengths:

The work published up to ≈1976 has been summarized in [76]. Measurements at 13 mm (23.4 GHz) with 80″ beam have been made of M 42, NGC 2024, NGC 7538, W 3 A, W 33, W 43, W 49 and W 51 [85]. See also 7.3.4.3.

H II regions in the galactic center:

For radio and infrared observations up to 1974, see Moorewood, A.F.M. (ed.): "H II Regions and the Galactic Center", ESRO SP-105, Nordwijk, Holland (1974). Radio recombination line and/or continuum observations at 5 and 10 GHz: [81a, 81b, 19a]; see also Table 2. Reviews on more recent work are given in: Shaver (see 7.3.1).

The thermal radio emission of the galactic center comes from extended, low-density ionized gas and a number of giant H II regions [76b].

Recent infrared and submillimeter work: [79, 83, 30, 62]. A modern review of infrared observations of the galactic center is given by Wollman [102b].

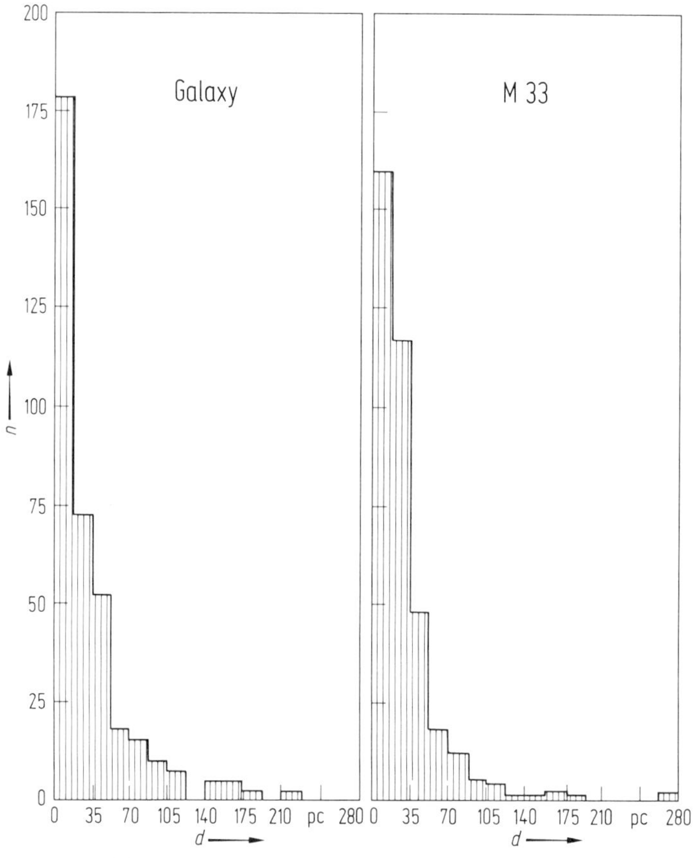


Fig. 1. Distribution of H II region diameters, $d$, in the Galaxy and in M 33 [35], see also [111]. $n$ = number of H II regions per interval of 17.5 pc width

The 5 largest H II regions in the Galaxy [35]:

| | |
|---|---|
| RCW 102, 104, 106 | $d$ = 216 pc |
| η Carina nebula = RCW 53 | = 175 pc |
| G 328°.5 − 1°.0 | = 175 pc |
| RCW 113 | = 164 pc |
| Orion loop | = 149 pc |

## 7.3.3 Classification

A size – mean density relation has been used to derive rough classification schemes for H II regions, which are believed to indicate an evolutionary sequence [16, 88]. Table 3 is mainly based on the proposal by Churchwell [16]. Another classification has been proposed in [128].

Table 3. A rough classification of H II regions [16, 88].
$d$ = diameter of the H II region or component
$N_e$ = electron density

| Class | Type of observed object | Evolutionary stage | Observational appearance | $d$ pc | $N_e$ $cm^{-3}$ | Examples |
|---|---|---|---|---|---|---|
| Probable first stage | some compact 20 μm emission sources | pre-main sequence O stars surrounded by an optically thick shell ("cocoon stars") | 20 μm continuum emission, but not at 2 μm; $H_2O$ masers; no observable radio continuum emission; optically obscured | $\lesssim 0.05$ | $> 5000$ | IRS 5 in W3 |
| 1 | compact H II regions (in groups) | earliest detectable H II regions | radio continuum emission; infrared continuum emission; usually $H_2O$ and/or OH masers; optically obscured or strongly reddened | 0.05···1 | $\gtrsim 1000$ | see Table 4a, b, e.g. components of DR 21 or NGC 7538 |
| 2 | dense complex H II region of intermediate size | expanded H II region | the whole range of radio, infrared and visible continuum and line emission (classes 2 and 3) | 1···100 | 10···1000 | M 42, M 17, RCW 38 |
| 3 | large diffuse H II region | very expanded H II region | | $> 100$ | $\lesssim 10$ | NGC 7000, IC 434, NGC 7822, IC 5070, η Carina |

## 7.3.4 The individual H II regions

### 7.3.4.1 Properties of selected objects

For Tables, see p. 92ff.

A comprehensive list of star forming regions in the Galaxy including many compact or extended H II regions is given in [128].

Giant H II regions:

These objects were defined by an "intrinsic flux density" $S_5 \cdot r^2 \geqq 400$ Jy $kpc^2$ (4 times that of the Orion nebula), where $S_5$ is the flux density at 5 GHz in [Jansky] and $r$ is the distance from the sun in [kpc].

Data for about 80 giant H II regions: [89]. An extensive list of the characteristics of H II regions including a quite homogeneous set of electron temperatures: [76a].

For electron temperatures, see also [5a, 71a, 99a].

Chemical composition of H II regions: [76a, 91a, 126, 127], see also 7.5.2.3 .

Table 4a. Selected optically visible evolved H II regions: position, distance, size, exciting stars.
Positions and angular diameter refer to the optical phenomenon seen on the Palomar Observatory Sky Survey prints [73] (exception: η Carina Nebula)
R.S. = associated radio source [98]  $r$ = distance
size = angular size (optical nebula)  $d$ = physical diameter (from radio observations)
Sp = spectral type

| Name of visible part | R.S. | $\alpha_{1950}$ | $\delta_{1950}$ | $l$ | $b$ | Size arc min | $r$ kpc | $d$ pc | Exciting stars: Name, HD/BD | Exciting stars: Sp | Ref. |
|---|---|---|---|---|---|---|---|---|---|---|---|
| NGC 7822 = S171 | W1 | $0^h01^m$ | 68°20′ | 118.°59 | 6.°18 | 20′ × 4′ | 0.9 | 30 | association Cep OB4 | | 73, 16, 33 |
| NGC 1491 = S206 | | 3 59 | 51 11 | 150.56 | − 0.99 | 6 × 9 | 3.0 | 2.6 (core); ≈ 90 (halo) | BD + 50°886 | O6 | 73, 16, 92 |
| S209 | | 4 07 | 51 02 | 151.6 | − 0.3 | 12 | 5.4 | 5 (core); ≈ 50 (halo) | | | 73, 16, 92 |
| NGC 1976 = M42, Orion Nebula | Ori A, W10 | 5 33 | − 5 30 | 209.13 | −19.35 | 90 × 60 | 0.5 | 0.6 (core) | HD 37022 = $\theta^1$ Ori C (+A, B, D) | O6p, (B0, B3, O9.5) | 73, 40, 41 |
| IC 434 | W12 | 5 39 | − 2 20 | 206.90 | −16.56 | 90 × 30 | 0.5 | 16 | σ Ori | O9.5V | 73, 16 |
| NGC 2237···46, Rosette N. | W 16 | 6 30 | 5 00 | 206.39 | − 1.87 | 80 × 60 | 1.6 | ≈ 50 | 6 members of NGC 2244 cluster | O5···O9V | 73, 16, 45, 148 |
| NGC 3372 = RCW 53, η Car N. | | 10 42 | −59 36 | 287.5 | − 0.9 | 180 × 120 | 2.5 | 175 | cluster Tr16 | | 73, 16, 33 |
| NGC 6514 = M20, Trifid N. | W28 | 17 59 | −23 00 | 6.99 | − 0.17 | 20 × 20 | 2.1 | 5 | HD 164492 | O7 | 73, 33, 14 |
| NGC 6523 = M8, Lagoon N. | W29 | 18 01 | −24 20 | 6.06 | − 1.23 | 45 × 30 | 1.4 | 3.5 | NGC 6530 cluster | O5···B0 | 73, 102, 33, 43 |
| NGC 6618 = M17, Omega N. | W38 | 18 10 | −14 30 | 15.67 | 1.74 | 20 × 15 | 2.2 | ≈ 5 | members of a cluster | O4···O8V | 73, 71, 15, 42 |
| NGC 6611 = M16 | W37 | 18 16 | −13 50 | 16.96 | 0.78 | 120 × 25 | 2.2 | 6 | members of star cluster NGC 6611 | O5···B0 | 73, 41, 44 |
| IC 1318b | W67 | 20 26 | 39 50 | 78.55 | 0.85 | 45 × 20 | 1.5 | ≈ 10 | | | 73, 16, 41 |
| IC 5067 | W80 (IC 5067, IC 5070, NGC 7000) | 20 49 | 44 10 | 84.59 | 0.13 | 25 × 10 | 1.2 (IC 5067, IC 5070, NGC 7000) | 54 (IC 5067, IC 5070, NGC 7000) | association Cyg OB2 (IC 5067, IC 5070, NGC 7000) | | 73, 16, 33 (IC 5067, IC 5070, NGC 7000) |
| IC 5070 | | 20 49 | 44 00 | 84.46 | 0.02 | 60 × 50 | | | | | |
| NGC 7000, North America N. | | 21 00 | 44 00 | 85.75 | − 1.48 | 120 × 30 | | | HD 199579 | O6 | |

Table 4b. Selected optically visible evolved H II regions: physical properties

$U$ = excitation parameter = $N_e^{2/3} R_S$ with $R_S$ = Strömgren radius
$T_e$ = mean electron temperature from optical and/or radio observations
$N_{Lyc}$ = Lyman continuum photon flux required to ionize the nebula
$E$ = mean emission measure = $\int N_e^2 \, dr$
$N_e$ = mean electron density
$\mathfrak{M}_{HII}$ = mass of ionized gas

| Name of visible part | $U$ pc cm$^{-2}$ | $T_e$ K | $N_{Lyc}$ photons s$^{-1}$ | $E$ pc cm$^{-6}$ | $N_e$ cm$^{-3}$ | $\mathfrak{M}_{HII}/\mathfrak{M}_\odot$ | Comments | Ref. |
|---|---|---|---|---|---|---|---|---|
| NGC 7822 | $\gtrsim 70$ | | $>1.2 \cdot 10^{49}$ | $<10^4$ | $<20$ | $>1.6 \cdot 10^3$ | 2° ring structure and condensations | 16 |
| NGC 1491 | $\gtrsim 80$ | 7800 | $6 \cdot 10^{49}$ | $9 \cdot 10^4$ (core) | $2 \cdot 10^2$ | $1 \cdot 10^2$ | | 16, 92, 76a |
| | | | | $<5 \cdot 10^3$ (halo) | $<14$ | $>2.8 \cdot 10^3$ | | |
| S 209 | $\gtrsim 110$ | 8400 | $2 \cdot 10^{49}$ | $1 \cdot 10^5$ (core) | $2 \cdot 10^2$ | $4.9 \cdot 10^2$ | | 16, 92, 76a |
| | | | | $<3 \cdot 10^3$ (halo) | $<8$ | $>9.6 \cdot 10^3$ | | |
| NGC 1976 | 50 | 8000 | $5 \cdot 10^{48}$ | $6 \cdot 10^6$ | $5 \cdot 10^3$ | $\approx 10$ | contour map of $T_e$: see [75a] | 41, 81, 76a |
| IC 434 | 30 | $\lesssim 3500$ | $1.9 \cdot 10^{48}$ | $5.8 \cdot 10^2$ | 6 | $3 \cdot 10^2$ | | 16 |
| NGC 2237···46 | 80 | 8000 | $1.5 \cdot 10^{49}$ | $\approx 3 \cdot 10^4$ | $\approx 16$ | $1.1 \cdot 10^4$ | | 16 |
| NGC 3372 | 150 | 7000 | $8 \cdot 10^{50}$ | $2.5 \cdot 10^5$ (centers) | $2 \cdot 10^2$ | $2 \cdot 10^3$ | two main sources only | 47, 63 |
| NGC 6514 | $\approx 50$ | 8000 | $\approx 4 \cdot 10^{49}$ | $5 \cdot 10^4$ | $1 \cdot 10^2$ | $2 \cdot 10^2$ | | 14, 43 |
| NGC 6523 | 80 | 8000 | $\approx 2 \cdot 10^{49}$ | $3.7 \cdot 10^5$ | $6 \cdot 10^2$ | $2 \cdot 10^2$ | | 41, 43, 81, 76a |
| NGC 6618 | 170 | 7700 | $2 \cdot 10^{50}$ | $3 \cdot 10^6$ | $5 \cdot 10^2$ | $6 \cdot 10^2$ | | 71, 41, 42, 81, 76a |
| NGC 6611 | 120 | $\approx 8000$ | $\approx 5 \cdot 10^{49}$ | $4 \cdot 10^5$ | $2 \cdot 10^2$ | $7 \cdot 10^2$ | | 41, 44 |
| IC 1318b | 90 | 7000 | $2.3 \cdot 10^{49}$ | $2.3 \cdot 10^4$ | 40 | $2 \cdot 10^3$ | | 16, 41, 46 |
| IC 5067, IC 5070, NGC 7000 | $\approx 100$ | $\approx 7000$ | $3.1 \cdot 10^{49}$ | $4.4 \cdot 10^3$ | $\approx 10$ | $1.8 \cdot 10^4$ | about 18 local maxima are embedded in the smooth component | 16, 46 |

Table 5a. Examples of compact H II regions in larger objects: results from radio observations.
Calculations of physical parameters are in most cases based on the assumption of the electron temperature $T_e = 10^4$ K.

size = mean angular diameter
$r$ = distance
$d$ = linear diameter
$E$ = emission measure (for definition, see Table 4b)
$N_e$ = electron density (rms)
$U$ = excitation parameter (for definition, see Table 4b)
$\mathfrak{M}_{HII}$ = mass of ionized hydrogen

| Extended source | Selected components | $\alpha_{1950}$ | $\delta_{1950}$ | Size arc sec | $r$ kpc | $d$ pc | $E$ $pc \cdot cm^{-6}$ | $N_e$ $cm^{-3}$ | $U$ $pc \cdot cm^{-2}$ | $\mathfrak{M}_{HII}/\mathfrak{M}_\odot$ | Ref. |
|---|---|---|---|---|---|---|---|---|---|---|---|
| W3 (with associated optical nebula | A1⋯A5 | $2^h21^m56\overset{s}{.}6$ | +61°52′43″ | 28⋯35″ | 3.1 | 0.44⋯0.54 | $2 \cdot 10^7$ | $2 \cdot 10^3 \cdots 1 \cdot 10^4$ | 83 | 10 | 102 |
| IC 1795) | C | 21 43.6 | 52 46 | 5 | | 0.071 | $3 \cdot 10^7$ | $2 \cdot 10^4$ | 25 | 0.09 | |
| | W3(OH), comp. A | 23 16.5 | 38 57 | 1.5 | | 0.024 | $1 \cdot 10^9$ | $2 \cdot 10^5$ | 54 | 0.08 | 102, 20, 101 |
| NGC 1976 = M42 | Ori A, G 209.0⋯19.4 | 5 32 50 | − 5 25 15 | 240 | 0.5 | 0.6 | $3 \cdot 10^6$ | $2 \cdot 10^3$ | 55 | 7 | 86 |
| NGC 6523 = M8 | A1 (ring) | 18 00 37.0 | −24 22 50 | 15 | 1.4 | 0.10 | $8 \cdot 10^6$ | $1 \cdot 10^4$ | 26 | 0.18 | 102 |
| | A4 | 36.3 | 52 | 46 | | 0.32 | $1 \cdot 10^6$ | $2 \cdot 10^3$ | 24 | 0.72 | |
| NGC 6618 = M17 | S (main) | 18 17 34 | −16 13 24 | 220 | 2.2 | 2.3 | $5 \cdot 10^6$ | $2 \cdot 10^3$ | 143 | 67 | 102, 71, 94 |
| | N | 39 | 10 30 | 280 | | 2.9 | | | 141 | | |
| | E | 51 | 11 30 | 190 | | 2.0 | | | 78 | | |
| W49A | B1 | 19 07 51.1 | + 9 01 22 | 13 | 13.8 | 0.9 | $1 \cdot 10^7$ | $3.5 \cdot 10^3$ | 105 | 32 | 102 |
| | C1 | 50.0 | 15 | 12 | | 0.8 | $1 \cdot 10^7$ | $3.6 \cdot 10^3$ | 95 | 23 | |
| | D1 | 52.5 | 13 | 19 | | 1.3 | $5 \cdot 10^6$ | $1.9 \cdot 10^3$ | 102 | 53 | |
| | K | 58.3 | 06 | 9 | | 0.34 | $8 \cdot 10^7$ | $1.5 \cdot 10^4$ | 106 | 7.3 | |
| W51 | G 49.5⋯0.4d | 19 21 22.3 | +14 25 15 | 11 | 7.3 | 0.4 | $>4 \cdot 10^7$ | $>8 \cdot 10^3$ | 116 | <20 | 20, 101, 10 |
| | e | 24.4 | 24 43 | 19 | | 0.7 | $7 \cdot 10^7$ | $1 \cdot 10^4$ | 170 | 50 | |
| W75 | DR 21A | 20 37 13.7 | +42 08 55 | 5.7 | 3.0 | 0.083 | $5.3 \cdot 10^7$ | $2.5 \cdot 10^4$ | 35.7 | 0.17 | 58 |
| | B | 14.0 | 09 03 | 4.0 | | 0.058 | $4.6 \cdot 10^7$ | $2.8 \cdot 10^4$ | 26.7 | 0.06 | |
| | C | 14.1 | 08 54 | 7.1 | | 0.101 | $9.1 \cdot 10^7$ | $3.0 \cdot 10^4$ | 48.8 | 0.37 | |
| | D | 14.2 | 09 15 | 4.2 | | 0.062 | $4.0 \cdot 10^7$ | $2.6 \cdot 10^4$ | 26.7 | 0.07 | |
| NGC 7538 = S158 | A1 | 23 11 30.3 | +61 12 56 | 110 | 2.5 | 1.4 | $7.5 \cdot 10^5$ | $5 \cdot 10^2 \cdots 1 \cdot 10^3$ | 60 | 33 | 102 |
| | A2 | 20.8 | 13 45 | 10 | | 0.12 | $1.6 \cdot 10^6$ | $4 \cdot 10^3$ | 14 | 0.07 | |
| | B | 36.7 | 12 00 | 12 | | 0.15 | $7.4 \cdot 10^6$ | $6 \cdot 10^3$ | 26 | 0.32 | |
| | C | 36.7 | 11 50 | 1.0 | | 0.013 | $1 \cdot 10^7$ | $1 \cdot 10^5$ | 12 | 0.002 | |

Table 5b. Examples of compact H II regions in larger objects: results from infrared observations.
$S(\lambda)$ = observed flux density (1 Jy = $10^{-26}$ W m$^{-2}$ Hz$^{-1}$)
$L_{IR}$ = total infrared luminosity in units of $L_{\odot} = 4 \cdot 10^{33}$ erg s$^{-1}$
$T_c$ = colour temperature (best fit to the observed spectral emittance by a black body of temperature $T_c$)

| Extended source | Selected components: radio | Selected components: infrared | $S(20\ \mu m)$ Jy | $\frac{S(20\ \mu m)}{S(6\ cm)}$ | $L_{IR}$ $L_{\odot}$ | Colour temperature: $\lambda$ μm | Colour temperature: $T_c$ K | Ref. |
|---|---|---|---|---|---|---|---|---|
| W3 | A1 – A5 | IRS 1 | $2 \cdot 10^3$ | 65 | $\approx 3 \cdot 10^5$ | 45···500 | 74 | 102, 20, 106, 68, 55 |
| | C | IRS 4 | $3 \cdot 10^3$ | 500 | | | | |
| | W3(OH) | IRS 8 | $2 \cdot 10^2$ | 320 | $\approx 2 \cdot 10^5$ | 45···500 | 77 | |
| NGC 1976 = M42 | G 209.0 – 19.4 | (Trapezium Nebula) IRe1 | $1.4 \cdot 10^5$ | 500 | $3.9 \cdot 10^5$ | 21, 350 | 100 | 70 |
| NGC 6523 = M8 | A1 – A4, G 6.0 – 1.2 | | $1.3 \cdot 10^3$ | 30 | $\approx 5 \cdot 10^4$ | 30···500 | 85 | 102, 43, 55 |
| NGC 6618 = M17 | S, G 15.0 – 0.7, IRe1 | | $5.1 \cdot 10^4$ | 220 | $5 \cdot 10^6$ | 5···30 | 240 | 102, 43, 55 |
| | N, G 15.1 – 0.7, IRe2a | | $3.1 \cdot 10^4$ | 141 | | 30···500 | 80 | |
| | E | | $2.4 \cdot 10^3$ | 65 | | | | |
| W49A | B1, C1, D1 (OH-1, W49NW) | | $\approx 10$ | $\approx$ 3 | $\approx 1.9 \cdot 10^7$ | 45···750 | 70 | 102, 100, 54, 6 |
| | K (OH-2, W49SE) | | $\approx 1 \cdot 10^2$ | $\approx$ 50 | | | | |
| W51 | G 495 – 0.4d | IRS 2 | $1.5 \cdot 10^3$ | 136 | $\approx 5 \cdot 10^6$ | 45···750 | 70 | 20, 10, 54 |
| | e | IRS 1 | | | | | | |
| W75 | DR 21D = DR 21N | | $\approx 1 \cdot 10^2$ | 46 | $6 \cdot 10^4$ | | | 20, 54 |
| NGC 7538 = S158 | B | IRS 2 | $7 \cdot 10^2$ | 500 | $\approx 3 \cdot 10^3 \cdots 3 \cdot 10^4$ (for 1···25 μm) | | | 102, 20, 107 |
| | C | IRS 1 | $2 \cdot 10^2$ | 1670 | | | | |

#### 7.3.4.2 Infrared brightness distribution

A summary of the work prior to 1973: [108]. Results for H II regions in the galactic central region: see 7.3.2.

Table 6. Recent observations of the infrared and submillimeter brightness distribution of selected H II regions (contour maps).

| Name of source | $\lambda$ μm | Angular resolution | Ref. |
|---|---|---|---|
| M 42 | 21 | 1′ | 70 |
| | 2.2; 10 | 7″5; 4″5 | 7 |
| | 33 | 10″ | 9 |
| | 69 | 1′ | 23 |
| | 91 | 2′2 | 55 |
| | 20; 30; 50; 100 | 1′; 20″ | 95, 96 |
| | 350 | 56″ | 37 |
| | 400 | 1′6; 3′0 | 90, 64, 121 |
| | 1000 | 1′6 | 59 |
| | 1000 | 1′ | 97 |
| | 1230 | 1′ | 17 |
| NGC 2024 | 2.2; 3.5 | 1′ | 27 |
| (OriB) | 8.4 | 0′6 | 48 |
| | 93 | 2′2 | 55 |
| | 400 | 1′6 | 64 |
| W 3 | 4.9; 8.0; 8.5; 9.7; 10.8; 11.8; 12.7; 20.0 | 8″ | 51 |
| | 69 | 1′ | 24 |
| | 40···350 | 5′5 | 29 |
| | 30; 50; 100 | 30″ | 145 |
| | 1000 | 1′ | 97 |
| M 17 | 21 | 1′ | 70a |
| | 12; 21; 100 | ≈2′5 (scans) | 57 |
| | 69 | 1′ | 25, 100a |
| | 91 | 2′2 | 69a |
| | 30; 50; 100 | 1′ | 30a |
| M 8 | 11 | 4″5 | 21 |
| | 69 | 1′ | 105 |
| M 20 | 69 | 3′ | 103 |
| W 31 | 69 | 1′ | 104 |
| W 49 | 53 | 20″ | 59a |
| | 1000 | 1′ | 97 |
| | 1230 | 1′ | 17 |
| W 51 | 1.65···20 | 10″ | 107 |
| | 74 | 30″ | 91a |
| | 1230 | 1′ | 17 |
| W 75 | 20 | 10″ | 107 |
| | 53 | 20″ | 59a |
| W 22, RCW 38, RCW 49, RCW 57, RCW 97 | 10 | 15″ | 28 |
| RCW 38 | 10 | 7″ | 22 |
| NGC 3372 (η Car) | 35; 53; 80; 100; 175 | 35···45″ | 60 |
| G 333.6−0.2, W 33, RCW 38, RCW 57, RCW 122 | 1000 | 65″ | 143 |

#### 7.3.4.3 Spectrum of H II regions

Optical and infrared emission lines:

Lines between the near ultraviolet and near infrared have been studied most extensively for the Orion nebula. A list of wavelengths and identifications is given in Table 7. A comprehensive catalogue of relative emission line intensities for many objects: [110]. Recently, lines of the middle- and far-infrared regions have been detected in several H II regions. These observations are compiled in Table 8.

Table 7. Emission line spectrum of the Orion nebula from 3187 to 10938 Å [2, 65, 69, 78].
$\lambda$ = laboratory wavelength if identification is given, otherwise observed wavelength
$I$ = relative observed intensity calibrated according to the system of Aller and Liller [2] (Hβ = 100)

| $\lambda$[Å] | Element | Transition | $I$ | $\lambda$[Å] | Element | Transition | $I$ |
|---|---|---|---|---|---|---|---|
| 3187.74 | He I | $2^3S–4^3P$ | 6.3 | 3682.81 | H | $2^2P–20^2D$ | 1.49 |
| 3286.2 | [Fe III] | $a^5D_3–a^3D_3$ | | 3686.83 | H | $2^2P–19^2D$ | 1.69 |
| 3296.79 | He I | $2^1S–8^1P$ | 0.25 | 3691.56 | H | $2^2P–18^2D$ | 1.90 |
| 3324.01 | S III | $3d^3P_1–4p^3P_2$ | | 3694.22 | Ne II | $3s^4P_{5/2}–3p^4P_{5/2}$ | 0.10 |
| 3324.87 | S III | $3d^3P_2–4p^3P_2$ | 0.18 | 3697.15 | H | $2^2P–17^2D$ | 2.11 |
| 3334.87 | Ne II | $3s^4P_{5/2}–3p^4D_{7/2}$ | 0.17 | 3703.86 | H | $2^2P–16^2D$ | 2.34 |
| 3334.90 | [Fe III] | $a^5D_2–a^3D_2$ | | 3705.00 | He I | $2^3P–7^3D$ | 1.12 |
| 3342.5 | [Ne III] | $^1D_2–^1S_0$ | 0.17 | 3709.64 | Ne II | $3s^4P_{3/2}–3p^4P_{1/2}$ | 0.10 |
| 3342.9 | [Cl III] | $^4S–^2P_{3/2}$ | | 3709.37 | S III | $3d^3P_1–4p^3D_2$ | |
| 3353.33 | [Cl III] | $^4S–^2P_{1/2}$ | 0.08 | 3711.97 | H | $2^2P–15^2D$ | 2.85 |
| 3354.55 | He I | $2^1S–7^1P$ | 0.35 | 3717.78 | S III | $4s^3P_2–4p^3S$ | 0.20 |
| 3356.6 | [Fe III] | $a^5D_1–a^3D_2$ | | 3721.94 | H | $2^2P–14^2D$ | 4.1 |
| 3360.63 | Ne II | $3s^4P_{1/2}–3p^4D_{3/2}$ | | 3721.69 | [S III] | $^3P_1–^1S$ | |
| 3370.38 | S III | $3d^3P_2–4p^3P_1$ | | 3726.05 | [O II] | $^4S–^2D_{3/2}$ | 127 |
| 3387.13 | S III | $3d^3P_1–4p^3P_0$ | 0.27 | 3728.80 | [O II] | $^4S–^2D_{5/2}$ | 127 |
| 3447.59 | He I | $2^1S–6^1P$ | 0.61 | 3732.86 | He I | $2^3P–7^3S$ | 0.13 |
| 3453.21 | He I | $2^3P–20^3D$ | | 3734.37 | H | $2^2P–13^2D$ | 3.5 |
| 3465.89 | He I | $2^3P–17^3D$ | | 3747.90 | S III | $3d^3P_0–4p^3D_1$ | |
| 3471.80 | He I | $2^3P–16^3D$ | 0.13 | 3750.15 | H | $2^2P–12^2D$ | 4.4 |
| 3478.97 | He I | $2^3P–15^3D$ | 0.13 | 3756.10 | He I | $2^1P–14^1D$ | 0.09 |
| 3487.72 | He I | $2^3P–14^3D$ | 0.18 | 3758.49 | — | — | 0.06 |
| 3497.34 | S III | — | 0.06 | 3770.63 | H | $2^2P–11^2D$ | 5.4 |
| 3498.64 | He I | $2^3P–13^3D$ | 0.22 | 3777.16 | Ne II | $3s^4P_{1/2}–3p^4P_{3/2}$ | 0.06 |
| 3502.4 | He I | $2^3P–13^3S$ | 0.1 | 3784.89 | He I | $2^1P–12^1D$ | 0.09 |
| 3512.51 | He I | $2^3P–12^3D$ | 0.27 | 3797.90 | H | $2^2P–10^2D$ | 7.8 |
| 3530.49 | He I | $2^3P–11^3D$ | 0.34 | 3805.77 | He I | $2^1P–11^1D$ | 0.10 |
| 3554.39 | He I | $2^3P–10^3D$ | 0.41 | 3813.43 | He II | $4^2F–19^2G$ | 1 |
| 3587.25 | He I | $2^3P–9^3D$ | 0.59 | 3819.61 | He I | $2^3P–6^3D$ | 1.88 |
| 3613.64 | He I | $2^1S–5^1P$ | 0.89 | 3829.77 | Ne II | $3p^2P_{3/2}–3d^2D_{5/2}$ | 0.05 |
| 3634.24 | He I | $2^3P–8^3D$ | 0.84 | 3831.85 | S III | $4s^3P_0–4p^3P_1$ | 0.08 |
| 3651.97 | He I | $2^3P–8^3S$ | 0.13 | 3833.57 | He I | $2^1P–10^1D$ | 0.13 |
| 3656.14 | H | $2^2P–38^2D$ | | 3835.39 | H | $2^2P–9^2D$ | 10.9 |
| 3656.67 | H | $2^2P–37^2D$ | 0.15 | 3837.80 | S III | $4s^3P_1–4p^3P_1$ | 0.10 |
| 3657.27 | H | $2^2P–36^2D$ | 0.12 | 3838.39 | N II | $3p^3P_2–4s^3P_2$ | 0.13 |
| 3657.93 | H | $2^2P–35^2D$ | 0.37 | 3838.32 | S III | $4s^3P_2–4p^3P_2$ | — |
| 3658.64 | H | $2^2P–34^2D$ | 0.42 | 3853.66 | Si II | $2p^2\ ^2D_{3/2}–4s^2P_{3/2}$ | 0.02 |
| 3659.42 | H | $2^2P–33^2D$ | 0.41 | 3856.02 | Si II | $2p^2\ ^2D_{5/2}–4s^2P_{3/2}$ | 0.34 |
| 3660.28 | H | $2^2P–32^2D$ | 0.44 | 3862.59 | Si II | $2p^2\ ^2D_{3/2}–4s^2P_{1/2}$ | 0.19 |
| 3661.22 | H | $2^2P–31^2D$ | 0.49 | 3867.48 | He I | $2^3P–6^3S$ | — |
| 3662.26 | H | $2^2P–30^2D$ | 0.59 | 3868.76 | [Ne III] | $^3P_2–^1D$ | 19.7 |
| 3663.41 | H | $2^2P–29^2D$ | 0.63 | 3871.82 | He I | $2^1P–9^1D$ | 0.16 |
| 3664.68 | H | $2^2P–28^2D$ | 0.70 | 3882.20 | O II | $3p^4D_{7/2}–3d^4D_{7/2}$ | |
| 3666.10 | H | $2^2P–27^2D$ | 0.70 | 3889.05 | H | $2^2P–8^2D$ | 18.1 |
| 3667.68 | H | $2^2P–26^2D$ | 0.79 | 3888.65 | He I | $2^3S–3^3P$ | 18.1 |
| 3669.47 | H | $2^2P–25^2D$ | 0.89 | 3918.98 | C II | $3^2P_{1/2}–4^2S$ | 0.13 |
| 3671.48 | H | $2^2P–24^2D$ | 0.96 | 3920.68 | C II | $3^2P_{3/2}–4^2S$ | 0.26 |
| 3673.76 | H | $2^2P–23^2D$ | 1.06 | 3926.53 | He I | $2^1P–8^1D$ | 0.22 |
| 3676.37 | H | $2^2P–22^2D$ | 1.18 | 3928.62 | S III | $3d^3D_3–4p^3P_2$ | 0.06 |
| 3679.36 | H | $2^2P–21^2D$ | 1.30 | 3933.66 | Ca II | $4^2S–4^2P_{3/2}$ | — |

continued

Table 7, continued

| λ[Å] | Element | Transition | I |
|---|---|---|---|
| 3935.9 | He I | $2^1P^0_1$–$8^1S_0$ | 0.1 |
| 3964.73 | He I | $2^1S$–$4^1P$ | 1.35 |
| 3967.47 | [Ne III] | $^3P_1$–$^1D_2$ | 34.4 |
| 3967.4 | O II | $3p^4P^0_{1/2}$–$3d^2P_{1/2}$ | 24.4 |
| 3970.07 | Hε | $2^2P$–$7^2D$ | 24.4 |
| 3982.61 | — | — | 0.02 |
| 3983.77 | S III | $3d^3D_2$–$4p^3P_1$ | 0.075 |
| 3985.97 | S III | $3d^3D_1$–$4p^3P_0$ | 0.075 |
| 3993.18 | — | — | 0.065 |
| 4009.27 | He I | $2^1P$–$7^1D$ | 0.28 |
| 4026.19 | He I | $2^3P$–$5^3D$ | 2.7 |
| 4068.60 | [S II] | $^4S_{3/2}$–$^2P_{3/2}$ | 1.49 |
| 4069.90 | O II | $3p^4D_{3/2}$–$3d^4F_{5/2}$ | 0.11 |
| 4069.64 | O II | $3p^4D_{1/2}$–$3d^4F_{3/2}$ | |
| 4072.16 | O II | $3p^4D_{5/2}$–$3d^4F_{7/2}$ | 0.12 |
| 4075.87 | O II | $3d^4D_{7/2}$–$3d^4F_{9/2}$ | |
| 4076.35 | [S II] | $^4S_{3/2}$–$^2P_{1/2}$ | 0.69 |
| 4087.16 | O II | $3d^4F_{3/2}$–$4f^4G_{5/2}$ | 0.015 |
| 4089.30 | O II | $3d^4F_{9/2}$–$4f^4G_{11/2}$ | 0.03 |
| 4097.31 | N III | $3^2S_{1/2}$–$3^2P_{3/2}$ | 0.06 |
| 4097.26 | O II | {$3pP_{1/2}$–$3d^4D_{3/2}$ / $3d^4F_{7/2}$–$4f^4G_{9/2}$ | — |
| 4099.48 | — | — | 0.05 |
| 4101.74 | Hδ | $2^2P$–$6^2D$ | 25 |
| 4104.74 | O II | $3p^4P_{3/2}$–$3d^4D_{5/2}$ | — |
| 4105.00 | O II | $3p^4P_{3/2}$–$3d^2D_{3/2}$ | — |
| 4110.80 | O II | $3p^4P_{3/2}$–$3d^4D_{1/2}$ | 0.05 |
| 4114.5 | [Fe II] | $a^4F_{9/2}$–$b^2H_{11/2}$ | 0.1 |
| 4119.22 | O II | $3p^4P_{5/2}$–$3d^4D_{7/2}$ | 0.06 |
| 4120.81 | He I | $2^3P$–$5^3S$ | } |
| 4121.0 | O II | $3p^4P^0_{5/2}$–$3d^4D_{3/2}$ | } 0.5 |
| 4121.5 | O II | $3p^4P^0_{1/2}$–$3d^4P_{1/2}$ | } |
| 4128.0 | Si II | $3^2D_{3/2}$–$4^2F^0_{5/2}$ | } |
| 4129.3 | O II | $3p^4P^0_{3/2}$–$3d^4P_{1/2}$ | } 0.1 |
| 4129.4 | [Fe III] | $a^5D_1$–$a^3G_3$ | } |
| 4132.81 | O II | $3p^4P_{1/2}$–$3d^4P_{3/2}$ | 0.06 |
| 4143.76 | He I | $2^1P$–$6^1D$ | 0.45 |
| 4153.30 | O II | $3p^4P_{3/2}$–$3d^4P_{5/2}$ | 0.08 |
| 4156.54 | O II | $3p^4P_{5/2}$–$3d^4P_{3/2}$ | 0.13 |
| 4168.97 | He I | $2^1P$–$6^1S$ | 0.07 |
| 4185.46 | O II | $3p^2F_{5/2}$–$3d^2G_{7/2}$ | 0.015 |
| 4189.79 | O II | $3p^2F_{7/2}$–$3d^2G_{9/2}$ | 0.03 |
| 4243.98 | [Fe II] | $a^4F_{9/2}$–$a^4G_{11/2}$ | 0.05 |
| 4253.74 | O II | $3d^2G_{9/2}$–$4f^2H_{11/2}$ | 0.06 |
| 4253.98 | O II | $3d^2G_{7/2}$–$4f^2H_{9/2}$ | |
| 4253.59 | S III | $4s^3P_2$–$4p^3D_3$ | |
| 4267.15 | C II | $3^3D$–$4^2F$ | 0.36 |
| 4276.83 | [Fe II] | $a^4F_{7/2}$–$a^4G_{9/2}$ | 0.03 |
| 4284.99 | S III | $4s^3P_1$–$4p^3D_2$ | 0.03 |
| 4287.40 | [Fe II] | $a^6D_{9/2}$–$a^6S_{5/2}$ | 0.11 |
| 4303.82 | O II | $3d^4P_{5/2}$–$4f^4D_{7/2}$ | 0.03 |
| 4305.90 | [Fe II] | $a^4F_{5/2}$–$a^4G_{5/2}$ | |
| 4317.14 | O II | $3s^4P_{1/2}$–$3p^4P_{3/2}$ | 0.06 |
| 4319.63 | O II | $3s^4P_{3/2}$–$3p^4P_{5/2}$ | 0.03 |
| 4319.62 | [Fe II] | $a^4F_{5/2}$–$a^4G_{7/2}$ | |
| 4332.71 | S III | $4s^3P_0$–$4p^3D_1$ | 0.03 |
| 4340.47 | Hγ | $2^2P$–$5^2D$ | 41 |
| 4345.56 | O II | $3s^4P_{3/2}$–$3p^4P_{1/2}$ | 0.11 |
| 4346.85 | [Fe II] | $a^4F_{7/2}$–$a^4G_{11/2}$ | 0.03 |
| 4347.43 | O II | $3s^2D_{3/2}$–$3p^2D_{3/2}$ | 0.03 |
| 4349.43 | O II | $3s^4P_{5/2}$–$3p^4P_{5/2}$ | 0.08 |
| 4351.27 | O II | $3s^2D_{5/2}$–$3p^2D_{5/2}$ | 0.015 |
| 4359.34 | [Fe II] | $a^6D_{7/2}$–$a^6S_{5/2}$ | 0.09 |
| 4363.21 | [O III] | $^1D$–$^1S$ | 1.55 |
| 4366.90 | O II | $3s^4P_{5/2}$–$3p^4P_{3/2}$ | 0.06 |
| 4368.1 | C II | $3d^4P^0$–$4f^4D$ | 0.1 |
| 4387.93 | He I | $2^1P$–$5^1D$ | 0.6 |
| 4409.12 | | | 0.03 |
| 4413.78 | [Fe II] | $a^6D_{5/2}$–$a^6S_{5/2}$ | 0.08 |
| 4414.91 | O II | $3s^2P_{3/2}$–$3p^2D_{5/2}$ | 0.06 |
| 4416.27 | [Fe II] | $a^6D_{9/2}$–$b^4F_{9/2}$ | 0.06 |
| 4416.98 | O II | $3s^2P_{1/2}$–$3p^2D_{3/2}$ | 0.05 |
| 4437.55 | He I | $2^1P$–$5^1S$ | 0.10 |
| 4447.03 | N II | $3p^1P$–$3d^1D$ | |
| 4452.38 | O II | $3s^2P_{3/2}$–$3p^2D_{3/2}$ | 0.06 |
| 4452.11 | [Fe II] | $a^6D_{3/2}$–$a^6S_{5/2}$ | — |
| 4457.95 | [Fe II] | $a^6D_{7/2}$–$b^4F_{7/2}$ | 0.05 |
| 4471.48 | He I | $2^3P$–$4^3D$ | 4.6 |
| 4590.97 | O II | $3s^2D_{5/2}$–$3p^2F_{7/2}$ | 0.04 |
| 4607.15 | N II | $3s^3P_0$–$3p^3P_1$ | 0.06 |
| 4607.0 | [Fe III] | $a^5D_4$–$a^3F_3$ | |
| 4621.39 | N II | $3s^3P_1$–$3p^3P_0$ | 0.03 |
| 4621.5 | [C I] | $^3P_1$–$^1S$ | |
| 4630.54 | N II | $3s^3P_2$–$3p^3P_2$ | 0.08 |
| 4634.16 | N III | $3^2P_{1/2}$–$3^2D_{3/2}$ | 0.04 |
| 4638.85 | O II | $3s^4P_{1/2}$–$3d^4D_{3/2}$ | 0.08 |
| 4640.64 | N III | $3^2P_{3/2}$–$3^2D_{5/2}$ | 0.08 |
| 4641.81 | O II | $3s^4P_{3/2}$–$3p^4D_{5/2}$ | 0.15 |
| 4649.14 | O II | $3s^4P_{5/2}$–$3p^4D_{7/2}$ | 0.25 |
| 4650.84 | O II | $3s^4P_{1/2}$–$3p^4D_{1/2}$ | 0.08 |
| 4658.1 | [Fe III] | $a^5D_4$–$a^3F_4$ | 1.2 |
| 4661.64 | O II | $3s^4P_{3/2}$–$3p^4D_{3/2}$ | 0.08 |
| 4676.23 | O II | $3s^4P_{5/2}$–$3p^4D_{5/2}$ | 0.06 |
| 4677.99 | — | — | |
| 4699.21 | O II | $3p^2D_{3/2}$–$3d^2F_{5/2}$ | 0.04 |
| 4701.5 | [Fe III] | $a^5D_3$–$a^3F_3$ | 0.24 |
| 4705.36 | O II | $3p^2D_{5/2}$–$3d^2F_{7/2}$ | 0.04 |
| 4711.34 | [A IV] | $^4S_{3/2}$–$^2D_{5/2}$ | 0.17 } 1.1 |
| 4713.14 | He I | $2^3P$–$4^3S$ | 1.05 } |
| 4733.9 | [Fe III] | $a^5D_2$–$a^3F_2$ | 0.09 |
| 4740.20 | [A IV] | $^4S_{3/2}$–$^2D_{3/2}$ | 0.17 |
| 4754.7 | [Fe III] | $a^5D_3$–$a^3F_4$ | 0.11 |
| 4769.4 | [Fe III] | $a^5D_2$–$a^3F_3$ | 0.07 |
| 4774.74 | [Fe II] | $a^4F_{9/2}$–$b^4F_{7/2}$ | |
| 4777.7 | [Fe III] | $a^5D_1$–$a^3F_2$ | 0.05 |
| 4814.55 | [Fe II] | $a^4F_{9/2}$–$b^4F_{9/2}$ | 0.10 |
| 4861.33 | Hβ | $2^2P$–$4^2D$ | **100** |
| 4864.13 | | | 0.08: |

continued

Table 7, continued

| λ[Å] | Element | Transition | I | λ[Å] | Element | Transition | I |
|---|---|---|---|---|---|---|---|
| 4881.0 | [Fe III] | $a^5D_4$–$a^3H_4$ | 0.31 | 7065.3 | He I | $2^3P$–$3^3S$ | ≈ 10 |
| 4921.93 | He I | $2^1P$–$4^1D$ | 1.5 | 7135.8 | [A III] | $^3P_2$–$^1D$ | |
| 4931.0 | [O III] | $^3P_0$–$^1D$ | 0.09 | 7155.1 | [Fe II] | $a^4F_{9/2}$–$a^2G_{9/2}$ | 0.6 |
| 4930.5 | [Fe III] | $a^5D_1$–$a^3P_0$ | | 7231.1 | C II | $3^2P_{1/2}$–$3^2D_{3/2}$ | 0.4 |
| 4958.92 | [O III] | $^3P_1$–$^1D$ | 113 | 7236.2 | C II | $3^2P_{3/2}$–$3^2D_{5/2}$ | 0.7 |
| 5006.85 | [O III] | $^3P_2$–$^1D$ | 342 | 7254.4 | O I | $3^3P$–$5^3S$ | 1.0 |
| 5015.68 | He I | $2^1S$–$3^1P$ | 2.24 | 7281.3 | He I | $2^1P$–$3^1S$ | ≈ 10 |
| 5041.1 | Si II | $4p^2P_{1/2}$–$4d^2D$ | 0 | 7319.9 | [O II] | $^2D_{5/2}$–$^2P$ | |
| 5046.1 | He I | $2^1P$–$4^1S$ | 0.4 | 7330.2 | [O II] | $^2D_{3/2}$–$^2P$ | |
| 5056.1 | Si II | $4p^2P_{3/2}$–$4d^2D$ | 0.4 | 7378.6 | | | 0.6 |
| 5191.4 | [A III] | $^1D$–$^1S$ | 0.2 | 7388.2 | [Fe II] | $a^4F_{5/2}$–$a^2G_{9/2}$ | 0 |
| 5198.5 | [N I] | $^4S$–$^2D_{5/2}$ | 0.6 | 7452.5 | [Fe II] | $a^4F_{7/2}$–$a^2G_{9/2}$ | 0 |
| 5200.7 | [N I] | $^4S$–$^2D_{3/2}$ | 0.4 | 7500 | | | 0.4 |
| 5262.1 | | | 0.4 | 7530.9 | [Cl IV] | $^3P_1$–$^1D$ | 0.4 |
| 5270.3 | [Fe III] | $a^5D_3$–$a^3P_2$ | 1.2 | 7751.0 | [A III] | $^3P_1$–$^1D$ | ≈ 5 |
| 5302.5 | | | 0.6 | 7774.2 | O I | $3^5S$–$3^5P$ | 0.2 |
| 5322.2 | [Cl IV] | $^1D$–$^1S$ | 0 | 7794.0 | | | 0.4 |
| 5487.0 | | | } 0.4 | 7816.2 | He I | $3^3S$–$7^3P$ | 0.4 |
| 5512.7 | O I | $3^3P$–$6^3D$ | | 7837.7 | | | 0.4 |
| 5517.7 | [Cl III] | $^4S$–$^2D_{5/2}$ | ≈ 10 | 8286.4 | H | $P_{30}$ | 1 *) |
| 5537.6 | [Cl III] | $^4S$–$^2D_{3/2}$ | ≈ 10 | 8292.3 | H | $P_{29}$ | 0 *) |
| 5554.9 | O I | $3^3P$–$7^3S$ | 0 | 8298.8 | H | $P_{28}$ | 1 *) |
| 5577.3 | [O I] | $^1D$–$^1S$ | ≈ 15 | 8306.8 | H | $P_{27}$ | 1 *) |
| 5679.6 | N II | $3s^3P_2$–$3p^3D_3$ | 0.4 | 8314.3 | H | $P_{26}$ | 1 *) |
| 5739 | | | 0.6 | 8323.4 | H | $P_{25}$ | 1 *) |
| 5754.6 | [N II] | $^1D$–$^1S$ | ≈ 15 | 8333.8 | H | $P_{24}$ | 1 *) |
| 5759 | | | 0.4 | 8345.6 | H | $P_{23}$ | 1 *) |
| 5875.6 | He I | $2^3P$–$3^3D$ | | 8359.0 | H | $P_{22}$ | 2 *) |
| 5957.6 | Si II | $4^2P_{1/2}$–$5^2S$ | 0.4 | 8374.5 | H | $P_{21}$ | 3 *) |
| 5958.5 | O I | $3^3P$–$5^3D$ | 0.4 | 8392.4 | H | $P_{20}$ | 3 *) |
| 5979.0 | Si II | $4^2P_{3/2}$–$5^2S$ | 0.4 | 8413.3 | H | $P_{19}$ | 2 *) |
| 6046.4 | O I | $3^3P$–$6^3S$ | 0.7 | 8438.0 | H | $P_{18}$ | 3 *) |
| 6258 | | | 0.4 | 8446.4 | O I | $3^3S$–$3^3P$ | 5 *) |
| 6300.3 | [O I] | $^3P_2$–$^1D$ | ≈ 10 | 8467.3 | H | $P_{17}$ | 3 *) |
| 6312.1 | [S III] | $^1D$–$^1S$ | ≈ 15 | 8502.5 | H | $P_{16}$ | 4 *) |
| 6347.5 | Si II | $4^2S$–$4^2P_{3/2}$ | 0.6 | 8545.4 | H | $P_{15}$ | 4 *) |
| 6363.8 | [O I] | $^3P_1$–$^1D$ | 1.2 | 8598.4 | H | $P_{14}$ | 3 *) |
| 6371.3 | Si II | $4^2S$–$4^2P_{1/2}$ | 0.4 | 8665.0 | H | $P_{13}$ | 1 *) |
| 6548.1 | [N II] | $^3P_1$–$^1D$ | | 9069.0 | [S III] | $3p^2\ ^3P$–$3p^2\ ^1D$ | 72 |
| 6562.8 | Hα | $2^2P$–$3^2D$ | 350 | 9229.0 | H | $3^2D$–$9^2F=P_9$ | 5.8 |
| 6583.4 | [N II] | $^3P_2$–$^1D$ | 55 | 9531.8 | [S III] | $3p^2\ ^3P$–$3p^2\ ^1D$ | 181 |
| 6678.1 | He I | $2^1P$–$3^1D$ | ≈ 15 | 9546.0 | H | $3^2D$–$8^2F=P_8$ | 8 |
| 6716.4 | [S II] | $^4S$–$^2D_{5/2}$ | ≈ 10 | 10049.4 | H | $3^2D$–$7^2F=P_7$ | 10 |
| 6730.8 | [S II] | $^4S$–$^2D_{3/2}$ | ≈ 10 | 10830.3 | He I | $2^3S$–$2^3P^0$ | 70 |
| 7002.1 | O I | $3^3P$–$4^3D$ | 0.7 | 10938.1 | H | $3D^2$–$6^2F=P_6$ | 20 |

*) Not the same scale as for the other intensities.

Recent near infrared line identifications in the spectra of M 42 and other H II regions: [122, 123, 124, 116]. Far ultraviolet spectrum of M 42 and NGC 7000: [142, 134].

Table 8. Observations of infrared line emission associated with H II regions.
$\lambda$ = laboratory wavelength

| $\lambda$ [μm] | Identification | | H II region | Ref. |
|---|---|---|---|---|
| 4.052 | H | Brackett α (5 − 4) | M 42 and others with compact IR sources | 49, 66, 88a, 131 |
| 2.166 | | Brackett γ (7 − 4) | Sgr A West, G 333.6 − 0.2 | 79, 133 |
| 4.654 | | Pfund β (7 − 5) | M 42 | 52 |
| 3.741 | | Pfund γ (8 − 5) | M 42 | |
| 2.058 | He I, | $2\,^1P - 2\,^1S$ | G 333.6 − 0.2 | 133 |
| 8.99 | [A III], | fine-structure line | G 45.1 + 0.1, G 298.2 − 0.3, G 333.6 − 0.2 | 61a, 131 |
| 10.52 | [S IV], | fine-structure line | | |
| 12.81 | [Ne II], | fine-structure line | G 333.6 − 0.2 | 1, 130, 131 |
| | | | Sgr A West, G 45.1 + 0.1 | 1a, 102b, 61a |
| 18.71 | [S III], | fine-structure line | M 42, M 17, W 51, NGC 2024 | 5, 77, 114, 115, 132 |
| 33 | [S III], | fine-structure line | M 17 | 133 |
| 51.8 | [O III], | fine-structure line | M 42, M 17, W 51, G 333.6 − 0.2 | 75, 115, 136 |
| 57 | [N III], | fine-structure line | M 17, NGC 6357 | 115, 136 |
| 63.2 | [O I], | fine-structure line | M 42, M 17 | 133, 136 |
| 88.35 | [O III], | fine-structure line | M 42, M 17, M 8, W 3, W 49, W 51, NGC 6557, NGC 6334, NGC 7538, G 333.6 − 0.2, Sgr A | 5, 75, 19, 93, 115, 136 |
| 157 | [C II], | fine-structure line | M 42, NGC 2024 | 144 |
| 1.958<br>2.034<br>2.122<br>2.223<br>2.407<br>2.413<br>2.424 | $H_2$, | $v = 1 \rightarrow 0$ vibration-rotation quadrupole spectrum | M 42 | 31, 8, 80, 67 |
| 12.28 | $H_2$, | $v = 0 - 0$, S(2) rotational transition | M 42 (OMC 1) | 112 |

Radio recombination lines:

Emission lines at radiofrequencies resulting from transitions between high quantum numbers $n$ of H and He atoms were discovered in 1964 (Dravskich, Z.V. et al., Sorochenko, R.L., and Borodich, E.V.: Reports to XII. General Assembly of the IAU 1964). Up to 1970 a large number of radio recombination lines of H, He, and C have been detected at wavelengths ranging from 8 mm (H 56α) to 75 cm (H 253α). For a general introduction, see e.g., Gordon (1974), quoted in 7.3.1. Recent review articles: Brown et al. (1978), Shaver (1980) (see 7.3.1).

Recent observations at 6 cm (H 109α) and 3.3 cm (H 91α): [16a, 141]; at shorter wavelengths: [71a, 99a, 91b].

Tables of the rest frequencies of radio recombination lines of H and He have been published in [72].

Table 9. Selected radio recombination line data for the Orion nebula derived with angular resolution of 1···3′. For more detailed data, see [137, 147].

nα corresponds to the transition from n+1 to n, nβ from n+2 to n etc.

$\nu_0$ = rest frequency

$\Delta v_L$ = line width at half-intensity points

$T_L$ = peak brightness temperature of the line

| Line | $\nu_0$ MHz | $T_L$ K | $\Delta v_L$ km s$^{-1}$ | Ref. |
|---|---|---|---|---|
| H 56α | 36466.32 | – | 30 | 90a |
| H 66α | 22364.17 | – | 25.8 | 99a |
| H 83β | 22196.47 | – | 25.5 | 99a |
| H 86α | 10161.30 | 2.222 | 27.5 | 71a |
| He 86α | 10165.44 | 0.240 | 17.9 | 71a |
| H 108β | 10157.63 | 0.645 | 29.6 | 71a |
| H 109α | 5008.92 | 7.633 | 30.3 | 16a |
| He 109α | 5010.96 | 0.818 | 23.2 | 16a |
| C 109α | 5011.3 | 0.37 | 4.8 | 16a |
| H 137β | 5005.03 | 1.587 | 34.3 | 16a |
| He 137β | 5007.07 | 0.122 | 24.5 | 16a |

Continuous spectrum:

In the UV and visible regions, continuous radiation results from the scattering of the light of exciting stars by dust grains inside and/or near the H II region (see 7.1.4). The infrared, mm-wave and radio spectrum is similar for all H II regions with only a few exceptions, and it has two main components: (1) free-free emission of the ionized gas and (2) thermal infrared emission of the dust, peaking at a wavelength of 50···100 μm. In the range 2···20 μm, the spectrum can be fitted by the power law $I_\nu \propto \nu^{-\gamma}$ with $\gamma \approx 3.5 \pm 0.5$.

Observational results for the infrared have been summarized by D. A. Harper in: Wilson and Downes (1975) p. 343 and by N. Panagia in: Fazio (1977) p. 43, cited in 7.3.1. For the mm-range, see e.g. [17, 73a]; for the radio continuum, e.g. Gordon (1974) or Terzian (1974) (see 7.3.1).

### 7.3.4.4 Molecular masers associated with H II regions

OH (Type I) and $H_2O$ maser sources occur quite frequently near compact H II regions. Furthermore, the probability of finding an infrared source within a few arc seconds of a maser is very high. Table 10 summarizes some observational results obtained mainly by recent extensive VLBI (= very-long-base-line interferometry) analysis. List of star forming regions with OH and/or $H_2O$ maser sources: [128]. Recent review article: [141a].

Table 10. Observed properties of OH and $H_2O$ masers associated with H II regions [11, 32, 149].

| Quantity | OH ($\lambda$ = 18 cm) | | $H_2O$ ($\lambda$ = 1.35 cm) |
|---|---|---|---|
| Probability of occurrence in H II regions | 1/3 | | 1/10 |
| Individual maser sources: | | | |
| size | | 1···100 AU | |
| line width, $\Delta v$ | | 0.5···2 km s$^{-1}$ | |
| brightness temperature | $\approx 10^{13}$ K | | $\approx 10^{15}$ K |
| polarization | circular | | none or linear (M 42) |
| luminosity | $10^{-5} \cdots 10^{-3} L_\odot$ | | $10^{-4} \cdots 1\, L_\odot$ |
| velocity relative to the H II region (or molecular cloud) | | $\approx 20$ km s$^{-1}$ | (large-scale outflow) |
| Clusters of individual masers ("centers of activity") | | | |
| size | | $\approx 2 \cdot 10^3$ AU | |
| Surrounding region of weak maser emission | | | |
| size | | $\approx 10^4$ AU | |
| velocity, $v$ | | $\pm 15 \cdots \pm 250$ km s$^{-1}$ | |

## 7.3.5 References for 7.3

1 Aitken, D.K., Jones, B.: Mon. Not. R. Astron. Soc. **167** (1974) 11 P.
1a Aitken, D.K., Jones, B., Penman, J.M.: Mon. Not. R. Astron. Soc. **169** (1974) 35 P.
2 Aller, L.H., Liller, W.: Astrophys. J. **130** (1959) 45.
3 Altenhoff, W.J., Downes, D., Goad, L., Maxwell, A., Rinehart, R.: Astron. Astrophys. Suppl. **1** (1970) 319.
4 Altenhoff, W.J., Downes, D., Pauls, T., Schraml, J.: Astron. Astrophys. Suppl. **35** (1979) 23.
5 Baluteau, J.-P., Bussoletti, E., Anderegg, M., Moorwood, A.F.M., Coron, N.: Astrophys. J. Lett. **210** (1976) L45.
5a Barker, T.: Astrophys. J. **227** (1979) 863.
6 Becklin, E.E., Neugebauer, G., Wynn-Williams, C.G.: Astrophys. Lett. **13** (1973) 147.
7 Becklin, E.E., Beckwith, S., Gratley, I., Matthews, K., Neugebauer, G., Sarazin, C., Werner, M.W.: Astrophys J. **207** (1976) 770.
8 Beckwith, S., Becklin, E.E., Neugebauer, G., Persson, S.E.: Bull. Am. Astron. Soc. **8** (1979) 564.
9 Beichmann, C.A., Dyck, H.M., Simon, T.: Astron. Astrophys. **62** (1978) 261.
10 Bieging, J. in: Wilson and Downes (see 7.3.1), p. 443.
11 Burke, B.F.: Int. Astron. Union Symp. **60** (1974) 267.
12 Caswell, J.L.: Austral. J. Phys. **25** (1972) 443.
13 Cederblad, S.: Medd. Lund Astron. Obs. Ser. 2, No. 119 (1946).
14 Chaisson, Y.P., Wilson, R.F.: Astrophys. J. **199** (1975) 647.
15 Chini, R.: Dissertation Heidelberg 1979.
16 Churchwell, E. in: Wilson and Downes (see 7.3.1), p. 245.
16a Churchwell, E., Smith, L.F., Mathis, J., Mezger, P.G., Huchtmeier, W.: Astron. Astrophys. **70** (1978) 719.
17 Clegg, P.E., Rowan-Robinson, M., Ade, P.A.R.: Astron. J. **81** (1976) 399.
18 Crampton, D., Georgelin, Y.M., Georgelin, Y.P.: Astron. Astrophys. **66** (1978) 1.
19 Dain, F.W., Gull, G.E., Melnick, G., Harwit, M.: Astrophys. J. Lett. **221** (1978) L17.
19a Downes, D., Goss, W.M., Schwarz, U.J., Wouterloot, J.G.A.: Astron. Astrophys. Suppl. **35** (1979) 1.
20 DeJong, T., Israel, F.P., Tielens, A. in: Wilson and Downes (see 7.3.1), p. 125.
21 Dyck, H.M.: Astron. J. **82** (1977) 129.
22 Epchtein, N., Turon, P.: Astron. Astrophys. **72** (1979) L4.
23 Fazio, G.G., Kleinmann, D.E., Noyes, R.W., Wright, E.L., Zeilik, M.: Astrophys. J. Lett. **192** (1974) L23.
24 Fazio, G.G., Kleinmann, D.E., Noyes, R.W., Wright, E.L., Zeilik, M., Low, F.J.: Astrophys. J. Lett. **199** (1975) L177.
25 Fazio, G.G. in: Infrared Astronomy (Setti, G., Fazio, G.G., eds.), Reidel, Dordrecht, Holland (1978) p. 33.
26 Felli, M., Churchwell, E.: Astron. Astrophys. Suppl. **5** (1972) 369.
26a Felli, M., Harten, R.H., Habing, H.J., Israel, F.P.: Astron. Astrophys. Suppl. **32** (1978) 423.
27 Frey, A., Lemke, D., Thum, C., Fahrbach, U.: Astron. Astrophys. **74** (1979) 133.
28 Frogel, J.A., Persson, S.E.: Astrophys. J. **192** (1974) 351.
29 Furniss, I., Jennings, R.E., Moorwood, A.F.M.: Astrophys. J. **202** (1975) 400.
30 Gatley, I., Becklin, E.E., Werner, M.W., Harper, D.A.: Astrophys. J. **220** (1978) 822.
30a Gatley, I., Becklin, E.E., Sellgren, K., Werner, M.W.: Astrophys. J. **233** (1979) 575.
31 Gautier III, T.N., Fink, U., Treffers, R.R., Larson, H.P.: Astrophys. J. Lett. **207** (1976) L129.
32 Genzel, R., et al.: Astron. Astrophys. **66** (1978) 13.
33 Georgelin, Y.P., Georgelin, Y.M.: Astron. Astrophys. **6** (1970) 349.
34 Georgelin, Y.P., Georgelin, Y.M.: Astron. Astrophys. Suppl. **3** (1970) 1.
35 Georgelin, Y.P.: Astron. Astrophys. **11** (1971) 414.
36 Georgelin, Y.M., Georgelin, Y.P., Roux, S.: Astron. Astrophys. **25** (1973) 337.
37 Gezari, D.Y., Joyce, R.R., Righini, G., Simon, M.: Astrophys. J. Lett. **191** (1974) L33.
38 Goss, W.M., Shaver, P.A.: Australian J. Phys., Astrophys. Suppl. **14** (1970) 1.
39 Goudis, C.: Astrophys. Space Sci. **35** (1975) 409.
40 Goudis, C.: Astrophys. Space Sci. **36** (1975) 79.
41 Goudis, C.: Astrophys. Space Sci. **38** (1975) 13.
42 Goudis, C.: Astrophys. Space Sci. **39** (1976) 273.
43 Goudis, C.: Astrophys. Space Sci. **40** (1976) 281.
44 Goudis, C.: Astrophys. Space Sci. **41** (1976) 105.
45 Goudis, C.: Astrophys. Space Sci. **43** (1976) 135.
46 Goudis, C.: Astrophys. Space Sci. **44** (1976) 281.
47 Goudis, C.: Astrophys. Space Sci. **47** (1977) 109.

47a Goudis, C.: Astrophys. Space Sci. **61** (1979) 417.
48 Grasdalen, G.L.: Astrophys. J. **193** (1974) 373.
49 Grasdalen, G.L.: Astrophys. J. Lett. **205** (1976) L83.
50 Gum, C.S.: Mem. R. Astron. Soc. **67** (1955) 155.
51 Hackwell, J.A., Gehrz, R.D., Smith, J.R., Briotta, D.: Astrophys. J. **221** (1978) 797.
52 Hall, D.N.B., Kleinmann, S.G., Ridgway, S.T., Gillett, F.C.: Astrophys. J. Lett. **223** (1978) L47.
53 Hase, V.F., Shain, G.A.: Izv. Krim. Astrofiz. Obs. **15** (1955) 1.
54 Harper, D.A., Low, F.J.: Astrophys. J. Lett. **165** (1971) L9.
55 Harper, D.A.: Astrophys. J. **192** (1974) 557.
56 Harper, D.A.: Astrophys. J. **192** (1974) 551.
57 Harper, D.A., Low, F.J., Rieke, G.H., Thronson, H.A.: Astrophys. J. **205** (1976) 136.
58 Harris, S.: Mon. Not. R. Astron. Soc. **162** (1973) 5 P.
59 Harvey, P.M., Gatley, I., Werner, M.W., Elias, J.H., Evans, N.J., Zuckerman, B., Morris, G., Sato, T., Litvak, M.M.: Astrophys. J. Lett. **191** (1974) L87.
59a Harvey, P.M., Campbell, M.F., Hoffmann, W.F.: Astrophys. J. **211** (1977) 786.
60 Harvey, P.M., Hoffmann, W.F., Campbell, M.F.: Astrophys. J. **227** (1979) 114.
61 Haynes, R.F., Caswell, J.L., Simons, L.W.J.: Australian J. Phys., Astrophys. Suppl. **45** (1978).
61a Hefele, H., Schulte i. d. Bäumen, J.: Astron. Astrophys. **66** (1978) 465.
62 Hildebrand, R.H., Whitcomb, S.E., Winston, R., Stiening, R.F., Harper, D.A., Moseley, S.H.: Astrophys. J. Lett. **219** (1978) L101.
63 Huchtmeier, W.K., Day, G.A.: Astron. Astrophys. **41** (1975) 105.
64 Hudson, H.S., Soifer, B.T.: Astrophys. J. **206** (1976) 100.
65 Johnson, H.M.: Stars and Stellar Systems VII (1968) p. 92.
66 Joyce, R.R., Simon, M., Simon, T.: Astrophys. J. **220** (1978) 156.
67 Joyce, R.R., Gezari, D.Y., Scoville, N.Z., Furenlid, I.: Astrophys. J. Lett. **219** (1978) L31.
68 Krügel, E., Mezger, P.G.: Astron. Astrophys. **42** (1975) 441.
68a Kuzmin, A.D., Levchenko, M.T., Noskova, R.I., Solomonvich, A.E.: Sovjet Astron. – Astron. J. **4** (1961) 909.
69 Lambrecht, H.: Landolt-Börnstein, NS., Vol. VI/1 (1965) p. 646.
69a Lada, C.J.: Astrophys. J. Suppl. **32** (1976) 603.
69b Large, M.I., Mathewson, D.S., Haslam, C.G.T.: Mon. Not. R. Astron. Soc. **123** (1961) 113, 123.
70 Lemke, D., Low, F.J., Thum, C.: Astron. Astrophys. **32** (1974) 231.
70a Lemke, D., Low, F.J.: Astrophys. J. Lett. **177** (1972) L53.
71 Lemke, D. in: Wilson and Downes (see 7.3.1), p. 245.
71a Lichten, S.M., Rodriguez, L.F., Chaisson, E.J.: Astrophys. J. **229** (1979) 524.
72 Lilley, A.E., Palmer, P.: Astrophys. J. Suppl. **16** (1968) 143.
73 Lynds, B.T.: Astrophys. J. Suppl. **12** (1965) 163.
73a Malkamäki, L., Sandell, G., Mattila, K., Gebler, K.-H.: Astron. Astrophys. **71** (1979) 198.
74 Marsalkova, P.: Astrophys. Space Sci. **27** (1974) 3.
74a Matthews, H.E., Goss, W.M., Winnberg, A., Habing, H.J.: Astron. Astrophys. **61** (1977) 261.
75 Melnick, G., Gull, G.E., Harwit, M.: Astrophys. J. Lett. **222** (1978) L137.
75a Mezger, P.G., Ellis, S.A.: Astrophys. Lett. **1** (1968) 159.
76 Mezger, P.G., Wink, J.E. in: Infrared and Submillimeter Astronomy (Fazio, G.G., ed.), Reidel, Dordrecht, Holland (1977) p. 58.
76a Mezger, P.G., Pankonin, V., Schmid-Burgk, J., Thum, C., Wink, J.: Astron. Astrophys. **80** (1979) L3.
76b Mezger, P.G., Pauls, T.: Int. Astron. Union Symp. **84** (1979) 357.
77 Moorwood, A.F.M., Baluteau, J.P., Anderegg, M., Coron, N., Biraud, Y.: Astrophys. J. **224** (1978) 101.
78 Morgan, L.A.: Mon. Not. R. Astron. Soc. **153** (1971) 393.
79 Neugebauer, G., Becklin, E.E., Matthews, K., Wynn-Williams, C.G.: Astrophys. J. **220** (1978) 149.
80 Ogden, P.M., Roesler, F.L., Reynolds, R.J., Scherb, F., Larson, H.P., Daehler, M.: Bull. Am. Astron. Soc. **10** (1978) 395.
81 Osterbrock, D.E.: Astrophysics of Gaseous Nebulae, Freeman, San Francisco 1974, p. 102, 105, 120.
81a Pauls, T., Mezger, P.G., Churchwell, E.: Astron. Astrophys. **34** (1974) 327.
81b Pauls, T., Downes, D., Mezger, P.G., Churchwell, E.: Astron. Astrophys. **46** (1976) 407.
82 Reifenstein III, E.C., Wilson, T.L., Burke, B.F., Mezger, P.G., Altenhoff, W.J.: Astron. Astrophys. **4** (1970) 357.
83 Rieke, G.H., Telesco, C.M., Harper, D.A.: Astrophys. J. **220** (1978) 556.
84 Rodgers, A.W., Campbell, C.T., Whiteoak, J.B.: Mon. Not. R. Astron. Soc. **121** (1960) 103.
85 Rodriguez, L., Chaisson, E.J.: Astrophys. J. **221** (1978) 816.
86 Schraml, J., Mezger, P.G.: Astrophys. J. **156** (1969) 269.

87 Sharpless, S.L.: Astrophys. J. Suppl. **4** (1959) 257.
88 Shaver, P.A. in: Topics in Interstellar Matter (van Woerden, H., ed.), Reidel, Dordrecht, Holland (1977) p. 58.
88a Simon, Th., Simon, M., Joyce, R.R.: Astrophys. J. **230** (1979) 127.
89 Smith, L.F., Biermann, P., Mezger, P.G.: Astron. Astrophys. **66** (1978) 65.
90 Soifer, B.T., Hudson, H.S.: Astrophys. J. Lett. **191** (1974) L83.
90a Sorochenko, R.L., Berulis, I.I.: Soviet Astron.–Astron. J. **14** (1971) 683.
91 Terzian, Y. in: Vistas in Astronomy **16** (1974) 279.
91a Thronson, H.A., Harper, D.A.: Astrophys. J. **230** (1979) 133.
91b Thum, C., Mezger, P.G., Pankonin, V.: Astron. Astrophys. **87** (1980) 269.
92 Walsmley, C.M., Churchwell, E.: Astron. Astrophys. **41** (1975) 121.
93 Ward, D.B., Dennison, B., Gull, G., Harwit, M.: Astrophys. J. Lett. **202** (1975) L31.
94 Webster, W.J., Altenhoff, W.J., Wink, J.E.: Astron. J. **76** (1971) 677.
95 Werner, M.W., Gatley, I., Harper, D.A., Becklin, E.E., Loewenstein, R.F., Telesco, C.M., Thronson, H.A.: Astrophys. J. **204** (1976) 420.
96 Werner, M.W., Becklin, E.E., Gatley, I., Neugebauer, G. in: Infrared and Submillimeter Astronomy (Fazio, G.G., ed.), Reidel, Dordrecht, Holland (1977) p. 89.
97 Westbrook, W.E., Werner, M.W., Elias, J.H., Gezari, D.Y., Hauser, M.G., Lo, K.Y., Neugebauer, G.: Astrophys. J. **209** (1976) 94.
98 Westerhout, G.: Bull. Astron. Inst. Netherl. **14** (1958) 215.
99 Wilson, T.L., Mezger, P.G., Gardner, F.F., Milne, P.K.: Astron. Astrophys. **6** (1970) 364.
99a Wilson, T.L., Bieging, J., Wilson, W.E.: Astron. Astrophys. **71** (1979) 205.
100 Wilson, T.L. in: Wilson and Downes (see 7.3.1), p. 424.
100a Wilson, T.L., Fazio, G.G., Jaffe, D., Kleinmann, D., Wright, E.L., Low, F.J.: Astron. Astrophys. **76** (1979) 86.
101 Wink, J.E., Altenhoff, W.J., Webster, W.J.: Astron. Astrophys. **22** (1973) 251.
102 Wink, J.E., Altenhoff, W.J., Webster, W.J.: Astron. Astrophys. **38** (1975) 109.
102a Wollman, E.R., Geballe, T.R., Lacy, J.H., Townes, C.H., Rank, D.M.: Astrophys. J. Lett. **205** (1976) L5; **218** (1977) L103.
102b Wollman, E.R.: Int. Astron. Union Symp. **84** (1979), p. 367.
103 Wright, E.L., Fazio, G.G., Low, F.J.: Astrophys. J. Lett. **208** (1976) L87.
104 Wright, E.L., Fazio, G.G., Low, F.J.: Astrophys. J. **217** (1977) 724.
105 Wright, E.L., Lada, C.J., Fazio, G.G., Kleinmann, D.E.: Astron. J. **82** (1977) 132.
106 Wynn-Williams, C.G., Becklin, E.E., Neugebauer, G.: Mon. Not. R. Astron. Soc. **160** (1972) 1.
107 Wynn-Williams, C.G., Becklin, E.E., Neugebauer, G.: Astrophys. J. **187** (1974) 473.
108 Wynn-Williams, C.G., Becklin, E.E.: Publ. Astron. Soc. Pac. **86** (1974) 5.
109 Dubout-Crillon, R.: Astron. Astrophys. Suppl. **25** (1976) 25.
110 Kaler, J.B.: Astrophys. J. Suppl. **31** (1976) 517.
111 Georgelin, Y.M., Georgelin, Y.P., Sivan, J.-P.: Int. Astron. Union Symp. **84** (1979) 65.
112 Beck, S.C., Lacy, J.H., Geballe, T.R.: Astrophys. J. Lett. **234** (1979) L213.
113 Storey, J.W.V., Watson, D.M., Townes, C.H.: Astrophys. J. **233** (1979) 109.
114 McCarthy, J.F., Forrest, W.J., Houck, J.R.: Astrophys. J. **231** (1979) 711.
115 Moorwood, A.F.M., Baluteau, J.-P., Anderegg, M., Coron, N., Biraud, Y., Fitton, B.: Astrophys. J. **238** (1980) 565.
116 Cosmovici, C.B., Strafella, F., Dirscherl, R.: Astrophys. J. **236** (1980) 498.
117 Sivan, J.-P.: Astron. Astrophys. Suppl. **16** (1974) 163.
118 Courtés, G., Sivan, J.-P., Saisse, M.: Astron. Astrophys. (1981) in press.
119 Lyngå, G., Hansson, N.: Astron. Astrophys. Suppl. **6** (1972) 327.
120 Rodgers, A.W., Campbell, C.T., Whiteoak, J.B., Bailey, H.H., Hunt, V.O.: An Atlas of H-alpha Emission in the Southern Milky Way. Mt. Stromlo Obs., Canberra (1960).
121 Smith, J., Lynch, D.K., Cudaback, D., Werner, M.W.: Astrophys. J. **234** (1979) 902.
122 Grandi, S.A.: Astrophys. J. Lett. **199** (1975) L43.
123 Hippelein, H., Münch, G.: Astron. Astrophys. **68** (1978) L7.
124 Hippelein, H., Münch, G.: Astron. Astrophys. **95** (1981) 100.
125 Reynolds, R.J.: Astrophys. J. **236** (1979) 153.
126 Hawley, S.A.: Astrophys. J. **224** (1979) 417.
127 Talent, D.L., Dufour, R.J.: Astrophys. J. **233** (1979) 888.
128 Habing, H.J., Israel, F.P.: Compact H II Regions and OB Star Formation. Annu. Rev. Astron. Astrophys. **17** (1979) 345.

129 Neckel, Th., Harris, A.W., Eiroa, C.: Astron. Astrophys. **92** (1980) L9.
130 de Vries, J.S., van der Wal, P.B., Andriesse, C.D.: Astron. Astrophys. **86** (1980) 248.
131 Rank, D.M., Dinerstein, H.L., Lester, D.F., Bregman, J.D., Aitken, D.K., Jones, B.: Mon. Not. R. Astron. Soc. **185** (1978) 179.
132 Greenberg, L.T., Dyal, P., Geballe, T.R.: Astrophys. J. Lett. **213** (1977) L71.
133 Wynn-Williams, C.G., Becklin, E.E., Matthews, K., Neugebauer, G.: Mon. Not. R. Astron. Soc. **183** (1978) 237.
134 Carruthers, G.R., Heckathorn, H.M., Gull, T.R.: Astrophys. J. **237** (1980) 438.
135 Parker, R.A.R., Gull, T.R., Kirshner, R.B.: An emissionline survey of the Milky Way, NASA SP-434, Washington D.C. (1979).
136 Moorwood, A.F.M., Salinari, P., Furiss, I., Jennings, R.E., King, K.J.: Astron. Astrophys. **90** (1980) 304.
137 Pankonin, V., Walmsley, C.M., Harwit, M.: Astron. Astrophys. **75** (1979) 34.
138 Pankonin, V., Payne, H.E., Terzian, Y.: Astron. Astrophys. **75** (1979) 365.
139 Donati Falchi, A., Felli, M., Tofani, G.: Astron. Astrophys. **89** (1980) 363.
140 van Gorkom, J.H., Goss, W.M., Shaver, P.A., Schwarz, U.J., Harten, R.H.: Astron. Astrophys. **89** (1980) 150.
141 Downes, D., Wilson, T.L., Bieging, J., Wink, J.: Astron. Astrophys. Suppl. **40** (1980) 379.
141a Downes, D., Genzel, R.: Int. Astron. Union Symp. **87** (1980) 565.
142 Bohlin, R.C., Stecher, T.P.: Bull. Amer. Astron. Soc. **7** (1975) 547.
143 Cheung, L.H., Frogel, J.A., Gezari, D.Y., Hauser, M.G.: Astrophys. J. **240** (1980) 74.
144 Russell, R.W., Melnick, G., Gull, G.E., Harwit, M.: Astrophys. J. Lett. **240** (1980) L99.
145 Werner, M.W., Becklin, E.E., Gatley, I., Neugebauer, G., Sellgren, K., Thronson, H.A., Harper, D.A., Loewenstein, R., Moseley, S.H.: Astrophys. J. **242** (1980) 601.
146 Martin, A.H.M., Gull, S.F.: Mon. Not. R. Astron. Soc. **175** (1976) 235.
147 Jaffe, D.T., Pankonin, V.: Astrophys. J. **226** (1978) 869.
148 Turner, D.G.: Astrophys. J. **210** (1976) 65.
149 Genzel, R., et al.: Astrophys. J. **244** (1981) 884.

# 7.4 Physics of interstellar dust

## 7.4.1 Optical properties of the grains

### 7.4.1.1 Definitions

Total extinction cross-section, $C_{ext}$, (depending on wavelength):

$$C_{ext}=C_{sca}+C_{abs}=\frac{\text{total radiant energy scattered and absorbed per unit time}}{\text{incident radiant energy per unit area per unit time}}$$

$Q$ = efficiency factor
$G$ = geometrical cross-section
$C=G\cdot Q$.

Extinction coefficient expressed in magnitudes per unit length (for a particular direction of polarization or for all directions of vibration):

$$a_\lambda=1.086\,G\cdot Q_{ext}\cdot N_d$$

$N_d$ = number density of the dust particles [$cm^{-3}$].

Albedo of the grain

$$\gamma=\frac{C_{sca}}{C_{ext}}=\frac{Q_{sca}}{Q_{ext}}=\frac{Q_{sca}}{Q_{sca}+Q_{abs}}$$

Asymmetry factor

$$g=\langle\cos\theta\rangle\,,$$

where $\theta$ denotes the scattering angle; it gives the degree of scattering in the forward direction $\theta=0$.

Force exerted by radiation pressure ($I_0$ = incident intensity):

$$K_{rad}=\frac{I_0}{c}G\cdot Q_{pr}\quad\text{with}\quad Q_{pr}=Q_{abs}+(1-g)Q_{sca}$$

### 7.4.1.2 Efficiency factors

$Q_{sca}$, $Q_{abs}$, and $Q_{ext}$ of a dust particle depend on its size, shape and complex index of refraction $m=m'-im''$. Exact solutions of the scattering problem are only available for concentric spheres (Mie theory) and concentric infinite circular cylinders. These have been applied numerically to a wide variety of indices of refraction. Results for spheroids have been derived on the basis of approximate theory. Scattering efficiencies for many types of non-spherical particles, including e.g. particles with rough surfaces, are available from experiments using the microwave-analogue method.

Table 1. Numerical calculations and analogous measurements of scattering efficiencies.

| Shape of particle | Method | Ref. |
|---|---|---|
| Homogeneous sphere of arbitrary size | Mie theory | 31, 8, 35 |
| Infinite cylinders (needles, whiskers) | theory | 19, 10, 35 |
| Prolate and oblate spheroids | approximate theory<br>microwave analogy | 10, 11, 41 |
| Composite spherical (core-mantle) grains | theory | 32, 10, 35 |
| Dielectric rough spheres, cubes, octahedra and irregular particles | microwave analogy | 39, 40, 44 |

### 7.4.1.3 Albedo and asymmetry factor derived from observations

The albedo of the interstellar dust particles is relatively high in the visible and UV, but has a minimum near $\lambda=2200$ Å. The angular distribution of scattered light seems to change gradually from a strongly forward-scattering form at visible wavelengths to a nearly isotropic form in the far UV.

Table 2. Estimates of albedo $\gamma$ and asymmetry factor $g$ of interstellar grains for visible and ultraviolet wavelengths from analysis of measurements of the diffuse galactic light, reflection nebulae and dark nebulae [2, 6, 18, 21, 22, 38].

| $\lambda$[Å] | $\gamma$ | $g$ |
|---|---|---|
| 1550 | 0.6 | 0.25 |
| 1900 | 0.4 | 0.35 |
| 2200 | 0.35 | 0.45 |
| 2400 | 0.4 | 0.5 |
| 3000 | 0.6 | 0.6 |
| 3300 | 0.7 | 0.6 |
| 4250 | 0.7 | 0.7 |
| 5500 | 0.7 | 0.75 |

### 7.4.1.4 Grain models

Table 3. Materials suggested as constituents of interstellar grains in recent works and summary of their optical properties characterized by the complex index of refraction: $m = m' - im''$.

| Material | Outline of optical properties | Ref. for detailed data |
|---|---|---|
| Ices: $H_2O$ pure or mixed with $CH_4$, $NH_3$ ("dirty ice") | $\lambda = 0.3 \cdots 1$ µm: approximately constant $m' \approx 1.33 \cdots 1.5$; $m'' \lessapprox 0.05$<br>$\lambda < 0.3$ µm: $m'' > 0.1$, absorption edges<br>$\lambda > 1$ µm: absorption at 3.1 µm and between 10 and 20 µm (peak at $\approx 12$ µm); $m'' \propto \lambda^{-1}$ in far IR | 10, 3, 12, 13 |
| Combination of molecules of O, C, N, and H with free radicals | — | 13 |
| Polyoxymethylene (polymerized formaldehyde) | absorption feature near $\lambda \approx 10$ µm | 36 |
| Carbon: graphite | $\lambda \gtrapprox 0.4$ µm: $\varepsilon = m^2 \approx 4 - 10i\lambda$; absorption peak between 0.2 and 0.3 µm.<br>IR: imaginary part decreases with increasing temperature | 27, 10, 29, 34, 20, 12, 13 |
| Silicates: enstatite (Fe, Mg)$SiO_3$ olivine $(Fe, Mg)_2SiO_4$ silicon dioxide $SiO_2$ hydrated silicates (as found in carbonaceous chondrite meteorites) | Visible region: approximately constant $m' \approx 1.66$; $m'' \approx 0.05$.<br>$\lambda < 0.3$ µm: substantial absorptivity.<br>IR: strong absorption at $\lambda \approx 10$ µm and at $\lambda \approx 20$ µm | 4, 14, 15, 24, 11, 26, 16, 12, 1 |
| Silicon carbide SiC | Visible region: approximately constant $m' \approx 2.6$; $m'' \approx 0$ | 28, 23, 7 |
| Magnetite $Fe_3O_4$ (as present in carbonaceous meteorites) | $\lambda \approx 0.3 \cdots 1$ µm: $m' \approx 2.5$; variable $m'' \approx 0.4 \cdots 0.9$ | 15 $\cdots$ 17 |
| Iron | Visible region: $m' \approx 3$; strong variation from UV to IR of $m'$ and $m''$; temperature dependence | 10, 34 |

Table 4. Constraints on dust models provided by the cosmic elemental abundances [13].
The entries at each particular material give the relative atomic abundances required for spherical and spheroidal (in parentheses, for non-sphericity of 5:1 or 1:5) grain models which can closely approximate the observed extinction curve by using that ingredient alone.
Assumed column density of hydrogen:
$\mathfrak{N}_H = 2.34 \cdot 10^{21}\, A_V$ with $\mathfrak{N}_H$ in [atoms cm$^{-2}$],
$A_V$ = visual absorption (see 7.4.1.5) in [mag]

| Material | Relative abundance $\times 10^4$ | | | | |
|---|---|---|---|---|---|
| | O/H | Si/H | (Mg, Fe)/H | Fe/H | C/H |
| Modified ice | 1.43<br>(1.62) | | | | 0.78<br>(0.88) |
| Graphite | | | | | 2.22<br>(2.10) |
| Enstatite | 1.95<br>(2.34) | 0.65<br>(0.78) | 0.65<br>(0.78) | | |
| Olivine | 1.84<br>(2.21) | 0.92<br>(1.10) | 1.84<br>(2.21) | | |
| Silicon carbide | | 0.89<br>(1.19) | | | 0.89<br>(1.19) |
| Magnetite | 1.25<br>(1.38) | | | 0.94<br>(1.03) | |
| Iron | | | | 1.07<br>(1.61) | |
| Cosmic abundances | 6.76 | 0.32 | 0.34 | 0.26 | 3.70 |

Table 5. Recent examples of grain models which can explain the main features of the observed interstellar extinction curves.
$a$ = radius of spherical grains
$a_c$ = core radius
$a_m$ = mantle radius
$\varepsilon$ = cylinder elongation = length/$2a_m$
$N_H$ = number density of H atoms [cm$^{-3}$]
$N_d$ = number density of grains [cm$^{-3}$]

| Material of components | Grain size [μm] | Size distribution | Relative contribution (number of grains) | Ref. |
|---|---|---|---|---|
| Spherical grains of | | | | 9, 42 |
| (1) graphite | $\bar{a}=0.020$ | each | 0.72 | |
| (2) meteoritic silicate | $\bar{a}=0.045$ | $N_d(a) \propto \exp\left[-0.69\left(\frac{a}{\bar{a}}\right)^{2.6}\right]$ | 0.25 | |
| (3) silicon carbide | $\bar{a}=0.075$ | | 0.03 | |
| Spherical grains of | | | | 20 |
| (1) graphite | 0.005⋯1.0 | $N_d(a) \propto a^{-3.5}$ | ≈0.5 | |
| (2) silicate (olivine or enstatite) | 0.025⋯0.25 | | ≈0.5 | |
| (1) cylindrical core-mantle grains with circular cross section | | single core size $a_c$; $N_d(a_m) \propto \exp\left[-5\,\frac{a_m - a_c}{a_i}\right]$ | $\approx 2 \cdot 10^{-4} \cdot \varepsilon^{-1}$ | 12 |
| core: silicate | $a_c=0.05$ | with $a_i=0.2$ μm | $(N_d \approx 10^{-12}\, N_H\, \varepsilon^{-1})$ | |
| mantle: modified ice | $\bar{a}_m=0.15$ | | | |
| (2) very small uncoated spherical silicate grains | $\bar{a}=0.005$ | single size | ≈1 $(N_d \approx 5 \cdot 10^{-9}\, N_H\, \varepsilon^{-1})$ | |

The interpretation of the observed wavelength-dependence of interstellar extinction – (1) in the visible and infrared, (2) at the UV peak (2200 Å) and (3) in the far UV (rise towards 1000 Å) – seems to require at least three different types of particles. Variations of the extinction curve from region to region in space might be the result of differences in the frequency distributions over various sizes and compositions of the grains. Theories of grain evolution lead to the expectation that part of the dust particles are coated with mantles of light-weight volatile substances of low index of refraction.

The observed wavelength-dependence of interstellar linear polarization requires non-spherical particles, e.g. cylinders or prolate spheroids, with (smallest) diameters comparable to the size values deduced from the extinction curve. Variation of the wavelength of maximum polarization, $\lambda_{max}$, from region to region again suggests spatial differences in the frequency distribution of grain sizes. Observations of circular polarization show a reversal of sign near 6000 Å which can only be interpreted by the assumption of mainly dielectric materials (with small imaginary part of the refractive index) for the "visual grain" component [41, 42, 43].

### 7.4.1.5 Extinction coefficient per unit mass and total amount of dust; dust-to-gas ratio

Symbols (see also 7.4.1.1)

$N_d$ number density of grains [$cm^{-3}$]
$\varrho_d$ mass density of the dust [$g\,cm^{-3}$]
$\varrho_g$ mass density of the gas [$g\,cm^{-3}$]
$V_d$ volume of a grain [$cm^3$]
$\varrho_s$ mass density of solid material within the grain [$g\,cm^{-3}$]
$k_\lambda^*$ extinction coefficient per unit mass of dust [$mag\,cm^2\,g^{-1}$]

The total extinction over a distance $r$ at the wavelength $\lambda$ by grains of a single type can be written

$$A(\lambda)=1.086\,C_{ext}(\lambda)\cdot N_d\cdot r=1.086\,\frac{C_{ext}(\lambda)}{V_d}\,\frac{\varrho_d}{\varrho_s}\cdot r$$

$$=1.086\,k_\lambda\cdot\varrho_d\cdot r=k_\lambda^*\cdot\varrho_d\cdot r$$

Values of the extinction coefficient $k_\lambda^*$ follow immediately from calculations of the "volume-extinction factor" $V_d/C_{ext}$ [13]. For different grain materials one obtains in the visual region

$$k_V^*=(3\cdots10)\cdot10^4\ \mathrm{mag\,cm^2\,g^{-1}}.$$

For the three-component model of Gilra [9] (see Table 5) the result is

$$k_V^*=4\cdot10^4\ \mathrm{mag\,cm^2\,g^{-1}}.$$

Using this value and $A_V/r=2.0$ mag kpc$^{-1}$ we obtain for the mean total mass density of the dust

$$\bar{\varrho}_d=\frac{A_V}{r}\cdot\frac{1}{k_V^*}=1.6\cdot10^{-26}\ \mathrm{g\,cm^{-3}}.$$

Estimating $C_{ext}(\lambda)/V_d$ from the Kramers-Kronig relationship (and using $A_V/r=1.8$ mag kpc$^{-1}$) Spitzer [26] finds

$$\bar{\varrho}_d=1.8\cdot10^{-26}\ \mathrm{g\,cm^{-3}}.$$

Dust-to-gas ratio: From the relation between column density $\mathfrak{N}$(H) and $E_{B-V}$ given in 7.1.2.1 one obtains with $A_V/E_{B-V}=3.1$ and $A_V=2.0$ mag kpc$^{-1}$ for a H/He ratio of 10 the mean gas density $\bar{\varrho}_g=2.9\cdot10^{-24}$ g cm$^{-3}$. This, with the value of $\bar{\varrho}_d$ given above, yields

$$\frac{\bar{\varrho}_d}{\bar{\varrho}_g}=0.6\cdot10^{-2}.$$

Dust opacity in the Lyman continuum [23]:
Total value: $\bar{\varkappa}\,(\lambda<912\,\text{Å})=5.7\cdot10^{-22}\,N_H\,\mathrm{cm^{-1}}$,
H-ionizing continuum ($504<\lambda<912$ Å): $\varkappa_H=3.5\cdot10^{-22}\,N_H\,\mathrm{cm^{-1}}$,
He-ionizing continuum ($\lambda<504$ Å): $\varkappa_{He}=1.5\cdot10^{-21}\,N_H\,\mathrm{cm^{-1}}$.
Total extinction coefficient per unit mass of interstellar hydrogen for $\lambda<912$ Å:

$$\bar{k}=\bar{\varkappa}/m_H=3.4\cdot10^2\ \mathrm{cm^2\,g^{-1}}$$

## 7.4.1.6 References for 7.4.1


1 Aannestad, P.A., Purcell, E.M.: Interstellar Grains, Annu. Rev. Astron. Astrophys. **11** (1973) 309.
2 Andriesse, C.D., Piersam, T.R., Witt, A.N.: Astron. Astrophys. **54** (1977) 841.
3 Bertie, J.E., Labbé, H.A., Whally, E.: J. Chem. Phys. **50** (1969) 4501.
4 Dorschner, J.: Astron. Nachr. **290** (1967) 171.
5 Field, G., Cameron, A. (eds.): The Dusty Universe, Neale Watson, New York (1975).
6 FitzGerald, M.P., Stephens, T.C., Witt, A.N.: Astrophys. J. **208** (1976) 709.
7 Friedemann, Chr.: Astron. Nachr. **291** (1969) 177.
8 Giese, R.H.: Z. Astrophys. **51** (1961) 119.
9 Gilra, D.P.: Nature **229** (1971) 237.
10 Greenberg, J.M.: Interstellar Grains, in: Stars and Stellar Systems VII (1968) Chap. 6, p. 221.
11 Greenberg, J.M. in: Planets, Stars and Nebulae (Gehrels, T., ed.), Univ. of Arizona Press, Tucson (1974) p. 107.
12 Greenberg, J.M. in: Infrared Astronomy (Setti, G., Fazio, G.G., eds.), Reidel, Dordrecht (1978) p. 51.
13 Greenberg, J.M. in: Cosmic Dust (McDonnell, J.A.M., ed.), Wiley, New York (1978) p. 187.
14 Huffman, D.R., Stapp, J.L.: Nature Phys. Sci. **229** (1971) 45.
15 Huffman, D.R., Stapp, J.L. in: Int. Astron. Union Symp. **52** (Interstellar Dust and Related Topics) (1973) p. 297.
16 Huffman, D.R.: Adv. Phys. **26** (1977) 129.
17 Knacke, R.F. in: Protostars and Planets (Gehrels, T., ed.), Univ. of Arizona Press, Tucson (1978) p. 112.
18 Lillie, C.F., Witt, A.N.: Astrophys. J. **208** (1976) 64.
19 Lind, A.C., Greenberg, J.M.: J. Appl. Phys. **37** (1966) 3195.
20 Mathis, J.S., Rumpl, W., Nordsieck, K.H.: Astrophys. J. **217** (1977) 425.
21 Mattila, K.: Astron. Astrophys. **9** (1970) 53.
22 Mattila, K.: Astron. Astrophys. **15** (1971) 292.
23 Panagia, N. in: Infrared and Submillimeter Astronomy (Fazio, G.G., ed.), Reidel, Dordrecht (1977) p. 43.
24 Philipp, H.R., Taft, E.A. in: Conference on Silicon Carbide, Pergamon Press, Boston (1960) p. 366.
25 Pollack, J.B., Toon, C.B., Khare, B.N.: Icarus **19** (1973) 372.
26 Spitzer, L.: Physical Processes in the Interstellar Medium, Wiley, New York (1978) p. 162.
27 Steyer, T.R., Day, K.L., Huffman, D.R.: Appl. Opt. **13** (1974) 1586.
28 Taft, E.A., Philipp, H.R.: Phys. Rev. **138** (1965) A 197.
29 Thibault, N.W.: American Mineral. **29** (1944) 327.
30 Tosatti, E., Bassani, F.: Nuovo Cimento **65** B (1970) 161.
31 Van de Hulst, H.C.: Light Scattering by Small Particles, Wiley, New York (1957).
32 Wickramasinghe, N.C.: Interstellar Grains, Chapman and Hall, London (1967).
33 Wickramasinghe, N.C., Nandy, K.: Mon. Not. R. Astron. Soc. **153** (1971) 205.
34 Wickramasinghe, N.C.: Interstellar Dust, in: Interstellar Matter (Wickramasinghe, N.C., Khan, F.D., Mezger, P.G., eds.), Swiss Society of Astronomy and Astrophysics, Second Advanced Course, Saas-Fee (1972) p. 209.
35 Wickramasinghe, N.C.: Light Scattering Functions for Small Particles, Hilger, London (1973).
36 Wickramasinghe, N.C.: Mon. Not. R. Astron. Soc. **170** (1975) 11 P.
37 Wickramasinghe, N.C., Morgan, D.J. (eds.): Solid State Astrophysics, Reidel, Dordrecht (1976).
38 Witt, A.N.: Publ. Astron. Soc. Pacific **89** (1977) 750.
39 Zerull, R., Giese, R.H. in: Planets, Stars and Nebulae (Gehrels, T., ed.), Univ. of Arizona Press, Tucson (1974) p. 901.
40 Zerull, R. in: Interplanetary Dust and Zodiacal Light (Elsässer, H., Fechtig, H., eds.), Lecture Notes in Physics, Springer, Berlin **48** (1976) p. 130.
41 Rogers, C., Martin, P.G.: Astrophys. J. **228** (1979) 450.
42 Mathis, J.S.: Astrophys. J. **232** (1979) 747.
43 Savage, B.D., Mathis, J.S.: Annu. Rev. Astron. Astrophys. **17** (1979) 73.
44 Schuerman, D. (ed.): Light Scattering by Irregularly Shaped Particles, Plenum Press, New York and London (1980).

## 7.4.2 Grain temperatures

### 7.4.2.1 Introduction

Symbols

| | |
|---|---|
| $B(\lambda, T)$ | Planck distribution per unit wavelength at the temperature $T$ |
| $I(\lambda)$ | intensity per unit wavelength of the environmental radiation field |
| $Q_{abs}(\lambda)$ | absorption efficiency factor (see 7.4.1.1) |
| $T_d$ | temperature of the dust grain |

In most cases, the temperature of the dust particles is entirely determined by the balance between absorbed and emitted radiative energy. Atom collisions and molecule formation usually make a negligible contribution (see 7.4.1.6: [12, 13, 26]). Applying the Kirchhoff relation between emissivity and absorptivity, the steady-state equation of equilibrium for spherical grains can be written

$$\int_0^\infty Q_{abs}(\lambda)\, I(\lambda)\, d\lambda = \int_0^\infty Q_{abs}(\lambda)\, B(\lambda, T_d)\, d\lambda\,.$$

$Q_{abs}(\lambda)$ depends on the radius of the dust particle. For non-spherical grains one may use absorption efficiency factors averaged over particle orientation.

In a very small grain with small low-temperature specific heat, the heat content may rise substantially with the absorption of a single photon and fall to some much lower value before the next photon is absorbed. Consequently, the grain temperature will fluctuate rapidly with time – if the photon density is relatively low. A single grain temperature is then not definable. This case may occur in tenuous and medium-density clouds. In H II regions the photon density is generally sufficiently high to reduce these fluctuations substantially (7.4.1.6: [12, 13]; for critical discussion, see [3]).

### 7.4.2.2 H I regions and cool dense clouds

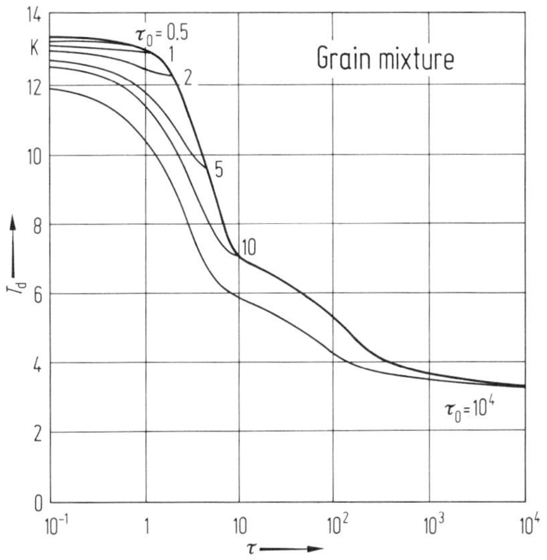


Fig. 1. Average grain temperature at optical depth $\tau$ within a typical dark cloud with a centrally condensed density distribution. Central optical depth $\tau_0$ at $\lambda = 5470$ Å is indicated by the number. Grain mixture model [6]:

| Grain | Radius | Fractional number density |
|---|---|---|
| SiC | $\bar{a} = 0.07$ μm | $9 \cdot 10^{-3}$ |
| graphite | $\bar{a} = 0.02$ μm, | $9 \cdot 10^{-2}$ |
| ice | $\bar{a} = 0.2$ μm, | $9 \cdot 10^{-4}$ |
| silicate-ice | $\left\{ \begin{array}{l} \bar{a}_c = 0.05\ \mu m \\ \bar{a}_m = 0.08\ \mu m \end{array} \right\}$ | $9 \cdot 10^{-1}$ |

For further explanations, see Fig. 2.

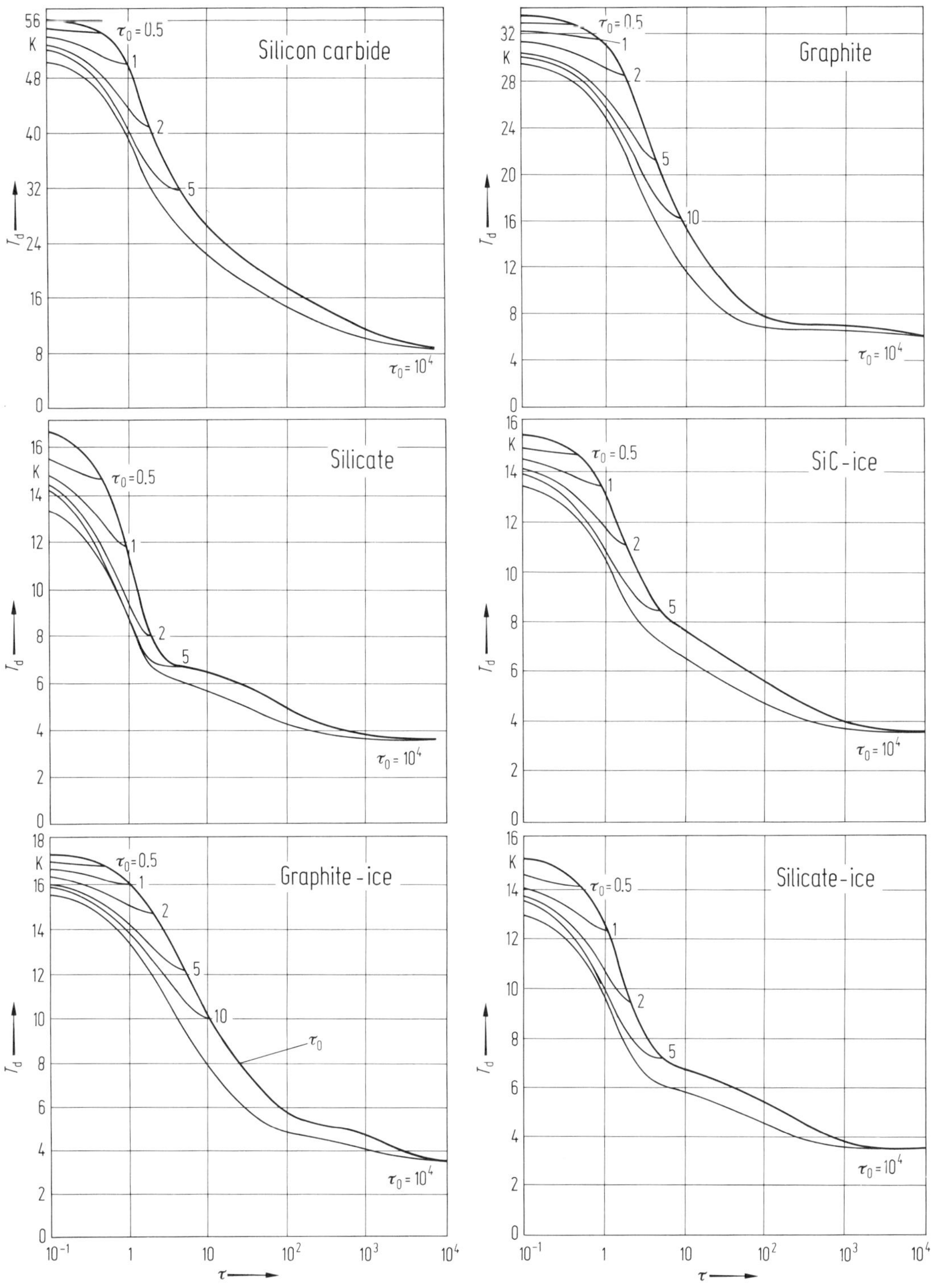
Silicon carbide
Graphite
Silicate
SiC-ice
Graphite-ice
Silicate-ice
$T_d$
K
$\tau$
$\tau_0 = 0.5$
$\tau_0 = 10^4$
$\tau_0$

Fig. 2.

Table 6. Grain temperatures, $T_d$, in H I regions. The radiation fields used are labeled by $I_1$, $I_2$, and $I_3$:
$I_1$ Planck function for $T=10000$ K with dilution factor $W=10^{-14}$
$I_2$ sum of two Planck functions for $T=12000$ and 5000 K combined with $W=4.3\cdot10^{-16}$ and $1.5\cdot10^{-13}$
$I_3$ sum of three Planck functions for $T=14500$, 7500, and 4000 K with $W=4\cdot10^{-16}$, $1.5\cdot10^{-14}$, and $1.5\cdot10^{-13}$ together with the undiluted 2.8 K background radiation, but with $\lambda<4250$ Å (observed values of Witt and Johnson, see 7.1.8)
The results given in the last column ($I_3$) are based on the distribution of grain radii $a$:

$$N_d(a)=Aa^{3/2}\exp\left[-\frac{1}{2}\left(\frac{a}{\bar{a}}\right)^3\right]$$

with $N_d$ in [cm$^{-3}$], $a$ in [μm]; the left-hand given values of the size of maximum frequency, $\bar{a}$, $\bar{a}_m$, and $\bar{a}_c$ for grain, mantle and core, respectively, keeping $a_m/a_c=$const.

| Radiation field | | $I_1$ [4] | $I_2$ [4] | $I_3$[6] |
|---|---|---|---|---|
| Material | Radius [μm] | | $T_d$[K] | |
| Grain type: | | | | |
| Black body | | 3.2 | 3.2 | — |
| Dirty ice spheres | $a=0.1$ | 14 | 13 | — |
| | $a=1.0$ | 13 | 12 | — |
| | $a=10$ | 8.2 | 8.0 | — |
| Dirty ice cylinders | $a=0.1$ | 15.5 | — | — |
| Graphite spheres | $a=0.05$ | 35 | 33 | 33.7 |
| Graphite spheres ice coated | $\bar{a}_c=0.05$, $\bar{a}_m=0.15$ | — | — | 17.3 |
| Silicate spheres | $\bar{a}=0.05$ | 9 | — | 16.9 |
| Silicate spheres ice coated | $\bar{a}_c=0.05$, $\bar{a}_m=0.15$ | — | — | 15.3 |
| SiC spheres | $\bar{a}=0.05$ | — | — | 56.7 |
| SiC spheres ice coated | $\bar{a}_c=0.05$, $\bar{a}_m=0.15$ | — | — | 15.6 |

Temperatures of silicate grains in dust clouds based on experimentally determined refractive indices for olivine and lunar basalt have been calculated by S. Aiello et al. [2].

◄

Fig. 2. Grain temperature, $T_d$, at optical depth $\tau$ within a spherical dust cloud of uniform density with central optical depth $\tau_0$ (at $\lambda$ 5470 Å), indicated by the number, calculated for different spherical grain models with the size distribution function

$$N_d(a)=Aa^{3/2}\exp\left[-\frac{1}{2}\left(\frac{a}{\bar{a}}\right)^3\right],$$

where $\bar{a}$ is the size of maximum frequency: pure grains of SiC, graphite and silicate with a characteristic radius $\bar{a}=0.05$ μm; ice-coated core-mantle grains of SiC, graphite and silicate with core radius $\bar{a}_c=0.05$ μm and mantle radius $\bar{a}_m=0.15$ μm. These grain temperatures are solely determined by the balance between radiation absorbed and radiation emitted, and it is assumed that the grains scatter almost isotropically.
Heavy line: temperature at the center of a cloud with optical depth $\tau_0$ [6].

## 7.4.2.3 H II regions

Sources of energy for heating the grains in H II regions are: (1) stellar radiation, mainly Lyman continuum photons with $\lambda < 912$ Å, and (2) strong diffuse Ly $\alpha$ radiation. Source (1) dominates very near to the exciting star; (2) is the most important source outside the central part of the H II region. In compact H II regions, stellar heating is important throughout.

Table 7. Temperatures of dust grains, $T_d$, near hot stars resulting from calculations using several simplifications. The stellar radiation is characterized by a colour temperature $T_c$ and is reduced only by the distance $r$ from the star: dilution factor $W = R_*^2/4r^2$, where $R_*$ is the stellar radius. In the case of Ly $\alpha$ heating $T_e$ and $N_H$ denote the electron temperature and the hydrogen density, respectively [5, 8].

| | | Stellar heating | | | | | | | | Ly $\alpha$ heating | |
|---|---|---|---|---|---|---|---|---|---|---|---|
| $T_c$ [K] | | 28200 | | 35500 | | 42400 | | 55300 | | $T_e = 8000$ K | |
| $R_*$ [$R_\odot$] | | 3.5 | | 4.4 | | 6.0 | | 11.8 | | | |
| | | | | | | | | | | $N_H = 10$ cm$^{-3}$ | $10^3$ cm$^{-3}$ |
| $r$ [pc] | | 0.1 | 0.5 | 0.1 | 0.5 | 0.1 | 0.5 | 0.1 | 0.5 | $\gtrsim 1$ | |
| Material | Radius [μm] | $T_d$[K] | | | | | | | | | |
| Grain type | | | | | | | | | | | |
| Ice [1]) | $a = 0.10$ | 50 | — | 70 | 40 | 100 | 50 | 200 | 80 | 20 | 45 |
| Graphite [2]) | $a = 0.05$ | — | — | — | — | — | — | 220 | 120 | 24 | 55 |
| Graphite core–ice mantle | $a_c = 0.05$, $a_m = 0.15$ | 50 | — | 65 | 30 | 80 | 50 | 170 | 70 | — | — |

[1]) Evaporation for $T_d > 100$ K.
[2]) Evaporation for $T_d > 1350$ K.

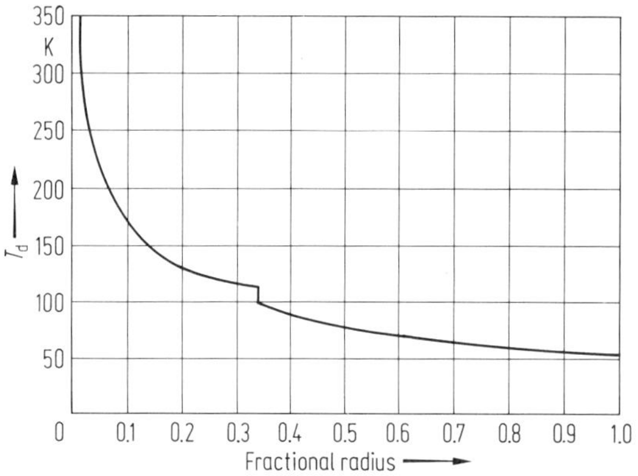


Fig. 3. Temperature variation with distance from the exciting star for olivine core – ice mantle grains in a spherical model of Orion A with uniform density $N_H = 10^3$ cm$^{-3}$ and total effective dust optical depth in the Lyman continuum $\tau_d = 1$. Grain core radius $a_c = 0.05$ μm, mantle radius $a_m = 0.15$ μm. In the inner region, where $T_d > 100$ K, ice mantles are evaporated while the olivine cores persist up to $T_d \approx 1000$ K. The distance from the center is expressed in units of the Strömgren radius $r_S = 0.6$ pc [1].

The distribution of grain temperature in molecular clouds with embedded O–B stars has been calculated by Leung [7].
Thermal emission from grains: [9] and 7.4.1.6: [13, 43]. Photoelectric yield of grains: [10, 11].

## 7.4.2.4 References for 7.4.2


1 Aannestad, P.A. in: Far Infrared Astronomy (Rowan-Robinson, M., ed.), Pergamon Press, Oxford (1976) p. 257.
2 Aiello, S., Mencaraglia, F., Blanco, A., Borghesi, A., Bussoletti, E.: Mon. Not. R. Astron. Soc. **180** (1977) 323.
3 Drapatz, S., Michel, K.W.: Astron. Astrophys. **56** (1977) 353.
4 Greenberg, J.M.: Astron. Astrophys. **12** (1971) 240.

5 Isobe, S.: Ann. Tokyo Astron. Obs., Second Ser., XII, No. 4 (1971) 286.
6 Leung, C.M.: Astrophys. J. **199** (1975) 340.
7 Leung, C.M.: Astrophys. J. **209** (1976) 75.
8 Spitzer, L.: Physical Processes in the Interstellar Medium, Wiley, New York (1978) p. 195.
9 Andriesse, C.D.: Vistas Astron. **21** (1977) 107.
10 Spitzer, L.: Physical Processes in the Interstellar Medium, Wiley, New York (1978) p. 145.
11 Gail, H.-P., Sedlmayr, E.: Astron. Astrophys. **86** (1980) 380.

# 7.5 Physics of the interstellar gas

## 7.5.1 List of symbols and definitions

| | |
|---|---|
| $h$ | Planck constant |
| $k$ | Boltzmann gas constant |
| $T$ | gas temperature [K] |
| $T_e$ | electron temperature [K] |
| $T_{exp}$ | experimental temperature [K] |
| $T_m$ | temperature divided by $10^m$ K, i.e. a dimensionless quantity |
| $Z$ | number of net positive charges on ion; $Z_0$: of nuclear charge |
| $\alpha_x$ | rate coefficient [$cm^3\,s^{-1}$], defined by: number of reactions (of type x) per $cm^3$ and sec $=\alpha_x N_1 N_2$, with |
| $N_i$ | number density of reaction partner i [$cm^{-3}$] |
| $N$ | particle density [$cm^{-3}$] |
| $\alpha_{rad}(i, T_e)$ | radiative recombination rate coefficient [$cm^3\,s^{-1}$] of recombining ionic species i at electron temperature $T_e$ |
| $\nu$ | frequency [$s^{-1}$] |

## 7.5.2 Components of the interstellar gas

### 7.5.2.1 Atoms and atomic ions

Energy levels [16, 159]; Grotrian diagrams [16, 129]; ionization potentials [3, 129]; for transition probabilities, see 4.3.4.4.3 and [g, 3], cf. also 7.5.3.7.

Summary of wavelengths and transition probabilities:

visual and UV interstellar absorption lines [161], updated (1978) [160]; lines between 1 and 250 Å [124, 192a], between 100 and 1000 Å [218a].

Particular recent results: for C I [53], C II [170b], C III [168], N I [143], O I [37, 188], Si II [166], S II [45], Fe I [23], Fe II [184], Fe III and VI [78, 169], N IV, O V, F V, F VI, and Ne VII [170]; for the Be and Mg isoelectronic sequences [137], for the B isoelectronic sequence [52], the C isoelectronic sequence [167]. Intercombination lines of C II, III, N III, IV, O IV, V referenced in [9].

Vacuum wavelengths of the fine-structure transitions in the ground terms are given in Table 1 (after [94] except for C I [64], C II [97a], S III [159a], Sc I···III, Ti I···IV, V I···V, and Fe I···VII [16, 78, 159, 169]).

For Table 1, see next page.

### 7.5.2.2 Molecules

Molecular parameters for all molecules detected up to April 1979 in interstellar space by microwave techniques (see Table 6 in 7.2) compiled and extensively referenced in [246].

Recommended rest frequencies with proper molecular and quantum mechanical labels of all known interstellar microwave transitions of these molecules plus NO (for NO see also [139]), $HC_4$ [95], and $HC_9N$ [25]: ordered and referenced in [142].

Table 1. Fine-structure transitions in the ground term, in [μm] [16, 64, 78, 94, 97a, 159, 159a, 169], see text of 7.5.2.1.
Notes: Transitions ordered by ascending energy of emitting level.
Transitions for B I and Al I isoelectronic sequences are $^2P_{3/2} \rightarrow {}^2P_{1/2}$, for C I and Si I sequences $^3P_1 \rightarrow {}^3P_0$, $^3P_2 \rightarrow {}^3P_1$, for O I and S I sequences $^3P_1 \rightarrow {}^3P_2$, $^3P_0 \rightarrow {}^3P_1$, for F I and Cl I sequences $^2P_{1/2} \rightarrow {}^2P_{3/2}$.
Last two lines: transition data and transitions for Fe; read $(^5D)3 \rightarrow 4$ as $^5D_3 \rightarrow {}^5D_4$ etc.
The F I value is the F 2→1 hyperfine transition component [94].

| Element | Ionization stage I | Ionization stage II | Ionization stage III |
|---|---|---|---|
| | ←(a)— | —(a)→ ←(b)→ | ←(c)— |
| B | 655.57 | none | none |
| C | 609.1333 ± 0.0008; 370.414 ± 0.0015 | 157.4 ± 0.2 | none |
| N | none | 203.5 ± 0.8; 121.7 ± 0.27 | 57.343 ± 0.003 |
| O | 63.185 ± 0.004; 145.6 ± 0.1 *) | none | 88.355 ± 0.008; 51.816 ± 0.003 |
| F | 24.745 ± 0.001 | 29.33; 67.16 | none |
| Ne | none | 12.8136 ± 0.0002 | 15.55; 36.02 |
| Na | none | none | 7.319 |
| Mg | none | none | none |
| Al | 89.237 ± 0.0025 | none | none |
| Si | 129.68; 68.473 | 34.814 | none |
| P | none | 60.68; 32.87 | 17.885 ± 0.007 |
| S | 25.2490 ± 0.0003; 56.309 ± 0.01 | none | 33.456 ± 0.001; 18.7129 ± 0.0004 |
| Cl | 11.333345 ± 0.000015 | 14.368; 33.281 | none |
| Ar | none | 6.9853 ± 0.0002 | 8.9910 ± 0.0002; 21.84 ± 0.1 |
| K | none | none | 4.617 ± 0.002 |
| Ca | none | none | none |
| Sc | 59.40 | 90.95; 147.8 | 50.60 |
| Ti | 58.78; 46.14 | 106.5; 76.03; 59.61 | 54.44; 41.98 |
| V | 72.79; 53.75; 43.55 | 277.4; 141.7; 97.79; 76.73 | 69; 51.5; 41.0 |
| Fe | 24.04; 34.71; 54.31; 111.18 | 25.99; 35.352; 51.28; 87.41 | 22.93; 33.04; 51.68; 105.4 |
| $J \rightarrow J'$ | $(^5D)3 \rightarrow 4$, $2 \rightarrow 3$, $1 \rightarrow 2$, $0 \rightarrow 1$ | $(^6D)7/2 \rightarrow 9/2$, $5/2 \rightarrow 7/2$, $3/2 \rightarrow 5/2$, $1/2 \rightarrow 3/2$ | $(^5D)3 \rightarrow 4$, $2 \rightarrow 3$, $1 \rightarrow 2$, $0 \rightarrow 1$ |

*) New values, added in proof: 63.17000 ± 0.00003; 145.52547 ± 0.000085 [249].

Measured values of electric dipole moments and some wavelengths of simple molecules compiled in [129].

Radiative lifetimes of excited electronic states in some molecular ions [144a].

Review (1977) of calculated and measured transition probabilities of diatomic molecules, and references to some small polyatomic molecule data [165]; compilation of transition frequencies and molecular constants for diatomics [194a].

Vibration-rotation transition probabilities of ground electronic state of $H_2$ [228], high rotational quadrupole transitions of $H_2$ referenced in [124b].

Additional physical data on molecules of possible interstellar importance:

| | | | | | |
|---|---|---|---|---|---|
| diatomic molecules | [104, 109] | NCO | [189] | $NH_2D$ | [229] |
| OH | [18] | HNC, $HN^{13}C$ | [77] | HDCO | [130] |
| SiH, $SiH^+$ | [34a, 211] | HNSi, HSiN | [125] | $C_2NC$, $HC_2NC$ | [243] |
| $CH^+$ | [63] | $HSiO^+$, $HOSi^+$ | [244] | $H_2CCO$ | [225] |
| $CO^+$ | [42, 107] | HDS | [99] | $CH_4$ | [13, 75] |
| CaO | [105] | $HCS^+$ | [245] | $SiH_4$ | [13] |
| $CN^+$, $HCN^+$, $HNC^+$ | [107] | $N_2D^+$ | [5] | $NH_2{}^{13}CHO$ | [133] |
| SO | [39a] | | | $CH_3SH$ | [137a] |
| | | | | $CH_3C_2H$ | [17a] |
| | | | | $CH_3NHD$ | [74] |

metastable isomers of small organic molecules ($H_2CC$, $H_3CN$, HCCNC, CCCNH, $H_2CCHOH$, and HCCOH) [88]; $H_2N \cdot CH_2 \cdot COOH$ [26]; the CO dimer $(CO)_2$ [229a].

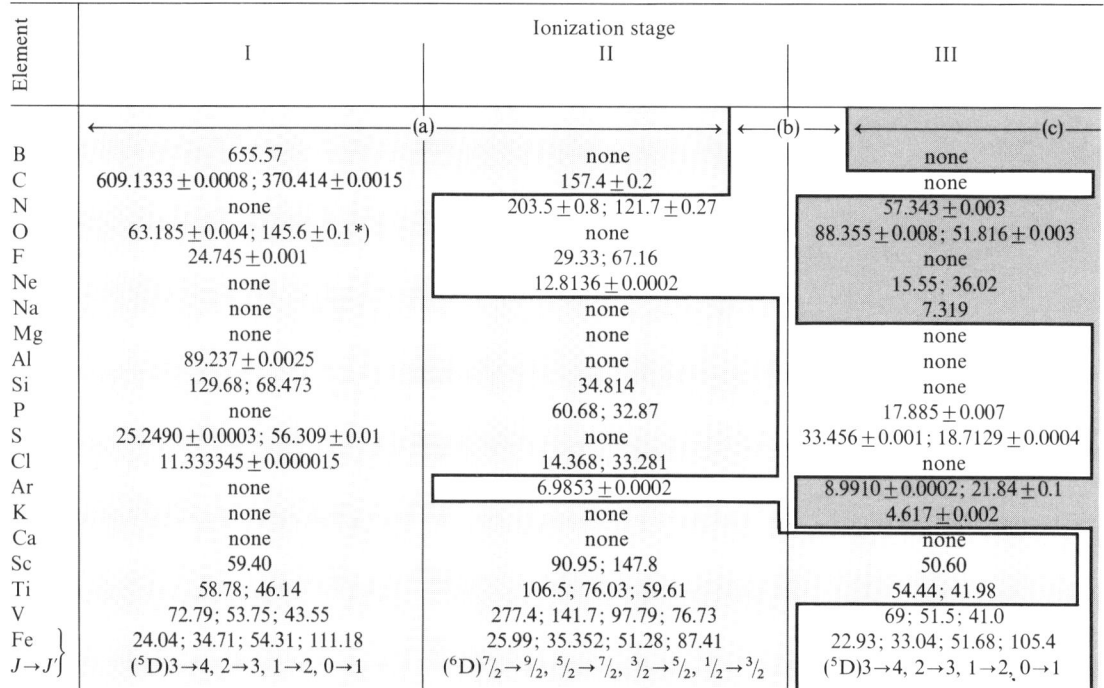

The boundaries in heavy lines separate atomic ions expected to be present in radiation fields (see 7.5.3.2.1) with:
(a) $h\nu < 13.598$ eV (H I regions)
(b) $h\nu < 24.587$ eV (H II/He I regions)
(c) $h\nu < 54.416$ eV (H II/He II regions)
(d) $h\nu > 54.416$ eV ("coronal" gas of the interstellar medium, high-excitation planetary nebulae, supernovae)

| Ionization stage: IV | V | VI | VII | Element |
|---|---|---|---|---|
| —(c)→ | ←—— | —(d)—— | ——→ | |
| none | none | none | none | B |
| none | none | none | none | C |
| none | none | none | none | N |
| 25.87 ± 0.02 | none | none | none | O |
| 44.25; 25.83 | 13.43 | none | none | F |
| none | 24.28 ± 0.02; 14.32 | 7.642 | none | Ne |
| 9.039; 21.29 | none | 14.33; 8.62 | 4.675 | Na |
| 4.488 | 5.609; 13.52 | none | 8.873; 5.519 | Mg |
| none | 2.905 | 3.66; 9.12 | none | Al |
| none | none | 1.965 ± 0.0025 | 2.481; 6.49 | Si |
| none | none | none | 1.376 | P |
| 10.5105 ± 0.0001 | none | none | none | S |
| 20.38 ± 0.02; 11.76 ± 0.01 | 6.708 ± 0.003 | none | none | Cl |
| none | 13.09; 7.903 | 4.53 | none | Ar |
| 5.983; 15.39 | none | 8.823; 5.575 | 3.191 | K |
| 3.207 ± 0.001 | 4.159; 11.48 | none | 6.154; 4.087 | Ca |
| none | 2.31182 ± 0.00004 | 2.989; 9.001 | none | Sc |
| 26.17 | none | 1.7155 ± 0.0004 | 2.206; 7.386 | Ti |
| 31.4; 24.3 | 16.1 | none | 1.3041 | V |
| none | 68.97; 36.50; 25.97; 20.79 | 19.57; 14.77; 12.30 | 9.55; 7.81 | Fe |
| | ($^5$D)1→0, 2→1, 3→2, 4→3 | ($^4$F)$^5/_2$→$^3/_2$, $^7/_2$→$^5/_2$, $^9/_2$→$^7/_2$ | ($^3$F)3→2, 4→3 | $J \to J'$ |

### 7.5.2.3 Chemical abundances in the interstellar medium

Reviews (1979) [39, 177, 178]. Gaseous abundance and depletion studies by interstellar absorption lines referenced in [138], UV results summarized in [k, 196, 199b, 215a]. For the S problem, see [164a].

Galactic abundance gradients in the interstellar matter recently estimated from H II region observations: optically for He, O, N and S [98, 179, 219], by radio techniques for He [221] and O [155]; from planetary nebulae: [55, 222]; see also [4]. For D in the Galaxy, see e.g. [180].

## 7.5.3 Particle processes

### 7.5.3.1 Interaction with photon fields: radiation

Interstellar radiation field (H I-shielded) [96, 101, 120], far-UV [101a], X-rays from hot interstellar plasma [207c] see also 7.1.8.

Photoionization of the interstellar medium by supernova explosions [216], supernova remnants [36], X-rays [44, 223, 240].

For H II regions: physical and spectral parameters of early-type stars [176]; line-blanketing effects see [126], attempts to incorporate non-solar abundances in atmospheres of early-type stars [10, 207].

Far-UV from central stars of planetary nebulae [164b].

## 7.5.3.2 Photoionization

### 7.5.3.2.1 Photoionization of atoms

For atomic ions expected to be present in radiation fields with $h\nu < 13.598$ eV (H I regions), $h\nu < 24.587$ eV (H II/He I regions), $h\nu < 54.416$ eV (H II/He II regions) and $h\nu \geqq 54.416$ eV ("coronal" gas of the interstellar medium, high-excitation planetary nebulae, supernovae), see Table 1.

Photo cross sections of atomic ions: review (1976) of computational methods [30]. For data, see 4.3.4.8 and [g]. Recent results and compilations: for ground-state configurations of the He to Ne isoelectronic sequences [118], for all ions with $Z_0 \leqq 30$ and $h\nu$ between 5 eV and 5 keV [193], for individual shells of Li-like ions [12], for O I [188] and for Ne and Ar [31]. Empirical fits to subshell cross sections for UV photons [242], for X-rays [241].

### 7.5.3.2.2 Photoionization and photodissociation of molecules

Discussion, and list of representative lifetimes against photodissociation in the galactic radiation field: unshielded by dust [m], accounting for variable dust extinction [156]; this process important for interstellar chemistry, but most rates very uncertain. Destruction by photoionization (rates [156]) usually less effective than photodissociation (possible exceptions CH and $CH_2$).

Data: ionization potentials [110, 129, 246]; dissociation energies [129]; cross sections (mostly estimates) for CO, OH, CH, $H_2CO$, and $NH_3$ referenced in [199], cross sections and rates of all H, C, N, O diatomic molecules plus $CH_2$, $H_2O$, and $C_2H$ in [14]. For CH also [232], for OH [215], for $CH^+$ [124a]. Detailed theoretical [73, 218b] and experimental [27] studies of $H_2$, HD, $D_2$ compared in [46]. Energy in dissociation products of $H_2$, HD and $D_2$ [218c]. Interaction of interstellar radiation with $H_2$ [21, 65, 207a], of X-rays with $H_2$ [44].

## 7.5.3.3 Interaction with cosmic rays and X-ray photoelectrons

Ionization, dissociation and heating of $H_2$ gas by cosmic rays [43, 81]; ionization of H, He [81], of C, N, O [156]; cosmic ray penetration into molecular clouds [35]. Heating and ionization by X-ray photoelectrons [207b].

## 7.5.3.4 Recombination of atomic ions

For general discussion, see 4.3.4.3.3 and 4.3.4.4.1. According to [1, 2], in the interstellar medium dielectronic recombination is not important relative to radiative recombination (i.e. rate ratio <1:5) below $T_e = 1.1 \cdot 10^4$ K for all ions of Table 2a (excluding Ar) except for Mg II. This ratio increases above 1:5 between $1.1 \cdot 10^4$ K and $2.9 \cdot 10^4$ K for all ionization stages of the recombining ion below C V, N VI, O VII, Si V, and S VII; at higher temperatures dielectronic recombination is often the dominant recombination mechanism. Simple expressions for both recombination rates, valid at $T_e \geqq 3 \cdot 10^5$ K, available for many ionic species [127].

### 7.5.3.4.1 Radiative recombination

Recent highly accurate calculations of rate coefficients $\alpha_{rad}$ [83]. In this subsection we use $T_m \equiv \frac{T_e[K]}{10^m\,K}$ for the reader's convenience. Then $\alpha_{rad}$ is approximated [83] in the range $0.3 \lessapprox T_m \lessapprox 3$ by

$$\alpha_{rad}(i, T_e) = A_m T_m^{-1/2}[f_m + (1-f_m)\phi(T_m)],$$

where $f_m$ = fraction of recombinations at $T_m = 1$ to the valence shell, and $\phi(T_m) = T_m^{-(\alpha_m + \beta_m T_m)}$. Note: the same function $\phi(T_m)$ is valid for several ionic species (see Table 2b which is based on [83]).

Alternative approximation [1, 2]:

$$\alpha_{rad}(i, T_e) = A_m T_m^{-\eta} \text{ (less accurate, larger range of } T_e).$$

Tables 2a and 2b contain data for m=2 (1. row) and m=4, from [83] as far as available, otherwise from [1, 2] (in italics).

For $T_m = 1$ follows $\phi(T_m) = 1$ and thus $\alpha_{rad} = A_m$.

Hydrogen-like ions:

$H^+$: $\alpha_{rad} = 4.18 \cdot 10^{-13} T_4^{-0.698}$ in [cm$^3$ s$^{-1}$], valid (±2%) for $0.12 \leqq T_4 \leqq 2$;

$\alpha_{rad}(i, T_e) = 9.3 \cdot 10^{-14} Z^2 T_4^{-1/2}(1.735 + \ln y + 1/6y)$ in [cm$^3$ s$^{-1}$], with $y \equiv 15.789\, Z^2/T_4$, valid (±2%) for $y \gtrapprox 1$ [82, 223]; for $y < 1$, see [j, 66a].

Tables 2a and 2b. Radiative recombination coefficients $A_m$, $f_m$, $\eta$, and $\phi(T_m)$.
From [1, 2, 83]; see text, $A_m$ in units $[10^{-13}\,\mathrm{cm^3\,s^{-1}}]$. All data for the case m = 4, except for rows marked * for which m = 2 ($\eta$ in these rows: interpolation after [83]). ($\eta$): uncertain. $Z$ is number of positive charges per ion before recombination. For $\phi(T_m)$ coefficients $\alpha_m$, $\beta_m$, see Table 2b (below, right).

Table 2a.

| | He | | | C | | | N | | | O | | | Ne | | |
|---|---|---|---|---|---|---|---|---|---|---|---|---|---|---|---|
| $Z$ | $A_m$ | $f_m$ | $\eta$ | $A_m$ | $f_m$ | $\eta$ | $A_m$ | $f_m$ | $\eta$ | $A_m$ | $f_m$ | $\eta$ | $A_m$ | $f_m$ | $\eta$ |
| 1* | 86.8 | 0.182 | 0.652 | 88.3 | 0.264 | 0.639 | 81.1 | 0.185 | 0.658 | 74.9 | 0.115 | 0.677 | 70.0 | 0.042 | 0.697 |
| 1 | 4.31 | 0.366 | *0.672* | 4.66 | 0.500 | *0.624* | 3.92 | 0.383 | *(0.608)* | 3.31 | 0.260 | *0.678* | 2.83 | 0.104 | *(0.759)* |
| 2 | | | | 24.5 | 0.474 | *0.645* | 22.8 | 0.425 | *0.639* | 20.5 | 0.346 | *0.646* | 17.1 | 0.193 | *0.693* |
| 3 | | | | 50.5 | 0.325 | *(0.770)* | 54.4 | 0.360 | *0.676* | 54.3 | 0.350 | *0.666* | 44.4 | 0.245 | *0.675* |
| 4 | | | | 84.5 | 0.193 | *0.817* | 95.5 | 0.272 | *(0.743)* | 103 | 0.314 | *0.670* | 98.1 | 0.274 | *0.668* |
| 5 | | | | *170* | *0.147* | *0.721* | *150* | *0.043* | *0.850* | *120* | *0.030* | *0.779* | *150* | *0.267* | *0.684* |
| 6 | | | | | | | *290* | *0.124* | *0.750* | *230* | *0.042* | *0.802* | *230* | *0.222* | *0.704* |
| 7 | | | | | | | | | | *410* | *0.115* | *0.742* | *280* | *0.024* | *0.771* |
| 8 | | | | | | | | | | | | | *500* | *0.036* | *0.832* |
| 9 | | | | | | | | | | | | | *860* | *0.088* | *0.769* |

| | Mg | | | Si | | | S | | | Ar | | |
|---|---|---|---|---|---|---|---|---|---|---|---|---|
| $Z$ | $A_m$ | $f_m$ | $\eta$ | $A_m$ | $f_m$ | $\eta$ | $A_m$ | $f_m$ | $\eta$ | $A_m$ | $f_m$ | $\eta$ |
| 1* | 68.4 | 0.144 | 0.709 | 105 | 0.426 | 0.605 | 86.1 | 0.294 | 0.634 | 72.6 | 0.156 | 0.671 |
| 1 | 2.61 | 0.300 | *(0.855)* | 6.48 | 0.677 | *0.601* | 4.65 | 0.527 | *0.630* | 3.30 | 0.309 | |
| 2 | 13.5 | 0 | *(0.838)* | 14.4 | 0.205 | *(0.786)* | 24.2 | 0.513 | *(0.686)* | 24.5 | 0.504 | |
| 3 | 41.9 | 0.123 | *0.734* | 36.5 | 0.197 | *0.693* | 35.3 | 0.140 | *(0.745)* | 43.1 | 0.250 | |
| 4 | 84.4 | 0.154 | *0.718* | 73.2 | 0 | *(0.821)* | 70.3 | 0.129 | *0.755* | 74.0 | 0.153 | |
| 5 | *140* | *0.193* | *0.716* | *120* | *0.083* | *0.735* | *120* | *0.208* | *0.701* | | | |
| 6 | *230* | *0.213* | *0.695* | *210* | *0.138* | *0.716* | *170* | *0.002* | *0.849* | | | |
| 7 | *320* | *0.206* | *0.691* | *300* | *0.153* | *0.702* | *270* | *0.063* | *0.733* | | | |
| 8 | *460* | *0.196* | *0.711* | *430* | *0.172* | *0.688* | *400* | *0.133* | *0.696* | | | |
| 9 | *580* | *0.021* | *0.804* | *580* | *0.171* | *0.703* | *550* | *0.129* | *0.711* | | | |
| 10 | *910* | *0.033* | *0.830* | *770* | *0.169* | *0.714* | *740* | *0.134* | *0.716* | | | |
| 11 | *1500* | *0.073* | *0.779* | *1200* | *0.015* | *0.855* | *920* | *0.130* | *0.714* | | | |
| 12 | | | | *1500* | *0.031* | *0.831* | *1400* | *0.129* | *0.755* | | | |
| 13 | | | | *2100* | *0.071* | *0.765* | *1700* | *0.015* | *0.832* | | | |
| 14 | | | | | | | *2500* | *0.024* | *0.852* | | | |
| 15 | | | | | | | *3300* | *0.061* | *0.783* | | | |

Table 2b.

For C, N, O, Ne:

| $Z$ | $\alpha_m$ | $\beta_m$ |
|---|---|---|
| 1* | 0.149 | 0.012 |
| 1 | 0.371 | 0.043 |
| 2 | 0.269 | 0.033 |
| 3 | 0.226 | 0.026 |
| 4 | 0.202 | 0.022 |

For Mg, Si, S, Ar:

| $Z$ | $\alpha_m$ | $\beta_m$ |
|---|---|---|
| 1* | 0.165 | 0.015 |
| 1 | 0.431 | 0.044 |
| 2 | 0.313 | 0.038 |
| 3 | 0.260 | 0.032 |
| 4 | 0.230 | 0.027 |

### 7.5.3.4.2 Dielectronic recombination

Review (1976) of rate coefficient calculations [204]: error of the general formula of Burgess [28] usually less than 40%. Approximate expressions for all ions of Table 2a except for Ar computed along lines of Burgess formula: [1, 2]. For high stages of ionization, results of [6] should be somewhat more accurate. Both these approximations not generally valid, however, for $kT_e \ll$ excitation energy of the stabilizing transition [97]. For He-like ions, see [29]. Papers on rate coefficients of Si II···XIII ($T_e \geqq 10^4$ K) [115], of Ne II···VIII, Mg II···X, and S II···VII ($T_e \geqq 10^4$ K) [116], and of Fe IX···XXVI ($T_e \geqq 4 \cdot 10^5$ K) [114] contain discussion of inadequacies of Burgess formula.

## 7.5.3.5 Radiative and dissociative recombination of molecules

General discussion [m, 11]. Dissociative recombination often dominates destruction of key molecular ions; radiative recombination usually much less important. Rate estimates [156]. Theory of dissociative recombination of polyatomic ions, applied to $NH_4^+$, $CH_3^+$, $H_3O^+$, and $HCNH^+$ [102]. Measured dissociative rate coefficients: Table 3. Note isotopic effects (H vs. D).

Table 3. Measured dissociative electron recombination coefficients, $\alpha_{de}$, for molecular ions, after [152] and measured at $T_{exp} = 120$ K unless marked otherwise. Alternative measurements and their references in brackets. Larger temperature range for $C_2H_n^+$ (with n = 0···3) [163a].
Usually $\alpha_{de} \propto T^{-1/2}$ (at least for 100···1000 K); for exceptions, see footnotes.

| Recombining ion | $\alpha_{de}$ $10^{-7}$ cm$^3$ s$^{-1}$ | $T_{exp}$ K | Ref. | Recombining ion | $\alpha_{de}$ $10^{-7}$ cm$^3$ s$^{-1}$ | $T_{exp}$ K | Ref. |
|---|---|---|---|---|---|---|---|
| $H_2^+ (v=0-2)$ | 0.8 | | | $H_2D^+$ | 5.4 | | |
| $H_2^+$ (all $v$) | 4 | | | $D_3^+$ | 4 | | |
| $HD^+$ (all $v$) | 4 | | | $H_2O^+$ | 12 | | |
| $D_2^+$ (all $v$) | 2.7 | | | $N_2H^+$, $N_2D^+$ | 15, (12) | 100, (120) | 163, (152) |
| $CH^+$ (ground state) | 3.1 ± 0.3, (4) | 120, (120) | 158, (152) [1] | $HCO^+$ | 3 | 205 | 135 [2] |
| $CH^+$ (excited) | 7.4 | | | $CO_2^+$ | 3.8 ± 0.5 | 300 | 11 |
| $NH^+$ | 1.7 | | | $H_3O^+$ | 11 | 100 | 163 |
| $OH^+$ | 1.3 | | | $D_3O^+$ | 9 | 100 | 163 |
| $He_2^+$ | < 0.1 | 300 | 11 | $CH_2^+$ | 8 | | |
| $C_2^+$ | 9.4 | | | $CH_3^+$ | 12 | | |
| $N_2^+$ | 3, (5.7) | 300, (120) | 11, (152) | $CH_4^+$ | 9.5 | | |
| $O_2^+$ | 2.1, (3.0) | 300, (120) | 11, (152) | $CH_5^+$ | 11 | | |
| $NO^+$ | 4.1, (3.7) | 300, (120) | 11, (152) | $C_2H^+$ | 9.4 | | 163a |
| $Ne_2^+$ | 2 | 300 | 11 | $C_2H_2^+$ | 9.4 | | |
| $H_3^+$ | 3 | 205 | 135 | $C_2H_3^+$ | 15 | | |
| $H_3^+$ $(v=0)$ | 6.6 | | | $H_5^+$ | 36 | 205 | 135 |
| | | | | $NH_4^+$ | 25 ± 10, (15 ± 3) | 200, (300) | 108, (108) |

[1]) $\alpha_{de} \propto T^{-(0.39 \pm 0.11)}$ indicated.
[2]) $\alpha_{de} \propto T^{-1}$ indicated.

## 7.5.3.6 Charge exchange

Potentially important process:

in partly ionized intercloud gas [k, 218]; in ionized nebulae [181a, 183]; in interstellar shocks [33b];

in interstellar matter near compact X-ray sources [151];

as the dominant recombination process for numerous ions when the neutral fraction of interstellar gas exceeds $\approx 10^{-3}$;

for initiating molecule formation, e.g. by ionizing D (→HD, subsequent D-enhancement in other molecules) and O (→OH) even at low proton densities [236].

Few rates known; cross sections strongly varying with ionic species. Rate coefficients for astrophysically important ions: Table 4.

Table 4. Rate coefficients $\alpha_{ex}$ for charge exchange with hydrogen, theoretical and empirical; radiationless processes unless marked otherwise. Empirical data are from laboratory measurements, planetary nebulae or H II regions. Additional LZA results in [32b].

| Initial state | $T$ K | $\alpha_{ex}$ (theoretical) $10^{-9}$ cm$^3$ s$^{-1}$ | Ref. | $\alpha_{ex}$ (empirical) $10^{-9}$ cm$^3$ s$^{-1}$ | Ref. | Notes |
|---|---|---|---|---|---|---|
| H II+H I | 100 | 0.5 | cf. 236 | | | |
| D II+H I | 10···300 | $1/(0.5+0.98\exp(-1.45\,T_2))$ | 237 | | | 1, 2 |
| He II+H I | 1···1000 | $\approx 1.9\cdot 10^{-6}(1+5/T_0)$ | 199a; cf. 122 | | | 1, 3, 7 |
| He III+H I | | $1.56\cdot 10^{-4}$ | 7 | | | 3 |
| C II+H I | $\approx 10^4$ | | | $\approx 7\cdot 10^{-4}$ | 217 | |
| C III+H I | 10···$10^5$ | $(1.28\cdots 2.44)\cdot 10^{-5}$ | 33 | } $\gtrsim 1.3\cdot 10^{-2}$ | 217 | 3 |
| C III+H I | $\approx 10^4$ | $\approx 10^{-3}$; $\approx 10^{-5}$? | 33a; 148 | | | |
| C IV+H I | 30···$3\cdot 10^4$ | 1.5···1.7 | 238 | $2.0\pm 0.6$ | 183 | |
| C IV+H I | $5\cdot 10^3\cdots 5\cdot 10^4$ | $3.58\,T_4^{0.26}$ | 33a, 49a | | | 1 |
| N I+H II | $\approx 10^4$ | $\approx 10^{-3}$ | cf. 47 | | | |
| N II+H I | $10^2\cdots 6\cdot 10^4$; $10^4$ | $(1\pm 0.4)\cdot 10^{-3}$; $1\cdot 10^{-3}$ | 32a | –; $2.2\cdot 10^{-3}$ | 181 | |
| N III+H I | $\approx 5\cdot 10^3\cdots 5\cdot 10^4$ | $0.86\,T_4^{0.18}$ | 33a | } $2\pm 1$ | 183 | |
| N III+H I | 10···$10^5$ | $(2.04\cdots 3.16)\cdot 10^{-5}$ | 33 | | | 3 |
| N IV+H I | 1···400 | 0.2···0.3 | 150, (238) | | | |
| N IV+H I | $10^3\cdots 3\cdot 10^4$ | $3.3\,T_4^{0.9}$ | 150, (238) | | | 1 |
| N IV+H I | $5\cdot 10^3\cdots 5\cdot 10^4$ | $2.93\,T_4^{0.8}$ | 33a | | | 1 |
| O I+H II | 300 | $\approx 1\exp(-2.3/T_2)$ | cf. 69, 236 | 0.38 | 66 | 1, 4 |
| O I+D II | 300 | | | 0.28 | 213a | |
| O II+H I | $\approx 100$; 300 | $\approx 1$; $\approx 1$ | 69; 236 | $\approx 0.5$; 0.68 | 22; 66 | 4 |
| O III+H I | $\approx 5\cdot 10^3\cdots 5\cdot 10^4$ | $0.77\,T_4^{0.45}$ | 33a | 0.8 | 181 | |
| O IV+H I | $5\cdot 10^3\cdots 5\cdot 10^4$ | $8.63\,T_4^{0.47}$ | 33a, 49a | | | 1 |
| Ne III+H I | $\approx 10^4$ | $\ll 0.1(\approx 10^{-5}?)$ | 32, 33a | 0.2 | 181 | 3 |
| Ne IV+H I | $5\cdot 10^3\cdots 5\cdot 10^4$ | $5.68\,T_4^{0.56}$ | 33a | | | 1 |
| Mg III+H I | $\approx 10^4$ | $1\cdot 10^{-5}$ | 17 | | | |
| Si III+H I | 10···$10^3$ | $\approx 1.7$ | 149 | | | |
| Si III+H I | $\approx 10^3\cdots 3\cdot 10^4$ | $5.16+1.34\ln T_4$ | 149 | | | 1, 8 |
| S III+H I | $\approx 10^4$ | | | $\ll 1$ | 183 | |
| S IV+H I | $\approx 10^4$ | | | $1.5\pm 0.3$ | 183 | |
| Fe I+H II | 300 | | | 7 | 195 | |
| $H_2$+H II | 300 | | | $<10^{-3}$ | 110 | |
| $O_2$+H II | 300 | | | $1.17\pm 0.06$ | 110 | |
| NO+H II | 300 | | | 1.9 | 66 | |
| $H_2O$+H II | 300 | | | $8.2\pm 0.4$ | 110 | |
| $H_2S$+H II | 300 | | | $7.6\pm 0.9$ | 110 | 5 |
| HCN+H II | 300 | | | $\ll 1$ | 110 | |
| $NH_3$+H II | 300 | | | $5.2\pm 0.2$ | 110 | |
| $CH_4$+H II | 300 | | | 1.5 | 110 | 6 |

Notes:

1) $T_m$ is temperature in [K], divided by $10^m$ K.
2) For inverse reaction, multiply $\alpha_{ex}$(theoretical) by $\exp(-0.43/T_2)$.
3) Radiative charge-exchange process.
4) Applicability of laboratory results [66] to interstellar gas uncertain [236].
5) Final state includes all reaction products.
6) Alternative reaction: $\rightarrow CH_3^+ + H_2$, for which $\alpha_{ex}$ (empirical)=2.3 [110].
7) This rate includes alternative channel $\rightarrow HeH^+ + h\nu$; rate coefficient for charge exchange alone is $\alpha_{ex}\approx 4.1\cdot 10^{-6}\,T_0^{-0.38}$ [199a].
8) For inverse reaction, multiply $\alpha_{ex}$ (theoretical) by $\frac{1}{3}\exp(-3.18/T_4)$ [10a].

#### 7.5.3.6.1 Charge exchange: atomic ions

Reviews (1978) [47, 236]. Some rates involving He I [33a, 236]. Discussion of Landau-Zener approximation (LZA, [162]) for ionization stages II···V of (He), C, N, O, Ne, Si, S, Ar, Fe, Mg exchanging with H, He [32b, 49]. LZA tends to be adequate at $T \gtrsim 10^4$ K, deficient at $T \cong 10^2$ K. Recent quantal calculations [33a].

Rate coefficients for exchange reactions with simultaneous photon emission expected to be often much lower ($\alpha = 10^{-15} \cdots 10^{-13}$ cm$^3$ s$^{-1}$) than corresponding radiationless process [49]. Double charge transfer may proceed faster than single transfer (e.g. C V + He I→C III + He III, [47]), but probably not important in ionized nebulae [220].

#### 7.5.3.6.2 Charge exchange: molecules

Discussion [m, 110]. Exothermic exchange processes involving molecules are expected to proceed often near collision rate (rate coefficient $\alpha \cong 10^{-9}$ cm$^3$ s$^{-1}$ near $T = 100$ K); important deviations from this trend for some small molecules (cf. Table 4 which contains the measurements now available for thermal proton reactions). Reactions slow compared to collision rate at $T \cong 300$ K may show significant temperature dependence; these rates at present unknown for $T$ between 10 K and ≈100 K.

For molecules, alternative reaction channels compete with charge transfer (noted in Table 4 where applicable).

Rate coefficients: measurements at $T = 300$ K [110], for $HC_3N$ [76]. Theoretical estimates [156].

### 7.5.3.7 Collisional excitation and de-excitation

Recent results for collision rate coefficients (often including information on corresponding radiative transition probabilities):

Atoms and atomic ions:

a) electron impact: H and He I referenced in [209]; He-like ions [187a];

He I $2^3S \rightarrow 2^3P$ [224];

[C I], [N I] and [O I] [182];

[N I] [56];

[O I] [134];

[N II], [Ne II], [O III] and [Ne III] [203];

[O II] [186];

[S II] [187];

O III [19];

[O IV] and O IV [72];

[Ne IV] [144];

S IV [18a]

C III and O V [58];

O V [145]; S V [229b];

[Ne V] [80];

[Fe II] and Fe II, estimates [41, 57], calculations [170a, 170c];

[Fe III] [78];

[Fe V] empirical [206];

[Fe VI] [78, 169];

references for collisional excitation of UV line transitions of C II···IV, N II···V, O III···VI, Mg II, Si IV [205]; of selected allowed and semi-forbidden UV transitions of some ions of C, N, O, Mg, Si, S, and Ca [175]; estimates for allowed transitions between 1 and 250 Å compiled [124]; for transitions between 100 and 1000 Å, see [218a]. Tabulation of collision strengths relevant for hot plasmas [207c].

b) proton impact: O V [145]; lowest-lying $^3P$ terms of the Be and Mg iso-electronic sequences [128], of the O and S sequences [127a], of N II, O III, Ne V, Na VI, and Mg VII [64a].

c) collisions with H I: [C I] and [O I] [131]; C I fine structure [248]; [C II] [132].

d) collisions with $H_2$: [C II] [71].

Molecules:

review on collision-induced transitions between rotational levels [171]. Approximate formulae for collisional excitation of rotations of linear molecules by spinless neutrals [230]. Rotational excitation of $H_2$ by H, e, photons [215b].

a) electron impact: $CH^+$ and $HeH^+$ [70]; rotational excitation of molecular ions [19a].

b) collisions with H I: $H_2$ [4a, 60, 93]; OH [123]; CO [92].

c) collisions with He and $H_2$ (usually He calculated, $H_2$ discussed): $H_2$ rotational [90]; $H_2$ vibrational [192]; $H_2$–$H_2$ [158a, 199c]; fits to $H_2$ data [208]; HD [38]; CO, CS, OCS, and $HC_3N$ [86]; CO [220a]; $H_2CO$ [87]; HCN [91]; $N_2H^+$ [84]; HCl [89]; $NH_3$ [85, 85a]; $NH_3$ measured [201], theory and measurement [52a]; $H_2O$ [85b].

Population ratios of fine-structure levels for O I, C I, C II, Si II, N II, and N III, as induced by collisions with e, p and H I and by the interstellar radiation field, tabulated for $10\ K \leqq T \leqq 15000\ K$ and $10^{-3}\ cm^{-3} \leqq N \leqq 10^3\ cm^{-3}$ [212]. Rotational level populations of CO, CS, and SiO in 3 K radiation field for $10\ K \leqq T \leqq 100\ K$ and $10^3\ cm^{-3} \leqq N \leqq 10^8\ cm^{-3}$ [231].

### 7.5.3.8 Chemical reactions in interstellar clouds

Reviews of molecule formation and destruction processes [m, 48, 103, 234, 236, 238a, 246]. Under interstellar conditions, gas-phase processes are mostly exothermic ion-molecule reactions like

a) proton transfer (e.g. $N_2H^+ + CO \rightarrow N_2 + HCO^+$); list of proton affinities of molecules [110], of $HC_3N$ [76],
b) hydrogen atom transfer ($OH^+ + H_2 \rightarrow OH_2^+ + H$), ($NH_3^+ + H_2 \rightarrow NH_4^+ + H$); for rates, see [143a],
c) $H^-$-abstraction ($CH_3^+ + H_2CO \rightarrow HCO^+ + CH_4$),
d) proton or hydrogen atom elimination ($C^+ + OH \rightarrow CO + H^+$),
e) deuterium exchange reactions ($H_3^+ + HD \rightarrow H_2D^+ + H_2$, important for ensuing concentration of D in molecules [0, 236]),
f) rare gas reactions ($He^+ + CO \rightarrow He + C^+ + O$); $He^+$ (due to high ionization potential of He I) very effective in breaking up abundant, unreactive molecules to provide $C^+$, $N^+$, $O^+$ [236]; for $He^+ + H_2$ rates, see [118a],
g) charge transfer ($HCO^+ + Ca \rightarrow HCO + Ca^+$), see 7.5.3.6.2; ions of low-ionization-potential metals (Na, Mg, Ca, Fe) unreactive under dense molecular cloud conditions, hence level of ionization of cloud codetermined by their low radiative recombination rates [173],
h) $H_2$ elimination and related reactions of $CH_n^+$ and hydrocarbons ($CH_3^+ + O \rightarrow HCO^+ + H_2$),
i) radiative association ($C^+ + H_2 \rightarrow CH_2^+ + h\nu$, [22, 143a]) and
j) negative ion reactions ($H + H^- \rightarrow H_2 + e$, [20]).

In hot gas behind shocks (e.g. around expanding H II regions) some endothermic ion-molecule reactions can proceed as well: [61, 103a] for $C^+ + H_2 \rightleftharpoons CH^+ + H - 0.4\ eV$.

Additional processes for interstellar chemistry:

k) neutral-neutral reactions (radiative association, attachment); rates at interstellar temperatures uncertain due to possible low energy barriers [234],
l) dissociative recombination: for polyatomics, relative probabilities for various possible reaction channels largely unknown,
m) radiative recombination: for l) and m), see 7.5.3.5,
n) photoionization and photodissociation: see 7.5.3.2,
o) collision-induced dissociation (in shocked regions [51]),
p) associative ionization (e.g. $Ti + O \rightarrow TiO^+ + e$ [174]),
q) interaction with cosmic rays: see 7.5.3.3, and
r) molecule formation on grain surfaces: e.g. [234].

Reaction rates

Collections of experimental data (mostly taken at $T = 300\ K$):
for processes a⋯h: including quantitative discussion of individual processes [110], mainly reactions involving a C-bearing molecular ion [200, 213], for $HC_3N$ + ion [76], for molecular ions + N, O, and NO [231a];

for i [m, 151a, 213], for $CH_3^+$ + neutral [214], for OH [215];

for j [50]; for k [112, 113]; for o [106].

Large library of reaction rates [190a]. References to recent low-temperature rate measurements [238a]. Compilations of mostly theoretical rate values and estimates, or references thereto: for processes a⋯n [14, 40, 100, 156]; for k and n [59]. Element- or molecule-specific rates, reaction sequences and networks, predicted abundances for small molecules of H, C, N, and O, see reviews above, for Si chemistry [227], S chemistry [58a, 172], Cl chemistry [120, 191], for destruction of NO [140], of $HC_3N$ [76], for synthesis of $HC_{2n}CN$ (n = 1⋯4) and related molecules [157, 200], for more complex species with emphasis on radiative association reactions [111]; for formation of $CO_2$ [190], $H_2CO$ [48], NCO [189], $HCS^+$ [147], $H_2C_2O$ [225], of isomers of small organic molecules [88]. Isotope fractionation, e.g. [235, 236], position of isotope in larger molecules ($^{13}C$ in $HC_3N$) [247], rate of $^{13}C^+ + ^{12}CO \rightleftharpoons ^{12}C^+ + ^{13}CO$ measured at low temperatures [214a].

## 7.5.4 Phases, bulk properties and processes of the interstellar gas

The range of physical parameters and processes of the interstellar medium may form a continuum of values; a clear-cut definition of individual phases is not always possible. The divisions in Table 5 follow recent trends: [f, 96a, 164, 198, 226]. Global theoretical models, based on dynamical interactions between supernovae and the interstellar medium (see e.g. [36, 209]), are available: [42a, 153]. For the physics of high-velocity gas, see [f].

Table 5. Phases of interstellar gas; lead references to gas parameters and to major physical processes. Definition of phases after [164, 198, 226]. Unreferenced quantities from [226]. For unreferenced topics, see head of respective column. Read series of three data as: lower limit (typical value) upper limit. a)···d) combine information within the same column. $A_v$ = visual absorption

| | "Coronal" gas [f, 117, 153] | Intercloud gas [8] | Diffuse interstellar clouds [j] |
|---|---|---|---|
| Subdivisions | a) hot coronal gas;<br>b) O VI gas | a) H I gas [226];<br>b) extended low-density H II regions [154, 198] | a) "standard";<br>b) large;<br>c) lukewarm [198] |
| Observational evidence | a) diffuse soft thermal X-rays ($h\nu \lesssim 1$ keV), O VII line emission<br>b) UV absorption lines of highly ionized species (mainly O VI) | a) broad, low-intensity emission features of 21-cm line<br>b) pulsar dispersion | UV, visual and 21-cm absorption lines; $A_v \leqq 1^m$ |
| Range of temperatures $T$ [K] | a) $\approx 5\cdot10^5$···few$\cdot10^6$ [f]<br>b) $2\cdot10^5$···$8\cdot10^5$ | a) 800···10000; $>10^4$?<br>b) $\approx 10^4$ [198] | a) 40(80)150<br>c) 100···1000 [198] |
| Range of number densities $N$ [cm$^{-3}$] | 0.005···0.03 | a) 0.1···1.0<br>b) $\lesssim 3$ [198], $<0.2$ [197] | a) 20(30)1000 |
| Range of sizes [pc] | scale height in Galaxy: $5\cdot10^3\ T_6$ [f] | | a) $\lesssim 1(5)$···<br>b) $\approx 35$ |
| Range of masses [$\mathfrak{M}_\odot$] | | | a) $\approx 400$ |
| Volume filling factor | 0.2···0.5(0.8) [164],<br>a) $\approx 0.5$; b) $\approx$ or $\ll 0.2$ [f] | a) $\approx 0.3$, or $\approx 1$? [164, 198]<br>b) $\geqq 0.01$ [194, 198] | a) 0.005 [198] |
| Pressure $p/k$ [K cm$^{-3}$] | | low: $\approx 10^3$···$10^4$ | |
| Fractional ionization; hydrogen largely | 1.0; H II | a) partially ionized, $<10^{-1}$?; H I<br>b) $\approx 1.0$; H II | $2\cdot10^{-4}$···$10^{-3}$; H I, for b) also $H_2$ [k] |
| Dominant source of ionization | collisional ionization [114,115,116] | a) soft X-rays from coronal gas and supernova remnants [153]<br>b) UV from OB stars, planetary nebulae [154, 198] | ionization of C I by starlight; of H I less important [236] |

Table 5

| "Isolated" dark clouds and globules [24, 146] | Massive dark molecular clouds with OBA-star formation [226] | H II regions [g, n, 96a] | |
|---|---|---|---|
| a) dark clouds; b) globules | a) with BA-star formation; b) with O-star formation; c) molecular cores of OBA-star-forming clouds; d) giant molecular clouds | a) (ultra) compact; b) dense; c) classical; d) giant | Subdivisions |
| molecular lines; not associated with early-type stars; $\approx 1^m \leqq A_v \leqq 25^m$ (possibly larger) | molecular lines, far IR and sub-mm continuum, often associated with emission or reflection nebulosity | forbidden and fine-structure lines from atomic ions, recombination lines, free-free continuum, IR from dust. a) OH, $H_2O$ masers | Observational evidence |
| a) 6(10)15 b) 7(10)14 | a)b)d) $\approx 10$; enhanced near embedded stars or shock fronts; c) 20(36)150 | 4000···15000 [155] | Range of temperatures $T$ [K] |
| a) 200(1000)7400; b) 2500(8000)14000 a) some contain denser cores ($\approx 10^4$) but without embedded stars | a)b)d) $\approx 10^3$, enhanced near embedded stars or shock fronts; c) $3\cdot10^3(4\cdot10^4)10^6$ | a) > 2000; b) 200···6000; c) 20···200; d) 6···100 [96a] | Range of number densities $N$ [$cm^{-3}$] |
| a) 0.2(0.9)2.3; b) 0.1(0.3)1.1 | a) 1(8)60; b) 3(30)170; c) 0.2(1)5.4; d) 30(60)170 | a) $\leqq 1$; b) 0.15···10; c) 1···30; d) > 10 [96a] | Range of sizes [pc] |
| a) 5(260)1300; b) 0.3(20)70 | a) $250(10^4)7.6\cdot10^5$; b) $1500(2\cdot10^4)2.5\cdot10^6$; c) $100(10^3)5\cdot10^4$; d) $7\cdot10^4(\geqq 10^5)2.5\cdot10^6$ | a) $\lessapprox 1$; b) $\approx 10$; c) 10···500; d) $\gtrsim 500$ [96a] | Range of masses [$\mathfrak{M}_\odot$] |
| total $\gtrsim 0.005$ [153], $\approx 0.02$ | b) $\cong 4\cdot10^{-4}$ [198] | | Volume filling factor |
| high: $> 10^4$ (not in pressure equilibrium with ISM) | | $\gg 10^4$ | Pressure $p/k$ [K $cm^{-3}$] |
| $3\cdot10^{-10}\cdots1\cdot10^{-8}$ [226]; $H_2$; a): the larger ones possibly shell of H I at $T \cong 30$ K [164] | $\lesssim 10^{-8}$ [236]; $H_2$ | 1.0; H II | Fractional ionization; hydrogen largely |
| cosmic rays [173, 233, 236]; decay of $^{40}K$? [34] | | UV radiation ($h\nu \geqq 13.6$ eV) from central star(s), see 7.5.3.1 | Dominant source of ionization |

continued

Table 5, continued

|  | "Coronal" gas [f, 117, 153] | Intercloud gas [8] | Diffuse interstellar clouds [j] |
|---|---|---|---|
| Heating processes [67] (for heating by X-ray photoelectrons, see [207b]; for mechanical heating [42a] | supernova explosions [f, 117, 153], OB stellar wind blast waves [239] | a) soft X-rays, thermal electron conduction [198], photo-electrons from dust [54, 57] <br> b) ionization of H by starlight [198] | electron ejection from grains by starlight ($hv < 13.6$ eV) [54, 57], cloud-cloud collisions [210]. C I ionization and cosmic rays less [185] |
| Cooling processes [67] | collisional excitation of X-ray lines [124, 218a], free-free emission less; thermal conduction to interfaces with cooler regions [f] | excitation of line radiation of H I, of neutral and ionized atoms by collisions with H I or electrons [j] | fine-structure line emission of C II [79, 121]; emission by $H_2$ (see [121])? |
| Dynamical state (for stirring by supernova and OB blast waves, see [153, 198]) | velocities [199b] | | Jeans mass not exceeded |
|  | possibly in pressure equilibrium [153, 164] | | |
| Star formation [c, d] | | | |
| Interaction with other phases; special processes | evaporation of embedded cool clouds [153]; O VI gas located near interfaces with clouds [f] | interacting supernova remnants [212a] | models [15]; chemical reactions [14] |

# 7.5.5 References for 7.5

## 7.5.5.1 General references

a   Balian, R., Encrenaz, P., Lequeux, J. (eds.): Atomic and Molecular Physics and the Interstellar Matter, North-Holland, Amsterdam (1975).

b   Burton, W.B. (ed.): The Large-Scale Characteristics of the Galaxy, Int. Astron. Union Symp. No. 84, Reidel, Dordrecht (1979).

c   De Jong, T., Maeder, A. (eds.): Star Formation, Int. Astron. Union Symp. No. 75, Reidel, Dordrecht (1977).

d   Gehrels, T. (ed.): Protostars and Planets, Univ. of Arizona Press, Tucson (1978).

e   Kaplan, S.A., Pikelner, S.B.: The Interstellar Medium, Harvard University Press, Cambridge (1970).

f   McCray, R., Snow, T.P.: Annu. Rev. Astron. Astrophys. 17 (1979) 213.

g   Osterbrock, D.E.: Astrophysics of Gaseous Nebulae, Freeman and Co., San Francisco (1974).

h   Pinkau, K. (ed.): The Interstellar Medium, Reidel, Dordrecht (1974).

i   Setti, G., Fazio, G.G. (eds.): Infrared Astronomy, Reidel, Dordrecht (1978).

j   Spitzer, L.: Physical Processes in the Interstellar Medium, J. Wiley, New York (1978).

k   Spitzer, L., Jenkins, E.B.: Annu. Rev. Astron. Astrophys. 13 (1975) 133.

l   Van Woerden, H. (ed.): Topics in Interstellar Matter, Reidel, Dordrecht (1977).

m   Watson, W.D.: Rev. Mod. Phys. 48 (1976) 513.

n   Wilson, T.L., Downes, D. (eds.): H II Regions and Related Topics, Springer, Berlin (1975).

| | | | |
|---|---|---|---|
| "Isolated" dark clouds and globules [24, 146] | Massive dark molecular clouds with OBA-star formation [226] | H II regions [g, n, 96a] | |
| cosmic rays [82]; ambipolar diffusion [82, 164] possibly, gravitational collapse heating [164] probably unimportant [226] | cosmic rays? Probably no collapse heating, but ambipolar diffusion important [164]; gas collisions with warm grains when $N > 10^4 \cdots 10^5\ cm^{-3}$ [82]; embedded (proto) stellar sources | photoeffect by stellar UV on H I and He I | Heating processes [67] |
| collisional excitation of rotational transitions of CO, $O_2$, $H_2O$ and of fine-structure emission of C I [79, 82] | cooling as in "isolated" dark clouds; heating and cooling in shocked molecular clouds [208] | fine-structure and forbidden line emission, mainly of ions O, N, S, Si, Mg | Cooling processes [67] |
| b) collapse? [146]<br>Jeans mass usually exceeded, gravitationally bound; free fall? Support by turbulence inefficient [226], by rotation [68], by magnetic field [226] | collapse? [141] | expansion; stellar wind blast wave from central star [239] | Dynamical state |
| only low-mass star formation (e.g. T Tau) [226] | b)d) O-star formation by stimulated collapse [62, 242a]? | a) [96a] | Star formation [c, d] |
| cloud-cloud collisions [202]; thermal equilibrium [40, 82]; line radiation transfer [136]; chemical evolution [112, 238a]; chemical evolution of contracting clouds [79]; chemistry in shocked clouds [103a, 113]; $H_2$ dissociation fronts [140a] | | shock expanding into molecular cloud [208]; "blister model", C II regions, masers [96a] | Interaction with other phases; special processes |

## 7.5.5.2 Special references


0 Adams, N., Smith, D.: preprint (1981).
1 Aldrovandi, S.M.V., Péquinot, D.: Astron. Astrophys. **25** (1973) 137.
2 Aldrovandi, S.M.V., Péquinot, D.: Astron. Astrophys. **47** (1976) 321.
3 Allen, C.W.: Astrophysical Quantities, 3rd ed. Athlone, London (1973).
4 Aller, L.H.: Publ. Astron. Soc. Pacific **88** (1976) 580.
4a Altkorn, R.I., Schatz, G.C.: J. Chem. Phys. **72** (1980) 3337.
5 Anderson, T.G., Dixon, T.A., Pittch, N.D., Saykally, R.J., Szanto, P.G., Woods, R.C.: Astrophys. J. **216** (1977) L 85.
6 Ansari, G.M.R., Elwert, G., Mucklich, P.: Z. Naturforsch. **25a** (1970) 1781.
7 Arthurs, A.M., Hyslop, J.: Proc. Phys. Soc. London **A 70** (1957) 849.
8 Baker, P.L. in: [b] p. 287.
9 Baldwin, J.A., Netzer, H.: Astrophys. J. **226** (1978) 1.
10 Balick, B., Sneden, C.: Astrophys. J. **208** (1976) 336.
10a Baliunas, S.L., Butler, S.E.: Astrophys. J. **235** (1980) L 45.
11 Bardsley, J.N., Biondi, M.A.: Adv. At. Mol. Phys. **6** (1970) 1.
12 Barfield, W.D.: Astrophys. J. **229** (1979) 856.
13 Barrett, A.H.: Astrophys. J. **220** (1978) L 81.
14 Barsuhn, J.: Astron. Astrophys. Suppl. **28** (1977) 453.
15 Barsuhn, J., Walmsley, C.M.: Astron. Astrophys. **54** (1977) 345.

16 Bashkin, S., Stoner, J.O.: Atomic Energy Levels and Grotrian Diagrams, North-Holland Publ. Co., Amsterdam I (1975) and II (1978).
17 Bates, D.R., Moiseivitsch, B.L.: Proc. Phys. Soc. London **A 67** (1954) 805.
17a Bauer, A., Boucher, D., Burie, J., Demaison, J., Dubrulle, A.: J. Phys. Chem. Ref. Data **8** (1979) 537.
18 Bekooy, J.P., Meerts, W.L., Dymanus, A.: Astrophys. J. **224** (1978) L 77.
18a Bhadra, K., Henry, R.J.W.: Astrophys. J. **240** (1980) 368.
19 Bhatia, A.K., Doschek, G.A., Feldmann, U.: Astron. Astrophys. **76** (1979) 359.
19a Bhattacharyya, S.S. and B., Narayan, M.V.: Astrophys. J. **247** (1981) 936.
20 Bienick, R.J., Dalgarno, A.: Astrophys. J. **228** (1979) 635.
21 Black, J.H., Dalgarno, A.: Astrophys. J. **203** (1976) 132.
22 Black, J.H., Dalgarno, A.: Astrophys. J. Suppl. **34** (1977) 405.
23 Blackwell, D.E., Ibbetson, P.A., Petford, A.D., Shallis, M.J.: Mon. Not. R. Astron. Soc. **186** (1979) 633.
24 Bok, B.J.: Publ. Astron. Soc. Pac. **89** (1977) 597.
25 Broten, N.W., Oka, T., Avery, L.W., MacLeod, J.M.: Astrophys. J. **223** (1978) L 105.
26 Brown, R.D., Godfrey, P.D., Storey, J.W.V., Bassez, M.P.: J. Chem. Soc. Chem. Comm. **1978** (1978) 547.
27 Browning, R., Fryar, J.: J. Phys. B **7** (1973) 364.
28 Burgess, A.: Astrophys. J. **141** (1965) 1588.
29 Burgess, A., Tworkowski, A.S.: Astrophys. J. **205** (1976) L 105.
30 Burke, P.G. in: Atomic Processes and Applications, (Burke, P.G., Moiseivitsch, B.L., eds.), North-Holland, Amsterdam (1976) 199.
31 Burke, P.G., Taylor, K.T.: J. Phys. B **8** (1975) 2620.
32 Butler, S.E., Bender, C.F., Dalgarno, A.: Astrophys. J. **230** (1979) L 59.
32a Butler, S.E., Dalgarno, A.: Astrophys. J. **234** (1979) 765.
32b Butler, S.E., Dalgarno, A.: Astrophys. J. **241** (1980) 838.
33 Butler, S.E., Guberman, S.L., Dalgarno, A.: Phys. Rev. A **16** (1977) 500.
33a Butler, S.E., Heil, T.G., Dalgarno, A.: Astrophys. J. **241** (1980) 442.
33b Butler, S.E., Raymond, J.C.: Astrophys. J. **240** (1980) 680.
34 Cameron, A.G.W.: Icarus **1** (1962) 13.
34a Carlson, T.A., Copley, J., Durić, N., Elander, N., Erman, P., Larsson, M., Lyyra, M.: Astron. Astrophys. **83** (1980) 238.
35 Cesarsky, C.J., Völk, H.J.: Astron. Astrophys. **70** (1978) 367.
36 Chevalier, R.A.: Annu. Rev. Astron. Astrophys. **15** (1977) 175.
37 Christensen, A.B.: Astrophys. J. **229** (1979) 448.
38 Chu, S.-I.: J. Chem. Phys. **62** (1975) 4089.
39 Churchwell, E. in: Radio Recombination Lines (Shaver, P., ed.), Reidel, Dordrecht (1980) 225.
39a Clark, W.W., de Lucia, F.C.: J. Molec. Spectrosc. **60** (1976) 332.
40 Clavel, J., Viala, Y.P., Bel, N.: Astron. Astrophys. **65** (1978) 435.
41 Collin-Souffrin, S., Joly, M., Heidmann, N., Dumont, S.: Astron. Astrophys. **72** (1979) 293.
42 Čonkić, L., Janjić, J.D., Pešić, D.S., Rakotoarijimy, D., Vujisić, B.R., Weniger, S.: Astrophys. J. **226** (1978) 1162.
42a Cox, D.P.: Astrophys. J. **245** (1981) 534.
43 Cravens, T.E., Dalgarno, A.: Astrophys. J. **219** (1978) 750.
44 Cruddace, R., Paresce, F., Bowyer, S., Lampton, M.: Astrophys. J. **187** (1974) 497.
45 Czyzak, S.J., Aller, L.H.: Mem. Soc. R. Sci. Liège **6.5** (1973) 179.
46 Dalgarno, A. in: Atomic and Molecular Processes in Astrophysics (Huber, M.C.E., Nussbaumer, H., eds.), Geneva (1975).
47 Dalgarno, A. in: Planetary Nebulae (Terzian, Y., ed.), Reidel, Dordrecht (1978) 139.
48 Dalgarno, A., Black, J.H.: Rept. Progr. Phys. **39** (1976) 573.
49 Dalgarno, A., Butler, S.E.: Comm. Atom. Mol. Phys. **7** (1978) 129.
49a Dalgarno, A., Heil, T.G., Butler, S.E.: Astrophys. J. **245** (1981) 793.
50 Dalgarno, A., McCray, R.A.: Astrophys. J. **181** (1973) 95.
51 Dalgarno, A., Roberge, W.G.: Astrophys. J. **233** (1979) L 25.
52 Dankwort, W., Trefftz, E.: J. Phys. B **10** (1977) 2541.
52a Davis, S.L., Boggs, J.E.: J. Chem. Phys. **69** (1978) 2355.
53 De Boer, K.S., Morton, D.C.: Astron. Astrophys. **71** (1979) 141.
54 De Jong, T.: Astron. Astrophys. **55** (1977) 137.
55 D'Odorico, S., Peimbert, M., Sabbadin, F.: Astron. Astrophys. **47** (1976) 341.
56 Dopita, M.A., Mason, D.J., Robb, W.D.: Astrophys. J. **206** (1976) 102.

57 Draine, B.T.: Astrophys. J. Suppl. **36** (1978) 595.
58 Dufton, P.L., Berrington, K.A., Burke, P.G., Kingston, A.E.: Astron. Astrophys. **62** (1978) 111.
58a Duley, W.M., Millar, T.J., Williams, D.A.: Mon. Not. R. Astron. Soc. **192** (1980) 945.
59 Elitzur, M., de Jong, T.: Astron. Astrophys. **67** (1978) 323.
60 Elitzur, M., Watson, W.D.: Astron. Astrophys. **70** (1978) 443.
61 Elitzur, M., Watson, W.D.: Astrophys. J. **222** (1978) L 141; **226** (1978) L 157.
62 Elmegreen, B.G., Lada, C.J.: Astrophys. J. **214** (1977) 725.
63 Erman, P.: Astrophys. J. **213** (1977) L 89.
64 Evenson, K.M.: private communication (1979). See Astrophys. J. **238** (1980) L 107.
64a Faucher, P., Masnou-Seeuws, F., Prudhomme, M.: Astron. Astrophys. **81** (1980) 137.
65 Federman, S.R., Glassgold, A.E., Kwan, J.: Astrophys. J. **227** (1979) 466.
66 Fehsenfeld, F.C., Ferguson, E.E.: J. Chem. Phys. **56** (1972) 3066.
66a Ferland, G.J.: Publ. Astron. Soc. Pac. **92** (1980) 596.
67 Field, G.B. in: [a] p. 467.
68 Field, G.B. in: [d] p. 243.
69 Field, G.B., Steigman, G.: Astrophys. J. **166** (1971) 59.
70 Flower, D.R.: Astron. Astrophys. **73** (1978) 237.
71 Flower, D.R., Launay, J.M.: J. Phys. B **10** (1977) L 229.
72 Flower, D.R., Nussbaumer, H.: Astron. Astrophys. **45** (1975) 145.
73 Ford, A.L., Docken, K.K., Dalgarno, A.: Astrophys. J. **195** (1975) 819.
74 Fourikis, N., Takagi, K., Saito, S.: Astrophys. J. **212** (1977) L 33.
75 Fox, K., Jennings, D.E.: Astrophys. J. **226** (1978) L 43.
76 Freeman, C.G., Harland, P.W., McEwan, M.J.: Mon. Not. R. Astron. Soc. **187** (1979) 441.
77 Frerking, M.A., Langer, W.D., Wilson, R.W.: Astrophys. J. **232** (1979) L 65.
78 Garstang, R.H., Robb, W.D., Rountree, S.P.: Astrophys. J. **222** (1978) 384.
79 Gerola, H., Glassgold, A.E.: Astrophys. J. Suppl. **37** (1978) 1.
80 Giles, K.: Mon. Not. R. Astron. Soc. **187** (1979) 49 P.
81 Glassgold, A.E., Langer, W.D.: Astrophys. J. **193** (1974) 73.
82 Goldsmith, P.F., Langer, W.D.: Astrophys. J. **222** (1978) 881.
83 Gould, R.J.: Astrophys. J. **219** (1978) 250.
84 Green, S.: Astrophys. J. **201** (1975) 366.
85 Green, S.: J. Chem. Phys. **64** (1976) 3463.
85a Green, S.: J. Chem. Phys. **70** (1979) 816.
85b Green, S.: Astrophys. J. Suppl. **42** (1980) 103.
86 Green, S., Chapman, S.: Astrophys. J. Suppl. **37** (1978) 169.
87 Green, S., Garrison, B.J., Lester, W.A., Miller, W.H.: Astrophys. J. Suppl. **37** (1978) 321.
88 Green, S., Herbst, E.: Astrophys. J. **229** (1979) 121.
89 Green, S., Monchick, L.: J. Chem. Phys. **63** (1975) 4198.
90 Green, S., Ramaswamy, R., Rabitz, R.: Astrophys. J. Suppl. **36** (1978) 483.
91 Green, S., Thaddeus, P.: Astrophys. J. **191** (1974) 653.
92 Green, S., Thaddeus, P.: Astrophys. J. **205** (1976) 766.
93 Green, S., Truhlar, D.G.: Astrophys. J. **231** (1979) L 101.
94 Greenberg, L.T.: private communication (1979).
95 Guélin, M., Green, S., Thaddeus, P.: Astrophys. J. **224** (1979) L 27.
96 Habing, H.J.: Bull. Astron. Inst. Neth. **19** (1968) 421.
96a Habing, H.J., Israel, F.P.: Annu. Rev. Astron. Astrophys. **17** (1979) 345.
97 Harrington, J.P., Lutz, J.H., Seaton, M.J., Stickland, D.J.: preprint (1979).
97a Harwit, M.: private communication (1980).
98 Hawley, S.A.: Astrophys. J. **224** (1978) 417.
99 Helminger, P., De Lucia, F.C., Kirchhoff, W.H.: J. Phys. Chem. Ref. Data **2** (1973) 215.
100 Henning, K.: Ph.D. Thesis, University of Hamburg 1979.
101 Henry, R.C.: Astrophys. J. Suppl. **33** (1977) 451.
101a Henry, R.C.: Astrophys. J. **244** (1981) L 69.
102 Herbst, E.: Astrophys. J. **222** (1978) 508.
103 Herbst, E., Klemperer, W.: Astrophys. J. **185** (1973) 505.
103a Herbst, E., Knudson, S.: Astrophys. J. **245** (1981) 529.
104 Herzberg, G.: Molecular Spectra and Molecular Structure I, Van Nostrand, New York (1950).

105 Hocking, W.H., Winnewisser, G., Churchwell, E., Percival, J.: Astron. Astrophys. **75** (1979) 268.
106 Hollenbach, D.J., McKee, C.F.: Astrophys. J. Suppl. **41** (1979) 555.
107 Hollis, J.M., Ulich, B.L.: Astrophys. J. **219** (1978) 74.
108 Huang, C.M., Biondi, M.A., Johnsen, R.: Phys. Rev. **A 14** (1976) 984.
109 Huber, K., Herzberg, G.: Spectroscopic Constants of Diatomic Molecules, Van Nostrand, Princeton (1977).
110 Huntress, W.T.: Astrophys. J. Suppl. **33** (1977) 495.
111 Huntress, W.T., Mitchell, G.F.: Astrophys. J. **231** (1979) 456.
112 Iglesias, E.: Astrophys. J. **218** (1977) 697.
113 Iglesias, E.R., Silk, J.: Astrophys. J. **226** (1978) 851.
114 Jacobs, V.L., Davis, J., Kepple, P.C., Blaha, M.: Astrophys. J. **211** (1977) 605.
115 Jacobs, V.L., Davis, J., Kepple, P.C., Blaha, M.: Astrophys. J. **215** (1977) 690.
116 Jacobs, V.L., Davis, J., Rogerson, J.E., Blaha, M.: Astrophys. J. **230** (1979) 627.
117 Jenkins, E.B.: Astrophys. J. **220** (1978) 107.
118 John, T.L., Morgan, D.J.: J. Quant. Spectr. Rad. Transfer **14** (1974) 777.
118a Johnsen, R., Chen, A., Biondi, M.A.: J. Chem. Phys. **72** (1980) 3085.
119 Jura, M.: Astrophys. J. **190** (1974) L 33.
120 Jura, M.: Astrophys. J. **191** (1974) 375.
121 Jura, M. in: [d] p. 165.
122 Jura, M., Dalgarno, A.: Astron. Astrophys. **14** (1971) 243.
123 Kaplan, H., Shapiro, M.: Astrophys. J. **229** (1979) L 91.
124 Kato, T.: Astrophys. J. Suppl. **30** (1976) 397.
124a Kirby, K., Roberge, W.G., Saxon, R.P., Liu, B.: Astrophys. J. **239** (1980) 855.
124b Knacke, R.F., Young, E.T.: Astrophys. J. **249** (1981) L 65.
125 Kroto, H.W., Murrell, J.N., Al-Derzi, A., Guest, M.F.: Astrophys. J. **219** (1978) 886.
126 Kurucz, R.L.: Astrophys. J. Suppl. **40** (1979) 1.
127 Landini, M., Monsignori Fossi, B.C.: Sol. Phys. **20** (1971) 322.
127a Landman, D.A.: Astrophys. J. **240** (1980) 709.
128 Landman, D.A., Brown, T.: Astrophys. J. **232** (1979) 636.
129 Lang, K.R.: Astrophysical Formulae, Springer, Berlin (1978).
130 Langer, W.D., Frerking, M.A., Linke, R.A., Wilson, R.W.: Astrophys. J. **232** (1979) L 169.
131 Launay, J.M., Roueff, E.: Astron. Astrophys. **56** (1977) 289.
132 Launay, J.M., Roueff, E.: J. Phys. B **10** (1977) 879.
133 Lazareff, B., Lucas, R., Encrenaz, P.: Astron. Astrophys. **70** (1978) L 77.
134 Le Dourneuf, M., Nesbet, R.K.: J. Phys. B **9** (1976) L 241.
135 Leu, M.T., Biondi, M.A., Johnsen, R.: Phys. Rev. **A 7** (1973) 292; **A 8** (1973) 413 and 420.
136 Leung, C.M.: Astrophys. J. **225** (1978) 427.
137 Lin, C.D., Laughlin, C., Victor, G.A.: Astrophys. J. **220** (1978) 734.
137a Linke, R.A., Frerking, M.A., Thaddeus, P.: Astrophys. J. **234** (1979) L 139.
138 Liszt, H.S.: Astrophys. J. **233** (1979) L 147.
139 Liszt, H.S., Turner, B.E.: Astrophys. J. **224** (1978) L 73.
140 Loew, G.H., Berkowitz, D.S., Chang, S.: Astrophys. J. **219** (1978) 458.
140a London, R.: Astrophys. J. **225** (1978) 405.
141 Loren, R.B.: Astrophys. J. **218** (1977) 716.
142 Lovas, F.J., Snyder, L.E., Johnson, D.R.: Astrophys. J. Suppl. **41** (1979) 451.
143 Lugger, P.M., York, D.G., Blanchard, T., Morton, D.C.: Astrophys. J. **224** (1978) 1059.
143a Luine, J.A., Dunn, G.H. in: Electronic and atomic collisions (Datz, S., ed.), (1981).
144 Lutz, J.H., Seaton, M.J.: Mon. Not. R. Astron. Soc. **187** (1979) 1 P.
144a Mahan, B.H., O'Keefe, A.: Astrophys. J. **248** (1981) 1209.
145 Malinovsky, M.: Astron. Astrophys. **43** (1975) 110.
146 Martin, R.N., Barrett, A.H.: Astrophys. J. Suppl. **36** (1978) 1.
147 McAllister, T.: Astrophys. J. **225** (1978) 857.
148 McCarroll, R., Valiron, P.: Astron. Astrophys. **44** (1975) 465.
149 McCarroll, R., Valiron, P.: Astron. Astrophys. **53** (1976) 83.
150 McCarroll, R., Valiron, P.: Astron. Astrophys. **78** (1979) 177.
151 McCray, R., Wright, C., Hatchett, S.: Astrophys. J. **211** (1977) L 29.
151a McEwan, M.J., Anichich, V.G., Huntress, W.T., Kemper, P.R., Bowers, M.T.: Chem. Phys. Lett. **75** (1980) 278.
152 McGowan, J.W., Mul, P.M., D'Angelo, V.S., Mitchell, J.B.A., Defrance, P., Froelich, H.R.: Phys. Rev. Lett. **42** (1979) 373.

153 McKee, C.F., Ostriker, J.P.: Astrophys. J. **218** (1977) 148.
154 Mezger, P.G.: Astron. Astrophys. **70** (1978) 565.
155 Mezger, P.G., Pankonin, V., Schmid-Burgk, J., Thum, C., Wink, J.: Astron. Astrophys. **80** (1979) L 3; **99** (1981) 400.
156 Mitchell, G.F., Ginsburg, J.L., Kuntz, P.J.: Astrophys. J. Suppl. **38** (1978) 39.
157 Mitchell, G.F., Huntress, W.T., Prasad, S.S.: Astrophys. J. **233** (1979) 102.
158 Mitchell, J.B.A., McGowan, J.W.: Astrophys. J. **222** (1978) L 77.
158a Monchick, L., Schaefer, J.: J. Chem. Phys. **73** (1980) 6153.
159 Moore, C.E.: Atomic Energy Levels, I (1949), II (1952) and III (1959), Nat. Bur. Stand., Washington.
159a Moorwood, A.F.M.: private communication (1980).
160 Morton, D.C.: Astrophys. J. **222** (1978) 863.
161 Morton, D.C., Smith, W.H.: Astrophys. J. Suppl. **26** (1973) 333.
162 Mott, N.F., Massey, H.S.W.: The Theory of Atomic Collisions, Oxford Univ. Press (1965).
163 Mul, P.M., McGowan, J.W.: Astrophys. J. **227** (1979) L 157.
163a Mul, P.M., McGowan, J.W.: Astrophys. J. **237** (1980) 749.
164 Myers, P.C.: Astrophys. J. **225** (1978) 380.
164a Natta, A., Panagia, N., Preite-Martinez, A.: Astrophys. J. **242** (1980) 596.
164b Natta, A., Pottasch, S.R., Preite-Martinez, A.: Astron. Astrophys. **84** (1980) 284.
165 Nicholls, R.W.: Annu. Rev. Astron. Astrophys. **15** (1977) 197.
166 Nussbaumer, H.: Astron. Astrophys. **58** (1977) 291.
167 Nussbaumer, H., Rusca, C.: Astron. Astrophys. **72** (1979) 129.
168 Nussbaumer, H., Storey, P.J.: Astron. Astrophys. **64** (1978) 139.
169 Nussbaumer, H., Storey, P.J.: Astron. Astrophys. **70** (1978) 37.
170 Nussbaumer, H., Storey, P.J.: Astron. Astrophys. **74** (1979) 244.
170a Nussbaumer, H., Storey, P.J.: Astron. Astrophys. **89** (1980) 308.
170b Nussbaumer, H., Storey, P.J.: Astron. Astrophys. **96** (1981) 91.
170c Nussbaumer, H., Pettini, M., Storey, P.J.: Astron. Astrophys. **102** (1981) 351.
171 Oka, T.: Adv. Atom. Mol. Phys. **9** (1973) 127.
172 Oppenheimer, M., Dalgarno, A.: Astrophys. J. **187** (1974) 231.
173 Oppenheimer, M., Dalgarno, A.: Astrophys. J. **192** (1974) 29.
174 Oppenheimer, M., Dalgarno, A.: Astrophys. J. **212** (1977) 683.
175 Osterbrock, D.E., Wallace, R.K.: Astrophys. Lett. **19** (1977) 11.
176 Panagia, N.: Astron. J. **78** (1973) 929.
177 Panagia, N.: Mem. Soc. Astron. Italiana **50** (1979) 79.
178 Peimbert, M. in: [b] p. 307.
179 Peimbert, M., Torres-Peimbert, S., Rayo, J.F.: Astrophys. J. **220** (1978) 516.
180 Penzias, A.A., Wannier, P.G., Wilson, R.W., Linke, R.A.: Astrophys. J. **211** (1977) 108.
181 Péquignot, D.: preprint (1979).
181a Péquignot, D.: Astron. Astrophys. **81** (1980) 356.
182 Péquignot, D., Aldrovandi, S.M.V.: Astron. Astrophys. **50** (1976) 141.
183 Péquignot, D., Aldrovandi, S.M.V., Stasińska, G.: Astron. Astrophys. **63** (1978) 313.
184 Phillips, M.M.: Astrophys. J. Suppl. **39** (1979) 377.
185 Pottasch, S.R., Wesselius, P.R., van Duinen, R.J.: Astron. Astrophys. **74** (1979) L 15.
186 Pradhan, A.K.: Mon. Not. R. Astron. Soc. **177** (1976) 31.
187 Pradhan, A.K.: Mon. Not. R. Astron. Soc. **184** (1978) 89 P.
187a Pradhan, A.K., Norcross, D.W., Hummer, D.G.: Astrophys. J. **246** (1981) 1031, and Phys. Rev. A **23** (1980) 619.
188 Pradhan, A.K., Saraph, H.E.: J. Phys. B. **10** (1977) 3365.
189 Prasad, S.S., Huntress, W.T.: Mon. Not. R. Astron. Soc. **185** (1978) 741.
190 Prasad, S.S., Huntress, W.T.: Astrophys. J. **228** (1979) 123.
190a Prasad, S.S., Huntress, W.T.: Astrophys. J. Suppl. **43** (1980) 1.
191 Quaiyum, A., Ansari, S.M.P.: Mon. Not. R. Astron. Soc. **186** (1979) 621.
192 Ramaswamy, R., Rabitz, H.: J. Chem. Phys. **66** (1977) 152 and 3021.
192a Raymond, J.C., Smith, B.W.: Astrophys. J. Suppl. **35** (1977) 419.
193 Reilman, R.F., Manson, S.T.: Astrophys. J. Suppl. **40** (1979) 815.
194 Reynolds, R.J.: Astrophys. J. **216** (1977) 433.
194a Rosen, B. et al.: Spectroscopic Data Relative to Diatomic Molecules, Pergamon Press, Oxford (1970).
195 Rutherford, J.A., Vroom, D.A.: J. Chem. Phys. **57** (1972) 3091.
196 Salpeter, E.E.: Annu. Rev. Astron. Astrophys. **15** (1977) 267.

197 Salpeter, E.E. in: Planetary Nebulae, Int. Astron. Union Symp. No. 76 (Terzian, Y., ed.), Reidel, Dordrecht (1978) p. 333.
198 Salpeter, E.E. in: [b] p. 245.
199 Sandell, G.: Astron. Astrophys. **69** (1978) 85.
199a Sando, K.M., Cohen, R., Dalgarno, A.: Abstracts of Papers, VII. International Conference on Electronic and Atomic Collisions (Govers, T.R., de Heer, F.J., eds.), (1972) 973.
199b Savage, B.D., de Boer, K.S.: Astrophys. J. **243** (1981) 460.
199c Schaefer, J., Meyer, W.: J. Chem. Phys. **70** (1979) 344.
200 Schiff, H.I., Bohme, D.K.: Astrophys. J. **232** (1979) 740.
201 Schwartz, P.R.: Astrophys. J. **229** (1979) L 45.
202 Scoville, N.Z., Solomon, P.M., Sanders, D.B. in: [b] p. 277.
203 Seaton, M.J.: Mon. Not. R. Astron. Soc. **170** (1975) 475.
204 Seaton, M.J., Storey, P.L. in: Atomic Processes and Applications (Burke, P.G., Moiseiwitsch, B.L., eds.), North-Holland Publ. Co., Amsterdam (1976) 133.
205 Shields, G.A.: Astrophys. J. **204** (1976) 330.
206 Shields, G.A.: Astrophys. J. **219** (1978) 559.
207 Shields, G.A., Searle, L.: Astrophys. J. **222** (1978) 821.
207a Shull, J.M.: Astrophys. J. **219** (1978) 877.
207b Shull, J.M.: Astrophys. J. **234** (1979) 761.
207c Shull, J.M.: Astrophys. J. Suppl. **46** (1981) 27.
208 Shull, J.M., Hollenbach, D.J.: Astrophys. J. **220** (1978) 525.
209 Shull, J.M., McKee, C.F.: Astrophys. J. **227** (1979) 131.
210 Silk, J.: Astrophys. J. **198** (1975) L 77.
211 Singh, P.D., Vanlandingham, F.G.: Astron. Astrophys. **66** (1978) 87.
212 Smeding, A.G., Pottasch, S.R.: Astron. Astrophys. Suppl. **35** (1979) 257.
212a Smith, B.W.: Astrophys. J. **211** (1977) 404.
213 Smith, D., Adams, N.G.: Astrophys. J. **217** (1977) 741.
214 Smith, D., Adams, N.G.: Astrophys. J. **220** (1978) L 87.
214a Smith, D., Adams, N.G.: Astrophys. J. **242** (1980) 424.
215 Smith, W.H., Zweibel, E.G.: Astrophys. J. **207** (1976) 758.
215a Snow, T.P., Weiler, E.J., Oegerle, W.R.: Astrophys. J. **234** (1979) 506.
215b Spitzer, L., Zweibel, E.G.: Astrophys. J. **191** (1974) L 127.
216 Stecher, T.P., Williams, D.A.: Astron. Astrophys. **67** (1978) 115.
217 Steigman, G.: Astrophys. J. **195** (1975) L 39.
218 Steigman, G.: Astrophys. J. **199** (1975) 642.
218a Stern, R., Wang, E., Bowyer, S.: Astrophys. J. Suppl. **37** (1978) 195.
218b Stephens, T.L., Dalgarno, A.: J. Quant. Spectr. Rad. Transf. **12** (1972) 569.
218c Stephens, T.L., Dalgarno, A.: Astrophys. J. **186** (1973) 165.
219 Talent, D.L., Dufour, R.J.: Astrophys. J. **233** (1979) 888.
220 Tarter, C.B., Weisheit, J.C., Dalgarno, A.: Astron. Astrophys. **71** (1979) 366.
220a Thomas, L.D., Kraemer, W.P., Diercksen, G.H.F.: Chem. Phys. **51** (1980) 131.
221 Thum, C., Mezger, P.G., Pankonin, V.: Astron. Astrophys. **87** (1980) 269.
222 Torres-Peimbert, S., Peimbert, M.: Rev. Mex. Astron. Astrofis. **2** (1977) 181.
223 Tucker, W.H.: Radiation Processes in Astrophysics, Mass. Inst. Technol. Press, Cambridge and London (1975).
224 Tully, J.A., Summers, H.P.: Astrophys. J. **229** (1979) L 113.
225 Turner, B.E.: Astrophys. J. **213** (1977) L 75.
226 Turner, B.E. in: [b] p. 257.
227 Turner, J.L., Dalgarno, A.: Astrophys. J. **213** (1977) 386.
228 Turner, J., Kirby-Docken, K., Dalgarno, A.: Astrophys. J. Suppl. **35** (1977) 281.
229 Turner, B.E., Zuckerman, B., Morris, M., Palmer, P.: Astrophys. J. **219** (1978) L 43.
229a Vanden Bout, P.A., Steed, J.M., Bernstein, L.S., Klemperer, W.: Astrophys. J. **234** (1979) 503.
229b Van Wyngaarden, W.L., Henry, R.J.W.: Astrophys. J. **246** (1981) 1040.
230 Varshalovich, D.A., Khersonsky, V.K.: Astrophys. Lett. **18** (1977) 167.
231 Varshalovich, D.A., Khersonsky, V.K.: Soviet Astron. Astron. J. **22** (1978) 667.
231a Viggiano, A.A., Howorka, F., Albritton, D.L., Fehsenfeld, F.C., Adams, N.G., Smith, D.: Astrophys. J. **236** (1980) 492.
232 Walker, T.E.H., Kelly, H.P.: Chem. Phys. Lett. **16** (1972) 511.

233 Walmsley, C.M.: Astron. Astrophys. **25** (1973) 129.
234 Watson, W.D. in: [a] p. 177.
235 Watson, W.D.: CNO Isotopes in Astrophysics (Audouze, J., ed.), Reidel, Dordrecht (1977).
236 Watson, W.D.: Annu. Rev. Astron. Astrophys. **16** (1978) 585.
237 Watson, W.D., Christensen, R.B., Deissler, R.J.: Astron. Astrophys. **69** (1978) 159.
238 Watson, W.D., Christensen, R.B.: Astrophys. J. **231** (1979) 627.
238a Watson, W.D., Walmsley, C.M. in: Star Formation near H II regions (Roger and Dudney, eds.), Reidel, Dordrecht (1981).
239 Weaver, R., McCray, R., Castor, J., Shapiro, P., Moore, R.: Astrophys. J. **218** (1977) 377.
240 Weisheit, J.C.: Astrophys. J. **185** (1973) 877.
241 Weisheit, J.C.: Astrophys. J. **190** (1974) 735.
242 Weisheit, J.C., Collins, L.A.: Astrophys. J. **210** (1976) 299.
242a Welter, G., Schmid-Burgk, J.: Astrophys. J. **245** (1981) 927.
243 Wilson, S.: Astrophys. J. **220** (1978) 363.
244 Wilson, S.: Astrophys. J. **220** (1978) 737.
245 Wilson, S.: Astrophys. J. **220** (1978) 739.
246 Winnewisser, G., Churchwell, E., Walmsley, C.M.: Advances in Microwave Spectroscopy (Chantry, G.W., ed.), Academic Press Ltd., London (1979).
247 Wolfsberg, M., Bopp, P., Heinzinger, K., Mallinson, P.D.: Astron. Astrophys. **74** (1979) 369.
248 Yau, A.W., Dalgarno, A.: Astrophys. J. **206** (1976) 652.
249 Saykally, R.J., Evenson, K.M.: J. Chem. Phys. **71** (1979) 1564.

# 7.6 Cosmic rays

Energetic particles observed in the energy range $10^5 \cdots 10^{21}$ eV. Below a few $10^8$ eV, three components can be distinguished:

solar cosmic rays
galactic cosmic rays (GCR)
the anomalous component.

## 7.6.1 Solar cosmic rays

Energetic particles produced in solar flares (see 3.1.2.7) in [1]. Their flux, energy spectrum and chemical composition show large variations from flare to flare. For details, see 3.3.5.3 in [1].

## 7.6.2 Galactic cosmic rays (GCR)

Energetic particles (nuclei, electrons and positrons) penetrating into the solar system against the outward streaming solar wind and the frozen-in magnetic fields; see also 3.3.5.3.1 in [1].

### 7.6.2.1 The total flux of GCR

The total integrated flux of cosmic rays defined as

$$J(>E) = \int_E^\infty J(E)\,dE \qquad [\text{particles cm}^{-2}\,\text{s}^{-1}\,\text{sr}^{-1}] \qquad (1)$$

is plotted in Fig. 1a, b for the energy range $10^{11} \cdots 10^{16}$ eV and $10^{16} \cdots 10^{20}$ eV, respectively. These Figures are taken from Iwai et al. [2] and Suga [3], respectively.

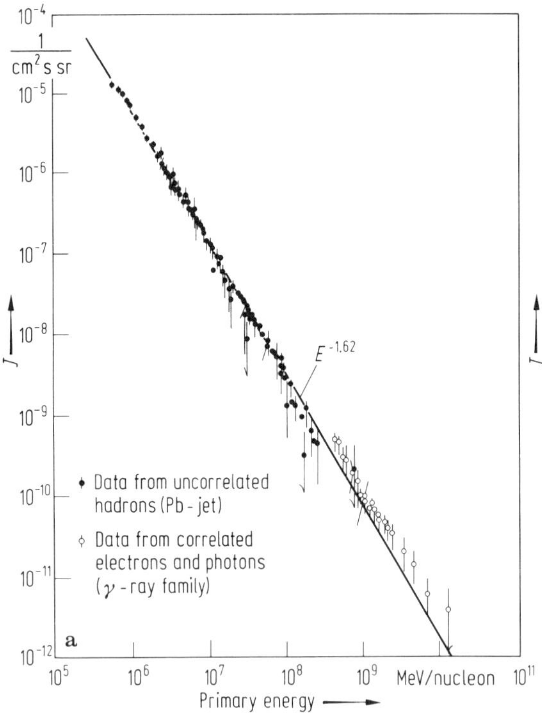


Fig. 1a. The integrated flux of cosmic rays, $J$, at medium energies. For further details, see [2].

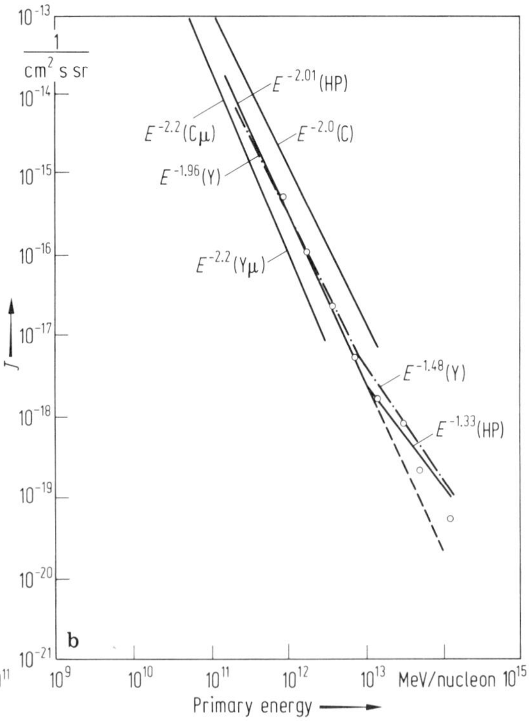


Fig. 1b. The integrated flux of cosmic rays, $J$, at very high energies. The different letters (C, Cμ, HP, Y, and Yμ) as well as the open circles refer to independent measurements at different laboratories and with different techniques. For further details, see [3].

The spectra can roughly be fitted to

$J(>E)=6.3\cdot10^{-6}\cdot E_{12}^{-1.62}$ particles $cm^{-2}\,s^{-1}\,sr^{-1}$ for $10^{11}\leqq E\leqq10^{14}$ (2a)
$J(>E)=3.0\cdot10^{-6}\cdot E_{12}^{-1.46}$ for $10^{14}\leqq E\leqq10^{16}$ (2b)
$J(>E)=5.7\cdot10^{-4}\cdot E_{12}^{-2.03}$ for $10^{16}\leqq E\leqq10^{19}$ (2c)
$J(>E)=2.8\cdot10^{-6}\cdot E_{12}^{-1.70}$ for $10^{19}\leqq E<10^{21}$ (2d)

where the particle energy is measured in [eV], with $E_{12}=E/10^{12}$ eV.

Obviously, the slope of the spectrum as described by the power law index is changing. One must bear in mind, however, that the experimental techniques also change with energy, and systematic effects can not fully be excluded. Recent measurements by Effimov et al. [4] in the range $10^{15}$ to $6\cdot10^{16}$ eV do support changes in the spectral slope at these energies. Previous measurements had indicated a kink at $10^{15}$ eV.

The total energy density of galactic cosmic rays in interstellar space can in principle be calculated from the spectra given above. Because of the steep slope of these spectra the result does, however, crucially depend on the minimum energy to which (2a) can be extrapolated. For $E_{min}\leqq10^{10}$ eV one obtains an energy density $u_{GCR}\geqq10^{-13}$ erg $cm^{-3}$, i.e. of the same order of magnitude as the energy density of star light and of the interstellar magnetic field.

## 7.6.2.2 Galactic cosmic-ray nuclei

### 7.6.2.2.1 Observed energy spectra of individual nuclei

In Fig. 2, observed energy spectra have been compiled for H, He, Be, B, C, N, O, Ne, Mg, Si, and the elements with $16\leqq Z\leqq30$. The individual plots have been taken from Meyer [5]. Note the large similarity in the shape of the spectra at energies above a few hundred MeV/nucleon. The deviation of the spectra from a power law at lower energies is largely caused by the solarwind modulation effect.

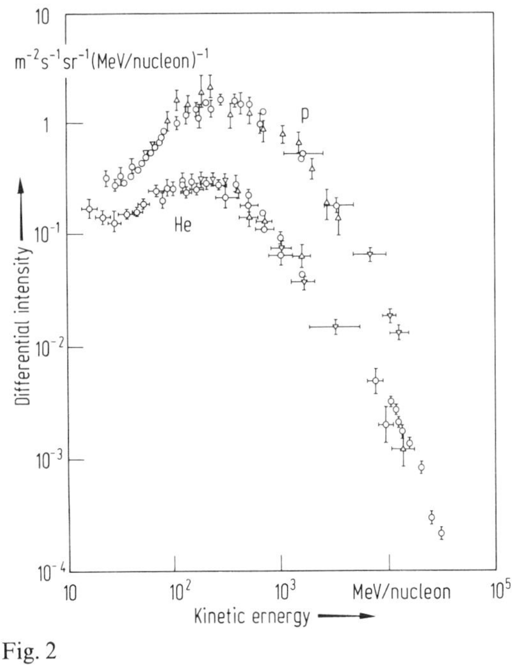


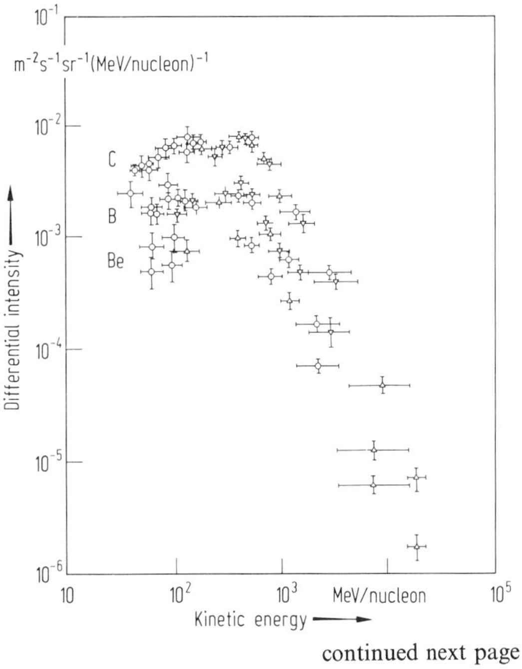


Fig. 2

continued next page

Fig. 2, continued

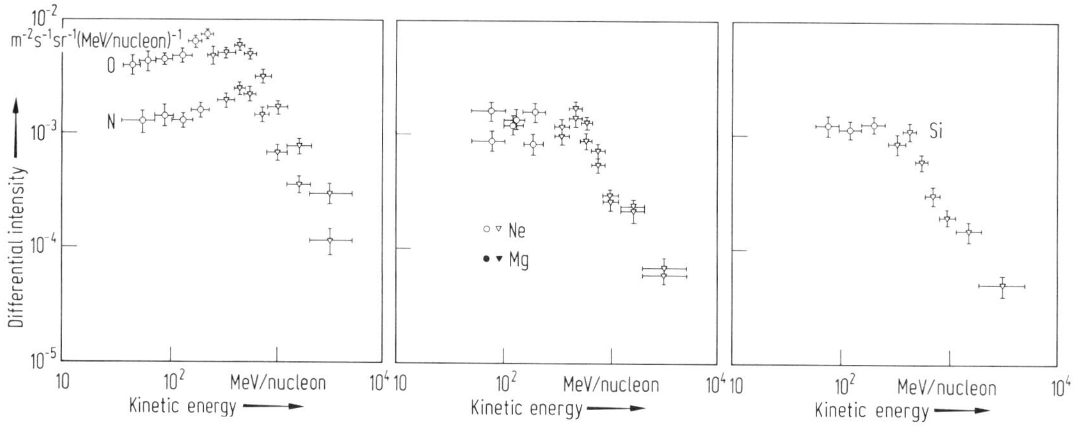


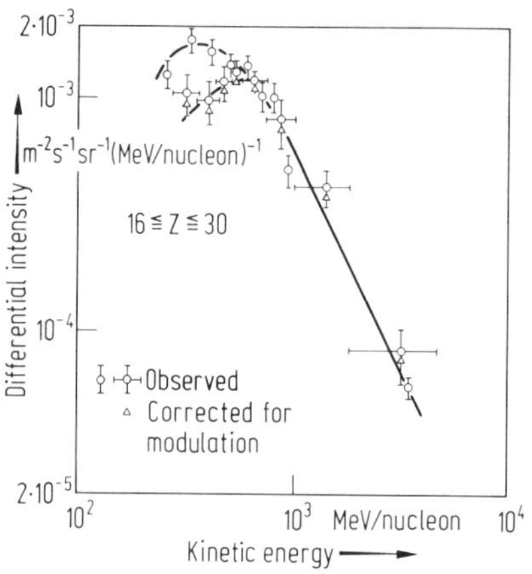


Fig. 2. The energy spectra of individual components of the primary cosmic radiation. These plots have been compiled from the review by Meyer [5] where further details can be found as well as the references to the original literature.

### 7.6.2.2.2 The isotopic composition of GCR nuclei

Data on isotopic ratios for several GCR nuclei have recently become available. By and large, these data have generally confirmed the similarity between the GCR composition and solar system material, if spallation effects and selective acceleration (see 7.6.2.2.4) are taken into account.

The present situation is illustrated in Fig. 3, taken from a recent review by Reeves [6].

For carbon, nitrogen and oxygen, spallation corrections which are large compared to the solar-system ratios must be applied. The resulting upper limits are not in contradiction with the solar-system ratios.

The only isotope ratio which seems to be different is that for neon (see e.g. [7, 8]). However, in this case the solar-system value is not known with certainty. Whereas the solar-wind ratio of $^{22}Ne/^{20}Ne \approx 0.08$ [9], this ratio is about 0.13 [10] for gas trapped in chondrites, which is similar to the value measured recently in solar flares [11]. The GCR $^{22}Ne/^{20}Ne$ ratio of roughly 0.33 is certainly different from either of these.

There is also some indication that the $^{25}Mg/^{24}Mg$ and the $^{26}Mg/^{24}Mg$ ratios in GCR and in the solar system differ, whereas the silicon and iron isotopic ratios are fully consistent with the meteoritic ratios (e.g. [12]).

The recent results have been summarized in [13], where the measurements made at different energies have also been compiled. Their theoretical implications are discussed in [6].

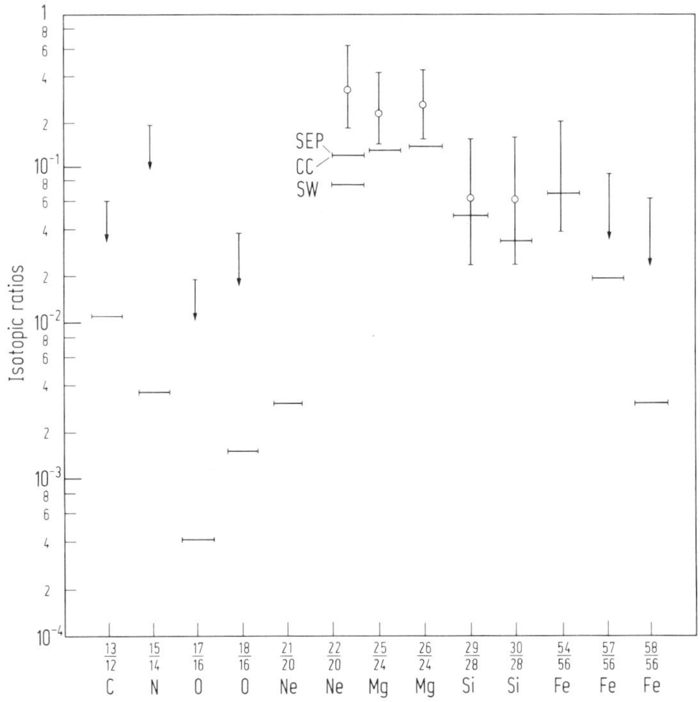


Fig. 3. Isotopic ratios in GCR or their upper limits (vertical bars) compared with the same ratios in the solar system (horizontal bars). SEP refers to solar energetic particles, C C to carbonaceous chondrites, and SW to the solar wind [6].

### 7.6.2.2.3 GCR with energies above $10^{13}$ eV

These are treated here seperately because the results are obtained by a different technique. The observations are made with ground-based instruments through studies of μ-mesons and, beyond $10^{14}$ eV, extensive airshowers.

The total energy spectrum has been measured up to energies of $10^{21}$ eV and the results have already been discussed in 7.6.2.1 .

All evidence on the chemical composition at these high energies is indirect, mostly from observations of the structure of extensive airshowers. It appears that most observations are consistent with a mixture of roughly 60% protons and 40% heavier nuclei. Among these the heavy primaries could be overabundant, but certainly not all these nuclei can be iron nuclei and a normal composition can not be excluded. For further details, see [3, 14, 15].

### 7.6.2.2.4 Source composition of GCR nuclei

In order to determine the cosmic-ray source composition which should provide some clues as to the origin of this radiation, one must correct the observed spectra for

the modulation effects caused by the outward streaming solar wind and the frozen-in magnetic field, and
for the effects of nuclear interactions during cosmic-ray propagation through interstellar space.

Modulation effects

For a detailed compilation of recent results, see [16] and the summary in 3.3.5.3 in [1], which also contains a large number of further references.

Nuclear interactions during propagation through interstellar space

It has been known for many years (see e.g. [17]) that nuclear interactions between cosmic-ray particles and interstellar matter seriously modify the chemical composition. In order to correct for this, one needs (i) a large number of nuclear cross sections since these reactions proceed through a correspondingly large number of reaction channels, and (ii) a model for the propagation of cosmic-ray nuclei through interstellar space. For a recent review which also discusses the pathlength distribution and lifetime of GCR, see [18].

Source composition

Fig. 4, taken from Meyer [13], shows the abundances of the elements in the GCR as measured by the cosmic-ray telescope on board the IMP-8 satellite from 70 to 280 MeV/nucleon. The results are normalized to carbon and compared to "solar-system" abundances [19], and local-Galactic abundances [20]. Note the high abundance e.g. of Li, Be, B due to spallation reactions and other abundance differences possibly caused by a charge dependence of the acceleration process.

In Fig. 5, also taken from Meyer [13], the relative abundances for ultraheavy (UH) nuclei up to Z=60 are plotted. Note the enhancement of elements around platinum, a feature that is expected if the r-process nucleosynthesis plays a dominant role in the production of these nuclei.

From the data of the IMP-8 satellite, together with what they consider the most reliable measurements at higher energies and a particular propagation model, Garcia-Munoz and Simpson [21] have recently determined a set of source abundances which is given in Table 1.

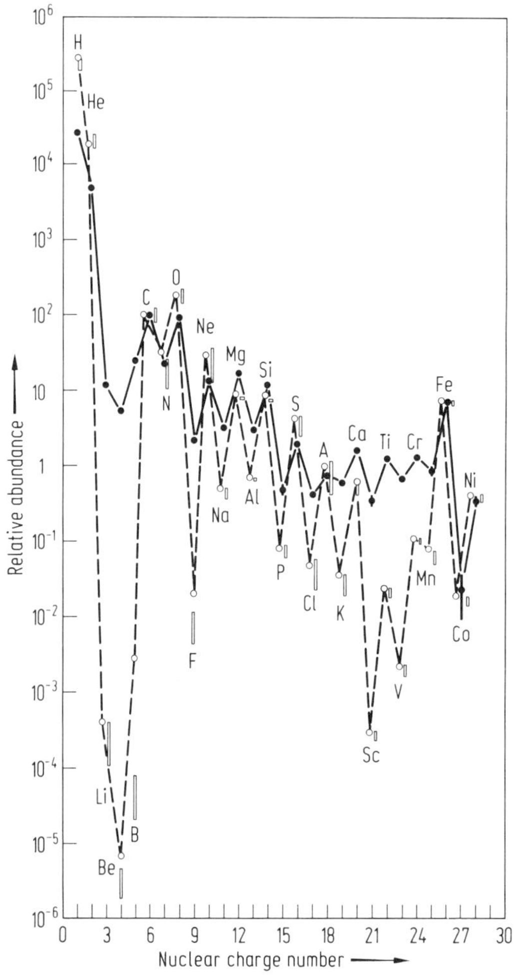


Fig. 4. The relative element abundances of the cosmic rays compared with the relative element abundances of the solar system (normalized to C; according to atomic numbers). Full circles represent measurements by the University of Chicago cosmic-ray telescope on board the IMP-8 satellite from 70 to 280 MeV/nucleon. Open circles denote solar-system abundances [19], open rectangles local-Galactic abundances [20]. For further comments, see [13].

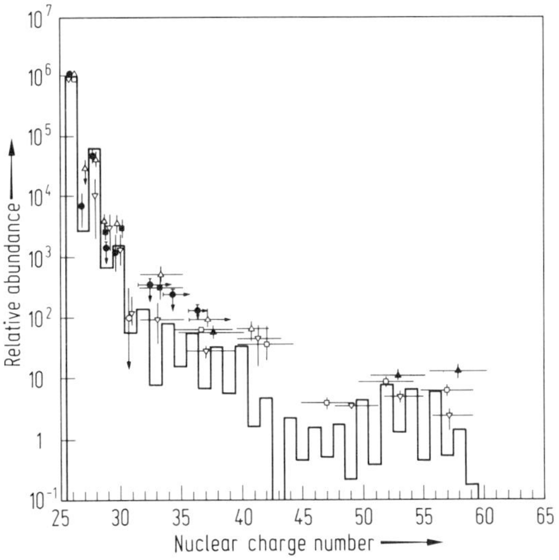


Fig. 5. Relative element abundances of ultraheavy nuclei between Fe and Nd compared with the relative element abundances in the solar system (solid curve). The different symbols represent independent measurements by different groups. For details, see [13].

Table 1. Cosmic-ray source abundances (normalized to C; according to atomic numbers).

| Element | Cosmic-ray source abundances according to different computations | | | Local-Galactic abundances | Ratio of GCR- to local-Galactic abundances (column 4/ column 5) |
|---|---|---|---|---|---|
| | [22] | [23] | [21] | [20] | |
| He | 2600 | 2980 ±60 | 3070 ±200 | $(2.08 \pm 0.46) \cdot 10^{4}$ | 0.148 ± 0.034 |
| C | 100 | 100 ± 2 | 100 ± 1 | 100 ±23 | 1.00 ±23 |
| N | 8 ±2 | 8 ± 2 | 6 ± 1 | 17.7 ± 7.7 | 0.34 ± 0.16 |
| O | 111 ±2 | 112 | 128 ± 3 | 177 ± 38 | 0.72 ± 0.16 |
| F | | 0.4 | 0.7 ± 0.5 | $7.15(1.6) \cdot 10^{-3}$ | |
| Ne | 15 ±2 | 16.0 ± 0.9 | 16.0 ± 2 | 20.8 (1.7) *) | 0.77 (1.7) *) |
| Na | 0.9 ± 0.4 | 0.7 ± 0.6 | 2.7 ± 1 | 0.43 ± 0.07 | 6.3 ± 2.6 |
| Mg | 24 ±2 | 23.6 ± 1.1 | 29.5 ± 2 | 8.08 ± 0.23 | 3.65 ± 0.27 |
| Al | 2.3 ± 1 | 3.0 ± 0.9 | 4.4 ± 1 | 0.65 ± 0.03 | 6.8 ± 1.57 |
| Si | 21 ±3 | 23.1 ± 1.3 | 28.0 ± 2 | 7.69 ± 0.23 | 3.64 ± 0.28 |
| P | $0.2^{+0.4}_{-0.2}$ | | 0.6 ± 0.4 | 0.074 ± 0.015 | |
| S | 3 ±0.6 | 3.7 ± 0.3 | 3.8 ± 0.6 | 3.46 ± 1 | 1.10 ± 0.36 |
| Cl | $0.1^{+0.5}_{-0.1}$ | | 0.3 ± 0.2 | 0.036 (1.6) *) | |
| A | 0.8 ± 0.5 | < 0.8 (Cl, A, K) | 0.8 ± 0.3 | 0.69 (1.7) *) | 1.16 (1.8) *) |
| K | $0.1^{+0.5}_{-0.1}$ | | 0.7 ± 0.3 | 0.028 ± 0.009 | |
| Ca | 2.4 ± 0.8 | 1.6 ± 0.3 | 3.7 ± 0.5 | 0.48 ± 0.06 | 7.71 ± 1.42 |
| Ti | $0.1^{+0.5}_{-0.1}$ | | 0 | | |
| Cr | 0.4 ± 0.3 | < 1.6 (Ti, Cr, Mn) | 0 | | |
| Mn | $0.1^{+0.5}_{-0.1}$ | | 0 | | |
| Fe | 22 ±3 | 26.5 ± 1.8 | 30.5 ± 2 | 6.77 ± 0.46 | 4.51 ± 0.43 |
| Ni | 0.8 ± 0.2 | 1.1 ± 0.2 | 1.7 ± 0.3 | 0.37 ± 0.046 | 4.59 ± 1.0 |

*) (1.6): within a factor 1.6.

It also contains the earlier results by Shapiro et al. [22], Lezniak and Webber [23], and Meyer [20]. This compilation may lead to the conclusion that the source composition is not too drastically different from the chemical abundances observed in the solar system and in many stars.

## 7.6.2.3 The lepton component of GCR

Electrons

Discovered in 1961 by Earl [24] and Meyer and Vogt [25]. Their energy spectrum has now been observed in the range 2 MeV···1000 GeV. For earlier references, see [5], for more recent results [26].

Fig. 6 shows the overall energy spectrum [5], Fig. 7 details of the measurements at low energies and an extrapolation of the galactic spectrum after correction of the observations for various loss processes [27]. Fig. 8 displays a compilation of the more recent results at high energies [26].

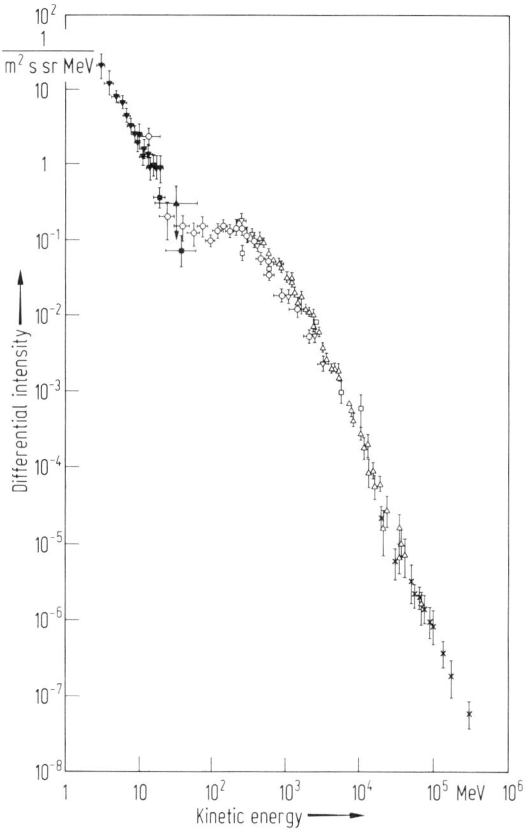


Fig. 6. The overall spectrum of GCR electrons [5]. The different symbols represent measurements by different groups.

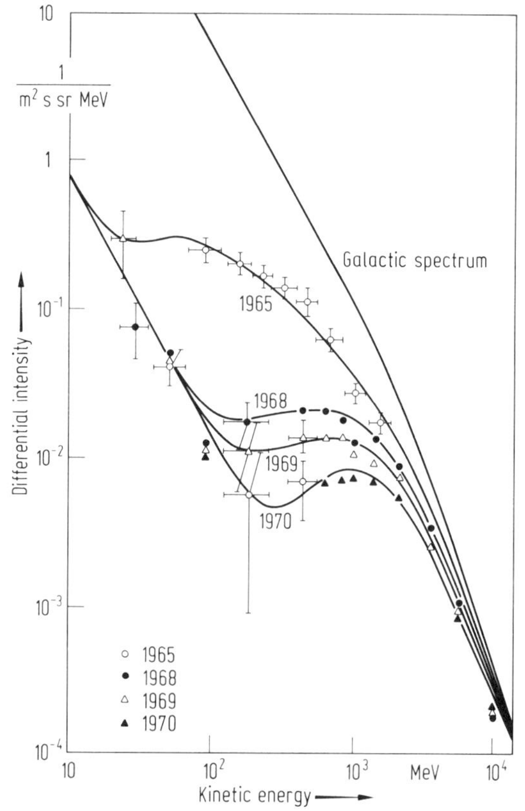


Fig. 7. The low energy end of the GCR electron spectrum as observed and corrected for loss processes [27]. The different symbols represent measurements in different years.

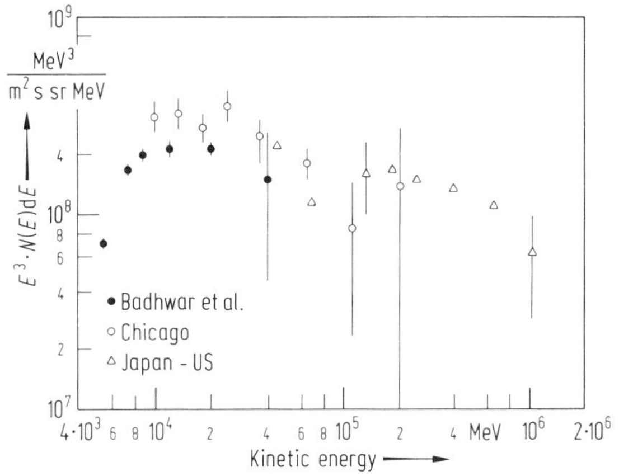


Fig. 8. The GCR electron spectrum at high energies as observed by different groups [26].

The electron to positron ratio

The relative abundance of electrons to positrons is of crucial importance in relation to the question of whether these particles are directly accelerated or whether they originate from nuclear interactions. Measurements are now available for the energy range 0.173···29 GeV. Earlier results are summarized in [5]. More recent results are quoted in [26]. The observations above 0.5 GeV are almost all consistent with $e^+/(e^+ + e^-) = 0.09 \pm 0.01$.

### 7.6.2.4 The origin of GCR

The bulk of the observed cosmic radiation must originate within our Galaxy. Two main potential sources have been considered in the past, (i) supernova explosions, and (ii) acceleration of particles from the interstellar medium. A recent discussion of the respective acceleration mechanisms is given by Axford [28]. This summary also contains a large number of references to the earlier work.

Irrespective of the details of the physics of the acceleration mechanism, some clue as to the origin of GCR is provided by the recently determined isotopic composition and the composition of the ultraheavy (UH) nuclei. Both these point toward the fact that the sample of material that constitutes the primary cosmic rays has features in its composition which distinguish it from solar system material, but also from interstellar material. The most convincing of these features are the over-abundance of $^{22}$Ne, an apparent peak in the Pt region, and the high abundance of the actinides, if that is substantiated (see 7.6.2.2.2 and [29, 30, 26]).

While evidence is mounting – corroborated by γ-ray observations – that low and intermediate energy cosmic rays are a galactic phenomenon, the highest energy cosmic rays must be an extragalactic phenomenon unless the current theories on particle confinement in the Galaxy and its vicinity are drastically revised. Several authors favor an origin of the high energy CR in the local supercluster, probably in the Virgo region. Some evidence for this comes from the distribution of the arrival directions as discussed in [3]. For further details, see [14, 15, 31].

## 7.6.3 The anomalous component

Discovered in 1972/73 through anomalies in the abundances and energy spectra of quiet-time cosmic ray He, N, O, and Ne nuclei. The anomalous abundances (e.g. $O/C \approx 10$, $He/H > 1$) in the energy range 1···30 MeV/nucleon, the positive radial gradient and the modulation effects all indicate a non-solar origin of these particles. They could be generated by a local acceleration mechanism acting within the heliosphere. However, a local galactic origin can not completely be excluded. For further details, see 3.3.5.2 in [1].

## 7.6.4 High-energy particles in extragalactic systems

Early in the 1950's, it was realized that the non-thermal radio emission that is observed in many extragalactic objects (normal galaxies, radiogalaxies, and quasi-stellar radio sources) must be produced by relativistic electrons radiating by the magneto-bremsstrahlung process. This implies the existence of cosmic ray electrons in these systems. By analogy one may then conclude that cosmic ray nuclei must also be present in these extragalactic system as they are in our own Galaxy.

## 7.6.5 References for 7.6


1 Landolt-Börnstein, NS, Vol. VI/2a (1981).
2 Iwai, J., Ogata, T., Ohta, I., Takahashi, Y., Yanagita, T. in: Proc. 16th Int. Cosmic Ray Conf., Vol. **8** (1979) 1.
3 Suga, K. in: Proc. 16th Int. Cosmic Ray Conf., Vol. **14** (1979) 325.
4 Effimov, N.N., Sokurov, V.F. in: Proc. 16th Int. Cosmic Ray Conf., Vol. **8** (1979) 153.
5 Meyer, P.: Annu. Rev. Astron. Astrophys. **7** (1969) 1.
6 Reeves, H. in: Origin of Cosmic Rays, Int. Astron. Union Symp. **94** (1981) 23.
7 Garcia-Munoz, M., Simpson, J.A., Wefel, J.P.: Astrophys. J. **232** (1979) L95.
8 Mewaldt, M.A., Spalding, J.D., Stone, E.C., Vogt, R.E.: Astrophys. J. **235** (1980) L95.
9 Geiss, J., Bochsler, P. in: Proc. 4th Solar Wind Conf. (1979).
10 Black, D.C., Pepin, R.O.: Earth and Planet. Sci. Lett. **6** (1969) 395.
11 Mewaldt, R.A. in: Proc. of the Conf. on the Ancient Sun (1980).
12 Garcia-Munoz, M., Simpson, J.A., Wefel, J.P. in: Proc. 16th Int. Cosmic Ray Conf., Vol. **1** (1979) 436.
13 Meyer, P. in: Origin of Cosmic Ray, Int. Astron. Union Symp. **94** (1981) 7.
14 Watson, A.A. in: Proc. 14th Int. Cosmic Ray Conf., Vol. **11** (1975) 4019.
15 Sreekantan, B.V. in: Proc. 16th Int. Cosmic Ray Conf., Vol. **14** (1979) 345.
16 Webber, W.R. in: Proc. 16th Int. Cosmic Ray Conf., Vol. **14** (1979) 253.
17 Shapiro, M.M., Silberberg, R.: Annu. Rev. Nucl. Sci. **20** (1970) 323.
18 Raisbeck, G.M. in: Proc. 16th Int. Cosmic Ray Conf., Vol. **14** (1979) 146.
19 Cameron, A.G.W.: Space Sci. Rev. **15** (1973) 121.
20 Meyer, J.P. in: Proc. 22nd Liège Int. Astrophys. Symp. (1978).
21 Garcia-Munoz, M., Simpson, J.A. in: Proc. 16th Int. Cosmic Ray Conf., Vol. **1** (1979) 270.
22 Shapiro, M.M., Silberberg, R., Tsao, C.H. in: Proc. 14th Int. Cosmic Ray Conf., Vol. **2** (1975) 532.
23 Lezniak, J.K., Webber, W.R.: Astrophys. Space Sci. **63** (1979) 35.
24 Earl, J.R.: Phys. Rev. Lett. **6** (1961) 125.
25 Meyer, P., Vogt, R.E.: Phys. Rev. Lett. **6** (1961) 193.
26 Balasubrahmanyan, V.K. in: Proc. 16th Int. Cosmic Ray Conf. Vol. **14** (1979) 121.
27 Urch, I.H., Gleeson, L.J.: Astrophys. Space Sci. **17** (1972) 426.
28 Axford, I. in: Origin of Cosmic Rays, Int. Astron. Union Symp. **94** (1981) 339.
29 Meyer, P.: Proc. 14th Int. Cosmic Ray Conf. Vol. **11** (1975) 3698.
30 Waddington, C.J. in: Proc. 15th Int. Cosmic Ray Conf., Vol. **10** (1977) 168.
31 Wolfendale, A.W.: Pramãna **12** (1979) 63.

# 7.7 Interstellar magnetic field

## 7.7.1 Symbols and units

| | |
|---|---|
| $B_\perp$ [Gauss] | magnetic field strength perpendicular to line of sight |
| $B_\parallel$ [Gauss] | magnetic field strength parallel to line of sight |
| $N$ [$cm^{-3}$] | number density of the particles relevant for the particular process in question (e.g. of H I atoms in case of 21-cm Zeeman splitting) |
| $N_e$ [$cm^{-3}$] | number density of free electrons |
| $\tilde{N}$ | quantity related (but not necessarily proportional) to the number density of the particles causing the process in question |
| $DM$ [pc $cm^{-3}$] | dispersion measure |
| $RM$ [rad $m^{-2}$] | rotation measure |
| $\langle f \rangle$ | $\int f(s)ds$ = integral along line of sight, from observer to infinity, over function $f$; $s$ [pc] = length-coordinate along line of sight. In case of non-negligible extinction, integrand $f(s)$ is to be multiplied by an appropriate attenuation factor. |

## 7.7.2 Methods of determination

Information on the galactic magnetic field is obtained from

(a) synchrotron emission by gyrating relativistic electrons, and in particular from its polarization: theory [23, 41, 42, 59], depolarization [9, 15, 64]; for definition of some observable quantities (e.g. polarization brightness temperature), see [6];

(b) Faraday rotation of linearly polarized radiation (from pulsars, radio galaxies or quasars) traversing ionized gas in magnetic fields [b, 57];

(c) polarization (usually linear) of optical light from reddened stars (or of infrared from protostars [19]), caused by preferential extinction by magnetically aligned interstellar grains [b, 1, 36, alignment mechanism also a];

(d) Zeeman splitting [e], in the H I 21-cm absorption line originating from H I gas in front of strong radio continuum sources [b] and in molecular (primarily OH maser) emission lines [14, 18a, 24].

More indirect information on the galactic fields may come from the shape of interstellar clouds [a] and from the cosmic ray spectrum [4, 28].

Processes (a) and (c) permit measurements of $\langle \tilde{N} B_\perp^\alpha \rangle$, the line-of-sight integral of a density-related quantity $\tilde{N}$ of the interacting agent (electrons, grains) weighted by a power $\alpha$ of $B_\perp$, and, in particular, derivation of the active-line-of-sight-averaged field orientation from the orientation of polarization. Note that the power $\alpha$ depends, in case of process (a), on the electron energy spectrum; for process (c), $\alpha = 2$, cf. [20]. The relation between $\tilde{N}$ and agent density $N$ may be complicated; cf. refs. under (a) and (c).

$\langle N B_\parallel \rangle$ is obtained from process (b) or (d), with density $N$ ($= N_e$ in process (b)) determined, e.g., from the dispersion measure $DM = \langle N_e \rangle$ as deduced from pulsar signal delays (see [b], cf. also 5.6.3.3).

The rotation measure $RM$ for process (b) is defined as $RM = \psi/\lambda^2 = 8.1 \cdot 10^5 \langle N_e B_\parallel \rangle$, with $\psi$ = total rotation angle in [rad], $\lambda$ = wavelength of observation in [m], and $RM$, $N_e$, $B_\parallel$ and $\langle\rangle$ in units as given in 7.7.1.

## 7.7.3 Observational results

Summary of properties of interstellar magnetic field: [c, in tabular form: e].

North polar spur: [27].

From individual processes (labeled here as in 7.7.2):

(a): [8, 28, 40 ($\nu < 10$ MHz, from satellite), 45, 64];

(b): extragalactic sources [53···55 ($RM$ data)] – for field parameters derived from them, see also [51], for fields in H II regions [26]; pulsars (mostly in galactic plane, up to several kpc from the sun) [34, 35], see also [50, 54];

(c): [3 (largest body of data), 10, 22, 30, 31, 37], field structure from stellar polarization data [20], Barnard's loop [2], RCrA dark cloud [61a];

(d): H I 21 cm (few detections only) [e, 60], upper limits [48]; OH (interpretation controversial [e, 66]; note that OH-masing regions are small compared to H I and SO regions) e.g. [13] (Orion; but see [66]), [17, 33, 38 (at 5 cm), 47a, 65 (V 1057 Cyg)]; SO (upper limits only) [16].

Alignment of dust clouds/H I filaments with optical polarization vectors: [a, 25, 52, 61], but see also [29].

Table 1. Some derived values of galactic magnetic field strengths $B$.

| Field region | $B$ [μG] ($B_{\parallel}$; $B_{\perp}$) | Comments | Method of determination | Ref. |
|---|---|---|---|---|
| Galaxy at large | ≈2 | regular component of large-scale galactic field | rotation measures of pulsars and extragalactic radio sources, polarization of galactic radio emission | c, e, 34 |
| | ≈2 | random component (with scale of variations ≈50 pc) | | |
| H I regions | 10···20; 50···70 | towards Cas A; Ori A | Zeeman splitting of 21-cm line | e |
| | ≲(4···10) | | in emission | 67 |
| | <0.8; <4.6 | towards Cas A; Cyg A | in absorption | 67 |
| Molecular clouds | ≃100 | R CrA dark cloud | polarization of starlight | 61a |
| | | | Zeeman splitting of OH lines | 17 |
| | < 50 | | in emission | 68 |
| | < 15···40 | | in absorption | 68 |
| H II regions | ≈10 | in S 232 | Faraday rotation of extra-galactic radio sources behind the H II regions | 26 |
| | ≈1···2; ≈20 | in S 117 and S 264; in S 119 | | 69 |
| OH maser sources (mostly associated with H II regions) | several 1000 | small (≲$10^{15}$ cm) high-density regions | Zeeman splitting of OH emission lines | 33, 65 |
| North polar spur | ≳1.2 | radio continuum shell | Faraday depolarization | 27 |
| | ≲6 | H I shell | Zeeman splitting of 21-cm line | 27 |

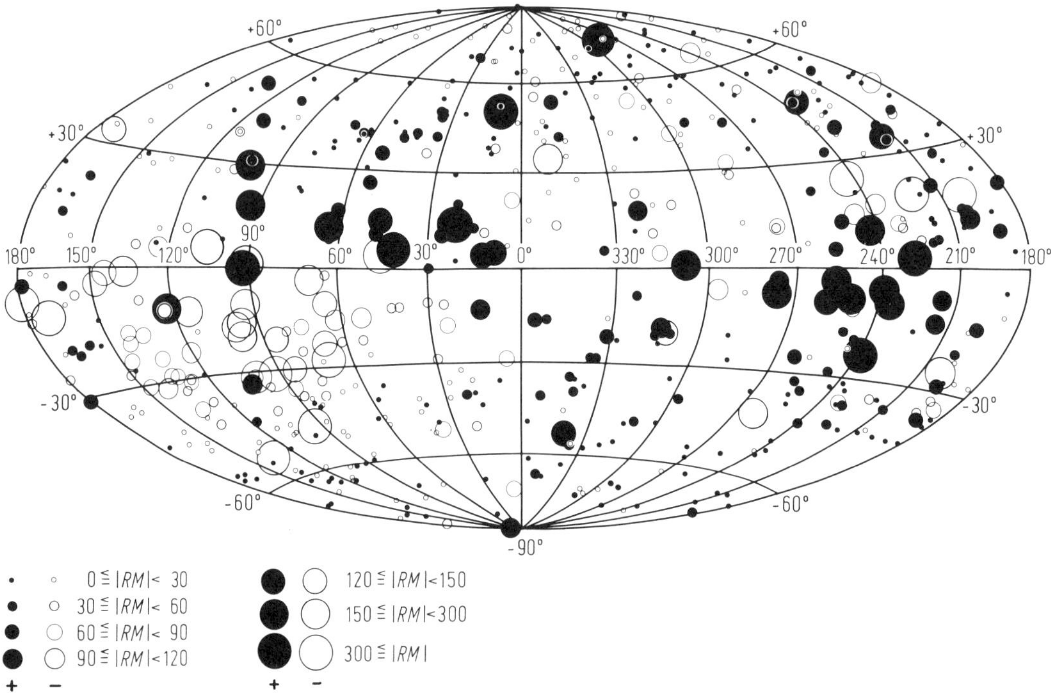


Fig. 1. Variation of rotation measures, $RM$ in [rad $m^{-2}$], of extragalactic sources over the sky. From P. P. Kronberg and M. Simard-Normandin, private communication 1981.

## 7.7.4 Theoretical aspects

Field models from synchrotron emission [7, 49, 62]; from rotation measures [54 (limited to large-scale field in galactic plane, and including thickness of magneto-ionic layer)]. Interaction of field with interstellar gas and cosmic rays [44, 46]. Hydromagnetic model of spiral arms (invoking intergalactic magnetic field) [47]. Origin of galactic magnetic field: seed fields from Poynting-Robertson effect [24a]; dynamo theory starting from small seed fields [43, 44, 58]; primordial origin (including comparison with dynamo theory) [46]. Galactic dynamo model [56, 63]. Cosmic rays: in field [10, 12], diffusion in turbulent galactic field [11], effects of convection, diffusion and radiation losses on electrons [32]. Spatial reversals of field direction: [43, 47, 58], observational evidence [54].

Dense clouds and star formation as result of local field structure in Orion [2]; relationship between field and cloud density [21]; flux-freezing in molecular clouds [a, 5, 39]; fields constraining motion of giant cloud complexes [21a].

## 7.7.5 References for 7.7

### General references


a Heiles, C.: Annu. Rev. Astron. Astrophys. **14** (1976) 1.
b Spitzer, L.: Physical Processes in the Interstellar Medium, Wiley, New York (1978).
c Spoelstra, T.A.Th.: Sov. Phys. Usp. **20** (1977) 336.
d Verschuur, G.L., Kellermann, K.I. (eds.): Galactic and Extragalactic Radio Astronomy, Springer, Berlin (1974) p. 179 seq.
e Verschuur, G.L.: Fundamentals of Cosmic Physics **5** (1979) 113.


### Special references


1 Aannestad, P.A., Purcell, E.: Annu. Rev. Astron. Astrophys. **11** (1973) 309.
2 Appenzeller, I.: Astron. Astrophys. **36** (1974) 99.
3 Axon, D.J., Ellis, R.S.: Mon. Not. R. Astron. Soc. **177** (1976) 177.
4 Badhwar, G.D., Stephens, S.A.: Astrophys. J. **212** (1977) 494.
5 Baker, P.L.: Astron. Astrophys. **75** (1979) 54.
6 Berkhuijsen, E.M.: Astron. Astrophys. **40** (1975) 311.
7 Brindle, C., French, D.K., Osborne, J.L.: Mon. Not. R. Astron. Soc. **184** (1978) 283.
8 Brouw, W.N., Spoelstra, T.A.Th.: Astron. Astrophys. Suppl. **26** (1976) 129.
9 Burn, B.J.: Mon. Not. R. Astron. Soc. **133** (1966) 67.
10 Carrasco, L., Strom, S.E., Strom, K.M.: Astrophys. J. **182** (1973) 95.
11 Carvalho, J.C., ter Haar, D.: Mon. Not. R. Astron. Soc. **187** (1979) 23.
12 Cesarsky, C.J.: Annu. Rev. Astron. Astrophys. **18** (1980) 289.
13 Chaisson, E.J., Beichman, C.A.: Astrophys. J. **199** (1975) L39.
14 Chaisson, E.J., Vrba, F.J. in: Protostars and Planets (Gehrels, T., ed.), Univ. of Arizona Press, Tucson (1978) p. 189.
15 Cioffi, D.F., Jones, T.W.: Astron. J. **85** (1980) 368.
16 Clark, F.O., Johnson, D.R., Heiles, C.E., Troland, T.H.: Astrophys. J. **226** (1978) 824.
17 Crutcher, R.M., Evans, N.J., Troland, T., Heiles, C.: Astrophys. J. **198** (1975) 91.
18 Daniel, R.R., Stephens, S.A.: Space Sci. Rev. **17** (1975) 45.
18a Davies, R.D. in: Galactic Radio Astronomy, Int. Astron. Union Symp. **60** (Kerr, F.J., Simonson, S.C., eds.), Reidel, Dordrecht (1974) p. 275.
19 Dyck, H.M., Beichman, C.A.: Astrophys. J. **194** (1974) 57.
20 Ellis, R.S., Axon, D.J.: Astrophys. Space Sci. **54** (1978) 425.
21 Elmegreen, B.G.: Astrophys. J. **225** (1978) L85.
21a Elmegreen, B.G.: Astrophys. J. **243** (1981) 512.
22 Gammelgaard, P.: Astron. Astrophys. **70** (1978) 195.
23 Ginzburg, V.L., Syrovatskij, S.I.: Annu. Rev. Astron. Astrophys. **3** (1965) 297.
24 Goldreich, P., Keeley, D.A., Kwan, J.Y.: Astrophys. J. **179** (1973) 111.
24a Harwit, M.: Astrophysical Concepts, Wiley, New York (1973) p. 415.
25 Heiles, C. in: Galactic Radio Astronomy, Int. Astron. Union Symp. **60** (Kerr, F.J., Simonson, S.C., eds.), Reidel, Dordrecht (1974) p. 13.
26 Heiles, C., Chu, Y.-H.: Astrophys. J. **235** (1980) L105.

27 Heiles, C., Chu, Y.-H., Reynolds, R.J., Yegingil, I., Troland, T.H.: Astrophys. J. **242** (1980) 533.
28 Higdon, J.C.: Astrophys. J. **232** (1979) 113.
29 Hopper, P.B., Disney, M.J.: Mon. Not. R. Astron. Soc. **168** (1974) 639.
30 Klare, G., Neckel, T., Schnur, G.: Astron. Astrophys. **11** (1971) 155.
31 Klare, G., Neckel, T., Schnur, G.: Astron. Astrophys. Suppl. **5** (1972) 239.
32 Lerche, I., Schlickeiser, R.: Astrophys. J. **239** (1980) 1089.
33 Lo, K.Y., Walker, R.J., Burke, B.F., Moran, J.M., Johnson, K.J., Ewing, M.S.: Astrophys. J. **202** (1975) 650.
34 Manchester, R.N.: Astrophys. J. **188** (1974) 637.
35 Manchester, R.N., Taylor, J.H.: Pulsars, Freeman, San Francisco (1977).
36 Martin, P.G.: Cosmic Dust, Clarendon, Oxford (1978).
37 Mathewson, D.S., Ford, V.L.: Mem. R. Astron. Soc. **74** (1970) 139.
38 Moran, J.M., Ried, M.J., Lada, C.J., Yen, J.L., Johnston, K.J., Spencer, J.H.: Astrophys. J. **224** (1978) L67.
39 Mouschovias, T.C.: Astrophys. J. **207** (1976) 141.
40 Novaco, J.C., Brown, L.W.: Astrophys. J. **221** (1978) 114.
41 Pacholczyk, A.G.: Radio Astrophysics, Freeman, San Francisco (1970).
42 Pacholczyk, A.G. in: Lecture Notes on Introductory Theoretical Astrophysics (Weymann, R.J. et al., eds.), Pachart Publ. House, Tucson (1976) p. 151.
43 Parker, E.N.: Astrophys. J. **163** (1971) 255.
44 Parker, E.N.: Cosmical Magnetic Fields , Clarendon, Oxford (1979) ch. 22.
45 Phillipps, S., Kearsey, S., Osborne, J.L., Haslam, C.G.T., Stoffel, H.: Astron. Astrophys. **98** (1981) 286.
46 Piddington, J.H.: Astrophys. Space Sci. **37** (1975) 183.
47 Piddington, J.H.: Astrophys. Space Sci. **59** (1978) 237.
47a Reid, M.J., Haschik, A.D., Burke, B.F., Moran, J.M., Johnston, K.J., Swenson, G.W.: Astrophys. J. **239** (1980) 89.
48 Reif, K., Booth, R., Mebold, U., Winnberg, A.: Astron. Astrophys. **70** (1978) 271.
49 Rockstroh, J.M., Webber, W.R.: Astrophys. J. **224** (1978) 677.
50 Ruzmaikin, A.A., Sokolov, D.D.: Astrophys. Space Sci. **52** (1977) 365; erratum: **56** (1978) 519.
51 Ruzmaikin, A.A., Sokolov, D.D., Kovalenko, A.V.: Soviet. Astron. **22** (1978) 395.
52 Schlosser, W., Schmidt-Kaler, Th. in: Planets, Stars and Nebulae (Gehrels, T., ed.), Univ. of Arizona Press, Tucson (1974) p. 972.
53 Simard-Normandin, M., Kronberg, P.P.: Nature **279** (1979) 115.
54 Simard-Normandin, M., Kronberg, P.P.: Astrophys. J. **242** (1980) 74.
55 Simard-Normandin, M., Kronberg, P.P., Button, S.: Astrophys. J. Suppl. **45** (1981) 97.
56 Soward, A.M.: Astron. Nachr. **299** (1978) 25.
57 Spitzer, L.: Physics of Fully Ionized Gases, Wiley, New York, 2nd ed. (1962).
58 Stix, M.: Astron. Astrophys. **42** (1975) 85; erratum: **68** (1978) 459.
59 Tucker, W.H.: Radiation Processes in Astrophysics, Mass. Inst. Technol. Press, Cambridge (1975).
60 Verschuur, G.L. in: [d] p. 192.
61 Vrba, F.J., Strom, S.E., Strom, K.M.: Astron. J. **81** (1976) 858.
61a Vrba, F.J., Coyne, G.V., Tapia, S.: Astrophys. J. **243** (1981) 489.
62 Webber, W.R., Simpson, G.A., Cane, H.V.: Astrophys. J. **236** (1980) 448.
63 White, M.P.: Astron. Nachr. **299** (1978) 209.
64 Wilkinson, A., Smith, F.G.: Mon. Not. R. Astron. Soc. **167** (1974) 593.
65 Winnberg, A., Graham, D., Walmsley, C.M., Booth, R.S.: Astron. Astrophys. **93** (1981) 79.
66 Zuckerman, B., Palmer, P.: Astrophys. J. **199** (1975) L35.
67 Troland, T.H., Heiles, C.: preprint (1981).
68 Crutcher, R.M., Troland, T.H., Heiles, C.: preprint (1981).
69 Heiles, C., Chu, Y.-H., Troland, T.H.: preprint (1981).

# 8 Our Galaxy

## 8.1 Positions, motions, parallaxes of stars

### 8.1.1 Star positions

#### 8.1.1.1 Introduction

Star positions are determined and catalogued

1. for identification. Star lists perform this task (see 8.1.1.3) without giving the exact position, also star maps and atlases (see 8.1.1.4). Occasionally exact positions of very faint objects are published for identification but normally finding charts are preferred for special faint stars.

2. to provide data for different scientific investigations. The exact measurements are the task of positional astronomy (see 8.1.1.5).

Normally, star positions are given in the equatorial system of right ascension and declination. In some cases, galactic longitudes and latitudes have been published for special investigations. Positions on photographic plates are measured and often given in rectangular coordinates.

Star positions are observed directly with transit circles (meridian circles), transit instruments, vertical circles, photographic zenith tubes (PZT), and also with astrolabes. Recently it has become possible to measure positions of radio sources by radio interferometric methods; a review on radio astrometry is given by Elsmore [2]. The observed positions are "apparent places" ([3] p. 149; see also 8.1.1.5).

Positions relative to objects in their close neighbourhood are determined visually by micrometer measurements, today mainly on photographic plates taken with long-focus instruments; purposes: double star positions, parallax determinations, proper motions of faint stars, and positions of minor planets or comets. See: Stars and Stellar Systems II, Astronomical Techniques (Hiltner, W.A., ed.) Chicago (1962) chs. 19···21.

#### 8.1.1.2 Constellations

The International Astronomical Union stipulated the boundaries of the constellations at its General Assemblies in 1925 and 1928. They are well-defined lines running along the hour circles and declination circles in the equatorial system of 1875.0 [1]. It was further agreed that Latin names for constellations should be used in the scientific literature. These names are normally shortened to three (seldom four) letters [11]. List of the constellations [9].

#### 8.1.1.3 Star lists and nomenclature of stars

Star lists are compilations which give characteristic data for each object. Usually the positions are given only approximately, but sufficiently for each star to be identified. Bright stars, if not named by Greek letters or numbers within their constellation, are marked by numbers in one of the star lists in Table 1. For faint stars the nomenclature is not systematic. Only the most important of the large number of lists used for denomination of faint objects have been included in Table 1. More than fifty references to the nomenclature of proper-motion stars are compiled by Giclas et al. [6]; see also Gliese and Jahreiss [7]. Standardization of the nomenclature of faint stars is under discussion (IAU). The following proposal is also used occasionally: Specification of a star by ten digits giving its rough position at a certain equinox; e.g. 14263–6228 = Proxima Centauri (the nearest star) at $14^h26^m.3$, $-62°28'$ (1950). Nomenclature and abbreviations of lists for special types (double stars, variables, white dwarfs, ...) are given in the chapters concerned.

Table 1. The most important star lists (used for nomenclature).
$m_{lim}$ = limiting of magnitude
$N$ = number of stars

| No. | Abbr. | Name | Author | Literature | $m_{lim}$ | $N$ | Equinox | Declinations |
|---|---|---|---|---|---|---|---|---|
| 1 | BS = HR | Catalogue of Bright Stars (Third revised edition) | Hoffleit, D. | New Haven, Yale, Univ. Obs. (1964) | $6^m.5$ | 9091 | 1900 2000 | $-90°\cdots+90°$ |

continued

Table 1, continued

| No. | Abbr. | Name | Author | Literature | $m_{lim}$ | $N$ | Equinox | Declinations |
|---|---|---|---|---|---|---|---|---|
| 2 | BD | Bonner Durchmusterung nördl. Teil | Argelander, F.W. | Astron. Beob. Bonn 3···5 (1859···1862) 3. Ed. Bonn 1951 | 9.5 | 324189 | 1855 | − 2°···+90° |
| | | südl. Teil | Schönfeld, E. | Astron. Beob. Bonn 8 (1886) 2. Ed. Bonn 1949 | 10 | 133659 | 1855 | −23°···−2° |
| 3 | CD CoD | Cordoba Durchmusterung | Thome, J., Perrine, C.D. | Result. Cordoba 16···18, 21 (1892···1932) | 10 | 613953 | 1875 | −22°···−90° |
| 4 | CPD | Cape Photographic Durchmusterung | Gill, D., Kapteyn, J.C. | Ann. Cape III···V (1896···1900) | 9.2 | 454875 | 1875 | −18°···−90° |
| 5 | HD | Henry Draper Catalogue | Cannon, A. | Ann. Harvard 91···99 (1918···1924) | 8.3 | 225300 | 1900 | −90°···+90° |
| | HDE | Henry Draper Extension | Cannon, A. | Ann. Harvard 100 (1925) | 8.5 | 47000 | 1900 | special regions |
| 6 | BPM | Bruce Proper Motion Survey | Luyten, W.J. | Univ. of Minnesota (1963) Vols. I, II | | 94263 | 1900 | −90°···+40° |
| 7 | G | Lowell Proper Motion Survey Northern Hem. | Giclas, H.L., Burnham, jr., R., Thomas, N.G. | The G numbered stars, Lowell Obs. (1971) | 16 (pg) | 8991 | 1950 | 0°···+90° |
| | | Southern Hem. | | Lowell Bull. **8** (1978) 89 | 17 (pg) | 2758 | 1950 | special regions −40°···0° |
| 8 | LFT | A Catalogue of 1849 Stars With Proper Motions Exceeding 0.″5/a | Luyten, W.J. | Minneapolis, Minn. (1955) | | 1849 | 1950 | −90°···+90° |
| 9 | LHS | LHS Catalogue, 2. Ed. | Luyten, W.J. | Minneapolis (1979) | 21 (pg) | 4045 | 2000 1950 | −90°···+90° |
| 10 | LTT | A Catalogue of 9867 Stars in the Southern Hemisphere with Proper Motions Exceeding 0.″2/a | Luyten, W.J. | Minneapolis, Minn. (1957) | | 9867 | 1950 | −90°··· 0° |
| | | A Catalogue of 7127 Stars in the Northern Hemisphere with Proper Motions Exceeding 0.″2/a | | (1961) | | 7127 | 1950 | 0°···+90° |
| | | First Supplement to the LTT Catalogues | | (1962) | | 1508 | 1950 | −90°···+90° |

Occasionally also numbers from the following zone catalogues (Table 3) are used for denomination of stars in the literature: Yale Zone Catalogues, Cape Photographic Zone Catalogues, AGK 3, Astrographic Catalogue (AC).

## 8.1.1.4 Sky charts and atlases

The currently used star charts and atlases are given in Table 2. Precession tables which are necessary for the comparison of star charts of different equinoxes or for the finding of objects are to be found in 8.1.1.6.

Table 2. Sky charts and atlases ($n$ = number of maps).

| No. | Name | Author | Publication | $n$ | Scale 1° ≙ | $m_{lim}$ | Equinox | Declinations |
|---|---|---|---|---|---|---|---|---|
| 1 | Himmelsatlas | Schurig-Götz | 8. Ed., Mannheim (1960) | 8 | 2.9 mm | $6^m\!.5$ | 1950 | −90°···+90° |
| 2 | Norton's Star Atlas and Reference Handbook | Norton, A.P. | Cambridge, Mass. (1973) | 8 | 3.2 | 6.3 | 1950 | −90°···+90° |
| 3 | Skalnate Pleso Atlas of the Heavens | Becvár, A. | Cambridge, Mass. (1969; 1976) | 16 | 5.3 | 7.75 | 1950 | −90°···+90° |
| 4 | Sky Atlas 2000.0 | Tirion, W. | Sky Publ. Corp. Cambridge, Mass. (1981) | 26 | 7.2 | 8.0 | 2000 | −90°···+90° |
| 5 | Bonner Durchmusterung nördl. Teil | Argelander, F.W. | 3. Ed., Bonn (1954) | 40 | 20 | 9.5 | 1855 | − 1°···+90° |
| | südl. Teil | Schönfeld, E. | 2. Ed., Bonn (1951) | 24 | 20 | 10 | 1855 | −23°··· − 1° |
| 6 | Cordoba Durchmusterung | Thome, J., Perrine, C.D. | Cordoba (1929) | 28 | 20 | 9.5 | 1875 | −21°···−90° |
| 7 | Atlas Stellarum | Vehrenberg, H. | Düsseldorf (1970) | 450 | 30 | 14 (pg) | 1950 | −90°···+90° |
| 8 | True Vis. Mag. Phot. Star Atlas, Vol. 1, 2 | Papadopoulos, C. | Oxford (1979) | 120<br>216 | 30 | 13.5 (pv) | | −90°···−30°<br>−30°···+30° |
| | Vol. 3 | Scovil, C. | Oxford (1980) | 120 | 30 | 13.5 (pv) | | +30°···+90° |
| 9 | Palomar Observ. Sky Survey (POSS) and Southern Extension | Mt. Palomar Observatory | Mt. Palomar (1954 and 1958) | 2×935[1] | 53.6 | 20···21 | | −27°···+90° and −33°···−27° |
| 10 | ESO/SRC Atlas of the Southern Sky ESO(R), SRC(J) | ESO SRC | in preparation, on film and on glass | 2×606 | 53.4 | 22 (R)<br>23 (J) | | −90°···−17° |
| | ESO(B) Atlas | ESO | on film and on glass (1973···1978) | 606 | 53.4 | 21.5 | | −90°···−17° |

[1]) For each area: one plate in blue and one plate in red.

## 8.1.1.5 Catalogues of star positions

The first lists of star positions were compiled more than 2000 years ago but catalogues produced before 1750 are of historical interest only. Usually, modern catalogues of star positions give the right ascensions to $0^s.01$ or $0^s.001$ and the declinations to $0''.1$ or $0''.01$. These data are obtained by direct observations (observational catalogues) or by combining various catalogues (compiled catalogues).

Comprehensive bibliographies of catalogues of star positions:

···1976 Servalić, B.M., Teleki, G., Szádeczky-Kardoss, G.: Bibliography of catalogues of star positions, Publ. Beograd **7** (1978) 69: 2087 catalogues.

1963···1981 Astronomisches Rechen-Institut, Heidelberg. (Manuscript, 1982): about 250 catalogues.

### Classification of position catalogues

In the past no unambiguous terminology of catalogue classes has been used. There is much confusion among the titles chosen by the authors and also in bibliographies. The following classification of position catalogues is recommended [8]:

Observational catalogues are those which are directly based on observations made usually with one instrument but occasionally with more than one. Later reductions of original observational data (e.g. Bessel's reduction of Bradley's observations) can still be regarded as giving an observational catalogue. (The use of the term "independent catalogue" for certain kinds of observational catalogues, [b] p. 380 and [a] p.102, is not recommended.)

Observational catalogues can be subdivided into

1. Absolute catalogues for which the reductions have been carried out without adopting any (previously known) star positions. For such a catalogue to be rigorously absolute, the zero points of the right ascension and declination systems must be independently determined. (The use of the term "fundamental" to describe such a catalogue should be avoided.)

2. Quasi- or semi-absolute catalogues are those which are absolute in most respects but for which the zero point of the right ascension system and/or the declination system has been assumed.

3. Relative catalogues in which the positions given were determined differentially from a number of reference stars whose positions were assumed. Relative catalogues restricted to stars within a narrow zone of declination are usually known as "zone catalogues" whilst those in which the positions given depend on the measurement of photographic plates are frequently referred to as "photographic catalogues".

There is no accepted terminology for describing the various types of catalogue in which the positions given depend on several instruments. If observations made at approximately the same epoch at two or more observatories or instruments are combined to form a single catalogue, this may be classified as being a "composite observational catalogue", while a catalogue derived by combining a number of observational catalogues together can be designated as a "compiled catalogue".

Proper motions are often derived and given in catalogues compiled from observational catalogues referring to different epochs but the designation of "complete" sometimes applied to such catalogues cannot be recommended.

Fundamental catalogues are compiled catalogues intended to define an inertial frame and must include positions, proper motions, precessional terms, epochs and error data for stars well distributed over the whole sky. The reliability of the system so defined depends very much on the number of absolute catalogues included in the compilation.

The term "general catalogue" is frequently applied to a collection of data for many objects for which no special criterion of inclusion has been applied; it cannot be strictly defined.

### Observational catalogues

Observational catalogues are included in the two bibliographies cited above. In many cases it is not possible to classify a catalogue by its title; study of the introduction and of the explanations of the methods of observation and of reduction is necessary. The vast majority of the observational catalogues have been used for deriving compiled catalogues or fundamental catalogues.

Positions of stars fainter than 7th or 8th magnitude usually have been observed relatively in declination zones, either visually or photographically. Lists and detailed descriptions of "systematic zone catalogues" are given by Eichhorn [a] ch. 5.

Table 3. Zone catalogues.
N = Number of stars

| No. | Catalogue | N | Zones | Epoch | Literature |
|---|---|---|---|---|---|
| 1 | AGK 1: Katalog d. Astronom. Gesellschaft I, II | 144218<br>43830 | − 2°···+80°<br>−23 − 2 | 1868···1908<br>1888···1905 | 15(I) and 5(II) zones; [a] Table V–1 |
| | South American Extensions: | | | | |
| | Cordoba A···D | 60542 | −22···−47 | 1891···1939 | |
| | La Plata A···E | 31443 | −47···−82 | 1913···1935 | |
| | Cordoba E | 1919 | −81···−90 | 1943···1945 | Result. Cordoba **39** (1973) |
| 2 | AGK 2A: Anhaltsterne f. AGK 2 | 13747 | − 5···+90 | 1928···1932 | Veröff. Kopernikusinst. **55** (1943) |
| | AGK 2: 2. Katalog d. Astron. Gesellsch. (photogr.) | 11322<br>117378 | +70···+90<br>+90···+20 | 1930···1932<br>1929···1930 | Publ. Pulkovo **60** (1947)<br>Bergedorf, Vols. 1···10 (1951···1954) |
| | | 66208 | +20···− 2 | 1928···1933 | Bonn, Vols. 11···15 (1957, 1958) |
| 3 | Yale Zone Catalogue Programs (photogr.) | 1031 | +90···+85 | 1951 | [a] Table V–2 |
| | | 16086 | +50···+60 | 1915···1917 | |
| | | 16554 | +50···+60 | 1946···1947 | |
| | | 7108 | − 2···+ 2 | 1913···1914 | |
| | | 128093 | −30···+30 | 1927···1940 | |
| | | 24996 | −30···−40 | 1956 | |
| | | 17373 | −40···−50 | 1942 | |
| | | ≈15000 | −60···−70 | 1941···1961 | not yet published |
| | | 18702 | −70···−90 | 1956 | |
| 4 | Cape Photographic Zone Catalogues | 20843 | −40···−52 | 1897···1910 | CZC (1923) |
| | | 20554 | −40···−52 | 1897···1910 | CFS (1939) |
| | | 24961 | −30···−40 | 1931···1937 | [a] Table V–4 |
| | | 43505 | −52···−90 | 1937···1955 | |
| 5 | AC: Astrographic Catalogue | | −90···+90 | 1891···1950 | [a] Table V–5 |
| 6 | AGK 3R: Reference stars for AGK 3 | 21499 | − 5···+90 | 1956···1962 | U.S. Naval Observatory, on tape |
| | AGK 3 | 183173 | − 2···+90 | 1956···1963 | Hamburg-Bergedorf, Vols. 1···8 (1975) |
| 7 | SRS: Southern Reference Star Program | 20495 | + 5···−90 | 1961···1974 | Reduction at Pulkovo and U.S. Naval Obs. in progress |

### Compiled catalogues

The goal of the compilation of several position catalogues observed at different epochs is, in nearly all cases, the computation of a compiled catalogue of positions and proper motions. From the data of such catalogues star positions can be deduced for a longer period of time.

Fundamental catalogues are the most important class of compiled catalogues. The purpose and history of these catalogues are described by Zverev [c], by Fricke in the Introduction to FK4 [5], and by Eichhorn [a] ch. 4. Table 4 gives the modern fundamental catalogues. Since the IAU (1935) has recommended [Trans. Int. Astron. Union **5** (1936) 281] that the ephemerides of stars in astronomical yearbooks should be transferred to the FK3 system, further work on the fundamental system was concentrated on the improvement and extension of FK3 and later on FK4. An improvement and extension of FK4 to fainter stars is in progress at the Astronomisches Recheninstitut, Heidelberg. The IAU recommends use of FK5 for the years 1984 onwards.

Apparent places of all FK4 stars are computed and published in yearly volumes "Apparent Places of Fundamental Stars" at the Astronomisches Recheninstitut, Heidelberg.

Table 4. Fundamental catalogues.
$N$ = number of stars
$m$ = magnitude

| No. | Abbr. | Author | Literature | $N$ | Equin. Epoch | $m$ |
|---|---|---|---|---|---|---|
| 1 | FK3 I Auwers stars | Kopff, A. et al. | 3. Fundamentalkatalog d. Berliner Astron. Jahrbuchs, Veröffentl. Kopernikusinst. **54** (1937) | 873 | 1925.0 1950.0 | $\cdots 7^m$ |
| | FK3 II Zusatzsterne | Kopff, A. | Abh. Preuß.Akad.Wiss., Phys.-Math. Kl. **3** (1938) | 662 | 1950.0 | $3^m \cdots 7^m$ |
| 2 | GC | Boss, B. | General Catalogue, Carnegie Inst., Wash., Publ. **468**, Vols. I···V (1936···1937) | 33342 | 1950.0 | $\cdots 9^m$ |
| 3 | N30 | Morgan, H.R. | Catalog of 5268 Standard Stars, 1950.0, based on the normal system N30, Astron. Papers Wash. **13**/3 (1952) | 5268 | 1950.0 | $\cdots 8^m$ |
| 4 | FK4 | Fricke, W., Kopff, A., et al. | Fourth Fundamental Catalogue, Veröffentl. Astron. Recheninst. Heidelberg **10** (1963) | 1535 | 1950.0 1975.0 | $\cdots 7^m$ |
| 5 | FK5 | Astron. Recheninst. Heidelberg | Fifth Fundamental Catalogue, in progress. See Fricke and Gliese (1979) [4] and Fricke (1980) [3a] | | | $\cdots 9^m$ |

Table 5. Compiled catalogues which do not represent a fundamental system.

| No. | Abbr. | Name and publication | Contents |
|---|---|---|---|
| 1 | FK4 Sup | Preliminary Supplement to the FK4, Veröff. Astron. Recheninst. Heidelberg **11** (1963) | positions and proper motions of 1987 non-fundamental stars in the FK4 system |
| 2 | SAO or SAOC | Smithsonian Astrophysical Observatory Star Catalog, Vols. 1···4, Washington (1966) | positions and proper motions of 258997 stars for the epoch and equinox 1950.0 |
| 3 | SKYMAP | Computer Sciences Corporation, Silver Spring, Maryland, USA (1978) (available on magnetic tape) | data for all stars brighter than 9.0 mag; about 255000 stars |
| 4 | PFKSZ | Predvaritel'nyj svodnyj katalog fundamentalnykh slabykh zvezd (Preliminary fundamental catalog of faint stars), Trudy Pulkovo (2) **72** (1958) | contains 587 stars $7 \cdots 9^m$ from $-20° \cdots +90°$. Approximately in the FK3 system |
| | PFKSZ–2 | PFKSZ re-reduced and supplemented, Kiev (1980) | 582 stars from $-20° \cdots +90°$ |
| 5 | KSV or KCB | Svodnyj Katalog Sluzhb Vremeni SSSR (General Catalogue of the USSR Time Services), Trudy Pulkovo (2) **78**, 59 (1971) | contains 807 stars brighter than 8th mag from $-13° \cdots +80°$; right ascensions only |
| 6 | KSZ | Katalog Slabykh Zvezd. (Catalogue of Faint Stars), Pulkovo (1956) (observations not yet finished) | list contains 15690 stars north of $-30°$ |

### 8.1.1.6 Precession tables

An approximate value of the precession is needed for the comparison of star charts of different equinoxes or for finding objects (e.g. comets, novae...) with a star chart which is of a different equinox than that in which the position of the object is given. The following tables give the annual precession in right ascension ($p_\alpha$) and declination ($p_\delta$) obtained by the formulae:

$$p_\alpha = m + n \sin\alpha \cdot \tan\delta$$
$$p_\delta = n \cdot \cos\alpha,$$

with the constants

$$\left.\begin{aligned} m &= 3\overset{s}{.}073\,27 \\ n &= 1\overset{s}{.}336\,17 = 20\overset{''}{.}042\,6 \end{aligned}\right\} 1950\,.$$

For these purposes it is sufficient to take $p_\alpha$ and $p_\delta$ from the tables and to multiply them by the time difference $t$ between both equinoxes ($t$ in years), to find the precession ($P_\alpha$, $P_\delta$) in $t$ years.

Table 6. Annual precession in right ascension $p_\alpha$.
For north (positive) declination, the argument $\alpha$ is on the left-hand side. For south (negative) declination, the argument $\alpha$ is on the right-hand side.

| α ↓ | δ → +80° | +70° | +60° | +50° | +40° | +30° | +20° | +10° | 0° | |
|---|---|---|---|---|---|---|---|---|---|---|
| 0$^h$ | +3$\overset{s}{.}$07 | +3$\overset{s}{.}$07 | +3$\overset{s}{.}$07 | +3$\overset{s}{.}$07 | +3$\overset{s}{.}$07 | +3$\overset{s}{.}$07 | +3$\overset{s}{.}$07 | +3$\overset{s}{.}$07 | +3$\overset{s}{.}$07 | 12$^h$ |
| 1 | 5.03 | 4.02 | 3.67 | 3.48 | 3.36 | 3.27 | 3.20 | 3.13 | 3.07 | 13 |
| 2 | 6.86 | 4.91 | 4.23 | 3.87 | 3.63 | 3.46 | 3.32 | 3.19 | 3.07 | 14 |
| 3 | 8.43 | 5.67 | 4.71 | 4.20 | 3.87 | 3.62 | 3.42 | 3.24 | 3.07 | 15 |
| 4 | 9.64 | 6.25 | 5.08 | 4.45 | 4.04 | 3.74 | 3.49 | 3.28 | 3.07 | 16 |
| 5 | +10.39 | +6.62 | +5.31 | +4.61 | +4.16 | +3.82 | +3.54 | +3.30 | +3.07 | 17 |
| 6 | 10.65 | 6.74 | 5.39 | 4.67 | 4.19 | 3.84 | 3.56 | 3.31 | 3.07 | 18 |
| 7 | 10.39 | 6.62 | 5.31 | 4.61 | 4.16 | 3.82 | 3.54 | 3.30 | 3.07 | 19 |
| 8 | 9.64 | 6.25 | 5.08 | 4.45 | 4.04 | 3.74 | 3.49 | 3.28 | 3.07 | 20 |
| 9 | 8.43 | 5.67 | 4.71 | 4.20 | 3.87 | 3.62 | 3.42 | 3.24 | 3.07 | 21 |
| 10 | + 6.86 | +4.91 | +4.23 | +3.87 | +3.63 | +3.46 | +3.32 | +3.19 | +3.07 | 22 |
| 11 | 5.03 | 4.02 | 3.67 | 3.48 | 3.36 | 3.27 | 3.20 | 3.13 | 3.07 | 23 |
| 12 | 3.07 | 3.07 | 3.07 | 3.07 | 3.07 | 3.07 | 3.07 | 3.07 | 3.07 | 0 |
| 13 | + 1.11 | 2.12 | 2.47 | 2.66 | 2.78 | 2.87 | 2.95 | 3.01 | 3.07 | 1 |
| 14 | − 0.72 | 1.24 | 1.92 | 2.28 | 2.51 | 2.69 | 2.83 | 2.95 | 3.07 | 2 |
| 15 | − 2.29 | +0.48 | +1.44 | +1.95 | +2.28 | +2.53 | +2.73 | +2.91 | +3.07 | 3 |
| 16 | 3.49 | −0.11 | 1.07 | 1.69 | 2.10 | 2.41 | 2.65 | 2.87 | 3.07 | 4 |
| 17 | 4.25 | 0.47 | 0.84 | 1.53 | 1.99 | 2.33 | 2.60 | 2.84 | 3.07 | 5 |
| 18 | 4.50 | 0.60 | 0.76 | 1.48 | 1.95 | 2.30 | 2.59 | 2.84 | 3.07 | 6 |
| 19 | 4.25 | 0.47 | 0.84 | 1.53 | 1.99 | 2.33 | 2.60 | 2.84 | 3.07 | 7 |
| 20 | − 3.49 | −0.11 | +1.07 | +1.69 | +2.10 | +2.41 | +2.65 | +2.87 | +3.07 | 8 |
| 21 | 2.29 | +0.48 | 1.44 | 1.95 | 2.28 | 2.53 | 2.73 | 2.91 | 3.07 | 9 |
| 22 | − 0.72 | 1.24 | 1.92 | 2.28 | 2.51 | 2.69 | 2.83 | 2.95 | 3.07 | 10 |
| 23 | + 1.11 | 2.12 | 2.47 | 2.66 | 2.78 | 2.87 | 2.95 | 3.01 | 3.07 | 11 |
| 24 | 3.07 | 3.07 | 3.07 | 3.07 | 3.07 | 3.07 | 3.07 | 3.07 | 3.07 | 12 |
| | −80° | −70° | −60° | −50° | −40° | −30° | −20° | −10° | 0° δ ← | α ↑ |

Table 7. Annual precession in declination $p_\delta$.
For the left-hand side, the argument $\alpha(\alpha=0^h \cdots 11^h)$ takes the upper sign, for the right-hand side, the argument $\alpha(\alpha=12^h \cdots 23^h)$ takes the lower sign.

| $\alpha$ | $0^m$ | $10^m$ | $20^m$ | $30^m$ | $40^m$ | $50^m$ | $60^m$ | $\alpha$ |
|---|---|---|---|---|---|---|---|---|
| $0^h$ | $\pm 20''.0$ | $\pm 20''.0$ | $\pm 20''.0$ | $\pm 19''.9$ | $\pm 19''.7$ | $\pm 19''.6$ | $\pm 19''.4$ | $12^h$ |
| 1 | $\pm 19.4$ | $\pm 19.1$ | $\pm 18.8$ | $\pm 18.5$ | $\pm 18.2$ | $\pm 17.8$ | $\pm 17.4$ | 13 |
| 2 | $\pm 17.4$ | $\pm 16.9$ | $\pm 16.4$ | $\pm 15.9$ | $\pm 15.4$ | $\pm 14.8$ | $\pm 14.2$ | 14 |
| 3 | $\pm 14.2$ | $\pm 13.5$ | $\pm 12.9$ | $\pm 12.2$ | $\pm 11.5$ | $\pm 10.8$ | $\pm 10.0$ | 15 |
| 4 | $\pm 10.0$ | $\pm 9.3$ | $\pm 8.5$ | $\pm 7.7$ | $\pm 6.9$ | $\pm 6.0$ | $\pm 5.2$ | 16 |
| 5 | $\pm 5.2$ | $\pm 4.3$ | $\pm 3.5$ | $\pm 2.6$ | $\pm 1.7$ | $\pm 0.9$ | 0.0 | 17 |
| 6 | 0.0 | $\mp 0.9$ | $\mp 1.7$ | $\mp 2.6$ | $\mp 3.5$ | $\mp 4.3$ | $\mp 5.2$ | 18 |
| 7 | $\mp 5.2$ | $\mp 6.0$ | $\mp 6.9$ | $\mp 7.7$ | $\mp 8.5$ | $\mp 9.3$ | $\mp 10.0$ | 19 |
| 8 | $\mp 10.0$ | $\mp 10.8$ | $\mp 11.5$ | $\mp 12.2$ | $\mp 12.9$ | $\mp 13.5$ | $\mp 14.2$ | 20 |
| 9 | $\mp 14.2$ | $\mp 14.8$ | $\mp 15.4$ | $\mp 15.9$ | $\mp 16.4$ | $\mp 16.9$ | $\mp 17.4$ | 21 |
| 10 | $\mp 17.4$ | $\mp 17.8$ | $\mp 18.2$ | $\mp 18.5$ | $\mp 18.8$ | $\mp 19.1$ | $\mp 19.4$ | 22 |
| 11 | $\mp 19.4$ | $\mp 19.6$ | $\mp 19.7$ | $\mp 19.9$ | $\mp 20.0$ | $\mp 20.0$ | $\mp 20.0$ | 23 |

More accurate values for precession can be taken from the tables by Schorr [10], or they can be computed with the following formulae for the reduction of equatorial coordinates from the mean equinox of $t_0$ to the mean equinox of $t$, or from $t$ to $t_0$:

$$\alpha = M + N \sin\tfrac{1}{2}(\alpha+\alpha_0)\tan\tfrac{1}{2}(\delta+\delta_0)$$
$$\delta = N \cos\tfrac{1}{2}(\alpha+\alpha_0),$$

where the right-hand sides are evaluated by successive approximation, if necessary. Here

$$M = \text{general precession in right ascension} = m_m(t-t_0)$$
$$N = \text{precession in declination} = n_m(t-t_0).$$

$m_m$, $n_m$ are the values of $m$ and $n$ at an epoch midway between the initial epoch $t_0$ and a subsequent epoch $t$, and they are derived from

$$m = 3^s.07234 + 0^s.00186\,T$$
$$n = 1^s.33646 - 0^s.00057\,T$$
$$= 20''.0468 - 0''.0085\,T,$$

where $T$ is measured in tropical centuries from 1900.0 [3].

From 1984 onwards the Int. Astron. Union recommends introduction of slightly corrected values for $m$ and $n$ ([9a]; see 2.3):

$$m = 3^s.07496 + 0^s.00186\,T$$
$$n = 1^s.33621 - 0^s.00057\,T$$
$$= 20''.0431 - 0''.0085\,T,$$

where $T$ is measured in Julian centuries from 2000 [9a].

## 8.1.1.7 References for 8.1.1

### General references

a Eichhorn, H.: Astronomy of Star Positions, New York (1974).
b Newcomb, S.: A Compendium of Spherical Astronomy, New York (1906).
c Zverev, M.S.: Fundamental Astrometry, Uspekhi astronomicheskikh nauk, Moscow, Acad. Sci. USSR, Vol. V (1950) and VI (1954).

### Special references

1 Delporte, E.: Délimitation scientifique des constellations and Atlas céleste; both: Cambridge University Press (1930).
2 Elsmore, B.: Int. Astron. Union Coll. No. 48, Modern Astrometry, (Prochazka, F.V., Tucker, R.H., eds.), Vienna (1979) 93.

3 Explanatory Supplement to the Astronomical Ephemeris and Nautical Almanac, London (1961).
3a Fricke, W.: Mitt. Astron. Ges. **48** (1980) 29.
4 Fricke, W., Gliese, W.: Int. Astron. Union Coll. No. 48, Modern Astrometry, (Prochazka, F.V., Tucker, R.H., eds.), Vienna (1979) 421.
5 Fricke, W., Kopff, A.: Veröffentl. Astron. Recheninst. Heidelberg **10** (1963).
6 Giclas, H.L., Burnham, R., jr., Thomas, N.G.: Lowell Obs. Bull. **8** (1978) 89.
7 Gliese, W., Jahreiss, H.: Astron. Astrophys. Suppl. **38** (1979) 423.
8 Gliese, W., Stoy, R.H., Tucker, R.H.: 1979 (not published).
9 Landolt-Börnstein, NS, Vol. VI/1 (1965) p. 255.
9a Lieske, J.H., Lederle, T., Fricke, W., Morando, B.: Astron. Astrophys. **58** (1977) 1.
10 Schorr, R.: Präzessions-Tafeln 1925.0, Bergedorf (1927).
11 Trans. Int. Astron. Union **4** (1933) 221.


## 8.1.2 Proper motions

### 8.1.2.1 Definition

Proper motion (p.m.): transverse (=angular) velocity of fixed stars measured in a spherical coordinate system which is assumed to be fixed in space with the sun as origin:

annual p.m. in [″/a]; centennial p.m. in [″/100 a], expressed as absolute value $\mu$ and position angle $P$ (or $\theta$) from N→E→S→W→N (0°···360°).

With known parallax $\pi$ in [″] the proper motion $\mu$ in [″/a] can be transformed to tangential velocity: $v_t = b\frac{\mu}{\pi}$ in [km s$^{-1}$] with $b=4.74$ [km a s$^{-1}$].

### 8.1.2.2 Components

a) Equatorial components

$$\mu''_\alpha = \mu \cdot \sin P \;[''/\mathrm{a}] \quad \text{or} \quad \mu^s_\alpha = \tfrac{1}{15}\sec\delta \cdot \mu \cdot \sin P \;[\mathrm{s/a}]$$
$$\mu''_\delta = \mu \cdot \cos P \;[''/\mathrm{a}]\ .$$

Reduction from epoch $t_0$ to epoch $t$: For stars with very large p.m. or with positions close to the poles the following terms must be added [b] p. 256:

$$\Delta\mu''_\alpha = 2\mu''_\alpha \cdot \mu''_\delta \sin 1'' \cdot \tan\delta \cdot (t-t_0)$$
$$\Delta\mu''_\delta = -\mu''^2_\alpha \sin 1'' \cdot \sin\delta \cos\delta \cdot (t-t_0)\,.$$

Stars with large proper motion $\mu$[″/a], large radial velocity $v_r$[km s$^{-1}$] and large parallax $\pi$[″] show observable perspective secular changes in p.m. ("foreshortening effect"):

$\Delta\mu_f[''/\mathrm{a}] = -0.205\cdot 10^{-5}\,\mu \cdot v_r \cdot \pi$ (applicable to $\mu_\alpha$ and $\mu_\delta$). The maximal known effect is $\Delta\mu_f = +125\cdot 10^{-5}$ [″/a] for Barnard's star [4].

Reduction of p.m. from equinox 1900.0 + $T_0$ to 1900.0 + $T_0$ + $T$ ($T_0$, $T$ in units of a tropical century): The rotation of the coordinate system by precession (see 2.3) leaves the value of the p.m. unchanged but the position angle varies [a, 1]: where

$$\mu = \text{const.}$$
$$P \to P + S\,,$$
$$\sin S = \sin\vartheta \sin a \sec\delta_{T_0}$$
$$a = \alpha_{T_0} + \zeta_0;\ \zeta_0 = (2304''\!.250 + 1''\!.396\ T_0)\ T + 0''\!.302\ T^2 + 0''\!.018\ T^3$$
$$\vartheta = (2004''\!.682 - 0''\!.853\ T_0)\ T - 0''\!.426\ T^2 - 0''\!.042\ T^3$$
$$\alpha_{T_0};\ \delta_{T_0} = \text{coordinates at } 1900.0 + T_0\,.$$

From 1984 onwards the Int. Astron. Union recommends introduction of slightly corrected values for $\zeta_0$ and $\vartheta$ ([3a]; see 2.3):

$$\zeta_0 = (2306''\!.218 + 1''\!.397\ T_0)T + 0''\!.302\ T^2 + 0''\!.018\ T^3$$
$$\vartheta = (2004''\!.311 - 0''\!.853\ T_0)T - 0''\!.427\ T^2 - 0''\!.042\ T^3$$

$\alpha_{T_0}$, $\delta_{T_0}$ = coordinates at 2000.0 + $T_0$; $T$, $T_0$ in units of a Julian century.

b) Transformation to galactic coordinates
Determination of $\mu_l$, $\mu_b$ from $\mu_\alpha$, $\mu_\delta$: see 8.4.1.2.1

c) $\upsilon$ and $\tau$ components
The $\upsilon$ component is the proper-motion component directed towards the antapex of the solar motion (see 8.4.1.2.2). It consists of the parallactic component in the direction to the antapex and a part due to the star's peculiar motion.
The $\tau$ component perpendicular to the direction of the solar motion results from the peculiar motion of the star only.

### 8.1.2.3 Determination of proper motions

a) Absolute proper motions. Proper motions in fundamental catalogues are called "absolute". Their system is derived from star positions in absolute catalogues (see 8.1.1.5) observed at different epochs $t_i$ (but reduced to a common equinox) by solving the set of equations for $\alpha_0$, $\mu_\alpha$; $\delta_0$, $\mu_\delta$:

$$\text{position}_i = \text{position}_0 + \mu\,(t_i - t_0)\,; \qquad (t_0 = \text{mean epoch})\,.$$

Very often different weights are given to the various catalogue positions (i=1, 2,..., n) used.
b) Relative proper motions are derived by measuring the differences between positions of the same object determined at different epochs relative to one or more reference stars. Photographically determined proper motions are relative.
Reduction from relative to absolute proper motions is possible by

1. referring to one or more stars with known absolute proper motions;
2. referring to a sample of faint distant stars whose mean "absolute" proper motions are estimated from statistics;
3. referring to extragalactic systems whose proper motions are assumed to be zero (see Table 10 Nos. 17···19).

c) Systematic terms in the observed proper motions

$$\mu_{\text{observed}} - \mu_{\text{peculiar}} = \mu_{\text{syst}}$$

are caused by the solar motion relative to the group of neighbouring stars, by errors in the adopted terms of precession, and by the galactic rotation. By an investigation of a statistically significant number of proper-motion stars these terms and corrections are determinable.

Equations for $(\mu_\alpha, \mu_\delta)_{\text{syst}}$: see Williams and Vyssotsky [6].

Equations for $(\mu_l, \mu_b)_{\text{syst}}$: see van de Kamp and Vyssotsky [5].

As corrections to the annual FK4 proper motions due to errors in the precessional constants, Fricke [2] recommends that one adds

$$(\Delta\mu_\alpha \cos\delta)'' = +0''\!.0020 \cos\delta - 0''\!.0044 \sin\alpha \sin\delta$$
$$\Delta\mu''_\delta = -0''\!.0044 \cos\alpha\,.$$

### 8.1.2.4 Numerical values and error estimates

Numbers in Table 8 are according to Luyten's catalogues (see 8.1.1.3 Table 1). Luyten has listed each visually or photographically distinguishable component of a multiple system as one separate object. The largest known proper motions are those of Barnard's star (10.3 ″/a) and of Kapteyn's star (8.7 ″/a) (see 8.2 Table 6).

Estimates of the accuracy of a proper-motion system are not very reliable. For FK4 the systematic errors are about the same size as the individual errors given here, probably somewhat smaller in declination proper motions. Furthermore, the accuracies of proper motions vary between the declination zones on the sphere. The errors in southern regions are about twice those in the northern hemisphere.

Table 8. Frequency distribution of large proper motions (1979).

| μ [″/a] | $N$ |
|---|---|
| ≧10.00 | 1 |
| 5.00···9.99 | 7 |
| 2.00···4.99 | 65 |
| 1.00···1.99 | 454 |
| 0.50···0.999 | 3085 |
| 0.2 ···0.499 | ≈60000 |

Table 9. Average mean internal errors of proper-motion components in some catalogues [3].

| Catalogue | Error [″/a] |
|---|---|
| FK3 | ±0.003 |
| FK4 | ±0.002 |
| GC | ±0.011 |
| N30 | ±0.005 |
| AGK3 | ±0.010 |
| SAOC | ±0.010 |
| Yale zones | ±0.012 |

## 8.1.2.5 Catalogues containing proper motions

1. Fundamental catalogues: see 8.1.1.5 Table 4.
2. Compiled catalogues: see 8.1.1.5 Table 5, Nos. 1···5.
3. Zone catalogues: see 8.1.1.5 Table 3, Nos. 3, 4, 6.
4. Proper-motion catalogues: see 8.1.1.3 Table 1 (star lists for nomenclature): Catalogues No. 6 (BPM), 7 (G numbered stars), 8 (LFT), 9 (LHS), 10 (LTT).

Table 10. Further proper-motion catalogues.
$N$ = number of stars
$m_{lim}$ = limiting of magnitude

| No. | Author | Publication | Zone | $N$ | $m_{lim}$ | μ[″/a] |
|---|---|---|---|---|---|---|
| 1 | Luyten, W.J. | The North Polar Cap, Minneapolis (1968) | +75°···+90° | 10592 | $21^m$pg | ≧0.04 |
| 2 | Luyten, W.J. | The zone +70° to +75°, Minneapolis (1970) | +70°···+75° | 7041 | 21 pg | ≧0.04 |
| 3 | Luyten, W.J. | Proper Motions for 1656 Stars between Declinations +68° and +70°, Minneapolis (1970) | +68°···+70° | 1656 | 21 pg | ≧0.04 |
| 4 | Luyten, W.J. | The +66° and +60° Zones, Minneapolis (1971) | +57°···+69° | 17418 | 21 pg | ≧0.04 |
| 5 | Luyten, W.J., La Bonte, A.E. | The South Galactic Pole, Minneapolis (1973) | $22^h0^m$···$3^h40^m$ −33°···+ 4° | 6878 | 21 pg | ≧0.180 |
| 6 | Luyten, W.J. | The South Galactic Pole Extension; Proper Motion Survey with the 48-Inch Schmidt Telescope, No. 47, Minneapolis (1976) | $21^h45^m$···$4^h45^m$ −45°···−33° | 3008 | 18 red | ≧0.180 |
| 7 | Luyten, W.J. | The North Galactic Pole, Minneapolis (1976) | $10^h$···$16^h$ − 5°···+60° | 10831 | 21 pg | ≧0.180 |
| 8 | Luyten, W.J. | Proper Motion Survey with the 48-Inch Schmidt Telescope, Nos. 1···53, 55, 59 Minneapolis (1963···1981) (LP, LDS stars) | special regions | | 21 pg | |
| 9 | Luyten, W.J. | NLTT Catalogue, 4 Vols., Minneapolis (1979···1980) | +90°···−90° | 58330 | 21 pg | ≧0.180 |
| 10 | Schorr, R. | Eigenbewegungslexikon, 2nd Ed. Bergedorf (1936) | +90°···−90° | 94731 | | |

continued

Table 10, continued

| No. | Author | Publication | Zone | $N$ | $m_{lim}$ |
|---|---|---|---|---|---|
| 11 | Van de Kamp, P., Vyssotsky, A.N., Williams, E.T.R. | McCormick Proper Motion Catalogues, Publ. McCormick **7** (1937) and **10** (1948) | $-30°\cdots+90°$ | 29000 | 11 |
| 12 | Spencer, Jones, H., Jackson, J. | Proper motions of stars in the Zone Catalogue, London (1936): see 8.1.1.5, Table 3, No. 4, CZC. | $-40°\cdots-52°$ | 20843 | |
| Proper motions in selected areas (=SA) | | | | | |
| 13 | Knox-Shaw, H., Scott-Barrett, H.G. | Radcliffe catalogue of proper motions, Oxford (1934) | SA 1···115 | 32408 | 15 pg |
| 14 | Deutsch, A.N. | Publ. Pulkovo (2) **55** (1940) | 74 SA from +75° to +15° | 18000 | 15 pg |
| 15 | Van Rhijn, P.J., Plaut, L. | Publ. Groningen **56** (1955) | SA at 0° | 11274 | 13 pg |
| 16 | Meurers, J., van Schewick, H., Stangenberg, B. | Veröffentl. Bonn **60** (1962) | 18 SA (north. Dec.) | 4914 | 14 pg |
| Programs for determination of absolute proper motions with respect to galaxies are in progress at several observatories. Preliminary results: | | | | | |
| 17 | Klemola, A.R., Vasilevskis, S., Shane, C.D., Wirtanen, C.A. | Publ. Lick **22**, part 2 (1971) | 83 fields | 8790 | 17 pg |
| 18 | Fatchikin, N.V. | Trudy Pulkovo (2) **81** (1974) 4 | 85 areas | 14600 | 15 pg |
| 19 | Rakhimov, A.G. | Tsirk. Tashkent **66** and in following Nos. (1976) | 41 areas | 10600 | 15 pg |

Proper motions between 0.09 and 0.18 [″/a] for about 200000 faint stars by Luyten will be sent to the NASA Data Center at Greenbelt, Maryland, U.S.A.

Proper motions of faint stars are determined photographically. Some of these catalogues and lists give "absolute" motions (reduced from relative to absolute proper motions by different methods), others give relative values. The differences between various measurements for the same faint object are often considerable ($\Delta\mu$ up to 0.1 ″/a).

Further literature: Astron. Jahresbericht (until 1958) § 92; 1959···1968 in § 102; Astron. Astrophys. Abstracts (from 1969 onwards) subject category 112.

## 8.1.2.6 References for 8.1.2

### General references


a Newcomb, S.: A Compendium of Spherical Astronomy, New York (1906) (=Dover Publications (1963) Part III).

b Smart, W.M.: Textbook on Spherical Astronomy, 6. ed. rev. by Green, R.M., Cambridge, London (1977) ch. XI.


### Special references


1 Explanatory Supplement to the Astronomical Ephemeris and Nautical Almanac, London (1961) 30.

2 Fricke, W.: Veröffentl. Astron. Recheninst. Heidelberg **28** (1977).

3 Lederle, T.: Bull. Inf. Centre Données Stellaires, Strasbourg **14** (1978) 62.

3a Lieske, J.H., Lederle, T., Fricke, W., Morando, B.: Astron. Astrophys. **58** (1977) 1.

4 van de Kamp, P.: Vistas Astron. **21** (1977) 289.

5 van de Kamp, P., Vyssotsky, A.N.: Astron. J. **45** (1936) 167.

6 Williams, E.T.R., Vyssotsky, A.N.: Astron. J. **53** (1948) 63.

## 8.1.3 Radial velocities

### 8.1.3.1 Definitions

The radial velocity $v_r$ is the component of the space velocity of an object (star, star group, part of a gas cloud, nebula, galaxy) in the line of sight. It is measured by means of spectrographs using the Doppler effect of the spectral lines and is expressed in [km s$^{-1}$], positive in the direction away from the observer. In most cases, the $v_r$ are given as heliocentric velocities, the observer having eliminated the orbital and rotational motion of the earth from the measured values:

The radial velocity correction for the earth's orbital motion about the sun is given by [16]

$$(\Delta v)_{orb} = -V_{♁} \cos\beta \sin(\lambda_s - \lambda) + V_{♁} e \sin(\Gamma - \lambda) \cos\beta ,$$

where $\lambda$ and $\beta$ are the ecliptic coordinates of the celestial object, $\lambda_s$ is the longitude of the sun at the epoch for which $\lambda$ and $\beta$ are determined, $\Gamma$ is the longitude of the sun at perigee given by

$$\Gamma = 281°13'15''.00 + 6''.18903\, T + 1''.63\, T^2 + 0''.012\, T^3 ,$$

where $T$ is the number of tropical centuries since 1900.0 (to the relevant epoch), $e = 0.0167$ is the eccentricity of the earth's orbit, and $V_{♁}$ is the mean orbital velocity of the earth, given by

$$V_{♁} = \frac{2\pi a}{P(1-e^2)^{1/2}} \approx 29.789 \text{ [km s}^{-1}] ,$$

where the semi-major axis of the earth's orbit is $a = 149\,597\,870$ km and the number of seconds in a sidereal year is $P = 31\,558\,150$ s.

The longitude of the sun, $\lambda_s$, for mean equinox of date is given for each day in the yearly volumes of "The Astronomical Ephemeris", London or "The American Ephemeris and Nautical Almanac", Washington. In the same volumes the mean longitude of perigee, $\Gamma$, at the beginning of each year was given before 1972. From 1972 onwards the mean longitude of perihelion of the earth, $\tilde{\omega} = \Gamma - 180°$, and its variation for 100 days is given for one date in the year under the heading "Inner Planets – mean elements for mean equinox and ecliptic of date".

The radial-velocity correction for the rotation of the earth about its axis is given by

$$(\Delta v)_{rot} = V_{eq} \sin h \cos\delta \cos\phi' ,$$

where the equatorial rotational velocity of the earth is $V_{eq} = 0.465$ [km s$^{-1}$], the hour angle $h$ of the celestial object is taken to be positive for objects east of the meridian and negative for those west of the meridian, $\delta$ is the declination of the celestial object, and $\phi'$ is the geocentric latitude of the observatory.

In special cases of high accuracy, the $v_r$ is referred to the center of gravity of the solar system. This requires additional corrections of up to 0.04 km s$^{-1}$.

### 8.1.3.2 Methods of observation

1. Prism and grating spectrographs [b]; improvement in optical design [14]; use of an image tube in conjunction with the spectrograph.

2. Objective-prism technique [8]; improved methods for the measurement of objective-prism spectrograms [9].

3. Photoelectric radial-velocity spectrometer [10]; CORAVEL [4a].

4. Fabry-Pérot interferometer [6], mainly suitable for radial-velocity determinations of diffuse objects such as planetary nebula, H II regions...

5. Radio spectrometers.

The spectrograms have been measured with a variety of devices [b]: measuring microscopes; projection micrometers, electronic scanners; recent increase in accuracy and speed by use of measuring machines such as the Grant machine [3, 6, 17].

### 8.1.3.3 Accuracy

a) Random errors. These result primarily from statistical irregularities of the photographic density, from local distortions of the emulsion and from uneven illumination of the spectrograph slit. Pure measuring errors (setting errors) contribute but little, except when the lines are very broad as they often are in the O to A spectra. But with modern measuring machines (profile-projection method) the setting on very broad lines is less problematic. The scatter of the $v_r$ must be determined by comparing different plates. The mean error attributable to one plate depends on the dispersion, line structure and width, number of measured lines, and on the reliability of the wavelength used in the reduction.

b) Systematic errors. These result chiefly from mechanical flexure effects, thermal deformations, from residual image defects, which act differently on absorption and emission lines (comparison spectra), and from errors in the adopted wavelengths. Virtually all stellar lines are either disturbed by neighbouring lines or form unresolvable groups (blended lines). Thus, their effective wavelengths change with spectral type, line structure (Stark effect, rotation), and contrast and density of the spectrograms. Using image tubes, additional errors are caused by geometrical distortion effects.

Wilson [21] has derived and listed corrections to the radial velocities of various observatories with respect to the Lick system which has been adopted as a reference system. In nearly all cases these corrections do not exceed 1.5 km $s^{-1}$; they concern long-term and routine programs. In modern research there is a growing tendency for each individual scientist to develop his own technique of measurement and his own method of reduction. Therefore, it is recommended that standard-velocity stars should be observed each night a spectrograph is used [a].

Table 11. Mean errors (m.e.) of radial velocities.

| Spectral class: | B | A | F···G |
|---|---|---|---|
| Spectrograms | m.e. [km $s^{-1}$] | | |
| Slit; dispersion [b] p. 76 | | | |
| 3.5 Å $mm^{-1}$ | – | – | ±0.4 |
| 11 | ±4 | ±2 | ±1 |
| 30 | ±5.5 | ±3 | ±2 |
| 90 | ±9 | ±6 | ±6 |
| Objective prism [9] | ±8 | ±8 | ±4.5 |
| With photoelectric spectrometer [10] | ±1 | ±1 | ±1 |

### 8.1.3.4 Standard-velocity stars

a) Primary standards are members of the solar system. Their radial velocities can be calculated quite independently of any spectroscopic observations: sun, moon, planets and, most suitable, minor planets which have no appreciable angular diameter.

b) Standard-velocity stars. Reliable $v_r$ of a series of standard stars are obtained by photographing their spectra under the same conditions as spectra of primary standards. These stars should be continuously observed along with the program stars. Commission 30 of the IAU has recommended the use of the stars given in Tables 3···5 of Vol. VI/1 of Landolt-Börnstein, New Series, p. 272/273:

Table 3. Standard-velocity stars brighter than $4^{m}.3$ [18]

Table 4. Standard-velocity stars fainter than $4^{m}.3$ [18]

Table 5. Proposed standard-velocity stars of spectral class B0···B9 [13].

Recent observations have raised questions about the constancy of the $v_r$ of the following objects, also listed in LB, NS, VI/1:

Table 3. ε Leo, β Vir, ε Peg,

Table 4. HD 26162 103095 144579
51250 114762 145001
80170 126053 184467
187691,

Table 5. HD 20365, 161573, and ADS 11593 A have variable $v_r$; ADS 11504 is suspected to have variable $v_r$; HD 170010 is in fact 170054[a].

The use of a list of 14 southern standard stars [19] should not be recommended because a possible zero-point error was reported by Maurice [12, a].

Table 12. 20 new IAU standard-velocity stars of $m_{pg} \approx 9$ [20]. HD 14969 has been omitted having shown a variable velocity [12]; Griffin [11a] has computed a spectroscopic orbit of HD 14969.
$m_{pg}$ = photographic apparent magnitude; $v_r$ = radial velocity; m.e. = mean error; Sp = spectral type; $n$ = number of observations

| HD or BD | 1950 α | 1950 δ | $m_{pg}$ | Sp | $v_r$ km s$^{-1}$ | m.e. km s$^{-1}$ | $n$ |
|---|---|---|---|---|---|---|---|
| 4388 | 00$^h$43$^m$.8 | +30°41′ | 8$^m$.80 | K3 III | −28.3 | ±0.6 | 10 |
| 12029 | 01 55.8 | +29 08 | 8.96 | K2 III | +38.6 | 0.5 | 12 |
| 23169 | 03 40.9 | +25 34 | 9.39 | G2 V | +13.3 | 0.2 | 13 |
| 32963 | 05 04.8 | +26 16 | 8.36 | G2 V | −63.1 | 0.4 | 13 |
| 42397 | 06 08.5 | +25 01 | 8.68 | G0 IV | +37.4 | 0.4 | 12 |
| 65934 | 07 59.1 | +26 47 | 8.87 | G8 III | +35.0 | 0.3 | 10 |
| 75935 | 08 50.9 | +27 06 | 9.35 | G8 V | −18.9 | 0.3 | 10 |
| 86801 | 09 58.7 | +28 48 | 9.48 | G0 V | −14.5 | 0.4 | 14 |
| 90861 | 10 27.1 | +28 50 | 8.36 | K2 III | +36.3 | 0.4 | 11 |
| 102494 | 11 45.3 | +27 37 | 8.26 | G8 IV | −22.9 | 0.3 | 18 |
| 112299 | 12 53.0 | +26 01 | 9.19 | F8 V | + 3.4 | 0.5 | 13 |
| 122693 | 14 00.5 | +24 48 | 8.74 | F8 V | − 6.3 | 0.2 | 12 |
| 132737 | 14 57.7 | +27 21 | 9.03 | K0 III | −24.1 | 0.3 | 13 |
| 140913 | 15 43.1 | +28 37 | 8.81 | G0 V | −20.8 | 0.4 | 13 |
| 149803 | 16 33.9 | +29 51 | 8.90 | F7 V | − 7.6 | 0.4 | 11 |
| 171232 | 18 30.6 | +25 27 | 8.66 | G8 III | −35.9 | 0.5 | 13 |
| 28°3402 | 19 33.0 | +28 59 | 9.55 | F7 V | −36.6 | 0.5 | 7 |
| 194071 | 20 20.5 | +28 05 | 9.06 | G8 III | − 9.8 | 0.1 | 12 |
| 213947 | 22 32.3 | +26 20 | 8.93 | K4 III | +16.7 | 0.3 | 13 |
| 223094 | 23 43.9 | +28 26 | 8.97 | K5 III | +19.6 | 0.3 | 13 |

## 8.1.3.5 Standard wavelengths for $v_r$ determinations

Literature for standard lines in stellar spectra and the standard lines of the iron arc are given in LB, NS, VI/1, p. 273, Tables 6 and 7 (correction in Table 7: for 4045.246 read 4045.815). Today also lines from He–Ar lamps and lines produced from Fe–Ar hollow cathodes are in use as standard lines.

Batten [a] gives a detailed report on difficulties in the choice of standard wavelengths: Each observatory has to determine its own wavelength system. In some cases it may even be necessary to derive different sets of wavelengths for use with spectrograms obtained on different emulsions with the same spectrograph.

**Gliese**

### 8.1.3.6 Catalogues of stellar radial velocities

Table 13. Catalogues.

| No. | Author | Publication | $N$ |
|---|---|---|---|
| 1 | Wilson, R.E. | General Catalogue of Stellar Radial Velocities (GCSRV), Carnegie Inst. Wash. Publ. No. 601 (1953) | 15106 |
| 2 | Evans, D.S. | Supplement to the GCSRV, Austin, Texas (in preparation): $\alpha = 0^h \cdots 21^h$ available on magnetic tape from "Centre de Données Stellaires" (= CDS) Strasbourg, France (1979) | |
| 3 | Abt, H.A., Biggs, E.S. | Bibliography of stellar radial velocities, Tucson, Arizona: New York (1972) | 25000 |
| | | Corrections to this Bibliography: | |
| | Bischoff, M. | Bull. Inf. CDS **12** (1977) 26 | |
| | Buscombe, W. | Bull. Inf. CDS **14** (1978) 19 | |
| | Duflot, M. | Bull. Inf. CDS **15** (1978) 105 | |
| | Hoffleit, D. | Bull. Inf. CDS **14** (1978) 17 | |
| 4 | Barbier, M. | Bibliography by Abt/Biggs updated to 1976, Marseille (supposed to be published) | |
| 5 | Abt, H.A. | Mount Wilson Data. Astrophys. J. Suppl. **19** (1970) 387 and **26** (1973) 365 | |
| 6 | NASA/Goddard Space Flight Center | Catalogue of Stellar Radial Velocities by Evans (see No. 2) will be updated and completed by the NASA/GSFC Data Center at Greenbelt, Md., USA | |

A systematic search for spectroscopic binaries among 135 solar-type stars is reported by Abt and Levy [1]. Further publications are cited in the volumes of "Astronomy and Astrophysics Abstracts", Heidelberg/Germany (subject categories 113 and 119; Spectroscopic Binaries).

### 8.1.3.7 Some statistical results

The bibliography by Abt and Biggs 8.1.3.6 gives the radial velocities $v_r$ of about 25000 stars available in June 1970. For 85% of the 9100 stars brighter than $m_v = 6.5$, $v_r$ are known. Radial velocities of very faint stars (down to $m = 13$) have been measured in the Magellanic Clouds [2, 4] and, with very low accuracy, for some nearby red dwarfs even down to $m_{pg} = 16$ [15]. Density distributions on the sphere for 4992 K stars, 4028 A stars, and 4423 B stars with known $v_r$ show a preference for the northern hemisphere [7].

Table 14. Distribution of high velocities among the radial velocities $v_r$ known in 1970. $N$ = number of stars

| $v_r$ [km s$^{-1}$] | > 500 | ··· 400 | ··· 300 | ··· 200 | ··· 100 | > ··· |
|---|---|---|---|---|---|---|
| $N$ | 1 | 3 | 40 | 100 | 340 | 24000 |

Largest measured $v_r$ of a star: +543 km s$^{-1}$ for CoD −29°2277 at $5^h26^m.9$, −29°56′ (1950), sd F6, $V = 11^m.59$, proper motion = 0.40 [″/a].

Table 15. Distribution of about 25000 stars with known radial velocity $v_r$ (see Table 13, No. 3) in various spectral types.

| Sp | W, O | B | A | F | G | K | M | N, R, S | Novae, Pec | Unknown |
|---|---|---|---|---|---|---|---|---|---|---|
| Frequency | 1 % | 20 | 18 | 15 | 14 | 18 | 8 | 1 | 1 | 4 |

Luminosity classes are known only for a fraction of these objects, mainly giants and dwarfs.

## 8.1.3.8 References for 8.1.3

### General references


a Batten, A.H.: Vistas Astron. **22** (1978) 265.

b Petrie, R.M.: Astronomical Techniques, = Stars and Stellar Systems II (Hiltner, W.A., ed.), Chicago (1962), ch. 3.

c Petrie, R.M.: Basic Astronomical Data, = Stars and Stellar Systems III (Strand, K.A., ed.), Chicago (1963), ch. 7.


### Special references


1 Abt, H.A., Levy, S.G.: Astrophys. J. Suppl. **30** (1976) 273.

2 Ardeberg, A., Brunet, J.-P., Maurice, E., Prévot, L.: Astron. Astrophys. Suppl. **6** (1972) 249.

3 Ardeberg, A., Maurice, E.: ESO Bull. **12** (1975) 25.

4 Ardeberg, A., Maurice, E.: Astron. Astrophys. Suppl. **30** (1977) 261.

4a Baranne, A., Mayor, M.: Optical and Infrared Telescopes for the 1990s, Kitt Peak Nat. Obs. Conf., I (Hewitt, A., ed.), Tucson, Arizona (1980) 395.

5 Courtés, G.: Publ. Haute Provence **5** (1960) No. 9.

6 Dennison, E.W.: Image Processing Techniques in Astronomy, = Astrophys. Space Sci. Library **54** (de Jager, C., Nieuwenhuijzen, N., eds.), Dordrecht (1975) 199.

7 Egret, D., Heck, A., Ochsenbein, F.: Astron. Astrophys. Suppl. **26** (1976) 65.

8 Fehrenbach, Ch.: Ann. Astrophys. **10** (1947) 257 and 306; **11** (1948) 35.

9 Gieseking, F.: Astron. Astrophys. **47** (1976) 43.

10 Griffin, R.F.: Astrophys. J. **148** (1967) 465.

11 Griffin, R.F.: Mon. Not. R. Astron. Soc. **171** (1975) 407.

11a Griffin, R.F.: Mon. Not. R. Astron. Soc. **190** (1980) 711.

12 Maurice, E.: Trans. Int. Astron. Union **16** B 1976 (1977) 214.

13 Petrie, R.M.: Publ. Victoria **9** (1953) 297.

14 Richardson, E.H.: J. R. Astron. Soc. Can. **62** (1968) 313.

15 Rodgers, A.W., Eggen, O.J.: Publ. Astron. Soc. Pac. **86** (1974) 742.

16 Schlesinger, F.: Astrophys. J. **10** (1899) 1.

17 Solf, J.: Sterne und Weltraum **16** (1977) 194.

18 Trans. Int. Astron. Union **9** 1955 (1957) 442.

19 Trans. Int. Astron. Union **13** B 1967 (1968) 170.

20 Trans. Int. Astron. Union **15** A (1973) 409.

21 Wilson, R.M.: General Catalogue of Stellar Radial Velocities, Carnegie Inst. Wash. Publ. No. 601 (1953).


## 8.1.4 Parallaxes

### 8.1.4.1 Introduction

A. The equatorial horizontal parallax $\pi$ of a star is defined as the angle which the equatorial radius of the earth $R_{\oplus}$ subtends at the star.

Since the value of $\pi$ is always very small, $\pi$ and the corresponding distance $r$ of the star are related by

$$\pi'' = 206265'' \frac{R_{\oplus}}{r}, \qquad \text{with}$$

$$R_{\oplus} = 6.3781 \cdot 10^8\ \text{cm}, \qquad \text{equatorial radius of the earth.}$$

With $\pi_{\odot} = 8\overset{''}{.}794$, parallax of the sun, we obtain the distance earth-sun

$$r_0 = 1\ \text{AU (astronomical unit)} = 1.49598 \cdot 10^{13}\ \text{cm}.$$

For $r_0$ compare 2.3 "Astronomical constants".

B. The annual parallax $\pi$ of a star is defined as the angle which the mean distance from the earth to the sun $r_0$ subtends at the star.

We get a similar relationship as in A:

$$\pi'' = 206265'' \frac{r_0}{r}$$

where $r$ is the distance of the star.

Normally the parsec (pc) is taken as unit of distance, defined as the distance of a star with a parallax $\pi$ of 1 second of arc, i.e.

$$r[\mathrm{pc}] = \frac{1}{\pi['']}$$

From the value of $r_0$ given above we get

$$1\ \mathrm{pc} = 3.0857 \cdot 10^{18}\ \mathrm{cm}\ .$$

Larger units are also used:

$$1\ \mathrm{kpc} = 10^3\ \mathrm{pc}; \quad 1\ \mathrm{Mpc} = 10^6\ \mathrm{pc}\,.$$

Another unit of distance is the light-year (l.y.), i.e. the distance travelled by a beam of light in one tropical year (see 2.2.3). Taking the velocity of light $c$ as

$$c = 2.99792 \cdot 10^{10}\ \mathrm{cm\,s^{-1}}$$

we have

$$1\ \mathrm{l.y.} = 0.94605 \cdot 10^{18}\ \mathrm{cm}; \quad 1\ \mathrm{pc} = 3.2616\ \mathrm{l.y.}$$

## 8.1.4.2 Determination of stellar parallaxes

### 8.1.4.2.1 Trigonometric parallaxes

The method of trigonometric parallaxes is the only direct method for deriving distances to objects outside the solar system. It remains the basic method for calibrating other methods. Trigonometric parallaxes are derived from the parallactic shifting of stellar positions caused by the yearly orbital motion of the earth. They are measured relative to a selected number of faint (probably distant) reference stars on photographic plates taken with long-focus refractors, astrometric reflectors and, recently, with Schmidt telescopes. Reduction to absolute values is made by adding an assumed value of the secular parallax of the reference stars.

Standard error of one determination of the order of $\pm 0\overset{''}{.}02$ has been reduced in modern series to about $\pm 0\overset{''}{.}01$ or even smaller (on the average $\pm 0\overset{''}{.}004$ for the results with the 61-inch reflector at Flagstaff). Possible further increase in accuracy by observations from space is discussed. Systematic differences which can be as large as $0\overset{''}{.}005$ appear to exist between parallaxes of the same stars measured at different observatories. The origin of these "observatory corrections" is not yet understood [10, 12, 22, 23].

### 8.1.4.2.2 Dynamical parallaxes

Dynamical parallaxes $\pi_d$ are derived from observations of visual double stars:

$$\pi'' = \frac{a''}{[P^2(\mathfrak{M}_1 + \mathfrak{M}_2)]^{1/3}}\,.$$

$a''$, the semi-major axis of the relative orbit is measured; $P$ = period in [a], $(\mathfrak{M}_1 + \mathfrak{M}_2)$ is the total mass of the system ($\mathfrak{M}_\odot = 1$).

Normally, the dynamical parallax is derived together with a computation of the orbit, but even before a complete period has been observed a value for $\pi_d$ can be derived. Various methods are possible depending on different assumptions on the masses of the components. A detailed review is given by Dommanget [4] who recommends a special terminology (orbital and non-orbital dynamical parallaxes; theoretical and hypothetical dynamical parallaxes).

In individual cases, only the use of orbital parallaxes is recommended; non-orbital parallaxes are of interest only for statistical investigations.

Angular separation determined from speckle interferometry of a spectroscopic binary whose elements are known permits the calculation of the parallax as a function of component masses [14].

The "radiation-energy parallaxes" for visual double stars with known orbit are very accurate; the masses are derived from the mass-luminosity relation and the knowledge of the spectral types and magnitudes [5].

### 8.1.4.2.3 Cluster parallaxes

a) Moving-cluster parallax: The classical method is the convergent-point method. Parallel space velocities are assumed for all cluster stars. Measured data are the proper motions $\mu$["/a] and at least one radial velocity $v_r$ [km s$^{-1}$]. When the convergent point has been determined from proper motions the parallax is

$$\pi'' = 4.74\,\mu''/v_r \tan\phi,$$

where $\phi$ is the angle between the position of the star and the convergent point. Normally $\pi''$ is averaged over all cluster stars having precise values of $\mu$ and $v_r$.

b) Another moving-cluster method is based on the rate of variation $\dot{\theta}$ of the angular diameter $\theta$ of the cluster due to its motion in the line of sight. It can be measured from proper motions and does not require the location of the convergent point. The distance $r$ is $r = -v_r\theta/\dot{\theta}$, where $\dot{\theta}/\theta = \mathrm{d}\mu_\alpha/\mathrm{d}\alpha = \mathrm{d}\mu_\delta/\mathrm{d}\delta$ can be determined separately from the proper-motion gradients in right ascension and declination [24].

c) Method of main-sequence fitting or "zero-age main-sequence fitting procedure": This method assumes basically that main-sequence stars have the same properties in all galactic clusters. When the distance of one o the clusters has been determined, all the clusters are calibrated by fitting together the main sequences in HR diagram. This procedure starts from the Hyades [2] p. 406.

### 8.1.4.2.4 Statistical parallaxes

Statistical parallaxes are to be considered mainly as a substitute where the precise individual trigonometric parallax fails, that is at about 20 or 25 pc. One of the principal applications is the calibration of luminosity criteria. Parallaxes of groups of stars are derived from proper motions plus radial velocities.

Secular parallaxes are derived from proper motions on the assumption that, on the average, the star motions must reflect the solar motion. If the angle between a star position and the apex of the solar motion relative to this group of stars is $\lambda$, the space velocity of the sun relative to the star group is $v_\odot$ [km s$^{-1}$], and $\upsilon$ ["/a] is the proper motion along the great circle connecting the star and the apex, $\tau$ ["/a] is the proper-motion component perpendicular to this circle

$$\overline{\pi''} = \frac{4.74\langle \upsilon \sin\lambda \rangle}{v_\odot \langle \sin^2\lambda \rangle}$$

and

$$\overline{\pi''} = \frac{4.74\langle |\tau| \rangle}{\langle |v_r + v_\odot \cos\lambda| \rangle} \quad [10, 20, 21]$$

($\langle \cdots \rangle$ indicates a mean over the number of observed stars).

The $\tau$-method requires the knowledge of the radial velocity $v_r$ of the stars.

Secular parallaxes play an important role for faint, distant star groups which are used as reference stars for reducing relative trigonometric parallaxes to absolute values [1, 15]. The results depend considerably on errors in absolute proper-motion systems [6], at least at distances exceeding 500 pc.

The maximum-likelihood method [11] uses proper motions and radial velocities and aims at finding a distribution of parallaxes of a given sample of stars (spectral type, luminosity class, apparent magnitude) so as to give the "best" fitting of proper motions to radial velocities (which means that radial velocities, on the one hand, and the tangential velocities, on the other hand, can be considered as resulting from the same kinematical model). For detailed description and literature, see [10].

In an extended sense also the use of a "reduced proper motion" $H = m + 5 + 5\log\mu$ by Luyten [13] for a very rough estimate of absolute magnitudes $M = a + bH$ of a group of stars from their proper motions is a statistical method of determining mean distances. Luyten has used this procedure for the derivation of a luminosity function in the region of low-luminosity stars for which neither trigonometric parallaxes nor radial velocities were known.

### 8.1.4.2.5 Spectroscopic and photometric parallaxes

The relation $\log\pi = -0.2\,(m - M + 5)$ permits the determination of the parallax of a star for which $m$ is measured and the absolute magnitude $M$ is derived from a correlation between intrinsic luminosity and certain characteristics in the stellar spectrum ("spectroscopic parallax") or $M$ is taken from a colour-luminosity relation ("photometric parallax"). The mean $M$ attached to the spectral types or colours respectively have been calibrated by trigonometric parallaxes, statistical parallaxes or, for very luminous objects, by cluster parallaxes. These methods are effective for individual stars. The errors in such parallaxes, caused by observational errors and by the suppression of any possible intrinsic dispersion, are between 10% and 60%.

Spectroscopic parallaxes: Visual absolute magnitudes $M_v$ for spectroscopic parallaxes are derived from spectral types plus luminosity classes (MK system, Mount Wilson types, Kuiper types of red dwarfs, ..., see 4.1.1) or from the measurement of luminosity-sensitive quantities in the spectra which have been calibrated versus absolute magnitude: line widths, line intensities, relative intensities, intensity variations in the continuous spectrum (see 4.1 and [2]). There is no limit to the distances measurable provided the stars are bright enough to have observable spectra.

Photometric parallaxes are based on colour-luminosity relations; in most cases their use presupposes knowledge of the luminosity class. For faint main-sequence stars broad-band *UBVRI* photometry (see 4.2) proves to be most effective. The use of other photometric systems (Strömgren, Genève, Vilnius..., see 4.1 and 4.2) seems to be favourable for bright stars in certain regions of the HR diagram. Luminosities and photometric parallaxes of white dwarfs have been determined [9] from multichannel observations; most useful colour: $1/\lambda = 2.12$ to $1.24\ \mu m^{-1}$.

At great distances these methods work with much greater reliability than the trigonometric method. However, a certain knowledge or reliable assumptions of the effects of interstellar reddening and interstellar absorption in the apparent magnitudes are necessary (see 4.2 and 7.4). The term $A_\lambda$, explained in 4.2, must be added:

$$\log \pi = -0.2(m - M - A_\lambda + 5)\,.$$

The period-luminosity relation for Cepheids (see 5.1.2.1):

Observed are the mean apparent magnitudes ($B$ and $V$) and the period $P$ in [d].

$$\langle M_V \rangle = -3.425 \log P + 2.52(B-V)_0 - 2.459$$

in the period interval $2^d < P < 42^d$ ([17] Table I, which gives $M_V$ and $M_B$ for $0.4 \leqq \log P \leqq 2.1$).

For RR Lyrae stars (see 5.1.2.2) $\langle M_V \rangle$ seems to be independent of $P$. $\langle M_V \rangle \approx +0^m\!.4$ [26], whereas Sandage [16] derived $\langle M_V \rangle = +0^m\!.75 \pm 0^m\!.23$; see also Clube and Dawe [3].

There are other methods to determine or estimate distances $r$ up to very distant extragalactic objects which are described by Sandage [16, 18]: magnitude variations of novae; globular clusters; sizes of H II regions in galaxies; the magnitudes of the brightest stars in an extragalactic system; and, finally, the parallax from the redshift $\Delta\lambda$ observed for extragalactic systems:

$$r = \frac{c}{H} \cdot \frac{\Delta\lambda}{\lambda} \quad \text{in [Mpc]},$$

where $c$ = velocity of the light, $H$ = the Hubble constant $\approx (50 \pm 7)\,\mathrm{km\,s^{-1}\,Mpc^{-1}}$ [18a]. However recent determinations summarized by Abell, G.O. [Trans. Int. Astron. Union **17** A-3 (1979) 203] and by Hodge [10a] yielded various values for $H$ up to $(110 \pm 10)\,\mathrm{km\,s^{-1}\,Mpc^{-1}}$.

A summary of the extragalactic distance indicators is compiled by Hodge [10a]. See also 9.7.5.

### 8.1.4.3 Parallax catalogues and lists

All stellar parallaxes measured up to 1935 (January) have been published in:

Schlesinger, F.: General Catalogue of Stellar Parallaxes, 2. Ed., Yale Univ. Obs. (1935).

The catalogue contains 7534 stars with trigonometric and/or spectroscopic parallaxes and 2444 stars with dynamical parallaxes. Errata and additions [19].

Trigonometric parallaxes are collected and compiled continuously at Yale Univ. Obs. Publications:

Jenkins, L.F.: General Catalogue of Trigonometric Stellar Parallaxes (1952), and

Supplement to the General Catalogue of Trigonometric Stellar Parallaxes (1963).

These catalogues give 5822 stars with parallaxes available in May 1950 and 577 further stars (end of 1962), respectively. A new edition is in preparation [25] at Yale Obs., publication probably 1982.

A bibliography of trigonometric parallax publications is given as an appendix to [8].

No general catalogue or list of publications with spectroscopic, photometric or dynamical parallaxes or distance moduli $m - M$ has been compiled. Only a summarizing report on the status in 1969 is available [7]. Even in the volumes of Astronomy and Astrophysics Abstracts, Heidelberg (two volumes for each year) such lists are cited under various categories. Literature concerning nearby stars is found in [8].

## 8.1.4.4 References for 8.1.4


1 Binnendijk, L.: Bull. Astron. Inst. Neth. **10** (1943) 9.
2 Blaauw, A.: Stars and Stellar Systems III = Basic Astronomical Data (Strand, K.A., ed.), Chicago (1963) 383.
3 Clube, S.V.M., Dawe, J.A.: Int. Astron. Union Symp. No. **80** (Philip, A.G.D., Hayes, D.S., eds.), Dordrecht (1978) 53.
4 Dommanget, J.: Ciel Terre **92** (1976) 65.
5 Franz, O.: Mitt. Univ. Sternw. Wien **8** (1956) 1.
6 Fricke, W.: Vistas Astron. **8** (1966) 205.
7 Gliese, W.: Publ. McCormick **16** (1971) 149.
8 Gliese, W., Jahreiss, H.: Astron. Astrophys. Suppl. **38** (1979) 423.
9 Greenstein, J.L.: Astron. J. **81** (1976) 323.
10 Heck, A.: Vistas Astron. **22** (1978) 221.
10a Hodge, P.W.: Annu. Rev. Astron. Astrophys. **19** (1981) 357.
11 Jung, J.: Astron. Astrophys. **4** (1970) 53.
12 Lutz, T.E.: Int. Astron. Union Coll. No. **48** (Prochazka, F.V., Tucker, R.H., eds.), Vienna (1978) 7.
13 Luyten, W.J.: Int. Astron. Union Symp. No. **54** (Hauck, B., Westerlund, B.E., eds.), Dordrecht (1973) 11.
14 McAlister, H.A.: Astrophys. J. **212** (1977) 459.
15 Murray, C.A., Clube, S.V.M.: Int. Astron. Union Coll. No. **7** (Luyten, W.J., ed.), Minneapolis (1970) 131.
16 Sandage, A.: Int. Astron. Union Symp. No. **15** (McVittie, G.C., ed.), New York (1962) 359.
17 Sandage, A.: Quart. J. R. Astron. Soc. **13** (1972) 202.
18 Sandage, A.: Quart. J. R. Astron. Soc. **13** (1972) 282.
18a Sandage, A., Tammann, G.A.: Astrophys. J. **256** (1982) 339.
19 Schlesinger, F., Jenkins, L.F.: Astron. J. **46** (1937) 104.
20 Smart, W.M.: Stellar Dynamics, Univ. Press Cambridge (1938) 189.
21 Smart, W.M.: Textbook on Spherical Astronomy, 6th ed. rev. by Green, R.M., Cambridge (1977).
22 Strand, K.A.: Stars and Stellar Systems III = Basic Astronomical Data (Strand, K.A., ed.), Chicago (1963) 55.
23 Upgren, A.R.: News Letters Astron. Soc. New York **1** (1979) No. 4.
24 Upton, E.K.L.: Astron. J. **75** (1970) 1097.
25 van Altena, W.: Int. Astron. Union Coll. No. **48** (Prochazka, F.V., Tucker, R.H., eds.), Vienna (1978) 79.
26 Woolley, Sir Richard: Quart. J. R. Astron. Soc. **13** (1972) 189.

# 8.2 The nearest stars

## 8.2.1 Introduction and data

Conventionally, objects at distances smaller than 20 or 25 pc are called "the nearest stars" as the observational errors become serious for trigonometric parallax measurements for $\pi < 0''.05$ or $\pi < 0''.04$. Lists of the known nearby stars together with their substantial data (positions, proper motions, radial velocities, parallaxes, spectral types, photometry and notes concerning duplicity, variability, and peculiarities) have been published in 1969 up to 22 pc [3], in 1970 to 25 pc [16], 1978 to 5.3 pc [11], and 1979 to 22 pc [6].

From trigonometric and spectroscopic/photometric distance determinations up to the end of 1978, 1578 stars – singles, doubles, and multiple systems – with altogether nearly 2000 components nearer than 22.5 pc ($\pi \geqq 0''.045$) were known. The increase in knowledge is shown by the following numbers of known stars:

| | 1957 Feb | 1968 Dec | 1978 Dec |
|---|---|---|---|
| $\pi \geqq 0''.050$ | 915 | 1049 | 1214 |
| $\pi \geqq 0''.045$ | | 1328 | 1578 |

Since 1970 new effective programs for trigonometric parallaxes have been started; preference has been given to faint, nearby objects. With the remarkable increase of photometric data in recent years the derivation of absolute magnitudes of many faint stars in the solar neighbourhood has become possible. For spectroscopic/photometric parallaxes given in the catalogues of stars nearer than 22 pc, mainly the calibrations by Gliese [4] have been used. Photometric parallaxes of large numbers of red, nearby dwarfs have been determined by Eggen [1, 2] and by Rodgers and Eggen [14]; photometric parallaxes of white dwarfs are given by Greenstein [7].

The stars in the solar neighbourhood are incompletely known; the space density of detected nearby stars decreases with growing distances from the sun (Table 1).

Table 1. Relative number density $n_{rel}$ of known stars within 22 pc ($n \triangleq 1$ within 4 pc).

| Interval [pc] | 0···4 | 4···5 | 5···10 | 10···15 | 15···20 | 20···22 |
|---|---|---|---|---|---|---|
| $n_{rel}$ | 1.00 | 0.80 | 0.54 | 0.43 | 0.32 | 0.29 |

## 8.2.2 Luminosity function in the solar neighbourhood

The luminosity functions $\Phi(M)$ derived in the vicinity of the sun by different methods are similar for the brighter stars (visual magnitude $M_v < +10$); in the region of the intrinsically fainter objects, however, some of the derivations show significant deviations. (For more details of luminosity function, see 8.3.2.1.)

Luyten's luminosity function is based on the frequency of large proper motions among the stars down to $m_{pg} = +21$ found in the proper-motion survey with the 48-inch Schmidt telescope [12]; it has a distinct maximum at $M_{pg}$ between +15 and +16. Down to $M_{pg} = 15$ this function is tolerably consistent with $\Phi(M)$ based on the observed numbers of nearby stars corrected for incompleteness [15]. The somewhat lower numbers $N_W$ between $M_{pg} \approx 7.5$ and 11 are essentially confirmed recently by Upgren and Armandroff [14a]. For $M_{pg} > +16$ the run of $\Phi(M)$ is very uncertain.

Table 2. Number of stars nearer than 10 pc according to the luminosity functions of Luyten, $N_L$ [12] and of Wielen, $N_W$ [15].
$N(M)$ = number of stars between $M-\frac{1}{2}$ and $M+\frac{1}{2}$
$M_{pg}$ = photographic magnitude

| $M_{pg}$ | $N_L$ | $N_W$ | $M_{pg}$ | $N_L$ | $N_W$ | $M_{pg}$ | $N_L$ |
|---|---|---|---|---|---|---|---|
| $0^M$ | | | $+8^M$ | | 11 | $+16^M$ | |
| +0.5 | 0.5 | | 8.5 | 20 | | 16.5 | 60 |
| 1 | | 1.4 | 9 | | 12 | 17 | |
| 1.5 | 1.3 | | 9.5 | 24.5 | | 17.5 | 37 |
| 2 | | 2.6 | 10 | | 17.5 | 18 | |
| 2.5 | 3.2 | | 10.5 | 30 | | 18.5 | 23 |
| 3 | | 3.0 | 11 | | 27 | 19 | |
| 3.5 | 4.8 | | 11.5 | 35 | | 19.5 | 13 |
| 4 | | 8 | 12 | | 36 | 20 | |
| 4.5 | 7.0 | | 12.5 | 42 | | 20.5 | 8 |
| 5 | | 10.5 | 13 | | 53 | 21 | |
| 5.5 | 9.5 | | 13.5 | 54 | | 21.5 | 4 |
| 6 | | 11.5 | 14 | | 53 | | |
| 6.5 | 13 | | 14.5 | 67 | | | |
| 7 | | 13 | 15 | | 75 | | |
| 7.5 | 16 | | 15.5 | 80 | | | |

From Luyten's function, the existence of about 550 stars nearer than 10 pc is expected (Table 2) but only 230 objects were known at the end of 1978. Most of the stars still undetected should be faint objects with proper motions smaller than 1 ["/a].

Table 3. Percentage frequency ($N$) of different spectral types of all stars in the solar neighbourhood based on Luyten's luminosity function.

| Type/class | Giants subgiants | Main sequence | | | | | Subdwarfs | White dwarfs |
|---|---|---|---|---|---|---|---|---|
| | | A | F | G | K | M | | |
| ($N$) % | <1 | 0.5 | 2 | 3 | 10 | 78*) | 1*) | 5*) |

*) M dwarfs, subdwarfs, and white dwarfs together make 84% but there is an uncertainty in splitting this number of low-luminosity star frequency into percentages of white dwarfs and of M dwarfs.

## 8.2.3 Star number density and density of matter in the solar neighbourhood

For more details, see 8.4.2.2.4

Table 4. Star number density and mass density in the solar neighbourhood.

| | | |
|---|---|---|
| Star number density, systems as one star | ≈0.13 stars/pc$^3$ | [12] |
| each component as one star | ≈0.2 | estimated |
| Density of matter in stars | 0.05···0.07 $\mathfrak{M}_\odot$/pc$^3$ | [12, 15, 8] |
| Total local mass density | 0.08···0.21 | recent investigations [10] |
| | 0.15 | [13] |
| Mass density due to interstellar hydrogen, He..., dust | 0.02···0.04 | |
| "Missing mass" (low-mass dark stars, more degenerates, higher percentage of multiple stars, black holes?) | 0.00···0.10 | |

The differences between the derived values for the density of matter in stars (mainly caused by uncertain values for the frequency and the masses of white dwarfs) yield discordant values for the percentage mass frequency (𝔐) % of different spectral types of all stars in the vicinity of the sun.

Table 5. Percentage mass of different spectral types in the solar neighbourhood.

| Types: | Giants | Main sequence | | | | | White dwarfs | Ref. |
|---|---|---|---|---|---|---|---|---|
| | | A | F | G | K | M | | |
| (𝔐) %: | 6 | | 9 | 9 | 16 | 31 | 29 | 8 |
| (𝔐) %: | 2 | 8 | 8 | 9 | 15 | 43 | 15 | 9 |

## 8.2.4 Colour-luminosity diagram of the nearest stars

For most of the stars nearer than 22 pc, broad-band *UBV* photometry is known but a representation of all these objects in one ($M_v$, $B-V$) diagram is far from showing the true distribution of the nearby stars in a colour-luminosity diagram. The systematic effects of the accidental errors in parallax measurements (or in $M_v$) near the lower parallax limit (0''.045) distort the distribution in the diagram significantly. Fig. 1 gives a reliable representation of the true distribution by restriction to objects with fairly accurately determined luminosities, standard error |s.e.| ≦ 0.30.

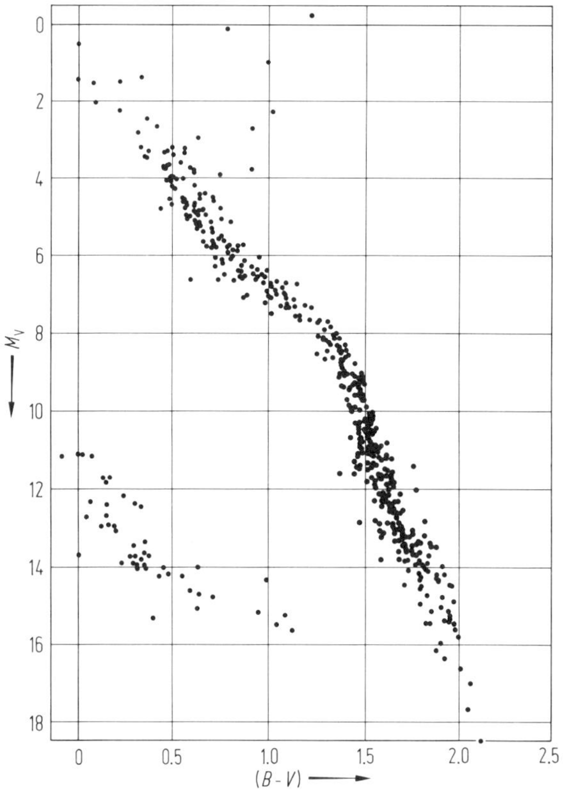


Fig. 1. Trigonometrically determined absolute visual magnitudes $M_v$ of 539 nearby stars ($\pi \geqq 0''.045$) vs. $(B-V)$ colours, limited to stars with s.e. $(M_v)$ not exceeding $\pm 0^m.30$ [5]; status 1980.

## 8.2.5 Stars within 5 pc

Table 6. Stars nearer than 5.25 pc (status 1979).
$\alpha, \delta$ = position for 1950.0
$\pi$ = trigonometric parallax ["] and its standard error
$\mu$ = proper motion ["/a]
$\theta$ = position angle of p.m.
$v_r$ = observed radial velocity [km s$^{-1}$]
Sp = spectral type (e denotes: H emission observed)
*UBV*, *RI* = photoelectric magnitudes and colours (if available)
$M_v$ = absolute visual magnitude
*L* = luminosity (in units of solar luminosity: $L_\odot = 1$)

| No. | Name | α (1950) | δ (1950) | π ± s.e. | μ "/a | θ | $v_r$ km s$^{-1}$ | Sp | V | B−V | U−B | R−I | $M_v$ | $L/L_\odot$ |
|---|---|---|---|---|---|---|---|---|---|---|---|---|---|---|
| 1 | Sun | | | | | | | G2 V | −26$^m$72 | 0$^m$65 | 0$^m$10 | | 4$^M$85 | 1.0 |
| 2 | Proxima Cen | 14$^h$26$^m$3 | −62°28′ | 0″772 ± 0″007 | 3″85 | 282° | − 16 | dM 5e | 11.05 | 1.97 | | 1$^m$65 | 15.49 | 0.00006 |
| | α Cen A | 14 36.2 | −60 38 | 0.750 ± 0.010 | 3.68 | 281 | − 22 | G2 V | − 0.01 | 0.68 | | 0.22 | 4.37 | 1.6 |
| | α Cen B | | | | | | | K0 V | 1.33 | 0.88 | | 0.24 | 5.71 | 0.45 |
| 3 | Barnard's star | 17 55.4 | + 4 33 | 0.545 ± 0.003 | 10.31 | 356 | −108 | M5 V | 9.54 | 1.74 | 1.29 | 1.25 | 13.22 | 0.00045 |
| 4 | Wolf 359 | 10 54.1 | + 7 19 | 0.421 ± 0.006 | 4.70 | 235 | + 13 | dM8 e | 13.53 | 2.01 | 1.54 | 1.85 | 16.65 | 0.00002 |
| 5 | BD +36°2147 | 11 00.6 | +36 18 | 0.397 ± 0.004 | 4.78 | 187 | − 84 | M2 V | 7.50 | 1.51 | 1.12 | 0.91 | 10.50 | 0.0055 |
| 6 | L 726−8 = A | 1 36.4 | −18 13 | 0.387 ± 0.012 | 3.36 | 80 | + 29 | dM6 e | 12.52 } | 1.85 | 1.09: | 1.6 | 15.46 | 0.00006 |
| | UV Cet = B | | | | | | + 32 | dM6 e | 13.02 } | | | | 15.96 | 0.00004 |
| 7 | Sirius A | 6 42.9 | −16 39 | 0.377 ± 0.006 | 1.33 | 204 | − 8 | A1 V | − 1.46 | 0.00 | −0.04 | 0.12 | 1.42 | 23.5 |
| | Sirius B | | | | | | | DA | 8.3: | −0.12: | −1.03: | | 11.2 | 0.003 |
| 8 | Ross 154 | 18 46.7 | −23 53 | 0.345 ± 0.012 | 0.72 | 104 | − 4 | dM5 e | 10.45 | 1.70 | 1.17 | 1.30 | 13.14 | 0.00048 |
| 9 | Ross 248 | 23 39.4 | +43 55 | 0.314 ± 0.004 | 1.60 | 176 | − 81 | dM6 e | 12.29 | 1.91 | 1.48 | 1.56 | 14.78 | 0.00011 |
| 10 | ε Eri | 3 30.6 | − 9 38 | 0.303 ± 0.004 | 0.98 | 271 | + 16 | K2 V | 3.73 | 0.88 | 0.58 | 0.30 | 6.14 | 0.30 |
| 11 | Ross 128 | 11 45.1 | + 1 06 | 0.298 ± 0.006 | 1.38 | 152 | − 13 | dM5 | 11.10 | 1.76 | 1.30 | 1.30 | 13.47 | 0.00036 |
| 12 | 61 Cyg A | 21 04.7 | +38 30 | 0.294 ± 0.006 | 5.22 | 52 | − 64 | K5 V | 5.22 | 1.17 | 1.11 | 0.47 | 7.56 | 0.082 |
| | 61 Cyg B | | | | | | | K7 V | 6.03 | 1.37 | 1.23 | 0.60 | 8.37 | 0.039 |
| 13 | ε Ind | 21 59.6 | −57 00 | 0.291 ± 0.010 | 4.70 | 123 | − 40 | K5 V | 4.68 | 1.05 | 1.00 | 0.40 | 7.00 | 0.14 |
| 14 | BD +43° 44A | 0 15.5 | +43 44 | 0.290 ± 0.006 | 2.90 | 82 | + 13 | M1 V | 8.08 | 1.56 | 1.24 | 0.88 | 10.39 | 0.0061 |
| | +43 44B | | | | | | + 20 | M6 V e | 11.06 | 1.80 | 1.40 | 1.22 | 13.37 | 0.00039 |
| 15 | L 789−6 | 22 35.7 | −15 36 | 0.290 ± 0.007 | 3.26 | 46 | − 60 | dM7 e | 12.18 | 1.96 | 1.54 | 1.66 | 14.49 | 0.00014 |
| 16 | Procyon A | 7 36.7 | + 5 21 | 0.285 ± 0.006 | 1.25 | 214 | − 3 | F5 IV−V | 0.37 | 0.42 | 0.03 | 0.14 | 2.64 | 7.65 |
| | Procyon B | | | | | | | DF | 10.7 | | | | 13.0 | 0.00055 |
| 17 | BD +59°1915 A | 18 42.2 | +59 33 | 0.282 ± 0.004 | 2.29 | 325 | 0 | dM4 | 8.90 | 1.54 | 1.11 | 1.07 | 11.15 | 0.0030 |
| | +59 1915 B | | | | 2.27 | 323 | + 10 | dM5 | 9.69 | 1.59 | 1.14 | 1.14 | 11.94 | 0.0015 |
| 18 | CD −36°15693 | 23 02.6 | −36 09 | 0.279 ± 0.024 | 6.90 | 79 | + 10 | M2 V | 7.35 | 1.48 | 1.18 | 0.85 | 9.58 | 0.013 |
| 19 | G 51−15 | 8 26.9 | +26 57 | 0.278 ± 0.004 | 1.27 | 242 | | | 14.81 | 2.06 | | 1.79 | 17.03 | 0.00001 |
| 20 | τ Cet | 1 41.7 | −16 12 | 0.277 ± 0.007 | 1.92 | 297 | − 16 | G8 V | 3.50 | 0.72 | 0.22 | 0.26 | 5.72 | 0.45 |
| 21 | BD +5°1668 | 7 24.7 | + 5 23 | 0.266 ± 0.006 | 3.77 | 171 | + 26 | dM5 | 9.82 | 1.56 | 1.12 | 1.19 | 11.94 | 0.0015 |

continued

Table 6, continued

| No. | Name | α (1950) | δ (1950) | π ± s.e. | μ ″/a | θ | $v_r$ km s$^{-1}$ | Sp | V | B−V | U−B | R−I | $M_v$ | $L/L_\odot$ |
|---|---|---|---|---|---|---|---|---|---|---|---|---|---|---|
| 22 | L 725−32 | 1ʰ09ᵐ9 | −17°16′ | 0″261 ± 0″012 | 1″32 | 62° | + 28 | dM5 e | 12ᵐ04 | 1ᵐ83 | 1ᵐ46 | 1ᵐ44 | 14ᴹ12 | 0.00020 |
| 23 | CD −39°14192 | 21 14.3 | −39 04 | 0.260 ± 0.012 | 3.46 | 251 | + 21 | M0 V | 6.66 | 1.40 | 1.20 | 0.69 | 8.74 | 0.028 |
| 24 | Kapteyn's star | 5 09.7 | −45 00 | 0.256 ± 0.010 | 8.72 | 131 | +245 | sdM0 pec | 8.84 | 1.56 | 1.05 | 0.77 | 10.88 | 0.0039 |
| 25 | Krüger 60 A | 22 26.2 | +57 27 | 0.253 ± 0.004 | 0.86 | 246 | − 26 | dM3 | 9.85 | 1.62 | 1.25 | 1.14 | 11.87 | 0.0016 |
| | Krüger 60 B | | | | | | | dM5 e | 11.3 | 1.8 | 1.3 | | 13.3 | 0.0004 |
| 26 | BD −12°4523 | 16 27.5 | −12 32 | 0.247 ± 0.007 | 1.18 | 183 | − 13 | dM5 | 10.11 | 1.60 | 1.16 | 1.20 | 12.07 | 0.0013 |
| 27 | Ross 614 A | 6 26.8 | − 2 46 | 0.246 ± 0.004 | 1.00 | 133 | + 24 | dM7 e | 11.10 } | 1.71 | 1.15 | 1.40 | 13.12 | 0.00049 |
| | Ross 614 B | | | | | | | | 14 } | | | | 16 | 0.00004 |
| 28 | Van Maanen's star | 0 46.5 | + 5 09 | 0.232 ± 0.004 | 2.99 | 155 | + 54: | DG | 12.37 | 0.56 | 0.02 | 0.16 | 14.20 | 0.00018 |
| 29 | Wolf 424 A | 12 30.9 | + 9 18 | 0.230 ± 0.006 | 1.76 | 279 | − 5 | dM6 e | 13.16 } | 1.80 | 1.18 | 1.62 | 14.97 | 0.00009 |
| | Wolf 424 B | | | | | | | dM6 e | 13.4 } | | | | 15.2 | 0.00007 |
| 30 | CD −37°15492 | 0 02.5 | −37 36 | 0.225 ± 0.012 | 6.11 | 112 | + 23 | M4 V | 8.56 | 1.46 | 1.03 | 0.92 | 10.32 | 0.0065 |
| 31 | L 1159−16 | 1 57.5 | +12 50 | 0.224 ± 0.004 | 2.09 | 149 | | dM8 e | 12.26 | 1.82 | 1.35 | 1.35 | 14.01 | 0.00022 |
| 32 | BD +50°1725 | 10 08.3 | +49 42 | 0.222 ± 0.010 | 1.45 | 250 | − 26 | K7 V | 6.59 | 1.36 | 1.28 | 0.60 | 8.32 | 0.041 |
| 33 | CD −46°11540 | 17 24.9 | −46 51 | 0.216 ± 0.012 | 1.06 | 147 | | dM4 | 9.37 | 1.53 | 1.21 | 1.03 | 11.04 | 0.0033 |
| 34 | G 158−27 | 0 04.2 | − 7 48 | 0.214 ± 0.007 | 2.04 | 204 | | dM | 13.74 | 1.95 | | 1.52 | 15.39 | 0.00006 |
| 35 | CD −49°13515 | 21 30.2 | −49 13 | 0.214 ± 0.010 | 0.81 | 184 | + 8 | M1 V | 8.67 | 1.46 | 1.05 | 0.93 | 10.32 | 0.0065 |
| 36 | CD −44°11909 | 17 33.5 | −44 17 | 0.213 ± 0.007 | 1.16 | 217 | | M5 | 10.96 | 1.65 | 1.20 | 1.26 | 12.60 | 0.00079 |
| 37 | BD +68°946 | 17 36.7 | +68 23 | 0.213 ± 0.006 | 1.31 | 196 | − 22 | M3.5 V | 9.15 | 1.50 | 1.08 | 1.10 | 10.79 | 0.0042 |
| 38 | G 208−44 = A | 19 52.3 | +44 18 | 0.211 ± 0.004 | 0.74 | 143 | | | 13.41 | 1.90 | | | 15.03 | 0.00008 |
| | G 208−45 = B | | | | | | | | 13.99 | 1.98 | | | 15.61 | 0.00005 |
| 39 | BD −15°6290 | 22 50.6 | −14 31 | 0.209 ± 0.007 | 1.14 | 124 | + 9 | dM5 | 10.17 | 1.60 | 1.15 | 1.22 | 11.77 | 0.0017 |
| 40 | $o^2$ (40) Eri A | 4 13.0 | − 7 44 | 0.207 ± 0.003 | 4.08 | 213 | − 42 | K1 V | 4.43 | 0.82 | 0.44 | 0.31 | 6.01 | 0.34 |
| | 40 Eri B | 4 13.1 | − 7 44 | | 4.07 | 212 | − 21 | DA | 9.52 | 0.03 | −0.68 | −0.10 | 11.10 | 0.0032 |
| | 40 Eri C | | | | | | − 45 | dM4 e | 11.17 | 1.66 | 0.83 | 1.31 | 12.75 | 0.00069 |
| 41 | BD +20°2465 | 10 16.9 | +20 07 | 0.206 ± 0.006 | 0.49 | 264 | + 11 | M4.5 V e | 9.43 | 1.54 | 1.06 | 1.12 | 11.00 | 0.0035 |
| 42 | L 145−141 | 11 43.0 | −64 33 | 0.206 ± 0.012 | 2.68 | 97 | | DC | 11.50 | 0.19 | −0.60 | 0.04 | 13.07 | 0.00052 |
| 43 | 70 Oph A | 18 02.9 | + 2 31 | 0.203 ± 0.006 | 1.12 | 167 | − 7 | K0 V | 4.22 } | 0.86 | 0.51 | 0.30 | 5.76 | 0.43 |
| | 70 Oph B | | | | | | | K5 V | 6.00 } | | | | 7.54 | 0.084 |
| 44 | BD +43°4305 | 22 44.7 | +44 05 | 0.200 ± 0.004 | 0.83 | 236 | − 2 | dM5 e | 10.2 | 1.6 | 1.1 | 1.15 | 11.7 | 0.0018 |
| 45 | Altair | 19 48.3 | + 8 44 | 0.198 ± 0.006 | 0.66 | 54 | − 26 | A7 IV, V | 0.76 | 0.22 | 0.08 | 0.02 | 2.24 | 11.1 |
| 46 | AC +79°3888 | 11 44.6 | +78 58 | 0.193 ± 0.007 | 0.89 | 57 | −119 | sdM4 | 10.80 | 1.60 | | 1.18 | 12.23 | 0.0011 |
| 47 | G 9−38 = A | 8 55.4 | +19 57 | 0.192 ± 0.004 | 0.89 | 267 | | m | 14.06 | 1.84 | | | 15.48 | 0.00006 |
| | LP426−40 = B | | | | | | | m | 14.92 | 1.93 | | | 16.34 | 0.000025 |
| 48 | BD +15°2620 | 13 43.2 | +15 10 | 0.192 ± 0.007 | 2.30 | 129 | + 15 | M4 V | 8.49 | 1.44 | 1.10 | 0.86 | 9.91 | 0.0095 |

Table 7. Duplicity and variability of the stars within 5 pc.
Nos. from Table 6
SB = spectroscopic binaries UV = UV Ceti stars
UC = unseen companions BY = BY Draconis stars

| No. | Duplicity | Variability | | |
|---|---|---|---|---|
| | | Component | Name | Type |
| 1 Sun | planets | | | |
| 2 A, B, Prox | visual triple | Prox | V 645 Cen | UV |
| 3 | UC (planets?) | | | |
| 4 | | | CN Leo | UV |
| 5 | UC | | | |
| 6 A, B | visual double | A | probably var. | UV |
| | | B | UV Cet | UV |
| 7 A, B | visual double | | | |
| 8 | | | V 1216 Sgr | UV |
| 9 | | | HH And | BY |
| 11 | | | FI Vir | UV |
| 12 A, B | vis. double; UC | | | |
| 14 A, B | visual double | A | GX And | UV |
| A | SB | B | GQ And | UV |
| 15 | | | EZ Aqr | UV |
| 16 A, B | visual double | | | |
| 17 A, B | visual double | A | | UV? |
| B | $v_r$ variable? | B | | UV? |
| 18 | | | var. suspect | BY? |
| 21 | UC | | | |
| 22 | | | YZ Cet | UV |
| 25 A, B | visual double | B | DO Cep | UV |
| 26 | SB | | | |
| 27 A, B | visual double | AB | V 577 Mon | UV |
| 29 A, B | visual double | AB | FL Vir | UV |
| 31 | | | TZ Ari | UV |
| 37 | UC (and SB?) | | | |
| 38 A, B | visual double | AB | V 1581 Cyg | UV |
| 40 A, B, C | visual triple | C | DY Eri | UV |
| 41 | UC? | | AD Leo | UV |
| 43 A, B | visual double | | | |
| 44 | UC | | EV Lac | UV |
| 47 A, B | visual double | | | |
| 48 | | | var. suspect | UV? |

Nearly half of the 48 nearest stars (Table 6) are in systems: visual doubles or triples, spectroscopic binaries SB, stars with unseen companions UC, or with UC suspected. Detailed notes on the duplicity of the nearest stars are given by Lippincott [11].

Most of the red dwarfs which show H emission in their spectra have proved to be flare stars (UV Cet type, see 5.1.3.10; a few others are variables of the BY Dra type (see 5.1.2.12). The majority of these variables are components of double (or triple) stars. For two of them (Nos. 41, 44) unseen companions have been determined. In a few cases (Nos. 27, 29, 38) it has not been observed which of the two components has flared up.

Space velocities $v$ are known for 40 of the 48 nearest stars. Certainly, the selection of objects for parallax programs has given preference to the high-velocity stars.

Table 8. Space velocity ($=v$) distribution of 40 stars nearer than 5.25 pc (standard solar motion eliminated).

| $v$ [km $s^{-1}$] | >100···65···40···20···0 |
|---|---|
| $N$ | 6 7 5 14 8 |

## 8.2.6 References for 8.2


1 Eggen, O.J.: Publ. Astron. Soc. Pac. **86** (1974) 697.
2 Eggen, O.J.: Astrophys. J. **204** (1975) 101.
3 Gliese, W.: Veröffentl. Astron. Recheninst. Heidelberg No. 22 (1969).
4 Gliese, W.: Veröffentl. Astron. Recheninst. Heidelberg No. 24 (1971).
5 Gliese, W.: Bull. Inf. CDS **20** (1981) 4.
6 Gliese, W., Jahreiss, H.: Astron. Astrophys. Suppl. **38** (1979) 423.
7 Greenstein, J.L.: Astron. J. **81** (1976) 323.
8 Hill, G., Hilditch, R.W., Barnes, J.V.: Mon. Not. R. Astron. Soc. **186** (1979) 813.
9 Jahreiss, H.: Private communication (1979).
10 Krisciunas, K.: Astron. J. **82** (1977) 195.
11 Lippincott, S.L.: Space Sci. Rev. **22** (1978) 153.
12 Luyten, W.J.: Mon. Not. R. Astron. Soc. **139** (1968) 221.
13 Oort, J.H.: Bull. Astron. Inst. Netherl. **15** (1960) 45.
14 Rodgers, A.W., Eggen, O.J.: Publ. Astron. Soc. Pac. **86** (1974) 742.
14a Upgren, A.R., Armandroff, T.E.: Astron. J. **86** (1981) 1898.
15 Wielen, R.: Highlights of Astronomy (Contopoulos, G., ed.), Vol. 3, Dordrecht (1974) 395.
16 Woolley, Sir Richard, Epps, E.A., Penston, M.J., Pocock, S.B.: R. Obs. Ann. 5 (1970).

# 8.3 Structure of the Galaxy

## 8.3.1 Apparent distribution of galactic objects on the celestial sphere

### 8.3.1.1 Galactic coordinates

The system of galactic coordinates ($l$, $b$) in use since $\approx 1960$ is defined by the equatorial coordinates of the north galactic pole:

$\alpha = 12^h\,49^m\!.0, \quad \delta = +27°\,24'\!.0$ (1950.0)

and the zero meridian of the galactic longitude ($l=0$): the great semicircle originating at the north galactic pole at the position angle of $\theta = 123°\!.0$ with respect to the equatorial system for 1950.0 (decision of the IAU 1959 [1]). $\theta$ is equal to the galactic longitude of the north pole ($\delta = +90°$).

Equatorial coordinates of the zero-point of the galactic system ($l=0$, $b=0$):

$\alpha = 17^h\,42^m\!.4, \quad \delta = -28°\,55'$ (1950.0).

Inclination of the galactic plane:

$i = 62°\,36'\!.0$

Ascending node of the galactic plane on the equator (1950.0):

$\alpha = 18^h\,49^m\!.0 = 282°\,15'$ corresponding to $l = 33°\!.0$.

From 1932 to 1960 the "old system", now labelled ($l^{\mathrm{I}}$, $b^{\mathrm{I}}$), was in general use. During a transition period the symbols ($l^{\mathrm{II}}$, $b^{\mathrm{II}}$) have been used for the "new system" or "IAU system" defined above. The new zero-point ($l = l^{\mathrm{II}} = 0$, $b = b^{\mathrm{II}} = 0$) corresponds to $l^{\mathrm{I}} = 327°\!.69$, $b^{\mathrm{I}} = -1°\!.40$ and coincides with the direction to the galactic center (see also 8.3.5.1).

Tables for conversion of equatorial coordinates (1950.0) into new galactic coordinates and vice versa, and tables for conversion of old into new galactic coordinates and vice versa: [6].

### 8.3.1.2 Distribution of surface brightness

Accurate surface brightness photometry of the Milky Way, including the integrated starlight and diffuse galactic light, is difficult from the earth's surface. Zodiacal light and airglow must be carefully subtracted from the observed quantity and corrections for atmospheric extinction and scattering are necessary. In recent measurements from extraterrestrial platforms the contribution of the zodiacal light only remains to be accounted for. A summary of modern photoelectric measurements in the visible and near-ultraviolet regions is given in Table 1. For measurements at shorter wavelengths, see 7.1.4.2 and 7.1.8. Maps of the infrared surface brightness of the Milky Way, mainly stellar radiation and thermal emission of dust particles, made using balloon-borne telescopes are given in Table 2.

Table 1. Surface brightness photometry of the Milky Way (MW) at visible and near-ultraviolet wavelengths $\lambda$. Field of view: area in [□°], diameter in [°].

| Spectral regions | Field of view | Region of MW | Ref. |
|---|---|---|---|
| P; V | 0.28; 0.61 □° | whole MW | Elsässer, H., Haug, U.: Z. Astrophys. **50** (1960) 121; [1]) |
| $\lambda = 5300$ Å, $\Delta\lambda = 30$ Å | 5° | whole MW, $\|b\| \leqq 20°$ | Smith, L.L., Roach, F.E., Owen, R.W.: Batelle Institute Report BNWL-1419-UC-2 (1970) |
| *U* | 1°.2 | MW except northern region Per to Cep/Cyg ($l \approx 90° \cdots 160°$) | Pfleiderer, J., Mayer, U.: Astron. J. **76** (1971) 691 |
| *B* | 1°.2 | MW except $l \approx 90° \cdots 160°$ | Classen, Ch.: Ph. D. Dissert. Bonn 1976 |
| *U*, *B*, *V* | 1° × 1° | Scorpius [2]) | Hanner, M., Leinert, C., Pitz, E.: Astron. Astrophys. **65** (1978) 245 |
| $\lambda = (4400 \pm 450)$ Å, $\lambda = (6400 \pm 500)$ Å | 2°.3 × 2°.3 | whole MW [3]) | Weinberg, J.L.: Sky and Telescope **61** (1981) 114 |
| *R* [$\lambda = (7100 \pm 1000)$ Å] | 2° | northern MW $\|b\| \leqq 20°$ | Zavarzin, Yu.M.: Astrophysics (transl. from russ.) **14**, No. 2 (1978) 168 |
| $\lambda = 3560$ Å | 0°.6 × 0°.6 | northern MW | Winkler, Chr., Schmidt-Kaler, Th., Schlosser, W.: Mitt. Astron. Ges. **52** (1981) 163 |

[1]) See also LB, NS, Vol. VI/1 p. 602.
[2]) From space probe Helios 1.
[3]) From space probe Pioneer 10.

Table 2. Surface brightness photometry of the Milky Way (MW) in the infrared with balloon-borne telescopes.

| $\lambda$ [μm] | Field of view | Region of MW | Ref. |
|---|---|---|---|
| 2.4 | 3° | $l = 23° \cdots 75°$ $\|b\| < 20°$ | Ito, K., Matsumoto, T., Uyama, W.: Publ. Astron. Soc. Jpn. **28** (1976) 427 |
| 2.4, 3.4 | 2° | $l = 350° \cdots 0° \cdots 10°$, $l = 30° \cdots 60°$ | Hofmann, W., Lemke, D., Thum, C.: Astron. Astrophys. **57** (1977) 111 |
| 2.4 | 1° | $l = 350° \cdots 0° \cdots 30°$ $\|b\| < 10°$ | Maihara, T., Oda, N., Sugiyama, T., Okuda, H.: Publ. Astron. Soc. Jpn. **30** (1978) 1 |
| 2.4; 3.4 | 0°.5, 0°.8, 1°.7; ≈2° | $l = 285° \cdots 0° \cdots 25°$ $\|b\| < 10°$ | Hayakawa, S., Matsumoto, T., Murakami, H., Uyama, K., Yamagami, T., Thomas, J.A.: Preprint DPNU-4-79 (1979) |
| 60···300 | 15′ | $l = 348° \cdots 0° \cdots 33°$ $b \approx 0°$ | Low, F.J., Wurtz, R.F., Poteet, W.M., Nishimura, T.: Astrophys. J. Lett. **214** (1977) L115 |
| 75···95, 115···196 | 0°.7 | $l = 36° \cdots 55°$ $b \approx 0°$ | Serra, G., Puget, J.L., Ryter, C.E., Wijnbergen, J.: Astrophys. J. Lett. **222** (1978) L21 |
| | | $l = 26° \cdots 40°$ $b \approx 0°$ | Serra, G., et al.: Astron. Astrophys. **76** (1979) 259 |
| 150 | 0°.7 | $l = 345° \cdots 0° \cdots 30°$ | Maihara, T., et al.: Int. Astron. Union Symp. **84** (1979) 109; Astron. Astrophys. **97** (1981) 139 |

Tables for the integrated starlight derived from star counts in the Selected Areas have been derived (1961) in photographic and visual magnitudes [4] – see also LB, NS, Vol. VI/1, p. 602 – and in *B* magnitude [5]. Comparison with modern measurements shows that these numbers may not be correct at lower galactic latitudes [2]. Therefore only results for $|b| \geqq 20°$ are reproduced in Table 3.

Table 3. Integrated starlight per square degree in $B$ magnitude for $|b| \geqq 20°$ [5].
$S_{10}(B)/\square°$ equivalent number of stars with $B=10^m$ per square degree

| $b$ | −80° | −70° | −60° | −50° | −40° | −30° | −20° | +20° | 30° | 40° | 50° | 60° | 70° | 80° |
|---|---|---|---|---|---|---|---|---|---|---|---|---|---|---|
| $l$ | $I_B\,[S_{10}(B)/\square°]$ | | | | | | | | | | | | | |
| 0° | 11 | 15 | 18 | 22 | 29 | 39 | 100 | 40 | 26 | 20 | 15 | 14 | 11 | 10 |
| 10 | 11 | 15 | 18 | 22 | 28 | 40 | 87 | 36 | 27 | 20 | 14 | 13 | 11 | 10 |
| 20 | 11 | 15 | 18 | 22 | 29 | 39 | 74 | 32 | 27 | 21 | 15 | 13 | 11 | 10 |
| 30 | 11 | 15 | 18 | 23 | 31 | 39 | 50 | 32 | 29 | 22 | 16 | 13 | 11 | 10 |
| 40 | 11 | 15 | 18 | 23 | 30 | 37 | 45 | 43 | 31 | 22 | 16 | 13 | 11 | 10 |
| 50 | 11 | 16 | 18 | 20 | 24 | 30 | 45 | 49 | 33 | 22 | 17 | 13 | 10 | 10 |
| 60 | 12 | 17 | 17 | 18 | 20 | 29 | 40 | 55 | 35 | 24 | 17 | 14 | 10 | 10 |
| 70 | 13 | 17 | 16 | 15 | 19 | 29 | 40 | 56 | 34 | 24 | 18 | 14 | 10 | 10 |
| 80 | 13 | 17 | 15 | 16 | 19 | 28 | 40 | 47 | 30 | 24 | 18 | 15 | 11 | 10 |
| 90 | 13 | 17 | 14 | 17 | 19 | 26 | 40 | 41 | 29 | 23 | 18 | 15 | 12 | 10 |
| 100 | 14 | 16 | 14 | 16 | 19 | 27 | 40 | 32 | 27 | 22 | 18 | 15 | 12 | 10 |
| 110 | 15 | 16 | 15 | 15 | 20 | 28 | 40 | 28 | 23 | 20 | 17 | 15 | 12 | 10 |
| 120 | 15 | 15 | 15 | 15 | 20 | 28 | 44 | 28 | 20 | 18 | 16 | 15 | 12 | 10 |
| 130 | 15 | 15 | 15 | 15 | 20 | 28 | 46 | 31 | 20 | 18 | 16 | 14 | 12 | 10 |
| 140 | 15 | 14 | 14 | 16 | 19 | 26 | 40 | 32 | 24 | 18 | 15 | 14 | 12 | 10 |
| 150 | 15 | 15 | 13 | 16 | 18 | 20 | 22 | 28 | 24 | 19 | 15 | 14 | 12 | 10 |
| 160 | 15 | 14 | 14 | 16 | 17 | 15 | 15 | 28 | 24 | 19 | 16 | 14 | 12 | 10 |
| 170 | 15 | 14 | 14 | 15 | 15 | 13 | 15 | 31 | 24 | 20 | 17 | 14 | 12 | 10 |
| 180 | 17 | 14 | 14 | 14 | 12 | 12 | 18 | 34 | 24 | 19 | 17 | 14 | 12 | 10 |
| 190 | 15 | 13 | 15 | 15 | 14 | 18 | 20 | 38 | 23 | 18 | 15 | 14 | 12 | 10 |
| 200 | 15 | 13 | 15 | 17 | 15 | 20 | 25 | 40 | 23 | 20 | 15 | 14 | 12 | 10 |
| 210 | 14 | 13 | 14 | 19 | 22 | 23 | 30 | 40 | 27 | 21 | 15 | 14 | 12 | 10 |
| 220 | 14 | 13 | 14 | 18 | 24 | 28 | 36 | 40 | 28 | 21 | 15 | 14 | 12 | 10 |
| 230 | 14 | 14 | 14 | 17 | 26 | 30 | 45 | 40 | 27 | 20 | 15 | 14 | 12 | 10 |
| 240 | 14 | 14 | 15 | 15 | 26 | 37 | 56 | 40 | 26 | 18 | 15 | 13 | 12 | 10 |
| 250 | 12 | 14 | 15 | 14 | 23 | 41 | 56 | 38 | 25 | 18 | 15 | 13 | 12 | 10 |
| 260 | 12 | 14 | 15 | 14 | 17 | 34 | 49 | 34 | 23 | 17 | 15 | 13 | 12 | 10 |
| 270 | 12 | 14 | 15 | 17 | 17 | 23 | 38 | 34 | 23 | 17 | 14 | 14 | 12 | 10 |
| 280 | 11 | 14 | 15 | 20 | 20 | 25 | 38 | 38 | 28 | 17 | 14 | 15 | 13 | 10 |
| 290 | 11 | 14 | 15 | 21 | 25 | 28 | 44 | 43 | 30 | 18 | 15 | 15 | 12 | 10 |
| 300 | 11 | 15 | 15 | 21 | 24 | 26 | 33 | 49 | 32 | 20 | 16 | 15 | 12 | 10 |
| 310 | 11 | 15 | 19 | 22 | 25 | 29 | 56 | 60 | 34 | 21 | 17 | 14 | 12 | 10 |
| 320 | 12 | 15 | 20 | 23 | 26 | 30 | 75 | 56 | 35 | 21 | 15 | 14 | 13 | 10 |
| 330 | 12 | 15 | 19 | 23 | 27 | 37 | 66 | 53 | 30 | 17 | 15 | 15 | 13 | 10 |
| 340 | 12 | 15 | 19 | 23 | 29 | 39 | 72 | 49 | 27 | 18 | 18 | 16 | 13 | 10 |
| 350 | 12 | 15 | 18 | 23 | 29 | 38 | 88 | 44 | 28 | 23 | 20 | 15 | 12 | 10 |

Wide-angle photographs of the Milky Way

A four-colour photographic atlas of the sky:

Schlosser, W., Schmidt-Kaler, Th., Hünecke, W.: Astronomisches Institut Bochum 1975. Separate publication. The atlas contains 22 super-wide-angle photographs in $U$, $B$, $V$, $R$; the Milky Way is shown over a span of 140° of galactic longitude. See also: Schlosser, W., Schmidt-Kaler, Th.: Vistas Astron. **21** (1977) 447; Sky and Telesc. **53** (1977) 436; Sterne und Weltraum **11** (1972) 98, 120.

Houck atlas of the Milky Way:

This atlas is intended to be a compilation of views of the Milky Way as obtained by University of Wisconsin astronomers. Eaton, J.A., Code, A.P.: Vistas Astron. **20** (1977) 319.

### 8.3.1.3 Apparent distribution of the different types of galactic objects

To obtain the structure of the galactic stellar system the spatial density distribution has to be deduced for each physical type of star separately. The large older systematic catalogues containing stars of all spectral types (see 8.1.3···8.1.5; 4.1, 4.2) do not include sufficient observational material to provide a basis for this kind of density analysis. Therefore special surveys have been performed. The most extensive of these are given in Table 4. For references to some further surveys in selected small regions of the Milky Way, see Table 9. Galactic belt surveys of M-, C-, and S-type stars have been summarized by Mavridis [3].

Table 4. Extended surveys and catalogues for selected spectral types in the Milky Way (MW).

| Spectral type | Region | Magnitude limit | Remarks | Ref. |
|---|---|---|---|---|
| O, B, AI···GI | northern MW | $13^m$(pg) | total number of OB stars 6243 | Luminous stars in the northern Milky Way. Hamburger Sternwarte – Warner & Swasey Observatory<br>I: Hardorp, J., Rohlfs, K., Slettebak, A., Stock, J., Hamburg-Bergedorf 1959<br>II: Stock, J., Nassau, J.J., Stephenson, C.B., Hamburg-Bergedorf 1960<br>III: Hardorp, J., Theile, I., Voigt, H.H., Hamburg-Bergedorf 1964<br>IV: Nassau, J.J., Stephenson, C.B., Hamburg-Bergedorf 1964<br>V: Hardorp, J., Theile, I., Voigt, H.H., Hamburg-Bergedorf 1965<br>VI: Nassau, J.J., Stephenson, C.B., MacConnell, D.J., Hamburg-Bergedorf 1965 |
| | southern MW | 13 (pg) | extension of I···VI (5132 stars) | Luminous stars in the southern Milky Way. Stephenson, C.B., Sanduleak, N.: Publ. Warner & Swasey Observatory **1** (1971) 100 |
| O, B, AI···GI | southern MW | 11.5 (pg) | >2000 stars | Finding lists of high-luminosity stars. Münch, L., Gonzáles, G., Iriarte, B., Chavira, E., Morgan, W.W.: Bol. Obs. Tonantzintla & Tacubaya, No. 5, 7–12, 14, 15 (1952···1956) |
| O, B | southern MW $l=233°\cdots0°\cdots20°$ | 11.5 ($B$) | 1660 stars | Klare, G., Szeidl, B.: OB-stars of the southern Milky Way (Heidelberg Catalogue). Veröff. Landessternwarte Heidelberg-Königstuhl Vol. **18** (1966) 9. Photometric Data: Klare, G., Neckel, T.: Astron. Astrophys. Suppl. **27** (1977) 215. |
| O, B, AI | $l=350°\cdots0°\cdots240°$ | 11.5 ($V$) | data for 1259 stars | Hiltner, W.A.: Astrophys. J. Suppl. **2** (1956) 389 |
| O | | | new edition of Hiltner's 1956 catalogue | A general O type stars catalogue. Goy, G.: Astron. Astrophys. Suppl. **12** (1973) 277; **26** (1976) 273 |
| O, BI··· MI | whole MW | | 699 supergiants of luminosity class Ib or brighter | Humphreys, R.M.: Astron. J. **75** (1970) 602 |

continued

Table 4, continued

| Spectral type | Region | Magnitude limit | Remarks | Ref. |
|---|---|---|---|---|
| O, B; M, N, R, S | $l = 237° \cdots 330°$ | 13 (*B*) | 13300 stars | Sundman, A., Lodén, K., Lodén, L.O., Nordström, B.: A spectral survey of the southern Milky Way. Astron. Astrophys. Suppl. **16** (1974) 445: $l = 306° \cdots 318°$; **21** (1975) 193: $l = 237° \cdots 280°$; **23** (1976) 283: $l = 280° \cdots 306°$ |
| Wolf-Rayet stars | whole sky | | 123 stars known up to 1962 | Roberts, M.S.: Astron. J. **67** (1962) 79 |
| | | | 107 stars | Smith, L.F.: Mon. Not. R. Astron. Soc. **138** (1968) 109 |
| Be | whole MW | | main catalogue; Be stars in clusters and associations | Catalogue and bibliography of B type emission line stars. Jaschek, C., Ferrer, L., Jaschek, M.: Obs. Astron. Univ. Nacional La Plata, Ser. Astron. Vol. **37** (1971) |
| Emission-line stars later than B | whole MW | | 1540 stars | Catalogue and bibliography of emission-line stars of types later than B. Bidelman, W.P.: Astrophys. J. Suppl. **1** (1954) 175. |
| M | whole MW | 10 (*I*) | all late M-type BD stars | The distribution of late M-type stars along the galactic equator. Nassau, J.J., Blanco, V.M., Cameron, D.M.: Astrophys. J. **124** (1956) 522; Neckel, H.: Astron. Astrophys. Suppl. **18** (1974) 169 |
| K···M dwarfs | northern sky | 10.5 (pg) | lists contain 900 M-type dwarfs | Vyssotsky, A.N.: Dwarf M stars found spectroscopically. Fourth List: Astron. J. **61** (1956) 201; supplementary list: Astron. J. **63** (1958) 211 |
| K···M dwarfs | southern hemisphere | 10.5 (pg) | 624 stars with spectral types K2 and later | Upgren, A.R., Grossenbacher, R., Penhallow, W.S., Mac Connell, D.J., Frye, R.L.: Astron. J. **77** (1972) 486 |
| Low-luminosity stars | whole sky | | 1055 stars | Luyten, W.J.: The stars of low luminosity. University of Minnesota, Minneapolis 1970, 48 pp., separate publication |
| White dwarfs | whole sky | | catalogue contains 2934 entries | Luyten, W.J.: White dwarfs, Univ. of Minnesota, Minneapolis, Minnesota 1970, separate publication |
| | | | 3513 stars | Luyten, W.J.: White dwarfs II, Univ. of Minnesota, Minneapolis, Minnesota 1977, separate publication. Further lists of degenerate stars: Greenstein, J.L.: Astrophys. J. Lett. **207** (1976) L119 Astrophys. J. Lett. **207** (1976) L119 |
| Carbon stars | whole MW | | 3219 stars | A general catalogue of cool carbon stars. Stephenson, C.B.: Publ. Warner & Swasey Observatory Vol. **1** No. 4 (1973) |
| | southern MW | | 1124 stars | A catalogue of carbon stars in the southern Milky Way. Westerlund, B.E.: Astron. Astrophys. Suppl. **4** (1971) 51 |
| S | whole MW | | 741 stars | A general catalogue of S-type stars. Stephenson, C.B.: Publ. Warner & Swasey Observatory Vol. **2**, No. 2 (1976) 21 |

Apparent distribution of star clusters and associations: see 6.2.

Apparent distribution of $\gamma$- and X-ray sources: see 5.7.

Apparent distribution of interstellar objects: for H II regions, see 7.3.2
for dark nebulae, see 7.2.2
for radio radiation, see 7.1.6 and 7.1.7.

Large surveys for infrared sources:

Two-Micron Sky Survey. Neugebauer, G., Leighton, R.B.: NASA SP-3047 (1969).
Measurements at the effective wavelength $\lambda = 2.2$ μm ($K$ magnitude) of over 5000 stars in the northern hemisphere.

Air Force Geophysics Laboratory Infrared Sky Survey. Revised and extended catalogue: Price, S.D., Walker, R.G.: AFCR-TR-73-0373 (1975).
Measurements of 2361 sources detected in one or more of the survey bands: 4.2 μm, 11.0 μm, 19.8 μm, 27.4 μm.

Steward Observatory Near Infrared Photographic Sky Survey (NIPSS). In progress since 1978, see for example: Horner, V.M., Craine, E.R.: Publ. Astron. Soc. Pacific **92** (1980) 209; Rossano, G.S., Craine, E.R.: Near infrared photographic sky survey: a field index, Pachart Publ. House, Tucson (1980).

## 8.3.1.4 References for 8.3.1

1 Blaauw, A., Gum, C.S., Pawsy, J.L., Westerhout, G.: Int. Astron. Union Inform. Bull. No. 1 (1959).
2 Hanner, M., Leinert, C., Pitz, E.: Astron. Astrophys. **65** (1978) 245.
3 Mavridis, L.M.: Space distribution of the late type stars, in: Structure and Evolution of the Galaxy (Mavridis, L.M., ed.), Reidel, Dordrecht, Holland (1971) p. 110.
4 Roach, F.E., Megill, L.R.: Astrophys. J. **133** (1961) 228.
5 Sharov, A.S., Lipaeva, N.A.: Soviet Astron.-Astron. J. **17** (1973) 69.
6 Torgård, I.: Ann. Lund **15**, **16**, **17** (1961).

# 8.3.2 The local star field

Symbols for 8.3.2 and 8.3.3

| | |
|---|---|
| $z$ | distance from the galactic plane |
| $R$ | distance from the galactic center |
| $R_0$ | distance of the sun from the galactic center |
| $r$ | distance from the sun |
| $D(r)$ | spatial star density at a distance $r$ for a fixed direction [number/volume] |
| $D_0$ | spatial star density in the neighbourhood of the sun |
| $D(z)$ | spatial star density as a function of distance $z$ from the galactic plane |
| $\varphi(M\|r)$ | fraction of the stars at the distance $r$ with absolute magnitudes in the range $M-\frac{1}{2}\cdots M+\frac{1}{2}$ |
| $\varphi_0(M)$ | $\varphi(M\|r)$ in the neighbourhood of the sun |
| $D_0\varphi_0(M)$ | luminosity function of the stars in the neighbourhood of the sun = number of stars per unit volume with absolute magnitudes in the range $M-\frac{1}{2}\cdots M+\frac{1}{2}$ |
| $D_0\varphi(M)$ | initial luminosity function = distribution of absolute magnitudes of newly born stars; related to the stellar birthrate. |
| $\varrho$ | stellar mass density in [$\mathfrak{M}_\odot$/volume] |
| $\varrho_0$ | stellar mass density in the solar neighbourhood |

### 8.3.2.1 Luminosity function

Results for the luminosity function in the vicinity of the sun are presented in Table 5. The second and third columns give the mean run deduced mainly from results obtained by van Rhijn [29], McCuskey [17], and Luyten [15]. The last column gives the luminosity function for $r \leqq 20$ pc derived from the material of Gliese's "Catalogue of nearby stars" (see 8.2). After transformation of the absolute magnitudes to the same colour system the results of Luyten (fourth column) and of Jahreiß and Wielen show good agreement. Both give the maximum of the luminosity function near $M_V \approx 13$ corresponding to $M_{pg} \approx M_B \approx 15$. Probably the numbers are incomplete for $M_V > 15$. For recent discussion of the luminosity function, fraction of stars of a given absolute magnitude on the Main Sequence, present-day and initial mass function, see [32].

The variation of the luminosity function with distance $z$ from the galactic plane has been investigated by W. Becker and his collaborators in Basle [7, see also Table 14]. Possible variations with position in the galactic plane have been discussed by McCuskey [18, 19] and in several of the papers listed in Table 9.

Table 5. Number of stars per $10^4$ $pc^3$ in an interval of absolute magnitude $\Delta M = 1$ mag for the star field in the immediate solar neighbourhood, $D_0\,\varphi_0\,(M)$. Absolute magnitudes for the photographic (pg) or the visual (V) spectral region.

| $M$ | $D_0\,\varphi_0\,(M)$ [number $\cdot 10^{-4}$ $pc^{-3}$ $mag^{-1}$] Van Rhijn, McCuskey, Luyten [1] pg | V | Luyten [15] pg | Jahreiß-Wielen [12, 13] V |
|---|---|---|---|---|
| $-6^M$ | $2\ \cdot 10^{-4}$ | $1\ \cdot 10^{-4}$ | | |
| $-5$ | $1.2 \cdot 10^{-3}$ | $6\ \cdot 10^{-4}$ | | |
| $-4$ | $4.3 \cdot 10^{-3}$ | $2.9 \cdot 10^{-3}$ | | |
| $-3$ | $1.6 \cdot 10^{-2}$ | $1.3 \cdot 10^{-2}$ | | |
| $-2$ | $6\ \cdot 10^{-2}$ | $5\ \cdot 10^{-2}$ | | |
| $-1$ | 0.20 | 0.25 | | |
| 0 | 1 | 1 | 1 | 1 |
| 1 | 2 | 3 | 3 | 4 |
| 2 | 5 | 6 | 8 | 7 |
| 3 | 10 | 12 | 11 | 13 |
| 4 | 15 | 19 | 17 | 23 |
| 5 | 22 | 34 | 23 | 32 |
| 6 | 30 | 42 | 31 | 36 |
| 7 | 34 | 35 | 38 | 30 |
| 8 | 41 | 42 | 48 | 39 |
| 9 | 59 | 54 | 58 | 47 |
| 10 | 65 | 78 | 72 | 73 |
| 11 | 80 | 98 | 84 | 102 |
| 12 | 93 | 107 | 100 | 153 |
| 13 | 102 | 117 | 129 | 178 |
| 14 | 115 | 129 | 160 | 102 |
| 15 | 126 | 125 | 191 | 102 |
| 16 | 120 | 120 | 143 | 64 |
| 17 | 107 | 107 | 88 | 64 |
| 18 | 89 | 83 | 55 | $>5$ |
| 19 | 63 | 50 | 31 | $>5$ |
| 20 | 40 | 30 | 19 | |
| 21 | 20 | 13 | 10 | |
| 22 | 8 | 5 | 6 | |
| 23 | | | 1 | |
| Total | 1247 | 1310 | 1327 | $>1080$ |

Table 6. Observed luminosity function $D_0\varphi_{MS}\,(M)$ including main-sequence stars only and initial luminosity function $D_0\varphi\,(M)$ [22]. $D_0\varphi_{MS}\,(M)$ and $D_0\varphi\,(M)$ in [number/$pc^3$].

| $M_V$ | $\log(D_0\varphi_{MS})+10$ | $\log(D_0\varphi)+10$ |
|---|---|---|
| $-6^M$ | 1.29 | 4.71 |
| $-5$ | 2.43 | 5.59 |
| $-4$ | 3.18 | 6.08 |
| $-3$ | 3.82 | 6.41 |
| $-2$ | 4.42 | 6.68 |
| $-1$ | 5.04 | 6.92 |
| 0 | 5.60 | 7.10 |
| 1 | 6.17 | 7.26 |
| 2 | 6.60 | 7.25 |
| 3 | 7.00 | 7.23 |
| 4 | 7.30 | 7.30 |
| 5 | 7.45 | 7.45 |
| 6 | 7.56 | 7.56 |
| 7 | 7.63 | 7.63 |

### 8.3.2.2 Star densities in the solar neighbourhood

Table 7. Spatial star density $D_0$ for different spectral types in the neighbourhood of the sun.

| Star types | $D_0$ [number $\cdot 10^{-3}$ pc$^{-3}$] | | | |
|---|---|---|---|---|
| Main sequence: | [28] | [4] | | |
| A0···A5 | 1.1 | 0.9 (A) | 0.5 | [8, 11] |
| A7···F2 | 1.2 } | | 0.9 | [8, 11] |
| F5 | 1.5 } | 4.5 (F) | 1.2 | [8, 11] |
| F8 | 2.2 } | | 1.4 | [8, 11] |
| G | 3.4 | 13 | (4) | [8, 12] |
| K | 2.9 } | 100 | (8) | [8, 12] |
| M | 39.0 } | | 30···60 | [27] |
| Giants: | | | | |
| G | $2\cdot 10^{-2}$ | 0.30 | 0.11 | [17] |
| K0 | 0.14 } | | – | |
| K1···K2 | 0.10 } | 0.09 | } 0.14 | [21] |
| K3...K5 | $6\cdot 10^{-3}$ } | | | |
| M | $3\cdot 10^{-3}$ | | $9\cdot 10^{-3}$ | [23] |
| G5···M | 0.27 | | 0.4 | [17] |
| White dwarfs | | | 20 | [6, 9, 30, 31] |

Table 8. Stellar mass density in the vicinity of the sun.

| Contributor | Mass density [$\mathfrak{M}_\odot$ pc$^{-3}$] | |
|---|---|---|
| | $r \leqq 20$ pc [12] | [11] |
| Main-sequence stars | | |
| $M_V$ range $<9^m.5$ | 0.020 } | |
| 9.5···13.5 | 0.014 } | 0.048 |
| $>13.5$ | 0.004 } | |
| Giants | 0.001 | |
| White dwarfs | 0.007 | 0.020 |
| Total mass density due to known stars | 0.046 | 0.068 |

### 8.3.2.3 Distribution of common stars near the galactic plane

Large programs for the investigation of the spatial distribution of different spectral types and based on extended surveys of many Milky Way fields have been carried out at the Observatories of Cleveland (Warner & Swasey), Basle, Abastumani, Stockholm, Uppsala and others. Some summaries of results have been given (McCuskey [18, 19], Becker [3], Sundman [25]). Recent work for selected regions of the Milky Way is presented in Table 9. For results on O and early B stars, see 8.3.3.2. Investigations of the distribution of late-type stars have been reviewed by Mavridis [16].

Table 9. Recent regional studies of spatial star density distribution at low galactic latitudes $|b| < 20°$ for common stars of spectral types B···M. (For OB stars, see 8.3.3.2).
For notation, see 4.2; MS = main sequence.

| Field (center or extension in *l*, *b*) *l* | *b* | Area □° | Number of stars | Observational data | Limiting magnitude | Spectral types | Ref. |
|---|---|---|---|---|---|---|---|
| *Galactic center direction* ($l = 350°\cdots0°\cdots10°$) | | | | | | | |
| 355° | + 2° | 34 | 380 | $V$, $B-V$, Sp | $12^m.75$ ($V$) | B0V···A0V | Poulakos, C.: Astron. Astrophys. **50**, (1976) 11 |
| 1.3 | − 7.1 | 0.16 | 3000 | $U$, $B$, $V$ | 16 ($V$) | K2III···K5III | Loibl, B.: Astron. Astrophys. **68** (1978) 107 |
| 359.6 | −10.1 | 0.14 | | | | | |
| 0.0 | −15.9 | 0.08 | | | | | |
| 0 | − 8 | | 2700 | $B$, $V$ | 20 ($V$) | red giants | Van den Bergh, S., Herbst, E.: Astron. J. **79** (1974) 603 |
| 0.9 | − 3.9 | | 557 | $U$, $B$, $V$ | 17.7 ($V$) | red giants | Van den Bergh, S.: Astron. J. **76** (1971) 1082 |
| 1.1 | − 1.1 | 0.14 | 2220 | $R$, $G$, $U$ | 16.5 ($G$) | MS: $M_G = 0^M\cdots+7^M$, late-type giants | Gschwind, P.: Astron. Astrophys. Suppl. **19** (1975) 281 |
| 1.2 | − 3.6 | 0.26 | 1400 | $R$, $G$, $U$ | 16.5 ($G$) | MS: $M_G = +2\cdots+7$, late-type giants | Fünfschilling, H.: Astron. Astrophys. **13** (1971) 454 |
| 6.5 | +10.3 | 0.19 | 1934 | $R$, $G$, $U$ | 17.9 ($G$) | MS: $M_G = 1\cdots3$; late-type giants | Becker, W.: Astron. Astrophys. Suppl. **38** (1979) 341 |
| 12.1 | − 0.8 | 0.065 | 1608 | $R$, $G$, $U$ | 17.2 ($G$) | MS: $M_G = 2\cdots8$, giants: 0···2 | Topaktas, L.: Astron. Astrophys. Suppl. **22** (1975) 207; Spaenhauer, A.M., Fenkart, R.P.: Astron. Astrophys. Suppl. **35** (1979) 249 |
| *Scutum-Aquila-Vulpecula* ($l = 10°\cdots60°$) | | | | | | | |
| 15° | 1°.5 | 68 | 1592 | $m_{pg}$, CI, Sp | 11···12.5 (pg) | B0···B3, B8···A0 | Grigoreva, N.B.: Sov. Astron.-A.J. **13** (1970) 807 |
| 38 | 1.2 | 36 | 974 | | | | |
| 58 | 0.6 | 28 | 316 | | | | |
| 25.4 | − 4.35 | 0.1 | 1648 | $R$, $G$, $U$ | 17.5 ($G$) | MS: $M_G = +3\cdots+6$; late-type giants | Schubarth, A.: Astron. Astrophys. **9** (1970) 130 |
| 26.2 | − 1.0 | 0.2 | 684 | $R$, $G$, $U$ | 15.5 ($G$) | MS: $M_G = -1.5\cdots+3$; late-type giants | Becker, W.: Z. Astrophys. **54** (1962) 155 |
| 26.2 | − 2.3 | 5.2 | 1608 | $R$, $G$, $U$ | 13.5 ($G$) | MS: $M_G = 2\cdots7$ giants: 0···2 | Karali, S.: Astron. Astrophys. Suppl. **35** (1979) 241 |

continued

Table 9, continued

| Field (center or extension in *l*, *b*) *l* | *b* | Area □° | Number of stars | Observational data | Limiting magnitude | Spectral types | Ref. |
|---|---|---|---|---|---|---|---|
| 26.70 | − 2.55 | 0.7 | 477 | *R*, *G*, *U* | 15.5 (*G*) | MS: $M_G = 0 \cdots +3$; late-type giants | Brodbeck, K.: Z. Astrophys. **58** (1963) 127 |
| 25.85 | − 3.65 | 0.7 | 499 | | | | |
| 36.0 | + 8.3 | 12 | 793 | *B*, *V*, Sp | 12.5 (*B*) | A2V···G0V gG5···gK5 | Wramdemark, S.: Astron. Astrophys. **58** (1977) 99 |
| 37.18 | 1.97 | 0.1 | 1 530 | *R*, *G*, *U* | 17.6 (*G*) | MS: $M_G = -1 \cdots +8$; late-type giants | Spaenhauer, A.: Astron. Astrophys. Suppl. **30** (1977) 63 |
| 55.3 | + 0.9 | 18.4 | 2045 | *B*, *V*, Sp | 12.5 (*V*) | B3···K3 | Guseva, N.G., Metreveli, M.D.: Bull. Abastumani Astrophys. Obs. **48** (1977) 116 |
| 59.4 | − 1 | 18.4 | 1 779 | *B*, *V*, Sp | 12.5 (*V*) | O···A9 | Metreveli, M.D., Kusnetsov, V.I.: Bull. Abastumani Astrophys. Obs. **48** (1977) 125 |
| *Cygnus-Cepheus-Cassiopeia-Perseus* ($l = 60° \cdots 150°$) | | | | | | | |
| 50°···150° | − 5°···+5° | | 56 190 | *V*, CI, Sp | 13 (*V*) | B8···A3 | McCuskey, S.W., Houk, N.: Astron. J. **76** (1971) 1117 |
| 62.7 | 16.0 | 19.5 | 524 | *U*, *B*, *V*, Sp | 13 (pg) | A1···F7 | Balázs, L.G.: Mitt. Sternwarte Budapest No. 68 (1975) |
| 65.7 | + 1.2 | 1 | 645 | *B*, *V*, Sp | 14.5 (*V*) | O···A9 | Voroshilov, V.I., Kalandadze, N.B., Kusnetsov, V.I.: Bull. Abastumani Astrophys. Obs. **43** (1972) 55, 67 |
| 112.8 | + 0.5 | 1 | 730 | | | | |
| 65.7 | + 1.2 | 18.3 | 4 500 | *B*, *V*, Sp | 12.5 (*V*) | O···G5 | Voroshilov, V.I., Kalandadze, N.B.: Astrometriya i Astrofizika Kiev **32** (1977) 46 |
| 75 | + 2.5 | | | | | M giants | Velghe, A.G.: Commun. Obs. Roy. Belg. Ser. A, No. 25 (1974) |
| 76.9 | + 0.6 | 1 | 264 | *B*, *V*, Sp | 15 (*B*) | B5···K | Kusnetsov, V.I.: Astrometriya i Astrofizika Kiev **12** (1971) 32, 40 |
| 76.9 | + 0.6 | 18 | 3 213 | *B*, *V*, Sp | 13 (*V*) | O···M8 | Kalandadze, N.B., Kolesnik, L.N.: Astrometriya i Astrofizika Kiev **32** (1977) 57 |

continued

Table 9, continued

| Field (center or extension in $l, b$) $l$ | $b$ | Area □° | Number of stars | Observational data | Limiting magnitude | Spectral types | Ref. |
|---|---|---|---|---|---|---|---|
| 90 | −1 | 0.19 | 677 | $U, B, V$ | 16 ($V$) | $M_V \leqq 4$, MS and late-type giants | Hartl, H.: Astron. Astrophys. **41** (1975) 321 |
| | | 0.24 | 705 | | | | |
| | | 0.57 | 300 | | | | |
| 80 | 4.0 | 30 | 186 | $m_{pg}$, CI, Sp | 11⋯12.5 (pg) | B0⋯B3 | Grigoreva, N.B.: Sov. Astron.-A.J. **13** (1970) 807; Metik, L.P.: Izv. Krim Astrophys. Obs. **27** (1962) 283; Brodskaya, E.S.: Izv. Krim Astrophys. Obs. **26** (1961) 382 |
| 85 | 1.3 | 42 | 3404 | | | B8⋯A0 | |
| 121 | −0.5 | 45 | 760 | | | partly later types | |
| 128 | −1.2 | 48 | 514 | | | | |
| 136 | −1.1 | 48 | 496 | | | | |
| 88 | 0 | 10 | 442 | $m_{ir}$, Sp | 15 ($V$) | M, C | Nandy, K., Smriglio, F.: Publ. R. Obs. Edinburgh **7** (1970) No. 1 |
| 102 | −1 | | 340 | $U, B, V$, H$\delta$,$\gamma$, Sp | 12.5 ($V$) | O⋯A3 | Barbier, M., Bernard, A., Bigay, J.H., Garnier, R.: Astron. Astrophys. **27** (1973) 421 |
| 102 | +5 | 19.5 | 843 | $U, B, V$, Sp | 13 ($B$) | B3⋯K | Kun, M.: Mitt. Sternw. Budapest No. 72 (1979) |
| 103 | 0 | 8 | 1305 | $V, B-V$, Sp | 13 ($V$) | B7⋯K3, M | Kubinec, W.R.: Publ. Warner & Swasey Obs. Vol. **1**, No. 3 (1973) |
| 112.6 | +0.5 | 18.3 | 4000 | $B, V$, Sp | 13.5 ($B$) | O⋯K | Voroshilov, V.I., Kalandadze, N.B.: Bull. Abastumani Astrophys. Obs. **48** (1977) 103 |
| 115 | −5.4 | | | | | M, C, S | Tsioumis, A.: Ph. D. Thesis, Univ. Thessaloniki, Greece |
| 120 | −2.8 | 20.9 | 170 | | | M | Poulakos, C.: Mem. Soc. Astron. Ital., Nuova Ser. **42** (1971) 421 |
| 133 | −1 | 21.3 | 445 | $B, V$, Sp | 12 ($V$) | B8⋯A3 | McCuskey, S.W.: Astron. J. **79** (1974) 107 |
| 133 | −1 | 25 | 99 | *uvby*, H$\beta$, Sp | 10 ($y$) | A | Pesch, P., McCuskey, S.W.: Astron. J. **79** (1974) 116 |
| 132⋯152 | −3⋯+3 | 130 | ~1850 | $m_{4400}$, CI, H$\gamma$,$\delta$, K, Sp etc. | 10.5 ($\lambda$ 4400) | A⋯F, gG, gK | Rydström, B.A.: Astron. Astrophys. Suppl. **32** (1977) 25 |

continued

Table 9, continued

| Field (center or extension in $l, b$) $l$ | $b$ | Area □° | Number of stars | Observational data | Limiting magnitude | Spectral types | Ref. |
|---|---|---|---|---|---|---|---|
| *Galactic anticenter direction* ($l = 150° \cdots 210°$) | | | | | | | |
| 161°.7 | − 0°.5 | 0.2 | 1743 | *R*, *G*, *U* | 18.0 (*G*) | MS: $M_G = +2 \cdots +7$; late-type giants | Becker, W., Fang, Ch.: Astron. Astrophys. **22** (1973) 187 |
| 177.8 | + 2.5 | 0.18 | 1234 | *R*, *G*, *U* | 17.7 (*G*) | MS: $M_G = +2 \cdots +7$; late-type giants | Becker, W., Svolopoulos, S.: Astron. Astrophys. Suppl. **23** (1976) 97; Spaenhauer, A.M., Fenkart, R.P.: Astron. Astrophys. Suppl. **35** (1979) 249 |
| 185.1 | + 1.7 | 0.27 | 2330 | *R*, *G*, *U* | 18.2 (*G*) | MS: $M_G = +2 \cdots +7$; late-type giants | Hersperger, Th.: Astron. Astrophys. **22** (1973) 195; Wick, Ch.: Z. Astrophys. **61** (1965) 110 |
| 185 | − 0.8 | 10 | 284 | $m_{ir}$, Sp | 15 (*V*) | M, C | Nandy, K., Smriglio, F.: Publ. R. Obs. Edinburgh **7** (1971) No. 6 |
| 186 | + 1 | 18.55<br>8 | 3621<br>493 } | *B*, *V*, Sp | 12.3 (*V*)<br>15···16 (*V*) | B5···M8<br>gM1···gM8 | McCuskey, S.W.: Astron. J. **72** (1967) 1199; **74** (1969) 807 |
| 186 | + 1.5 | 2.32 | 2400 | *R*, *G*, *U* | 15.5 | MS: $M_G = 2 \cdots 7$; late-type giants | Topaktas, L.: Astron. Astrophys. Suppl. **20** (1975) 269 |
| 186.6 | + 1 | 1 | 525 | *B*, *V*, Sp | 15.4 (*V*) | O···G8 | Kalandadze, N.B., Kolesnik, L.N., Kuznetsov, V.I.: Bull. Abastumani Astrophys. Obs. **40** (1971) 39 |
| 186.6 | + 1 | 18 | 2047 | *B*, *V*, Sp | 13 (*V*) | O···M | Kalandadze, N.B., Kolesnik, L.N.: Astrometriya i Astrofizika Kiev **34** (1978) 19 |
| 203 | + 2 | 13 | 825 | $m_{4400}$, CI, Sp | 13 (λ4400) | B···F | Karlsson, B.: Reports Obs. Lund No. 1 (1969); summary in: Medd. Lund Ser. I, Nr. 246 (1969) |
| 203 | + 2 | 13 | 825 | $m_{4400}$, CI, Hγ,δ, K, Sp | 13 (λ4400) | B···F7 | Karlsson, B.: Astron. Astrophys. Suppl. **7** (1972) 35 |
| *Puppis-Vela-Carina-Crux* ($l = 210° \cdots 300°$) | | | | | | | |
| 245° | − 0°.2 | 8.43 | 706 | *B*, *V*, Sp | 12.7 (*V*) | O···B9 | Wilson, W.J.F., FitzGerald, M.P.: J. R. Astron. Soc. Canada **66** (1972) 254 |

continued

Table 9, continued

| Field (center or extension in $l, b$) $l$ | $b$ | Area $\square^\circ$ | Number of stars | Observational data | Limiting magnitude | Spectral types | Ref. |
|---|---|---|---|---|---|---|---|
| 246 | − 0.6 | 12.2 | 370 | $V, I$, Sp | 11.7 ($I$) | M giants, C | Kirton, K.A., FitzGerald, M.P.: J. R. Astron. Soc. Canada **68** (1974) 154 |
| 268.5 | − 0.3 | 15.24 | 470 | $V, B-V$, Sp | 12.5 ($V$) | O···A0 | Stegman, J.E., FitzGerald, M.P.: J. R. Astron. Soc. Canada **66** (1972) 303 |
| 293.5 | + 0.8 | 2 | 290 | $R, I$, Sp | 13 ($I$) | M giants, C | Westerlund, B.E.: Mon. Not. R. Astron. Soc. **130** (1965) 45 |
| 270···303 | − 5···+5 | | 21575 | $V$, CI, Sp | 13 ($V$) | B8···A3 | McCuskey, S.W., Lee, S.G.: Astron. J. **81** (1976) 604 |
| 277 | 0 | 7.8 | 756 | $V, B-V$, Sp | 12.5 ($V$) | O···B3, B5···A0 | Moore, J.H., FitzGerald, M.P.: J. R. Astron. Soc. Canada **67** (1973) 291 |
| 281 | + 3.9 | 17.1 | 6195 | $V, B-V$, Sp | 12.5 ($V$) | B4···K4 | Wooden, W.H.: Publ. Warner & Swasey Obs., Vol. **1**, No. 2 (1971) |
| 298 | + 1.4 | 9 | 406 | $R, I$, Sp | 12.5 ($I$) | M giants | Vleeming, G.: Astron. Astrophys. Suppl. **19** (1974) 21 |
| 281 | + 3.9 | 9 | 284 | $R, I$, Sp | 12.5 ($I$) | M giants | Vleeming, G., Thé, P.S.: Astron. Astrophys. Suppl. **21** (1975) 33 |
| 294.8 | − 1.7 | | 125 | $B, V$, Sp | 12.3 ($B$) | B, A | Ardeberg, A., Maurice, E.: Astron. Astrophys. Suppl. **39** (1980) 325 |
| *Centaurus-Circinus-Norma-Scorpius* ($l=300°\cdots350°$) | | | | | | | |
| 280°···319° | −4°···+4° | | 13300 | $m_{4400} \approx B$, Sp | 13 ($B$) | O···A2; M; C | Sundman, A.: Stockholms Obs. Report No. 2 (1974); Mavridis, L.N. (ed.): Stars and the Milky Way System, Springer Berlin 1974, p. 76 |
| 307 | 0 | 5 | 23098 | $U, B, V$ | 16.5 ($V$) | O···K | McGruder III, C.H.: Astron. Astrophys. **43** (1975) 51 |
| 329.8 | − 2.2 | 20.3 | 1292 | $V, B-V$, Sp | 11.5 ($V$) | B0···G6; K; M | Drilling, J.S.: Astron. J. **73** (1968) 590 |
| 329.8 | − 2.2 | 6 | 600 | $R, I$, Sp | 12.5 ($I$) | M giants | Thé, P.S., Staller, R.F.A., Meurs, E.J.A.: Astron. Astrophys. Suppl. **15** (1974) 141 |
| 330.1 | − 2.3 | 0.12 | 2164 | $R, G, U$ | 16.0 ($G$) | MS: $M_G=2\cdots7$ late-type giants | Topaktas, L.: Astron. Astrophys. Suppl. **26** (1976) 19 |
| 331.7 | − 2.3 | 0.2 | 1500 | $U, B, V$ | 14.5 ($V$) | AV, FV; GIII; KIII | Schnur, G.: Astron. Astrophys. **5** (1970) 431 |

## 8.3.2.4 Distribution of common stars perpendicular to the galactic plane

Table 10. Relative spatial stellar density $D(z)/D(z_0)$ as a function of distance from the galactic plane, $z$, at the position of the sun. The values given for M giants have been calculated by the formula $D(z) \propto \exp(-0.0030\, z)$ [23]. Note that these spatial densities do not differentiate between the different population groups. For distributions of disk and halo stars, see 8.3.3.3, Table 13. Discussion of several types of red giants: [33]. Scale heights for stars of different absolute magnitudes $M_V$: [32].

| | Main sequence [11] | | | | Giants [10, 23, 24, 28] | | | | | |
|---|---|---|---|---|---|---|---|---|---|---|
| $z$ [pc] | A0···A5 | A7···F2 | F5 | F8 | G5 | G8···G9 | K0 | K1···K2 | G4, K4 | M |
| 25 | 1.00 | 1.00 | 1.00 | 1.00 | | | | | 2.45 | 1.69 |
| 50 | 1.00 | 0.89 | 0.85 | 0.98 | 1.00 | | | | | 1.57 |
| 100 | (0.50) | 0.28 | 0.50 | 0.76 | 1.00 | | | | 1.66 | 1.35 |
| 200 | 0.19 | 0.069 | 0.22 | 0.38 | 1.00 | 1.00 | 1.00 | 1.00 | 1.00 | 1.00 |
| 300 | 0.047 | 0.025 | 0.16 | 0.11 | 0.9 | 1.0 | 0.5 | 0.5 | | 0.74 |
| 400 | 0.023 | 0.015 | 0.048 | 0.058 | 0.8 | 1.0 | 0.3 | 0.3 | 0.44 | 0.55 |
| 500 | 0.012 | 0.009 | 0.027 | 0.031 | 0.7 | 0.5 | 0.2 | 0.3 | | 0.41 |
| 600 | | | | | 0.7 | 0.5 | 0.2 | 0.2 | 0.22 | 0.30 |
| 800 | | | | | 0.6 | 0.4 | 0.1 | 0.15 | 0.13 | 0.17 |
| 1000 | | | | | 0.6 | 0.3 | 0.1 | 0.1 | | 0.09 |

## 8.3.2.5 References for 8.3.2


1 Allen, C.W.: Astrophysical Quantities, Athlone Press, London, Third Edition (1973) p. 248.
2 Becker, W.: Z. Astrophys. **62** (1965) 54.
3 Becker, W.: Q. J. R. Astron. Soc. **13** (1972) 226; Mitt. Astron. Ges. **43** (1978) 21.
4 Borzov, G.G.: Sov. Astron.-A.J. **17** (1974) 659.
5 Chiu, L.G.: Publ. Astron. Soc. Pac. **89** (1977) 614.
6 Eggen, O.J., Bessell, M.S.: Astrophys. J. **226** (1978) 411.
7 Fenkart, R.P.: Astron. Astrophys. **56** (1977) 91.
8 Gliese, W.: Veröff. Astron. Rechen-Inst. Heidelberg Nr. 22 (1969).
9 Green, R.F.: Ph. D. Thesis, California Inst. of Technology (1977).
10 Grenon, M. in: Highlights of Astronomy (Müller, E.A., ed.), **4**, Part II (1977) p. 55.
11 Hill, G., Hilditch, R.W., Barnes, J.V.: Mon. Not. R. Astron. Soc. **186** (1979) 813.
12 Jahreiß, H.: Dissertation Heidelberg (1974).
13 Jahreiß, H., Wielen, R.: Astron. Rechen-Inst. Heidelberg, Mitt. Ser. A, No. 79 (1974) = Mitt. Astron. Ges. **35** (1974) 212.
14 Liebert, J.: Astron. Astrophys. **70** (1978) 125.
15 Luyten, W.J.: Mon. Not. R. Astron. Soc. **139** (1968) 221.
16 Mavridis, L.N.: Structure and Evolution of the Galaxy, Reidel, Dordrecht (1971) p. 110.
17 McCuskey, S.W.: Astrophys. J. **123** (1956) 458.
18 McCuskey, S.W.: Stars and Stellar Systems **V** (1965) 1, (see 8.3.3.4 [a]).
19 McCuskey, S.W.: Vistas in Astronomy **7** (1966) 155.
20 McCuskey, S.W.: Int. Astron. Union Symp. **38** (1970) 189.
21 McRae, D.A.: The Obs. Handb. R. Soc. Canada 1958.
22 Sandage, A.R. in: Stellar Populations (O'Connell, D.J.K., ed.), North Holland Publ. Comp., Amsterdam and Interscience Publ. Inc., New York (1958) p. 78 = Specola Vaticana Ric. Astron. **5** (1958).
23 Sanduleak, N. in: Highlights of Astronomy **4**, Part II (1977) 35.
24 Sturch, C.R., Helfer, H.L.: Astron. J. **77** (1972) 726.
25 Sundman, A., Lodén, L.O., Nordström, B.: Astron. Astrophys. Suppl. **16** (1974) 445.
26 Sundman, A. in: Stars and the Milky Way System (Mavridis, L.N., ed.), Springer, Berlin (1974) p. 76.
27 Thé, P.S., Staller, R.F.A.: Astron. Astrophys. **36** (1974) 155.
28 Upgren, A.R.: Astron. J. **67** (1962) 37; **68** (1963) 194, 475.
29 van Rhijn, P.J.: Publ. Groningen No. 47 (1936).
30 Weidemann, V.: Z. Astrophys. **67** (1967) 286.
31 Weidemann, V. in: Low-luminosity stars (Kumar, S., ed.), Gordon and Breach, New York (1969) p. 311.
32 Miller, G.E., Scalo, J.M.: Astrophys. J. Suppl. **41** (1979) 513.
33 Scalo, J.M., Miller, G.E.: Astrophys. J. **233** (1979) 596.

## 8.3.3 Large-scale distribution of the stars

For symbols, see 8.3.2, p. 180

### 8.3.3.1 Subsystems of the Galaxy, stellar populations

Stellar objects of different physical type (stars of the different spectral classes, open and globular star clusters, classes of variable stars, etc.) show widely differing large-scale spatial distributions and kinematics. Extreme cases are the very flat distribution of the OB stars and the nearly spherical volume occupied by the globular clusters. These facts led to the conception of distinct galactic subsystems with the observed main characteristics summarized in Table 11.

Table 11. Characteristic data of galactic subsystems.
For symbols, see also 8.3.2.

$\Sigma D$ = rough estimate of the total number of objects in the Galaxy

$\langle|z|\rangle$ = mean absolute distance from the galactic plane in the vicinity of the sun

$\Sigma_W$ = velocity dispersion perpendicular to the galactic plane ($W$-component)

$\left(\frac{\partial \log D}{\partial \log R}\right)_{R_0}$ = radial logarithmic density gradient in the vicinity of the sun

| Objects | $D_0$ $N/10^3$ pc$^3$ | $\varrho_0$ $\mathfrak{M}_\odot/10^3$ pc$^3$ | $\Sigma D$ | $\langle\lvert z\rvert\rangle$ pc | $\Sigma_W$ km s$^{-1}$ | $\left(\frac{\partial \log D}{\partial \log R}\right)_{R_0}$ | Ref. |
|---|---|---|---|---|---|---|---|
| O···B2 (OB) ($M_V \lesssim -3$) | $1\cdot10^{-4}$ | $5\cdot10^{-4}$ | $3\cdot10^4$ | 65 | < 10 [1]) | 0 | 3, 5, 18 |
| Cepheids type I | $1\cdot10^{-4}$ | $5\cdot10^{-4}$ | $3\cdot10^4$ | 70 | 5 | 0 | 4, 19, 33 |
| Open clusters O···B6 | $4\cdot10^{-4}$ | 0.1 | $3\cdot10^4$ | 50 | 6 | 0 | 19, see also 6.2.2 |
| Open clusters B7···F | | | | 80 | | | |
| Main sequence (V) A | 1 | 48 | $10^{11}$ | 90 | 5 | 0 | 4, 16; Table 7, Table 8 |
| F | 3 | | | 130 | 15 | | |
| G | 100 | | | 180 | 20 | $-1\cdots-2$ | |
| K | | | | 270 | | | |
| M | | | | 270 | | | |
| Giants (III, IV) G | 1 | 1 | | 400 | 15 | $-1.5$ | 16, 19; Table 7, Table 8 |
| K | | | | 270 | | | |
| M | | | | 290 | | $-2$ | |
| White dwarfs | 20 | 20 | $10^{10}$ | (270) | 20 | $-2$ | see Table 7 and 8 |
| Pulsars | $5\cdot10^{-5}$ | $10^{-4}$ | $10^5$ | 250 | 100 | $(-1.5)$ | 31a |
| Planetary nebulae | $5\cdot10^{-5}$ | $5\cdot10^{-5}$ | $5\cdot10^4$ | 140 | – | $-1.5$ | 6 |
| Long-period variables (Mira-type) $P=350\cdots400$ d | $10^{-3}$ | $10^{-3}$ | $10^6$ | 250 | 20 | $-1$ | 25, 19, 31 |
| $300\cdots350$ d | | | | 300 | 35 | $-2$ | |
| $250\cdots300$ d | | | | 400 | 30 | | |
| RR Lyrae variables $\Delta S<5$, $P<0.4$ d | $10^{-5}$ | $10^{-5}$ | $5\cdot10^4$ | (400) | 30 | $-2$ | 25, 31 |
| Novae | | | 25···50 p.a. | (400) | (20) | $-2$ | 26, 19 |
| Long-period variables $P=200\cdots250$ d | $10^{-3}$ | $10^{-3}$ | $10^6$ | 600 | 50 | $-3$ | 24, 25, 29 |
| $140\cdots200$ d | | | | 900 | 60 | | |
| Halo stars (metal-poor) | 1 | 0.3 | $10^{10}$ | 700 | | $-3$ | 10, 28, 35 |
| RR Lyrae variables $\Delta S\geqq5$, $P>0.4$ d | $10^{-5}$ | $10^{-5}$ | $5\cdot10^4$ | (2000) [2]) | 70···90 | $-3$ [2]) | 17, 23, 26, 19, 31, 34 |
| Subdwarfs $\delta(U-B)>0.15$ | 1 | 1 | $10^{10}$ | 2000 [2]) | 90 | $-3.5$ [2]) | 4, 8, 16, 24 |
| Globular clusters Sp>G1 | $10^{-8}$ | $1\cdot10^{-3}$ | $2\cdot10^2$ | 2000 [2]) | 100 [2]) | $-4.5$ [2]) | 4, 12, 34 |
| halo-type (≦G1) | | | | 6000 [2]) | | $-3.4$ [2]) | |

[1]) Local OB stars are mainly grouped in associations with internal velocity dispersion <3 km s$^{-1}$ in one component; velocities of associations as a whole are in the order of 10 km s$^{-1}$.

[2]) These numbers refer to the subsystem as a whole.

The characteristics of the density and velocity distribution in the subsystems are correlated with the age and metal content of the member stars. The observed differences in the colour-magnitude diagrams of globular and open clusters, which result from different age and chemical composition in these classes of objects, led W. Baade [1] to propose two distinct stellar populations: the young, metal-rich population I which is strongly concentrated in the galactic plane, and the old, metal-poor population II including subsystems of more or less spherical shape.

Later, five subdivisions were introduced [22, 4]. Table 12 gives their main properties.

Table 12. Schema of stellar populations proposed at a conference of astronomers in 1957.
$\langle|z|\rangle$ = mean absolute distance from the galactic plane
$\Sigma_W$ = velocity dispersion perpendicular to the galactic plane (W component)
$\left(\frac{\partial \log D}{\partial \log R}\right)_{R_0}$ = radial logarithmic density gradient in the vicinity of the sun
M/H = metal to hydrogen ratio in the stellar atmospheres relative to extreme population I

| | Halo population II | Intermediate population II | Disk population | Older population I | Extreme population I |
|---|---|---|---|---|---|
| Typical objects | globular clusters subdwarfs, RR Lyrae variables with $P > 0.4$ d | high-velocity stars ($\lvert W\rvert > 30$ km s$^{-1}$), long-period variables with $P < 250$ d (Sp < M5) | stars with weak metallic lines ("weak-line stars"), planetary nebulae, novae, RR Lyrae variables with $P < 0.4$ d | sun, stars with normal (strong) metallic lines ("strong-line stars"), A stars, Me dwarfs | O stars, supergiants, Type I Cepheids, T Tau stars |
| Properties: | | | | | |
| $\langle\lvert z\rvert\rangle$ [pc] | 2000 | 700 | 400 | 150 | 70 |
| $\Sigma_W$ [km s$^{-1}$] | 80···100 | 40···60 | 20···40 | 10···20 | 5···10 |
| $\left(\frac{\partial \log D}{\partial \log R}\right)_{R_0}$ | −3.5 | −3 | −2 | −1 | 0 |
| Concentration to galactic center | strong | strong | moderate | little | none |
| Distribution | smooth | smooth | smooth | patchy | extremely patchy |
| M/H | 0.01···0.1 | 0.01···0.1 | 0.03···0.3 | 0.3···1 | 1 |
| Age [a] | $\approx 10^{10}$ | $\approx 10^{10}$ | $2\cdot10^9\cdots1\cdot10^{10}$ | $5\cdot10^8\cdots5\cdot10^9$ | $<5\cdot10^8$ |

## 8.3.3.2 Distribution of stars in the galactic disk

Information about the large-scale structure of the stellar system can be derived in three ways:

1. Determination of direction and distance for a sufficient number of individual stellar objects.
2. Statistical analysis of star counts [numbers of stars in the magnitude range $(m-\frac{1}{2}\Delta m)\cdots(m+\frac{1}{2}\Delta m)$ per unit area of the sky] for distinct spectral types or luminosity groups. It results in spatial densities of the stars under consideration.
3. Analysis of the surface brightness distribution over the sky. This yields the spatial distribution of stellar emission per unit volume.

The existence of a limiting observable magnitude and the extinction of starlight in the interstellar space have the consequence that by using methods (1) and (2) the stellar distribution near the galactic plane can be obtained for distances greater than about 1 kpc only for the most luminous objects, e.g., OB stars, supergiants, Be stars, very young open clusters, classical Cepheids [15]. In other galaxies objects of these types are observed to be well correlated with spiral arms. Some results derived for the distribution of such optical indicators of spiral structure in our own Galaxy are given in Figs. 1, 2, and 3; for distribution of young open clusters, see also 6.2.2. These distributions show several condensations which seem to form three arms often denoted by the symbols +I (Perseus arm), O (local arm or Orion arm) and −I (Sagittarius arm). The considerable scatter of the positions of individual objects and the limited distance range covered make it difficult to trace the true large-scale spiral structure of our Galaxy in this way at the moment.

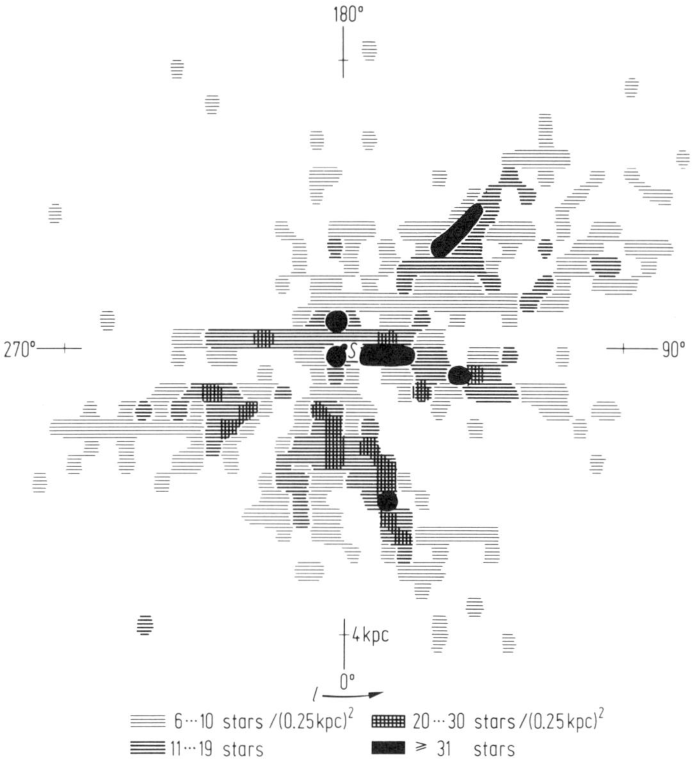


Fig. 1. Distribution of OB stars in the galactic plane up to distances from the sun of $r \approx 4$ kpc [18].

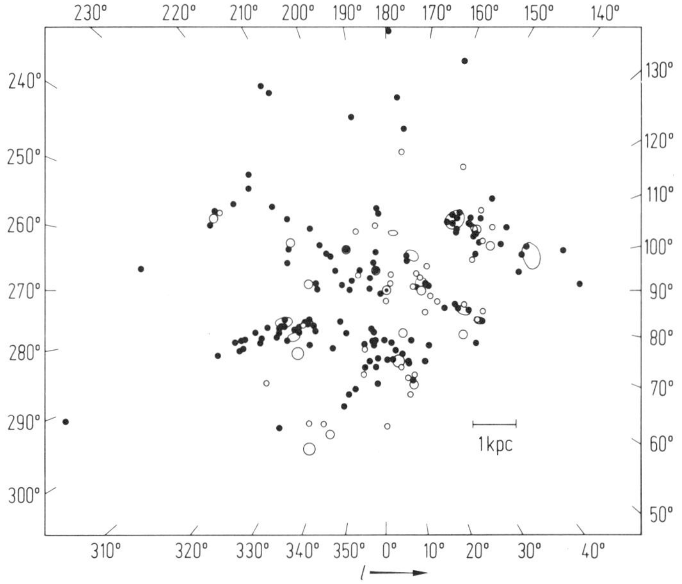


Fig. 2. Distribution of stellar associations (open circles) and young open clusters (full circles) projected onto the galactic plane [15].

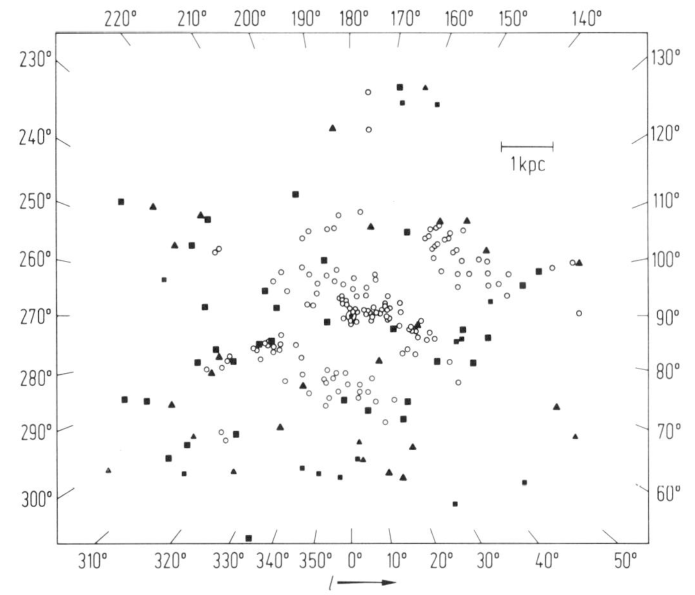


Fig. 3. Distribution of classical Cepheids with period $P>15$ d (squares) and $15\,\mathrm{d}>P>11.25$ d (triangles) projected onto the galactic plane. Open circles represent young open clusters and H II regions for comparison [30].

The distributions perpendicular to the galactic plane of OB stars and associations and of Cepheids for different distances from the galactic center have been compiled in [11a].

General star counts (including all spectral types) for limiting magnitudes $m_V \approx 24\cdots30$ derived by ground based observations with large telescopes aided by computer analysis [36, 37, 38] and by future Space Telescope observations will permit to prove global Galaxy models. Calculations of the expected star counts for $m_V \leqq 30$ have been provided by Bahcall and Soneira [39, 40]. The models used consist of disk and spheroidal stellar distributions. An additional large and massive halo is also discussed. For discussion, see [41].

Models of the distribution of stellar emissivity per unit volume in the galactic disk that can reproduce the observed surface brightness distribution have been derived for visible and near-infrared wavelengths. Some results are given in Table 13 and Figs. 4 and 5. For two-component (disk + spheroid) models of the Galaxy as a whole, see de Vaucouleurs and Pence [6a].

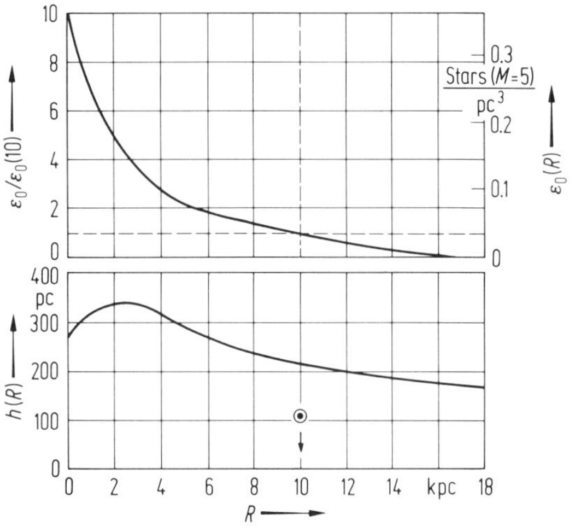


Fig. 4. Model of the galactic disk derived from the observed surface brightness distribution in the visual and blue spectral regions. For explanation, see Table 13 [21].

Table 13. Models of the galactic stellar disk derived from surface photometry.
For symbols, see 8.3.2
$R$ = distance of a general point from the galactic center projected onto the galactic plane
$\varepsilon$ = stellar emission per unit volume in the spectral region considered

| Spectral region | Representation of the emissivity | Unit of $\varepsilon$ | $\varepsilon(R_0, 0)$ | Main contributors | Ref. |
|---|---|---|---|---|---|
| Visual | $\varepsilon(R,z)=\varepsilon_0(R)\left[1+\left(\frac{z}{h(R)}\right)^2\right]^{-1}$<br>For $\varepsilon_0(R)/\varepsilon_0(R_0, 0)$ and $h(R)$ ("half-thickness"), see Fig. 4 (contribution of OB stars excluded) | number of stars per pc$^3$ with $M_V = 5^m$ | 0.036 | F···K stars | 21 |
| $\lambda = 2.4\ \mu$m | $\varepsilon(R,z)=1.4\cdot 10^{35}\exp(-0.45\,a)$<br>with $a=(R^2+\eta^{-2}z^2)^{1/2}$,<br>$R$ and $z$ in [kpc] and<br>$\eta$ = axial ratio = 0.035. | W µm$^{-1}$ kpc$^{-3}$ | $1.6\cdot 10^{33}$ | K and M stars | 20 |
| | The contours shown in Fig. 5 refer to a model which also represents the central bulge with semi-axes 2.5 kpc and 1.3 kpc and an emissivity concentration in a 5-kpc ring. | | $\varepsilon$(5 kpc, 0) $\approx 4\cdot 10^{34}$ | K and M giants | 27 |
| | Spheroidal component:<br>$\varepsilon_s(R,z)=1.6\cdot 10^{35}\exp(-0.488a)$<br>$\eta = 0.038$<br>Two rings ($k$=1,2): For $R > R_k$<br>$\varepsilon_r(R,z)=\varepsilon_r(R_k)\exp[-3.0(R-R_k)]\times\exp(-20z)$<br>$\varepsilon_r(R_1=4.6)=1.6\cdot 10^{35}$<br>$\varepsilon_r(R_2=7.2)=1.6\cdot 10^{34}$<br>$\varepsilon_r = 0$ for $R < R_k$ | W µm$^{-1}$ kpc$^{-3}$ | $1.2\cdot 10^{33}$ | – | 42 |

While the brightness observed at 2.4 µm near the galactic equator between $l \approx 50°$ and the anticenter is in agreement with that expected from K and M star counts, the values observed for $l \approx 5°\cdots 50°$ are about one order of magnitude higher than predicted. This excess has been attributed to massive red giants associated with regions of star formation [27]. The diffuse emission observed in the far infrared comes mainly from dust heated by OB and possibly other stars, and which is located in radio HII regions, extended low-density HII regions and neutral intercloud medium [20a, 7, 27], see also 8.3.4.2.

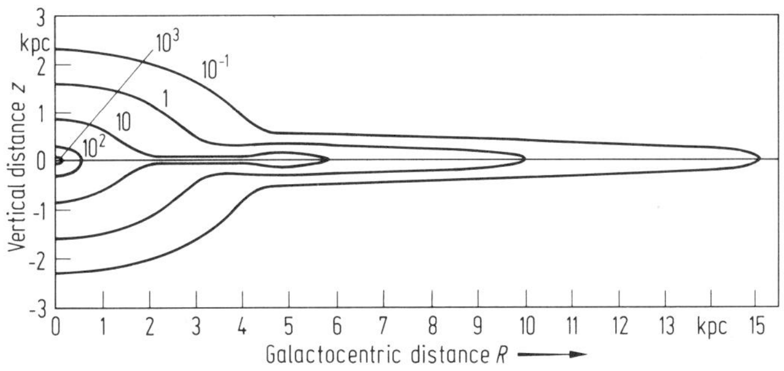


Fig. 5. Contours of the 2.4 µm emissivity distribution derived from surface brightness measurements [20].

### 8.3.3.3 Distribution of stars in the galactic halo

For characteristic data of the spatial distribution of globular clusters, isolated RR Lyrae variables and subdwarfs, see Table 11. For more detailed results for globular clusters, see 6.2.1.

The stellar mass content of the galactic halo is dominated by non-variable, metal-poor, old stars of relatively low absolute magnitude ($M_V > +3$) situated outside of globular clusters. These halo stars can be separated from the disk population by their UV excesses. Space-density distributions up to distances of almost 15 kpc have been derived for several directions by using the three-colour method of W. Becker (Table 14). Results for the space-density distribution perpendicular to the galactic plane are given in Table 15; for more detailed results, see [43].

Table 14. Spatial density distributions of halo stars in several test-directions, determined by *RGU* photometry (Basle halo-program). The first column gives the number of the Kapteyn Selected Area (SA) containing the test-field or the designation of a nearby star cluster or bright galaxy.

| SA | $l$ | $b$ | Area □° | Number of stars in this area | Limiting magnitude ($G$) | Ref. |
|---|---|---|---|---|---|---|
| 51 | 189° | +21° | 0.45 | | $19^m.1$ | Becker, W.: Z. Astrophys. **62** (1965) 54; Z. Astrophys. **66** (1966) 404 |
| 57 | 67 | +85 | 2.61 | | 19.1 | Fenkart, R.P.: Z. Astrophys. **66** (1967) 390 |
| 54 | 199 | +59 | 2.56 | | 18.5 | Fenkart, R.P.: Z. Astrophys. **68** (1968) 87 |
| 141 | 249 | −86 | 1.92 | 672 | 17.5 | Fenkart, R.P.: Astron. Astrophys. **3** (1969) 228 |
| M 3 | 42 | +79 | 2.72 | | 17.4 | Yilmaz, F.: Istanbul Univ. Fen. Fak. Mec. Seri, C., **37**, **38** (1974) 93 |
| 71 | 167 | −35 | 2.97 | 2022 | 18.5 | Fenkart, R.P., Wagner, R.: Astron. Astrophys. **19** (1972) 1 |
| 94 | 176 | −49 | 2.03 | 1350 | 19.0 | Schaltenbrand, R.: Astron. Astrophys. Suppl. **18** (1974) 27 |
| M 13 | 58 | +40 | 1.10 | 1919 | 18.5 | Fenkart, R.P., Wagner, R.: Astron. Astrophys. **41** (1975) 315 |
| NGC 4147 | 256 | +78 | 3.46 | 1333 | 18.5 | Fenkart, R.P., Schaltenbrand, R.: Astron. Astrophys. Suppl. **27** (1977) 409 |
| 82 | 6 | +66 | 1.20 | 988 | 19.0 | Becker, W., Steppe, H.: Astron. Astrophys. Suppl. **28** (1977) 377 |
| 107 | 6 | +41 | 0.95 | 1693 | 19.0 | (same as above) |
| 133 | 7 | +10 | 0.19 | 1934 | 17.9 | Becker, W.: Astron. Astrophys. Suppl. **38** (1979) 341 |
| NGC 6171 | 3 | +23 | | | | Wiedeman, D.: 1979 (from [43]) |
| M 5 | 4 | +47 | 1.05 | | 19.0 | Steinlin, U.W.: 1979 (from [43]) |

Table 15. Spatial star-density distributions perpendicular to the galactic plane for disk and halo stars of absolute magnitude intervals $M_G = 4^M.0 \cdots 4^M.9$ and $M_G = 3^M.0 \cdots 8^M.0$. [9, 10].
For symbols, see p. 180.

| $z$ [kpc] | log $D(z)+10$ | | | |
|---|---|---|---|---|
| | $M_G = 4^M \cdots 5^M$ | | $M_G = 3^M \cdots 8^M$ | |
| | disk | halo | disk | halo |
| 0.0 | 7.3 | 6.7 | 8.2 | 7.2 |
| 0.5 | 6.0 | 6.1 | 7.2 | 6.8 |
| 1.0 | 5.0 | 5.7 | 6.3 | 6.5 |
| 1.5 | 4.3 | 5.4 | 5.4 | 6.1 |
| 2.0 | 3.8 | 5.0 | 4.4 | 5.8 |
| 2.5 | | 4.7 | | 5.5 |
| 3.0 | | 4.5 | | 5.3 |
| 4.0 | | 4.2 | | 4.8 |
| 5.0 | | 4.1 | | 4.4 |
| 6.0 | | 4.0 | | 4.1 |

## 8.3.3.4 References for 8.3.3

### General references


a Galactic Structure (Blauuw, A., Schmidt, M., eds.), Stars and Stellar Systems V, Univ. of Chicago Press, Chicago (1965).

b The Spiral Structure of the Galaxy (Becker, W., Contopoulos, G., eds.), Int. Astron. Union Symp. **38**, Reidel, Dordrecht (1970).

c The Large-Scale Characteristics of the Galaxy (Burton, W.B., ed.), Int. Astron. Union Symp. **84**, Reidel, Dordrecht (1979).


### Special references


1 Baade, W.: Astrophys. J. **100** (1944) 137.

2 Becker, W., Fenkart, R.P.: Int. Astron. Union Symp. **38** (1970), Astron. Astrophys. Suppl. **4** (1971) 241.

3 Bertiau, F.C.: Astrophys. J. **128** (1958) 533.

4 Blaauw, A.: Stars and Stellar Systems **V** (1965) 435.

5 Blaauw, A.: Int. Astron. Union Symp. **38** (1970) 199.

6 Cahn, J.H., Wyatt, S.P.: Int. Astron. Union Symp. **76** (1978) 3.

6a de Vaucouleurs, G., Pence, W.D.: Astron. J. **83** (1978) 1163.

7 Drapatz, S.: Astron. Astrophys. **75** (1979) 26.

8 Eggen, O.J.: Vistas in Astronomy **12** (1970) 409.

9 Fenkart, R.P.: Z. Astrophys. **66** (1967) 390.

10 Fenkart, R.P.: Astron. Astrophys. **56** (1977) 91.

11 Gliese, W.: Z. Astrophys. **39** (1965) 1.

11a Guibert, J., Lequeux, J., Viallefond, F.: Astron. Astrophys. **68** (1978) 1.

12 Harris, W.A.: Astron. J. **81** (1976) 1095.

13 Hofmann, W., Lemke, D., Thum, C.: Astron. Astrophys. **57** (1977) 111.

14 Humphreys, R.M.: Publ. Astron. Soc. Pac. **88** (1976) 647.

15 Humphreys, R.M.: Int. Astron. Union Symp. **84** (1979) 93.

16 Jahreiß, H.: Dissertation Heidelberg (1974).

17 Kinman, T.D., Wirtanen, G.A., Janes, K.A.: Astrophys. J. Suppl. **13** (1966) 379.

18 Klare, G., Neckel, T.: Z. Astrophys. **66** (1967) 45.

19 Landolt-Börnstein, NS, Vol. VI/1 (1965) Tab. 18, p. 619.

20 Maihara, T., Oda, N., Sugiyama, T., Okuda, H.: Publ. Astron. Soc. Japan **30** (1978) 1.

20a Mezger, P.G., Mathis, J.S., Panagia, N.: Astron. Astrophys. **105** (1982) 372.

21 Neckel, T.: Z. Astrophys. **69** (1968) 112.

22 Oort, J.H. in: Stellar Populations (O'Connell, D.J.K., ed.), North Holland Publ. Comp. Amsterdam (1958) p. 415 = Specola Vaticana Ric. Astron. **5** (1958).

23 Oort, J.H.: Stars and Stellar Systems **V** (1965) 483.

24 Oort, J.H., Plaut, L.: Astron. Astrophys. **41** (1975) 71.

25 Plaut, L.: Bull. Astron. Inst. Neth. **17** (1963) 81.

26 Plaut, L.: Stars and Stellar Systems **V** (1965) 267.

27 Puget, J.L., Serra, G., Ryter, C.: Int. Astron. Union Symp. **84** (1979) 105.

28 Schmidt, M. in: The Galaxy and the Local Group (Dickens, R.J., Perry, J.E., eds.), R. Greenwich Obs. Bull. No. **182** (1976) 12.

29 Smak, J.I.: Annu. Rev. Astron. Astrophys. **4** (1966) 26.

30 Tammann, G.A.: Int. Astron. Union Symp. **38** (1970) 236.

31 Tammann, G.A. in: Conference on optical observing programms on galactic structure and dynamics (Schmidt-Kaler, Th., ed.), Bochum (1975) p. 1.

31a Taylor, J.H., Manchester, R.N.: Astrophys. J. **215** (1977) 885.

32 Van Herk, G.: Bull. Astron. Inst. Neth. **18** (1965) 71.

33 Wielen, R.: Astron. Astrophys. Suppl. **15** (1974) 1.

34 Woltjer, L.: Astron. Astrophys. **42** (1975) 109.

35 Woltjer, L. in: Astronomical Papers dedicated to Bengt Strömgren (Reiz, A., Andersen, T., eds.), Copenhagen (1978) p. 349.

36 Kron, R.G.: Ph. D. thesis, University of California, Berkeley (1978).

37 Tyson, J.A., Jarvis, J.F.: Astrophys. J. Lett. **230** (1979) L153.

38 Peterson, B.A., Ellis, R., Kibblewhite, E.J., Bridgeland, M.T., Hooley, T., Horne, D.: Astrophys. J. Lett. **233** (1979) L109.

39 Bahcall, J.N., Soneira, R.M.: Astrophys. J. Suppl. **44** (1980) 73.
40 Bahcall, J.N., Soneira, R.M.: Astrophys. J. Lett. **238** (1980) L17.
41 Spaenhauer, A.: Milchstraßenmodelle. In: Galaktische Struktur und Entwicklung (Buser, R., ed.), Preprint of the Astronomical Institute of the University of Basel, No. 2 (1981).
42 Hayakawa, S., Matsumoto, T., Murakami, H., Uyama, K., Thomas, J.A., Yamagami, T.: Astron. Astrophys. **100** (1981) 116.
43 Becker, W.: Astron. Astrophys. **87** (1980) 80.

## 8.3.4 Large-scale distribution of interstellar matter

### 8.3.4.1 General remarks

Observations have led to a description of the widespread interstellar matter in terms of several different components or "transient" phases: hot "coronal" gas, warm intercloud gas, diffuse clouds and a few types of denser cloud, the coolest of which consist mainly of molecules and dust (for cloud-structure of the interstellar matter, see 7.2.3 and 7.2.4). The expanding shells of supernovae appear to have generated a network of overlapping cavities containing the hot, rarefied "corona" gas and seem – rather than stellar winds from O stars – to provide most of the kinetic energy for "stirring" the matter. The warm component may be a kind of envelope around the clouds [14, 17]. Physical parameters and processes: see 7.5.4.

Table 16. General characteristics of the spatial distribution of different interstellar components in the Galaxy.
$R$ = distance from the galactic center (sun: $R = R_0$)
$T$ = temperature
$T_e$ = electron temperature
$N$ = number density of interstellar particles in [$cm^{-3}$]
$\Delta\bar{N}$ = overall contribution to mean number density
$N_e$ = electron number density
$H$ = effective half-thickness of the layer at $R = R_0$, defined as half of the total mass per $cm^2$ divided by the density in the galactic plane

| Component | $\Delta\bar{N}$ $cm^{-3}$ | Filling factor | Radial extent | $H$ pc | Total mass $\mathfrak{M}_\odot$ | Ref. |
|---|---|---|---|---|---|---|
| Hot gas (at least two subcomponents) ($T \approx 0.3 \cdots 1 \cdot 10^6$ K, $N \approx 0.5 \cdots 3 \cdot 10^{-3}$ $cm^{-3}$) | $\approx 10^{-3}$ | $\approx 0.5 \cdots 0.8$ | up to $R \approx 10$ kpc | $\sim 4000$ | $\approx 5 \cdot 10^6$ | 17, 18, 21, 22, 7a, 27 |
| Extended low-density H II region ($T_e \approx 5000$ K, $N_e \lesssim 1$ $cm^{-3}$) | $\approx 1 \cdot 10^{-2}$ ($N_e$) | $\gtrsim 0.1$ | $4 \lesssim R \lesssim 10$ kpc (bulk) | 120 | $\approx 10^8$ | 16, 17, 29 |
| Layer of electron-ion gas from pulsar dispersion measurement ($\langle N_e \rangle \approx 0.03$ $cm^{-3}$) | | | | $\lesssim 1000$ | | 8, 16, 25 |
| "Warm", predominantly neutral (weakly ionized), intercloud medium ($\langle T \rangle \approx 6000$ K, $\langle N \rangle \approx 0.1$ $cm^{-3}$) | $\approx 0.2$ | $\approx 0.1$ | ($R \approx 15$ kpc) | 200 [1] | $1.2 \cdot 10^9$ | 1, 15, 17 |
| Diffuse clouds including "standard" H I clouds ($\langle T \rangle \approx 80$ K, $\langle N \rangle \approx 40$ $cm^{-3}$) and "lukewarm" clouds ($T \approx 10^2 \cdots 10^3$ K) | 0.3 | $\approx 1 \cdot 10^{-2}$ | up to $R \approx 15$ kpc | 100 [1] | $1.4 \cdot 10^9$ | 1, 17 |
| Molecular clouds (typical internal values: $T \approx 10$ K, $N \approx 10^3$ ($H_2$) $cm^{-3}$) | 0.4 | $\approx 5 \cdot 10^{-4}$ | $4 \lesssim R \lesssim 10$ kpc; max. at $\approx 5.5$ kpc and in the center | 65 | $2 \cdot 10^9$ | 3, 5, 9, 10, 17, 20 |

[1]) Increases for $R > R_0$.

### 8.3.4.2 Interstellar dust

Distribution of dust in the galactic plane for distances from the sun $r \lesssim 3$ kpc: see 7.1.2.4. Comparison with the distribution of spiral tracers (OB stars, young star clusters, medium-dense H II regions) shows no obvious positional correlations [13].

Scale height of the dust layer above the galactic plane: see 7.1.2.1.

Mean mass density of the dust and average dust-to-gas ratio in the galactic plane: see 7.4.1.5.

Since interstellar grains partly transform the stellar radiation into far-infrared radiation, observations of the diffuse galactic emission at wavelengths of $\lambda > 30\,\mu$m may provide information on the large-scale distribution of the dust. Model calculations, however, show that the dust density is of moderate influence compared with that of early-type stars [16, 7]. Interpretation of recent observations in the far infrared: see [28, 33].

### 8.3.4.3 Interstellar gas in the galactic disk

Relation of gas column density to colour excess or visual extinction: see 7.1.2.1 and 7.2.4 respectively.

The overall radial distributions in the galactic plane of a) the cold clouds of molecular hydrogen and other molecules, b) low-density regions of ionized hydrogen c) the medium density H II regions, d) supernovae remnants and e) regions of emissivity of $\gamma$-radiation look quite similar: Outside the galactic center, these constituents of the interstellar matter are concentrated mainly at distances from the galactic center of $R \approx 4 \cdots 8$ kpc. In contrast, the disk formed by the neutral atomic hydrogen, mainly in low density clouds, extends to $R > 15$ kpc. An exponential model has been proposed for $R > R_0$ (= the sun's distance to the galactic center) with scale length $0.4\,R_0$ [31].

The thickness of the H I layer increases in the outer Galaxy with the distance $R$ from the galactic center and is warped so that the maximum density is reached at non-zero distances $z_0$ above or below the plane defined by $b = 0°$. A simple description of the $z$ dependence of H I densities, $N$, using a Gaussian distribution

$$N(z) = N(z_0)\exp\left[-\frac{1}{2}\left(\frac{z-z_0}{h}\right)^2\right]$$

with "scale height" $h$ and mean height of the layer $z_0$ is given by the relations [1, 2]

$$h = \begin{cases} 0.12\ \text{kpc} & \text{for } R < 9.5\ \text{kpc} \\ 0.12 + 0.023\,(R - 9.5)\ \text{kpc} & \text{for } R > 9.5\ \text{kpc} \end{cases}$$

$$z_0 = \begin{cases} 0 & \text{for R} < 9.5\ \text{kpc} \\ 0.12\,(R - 9.5)\cos(\phi - 85°)\ \text{kpc} & \text{for } R > 9.5\ \text{kpc} \end{cases}$$

$\phi$ is the galactocentric longitude (azimuth) measured in the direction of galactic rotation from the sun-center line; $h$ is related to the effective half-thickness $H$ as defined in Table 16 by $H = (\pi/2)^{1/2} h = 1.253\,h$. For $|z| \gtrsim 3h$ the distribution $N(z)$ shows extended exponential wings [6].

Recent observational results on the warp of the galactic H I layer which include a chart of the distribution of $z_0$ over the galactic plane for $R > 10$ kpc have been presented by Henderson (1979) [11].

The distributions perpendicular to the galactic plane of H II regions and supernova remnants for different $R$ have been compiled by Guibert et al. [10a].

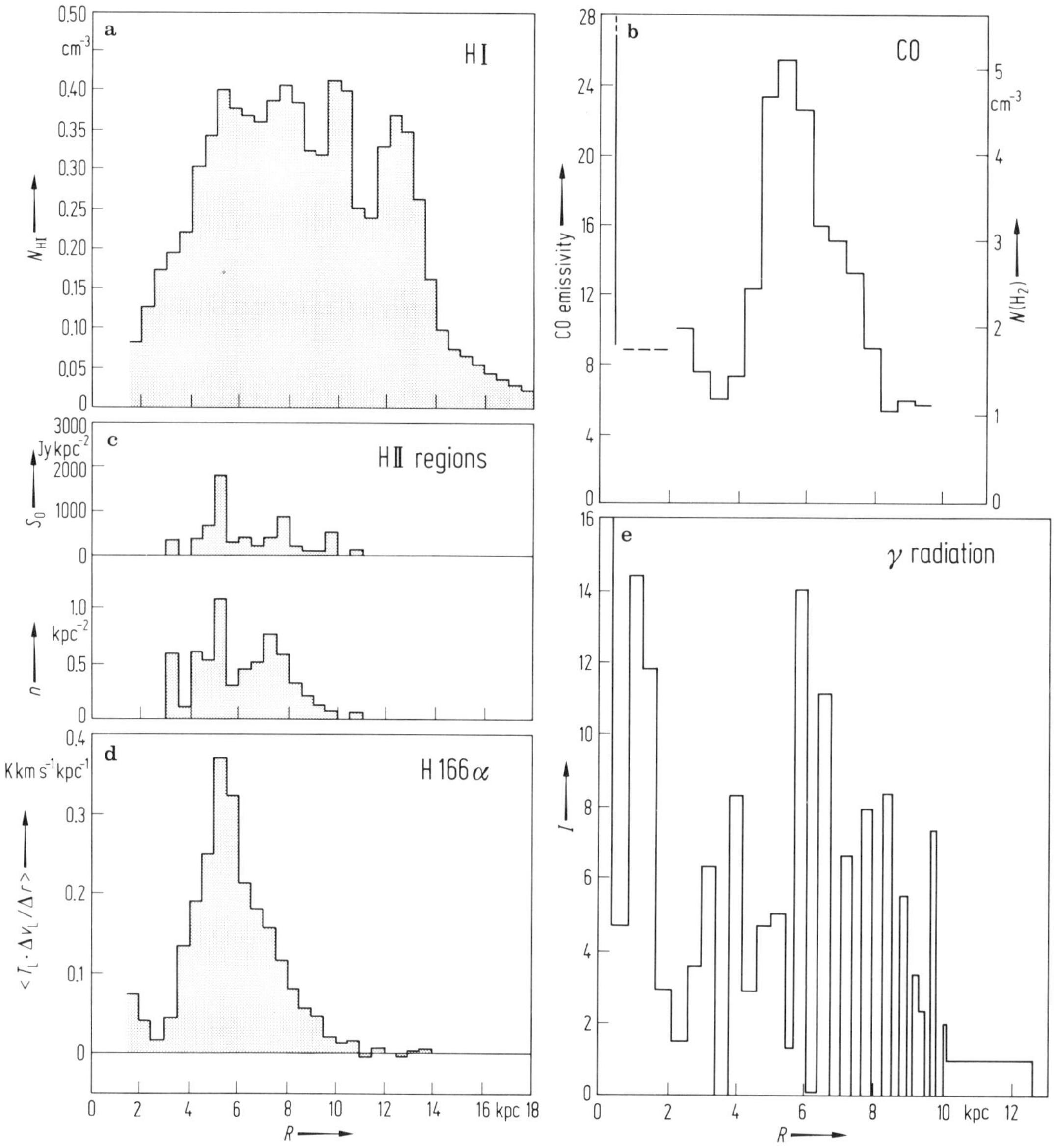


Fig. 6. Radial distributions of several constituents of the interstellar matter in the galactic plane.

a) HI: Number density $N$ [$cm^{-3}$] of neutral hydrogen from Burton and Gordon (1978) [4a].

b) CO: Emissivity of the $^{12}C^{16}O$ molecule (left scale) and number density of $H_2$ molecules (right scale; it may be overestimated by a factor of five) from Solomon et al. (1979) [20].

c) HII regions: Surface density of absolute flux $S_0$ [Jy $kpc^{-2}$] and of number $n$ [$kpc^{-2}$] for intrinsically luminous HII regions ($S_0 \geq 100$ Jy $kpc^{-2}$) derived under the assumption of circular galactic rotation by Lockman (1979) [12b].

d) H 166 α: Average power of line emission $\langle T_L \cdot \Delta V_L / \Delta r \rangle$ which comes mainly from diffuse low-density ionized hydrogen [K km $s^{-1}$ $kpc^{-1}$], and derived under the assumption of circular galactic rotation by Lockman (1976) [12a]. $T_L$ = peak brightness temperature of the line, $\Delta V_L$ = line width at half-intensity points.

e) γ radiation: Emissivity of photons with energy > 100 MeV deduced from measurements in the northern Milky Way assuming the effective half-thickness of the emitting layer $H = 60$ pc and emissivity near the sun of $2.4 \cdot 10^{-25}$ photons (> 100 MeV) $cm^{-3}$ $s^{-1}$ by Caraveo and Paul (1979) [5a].

Table 17. Radial galactic distribution of neutral atomic and molecular hydrogen and of the ionized gas in the low-density H II region. The numbers given are mean values in an annulus referred to a layer of thickness $2H = 260$ pc [16] (see also [4]).

$R$ = distance from the galactic center
$N$(H I) = mean number density of neutral atomic hydrogen
$2N(H_2)$ = mean number density of hydrogen in molecular form
$\langle N_e^2 \rangle^{1/2}$ = root-mean-square electron density in the extended low-density H II region
$\mathfrak{M}(H + H_2)$ = total mass of neutral gas, including the contribution of $^4$He(Y = 0.28) and heavier elements (Z = 0.02)
$\mathfrak{M}_{ion}$ = total mass of ionized gas for clumping factor $\langle N_e^2 \rangle / \langle N_e \rangle^2 = 70$

| $R$ interval kpc | $N$(H I) $cm^{-3}$ | $2N(H_2)$ $cm^{-3}$ | $\langle N_e^2 \rangle^{1/2}$ $cm^{-3}$ | $\mathfrak{M}(H+H_2)$ $10^6\,\mathfrak{M}_\odot$ | $\mathfrak{M}_{ion}$ $10^6\,\mathfrak{M}_\odot$ |
|---|---|---|---|---|---|
| 0.8 ··· 4 | 0.18 | 0.67 | 0.20 | 352 | 5.2 |
| 4 ··· 5 | 0.32 | 1.56 | 0.72 | 475 | 41.0 |
| 5 ··· 6 | 0.38 | 1.82 | 0.62 | 681 | 37.2 |
| 6 ··· 7 | 0.35 | 1.21 | 0.44 | 572 | 22.1 |
| 7 ··· 8 | 0.43 | 1.13 | 0.34 | 664 | 15.4 |
| 8 ··· 9 | 0.39 | 0.51 | 0.28 | 417 | 11.4 |
| 9 ··· 10 | 0.40 | 0.39 | 0.26 | 418 | 11.2 |
| 10 ··· 11 | 0.44 | 0.24 | 0.17 | 396 | 5.3 |
| 11 ··· 12 | 0.44 | 0.11 | 0.14 | 354 | 3.9 |
| 12 ··· 13 | 0.62 | 0.02 | 0.12 | 456 | 3.2 |
| 4 ··· 13 | mean 0.40 | 0.77 | 0.34 | total 4785 | 155.9 |

Spiral structure: No conclusive results on a large-scale spiral structure of the interstellar gas have been presented so far. The reason is that the localization of the gas not associated with stellar objects must be based on its large-scale velocity field, which is not known with sufficient accuracy (see also 8.4.1.3).

Distances of H II regions derived from spectrophotometric data of the exciting stars are mostly restricted to $r < 4$ kpc (recent work [16a]). An attempt to provide information on the spatial distribution of H II regions including more distant objects by combining optical and kinematical (radio recombination line) data has been made by Georgelin and Georgelin (1976) [8a]. The resulting "spiral-arm model" obtained for giant H II regions (see

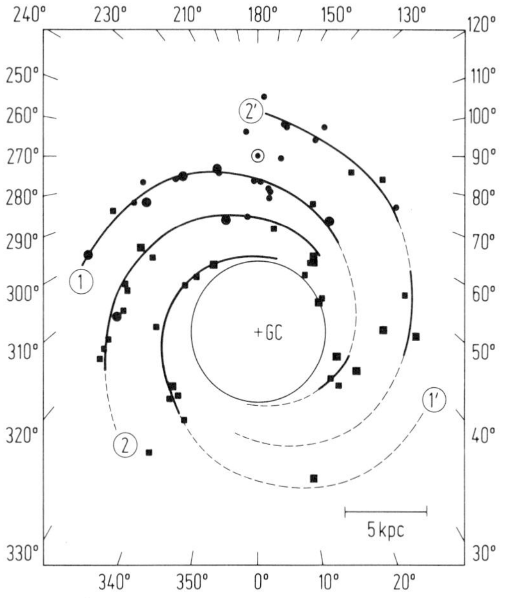


Fig. 7. Large-scale distribution of 60 giant H II regions in the galactic plane tentatively represented by a four-arm spiral pattern (1 Sagittarius-Carina arm, 2 Scutum-Crux arm, 1' Norma arm, 2' Perseus arm). Circles: optical H II regions; squares: radio H II regions. Large symbols: excitation parameter $U > 200$ pc $cm^{-2}$; small symbols $U = 100 \ldots 200$ pc $cm^{-2}$ [8 b].

7.3.4.1) is reproduced in Fig. 7. Each pair of spirals may be approximated by a logarithmic spiral in a galactocentric cylindrical coordinate system ($R$, $\varphi$, $z$) by $R = R^* \exp(k\varphi)$ with $R^* = 3$ and 4 kpc, and $k = 0.24$, corresponding to the pitch angle $i = \arctan k = 13.^\circ 5$ [7]. For molecular clouds as tracers of spiral structure, see [32].

Distribution of cosmic-ray particles in the galactic disk (see also 7.6): Recent observations of galactic $\gamma$-radiation appear to rule out a quasi-spherical cosmic-ray halo. Significant trapping of cosmic-ray electrons and nucleons seems to be confined to a disk of half-thickness $H < 3$ kpc [19, 23, 24]. A weak nonthermal radio halo observed at meter wavelengths may be generated by synchrotron emission of cosmic-ray electrons escaping from the disk [26]. Recent review articles: see [30].

## 8.3.4.4 References for 8.3.4

### General reference

a The large-scale characteristics of the Galaxy (Burton, W.B., ed.), Int. Astron. Union Symp. **84**, Reidel, Dordrecht (1979).

### Special references

1 Baker, P.L., Burton, W.B.: Astrophys. J. **198** (1975) 281.
2 Burton, W.B.: Annu. Rev. Astron. Astrophys. **14** (1976) 275.
3 Burton, W.B., Gordon, M.A.: Astrophys. J. Lett. **207** (1976) L189.
4 Blitz, L., Shu, F.H.: Astrophys. J. **238** (1980) 148.
4a Burton, W.B., Gordon, M.A.: Astron. Astrophys. **63** (1978) 7.
5 Burton, W.B., Gordon, M.A.: Int. Astron. Union Symp. **84** (1979) 271.
5a Caraveo, P.A., Paul, J.A.: Astron. Astrophys. **75** (1979) 340.
6 Celnik, W., Rohlfs, K., Braunsfurth, E.: Astron. Astrophys. **76** (1979) 24.
7 Drapatz, S.: Astron. Astrophys. **75** (1979) 26.
7a Dwek, E., Scalo, J.M.: Astrophys. J. Lett. **233** (1979) L81.
8 Falgarone, E., Lequeux, J.: Astron. Astrophys. **25** (1973) 253.
8a Georgelin, Y.M., Georgelin, Y.P.: Astron. Astrophys. **49** (1976) 57.
8b Georgelin, Y.M., Georgelin, Y.P., Sivan, J.-P.: Int. Astron. Union Symp. **84** (1979) 15.
9 Gordon, M.A., Burton, W.B.: Astrophys. J. **208** (1976) 346.
10 Gordon, M.A., Burton, W.B.: Int. Astron. Union Symp. **84** (1979) 271.
10a Guibert, J., Lequeux, J., Viallefond, F.: Astron. Astrophys. **68** (1978) 1.
11 Henderson, A.P.: Int. Astron. Union Symp. **84** (1979) 493.
12 Jackson, P.D., Kellman, S.A.: Astrophys. J. **190** (1975) 53.
12a Lockman, F.J.: Astrophys. J. **209** (1976) 429.
12b Lockman, F.J.: Astrophys. J. **232** (1979) 761.
13 Lynga, G.: Int. Astron. Union Symp. **84** (1979) 87.
14 McCray, R., Snow, T.P.: Annu. Rev. Astron. Astrophys. **17** (1979) 213.
15 Mebold, U., Hills, D.L.: Astron. Astrophys. **42** (1975) 187.
16 Mezger, P.G.: Astron. Astrophys. **70** (1978) 565.
16a Moffat, A.F.J., FitzGerald, M.P., Jackson, P.D.: Astron. Astrophys. Suppl. **38** (1979) 197.
17 Salpeter, E.E.: Int. Astron. Union Symp. **84** (1979) 245.
18 Savage, B.D., de Boer, K.S.: Astrophys. J. Lett. **230** (1979) L77.
19 Schlickeiser, R., Thielheim, K.O.: Astrophys. Space Sci. **47** (1977) 415.
20 Solomon, P.M., Sanders, D.B., Scoville, N.Z.: Int. Astron. Union Symp. **84** (1979) 35.
21 Spitzer, L., Jenkins, E.B.: Annu. Rev. Astron. Astrophys. **13** (1975) 133.
22 Spitzer, L.: Physical Processes in the Interstellar Medium, Wiley, New York (1978).
23 Stecker, F.W.: Astrophys. J. **212** (1977) 60.
24 Stecker, F.W.: Int. Astron. Union Symp. **84** (1979) 475.
25 Terzian, Y., Davidson, K.: Astrophys. Space Sci. **44** (1976) 479.
26 Webster, A.: Mon. Not. R. Astron. Soc. **171** (1975) 243.
27 York, D.G.: Astrophys. J. **213** (1977) 43.
28 Maihara, T., Oda, N., Shibai, H., Okuda, H.: Astron. Astrophys. **97** (1981) 139.
29 Bromage, G.E., Gabriel, A.H., Sciama, D.W.: 2nd European IUE Meeting (1980).
30 "Origin of cosmic rays", Int. Astron. Union Symp. **94** (1981).
31 Knapp, G.R., Tremaine, S.D., Gunn, J.E.: Astron. J. **83** (1978) 1585.
32 Cohen, R.S., Cong, H., Dame, T.M., Thaddeus, P.: Astrophys. J. Lett. **239** (1980) L53.
33 Serra, G., Puget, J.L., Ryter, C.E.: Astron. Astrophys. **84** (1980) 220.

**Scheffler**

# 8.3.5 The galactic center

## 8.3.5.1 Position and distance of the nucleus

Table 18. Equatorial coordinates of the galactic nucleus.

| | $\alpha_{1950}$ | $\delta_{1950}$ | Ref. |
|---|---|---|---|
| Ultracompact core of the radio source Sgr A West... | $17^h42^m29\overset{s}{.}291 \pm 0\overset{s}{.}005$ | $-28°59'17\overset{''}{.}6 \pm 0\overset{''}{.}1$ | 11 |
| Centroid of the near-infrared cluster in the nuclear region | $17^h42^m29\overset{s}{.}3 \pm 0\overset{s}{.}15$ | $-28°59'18'' \pm 3''$ | 6 |

Galactic coordinates of the nucleus [28]:

$l = -3\overset{'}{.}34, b = -2\overset{'}{.}75$

Table 19. Recent determinations of the sun's distance $R_0$ from the galactic center.

| Method | $R_0$ [kpc] | Ref. |
|---|---|---|
| Kinematical and photometric data of OB stars and Cepheids | $9 \pm 1$ | 23 |
| Kinematical and photometric data of OB stars | $9.0 \pm 1.6$ | 3 |
| Calibration of kinematic distances of hydrogen features by photometric distances of OB stars | $8.5 \pm 0.5$ | 44 |
| Galactic disk model | $9.0 \pm 1.0$ | 36, 31 |
| Main-sequence stars in the central bulge | $\begin{cases} 9.0 \pm 2.0 \\ 9.2 \pm 2.2 \end{cases}$ | 38<br>39 |
| RR Lyrae stars in the central region | $8.7 \pm 0.6$ | 27 |
| Globular cluster system | $8.5 \pm 1.6$ | 13 |
| Best value | $8.7 \pm 0.6$ | |

In most investigations published since ≈1963 the "standard distance" of $R_0 = 10$ kpc has been used following a recommendation of the IAU [15].

## 8.3.5.2 Gas distribution in the central region

From observations of atomic hydrogen (HI) two different phenomenological models have been deduced. The first one is referred to here as "Rougoor-Oort model"; its further development is reviewed by Oort [28, 28a]. Fig. 8 gives a sketch of the main features of this model for distances from the center up to about $R = 4$ kpc. Outside a rotating "nuclear disk" (and "ring") an expanding arm, the so-called "3 kpc-arm", occurs; it has a possible counterpart on the opposite side of the center. The 3 kpc-arm moves away from the center at 53 km $s^{-1}$; the distance of 3.0 kpc has been derived on the assumption that the sun's distance $R_0 = 8.2$ kpc. Furthermore, a number of observed spectral features were identified as isolated condensations, usually considered as ejecta produced by violent activity of the galactic nucleus (see Oort's review loc. cit.). For the inner region, $R \lessapprox 1.5$ kpc, an alternative "tilted-disk model" has been proposed recently by Burton and Liszt [9, 45] (see also Cohen and Davis [50]). It appears to account in a simple way for various anomalous spectral features quoted above (Fig. 9).

Table 20. Parameters of the Rougoor-Oort model [28, 8].

| | Nuclear disk | | Ring [1]) | 3 kpc-arm |
|---|---|---|---|---|
| Radius or distance from the center | 300 pc (effective) | | 600···900 pc | 3.7 kpc (for $R_0 = 10$ kpc) |
| H I density | 4··· 0.3 $cm^{-3}$ | for $R = 100 \cdots 500$ pc | $\approx 1$ $cm^{-3}$ | $\approx 2$ $cm^{-3}$ |
| Effective half-thickness $H$ ($H = 1.253 \times$ scale height used in 8.3.4.3) | 25···40 pc | | 100 pc | 100 pc |
| Inclination to the galactic plane | | | $\approx 8°$ | |
| Rotation velocity | $\approx 200$ km $s^{-1}$ | | 240 km $s^{-1}$ | $\approx 200$ km $s^{-1}$ |
| Expansion | 0 | | 0 | 53 km $s^{-1}$ |
| Total mass of H I | | $\approx 4 \cdot 10^6$ $\mathfrak{M}_\odot$ | | $\approx 3 \cdot 10^7$ $\mathfrak{M}_\odot$ |

[1]) Recent observations appear to give no evidence for the "ring" [10].

Table 21. Parameters of the tilted-disk models of Burton and Liszt [9, 45].
$R_d$, $a_d$, $b_d$, $\Pi_d$, $\theta_d$, and $v_d$ refer to the plane of the tilted disk.

| Parameter | Circular disk | Elliptical disk ("bar") |
|---|---|---|
| Radius $R_d$/axes $a_d$, $b_d$ of the disk/"bar" | 1.5 kpc | 1.9 kpc, 0.6 kpc |
| Effective half-thickness $H_d$ ($H = 1.253 \times$ scale height used in 8.3.4.3) | 125 pc | 125 pc |
| H I density | 0.33 $cm^{-3}$ | 0.33 $cm^{-3}$ |
| Angle between the plane of the tilted disk and the galactic plane | 25° | 24° |
| Angle in the plane of the sky between the disk and the galactic equator [1]) | 22° | 13.5° |
| Inclination angle between the plane of the disk and the plane of the sky | 78° | 70° |
| Velocity of expansion $\Pi_d(R_d)$ [km $s^{-1}$] | $170\left[1-\exp\left(-\frac{R_d}{0.07}\right)\right]$ | 0 |
| Velocity of rotation $\theta_d(R_d)$ [km $s^{-1}$] | $180\left[1-\exp\left(-\frac{R_d}{0.07}\right)\right]$ for $R_d \leqq 0.85$ kpc | motion along closed elliptical paths |
| | $180\left[1-\exp\left(-\frac{1.7-R_d}{0.20}\right)\right]$ for $R_d > 0.85$ kpc | "Rotational" speed at the minor axes: $v_d(b_d) = 360\left[1-\exp\left(-\frac{b_d}{0.10}\right)\right]$; $v_d$ in [km $s^{-1}$] |

[1]) In the bar-like model the long axis makes an angle of 41°.5 with respect to the plane of the sky measured in the plane of the disk.

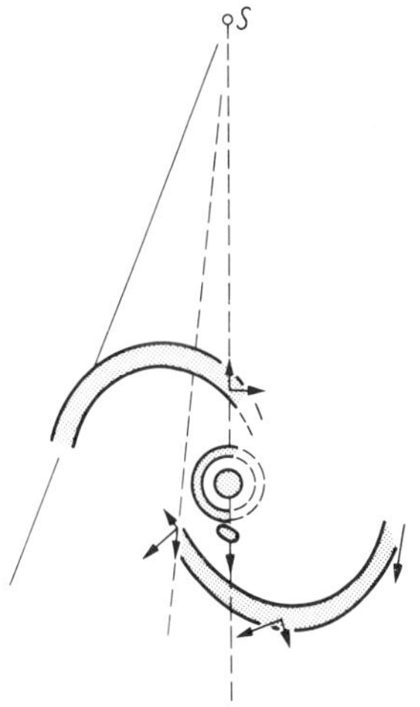


Fig. 8. Sketch of the Rougoor-Oort model of the galactic central region. S = sun. For explanation, see Table 20 [30a].

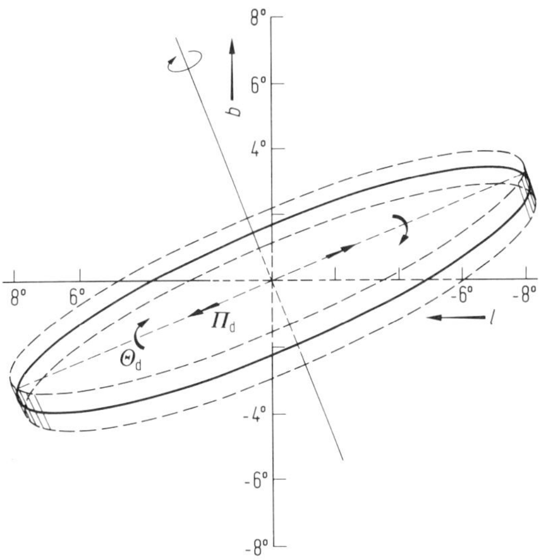


Fig. 9. Tilted-disk model projected onto the plane of the sky. For explanation, see Table 21 [9, 45].

The distribution of dense clouds derived mainly from observations of interstellar molecules, especially CO, OH, and $H_2CO$, shows a general deficiency between $R \approx 1$ and 3.5 kpc. Inward of a few hundred pc the mean density increases strongly, and the hydrogen is dominantly in the molecular state:

$\mathfrak{M}(H_2)/\mathfrak{M}(HI) > 10$. The molecular clouds lie close to the galactic plane ($|z| \lesssim 100$ pc) and their thickness (in $z$-direction) is quite small ($\lesssim 30$ pc).

The absorption spectra of OH and $H_2CO$ led some investigators to suggest an "expanding molecular ring" of radius $\approx 250$ pc, velocity of expansion $\approx 135\ \mathrm{km\,s^{-1}}$ and velocity of rotation $\approx 50\ \mathrm{km\,s^{-1}}$ [16, 35]. A similar model has been deduced from the more recent observations of CO emission [4]. An alternative explanation of the spectral CO-emission features is given by the tilted-disk model described in Table 21 when using a mean molecular density $N(H_2) \approx 10^2\ \mathrm{cm^{-3}}$ [19, 48]. Interpretation of recent measurements of the 6 cm $H_2CO$ line: [49].

Table 22. Mean density and mass of molecular gas in the central region.
$\langle N(H_2) \rangle$ = mean number density of molecular hydrogen
$\mathfrak{M}(H_2)$ = total mass of molecular hydrogen

| | $\langle N(H_2) \rangle$ $\mathrm{cm^{-3}}$ | $\mathfrak{M}(H_2)$ $\mathfrak{M}_\odot$ | Ref. |
|---|---|---|---|
| Expanding molecular ring ($R < 0.3$ kpc) | $\approx 3 \cdot 10^2$ | $\lesssim 3 \cdot 10^8$ | 28, 4 |
| Tilted-disk model ($R < 1.5$ kpc) | $\gtrless 25$ | $\gtrsim 10^9$ | 19,51 |
| 3 kpc-arm (90° galactocentric segment) | $\approx 1$ | $\approx 2 \cdot 10^7$ | 5,48 |
| $0 \leqq R \leqq 4$ kpc | $\approx 3$ | $\gtrsim 10^9$ | 28, 19 |

Ionized hydrogen detected by its thermal radio emission is strongly concentrated within a region of $R \approx 150$ pc which has an effective half-thickness of $H \approx 60$ pc. It consists of "extended low-density HII" and a number of "giant HII regions". The extended low-density HII appears to be a superposition of many evolved, widely expanded HII regions [25]. A representation of its distribution by using the sum of two spheroidal components of constant density has been proposed by J. Schmidt [34] (Table 23). Giant H II regions inside $R \approx 150$ pc are listed in Table 24. Nonthermal radiation, probably produced by supernova remnants, comes from a region with bounding diameters of $\approx 900$ pc and $\approx 350$ pc [28].

Table 23. Representation of the extended low-density H II region in the galactic center by two spheroids of constant density [25].
$T_e$ = electron temperature $N_e$ = electron number density
$E$ = emission measure $(=\int(N_e)^2 dr)$ $\mathfrak{M}$(H II) = total mass of ionized hydrogen

| | Apparent dimensions | True dimensions | $T_e$ K | $E$ cm$^{-6}$ pc | $N_e$ cm$^{-3}$ | $\mathfrak{M}$(H II) $\mathfrak{M}_\odot$ |
|---|---|---|---|---|---|---|
| Spheroid I | 90′ × 36′ | 300 pc × 120 pc | 5000 | $1.4 \cdot 10^4$ | 7 | $1.4 \cdot 10^6$ |
| Spheroid II | 38′ × 22′ | 130 pc × 80 pc | 5000 | $3.1 \cdot 10^4$ | 16 | $3.7 \cdot 10^5$ |
| | | | Total mass | | | $\approx 2 \cdot 10^6$ |

Table 24. Giant H II regions in the galactic center ($R \lesssim 150$ pc). Galactic coordinates of the sources from observations at 5 GHz [1]. $E$, $N_e$, and $\mathfrak{M}$ (H II) from [24, 29].
$S_5$ (max) = peak flux of the source at 5 GHz after the background has been subtracted [1]
For symbols, see Table 23.

| Source | $S_5$ (max) Jy | $E$ cm$^{-6}$ pc | $N_e$ cm$^{-3}$ | $\mathfrak{M}$(H II) $\mathfrak{M}_\odot$ | Remarks |
|---|---|---|---|---|---|
| G 359.429 − 0.090 (Sgr C) | 4.90 | – | – | – | probably a supernova remnant [25] |
| G 359.718 − 0.042 | 2.0 | – | – | – | |
| G 359.945 − 0.055 (Sgr A West) | 66.0 | $6.5 \cdot 10^6$ | $1.4 \cdot 10^3$ | $0.6 \cdot 10^3$ | detailed structure at 5 GHz see [11] |
| G 0.068 + 0.014 | 10.0 | $2.8 \cdot 10^5$ | $1 \cdot 10^2$ | $2 \cdot 10^4$ | main sources of the so-called "Arc" |
| G 0.178 − 0.050 | 11.0 | $2.3 \cdot 10^5$ | $1 \cdot 10^2$ | $5 \cdot 10^4$ | |
| G 0.510 − 0.051 | 10.24 | $1.9 \cdot 10^5$ | $0.8 \cdot 10^2$ | $1 \cdot 10^4$ | |
| G 0.670 − 0.036 (Sgr B2) | 18. | $5.6 \cdot 10^5$ | $2 \cdot 10^2$ | $0.6 \cdot 10^4$ | |

The radio source Sgr A is associated with the galactic nucleus. Sgr A consists of four components: a thermal source, Sgr A West: a nonthermal source, Sgr A East, probably a supernova remnant; a "halo" of ≈6′ diameter (thermal and nonthermal emission) and an extremely compact core in Sgr A West [11]. Observations of Ne II infrared line radiation at 12.8 µm have shown that the ionized gas of Sgr A West is strongly concentrated in an area of about 1.5 pc diameter (≈30″), the center of which is the compact core of Sgr A West [42].

Cosmic rays in the galactic center: [46].

### 8.3.5.3 Central stellar bulge and nucleus

Nearly all the luminosity and mass of the galactic central region is accounted for by a stellar bulge. Because of the contribution of red giants, it can best be observed at near-infrared wavelengths $\lambda \approx 2$ µm [43]. Distribution of the emissivity from observations with low spatial resolution: see 8.3.3.2, Fig. 5 and Table 13. A detailed model using specific radial distributions of O stars, giants with a black-body spectrum of 4000 K and cool supergiants of 2500 K has been discussed by Krügel and Tutukov (1978) [18].

The inner region of about 1.5 pc diameter, centred on the compact core of Sgr A West (see 8.3.5.2), contains an agglomeration of compact near-infrared sources [7, 26, 30, 41], most of which are probably high-luminosity stars. One of the sharp maxima coincides with the unresolved core in Sgr A West observed at cm wavelengths (see 8.3.5.1). This source is probably the actual nucleus of the Galaxy [28]. Very-long-baseline-interferometry indicates an overall diameter of ≈0″.01 at cm-wavelengths, and about 25% of the emission comes from a core diameter ≲0″.001 corresponding to ≲ 10 AU [17, 2].

The mean dust density around the nucleus is relatively small (visual extinction ≲ $3^m$ across the central 10 pc) in spite of high gas density [12, 25]. Strong middle- and far-infrared radiation observed from the region inside $R \approx 200$ pc appears to be thermal radiation of dust heated by hot stars of high luminosity and by population II stars. Most of this inner volume is filled by ionized gas while the massive molecular clouds occupy only a few percent of it [28, 25].

Table 25. Global data for the galactic center.

| Data | | Ref. |
|---|---|---|
| Total luminosity $L_{bol}$: | | |
| $R<$ 1 pc | $3\cdot10^6\ L_\odot$ | 12 |
| $R<150$ pc | $2\cdot10^8\ L_\odot$ | 18, 37 |
| central bulge ($R<2.5$ kpc) | $2\cdot10^{10}\ L_\odot$ | 22, 14, 21 |
| Total mass of the gas: | | |
| $R<150$ pc | $2\cdot10^7\ \mathfrak{M}_\odot$ | 28, 25, 37 |
| $R<300$ pc | $1.3\cdot10^8\ \mathfrak{M}_\odot$ | 46 |
| central disk ($R<1.5$ kpc) | $\approx10^9\ \mathfrak{M}_\odot$ | 20 |
| Total mass of dust: | | |
| $R<300$ pc | $\approx5\cdot10^5\ \mathfrak{M}_\odot$ | 46 |
| Total stellar mass: | | |
| nucleus $R<0.4$ pc | $\approx5\cdot10^6\ \mathfrak{M}_\odot$ | 28, 43 |
| $R<150$ pc | $1\cdot10^9\ \mathfrak{M}_\odot$ | 32, 25 |
| central bulge ($R<2.5$ kpc) | $\approx4\cdot10^{10}\ \mathfrak{M}_\odot$ | 28, 14 |
| Mass-to-luminosity ratio $\mathfrak{M}/L_V$ | $\approx15$ | 40 |
| $\mathfrak{M}/L_V$ | $\approx$ 7.5 | 22a |
| $\mathfrak{M}/L_{bol}$ | $\approx$ 2 | 47 |
| ($\mathfrak{M}$ and $L$ in solar units) | | |

Table 26. Energy balance for the galactic central region $R<100$ pc [37].

| Source | Energy input $L_\odot$ | Ionization input photons$\cdot$s$^{-1}$ | X-ray luminosity $L_\odot$ | Kinetic energy $L_\odot$ |
|---|---|---|---|---|
| OB stars | $10^8$ | $10^{52}$ | | $10^6$ |
| Cool giants | $10^8$ | | | |
| Planetaries | $10^6$ | $10^{50}$ | | |
| Supernovae | $5\cdot10^7$ | | $10^5$ | $2\cdot10^6$ |
| Stellar wind | $2\cdot10^6$ | | $10^4$ | $2\cdot10^5$ |
| X-ray binaries | | | $10^5$ | |
| Dissipation of gas clouds | $7\cdot10^7$ | | | |
| Total | $2.5\cdot10^8$ | $10^{52}$ | $\approx10^5$ | |
| Observed | $2\ \cdot10^8$ | $10^{52}$ | $3\cdot10^3$ | |

## 8.3.5.4 References for 8.3.5

### General reference

a The Large-Scale Characteristics of the Galaxy (Burton, W.B., ed.), Int. Astron. Union Symp. **84**, Reidel, Dordrecht (1979).

### Special references

1 Altenhoff, W.J., Downes, D., Pauls, T., Schraml, J.: Astron. Astrophys. Suppl. **35** (1979) 23.
2 Backer, D.B.: Astrophys. J. Lett. **222** (1978) L9.
3 Balona, L.A., Feast, M.W.: Mon. Not. R. Astron. Soc. **167** (1974) 621.
4 Bania, T.M.: Astrophys. J. **216** (1977) 381.
5 Bania, T.M.: Int. Astron. Union Symp. **84** (1979) 351.

6 Becklin, E.E., Neugebauer, G.: Astrophys. J. Lett. **200** (1975) L71.
7 Becklin, E.E., Matthews, K., Neugebauer, G., Willner, S.P.: Astrophys. J. **219** (1978) 121.
8 Burton, W.B. in: Galactic and Extra-Galactic Radio Astronomy (Verschuur, G.L., Kellermann, K.I., eds.), Springer, Berlin (1974) p. 82.
9 Burton, W.B., Liszt, H.S.: Astrophys. J. **225** (1978) 815.
10 Cohen, R.J.: Int. Astron. Union Symp. **84** (1979) 337.
11 Ekers, R.D., Downes, D., Goss, W.M., Rogstad, D.H., Schwarz, U.J.: Astron. Astrophys. **43** (1975) 159.
12 Gatley, I., Becklin, E.E., Werner, M.W., Wynn-Williams, C.G.: Astrophys. J. **216** (1977) 277.
13 Harris, W.E.: Astron. J. **81** (1976) 1095.
14 Hofmann, W., Lemke, D., Thum, C.: Astron. Astrophys. **57** (1977) 111.
15 Int. Astron. Union Inform. Bull. No. 11 (1963).
16 Kaifu, N., Wato, T., Iguchi, T.: Nature Phys. Sci. **238** (1972) 105.
17 Kellermann, K.I., Shaffer, D.B., Clark, B.G., Geldzahler, B.J.: Astrophys. J. Lett. **214** (1977) L61.
18 Krügel, E., Tutukov, A.V.: Astron. Astrophys. **63** (1978) 375.
19 Liszt, H.S., Burton, W.B.: Astrophys. J. **226** (1978) 790.
20 Liszt, H.S., Burton, W.B.: Int. Astron. Union Symp. **84** (1979) 343.
21 Low, F.J., Kurtz, R.F., Poteet, W.M., Nishimura, T.: Astrophys. J. Lett. **214** (1977) L115.
22 Maihara, T., Oda, N., Sugiyama, T., Okuda, H.: Publ. Astron. Soc. Japan, **30** (1978) 1.
22a Maihara, T.: Int. Astron. Union Symp. **84** (1979) 381.
23 Martin, P.G.: Mon. Not. R. Astron. Soc. **153** (1971) 251.
24 Mezger, P.G., Churchwell, E., Pauls, T. in: Stars and the Milky Way (Mavridis, L.N., ed.), Springer, Berlin 1974.
25 Mezger, P.G., Pauls, T.: Int. Astron. Union Symp. **84** (1979) 357.
26 Neugebauer, G., Becklin, E.E., Matthews, K., Wynn-Williams, C.G.: Astrophys. J. **220** (1978) 149.
27 Oort, J.H., Plaut, L.: Astron. Astrophys. **41** (1975) 71.
28 Oort, J.H.: Annu. Rev. Astron. Astrophys. **15** (1977) 295.
28a Oort, J.H.: Physica Scripta **17** (1978) 175.
29 Pauls, T., Mezger, P.G., Churchwell, E.: Astron. Astrophys. **34** (1974) 327.
30 Rieke, G.H., Telesco, C.M., Harper, D.A.: Astrophys. J. **220** (1978) 556.
30a Rougoor, G.W., Oort, J.H.: Proc. Natl. Acad. Sci. USA **46** (1960) 1.
31 Rybicki, G., Lecar, M., Schaefer, M.: Bull. American Astron. Soc. **6** (1974) 453.
32 Sanders, R.H., Lowinger, T.: Astron. J. **77** (1972) 292.
33 Schmidt, M.: Stars and Stellar Systems **V** (Galactic Structure) (1965) p. 513.
34 Schmidt, J.: Dissertation, Universität Bonn 1978.
35 Scoville, N.Z.: Astrophys. J. Lett. **175** (1972) L127.
36 Toomre, A.: Quart. J. R. Astron. Soc. **13** (1972) 241.
37 Tutukov, A., Krügel, E.: Astron. Astrophys. **67** (1978) 437.
38 Van den Bergh, S.: Astrophys. J. **188** (1974) L9.
39 Van den Bergh, S., Herbst, E.: Astron. J. **79** (1974) 603.
40 Van den Bergh, S.: Annu. Rev. Astron. Astrophys. **13** (1975) 217.
41 Willner, S.P.: Astrophys. J. **219** (1978) 870.
42 Wollman, E.R., Geballe, T.R., Lacy, J.H., Townes, C.H., Rank, D.M.: Astrophys. J. **218** (1977) 103.
43 Wollman, E.R.: Int. Astron. Union Symp. **84** (1979) 367.
44 Quiroga, R.J.: Astron. Astrophys. **92** (1980) 186.
45 Liszt, H.S., Burton, W.B.: Astrophys. J. **236** (1980) 779.
46 Audouce, J., Lequeux, J., Masnou, J.-L., Puget, J.-L.: Astron. Astrophys. **80** (1979) 276.
47 Okuda, H., Maihara, T., Oda, N., Sugiyama, T.: Int. Astron. Union Symp. **84** (1979) 377.
48 Bania, T.M.: Astrophys. J. **242** (1980) 95.
49 Güsten, R., Downes, D.: Astron. Astrophys. **87** (1980) 6.
50 Cohen, R.J., Davis, R.D.: Mon. Not. R. Astron. Soc. **186** (1979) 453.
51 Linke, R.A., Stark, A.A., Frerking, M.A.: Astrophys. J. **243** (1981) 147.

## 8.3.6 Properties of the Galaxy as a whole

Table 27. Summary of global parameters for an external observer.

| | Data | Ref. |
|---|---|---|
| Type of galaxy: Hubble-van den Bergh system | Sb ($-$Sb$^+$) I–II | 8, 10, 11 |
| de Vaucouleur's system | SAB (rs) bc II | 8 |
| Morgan's system | gkS 7 | 6, 8 |
| Face-on total absolute magnitude $M$ [mag] | $-20.5$ (P), $-21.1$ ($V$) | 7 |
| | $-19.9$ (pg) | 8 |
| | $-21.2$ (pg) | 9 |
| | $-20.1$ ($B$), $-20.6$ ($V$) | 4 |
| Colour index $B-V$ [mag] | $+0.65$ | 8 |
| | $+0.53$ | 4 |
| Luminosity $L$ [$L_\odot$] | $4\cdot10^{10}$ ($V$) | 7 |
| | $4\cdot10^{10}$ (pg) | 9 |
| | $1.6\cdot10^{10}$ ($B$) | 4 |
| Distance of the sun from the galactic center $R_0$ [kpc] | 8.7 | see Table 19 |
| Holmberg radius [kpc] (see 9.1.2.2) (isophote of the face-on surface brightness of 26.5 mag (pg) arc s$^{-2}$) | 17 | 4, 9 |
| Standard face-on isophotal radius [kpc] (at the surface brightness of 25.0 mag ($B$) arc s$^{-2}$) | 11.5 | 4 |
| H I radius [kpc] | 21 | 3, 9 |
| H II radius [kpc] | 13 | 2, 9 |
| Total mass $\mathfrak{M}_T$ [$\mathfrak{M}_\odot$] | $1.8\cdot10^{11}$ | 9, see also 8.4. |
| Mass-luminosity ratio $\mathfrak{M}/L_B$ [solar units] | $\approx 10$ | 4, 7 |
| Total mass of the gas $\mathfrak{M}_G$ [$\mathfrak{M}_\odot$] | $\approx 8\cdot10^9$ | [1]) |
| Fractional mass of the gas $\mathfrak{M}_G/\mathfrak{M}_T$ | $\approx 4\%$ | |
| Total mass of neutral atomic hydrogen $\mathfrak{M}_G$ (H I) [$\mathfrak{M}_\odot$] | $2.5\cdot10^9$ | 1, 4 |
| H I-luminosity ratio $\mathfrak{M}_G$ (H I)$/L_B$ [solar units] | 0.15 | 4, 9 |
| Total mass of the stellar halo [$\mathfrak{M}_\odot$] | $2\cdots4\cdot10^{10}$ | [2]) |
| | $\approx 1\cdot10^{11}$ | 12, 13 |
| Halo-disk stellar mass ratio | $0.10\cdots0.25$ | 5 |

[1]) $5\cdot10^9\,\mathfrak{M}_\odot$ for $R>0.8$ kpc (Table 17) plus $1\cdot10^9\,\mathfrak{M}_\odot$ for $R\lesssim0.8$ kpc (Table 25) multiplied by 1.3 (allowance for heavier elements).

[2]) Corresponding to the halo-disk ratio given in the last line.

### References for 8.3.6


1 Baker, P.L., Burton, W.B.: Astrophys. J. **198** (1975) 281.
2 Burton, W.B., Gordon, M.A., Bania, T.M., Lockman, F.J.: Astrophys. J. **202** (1975) 30.
3 Burton, W.B.: Annu. Rev. Astron. Astrophys. **14** (1976) 275.
4 de Vaucouleurs, G., Pence, W.D.: Astron. J. **83** (1978) 1163; Int. Astron. Union Symp. **84** (1979) 203.
5 Fenkart, R.P.: Astron. Astrophys. **56** (1977) 91.
6 Morgan, W.W.: Publ. Astron. Soc. Pac. **70** (1958) 364.
7 Neckel, T.: Z. Astrophys. **69** (1968) 112.
8 Schmidt-Kaler, Th., Schlosser, W.: Astron. Astrophys. **29** (1973) 409.
9 Tammann, G.A.: Proc. Conf. on optical observing programs on Galactic Structure and Dynamics (Schmidt-Kaler, Th., ed.), Bochum (1975) p. 1.
10 Van den Bergh, S.: J. R. Astron. Soc. Canada **62** (1968) 149.
11 Van den Bergh, S.: Astron. Astrophys. **20** (1972) 469.
12 Woltjer, L.: Astron. Astrophys. **42** (1975) 109.
13 Woltjer, L. in: Astronomical Papers dedicated to Bengt Strömgren (Reiz, A., Andersen, T., eds.), Copenhagen Univ. Obs. (1978) p. 349.

# 8.4 Kinematics and dynamics

## 8.4.0 Notation

| | |
|---|---|
| $\boldsymbol{r}, \boldsymbol{v}$ | vectors of position and velocity |
| $t$ | time |
| $\tau$ | age |
| $M_v$ | absolute visual magnitude |
| $B-V$ | colour index |
| $P$ | period of a variable star |
| $\Delta S$ | Preston's metallicity parameter for RR Lyrae variables |
| $\alpha, \delta$ | equatorial coordinates |
| $l, b$ | galactic coordinates (IAU 1960) |
| $\alpha_n, l_n$ | position of the ascending node of the galactic equator |
| $\varepsilon$ | inclination of the galactic plane with respect to the equatorial plane |
| $\mu_\alpha, \mu_\delta, \mu_l, \mu_b$ | proper motions (see 8.1.2) |
| $\varphi$ | parallactic angle |
| $v_r, v_t$ | radial and tangential velocities with respect to the sun |
| $v$ | modulus of the space velocity |
| $U, V, W$ | galactic components of the space velocity (see 8.4.1.2.1) |
| $U_\odot, V_\odot, W_\odot$ | galactic components of the solar motion |
| $\sigma_v; \sigma_U, \sigma_V, \sigma_W$ | velocity dispersions (total; in $U, V, W$) |
| $\psi$ | vertex deviation |
| $r$ | distance from the sun |
| $z$ | height above the galactic plane |
| $R$ | distance from the galactic center (if $z=0$) or, in general, from the galactic axis of rotation |
| $R_0$ | distance of the sun from the galactic center |
| $\theta$ | angular coordinate in a galactocentric system |
| $A, B$ | Oort's constants of galactic rotation |
| $v_c$ | circular velocity |
| $\omega_c$ | angular frequency of circular rotation |
| $\omega_{c,0}$ | local frequency of galactic rotation ($=\omega_c(R_0, z=0)$) |
| $\varkappa$ | epicyclic frequency of oscillations parallel to the galactic plane |
| $\varkappa_0$ | local epicyclic frequency ($=\varkappa(R_0)$) |
| $\omega_z$ | oscillation frequency perpendicular to the galactic plane |
| $C=\omega_{z,0}$ | local frequency of $z$-oscillations ($=\omega_z(R_0, z=0)$), Kuzmin constant |
| $T_c$ | period of galactic rotation |
| $T_\varkappa$ | epicyclic period |
| $T_z$ | period of $z$-oscillation |
| $\mu$ | surface density (mass) |
| $\varrho$ | space density (mass) |
| $\varrho_0$ | local mass density |
| $e$ | numerical eccentricity |
| $a$ | semi-major axis |
| $M_R$ | mass within a sphere of radius $R$ |
| $G$ | gravitational constant |
| $\mathfrak{M}_\odot$ | solar mass |
| $\Psi$ | galactic gravitational potential (per unit mass) |
| $\boldsymbol{K}$ | galactic gravitational force (per unit mass) |
| $K_R, K_\theta, K_z$ | galactic components of $\boldsymbol{K}$ |
| $\Omega_p$ | pattern speed of the density wave |
| $\Psi_1$ | gravitational potential (per unit mass) due to the density wave |
| $A_1$ | amplitude of the potential $\Psi_1$ |
| $f_{R,0}$ | local relative amplitude of the radial force of the density wave |
| $\Phi$ | radial phase of the density wave |
| $i$ | pitch angle of the spiral arm with respect to a circle |

## 8.4.1 Kinematics

### 8.4.1.1 Basic concepts of galactic kinematics

Galactic kinematics describe the present velocities of objects in our Galaxy. Kinematical data are observationally based on radial velocities (see 8.1.3), proper motions (see 8.1.2), distances (see 8.1.4), and positions (see 8.1.1). From these observed properties, three-dimensional space velocities $\boldsymbol{v}$ may be obtained, which are usually more suitable for theoretical discussions.

In most cases, the kinematics of rather homogeneous groups of objects (defined by common spectral type, common age, etc.) are studied. At any position in space, $\boldsymbol{r}$, we can obtain the mean motion of this group, $\boldsymbol{v}_{\text{mean}}$. This mean motion depends usually on position, $\boldsymbol{v}_{\text{mean}}(\boldsymbol{r})$, and represents a velocity field. The velocity field of objects of the galactic disk is mainly determined by the differential galactic rotation, thus reflecting the circular velocities in the Galaxy. Deviations from an axisymmetric velocity field are probably correlated with the spiral structure of our Galaxy. The mean motion of halo objects is significantly smaller than the circular velocity.

The difference between the individual velocity of an object, $\boldsymbol{v}_i$, and a reference velocity, $\boldsymbol{v}_{\text{ref}}(\boldsymbol{r}_i)$, at its present position, $\boldsymbol{r}_i$, is called the peculiar motion of this object. As the reference velocity, we may choose either the observable mean group motion, $\boldsymbol{v}_{\text{mean}}$, or the circular velocity $\boldsymbol{v}_c(\boldsymbol{r}_i)$. The circular velocity is the proper reference velocity for theoretical studies, but lacks direct observability. The peculiar motion of the sun with respect to the local mean motion of a group of objects is called the solar motion. Since the mean motions of various groups usually differ, the solar motion depends on the group chosen, thus reflecting the differences in the mean motions of various groups of objects.

Significant observational information on peculiar motions can be derived usually for nearby objects only. For disk stars, the distribution of the peculiar motions can be approximated by a Gaussian velocity distribution with anisotropic velocity dispersions (Schwarzschild distribution). Formally, it is always possible to obtain the velocity dispersions, $\sigma$, as the root-mean-square peculiar velocities of a group of objects. The velocity dispersions in the three directions of velocity space define, together with the corresponding correlation terms, a velocity ellipsoid. The angular deviation of the major axis of this ellipsoid from the direction towards the galactic center is called the vertex deviation and is probably connected with the spiral structure. The velocity dispersions of disk stars increase strongly with the age of the stars. Halo objects show the largest velocity dispersions, thus contributing most of the high-velocity objects. With increasing velocity dispersion (and hence with age), the difference between the circular velocity and the mean velocity of a group, $\boldsymbol{v}_c - \boldsymbol{v}_{\text{mean}}$, called the asymmetric drift, increases too.

Reviews on galactic kinematics: [13, 63, 89, 65, 4].

### 8.4.1.2 Velocities

#### 8.4.1.2.1 Space velocities

Computation of space velocities:

Convert the proper motion $\mu_\alpha$ from [s/a] into [″/a] by the relation $\mu_\alpha[''/a] = 15\,\mu_\alpha[s/a]$. Published proper motions usually do not refer to an inertial system but (usually) to the FK4 system. They have to be corrected for the unphysical rotation of the FK4 system before they can be used for kinematical investigations. According to Fricke [37], the following precessional corrections should be added to the proper motions in the FK4 system (both in [″/a]):

$$\Delta(\mu''_\alpha \cos\delta) = +0\overset{''}{.}0020\cos\delta - 0\overset{''}{.}0044\sin\alpha\cos\delta$$

$$\Delta\mu''_\delta = -0\overset{''}{.}0044\cos\alpha\,.$$

In some cases, it is even necessary to correct the published radial velocities by eliminating a "constant $K$ term" if this term is either due to systematic errors in the radial velocity system or due to systematic motions in the stellar atmospheres.

Transform $\alpha$[h, m, s] and $\delta$[°, ′, ″] into $\alpha$[°] and $\delta$[°]. Determine the galactic coordinates $l, b$ and the parallactic angle $\varphi$ from $\alpha, \delta$:

$$\cos b \cos(l - l_n) = \cos\delta \cos(\alpha - \alpha_n)$$
$$\cos b \sin(l - l_n) = \cos\delta \sin(\alpha - \alpha_n) \cos\varepsilon + \sin\delta \sin\varepsilon$$
$$\sin b = -\cos\delta \sin(\alpha - \alpha_n) \sin\varepsilon + \sin\delta \cos\varepsilon$$
$$\cos b \sin\varphi = \sin\varepsilon \cos(\alpha - \alpha_n)$$
$$\cos b \cos\varphi = \sin\varepsilon \sin\delta \sin(\alpha - \alpha_n) + \cos\varepsilon \cos\delta$$

(for 1950.0: $\alpha_n = 282.°25$, $l_n = 33.°0$, $\varepsilon = 62.°6$; for 1900.0: $\alpha_n = 281.°65$, $l_n = 33.°07$, $\varepsilon = 62.°33$).

Determine $\mu_l$, $\mu_b$ from corrected $\mu_\alpha$, $\mu_\delta$:

$$(\mu_l \cos b) = \cos\varphi(\mu_\alpha \cos\delta) + \sin\varphi \mu_\delta$$
$$\mu_b = -\sin\varphi(\mu_\alpha \cos\delta) + \cos\varphi \mu_\delta .$$

Determine the galactic components $U$, $V$, $W$ of space velocities in [km s$^{-1}$]:

$$U = -\gamma r \mu_l \cos b \sin l - \gamma r \mu_b \cos l \sin b + v_r \cos l \cos b$$
$$V = +\gamma r \mu_l \cos b \cos l - \gamma r \mu_b \sin l \sin b + v_r \sin l \cos b$$
$$W = +\gamma r \mu_b \cos b + v_r \sin b$$

($\gamma = 4.74$ km s$^{-1}$/(″/a)).

By definition, the positive $U$-axis points towards the galactic center ($l = 0°$, $b = 0°$), the $V$-axis in the direction of galactic rotation ($l = 90°$, $b = 0°$), and the $W$-axis towards the north galactic pole ($b = +90°$). This ($U$, $V$, $W$)-system is a right-handed system. Caution! Some authors define the positive $U$-axis towards the anticenter ($l = 180°$, $b = 0°$). In many papers, the symbols $\Pi = -U$, $\Theta = V$, $Z = W$ are used.

Since $v_r$ and the corrected $\mu_\alpha$, $\mu_\delta$ refer to an inertial system which is comoving with the sun, the space velocity components $U$, $V$, $W$ are primarily determined relative to the sun also. They may, however, be transformed to other velocity systems by adding the corresponding solar motion.

Catalogues of space velocities:

| | | | |
|---|---|---|---|
| Nearby stars | [40, 76, 116] | Classical Cepheids | [106] |
| K and M dwarfs | [20, 22, 25, 98] | G, K, and M giants | [19, 24, 118] |
| Bright stars | [14, 15] | Subdwarfs | [83] |
| A stars | [16, 23] | RR Lyrae stars | [100] |
| Early-type stars | [26, 27, 28, 82] | High-velocity (halo) stars | [2, 17, 29, 81] |
| Gould's belt | [54] | High-proper-motion stars | [30] |

Most papers of O.J. Eggen contain data on space velocities. "Space velocities" of white dwarfs are often computed without taking the radial velocity into account.

### 8.4.1.2.2 Solar motions

The mean motion $\boldsymbol{v}_{\text{mean}}$ of a group of objects with individual space velocities $\boldsymbol{v}_i$ (i = 1, …, N), relative to the sun, is given by:

$$\boldsymbol{v}_{\text{mean}} = \frac{1}{N} \sum_{i=1}^{N} \boldsymbol{v}_i .$$

The solar motion $\boldsymbol{v}_\odot$ relative to the mean motion of this group is given by:

$$\boldsymbol{v}_\odot = -\boldsymbol{v}_{\text{mean}} .$$

The direction of the solar motion is called the apex. The solar motion can also be derived from radial velocities alone, and the apex from proper motions alone [63, 89].

The most complete list of solar motions and other kinematical data for various groups of stars was compiled by Delhaye [13] Table 1, p. 64 from data published up to 1961. From this compilation, Delhaye selected the kinematical data for stars of various spectral types given in Table 1. In Table 2, corresponding data for groups of selected objects are listed. Representative kinematical data based on nearby stars are given in 8.4.1.2.5.

Table 1. Solar motion and velocity dispersion for stars of various spectral types [13].
$\psi$ = vertex deviation in [°]

| Spectral type | Solar motion [km s$^{-1}$] | | | Velocity dispersion [km s$^{-1}$] | | | | |
|---|---|---|---|---|---|---|---|---|
| | $U_\odot$ | $V_\odot$ | $W_\odot$ | $\sigma_U$ | $\sigma_V$ | $\sigma_W$ | $\sigma_\upsilon$ | $\psi$ |
| Supergiants: | | | | | | | | |
| cO – cB5 | + 9.0 | +13.4 | + 3.7 | 12 | 11 | 9 | 19 | +36° |
| cF – cM | + 7.9 | +11.7 | + 6.5 | 13 | 9 | 7 | 17 | +18 |
| Giants: | | | | | | | | |
| gA | +13.4 | +11.6 | +10.3 | 22 | 13 | 9 | 27 | +27 |
| gF | +19.7 | +18.5 | + 9.5 | 28 | 15 | 9 | 33 | +14 |
| gG | + 7.2 | +11.1 | + 6.9 | 26 | 18 | 15 | 35 | +12 |
| gK0 | +10.6 | +18.6 | + 6.5 | 31 | 21 | 16 | 41 | +21 |
| gK3 | + 9.0 | +17.6 | + 6.4 | 31 | 21 | 17 | 41 | +14 |
| gM | + 4.5 | +18.3 | + 6.2 | 31 | 23 | 16 | 42 | + 7 |
| Main sequence: | | | | | | | | |
| B0 | + 9.6 | +14.5 | + 6.7 | 10 | 9 | 6 | 15 | −50 |
| dA0 | + 7.3 | +13.7 | + 7.2 | 15 | 9 | 9 | 20 | +15 |
| dA5 | + 8.5 | + 7.8 | + 7.4 | 20 | 9 | 9 | 24 | +19 |
| dF0 | +11.2 | +10.8 | + 7.3 | 24 | 13 | 10 | 29 | +21 |
| dF5 | +10.1 | +12.3 | + 6.2 | 27 | 17 | 17 | 36 | +13 |
| dG0 | +14.5 | +21.1 | + 6.4 | 26 | 18 | 20 | 37 | + 2 |
| dG5 | + 8.1 | +22.1 | + 4.3 | 32 | 17 | 15 | 39 | +14 |
| dK0 | +10.8 | +14.9 | + 7.4 | 28 | 16 | 11 | 34 | + 3 |
| dK5 | + 9.5 | +22.4 | + 5.8 | 35 | 20 | 16 | 43 | +11 |
| dM0 | + 6.1 | +14.6 | + 6.9 | 32 | 21 | 19 | 43 | + 8 |
| dM5 | + 9.8 | +19.3 | + 8.6 | 31 | 23 | 16 | 42 | − 7 |

Table 2. Solar motion and velocity dispersion for groups of selected objects.

| Objects | Solar motion [km s$^{-1}$] | | | Velocity dispersion [km s$^{-1}$] | | | | Ref. |
|---|---|---|---|---|---|---|---|---|
| | $U_\odot$ | $V_\odot$ | $W_\odot$ | $\sigma_U$ | $\sigma_V$ | $\sigma_W$ | $\sigma_\upsilon$ | |
| Interstellar H I | +12 | + 15 | + 9 | ( | 5.7 | ) | 10 | 10 |
| Interstellar Ca II | +11 | + 14 | + 8 | ( | 6 | ) | 10 | 3 |
| Classical Cepheids | +11 | + 12 | +10 | 8 | 7 | 5 | 12 | 106 |
| Carbon stars | +10 | + 12 | + 5 | 30 | 20 | 14 | 39 | 12 |
| White dwarfs | +10 | + 15 | + 7 | 42 | 22 | 18 | 50 | 47, 107 |
| Planetary nebulae | | | | | | | | |
| class B | +15 | + 21 | – | 28 | 28 | – | – | 11 |
| class C | +11 | + 22 | – | 56 | 28 | 30 | 64 | 11 |
| Semi-regular red variables | +20 | + 27 | +12 | 42 | 42 | 34 | 68 | 35 |
| Long-period variables | | | | | | | | 34 |
| $P$ [days] | | | | | | | | |
| <150 | + 6 | + 32 | +36 | 46 | 30 | 34 | 65 | |
| 150···200 | +44 | +119 | − 1 | 94 | 91 | 61 | 144 | |
| 200···300 | + 6 | + 46 | +11 | 52 | 45 | 40 | 80 | |
| >300 | +15 | + 26 | + 2 | 43 | 38 | 30 | 65 | |

continued

Table 2, continued

| Objects | Solar motion [km s$^{-1}$] | | | Velocity dispersion [km s$^{-1}$] | | | | Ref. |
|---|---|---|---|---|---|---|---|---|
| | $U_\odot$ | $V_\odot$ | $W_\odot$ | $\sigma_U$ | $\sigma_V$ | $\sigma_W$ | $\sigma_v$ | |
| RR Lyrae variables[1]) | | | | | | | | 43 |
| a-type | | | | | | | | |
| *P* [days] | | | | | | | | |
| 0.2···0.42 | +33 | + 15 | −26 | | 35 | 23 | | |
| 0.42···0.5 | +37 | +160 | +31 | 180 | 140 | 72 | 239 | |
| 0.5···0.6 | +23 | +169 | −12 | 151 | 119 | 90 | 212 | |
| ≧0.6 | +13 | +249 | +16 | 267 | 178 | 182 | 369 | |
| $\Delta S$=0···3 | + 3 | +109 | + 7 | 131 | 141 | 88 | 212 | |
| 4···6 | +42 | +228 | + 5 | 209 | 146 | 102 | 275 | |
| 7···10 | +30 | +199 | +10 | 210 | 133 | 117 | 275 | |
| all a-type | +24 | +163 | + 7 | 170 | 137 | 99 | 240 | |
| c-type | +19 | +127 | +47 | 80 | 68 | 23 | 107 | |
| Subdwarfs | | | | | | | | 83 |
| UV excess δ (0.6) [2]) | | | | | | | | |
| $0^{m}.00\cdots0^{m}.09$ | +15 | + 45 | + 9 | 76 | 35 | 19 | 86 | |
| 0.10···0.15 | +14 | + 62 | + 9 | 75 | 34 | 38 | 91 | |
| 0.16···0.20 | +11 | +149 | − 4 | 146 | 103 | 64 | 190 | |
| 0.21···0.31 | +48 | +204 | + 2 | 192 | 99 | 116 | 245 | |

[1]) See 5.1.2.2.

[2]) Normalized value at $B-V=0^{m}.6$, to account for the variation of δ with $B-V$ for stars of the *same* metal abundance [83].

#### 8.4.1.2.3 Local standard of rest

For text, see next page.

Table 3. Solar motion with respect to the local standard of rest (LSR). **Bold face**: primary values

| Designation | Remarks | Solar motion [km s$^{-1}$] | | | | Apex [°] | | | | Ref. |
|---|---|---|---|---|---|---|---|---|---|---|
| | | $U_\odot$ | $V_\odot$ | $W_\odot$ | $v_\odot$ | α | δ | *l* | *b* | |
| Kinematical definition: | | | | | | | | | | |
| Standard solar motion | Conventionally adopted value, especially in radio astronomy, based on the mean motions of nearby objects. | +10.3 | +15.3 | +7.7 | **20** | **270** **(1900)** 270.5 (1950) | **+30** +30.0 | 56.2 | +22.8 | 99 |
| Basic solar motion | LSR defined by the maximum in the velocity distribution of nearby stars. | **+ 9** | **+11** | **+6** | 15.4 | 267.4 (1950) | +25.7 | 50.7 | +23.9 | 13 |
| Dynamical definition: | | | | | | | | | | |
| Peculiar solar motion | Derived from the observed asymmetric drift as a function of velocity dispersion and a theoretical extrapolation to zero velocity dispersion. | **+ 9** | **+12** | **+7** | 16.6 | 267.0 (1950) | +28.1 | 53.1 | +25.0 | 13 |

The local standard of rest (LSR) is the origin of a velocity system which is corrected for the peculiar motion of the sun. The LSR may be defined either by the mean motion of common nearby stars (kinematical definition) or, using theoretical concepts, by the local circular velocity (dynamical definition). Both definitions differ slightly because of the asymmetric drift and other non-circular motions. The LSR is usually defined implicitly by giving the solar motion with respect to the LSR (Table 3).

Because of non-circular motions, especially of young objects, the conventional LSR (Table 3) may differ from the dynamically defined local circular velocity by a significant amount. For example, Mayor [62] proposed $V_{\odot} = +6\,\mathrm{km\,s^{-1}}$, and Wielen [114] suggested $U_{\odot} = +4\,\mathrm{km\,s^{-1}}$.

### 8.4.1.2.4 Velocity dispersions

For a group of objects with individual space-velocity components $U_i$, $V_i$, $W_i$ (i=1, ..., N), the velocity dispersions $\sigma_U$, $\sigma_V$, $\sigma_W$ are given by

$$\sigma_U^2 = \frac{1}{\mathrm{N}-1} \sum_{i=1}^{\mathrm{N}} (U_i - U_{\mathrm{mean}})^2, \text{ etc. for } V \text{ and } W.$$

The total velocity dispersion $\sigma_v$ is given by

$$\sigma_v^2 = \sigma_U^2 + \sigma_V^2 + \sigma_W^2 .$$

The vertex deviation $\psi$, i.e. the angle between the major axis of the velocity ellipsoid in the $UV$ plane and the $U$ axis, is given by

$$\frac{1}{2}\tan 2\psi = \frac{\sum_i (U_i - U_{\mathrm{mean}})(V_i - V_{\mathrm{mean}})}{\sum_i (U_i - U_{\mathrm{mean}})^2 - \sum_i (V_i - V_{\mathrm{mean}})^2} .$$

(The vertex deviations in the $UW$ and $VW$ planes are usually small.)

Observed velocity dispersions and vertex deviations are listed in Tables 1, 2, 4, and 5 (see also [13]).

For stellar objects in the galactic disk, the velocity dispersions are strongly correlated with the age $\tau$ of the objects. The ratios between the velocity dispersions in the different directions remain rather constant. For theoretical interpretations, see 8.4.2. Convenient approximations [111] are:

$$\sigma_v^2 = \sigma_{v,0}^2 + C_v \tau$$

with $\sigma_{v,0} = 10\,\mathrm{km\,s^{-1}}$; $C_v = 6.0 \cdot 10^{-7}\,(\mathrm{km/s})^2/\mathrm{a}$ and

$$\sigma_v : \sigma_U : \sigma_V : \sigma_W = 1 : 0.77 : 0.49 : 0.41 .$$

### 8.4.1.2.5 Representative nearby stars

It is a very severe problem in galactic kinematics to ensure that a chosen group of stars is not biased by selection effects but is really representative for the kinematical behaviour of the desired class of objects. For example, in many cases objects with large velocities are more easily detected or are specifically searched for, thus being overrepresented in many astronomical catalogues.

For nearby stars, the completeness and/or homogeneity is usually greatest. For stars with $M_v$ brighter than $+7^M$, Gliese's catalogue of nearby stars [40] is almost complete. In Table 4 unbiased kinematical data for groups of stars within 20 pc are given [47, 107].

For fainter stars, even the catalogues of nearby stars are strongly biased by kinematical selection effects [107]. Probably the most representative group for the kinematical behaviour of "common stars in the solar neighbourhood" are the nearby K and M dwarfs discovered at the McCormick Observatory [101] on objective prism plates. Data for these stars are listed in Table 5, and velocity distributions are shown in Figs. 1···3 [47, 107]. Ages for these stars have been derived from Wilson's estimates of the emission intensity (HK) in the CaII H and K lines [115].

Due to the larger velocity dispersion, $\sigma_W$, and hence the larger average distances, $z$, from the galactic plane, older stars are underrepresented in a volume of $r \leq 20$ pc around the sun, at $z \approx 0$, with respect to a more representative cylinder perpendicular to the galactic plane. The last line of Table 5 has been derived for such a cylinder by weighting each nearby star with its $W$ velocity component [107], see also [117]. These kinematical data should be considered as being the most representative for common nearby stars.

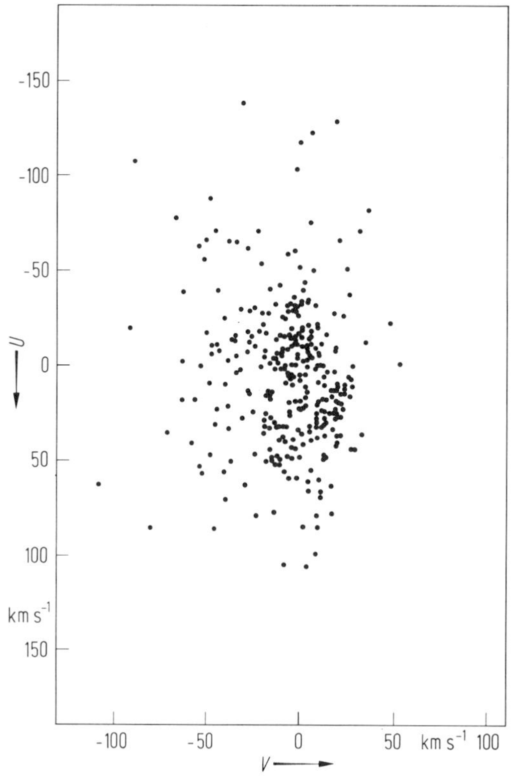


Fig. 1. Velocity distribution of nearby K and M dwarfs ($UV$ plane). The velocities have been corrected for the peculiar solar motion (Table 3).

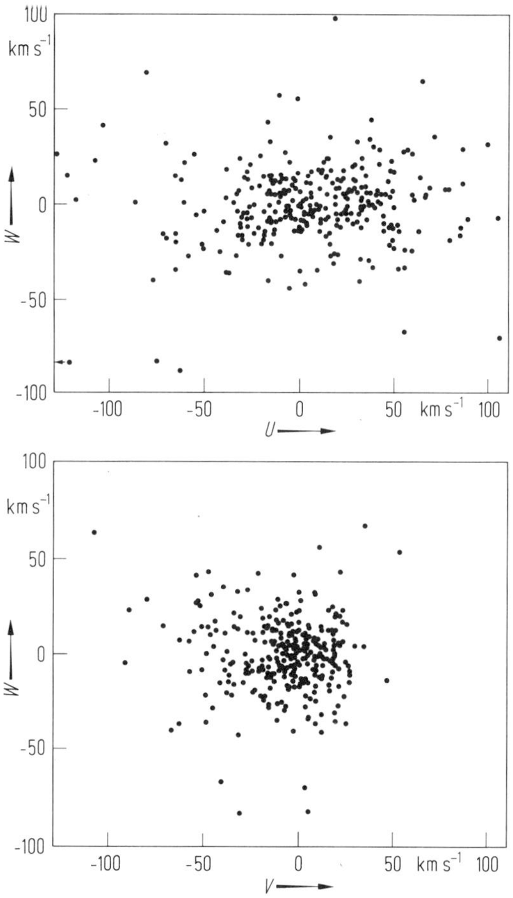


Fig. 2. Velocity distribution of nearby K and M dwarfs (upper part: $UW$ plane; lower part: $VW$ plane). The velocities have been corrected for the peculiar solar motion (Table 3).

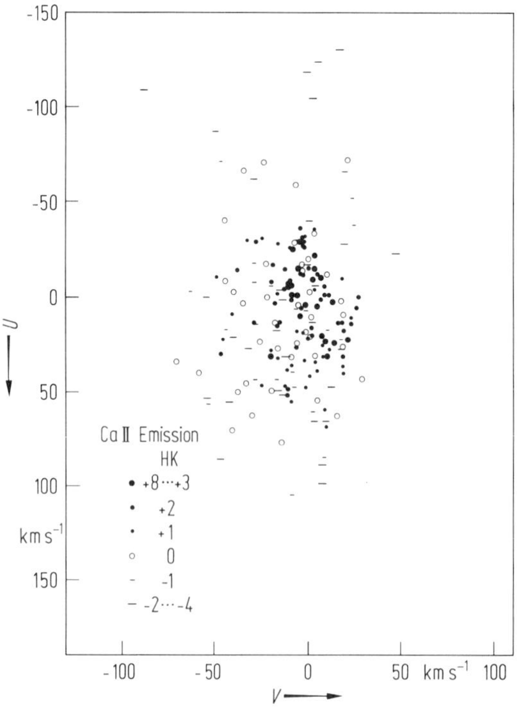


Fig. 3. Velocity distribution ($UV$ plane) of nearby K and M dwarfs with different Ca II emission intensities HK. The velocities have been corrected for the peculiar solar motion (Table 3).

Table 4. Galactic components of the solar motion, $U_{\odot}$, $V_{\odot}$, $W_{\odot}$, and the velocity dispersions, $\sigma_U$, $\sigma_V$, $\sigma_W$, $\sigma_v$, of stars within $r \leqq 20$ pc on or near the main sequence (without classified subdwarfs) [47, 107].

$B-V$ = range in colour
Sp. type = range in spectral type
$N$ = number of stars
$\psi$ = vertex deviation in [°]
$\langle\tau\rangle$ = mean age in [$10^9$a]
colour groups 3···6d according to [107]

| Group | $B-V$ | Sp. type | $N$ | $U_{\odot}$ (km s⁻¹) | $V_{\odot}$ (km s⁻¹) | $W_{\odot}$ (km s⁻¹) | $\sigma_U$ (km s⁻¹) | $\sigma_V$ (km s⁻¹) | $\sigma_W$ (km s⁻¹) | $\sigma_v$ (km s⁻¹) | $\psi$ | $\langle\tau\rangle$ $10^9$a |
|---|---|---|---|---|---|---|---|---|---|---|---|---|
| 6d | $\leqq 0^m.05$ | $\leqq$A1 | 6 | +7 | +5 | +7 | 14 | 8 | 4 | 16 | +23° | 0.2 |
| 6c | 0.05···0.19 | A2···A6 | 14 | +3 | +7 | +9 | 17 | 7 | 4 | 19 | +20 | 0.4 |
| 6b | 0.20···0.34 | A7···F1 | 16 | +12 | +7 | +6 | 14 | 11 | 8 | 19 | +32 | 0.9 |
| 6a | 0.35···0.49 | F2···F6 | 47 | +16 | +11 | +10 | 27 | 18 | 11 | 34 | +12 | 2 |
| 5 | 0.50···0.64 | F7···G3 | 83 | +17 | +17 | +5 | 34 | 21 | 21 | 45 | +13 | 4 |
| 4 | 0.65···0.79 | G4···G9 | 84 | +19 | +27 | +7 | 34 | 27 | 20 | 48 | +18 | 5 |
| 3 | 0.80···1.09 | K0···K4 | 138 | +11 | +19 | +7 | 36 | 27 | 17 | 48 | +20 | 5 |

Table 5. Solar motion and velocity dispersion of nearby McCormick K and M dwarfs. For explanation of symbols, see Table 4.

| Group | $N$ | $U_{\odot}$ (km s⁻¹) | $V_{\odot}$ (km s⁻¹) | $W_{\odot}$ (km s⁻¹) | $\sigma_U$ (km s⁻¹) | $\sigma_V$ (km s⁻¹) | $\sigma_W$ (km s⁻¹) | $\sigma_v$ (km s⁻¹) | $\psi$ | $\langle\tau\rangle$ $10^9$a |
|---|---|---|---|---|---|---|---|---|---|---|
| HK emission intensity | | | | | | | | | | |
| +8 to +3 | 23 | +9 | +11 | +4 | 18 | 10 | 8 | 22 | +15° | 0.3 |
| +2 | 40 | +8 | +15 | +7 | 21 | 16 | 13 | 30 | +8 | 1.4 |
| +1 | 36 | −8 | +18 | +4 | 29 | 17 | 15 | 37 | +14 | 3.0 |
| 0 | 41 | 0 | +25 | +6 | 38 | 23 | 20 | 48 | −4 | 5.2 |
| −1 | 24 | −7 | +21 | +4 | 40 | 27 | 26 | 55 | +6 | 7.2 |
| −2 to −5 | 31 | +14 | +25 | +2 | 66 | 27 | 23 | 75 | −0 | 9.0 |
| All K+M dwarfs at $z \approx 0$ | 317 | +5 | +19 | +6 | 39 | 23 | 20 | 49 | +4 | |
| All K+M dwarfs averaged over $z$ | 317 | +8 | +23 | +6 | 48 | 29 | 25 | 62 | −2 | 5.0 |

### 8.4.1.2.6 Moving groups

The velocity distribution of stars is in certain cases not a smooth (random) distribution but is "clumped", indicating that some of the stars have a tendency to share a common space velocity (star streams, moving groups or moving clusters). The similarity of space velocities is in general an indication for the common origin of these stars in space and time. The stars may form either a gravitationally bound system (open star cluster) or may slowly disperse in space (associations, moving groups). In many cases, an open star cluster is surrounded by, or is a part of, a moving group of stars which has a much larger spatial extent than the cluster and has a higher internal velocity dispersion than the cluster. In Table 6, the mean space velocities of some prominent moving groups are listed. Eggen [18] (and literature cited there) has identified many moving groups, even among old disk and halo stars. It is often difficult to assess the reality of a moving group.

Table 6. Mean space velocity (relative to the sun) of some moving groups.

| Name | $U$ (km s⁻¹) | $V$ (km s⁻¹) | $W$ (km s⁻¹) | Ref. |
|---|---|---|---|---|
| Ursa Major group[1]) | +14 | +3 | −8 | 112 |
| Hyades group | −40 | −17 | −3 | 102 |
| Pleiades group | −9 | −27 | −12 | 21 |

[1]) Called the Sirius group by Eggen.

### 8.4.1.3 Galactic rotation

To a first approximation, the velocity field of disk objects can be represented by purely circular motion: The streamlines are circles, parallel to the galactic plane, centred on the galactic rotation axis (through the galactic center, perpendicular to the galactic plane). The linear circular rotational velocity, $v_c(R)$, measured in [km s$^{-1}$], as a function of the galactocentric distance, $R$, defines the rotation curve. The corresponding rotational frequency, $\omega_c(R)=v_c(R)/R$, is usually measured in [(km s$^{-1}$/kpc]. Equivalent to this unit are

$$1\,(\mathrm{km\,s^{-1}})/\mathrm{kpc} = 2.109\cdot 10^{-4}\,''/\mathrm{a} = 1.023\cdot 10^{-9}\,\mathrm{rad/a}.$$

Let an observer at the position of the sun ($z\approx 0$) move with the local circular velocity $v_c(R_0)$ (see 8.4.1.2.3 for the reduction to the LSR). For an object which moves with the circular velocity at the galactocentric distance $R$, and is located at the distance $r$ from the sun in the direction $l$ in the galactic plane ($b=0$), he would measure the following (relative) radial and tangential velocities:

$$v_r = R_0(\omega_c(R)-\omega_c(R_0))\sin l$$

$$v_t = R_0(\omega_c(R)-\omega_c(R_0))\cos l - \omega_c(R)r.$$

For objects with $r \ll R_0$, a linear approximation gives:

$$v_r = Ar\sin 2l$$

$$v_t = Ar\cos 2l + Br,$$

where $A$ and $B$ are the local Oort constants of galactic rotation:

$$A = +\frac{1}{2}\left(\frac{v_c}{R} - \frac{dv_c}{dR}\right)_0 = -\frac{1}{2}R_0\left(\frac{d\omega_c}{dR}\right)_0$$

$$B = -\frac{1}{2}\left(\frac{v_c}{R} + \frac{dv_c}{dR}\right)_0 = -\frac{1}{2}R_0\left(\frac{d\omega_c}{dR}\right)_0 - \omega_c(R_0).$$

We also have the relations:

$$\omega_c(R_0) = A - B; \qquad \left(\frac{d\omega_c}{dR}\right)_0 = -2A/R_0$$

$$v_c(R_0) = R_0(A-B); \qquad \left(\frac{dv_c}{dR}\right)_0 = -(A+B).$$

#### 8.4.1.3.1 Constants of galactic rotation

The quantities $A$, $B$, and $R_0$, and the derived value of $v_c(R_0)$, are the (local) "constants of galactic rotation". The distance $R_0$ of the sun from the galactic center is most accurately determined from the spatial distribution of halo objects (RR Lyrae stars [71], globular clusters [1, 41]). The Oort constants $A$ and $B$ are usually determined from the radial velocities and proper motions of stars within a few kpc, for which the linearized formulae for $v_r$ and $v_t$ are adequate. The constant $A$ can be determined both from radial velocities and from proper motions, while the direct determination of the constant $B$ rests on proper motions alone. Using galactic dynamics (8.4.2.3.2), however, it is possible to derive $B$ indirectly from $A$ and the ratio between the stellar velocity dispersions, $\sigma_U/\sigma_V$:

$$B = -A\bigg/\left(\left(\frac{\sigma_U}{\sigma_V}\right)^2 - 1\right).$$

In Table 7 we list (a) the conventional values adopted by the IAU [68, 86], which should be used in general to ensure conformity, and (b) the currently most probable values according to Oort and Plaut [71], which may be used if strict accuracy is really necessary. Here, we shall always use the IAU values.

Table 7. Local constants of galactic rotation.

| Quantity | Unit | Conventional IAU values [68, 86] | Present most probable value (with mean error) [71] |
|---|---|---|---|
| Primary constants | | | |
| A | $(\mathrm{km\,s^{-1}})/\mathrm{kpc}$ | + 15 | + 16.9 (±0.9) |
| B | $(\mathrm{km\,s^{-1}})/\mathrm{kpc}$ | − 10 | − 9.0 (±1.5) |
| $R_0$ | kpc | 10 | 8.7 (±0.6) |
| Derived quantities | | | |
| $v_c(R_0)$ | $\mathrm{km\,s^{-1}}$ | 250 | 225 (±22) |
| $\left(\frac{dv_c}{dR}\right)_0$ | $(\mathrm{km\,s^{-1}})/\mathrm{kpc}$ | − 5 | − 7.9 |
| $\omega_c(R_0)$ | $(\mathrm{km\,s^{-1}})/\mathrm{kpc}$ | 25 | 25.9 |
| $\omega_c(R_0)$ | ″/a | 0.00527 | 0.00546 |
| $\left(\frac{d\omega_c}{dR}\right)_0$ | $(\mathrm{km\,s^{-1}})/\mathrm{kpc}^2$ | − 3 | − 3.9 |
| $T_{c,0}=2\pi/\omega_c(R_0)$ | a | $2.46\cdot10^8$ | $2.37\cdot10^8$ |
| $\varkappa_0=\sqrt{-4B(A-B)}$ | $(\mathrm{km\,s^{-1}})/\mathrm{kpc}$ | 31.6 | 30.5 |
| $T_{\varkappa,0}=2\pi/\varkappa_0$ | a | $1.94\cdot10^8$ | $2.01\cdot10^8$ |

### 8.4.1.3.2 Galactic rotation curve

While the local constants (see 8.4.1.3.1) describe the galactic rotation for $R\approx R_0$, we now consider the galactic rotation curve, $v_c(R)$, for "all" values of $R$. The derivation of the rotational velocities $v_c(R)$ are based on the first (general, not linearized) equation for $v_r$ in 8.4.1.3. We assume that the local constants, especially $\omega_c(R_0)$ and $R_0$, are already known (8.4.1.3.1). Within the limits of optical observations, i.e. $|R-R_0|\leqq$ a few kpc, it is possible to derive the rotation curve from individual objects, such as classical Cepheids [51] or open star clusters [46] or H II regions [39], for which the distance $r$ (and hence $R$) and the radial velocity $v_r$ can be measured.

$$\omega_c(R)-\omega_c(R_0)=v_r/(R_0\sin l).$$

For a larger range in $R$, it is necessary to rely on 21-cm line observations of H I. Since it is not possible to directly measure distances of H I, the tangential-point method is used:

From observations of other galaxies and theoretical considerations, it is known that, in most cases, $\omega_c(R)$ is a monotonic decreasing function of $R$. Then, viewing along a given line-of-sight in the direction $l$ the (absolutely) largest radial velocity, $v_{r,max}(l)$, occurs for the smallest value of $R$,

$$R_{min}(l)=R_0\sin l<R_0,$$

i.e. at the "tangential point". Since $v_{r,max}(l)$ can be directly determined from the "edge" of the 21-cm line profile, the rotation curve can be obtained, at least for $R<R_0$, from H I observations at various longitudes $l$ [$0°<l<90°$: (mainly) northern hemisphere observations; $270°<l<360°$; southern hemisphere observations]:

$$\omega_c(R_{min}(l))-\omega_c(R_0)=v_{r,max}(l)/R_{min}(l).$$

In Fig. 4, we reproduce the galactic rotation curve for $R<R_0$, obtained by Kerr, using the tangential-point method for 21-cm profiles of H I.

The rotation curve is rather uncertain for $R>R_0$. Similar to other galaxies, the galactic rotation curve may be rather flat ($v_c=$const.) even for very large values of $R$, instead of approaching the Keplerian asymptote, $v_c(R)\propto 1/\sqrt{R}$. Such behaviour can be explained by the existence of a large, massive "corona" of our Galaxy [73, 74, 32, 42].

In Table 8 we list the rotation curve according to [86] and [9].

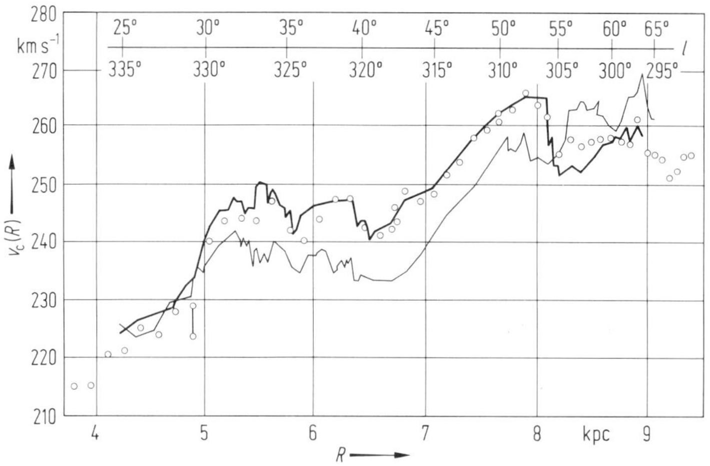


Fig. 4. Galactic rotation curve according to Kerr [49, 36]. Northern hemisphere observations ($l<90°$): heavy line; southern hemisphere observations ($l>270°$): thin line.

Table 8. Galactic rotation curve.
$R$ = distance from the galactic center
$v_c$ = circular velocity
[86]: mass model of M. Schmidt
[9]: analytic approximation of the galactic gravitational force by Contopoulos and Strömgren
Both the model and the analytic approximation are explained in section 8.4.2.2 and are based essentially on HI data similar to those shown in Fig. 4.

| $R$ [kpc] | $v_c(R)$ [km s$^{-1}$] | | $R$ [kpc] | $v_c(R)$ [km s$^{-1}$] | |
|---|---|---|---|---|---|
| | [86] | [9] | | [86] | [9] |
| 1 | 200 | | 11 | 244 | 244 |
| 2 | 187 | | 12 | 238 | 235 |
| 3 | 198 | 202 | 13 | 231 | 228 |
| 4 | 213 | 211 | 14 | 225 | 224 |
| 5 | 227 | 225 | 15 | 218 | (228) |
| 6 | 238 | 238 | 16 | 213 | |
| 7 | 247 | 247 | 17 | 207 | |
| 8 | 252 | 252 | 18 | 202 | |
| 9 | 253 | 253 | 19 | 197 | |
| 10 | 250 | 250 | 20 | 193 | |

### 8.4.1.3.3 Deviations from circular motion

The apparent rotation curve (Fig. 4) shows wavy irregularities ($\lesssim 10$ km s$^{-1}$). These are probably due to non-circular motions connected with the spiral structure of the Galaxy (see 8.4.2.4). Some authors have tried to fit the non-circular motions of H I or of young stars by simple sinusoidal expressions, based on the linear version of the density-wave theory. Such a procedure is, however, inadequate from a theoretical point of view [114, 57], since the motion of the gas is probably highly non-linear, including shock fronts [77], while young objects still reflect mainly the special conditions at their birth instead of the gravitational effects of a density wave.

The northern and southern data for the rotation curve (Fig. 4) differ systematically [50]. Probable explanations for this systematic difference are either large-scale perturbations of the galactic velocity field, e.g. by companion galaxies, or radial expansions of the Galaxy, or an error in the adopted LSR (see 8.4.1.2.3). Expanding motions are evident in the inner parts of the Galaxy, $R<4$ kpc, [70]. In the solar neighbourhood, expanding motions (positive "$K$ term") have been found for the constituents of Gould's belt [54, 59, 38, 97].

## 8.4.2 Dynamics

### 8.4.2.1 Basic concepts of galactic dynamics

Galactic dynamics aims at explaining the observed spatial distribution and kinematics of objects in the Galaxy by the action of gravitational forces. Magnetic and other forces are negligible in most cases. Our knowledge of the dynamical evolution of the Galaxy is still very limited.

The gravitational forces cannot be measured directly. They are usually derived indirectly from the observed kinematics and observed spatial distribution of suitable tracers, using theoretical concepts for the interpretation of these data. By such a procedure, a mass model of the Galaxy can be constructed which gives both the (total) density distribution $\varrho(\boldsymbol{r})$ and the corresponding gravitational force $\boldsymbol{K}(\boldsymbol{r})$ in the Galaxy.

It is customary in galactic dynamics to talk of "forces per unit mass" instead of accelerations and even to drop for brevity the words "per unit mass"!

If the gravitational forces are known, the galactic orbits of objects can be calculated (e.g. numerically) from the equations of motion. For those orbits which differ only slightly from a circular one, the epicyclic theory offers an adequate analytic description. The regular orbits of stars are perturbed by local irregularities in the gravitational field. The relaxation time for two-body encounters between stars is much larger than the age of the Galaxy. Other gravitational perturbations, e.g. due to large interstellar cloud complexes, are probably important even over periods of one or a few galactic rotations, thus explaining the observed increase of the stellar velocity dispersions with age.

In a statistical sense, a stellar system such as the Galaxy can be described by a distribution function $f(\boldsymbol{r}, \boldsymbol{v}, t)$ of matter in the six-dimensional phase space $\boldsymbol{r} \otimes \boldsymbol{v}$. Our empirical knowledge about the velocity distribution of common stars is up to now restricted to the small neighbourhood of the sun. Therefore, a reliable dynamical model, giving $f(\boldsymbol{r}, \boldsymbol{v})$ for our Galaxy is at present not available.

The spiral structure is probably caused by a density wave in the galactic disk, which is coincident with a wave in the gravitational potential of the Galaxy. The interstellar gas reacts strongly, with the occurrence of global shock fronts, to even small waves in the potential. The large increase in gas pressure and density should trigger star formation at the shock front very effectively, thus explaining the extreme concentration of young objects towards the spiral arms.

Reviews on galactic dynamics: [69, 63, 36, 65, 4].

### 8.4.2.2 Mass models and gravitational forces

#### 8.4.2.2.1 Mass models

A galactic mass model describes primarily the total mass density as a function of position, $\varrho(\boldsymbol{r})$. Some models are composed of separate components, representing different stellar populations. Usually, it is assumed that the model has rotational symmetry, $\varrho(R, z)$, and that the surfaces of constant density, at least in each component, are oblate spheroids of constant numerical eccentricity $e$. In this case, the density depends only on one variable, $a$,

Table 9. M. Schmidt's (1965) mass model of the Galaxy [86].

| | |
|---|---|
| Total mass of the Galaxy | $1.82 \cdot 10^{11}\,\mathfrak{M}_\odot$ |
| Mass of the central mass point | $0.07 \cdot 10^{11}\,\mathfrak{M}_\odot$ |
| Total mass of the spheroid | $1.75 \cdot 10^{11}\,\mathfrak{M}_\odot$ |
| mass of the inner part of the spheroid ($a < a_s$) | $0.82 \cdot 10^{11}\,\mathfrak{M}_\odot$ |
| mass of the outer part of the spheroid ($a > a_s$) | $0.93 \cdot 10^{11}\,\mathfrak{M}_\odot$ |
| Position of the sun | $R_0 = 10$ kpc, $z \approx 0$ |
| Flattening of the spheroid | $c/a = \sqrt{1-e^2} = 0.05$ |
| Eccentricity of the spheroid | $e = 0.998749$ |

Density in the spheroid
for $a < a_s$: $\varrho(a) = 3.930\,a^{-1} - 0.02489\,a$,
for $a > a_s$: $\varrho(a) = 1449.2\,a^{-4}$,
where $a_s e = 9.72$ kpc ($a_s = 9.73217$ kpc);
$a$ (which equals $R$ for $z=0$) measured in [kpc],
$\varrho$ in [$\mathfrak{M}_\odot/\text{pc}^3$].

Radial force in the galactic plane for $R < a_s$:
$K_R(R, z=0) = -30000\,R^{-2} - 10120.2 + 41.722\,R$
$R$ measured in [kpc], $K_R$ in [$(\text{km/s})^2/\text{kpc}$].

Table 10. Galactic quantities according to M. Schmidt's (1965) mass model.
$V_{esc}$=escape velocity; $\sigma_{V,\,crit}$=Toomre's critical velocity dispersion [94]; $\omega_c - \frac{\varkappa}{2}, \omega_c + \frac{\varkappa}{2}$=Lindblad resonance frequencies. (For other symbols, see 8.4.0)

| $R$ kpc | $\varrho$ $\mathfrak{M}_\odot$/pc$^3$ | $\mu$ $\mathfrak{M}_\odot$/pc$^2$ | $\Psi$ $10^4$ (km/s)$^2$ | $-K_R$ (km/s)$^2$kpc | $V_{esc}$ km s$^{-1}$ | $\sigma_{V,\,crit}$ km s$^{-1}$ |
|---|---|---|---|---|---|---|
| 3 | 1.235 | 646 | −13.75 | 13078 | 524 | 91 |
| 4 | 0.883 | 521 | −12.54 | 11328 | 501 | 89 |
| 5 | 0.662 | 421 | −11.45 | 10277 | 478 | 84 |
| 6 | 0.506 | 338 | −10.47 | 9452 | 458 | 78 |
| 7 | 0.387 | 267 | − 9.56 | 8688 | 437 | 71 |
| 8 | 0.292 | 206 | − 8.74 | 7919 | 418 | 64 |
| 9 | 0.213 | 155 | − 7.98 | 7111 | 400 | 57 |
| 10 | 0.145 | 114 | − 6.80 | 6250 | 369 | 52 |
| 11 | 0.099 | 86 | − 6.21 | 5419 | 352 | 47 |
| 12 | 0.070 | 66 | − 5.71 | 4704 | 337 | 42 |
| 13 | 0.051 | 52 | − 5.27 | 4104 | 325 | 37 |
| 14 | 0.038 | 41 | − 4.89 | 3602 | 312 | 34 |
| 15 | 0.029 | 34 | − 4.54 | 3181 | 301 | 31 |
| 16 | 0.022 | 28 | − 4.25 | 2827 | 291 | 28 |
| 17 | 0.017 | 23 | − 3.98 | 2527 | 282 | 26 |
| 18 | 0.014 | 20 | − 3.74 | 2271 | 273 | 24 |
| 19 | 0.011 | 17 | − 3.52 | 2051 | 265 | 22 |
| 20 | 0.009 | 14 | − 3.33 | 1861 | 257 | 20 |

Note: All quantities for $z=0$. For $v_c(R)$, see Table 8.

the semi-major axis of each surface, given by:

$$a^2 = R^2 + (1-e^2)^{-1} z^2 .$$

$a$ is equal to $R$ for points in the galactic plane. The density $\varrho(a)$ is normally derived empirically from the observed galactic rotation curve, $v_c(R)$, via the radial gravitational force in the galactic plane, $K_R(R, z=0) = -v_c^2(R)/R$:

$$v_c^2(R) = 4\pi G \sqrt{1-e^2} \int_0^R \frac{\varrho(a)a^2}{\sqrt{R^2 - a^2 e^2}}\, da .$$

Various techniques are used to invert this integral equation in order to find $\varrho(a)$ from the observed run of $v_c(R)$.

Mathematics of mass models: [75, 5, 85, 93, 31].

Mass models of the Galaxy: [86, 72, 45, 64].

The most widely used mass model of our Galaxy was constructed by M. Schmidt in 1965 [86]. It consists of a single inhomogeneous spheroid plus a mass point at the galactic center. Table 9 lists the basic properties of this model, while Table 10 gives derived numerical data for this model, mainly due to [7]. The main advantages of this model are its relative simplicity, its reasonably accurate reproduction of the observed rotation curve for a large range of $R$ and its acceptance by many authors which facilitates the comparison of their work. The main disadvantage of the model is that it ignores the multi-component structure of the Galaxy (disk, halo, perhaps a corona).

### 8.4.2.2.2 Galactic gravitational force $K_R$

$K_R = -\dfrac{\partial \Psi}{\partial R}$ is the radial component of the galactic gravitational force (per unit mass). The other component parallel to the galactic plane, $K_\theta = -\dfrac{1}{R}\dfrac{\partial \Psi}{\partial \theta}$, is zero for a strictly axisymmetric galaxy. Close to the galactic

Table 10, continued

| $\omega_c$ | $\varkappa$ (km s$^{-1}$)/kpc | $\omega_z$ | $T_c$ | $T_\varkappa$ $10^6$a | $T_z$ | $\omega_c - \frac{\varkappa}{2}$ (km s$^{-1}$)/kpc | $\omega_c + \frac{\varkappa}{2}$ | $R$ kpc |
|---|---|---|---|---|---|---|---|---|
| 66.02 | 102.98 | 254.72 | 93 | 60 | 24 | 14.53 | 117.51 | 3 |
| 53.21 | 84.99 | 214.85 | 115 | 72 | 29 | 10.72 | 95.71 | 4 |
| 45.33 | 72.58 | 186.00 | 136 | 85 | 33 | 9.04 | 81.62 | 5 |
| 39.68 | 62.82 | 162.89 | 155 | 98 | 38 | 8.27 | 71.09 | 6 |
| 35.22 | 54.44 | 142.98 | 174 | 113 | 43 | 8.00 | 62.44 | 7 |
| 31.46 | 46.74 | 124.84 | 195 | 131 | 49 | 8.09 | 54.83 | 8 |
| 28.10 | 39.20 | 107.41 | 219 | 157 | 57 | 8.50 | 47.70 | 9 |
| 25.00 | 31.62 | 89.90 | 246 | 194 | 68 | 9.19 | 40.81 | 10 |
| 22.19 | 26.49 | 75.05 | 277 | 232 | 82 | 8.95 | 35.44 | 11 |
| 19.79 | 22.81 | 63.56 | 310 | 269 | 97 | 8.39 | 31.20 | 12 |
| 17.76 | 19.97 | 54.54 | 346 | 308 | 113 | 7.78 | 27.75 | 13 |
| 16.04 | 17.70 | 47.33 | 383 | 347 | 130 | 7.19 | 24.89 | 14 |
| 14.56 | 15.85 | 41.47 | 422 | 388 | 148 | 6.64 | 22.49 | 15 |
| 13.29 | 14.31 | 36.66 | 462 | 429 | 168 | 6.14 | 20.45 | 16 |
| 12.19 | 13.01 | 32.65 | 504 | 472 | 188 | 5.69 | 18.70 | 17 |
| 11.23 | 11.90 | 29.27 | 547 | 516 | 210 | 5.28 | 17.18 | 18 |
| 10.39 | 10.94 | 26.41 | 591 | 562 | 233 | 4.92 | 15.86 | 19 |
| 9.64 | 10.10 | 23.95 | 637 | 608 | 257 | 4.59 | 14.69 | 20 |

plane, $K_R$ is nearly independent of $z$, since by symmetry: $(\partial K_R/\partial z)_{z=0}=0$. Hence $K_R(R, z=0)$ is often sufficient for studying the orbits of disk stars in the $(R, \theta)$-plane, as long as $|z|$ remains reasonably small, say less than a few hundred parsecs.

$K_R(R, z=0)$ can be determined from the observed rotation curve, $v_c(R)$:

$$K_R(R, z=0) = -v_c^2(R)/R.$$

Linear case:

For $|R-R_0| \ll R_0$, we can use a linear approximation:

$$K_R(R, z=0) = K_{R,0} + \left(\frac{\partial K_R}{\partial R}\right)_0 (R-R_0).$$

Both quantities $K_{R,0}$ and $\left(\frac{\partial K_R}{\partial R}\right)_0$ are fixed by the Oort constants of galactic rotation (IAU values adopted):

$$K_{R,0} = -R_0(A-B)^2 = -6250\,(\text{km/s})^2/\text{kpc}$$

$$\left(\frac{\partial K_R}{\partial R}\right)_0 = (A-B)(3A+B) = +\ 875\,(\text{km/s})^2/\text{kpc}^2.$$

The quality of this approximation can be judged from Fig. 5 (next page).

General case:

Contopoulos and Strömgren [9] gave a simple approximation formula for $K_R(R, z=0)$ which fits the observed rotation curve quite well in the range $3\,\text{kpc} < R < 14\,\text{kpc}$:

$$K_R(R, z=0) = -73340\,R^{-2} + 1581.8 - 3442.03\,R + 402.621\,R^2 - 12.9402\,R^3$$

$$(K_R \text{ measured in } [(\text{km/s})^2/\text{kpc}],\ R \text{ in } [\text{kpc}]).$$

For M. Schmidt's (1965) mass model, a corresponding formula is given in Table 9. Numerical values for $K_R(R, z=0)$ are shown in Fig. 5 and listed in Table 10.

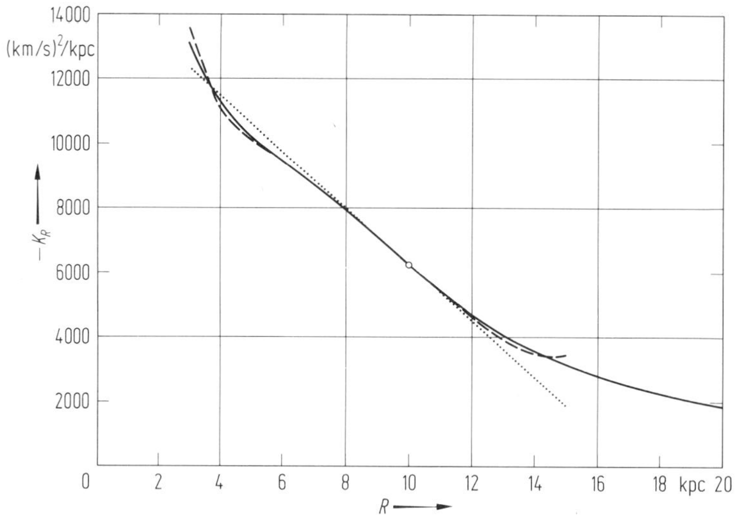


Fig. 5. Radial component of the galactic gravitational force, $K_R(R, z=0)$. Full line: M. Schmidt's mass model [86]; dashed line: formula of Contopoulos and Strömgren [9]; dotted line: local linear approximation $K_R = K_{R,0} + (\partial K_R/\partial R)_0 (R - R_0)$.

### 8.4.2.2.3 Galactic gravitational force $K_z$

$K_z = -\dfrac{\partial \Psi}{\partial z}$ is the vertical component of the galactic gravitational force (per unit mass). $K_z$ depends strongly on $z$, but also significantly on $R$.

Local $K_z$ for $R \approx R_0$:

The function $K_z(R_0, z)$ can be empirically determined from observations of adequate tracers, such as A dwarfs or K giants. It is necessary to observe their density along a line of sight towards the galactic poles, $\varrho(R_0, z)$, and the distribution of their $W$ velocities at $z=0$. For theory, see [69, 63]. In Table 11, we list the values of $K_z(R_0, z)$ derived by Oort [67], which are most widely used by other authors.

Table 11. Vertical component of the galactic gravitational force, $K_z$ (per unit mass) [67].

| $z$ [kpc] | 0.000 | 0.050 | 0.100 | 0.200 | 0.300 | 0.400 | 0.600 | 0.800 | 1.000 |
|---|---|---|---|---|---|---|---|---|---|
| $-K_z(R_0, z)$ [(km/s)$^2$/kpc] | 0 | 423 | 771 | 1321 | 1666 | 1904 | 2191 | 2360 | 2484 |

Linear approximation:

For small values of $|z|$, a linear approximation for $K_z$ can be used:

$$K_z(R, z) = -\omega_z^2(R) z .$$

The quantity $\omega_z(R)$ is directly connected with the total mass density $\varrho(R, z=0)$. From an application of the Poisson equation, $\Delta\Psi = 4\pi G\varrho$, we get:

$$\omega_z^2(R) = 4\pi G \varrho(R, z=0) + \frac{1}{R}\left(\frac{\partial}{\partial R}(R K_R)\right)_{z=0} .$$

For $R = R_0$, the last term is given by $2(A^2 - B^2)$. The run of $\omega_z(R)$ for M. Schmidt's (1965) mass model is given in Table 10. The local value of $\omega_z$ which corresponds to the IAU values of $A$ and $B$ and Oort's value of $\varrho_0$ (see 8.4.2.2.4) is

$$\omega_z(R_0) = C = 91.4\,(\mathrm{km\,s^{-1}})/\mathrm{kpc} .$$

Soviet astronomers [31] have proposed calling $C$ the "Kuzmin constant".

Locally, the force gradient in the $z$ direction is about ten times stronger than the radial force gradient:

$$|\mathrm{d}K_R/\mathrm{d}R|_0 : |\mathrm{d}K_z/\mathrm{d}z|_0 = (A - B)(3A + B) : C^2 = 1 : 9.5 .$$

### 8.4.2.2.4 Local mass density

The local mass density, $\varrho(R=R_0, z=0)=\varrho_0$, can be derived from the observed gradients of the galactic forces by applying Poisson's equation:

$$\Delta\Psi = -\frac{1}{R}\frac{\partial}{\partial R}(RK_R) - \frac{1}{R}\frac{\partial K_\theta}{\partial \theta} - \frac{\partial K_z}{\partial z} = 4\pi G\varrho .$$

For $R=R_0$, $z=0$ and $K_\theta=0$, we find (see 8.4.2.2.2/3):

$$4\pi G\varrho_0 = -\left(\frac{\partial K_z}{\partial z}\right)_0 - 2(A^2-B^2).$$

The first term on the right-hand side is much larger than the second one. Hence $\varrho_0$ is essentially determined by the local gradient of the vertical force. From the observed vertical force $K_z$ (see 8.4.2.2.3), Oort [67] derived

$$\varrho_0 = 0.15\,\mathfrak{M}_\odot/\text{pc}^3 .$$

This value is widely used in galactic dynamics. Other authors [44, 52] have found values for $\varrho_0$ between 0.08 and 0.21 $\mathfrak{M}_\odot/\text{pc}^3$.

In Table 12, we compare the gravitationally determined total mass density $\varrho_0$ of all local material with the contributions of known objects. The difference is probably significant (local "missing mass"). The missing mass is very probably not due to "low-velocity M dwarfs" [33, 104]. It cannot be excluded that hitherto undetected "black dwarfs", i.e. cool, degenerate, low-mass objects which never started hydrogen burning, may account for most of the missing mass.

The local mass density of halo stars is rather uncertain [87, 103, 109, 110, 30].

For the local value of the surface mass density, $\mu(R=R_0)=\mu_0$, the Schmidt model (Table 10) gives $\mu_0=114\,\mathfrak{M}_\odot/\text{pc}^2$, but see Toomre [95].

Table 12. Local mass density $\varrho$.

| Objects | Density $\varrho$ [$\mathfrak{M}_\odot/\text{pc}^3$] | Remarks | Ref. |
|---|---|---|---|
| Stars on or near the main sequence: | | Based on the luminosity function derived by Wielen for nearby stars with $M_v<13.5$ and Luyten's luminosity function for fainter dwarfs. | 107, 47 |
| $M_v< 9.5$ | 0.020 | | |
| $9.5\leqq M_v<13.5$ | 0.014 | | |
| $13.5\leqq M_v$ | 0.007 | | |
| Sum of main sequence: | 0.041 | | |
| Giants and subgiants | 0.001 | 27 stars within 20 pc | 107, 47 |
| White dwarfs (known) | 0.007 | 5 stars within 5 pc | 107, 47 |
| Sum of stars: | $\varrho_{\text{stars}}=0.049$ | | 107, 47 |
| HI | 0.021 | from 21-cm observations | 84 |
| HII | $\approx 0$ | | |
| $H_2$ | 0.007 | from UV observations of interstellar $H_2$ absorption | 84 |
| Other elements | 0.010 | from cosmic abundance | 44 |
| Dust | $\lesssim 0.001$ | $\approx 1\%$ of gas | |
| Sum of interstellar matter (i.m.): | $\varrho_{\text{i.m.}}=0.039$ | | |
| Sum of stars and interstellar matter: | $\varrho_{\text{stars}}+\varrho_{\text{i.m.}}=0.088$ | | |
| Gravitationally determined total density: | $\varrho_0=0.150$ | | 67 |
| "Missing mass" | $\varrho_0-(\varrho_{\text{stars}}+\varrho_{\text{i.m.}})=0.062$ | | |

#### 8.4.2.2.5 Galactic center

The interpretation of the observed spatial distributions and kinematics of stars and gas in the galactic center region is at present still rather uncertain. In a summary and review, Oort [70] has prepared Table 13, which lists important properties of the central regions of the Galaxy.

Table 13. Galactic center region [70].
All quantities for $z=0$. The values for $R=0.1$ and 1.0 pc may be too conservative.
$R$ = distance from center; $v_c$ = circular velocity
$M_R$ = mass within the sphere of radius $R$; $T_c$ = circular period (rotation)
$\Psi$ = gravitational potential; $\varrho$ = volume density

| $R$ pc | $M_R$ $10^6\,\mathfrak{M}_\odot$ | $\Psi(R)-\Psi(0)$ $10^4\,(\mathrm{km/s})^2$ | $v_c$ km s$^{-1}$ | $T_c$ $10^6$ a | $\varrho$ $\mathfrak{M}_\odot/\mathrm{pc}^3$ |
|---|---|---|---|---|---|
| 0.1 | 0.3 | 0.64 | 114 | 0.006 | $3\cdot10^7$ |
| 1.0 | 4.4 | 4.4 | 138 | 0.045 | $4\cdot10^5$ |
| 5 | 30 | 8.2 | 161 | 0.20 | $2\cdot10^4$ |
| 10 | 70 | 10.0 | 174 | 0.36 | $7\cdot10^3$ |
| 20 | 160 | 12.2 | 186 | 0.67 | $1.9\cdot10^3$ |
| 50 | 490 | 15.6 | 206 | 1.5 | |
| 100 | 920 | 18.6 | 200 | 3.1 | $1.1\cdot10^2$ |
| 200 | 2500 | 21.9 | 233 | 5.4 | |
| 500 | 8300 | 27.8 | 268 | 12 | 6 |
| 1000 | 16400 | 32.8 | 266 | 24 | |

### 8.4.2.3 Stellar orbits

#### 8.4.2.3.1 Unrestricted orbits

The orbit of a star can be obtained from its equations of motion:

$$\ddot{\mathbf{r}} = -\operatorname{grad}\Psi.$$

In cylindrical coordinates, measured in an inertial system comoving with the galactic center, the equations of motion are given by:

$$\ddot{R} - R\dot{\theta}^2 = -\frac{\partial\Psi}{\partial R} = K_R,$$

$$R\ddot{\theta} + 2\dot{R}\dot{\theta} = \frac{1}{R}(R^2\dot{\theta})^{\cdot} = -\frac{1}{R}\frac{\partial\Psi}{\partial\theta} = K_\theta,$$

$$\ddot{z} = -\frac{\partial\Psi}{\partial z} = K_z.$$

For an axisymmetric and stationary gravitational potential, $\Psi=\Psi(R,z)$, two integrals of motion along each individual stellar orbit are analytically known: energy (per unit mass)

$$E = \tfrac{1}{2}(\dot{R}^2 + (R\dot{\theta})^2 + \dot{z}^2) + \Psi = \text{const.}$$

$z$-component of the angular momentum (per unit mass)

$$J_z = R^2\dot{\theta} = \text{const.}$$

Theoretical investigations and numerical calculations of stellar orbits have shown that a "third integral" is important for the motions of essentially all disk stars and even many halo stars. Literature on the "third integral" of stellar orbits: [8, 61, 65].

Orbits in the general case can only be obtained by numerical integration of the equations of motion [66, 63]. For plane galactic orbits which pass at present through the position of the sun, extensive tables have been published [9]. In Table 14, we list the maximum and minimum distances from the galactic center of plane orbits ($z\equiv0$, $W\equiv0$) of stars which pass now with the velocity components $U$ and $V$ (measured relative to the local circular velocity, see 8.4.1.2.3) through the position of the sun ($R_0 = 10$ kpc). The data are taken from [9] and are based on the radial force $K_R$ given by Contopoulos and Strömgren [9], see 8.4.2.2.2.

Table 14. Maximum and minimum distance $R$ from the galactic center of various stellar orbits as a function of the present space velocity (with respect to local circular velocity!) [9].
$U$, $V$ = galactic components of space velocity, see 8.4.1.2.1

| $\|U\|$ [km s$^{-1}$] / $V$ [km s$^{-1}$] ($R_{max}$ [kpc] / $R_{min}$ [kpc]) | 0 | 10 | 20 | 30 | 40 | 50 | 60 | 70 | 80 | 90 | 100 |
|---|---|---|---|---|---|---|---|---|---|---|---|
| −60 | 10.00 | 10.02 | 10.08 | 10.17 | 10.30 | 10.47 | 10.67 | 10.91 | 11.19 | 11.52 | 11.88 |
| | 6.06 | 6.06 | 6.03 | 6.00 | 5.95 | 5.90 | 5.83 | 5.76 | 5.68 | 5.60 | 5.52 |
| −50 | 10.00 | 10.02 | 10.09 | 10.20 | 10.35 | 10.54 | 10.77 | 11.04 | 11.36 | 11.72 | 12.13 |
| | 6.56 | 6.55 | 6.52 | 6.47 | 6.41 | 6.34 | 6.26 | 6.17 | 6.08 | 5.98 | 5.89 |
| −40 | 10.00 | 10.03 | 10.11 | 10.24 | 10.42 | 10.64 | 10.91 | 11.22 | 11.58 | 11.99 | 12.45 |
| | 7.09 | 7.08 | 7.04 | 6.98 | 6.90 | 6.81 | 6.71 | 6.60 | 6.49 | 6.37 | 6.26 |
| −30 | 10.00 | 10.04 | 10.14 | 10.30 | 10.52 | 10.79 | 11.10 | 11.47 | 11.88 | 12.35 | 12.88 |
| | 7.68 | 7.66 | 7.61 | 7.52 | 7.41 | 7.29 | 7.16 | 7.03 | 6.90 | 6.76 | 6.63 |
| −20 | 10.00 | 10.05 | 10.20 | 10.41 | 10.69 | 11.02 | 11.39 | 11.82 | 12.30 | 12.84 | 13.44 |
| | 8.34 | 8.31 | 8.22 | 8.09 | 7.94 | 7.78 | 7.62 | 7.46 | 7.30 | 7.14 | 6.99 |
| −10 | 10.00 | 10.09 | 10.32 | 10.63 | 10.98 | 11.38 | 11.83 | 12.34 | 12.90 | 13.51 | 14.17 |
| | 9.10 | 9.02 | 8.86 | 8.66 | 8.45 | 8.25 | 8.05 | 7.86 | 7.68 | 7.51 | 7.34 |
| 0 | 10.00 | 10.33 | 10.69 | 11.08 | 11.51 | 11.99 | 12.52 | 13.10 | 13.72 | 14.37 | 15.01 |
| | 10.00 | 9.70 | 9.41 | 9.15 | 8.90 | 8.66 | 8.44 | 8.23 | 8.03 | 7.85 | 7.66 |
| +10 | 11.14 | 11.26 | 11.55 | 11.95 | 12.42 | 12.95 | 13.53 | 14.13 | 14.74 | | |
| | 10.00 | 9.91 | 9.71 | 9.47 | 9.23 | 9.00 | 8.77 | 8.55 | 8.35 | | |
| +20 | 12.66 | 12.74 | 12.98 | 13.33 | 13.77 | 14.26 | 14.76 | 15.26 | 15.74 | | |
| | 10.00 | 9.95 | 9.83 | 9.65 | 9.45 | 9.24 | 9.02 | 8.81 | 8.61 | | |

### 8.4.2.3.2 Epicyclic orbits

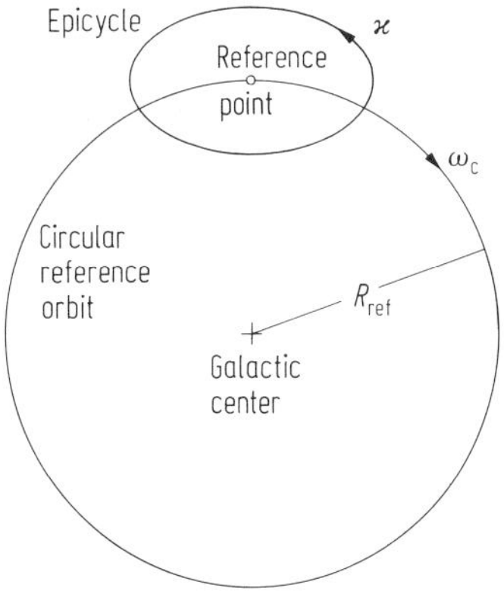


Fig. 6. Elliptic epicycle and circular reference orbit (schematically).

The orbits of most disk stars deviate only slightly from a circular one. In this case, the epicyclic theory [58, 69, 63] gives a rather accurate, analytical description of the orbit.

According to the epicyclic theory, a star carries out (small) harmonic oscillations around its "reference point" which moves on a suitably chosen circular orbit of radius $R_{ref}$ (Fig. 6). The radius $R_{ref}$ is chosen such that the angular momenta $J_z$ (and hence the periods in $\theta$) of the reference point and of the actual stellar orbit are equal,

$R_{\rm ref}^2\omega_c(R_{\rm ref})=R^2\dot\theta$. The harmonic oscillations in $\theta$ and $R$ have the common epicyclic frequency $\varkappa$, given by

$$\varkappa(R)=2\omega_c\sqrt{1+\frac{1}{2}\frac{R}{\omega_c}\frac{d\omega_c}{dR}}\,.$$

$\varkappa$ is usually larger than $\omega_c$ (see Table 10). The local value of $\varkappa$ is

$$\varkappa_0=\sqrt{-4B(A-B)}=31.6\,(\mathrm{km\,s^{-1}})/\mathrm{kpc}\,.$$

The axial ratio of the elliptic epicycle in the $(R,\theta)$-plane is $\varkappa/(2\omega_c)$, or locally $\sqrt{-B/(A-B)}=0.63=1/1.58$.

We consider a star which passes now ($t=0$) through the position of the sun ($R=R_0$, $\theta=0$) with the velocities $U$, $V$, $W$ (measured relative to the local circular velocity, see 8.4.1.2.3). Its epicyclic orbit is then given by:

$$R(t)-R_0=\frac{V}{-2B}-\frac{U}{\varkappa_0}\sin\varkappa_0 t-\frac{V}{-2B}\cos\varkappa_0 t$$

$$R_0(\theta(t)-\omega_c(R_0)t)=-\frac{A}{-B}Vt+\frac{U}{-2B}(1-\cos\varkappa_0 t)+\frac{\varkappa_0}{4B^2}V\sin\varkappa_0 t$$

$$z(t)=\frac{W}{\omega_{z,0}}\sin\omega_{z,0}t\,.$$

The velocity components $\tilde U$, $\tilde V$, $\tilde W$, measured with respect to the circular velocity at the actual position of the star at time $t$, are given by:

$$\tilde U(t)=U\cos\varkappa_0 t-\frac{\varkappa_0}{-2B}V\sin\varkappa_0 t$$

$$\tilde V(t)=\frac{-2B}{\varkappa_0}U\sin\varkappa_0 t+V\cos\varkappa_0 t$$

$$\tilde W(t)=W\cos\omega_{z,0}t\,.$$

The radius of the reference orbit in [kpc] is

$$R_{\rm ref}=R_0+\frac{V}{-2B}$$

$$=\left(R_0+\frac{V}{20}\right),\ \text{with } V \text{ in } [\mathrm{km\,s^{-1}}]$$

The semi-minor axis of the epicycle in [pc] is

$$a_R=\sqrt{\left(\frac{U}{\varkappa_0}\right)^2+\left(\frac{V}{2B}\right)^2}$$

$$=31.6\,U\sqrt{1+2.5\left(\frac{V}{U}\right)^2},\quad \text{with}\quad U,V \text{ in } [\mathrm{km\,s^{-1}}]\,.$$

The semi-major axis of the epicycle is

$$a_\theta=a_R\sqrt{(A-B)/(-B)}=1.58\,a_R\,.$$

The amplitude of the $z$-motion in [pc] is

$$a_z=|W|/\omega_{z,0}=10.9\,|W|\,,\quad \text{with } W \text{ in } [\mathrm{km\,s^{-1}}]\,.$$

The largest velocities along the epicycle are

$$\tilde U_{\max}=\varkappa_0 a_R\,,$$

$$\tilde V_{\max}=\sqrt{-B/(A-B)}\,\tilde U_{\max}=-2Ba_R\,,$$

$$\tilde W_{\max}=W\,.$$

For example, the epicyclic orbit of the sun, using $U=+9\,\mathrm{km\,s^{-1}}$, $V=+12\,\mathrm{km\,s^{-1}}$, $W=+7\,\mathrm{km\,s^{-1}}$ (see 8.4.1.2.3) and $R_0=10\,\mathrm{kpc}$, $A=+15\,(\mathrm{km\,s^{-1}})/\mathrm{kpc}$, $B=-10\,(\mathrm{km\,s^{-1}})/\mathrm{kpc}$, $\omega_{z,0}=91.4\,(\mathrm{km\,s^{-1}})/\mathrm{kpc}$ (see 8.4.1.3.1 and 8.4.2.2.3), is characterized by:

$$\begin{aligned} R_{\mathrm{ref}} &= 10\,\mathrm{kpc}+0.600\,\mathrm{kpc} & R_{\max} &= 11.264\,\mathrm{kpc}\\ &= 10.600\,\mathrm{kpc} & R_{\min} &= 9.936\,\mathrm{kpc}\\ a_R &= 0.664\,\mathrm{kpc} & \tilde{U}_{\max} &= 21.0\,\mathrm{km\,s^{-1}}\\ a_\theta &= 1.050\,\mathrm{kpc} & \tilde{V}_{\max} &= 13.3\,\mathrm{km\,s^{-1}}\\ a_z &= 0.077\,\mathrm{kpc} & \tilde{W}_{\max} &= 7.0\,\mathrm{km\,s^{-1}}. \end{aligned}$$

The accuracy of epicyclic orbits, in comparison to numerically integrated orbits which do not use the epicyclic approximation $a_R \ll R_0$, has been discussed in [9]. Table 15 gives such a comparison for $V=W=0$ and various values of $U$, taken from [9]. In Table 16, a similar comparison is given for the $z$-motions of stars with $U=V=0$ for various values of $W$, taken from [69]; in this case, however, the epicyclic theory uses $K_z(R_0, z) = -\omega_{z,0}^2 z$, while the numerically obtained $z$-motions use $K_z$ according to Table 11.

Table 15. Comparison between results of the epicyclic theory and numerically integrated orbits for $V=W=0$ [9].
$U$ = galactic component of space velocity towards the galactic center (see 8.4.1.2.1)
$P_R$ = period of radial oscillation
epi = epicyclic theory, num = numerical integration

| $\|U\|$ km s$^{-1}$ | $R_{\max}-R_0$ kpc | | $R_{\min}-R_0$ kpc | | $P_R$ $10^6$a | |
|---|---|---|---|---|---|---|
| | epi | num | epi | num | epi | num |
| 0 | 0.00 | 0.00 | −0.00 | −0.00 | 194 | 194 |
| 10 | 0.32 | 0.33 | −0.32 | −0.30 | 194 | 195 |
| 20 | 0.63 | 0.69 | −0.63 | −0.59 | 194 | 197 |
| 30 | 0.95 | 1.08 | −0.95 | −0.85 | 194 | 200 |
| 40 | 1.27 | 1.51 | −1.27 | −1.10 | 194 | 204 |
| 50 | 1.58 | 1.99 | −1.58 | −1.34 | 194 | 210 |
| 60 | 1.90 | 2.52 | −1.90 | −1.56 | 194 | 216 |

Table 16. Comparison between results of the epicyclic theory with numerically integrated orbits for $U=V=0$ [69].
$W$ = galactic component of space velocity vertical to the galactic plane
$P_z$ = period of vertical oscillation
epi = epicyclic theory, num = numerical integration

| $\|W\|$ km s$^{-1}$ | $z_{\max}$ kpc | | $P_z$ $10^6$ a | |
|---|---|---|---|---|
| | epi | num | epi | num |
| 0 | 0.000 | 0.000 | 67 | 67 |
| 10 | 0.109 | | 67 | 68 |
| 20 | 0.219 | | 67 | 79 |
| 30 | 0.328 | | 67 | |
| 40 | 0.438 | | 67 | 92 |
| 50 | 0.547 | | 67 | |
| 60 | 0.656 | 1.000 | 67 | |

### 8.4.2.3.3 Relaxation and diffusion

Beside the regular gravitational field of the Galaxy, described by $\boldsymbol{K}(\boldsymbol{r})$ and caused by the smoothed-out distribution of matter $\varrho(\boldsymbol{r})$, there must be an irregular field which causes relaxation effects and a diffusion of the regular stellar orbits. The importance and the source of this irregular field are at present not well established.

The relaxation due to two-body encounters between stars is negligible in galaxies [6, 63], because the corresponding relaxation time is about $10^{14}$ years and hence much longer than the age of the universe ($\approx 10^{10}$ years). Gravitational encounters between stars and large interstellar (molecular) cloud complexes may be important [90, 91, 92]. Assuming stochastic perturbations of the regular orbits of stars, it is possible to derive empirically the corresponding diffusion coefficient in the Galaxy from the observed age-dependence of the stellar velocity dispersion [111]. Table 17 lists the expected root-mean-square differences between the irregular and regular orbits, based on a constant diffusion coefficient in each velocity component, $D_U=D_V=D_W=2.0\cdot10^{-7}\,(\mathrm{km/s})^2/\mathrm{a}$ [111]. These differences describe the uncertainty of galactic orbits of stars due to the irregular gravitational field. Other explanations proposed for the increase of the stellar velocity dispersion with age [53] would lead to a much smaller diffusion coefficient.

Table 17. Diffusion of stellar orbits [111].

| Age $\tau$ | Expected rms difference between the irregular and regular stellar orbits | | | | | | | |
|---|---|---|---|---|---|---|---|---|
| a | $\Delta U$ | $\Delta V$ | $\Delta W$ | $\Delta \boldsymbol{v}$ | $\Delta R$ | $R\Delta\theta$*) | $\Delta z$ | $\Delta \boldsymbol{r}$*) |
| | $\mathrm{km\,s^{-1}}$ | | | | kpc | | | |
| $1\cdot10^{7}$ | 1.5 | 1.4 | 1.2 | 2.4 | 0.008 | 0.008 | 0.008 | 0.014 |
| $2\cdot10^{7}$ | 2.2 | 1.9 | 1.3 | 3.2 | 0.025 | 0.023 | 0.017 | 0.038 |
| $5\cdot10^{7}$ | 4.2 | 2.7 | 2.3 | 5.5 | 0.108 | 0.091 | 0.025 | 0.143 |
| $1\cdot10^{8}$ | 5.9 | 3.8 | 3.2 | 7.7 | 0.300 | 0.392 | 0.035 | 0.495 |
| $2\cdot10^{8}$ | 8.4 | 5.4 | 4.5 | 10.9 | 0.410 | 1.356 | 0.050 | 1.418 |
| $5\cdot10^{8}$ | 13.3 | 8.5 | 7.1 | 17.3 | 0.67 | 4.47 | 0.078 | 4.52 |
| $1\cdot10^{9}$ | 18.8 | 12.0 | 10.1 | 24.5 | 0.92 | 12.7 | 0.111 | 12.8 |
| $2\cdot10^{9}$ | 26.7 | 16.9 | 14.3 | 34.7 | 1.30 | 35.8 | 0.156 | 35.8 |
| $5\cdot10^{9}$ | 42.2 | 26.7 | 22.5 | 54.8 | 2.09 | 141 | 0.247 | 141 |
| $1\cdot10^{10}$ | 59.7 | 37.7 | 31.9 | 77.5 | 2.94 | 399 | 0.349 | 399 |

*) Large values of $R\Delta\theta$ and $\Delta \boldsymbol{r}$ indicate an uncertainty in the number of revolutions around the galactic center ($2\pi R_0 = 62.8$ kpc).

### 8.4.2.4 Density-wave theory of the spiral structure

Density waves are probably the most general phenomenon producing spiral structure in disk galaxies. The density-wave theory is able to give a fairly successful interpretation of the observed spiral structure and of the related kinematics in external galaxies and in our Galaxy. The density-wave theory (in its present form mainly due to C.C. Lin and his coworkers) and its application are described in many recent reviews [108, 113, 114, 55, 56, 88, 78, 48, 60, 96, 79, 80] and further literature cited in these reviews. For example, [108] is a review on an introductory level; [96] gives a critical discussion of the theory; [56] is most detailed in giving equations and tabular material.

For our Galaxy, it is rather difficult to determine accurately the parameters describing the density wave. The following simple model of the galactic density wave [119, 105, 114] has often been used for studying the stellar orbits or gaseous streamlines in our Galaxy.

Pattern speed (rotational frequency of the rigidly rotating wave):

$$\Omega_p = 13.5\,(\mathrm{km\,s^{-1}})/\mathrm{kpc}\,.$$

This pattern speed corresponds to the following resonances, using M. Schmidt's (1965) mass model:

| | | | |
|---|---|---|---|
| Inner Lindblad resonance | $\left(\Omega_p = \omega_c - \frac{\varkappa}{2}\right)$ | at | $R = 3.2$ kpc |
| Corotation resonance | $(\Omega_p = \omega_c)$ | at | $R = 15.8$ kpc |
| Outer Lindblad resonance | $\left(\Omega_p = \omega_c + \frac{\varkappa}{2}\right)$ | at | $R = 21$ kpc. |

Local kinematical data:

| | | |
|---|---|---|
| Relative linear speed | $R_0(\Omega_p - \omega_c(R_0))$ | $= -115\ \mathrm{km\,s^{-1}}$ |
| Relative frequency | $\nu_0 = 2(\Omega_p - \omega_c(R_0))/\varkappa_0$ | $= -0.73$ |
| Period of wave passage | $\pi/(\Omega_p - \omega_c(R_0))$ | $= 2.7\cdot10^{8}$ a. |

Gravitational potential of the density wave:

$$\Psi_1(R, \theta, z=0) = A_1 \cos(2(\Omega_p t - \theta) + \Phi(R))\,.$$

In the linear theory, the density response $\varrho_1$ is proportional to $-\Psi_1$, i.e. minima of $\Psi_1$ correspond to maxima of $\varrho_1$.

The spiral pattern is described by the radial phase function:

$$\Phi(R) = -2\cot i \ln(R/R_0) + \Phi_0,$$

i.e. a two-armed logarithmic spiral with a constant pitch angle

$$i = 6\overset{\circ}{.}2.$$

The phase constant

$$\Phi_0 = -0.3777\,\text{rad} = -21\overset{\circ}{.}64$$

corresponds to a minimum in the potential (spiral arm) at $R = 8.26$ kpc and $l = 0$, if our sun is at present ($t = 0$) located at $R_0 = 10$ kpc and $\theta_0 = 0$. The constant amplitude $A_1$ of the wave in the potential can be expressed as

$$A_1 = \tfrac{1}{2} f_{R,0} v_c^2(R_0) \tan i = 169.7\,(\text{km/s})^2,$$

where

$$f_{R,0} = 0.05$$

is the ratio between the amplitude of the radial force due to the wave and the mean force at $R = R_0$.

The radial and tangential forces, $K_{R,1}$ and $K_{\theta,1}$, produced by the potential $\Psi_1$ of the density wave are given by

$$K_{R,1} = -\frac{\partial \Psi_1}{\partial R} = -f_{R,0} v_c^2(R_0) \frac{1}{R} \sin(2(\Omega_p t - \theta) + \Phi(R))$$

$$K_{\theta,1} = -\frac{1}{R}\frac{\partial \Psi_1}{\partial \theta} = K_{R,1} \tan i.$$

These forces may be used in the equations of motion for stellar orbits (see 8.4.2.3.1) in addition to the axisymmetric forces, $K_R$ (given in 8.4.2.2.2) and $K_\theta = 0$.

## 8.4.3 References for 8.4


0 Galactic Structure (Blaauw, A., Schmidt, M., eds.) = Stars and Stellar Systems **5**, The University of Chicago Press (1965).
1 Arp, H.C. in: [0] p. 401.
2 Augensen, H.J., Buscombe, W.: Astrophys. Space Sci. **59** (1978) 35.
3 Blaauw, A.: Bull. Astron. Inst. Neth. **11** (1952) 459.
4 Bok, B.J. in: Galactic Astronomy (Chiu, H.-Y., Muriel, A., eds.), Gordon and Breach, New York Vol. 1 (1970) p. 1.
5 Burbidge, E.M., Burbidge, G.R.: Stars and Stellar Systems **9** (1975) 81; cf. [0].
6 Chandrasekhar, S.: Principles of Stellar Dynamics, Dover Publ., New York (1960) (Reprint of edition 1942).
7 Contopoulos, G. in: Dynamical Structure and Evolution of Stellar Systems (Martinet, L., Mayor, M., eds.), Geneva Observatory (1973) 51.
8 Contopoulos, G.: Int. Astron. Union Symp. **69** (1975) 209.
9 Contopoulos, G., Strömgren, B.: Tables of Plane Galactic Orbits, Goddard Space Flight Center, NASA (1965).
10 Crovisier, J.: Astron. Astrophys. **70** (1978) 43.
11 Cudworth, K.M.: Astron. J. **79** (1974) 1384.
12 Dean, C.A.: Astron. J. **81** (1976) 364.
13 Delhaye, J. in: [0] p. 61.
14 Eggen, O.J.: R. Obs. Bull. No. 41 (1961).
15 Eggen, O.J.: R. Obs. Bull. No. 51 (1962).
16 Eggen, O.J.: Astron. J. **68** (1963) 697.
17 Eggen, O.J.: R. Obs. Bull. No. 84 (1964).
18 Eggen, O.J. in: [0] p. 111.
19 Eggen, O.J.: R. Obs. Bull. No. 125 (1966).
20 Eggen, O.J.: Astrophys. J. Suppl. **16** (1968) 49.
21 Eggen, O.J.: Vistas Astron. **12** (1970) 367.
22 Eggen, O.J.: Astrophys. J. Suppl. **22** (1971) 389.

23 Eggen, O.J.: Publ. Astron. Soc. Pac. **84** (1972) 757.
24 Eggen, O.J.: Publ. Astron. Soc. Pac. **86** (1974) 129.
25 Eggen, O.J.: Publ. Astron. Soc. Pac. **86** (1974) 697.
26 Eggen, O.J.: Publ. Astron. Soc. Pac. **87** (1975) 37.
27 Eggen, O.J.: Publ. Astron. Soc. Pac. **89** (1977) 187.
28 Eggen, O.J.: Publ. Astron. Soc. Pac. **89** (1977) 205.
29 Eggen, O.J.: Astrophys. J. **229** (1979) 158.
30 Eggen, O.J.: Astrophys. J. **230** (1979) 786.
31 Einasto, J. in: Stars and the Milky Way System (Mavridis, L.N., ed.) Berlin: Springer (1974) 291.
32 Einasto, J., Kaasik, A., Saar, E.: Nature **250** (1974) 309.
33 Faber, S.M., Burstein, D., Tinsley, B.M., King, I.R.: Astron. J. **81** (1976) 45.
34 Feast, M.W.: Mon. Not. R. Astron. Soc. **125** (1963) 367.
35 Feast, M.W., Woolley, R., Yilmaz, N.: Mon. Not. R. Astron. Soc. **158** (1972) 23.
36 Freeman, K.C.: Stars and Stellar Systems **9** (1975) 409; cf. [0].
37 Fricke, W.: Veröffentl. Astron. Rechen-Inst. Heidelberg No. 28 (1977).
38 Frogel, J.A., Stothers, R.: Astron. J. **82** (1977) 890.
39 Georgelin, Y.M., Georgelin, Y.P.: Astron. Astrophys. **49** (1976) 57.
40 Gliese, W.: Veröffentl. Astron. Rechen-Inst. Heidelberg No. 22 (1969).
41 Harris, W.E.: Astron. J. **81** (1976) 1095.
42 Hartwick, F.D.A., Sargent, W.L.W.: Astrophys. J. **221** (1978) 512.
43 Hemenway, M.K.: Astron. J. **80** (1975) 199.
44 Hill, G., Hilditch, R.W., Barnes, J.V.: Mon. Not. R. Astron. Soc. **186** (1979) 813.
45 Innanen, K.A.: Astrophys. Space Sci. **22** (1973) 393.
46 Jackson, P.D., Fitzgerald, M.P., Moffat, A.F.J.: Int. Astron. Union Symp. **84** (1979) 221.
47 Jahreiß, H.: Thesis, Univ. Heidelberg (1974).
48 Kalnajs, A.J.: Int. Astron. Union Symp. **77** (1978) 113.
49 Kerr, F.J.: Int. Astron. Union Symp. **20** (1964) 81.
50 Kerr, F.J., Westerhout, G. in: [0] p. 167.
51 Kraft, R.P., Schmidt, M.: Astrophys. J. **137** (1963) 249.
52 Krisciunas, K.: Astron. J. **82** (1977) 195.
53 Larson, R.B.: Mon. Not. R. Astron. Soc. **186** (1979) 479.
54 Lesh, J.R.: Astrophys. J. Suppl. **17** (1968) 371.
55 Lin, C.C. in: Structure and Evolution of Galaxies (Setti, G., ed.), Reidel Publ. Co., Dordrecht (1975) 119.
56 Lin, C.C., Shu, F.H. in: Galactic Astronomy (Chiu, H.-Y., Muriel, A., eds.), Gordon and Breach, New York Vol. 2 (1970) p. 1.
57 Lin, C.C., Yuan, C., Roberts, W.W.: Astron. Astrophys. **69** (1978) 181.
58 Lindblad, B.: Hdb. Physik **53** (1959) 21.
59 Lindblad, P.O.: Highlights of Astronomy **3** (1974) 381.
60 Lindblad, P.O.: Int. Astron. Union Symp. **58** (1974) 339.
61 Martinet, L., Mayer, F.: Astron. Astrophys. **44** (1975) 45.
62 Mayor, M.: Astron. Astrophys. **32** (1974) 321.
63 Mihalas, D., Routly, P.M.: Galactic Astronomy, W.H. Freeman and Co., San Francisco and London (1968).
64 Miyamoto, M., Nagai, R.: Publ. Astron. Soc. Jpn. **27** (1975) 533.
65 Ogorodnikov, K.F.: Dynamics of Stellar Systems, Pergamon Press, Oxford (1965).
66 Ollongren, A. in: [0] p. 501
67 Oort, J.H.: Bull. Astron. Inst. Neth. **15** (1960) 45.
68 Oort, J.H.: Int. Astron. Union Symp. **20** (1964) 1 [see also Int. Astron. Union Inform. Bull. No. 11 (1963) 11].
69 Oort, J.H. in: [0] p. 455.
70 Oort, J.H.: Annu. Rev. Astron. Astrophys. **15** (1977) 295.
71 Oort, J.H., Plaut, L.: Astron. Astrophys. **41** (1975) 71.
72 Ostriker, J.P., Caldwell, J.A.R.: Int. Astron. Union Symp. **84** (1979) 441.
73 Ostriker, J.P., Peebles, P.J.E.: Astrophys. J. **186** (1973) 467.
74 Ostriker, J.P., Peebles, P.J.E., Yahil, A.: Astrophys. J. **193** (1974) L1.
75 Perek, L.: Advances in Astron. Astrophys. **1** (1962) 165.
76 Perrin, M.-N., Hejlesen, P.M., Cayrel de Strobel, G., Cayrel, R.: Astron. Astrophys. **54** (1977) 779.
77 Roberts, W.W.: Astrophys. J. **158** (1969) 123.

78 Roberts, W.W.: Vistas Astron. **19** (1977) 91.
79 Rohlfs, K.: Lectures on Density Wave Theory, Lecture Notes in Physics, Berlin: Springer Vol. 69 (1977).
80 Rohlfs, K.: Mitt. Astron. Ges. **43** (1978) 48.
81 Roman, N.G.: Astrophys. J. Suppl. 2 (1955) 195 (with a correction in [0] p. 353).
82 Rubin, V.C., Burley, J., Kiasatpoor, A., Klock, B., Pease, G., Rutscheidt, E., Smith, C.: Astron. J. **67** (1962) 491.
83 Sandage, A.: Astrophys. J. **158** (1969) 1115.
84 Savage, B.D., Bohlin, R.C., Drake, J.F., Budich, W.: Astrophys. J. **216** (1977) 291.
85 Schmidt, M.: Bull. Astron. Inst. Neth. **13** (1956) 15.
86 Schmidt, M. in: [0] p. 513.
87 Schmidt, M.: Astrophys. J. **202** (1975) 22.
88 Shu, F.H.: American Scientist **61** (1973) 524.
89 Smart, W.M.: Stellar Kinematics, Longmans and Co., London (1968).
90 Spitzer, L., Schwarzschild, M.: Astrophys. J. **114** (1951) 385.
91 Spitzer, L., Schwarzschild, M.: Astrophys. J. **118** (1953) 106.
92 Stark, A.A., Blitz, L.: Astrophys. J. **225** (1978) L15.
93 Toomre, A.: Astrophys. J. **138** (1963) 385.
94 Toomre, A.: Astrophys. J. **139** (1964) 1217.
95 Toomre, A.: Highlights of Astronomy **3** (1974) 457.
96 Toomre, A.: Annu. Rev. Astron. Astrophys. **15** (1977) 437.
97 Tsioumis, A., Fricke, W.: Astron. Astrophys. **75** (1979) 1.
98 Upgren, A.R.: Astron. J. **83** (1978) 626.
99 van de Hulst, H.C., Muller, C.A., Oort, J.H.: Bull. Astron. Inst. Neth. **12** (1954) 117.
100 van Herk, G.: Bull. Astron. Inst. Neth. **18** (1965) 71.
101 Vyssotsky, A.N.: Stars and Stellar Systems **3** (1963) 192; cf. [0].
102 Wayman, P.A., Symms, L.S.T., Blackwell, K.C.: R. Obs. Bull. No. 98 (1965).
103 Weistrop, D.: Astron. J. **80** (1975) 303.
104 Weistrop, D.: Highlights of Astronomy **4**, part II (1977) p. 31.
105 Wielen, R.: Astron. Astrophys. **25** (1973) 285.
106 Wielen, R.: Astron. Astrophys. Suppl. **15** (1974) 1.
107 Wielen, R.: Highlights of Astronomy **3** (1974) 395.
108 Wielen, R.: Publ. Astron. Soc. Pac. **86** (1974) 341.
109 Wielen, R.: Mitt. Astron. Ges. **38** (1976) 254.
110 Wielen, R.: Proc. Third European Astron. Meeting (Kharadze, E.K., ed.) Tbilisi (1976) 402.
111 Wielen, R.: Astron. Astrophys. **60** (1977) 263.
112 Wielen, R.: Mitt. Astron. Ges. **43** (1978) 261.
113 Wielen, R. in: Stars and Star Systems (Westerlund, B.E., ed.) Reidel Publ. Co., Dordrecht (1979) 81.
114 Wielen, R.: Int. Astron. Union Symp. **84** (1979) 133.
115 Wilson, O., Woolley, R.: Mon. Not. R. Astron. Soc. **148** (1970) 463.
116 Woolley, R., Epps, E.A., Penston, M.J., Pocock, S.B.: R. Obs. Ann. No. 5 (1970).
117 Woolley, R., Pocock, S.B., Epps, E.A., Flinn, R.: R. Obs. Bull. No. 166 (1971).
118 Yoss, K.M., Lutz, T.E.: Mem. R. Astron. Soc. **75** (1971) 21.
119 Yuan, C.: Astrophys. J. **158** (1969) 889.

# 9 Galaxies and universe

## 9.1 General information and integral properties of galaxies

### 9.1.1 Catalogues, atlases, positions

#### 9.1.1.1 General catalogues of non-stellar objects

M = Messier's (1784) catalogue of nebular objects [13], modern version [8]
GC = General Catalogue by Herschel 1864 [9]
NGC = New General Catalogue by Dreyer [5]
IC = Index Catalogue of Dreyer [6]
RNGC = Revised New General Catalogue by Sulentic and Tifft [19]

Of 110 entries in modern versions [8] of Messier's catalogue, 39 are now known to be external galaxies. A similarly large proportion of the entries of GC, NGC, and IC are also extragalactic as evidenced by the early re-examination of the GC [70] and the modern treatment RNGC [19]. This latter study is based upon an inspection of the objects as they appear on the POSS-photographs [24] and so descriptions are limited to northern objects, but a southern extension is planned.

Only as the true nature of the various classes of "nebulae" became known in the early decades of the twentieth century, external galaxies laid claim to being catalogued in their own right. Nevertheless, in the tradition of the NGC, the "Quick Blue" European Southern Observatory (ESO) Sky Survey recently became available with extensive listing and descriptions of all non-stellar southern objects [10] and makes interesting new galactic and extragalactic objects immediately available to the large reflectors being completed in the southern hemisphere. Collectively, the compilations cited above contain most of the apparently brightest galaxies in the northern and southern hemispheres in their files.

#### 9.1.1.2 Galaxy catalogues

One of the earliest and most useful lists containing galaxies only is that of Shapley and Ames:

SA = Shapley-Ames Catalogue [18]: rough magnitudes; sizes; positions; descriptions; 1249 galaxies; $m_{\text{lim}} \approx 13\overset{\text{m}}{.}5$.

RSA = Revised Shapely-Ames Catalogue (Sandage, Tammann [16]): based on large-scale reflector plates.

Following production of the POSS [24] several new catalogues of faint galaxies became available:

CGCG = Catalogue of Galaxies and Clusters of Galaxies (Zwicky et al. [23]): positions; magnitudes; available redshifts; $m_{\text{lim}} \approx 15\overset{\text{m}}{.}5$.

MCG = Morphological Catalogue of Galaxies (Vorontsov-Vel'yaminow et al. [21]): positions; rough magnitudes; diameters; inclination classes; coded descriptions; about 29000 galaxies; $m_{\text{lim}} \approx 15\overset{\text{m}}{.}0$; identification charts.

UGC = Uppsala General Catalogue of Galaxies (Nilson [14]): 12912 galaxies ($\delta > -2°30'$) with diameter $> 1'$ or $m_{\text{pg}} \leqq 14\overset{\text{m}}{.}5$; diameters; position angles; magnitudes; Hubble types; luminosity classes; cross references to MCG, NGC, IC, CGCG and other lists.

RCBG = Reference Catalogue of Bright Galaxies (de Vaucouleurs [3]): state-of-the-art data for 4364 well-studied galaxies; catalogue does not claim completeness but contains all galaxies with diameter $> 30''$, $m_{\text{pg}} < 16^{\text{m}}$, or with redshift $< 15000$ km s$^{-1}$.

RC2 = Second Reference Catalogue of Bright Galaxies (de Vaucouleurs et al. [4]): same material as RCBG; diameters, magnitudes, colours and redshifts on a basis as uniform as possible; references for radio continuum and 21-cm data; classifications.

Further special catalogues:

CPGA = Catalogue of Peculiar Galaxies and Associations (Arp, Madore [2]); samples are given in [30].

Catalogue of Selected Compact Galaxies and Post-Eruptive Galaxies (Zwicky [22]): positions and descriptions of over 36000 compact and related galaxies;
Smaller lists of compact galaxies [46, 47, 75, 79, 32, 28];
Catalogues of low surface-brightness dwarf galaxies [77, 56];
Dwarf galaxies in the Virgo cluster [67, 68],
in the NGC 5128 group [78],
in the Fornax cluster [53];
Irregular and dwarf galaxies in two fields centered on M 81 and M 101 [54],
in the Leo group [52],
in the NGC 1023 group [15];
Late-type spirals (including dwarfs) [35].

### 9.1.1.3 Optical identification

#### 9.1.1.3.1 Pictures and photographic atlases

Photographs of northern galaxies:
Lick Observatory [57, 36],
Mt. Wilson 60-inch reflector [65].
Photographs of southern galaxies:
Reynolds 30-inch reflector (Australia) [37],
Cape photographic atlas of southern galaxies (Radcliffe 74-inch reflector) [7],
furthermore [42, 45, 39].

Illustrations to support classification schemes [64, 63, 40].

The Hubble Atlas of Galaxies [15], the standard reference for galaxy illustrations, contains pictures of 176 galaxies photographed with the Hale Observatories' reflectors. This atlas also provides the definitive description of the Hubble-Sandage classification scheme (see 9.1.3.1). A mini-atlas illustrating a general discussion of galaxy classification is provided by Sandage [71].

The Atlas de Galaxias Australes [17] shows photographs and isodensity tracings for 59 southern galaxies.

More recently the first photographs from the Las Campanas 2.5- and 1-m reflectors have been published [44, 72], in preparation of publication of RSA [16].

Image tube plates have been used for:
an atlas of 41 galaxies designed to illustrate the dust distribution in galaxies [60], see also [39],
a survey of galactic structure underlying Seyfert galaxies [26],
a survey of dwarf galaxies in the vicinity of the M 81 and Local Group galaxies [49].

Interacting and peculiar galaxies:
Atlas and Catalogue of interacting galaxies [20], with objects found and reproduced from POSS paper prints;
Atlas of peculiar galaxies [1, 29], based on original large-scale reflector plates;
Isodensity tracings and small-scale reproduction of 91 peculiar galaxies [73].

Of course, photographic reproductions of galaxies of all types and descriptions can be found on prints and glass copies of the POSS [24] ($\delta > -45°$) and the ESO/SRC (= Science Research Council) sky survey ($\delta < -25°$).

#### 9.1.1.3.2 Positional identification

(For abbreviations, see 9.1.1.1, 9.1.1.2.) Most of the NGC/IC objects are identified on the Smithonian Astrophysical Observatory (SAO) charts. For northern NGC Galaxies $(x, y)$ mm positions on the POSS prints are calculated from nominal $(\alpha, \delta)$ in [58]. Sulentic and Tifft [19] also give rectangular coordinates measured directly from the POSS. In CGCG [23], galaxies are identified with respect to bright stars and clusters of galaxies on individual charts for each POSS field. For the ESO/SRC survey, the tabulations of [10] also give $(x, y)$ positions for the "Quick Blue" survey, as does the CPGA [2] for the ESO/SRC J-survey.

Finding charts for galaxies with ultraviolet continua can be found in [11, 12, 76, 62, 61, 31].

Fragmentary lists of peculiar and interacting galaxies are given in [27, 74], containing positions and selected photographs.

High-precision positions

With the advent of high-precision radio source positions, similarly accurate optical positions are needed. To meet this need a number of observers have remeasured positions of galaxies with respect to AGK stars (see 8.1.1.5) either directly or by using the intermediate step of producing transparent overlays. Extensive lists of positions are given in [51, 33, 50, 66, 71, 48, 43, 59]. The new positions published up to 1975 are incorporated into RC2 [4], and, depending on the method used and the definition of the galaxy's nucleus, the accuracies quoted for the best cases are of the order of $\pm 10''$.

### 9.1.1.4 Named galaxies

A number of galaxies are known by name; they are familiar but their positions are not necessarily easy to obtain. Table 1 contains the names, cross references and positions for a number of these galaxies.

Table 1. Positions for named extragalactic systems.
For cross references, see 9.1.1.1; furthermore:
Pal = Palomar Observations, Abell [25]
Arp = Arp, Atlas of peculiar galaxies [1]
DDO = David Dunlap Observatory, van den Bergh [77]
VV = Atlas and Catalogue of interacting galaxies, Vorontsov-Vel'yaminov [20]
cf. also "Local Group" in 9.3.3.3

| Name | α (1950) | δ |
|---|---|---|
| Andromeda galaxy = M31 = NGC 224 | $00^h40^m.0$ | $+41°00'$ |
| Andromeda I | 00 43.0 | +37 44 |
| Andromeda II | 01 13.5 | +33 09 |
| Andromeda III | 00 32.6 | +36 14 |
| Andromeda IV | 00 39.8 | +40 18 |
| BL Lac | 22 01.9 | +42 11 |
| Capricorn dwarf = Pal 13 *) | 21 44.0 | −21 29 |
| Caraffe galaxy | 04 26.6 | −48 01 |
| Carina dwarf | 06 45.1 | −51 00 |
| Cartwheel galaxy | 00 35.0 | −34 01 |
| Centarus A = NGC 5128 = Arp 135 | 13 22.5 | −42.46 |
| Circinus galaxy | 14 09.3 | −65 06 |
| Copeland Septet = NGC 3745/54 = Arp 320 | 11 35.1 | +22 18 |
| Cygnus A | 19 57.7 | +40 36 |
| Draco dwarf = DDO 208 | 17 19.2 | +57 58 |
| Fath 703 | 15 11.0 | −15 17 |
| Fornax A = NGC 1316 | 03 20.8 | −37 23 |
| Fornax dwarf | 02 37.8 | −34 44 |
| Fourcade – Figueroa object | 13 32.4 | −33 38 |
| GR8 = DDO 155 | 12 56.2 | +14 29 |
| Hardcastle Nebula | 13 10.2 | −32 26 |
| Hercules A | 16 48.7 | +05 06 |
| Holmberg I = DDO 63 | 09 36.0 | +71 25 |
| Holmberg II = DDO 50 = Arp 268 | 08 13.7 | +70 52 |
| Holmberg III | 09 09.6 | +74 26 |
| Holmberg IV = DDO 185 | 13 52.8 | +54 09 |
| Holmberg V | 13 38.8 | +54 35 |
| Holmberg VI = NGC 1325 A | 03 22.6 | −21 31 |
| Holmberg VII = DDO 137 | 12 32.2 | +06 35 |
| Holmberg VIII = DDO 166 | 13 11.0 | +36 29 |
| Holmberg IX = DDO 66 | 09 53.5 | +69 17 |
| Hydra A | 09 15.7 | −11 53 |

continued

Table 1, continued

| Name | α (1950) | δ |
|---|---|---|
| Large Magellanic Cloud | $05^h24^m.0$ | −69°48′ |
| Leo I = Harrington-Wilson No. 1 = Regulus Dwarf = DDO 74 | 10 05.8 | +12 33 |
| Leo II = Harrington-Wilson No.2 = Leo B = DDO 93 | 11 10.8 | +22 26 |
| Leo A = Leo III = DDO 69 | 09 56.5 | +30 59 |
| Lindsay-Shapley ring | 06 44.4 | −74 11 |
| Maffei I | 02 32.6 | +59 26 |
| Maffei II | 02 38.1 | +59 23 |
| Mayall's Object = Arp 148 = VV 32 | 11 01.1 | +41 07 |
| Mice = NGC 4676 = Arp 242 | 12 44.7 | +30 54 |
| Pegasus dwarf = DDO 216 | 23 26.0 | +14 28 |
| Perseus A = NGC 1275 | 03 16.5 | +41 20 |
| Reticulum dwarf | 04 35.4 | −58 56 |
| Reinmuth 80 = NGC 4517 A | 00 57.6 | −33 58 |
| Seashell galaxy | 13 44.5 | −30 10 |
| Serpens dwarf | 15 13.5 | +00 03 |
| Seyfert's Sextet = NGC 6027 A–D | 15 57.0 | +20 54 |
| Sextans A = DDO 75 | 10 08.6 | −04 28 |
| Sextans B = DDO 70 | 09 57.4 | +05 34 |
| Sextans C | 10 03.0 | +00 19 |
| Small Magellanic Cloud | 00 51.0 | −73 06 |
| Sombrero galaxy = M104 = NGC 4594 | 12 37.6 | −11 21 |
| Stephan's Quintet = NGC 7317–20 = Arp 319 | 22 33.7 | +33 42 |
| Triangulum galaxy = M33 = NGC 598 | 01 31.0 | +30 24 |
| Ursa Minor dwarf = DDO 199 | 15 08.2 | +67 23 |
| Virgo A = M87 = NGC 4486 = Arp 152 | 12 28.3 | +12 40 |
| Whirlpool galaxy = M51 = NGC 5194 | 13 27.8 | +47 27 |
| Wild's Triplet = Arp 248 | 11 44.2 | −03 33 |
| Wolf-Lundmark-Melotte object = DDO 221 | 23 59.4 | −15 44 |
| Zwicky No. 2 = DDO 105 | 11 55.9 | +38 21 |
| Zwicky's Triplet = Arp 103 | 16 48.0 | +45 33 |

*) Probably a distant globular cluster.

## 9.1.1.5 References for 9.1.1

### Catalogues and Atlases


1 Arp, H.C.: Atlas of Peculiar Galaxies, Calif. Inst. Techn. Pasadena (1966), see also [29].

2 CPGA Arp, H.C., Madore, B.F.: A Catalogue of Peculiar Galaxies and Associations, (1983) in preparation. (Samples are given in [30].)

3 RCBG De Vaucouleurs, G., de Vaucouleurs, A.: Reference Catalogue of Bright Galaxies, Univ. Texas Press, Austin (1964).

4 RC2 De Vaucouleurs, G., de Vaucouleurs, A., Corwin, H.: Second Reference Catalogue of Bright Galaxies, Univ. Texas Press, Austin (1976).

5 NGC Dreyer, J.L.E.: New General Catalogue of Nebulae and Clusters, Mem. R. Astron. Soc. **49** (1888) pt. 1.

6 IC Dreyer, J.L.E.: Index Catalogue, Mem. R. Astron. Soc. **51** (1895) 185; **59** (1908) pt. 2, p. 105.

7 Evans, D.S.: Cape Atlas of Southern Galaxies, Cape Obs. (1957).

8 M Glyn Jones, K.: Messier's Nebulae and Star Clusters, Faber & Faber, London (1968).

9 GC Herschel, J.: General Catalogue, Phil. Trans. R. Soc. (1864) pt. I.

10 Holmberg, E.B., Lauberts, A., Schuster, H.-E., West, R.M.: The ESO/Uppsala Survey of the ESO(B) Atlas of the Southern Sky, Astron. Astrophys. Suppl. **18** (1974) 463 (I); **18** (1974) 491 (II); **22** (1975) 327 (III); **27** (1977) 295 (IV); **31** (1978) 15 (V); **34** (1978) 285 (VI); **39** (1980) 173 (VII).

11 Markarian, B.E.: Galaxies with strong UV continuum, finding charts, Astrofizika **3** (1967) 55 (Engl. **3**, 24); **5** (1969) 443 (Engl. **5**, 206); **5** (1969) 581 (Engl. **5**, 286); **8** (1972) 628 (Engl. **8**, 370); **8** (1972) 165 (Engl. **8**, 100).

12 Markarian, B.E., Lipovetsky, V.A.: Galaxies with strong UV continuum, finding charts, Astrofizika **7** (1971) 511 (Engl. **7**, 299); **8** (1972) 155 (Engl. **8**, 89); **9** (1973) 487 (Engl. **9**, 283); **10** (1974) 307 (Engl. **10**, 185); **12** (1976) 389 (Engl. **12**, 241); **12** (1976) 657 (Engl. **12**, 429).

12a Markarian, B.E., Lipovetsky, V.A., Stepanyan, Dzh. A.: Astrofizika **13** (1977) 225 (Engl. **13**, 116); **13** (1977) 397 (Engl. **13**, 215); **15** (1979) 201 (Engl. **15**, 130); **15** (1979) 363 (Engl. **15**, 235).

13 M Messier, C.: Connaissance des Temps pour 1787, Paris (1784) 238.

14 UGC Nilson, P.N.: Uppsala General Catalogue of Galaxies, Uppsala Obs. Ann. **6** (1973).

15 Sandage, A.R.: The Hubble Atlas of Galaxies, Carnegie Institution, Washington, D.C. No. 618 (1961).

16 RSA Sandage, A.R., Tammann, G.A.: Revised Shapley-Ames Catalogue, Carnegie Institution, Washington (1981).

17 Sersic, J.L.: Atlas de Galaxias Australes, Cordoba, Argentina (1968).

18 SA Shapley, H., Ames, A.: Shapley-Ames-Catalogue, Harvard Ann. **88** (1932) No. 2.

19 RNGC Sulentic, J.W., Tifft, W.G.: The Revised New General Catalogue of Non-Stellar Objects, Univ. Arizona Press, Tucson (1973).

20 Vorontsov-Vel'yaminov, B.A.: Atlas and Catalogue of Interacting Galaxies, Sternberg Inst. Moscow (1959).

21 MCG Vorontsov-Vel'yaminov, B.A., Krasnogorskaya, A.A.: Morphological Catalogue of Galaxies **1**, Sternberg Inst. Moscow (1962);
Vorontsov-Vel'yaminov, B.A., Arkhipova, V.P.: Morphological Catalogue of Galaxies, Sternberg Inst. Moscow, **2** (1963); **3** (1964); **4** (1968); **5** (1974).

22 Zwicky, F.: Catalogue of Selected Compact Galaxies and Post-Eruptive Galaxies, Speich, Zürich (1971).

23 CGCG Zwicky, F. et al.: Catalogue of Galaxies and Clusters of Galaxies, Cal. Inst. Techn. Pasadena:
Zwicky, F., Herzog, E., Wild, P.: **1** (1960);
Zwicky, F., Herzog, E.: **2** (1963); **3** (1966); **4** (1968);
Zwicky, F., Karpowicz, M., Kowal, C.: **5** (1965);
Zwicky, F., Kowal, C.: **6** (1968).

24 POSS Palomar Observatory Sky Survey, National Geographic Soc., Calif. Inst. Techn., Pasadena (1954); User's guide: Lund, J.M., Dixon, R.S.: Publ. Astron. Soc. Pacific **85** (1973) 230.


## Special references


25 Abell, M.D.: Publ. Astron. Soc. Pac. **67** (1955) 258.

26 Adams, T.F.: Astrophys. J. Suppl. **33** (1977) 19.

27 Aguero, E.L.: Publ. Astron. Soc. Pac. **83** (1971) 310.

28 Allen, D.A., Longmore, A.J., Hawarden, T.G., Cannon, R.D., Allen, C.J.: Mon. Not. R. Astron. Soc. **184** (1978) 303.

29 Arp, H.C.: Astrophys. J. Suppl. **14**, No. 123 (1966) 1.

30 Arp, H.C., Madore, B.F.: Quarterly J. R. Astron. Soc. **18** (1977) 234.

31 Bohuski, T.J., Fairall, A.P., Weedman, D.W.: Astrophys. J. **221** (1978) 776.

32 Borngen, F., Kalloglyan, A.T.: Astrofizika **14** (1978) 613 = Astrophysics **14** (1978) No. 4.

33 Cameron, M.J.: Cornell-Sydney Univ. Astr. Center Report No. 219 (1970).

34 Corwin, H.C.: Mon. Not. R. Astron. Soc. **191** (1980) 1.

35 Corwin, H.C., de Vaucouleurs, G., de Vaucouleurs, A.: Astron. J. **83** (1978) 1566.

36 Curtis, H.C.: Lick Obs. Publ. **13** (1918) 43.

37 De Vaucouleurs, G.: Mem. Commonwealth Obs. **3**, No. 13 (1956).

38 De Vaucouleurs, G.: Ann. Obs. Houga **2**, No. 1 (1957).

39 De Vaucouleurs, G.: Astrophys. J. **127** (1958) 487.

40 De Vaucouleurs, G.: Handb. Phys. **53** (1959) 275.

41 De Vaucouleurs, G.: Astrophys. J. **130** (1959) 728.

42 De Vaucouleurs, G., de Vaucouleurs, A.: Mem. R. Astron. Soc. **68** (1961) 69.

43 Dressel, L.L., Condon, J.J.: Astrophys. J. Suppl. **31** (1976) 187.

44 Dressler, A., Sandage, A.R.: Publ. Astron. Soc. Pac. **90** (1978) 5.

45 Evans, D.S.: Vistas Astron. **2** (1956) 1553.

46 Fairall, A.P.: Mon. Not. Astron. Soc. South Africa **27** (1968) 67; **29** (1970) 52.
47 Fairall, A.P.: Compact Galaxies, Thesis, Univ. Texas, Austin (1971).
48 Fisher, J.R., Tully, R.B.: Astron. Astrophys. **44** (1975) 151.
49 Fisher, J.R., Tully, R.B.: Astron. J. **84** (1979) 62.
50 Gallouet, L., Heidmann, J., Dampierre, F.: Astron. Astrophys. Suppl. **3** (1971) 325; **12** (1973) 89; **19** (1975) 1.
51 Glanfield, L.R., Cameron, M.J.: Australian J. Physics **20** (1967) 613.
52 Hodge, P.: Astron. J. **69** (1964) 438; Sci. American **210**, No. 5 (1964) p. 78.
53 Hodge, P., Pyper, D., Webb, C.J.: Astron. J. **70** (1965) 559.
54 Holmberg, E.: Medd. Lund Obs. Ser. II, No. 128 (1950).
55 Karachentseva, V.E.: Soobshch Byurakan Obs. **39** (1968) 61.
56 Karachentseva, V.E.: Astron. Tsirk. No. 723 (1972) 1; Astrofiz. Issled. Izv. Spets. Astrofiz. Obs., Akad. Nauk. SSSR No. 5 (1973) 10.
57 Keeler, J.E.: Lick Obs. Publ. **8** (1908) 1.
58 King, I.R., Setteducati, A.F.: Astrophys. Suppl. **14**, No. 129 (1967) 239.
59 Kojoian, G., Elliot, K., Tovmassian, H.M.: Astron. J. **83** (1978) 1545.
60 Lynds, B.T.: Astrophys. J. Suppl. **28**, No. 267 (1974) 391.
61 McAlpine, C.M., Lewis, D.W., Smith, S.B.: Astrophys. J. Suppl. **35** (1977) 203.
62 McAlpine, C.M., Smith, S.B., Lewis, D.W.: Astrophys. J. Suppl. **34** (1977) 95; **35** (1977) 197.
63 Morgan, W.W.: Publ. Astron. Soc. Pac. **70** (1958) 364.
64 Morgan, W.W., Mayall, N.U.: Publ. Astron. Soc. Pac. **69** (1957) 291.
65 Pease, F.G.: Astrophys. J. **46** (1917) 24; **51** (1920) 276.
66 Peterson, S.D.: Astron. J. **78** (1973) 811.
67 Reaves, G.: Astron. J. **61** (1956) 69.
68 Reaves, G.: Publ. Astron. Soc. Pac. **74** (1962) 392.
69 Reaves, G.: Publ. Astron. Soc. Pac. **89** (1977) 620.
70 Reinmuth, K.: Veröff. Sternw. Heidelberg **9** (1926) 1.
71 Sandage, A.R.: Astrophys. J. **202** (1975) 563.
72 Sandage, A.R., Brucato, R.: Astron. J. **84** (1979) 472.
73 Schamberg, B.C.: Astrophys. J. Suppl. **26**, No. 230 (1973) 115.
74 Sersic, J.L.: Astrophys. Space Sci. **28** (1974) 365.
75 Sharov, A.S.: Astron. Zhurn. **50** (1973) 1023 = Soviet Astron. **17**, No. 5 (1973).
76 Smith, M.G., Aguirre, C., Zemelman, M.: Astrophys. J. Suppl. **32** (1976) 217.
77 Van den Bergh, S.: Publ. David Dunlap Obs. **2**, No. 5 (1959) 147.
78 Webster, B.L., Goss, W.H., Hawarden, R.G., Longmore, A.J., Mebold, V.: Mon. Not. R. Astron. Soc. **186** (1979) 31.
79 Zwicky, F., Sargent, W.L.W., Kowal, C.T.: Astron. J. **80** (1975) 545.


## 9.1.2 Apparent integral properties of galaxies

### 9.1.2.1 Magnitudes

For abbreviations of catalogues, see 9.1.1.

For bright galaxies, probably the most widely used system of magnitudes is the *B*(0) system [15]. *B*(0) is the integrated magnitude derived from photoelectric or modern photographic magnitudes in the Johnson *B* system (see 4.2.5.12) or magnitudes transformed to this system. *B*(0) magnitudes are given for 873 objects in RCBG (= [3] in 9.1.1.5). To incorporate the rich data base of galaxy magnitudes given by Shapley and Ames [50], systematic errors in these so-called Harvard magnitudes, $m_H$, have been investigated in [9, 10, 30]. These magnitudes have been corrected for surface-brightness and luminosity-gradient dependent errors and transformed to the *B*(0) system as corrected Harvard magnitudes $m_H^c$ in RC2 (= [4] in 9.1.1.5). A comprehensive source list of *B* magnitudes is given in RCBG (Table 10) and RC2 (Tables 10 and 11), but some of the more important individual sources of magnitudes of galaxies in the *UBV* system are [12, 13, 15, 21, 5, 6, 45].

For fainter galaxies, the most extensive photometry is that contained in the CGCG (= [23] in 9.1.1.5) and derived from "schraffierkassette" photographic photometry down to an estimated limit of $m_{pg} \approx 15\overset{m}{.}5$. The MCG (= [21] in 9.1.1.5) also contains magnitude estimates down to $m_{pg} \approx 15\overset{m}{.}0$. Magnitude errors in the CGCG have been investigated in [32, 47] and more extensively by Huchra [33], who shows that the CGCG magnitudes differ by $0\overset{m}{.}12$ per mag with respect to *B*(0) magnitudes and have a scatter of $\pm 0\overset{m}{.}35$.

Asymptotic/total magnitudes, $B_{\text{total}}$

In order to transform aperture-photometry of galaxies out to "infinite" radius (or zero surface brightness), magnitude-aperture curves for various galaxy types are given in RC2 (= [4] in 9.1.1.5). These curves were derived from published standard total magnitudes [1⋯4, 23, 11⋯13, 15⋯19] and applied to the data in RC2.

### 9.1.2.2 Dimensions of galaxies

Isophotal major-axis diameters

Redman [41] suggested that galaxy diameters should be operationally defined by the $\mu_B = 25^{\text{m}}.0/\square''$ isophote. This corresponds to a surface brightness about one-tenth of the night-sky brightness and is very nearly the maximum detectible diameter on blue prints of the POSS (= [24] in 9.1.1.5). Holmberg [30], however, chose to define his diameters at a few percent of the sky brightness, $\mu_{\text{pg}} \simeq 26^{\text{m}}.5/\square''$. The choice is arbitrary, but as fainter isophotal diameters are chosen the constraints on the data become more demanding and, as a result, the diameters in general are probably less well defined.

Major sources for diameters of galaxies are [42, 50, 49, 28, 43, 9, 30, 55, 38] and others as listed in Table 8 of RCBG (= [3] in 9.1.1.5). In addition, diameters are listed in the Atlas de Galaxias Australes (= [17] in 9.1.1.5), UGC (= [14] in 9.1.1.5) and MCG (= [21] in 9.1.1.5) which in turn have been transformed to a standard diameter, $D_{25}$, in RC2 (= [4] in 9.1.1.5).

Axial ratios and "face-on" diameters

The measured apparent major-axis diameter is a function of inclination [29, 57, 9, 10]. Galaxies viewed more edge on have larger apparent diameters due to finite thickness of the disk resulting in a greater optical path, and thus an increased surface brightness at fixed radius. Corrections to a "face-on" isophotal diameter $D(0)_{25}$ have been investigated [10], revised [27] and applied to galaxies in the RC2 (= [4] in 9.1.1.5). However, see also [53, 54]. Axial ratios, necessary for correcting the diameters, are taken principally from the MCG ([21] in 9.1.1.5), UGC ([14] in 9.1.1.5) or [30], and are transformed to the standard $25^{\text{m}}/\square''$ isophotal ratio, $R_{25}$, in RC2 ([4] in 9.1.1.5).

Mean surface brightness

The RC2 ([4] in 9.1.1.5) tabulates two average surface brightnesses: $M'_{25}$, the average surface brightness contained within an ellipse of diameter $D_{25}$ and axial ratio $R_{25}$, and $M'_{\text{e}}$, the average surface brightness within an aperture $A_{\text{e}}$ that contains half of the total light (i.e. containing half of the asymptotic magnitude $B_{\text{total}}$). The first definition is relatively straightforward to compute, the latter is more complicated and somewhat model dependent.

### 9.1.2.3 Colours

In the RC2 ([4] in 9.1.1.5), 959 galaxies have total $(B-V)$ colour indices derived from multi-aperture data; 682 have $(U-B)$ colours listed. These colours have been transformed either from original *UBV* observations or from other similar colour systems as enumerated in Table 11 of RC2. Recent *UBV* photometry for additional galaxies: [7, 8, 25, 26, 40, 56].

### 9.1.2.4 Redshifts

Stromberg's original list [51] of optical radial velocities contained 41 galaxies. Some forty years later optical redshifts for 583 galaxies were published [34]. In the RCBG and RC2 ([3, 4] in 9.1.1.5) now 280 references to sources of optical velocity data are listed (Tables 12 and 13, respectively). Velocity data is no longer restricted to the optical measurements but includes derivations from 21-cm radio observations (see for instance [52] and Table V 2 in RC2).

A more recent compilation of velocity data: [24]; a study of a complete sample of field galaxies: [36]; cross-correlation redshifts are discussed for 59 galaxies in [35].

Accidental and systematic errors in various sets of velocity measurements are discussed in [31, 39, 20, 44, 37]. A total of 2713 velocities are reduced to a standard system in RC2.

Corrections for the solar motion

Radial velocities $v$ referred to the galactic centre must be corrected for the orbital motion of the sun. Most researchers in the field (e.g. [46], or RC2) use the same correction

$$\Delta v = 300 \sin l \cos b \quad \mathrm{km\,s^{-1}},$$

where it is assumed that the sun is moving at 300 $\mathrm{km\,s^{-1}}$ toward $l=90°$, $b=0°$, which is in good agreement with the estimate of $V_{\odot}=(315\pm15)$ $\mathrm{km\,s^{-1}}$ toward $l=95°\pm6°$, $b=-8°\pm3°$ for the Local-Group velocity centroid [22]. For definition of galactic coordinates $l$ and $b$, see 8.3.1.1.

## 9.1.2.5 References for 9.1.2


1 Ables, H.D.: U. S. Naval Obs., Publ. **20** (1971) pt. IV.
2 Arp, H.C., Bertola, F.: Astrophys. Lett. **4** (1969) 23.
3 Bertola, F.: Contr. Asiago, No. 186 (1966).
4 Bertola, F.: Contr. Asiago, No. 197 (1967).
5 Bigay, J.H.: Publ. Obs. Haute Prov. **7**, No. 10 (1964).
6 Bigay, J.H.: Publ. Obs. Haute Prov. **7**, No. 12 (1964) = Ann. d'Astrophys. **27** (1964) 170.
7 Buchnell, M.J., Peach, J.V.: Observatory **96** (1976) 61.
8 Corvin, H.C.: Mon. Not. R. Astron. Soc. **191** (1980) 1.
9 De Vaucouleurs, G.: Astron. J. **61** (1956) 430.
10 De Vaucouleurs, G.: Publ. Astron. Soc. Pac. **69** (1957) 252.
11 De Vaucouleurs, G.: Astrophys. J. **128** (1958) 465.
12 De Vaucouleurs, G.: Lowell Obs. Bull. **4**, No. 97 (1959).
13 De Vaucouleurs, G.: Lowell Obs. Bull. **4**, Nr. 98 (1959).
14 De Vaucouleurs, G.: Astrophys. J. **131** (1961) 405.
15 De Vaucouleurs, G.: Astrophys. J. Suppl. **5**, No. 48 (1961) 233.
16 De Vaucouleurs, G.: Astrophys. J. **134** (1961) 666; Astrophys. J. Suppl. **6** (1961) 213.
17 De Vaucouleurs, G.: Astrophys. J. **136** (1963) 107; **137** (1963) 720; **138** (1964) 924; **139** (1964) 899; **181** (1973) 31; Astrophys. Lett. **4** (1969) 17; Astrophys. J. Suppl. **29**, No. 284 (1975) 193.
18 De Vaucouleurs, G., Ables, H.D.: Publ. Astron. Soc. Pac. **77** (1965) 272; Astron. J. **151** (1968) 105.
19 De Vaucouleurs, G., Aguero, E.: Publ. Astron. Soc. Pac. **85** (1973) 150.
20 De Vaucouleurs, G., de Vaucouleurs, A.: Astron. J. **68** (1963) 96.
21 De Vaucouleurs, G., de Vaucouleurs, A.: Mem. R. Astron. Soc. **77** (1972) pt. 1.
22 De Vaucouleurs, G., Peters, W.L.: Nature **220** (1968) 268.
23 Fraser, C.: Astron. Astrophys. Suppl. **29** (1977) 161.
24 Gisler, G.R., Friel, E.D.: Index of Galaxy Spectra, Pachart, Tucson (1979).
25 Green, M.R., Dixon, K.L.: Observatory **98** (1978) 166.
26 Graham, J.A.: Astron. J. **81** (1976) 681.
27 Heidmann, J., Heidmann, N., de Vaucouleurs, G.: Mem. R. Astron. Soc. **75**, pt. 4···6 (1971) 85.
28 Holmberg, E.: Ann. Lund Obs. No. 6 (1937) 1.
29 Holmberg, E.: Medd. Lund Obs. Ser. II, No. 117 (1946).
30 Holmberg, E.: Medd. Lund Obs. Ser. II, No. 136 (1958).
31 Holmberg, E.: Medd. Uppsala Obs. Ser. II, No. 138 (1961).
32 Holmberg, E.: Arkiv Astron. **5** (1969) 305.
33 Huchra, H.: Astron. J. **81** (1976) 952.
34 Humason, M.L., Mayall, N.U., Sandage, A.R.: Astron. J. **61** (1956) 97.
35 Kelton, P.W.: Astron. J. **85** (1980) 89.
36 Kirshner, R.P., Oemler, A., Schechter, P.L.: Astron. J. **83** (1978) 1549.
37 Lewis, B.M.: Mem. R. Astron. Soc. **78** (1975) 75.
38 Liller, M.H.: Astrophys. J. **132** (1960) 306; **146** (1966) 28.
39 Page, T.: Proc. Fourth Berkeley Symp. Math. Statistics & Probability (1961) 3, 277.
40 Persson, S.E., Frogel, J.A., Aaronson, M.: Astrophys. J. Suppl. **39** (1979) 61.
41 Redman, R.O.: Mon. Not. R. Astron. Soc. **96** (1963) 588; **98** (1963) 613.
42 Reinmuth, K.: Veröff. Sternw. Heidelberg **9** (1926) 1.
43 Reiz, A.: Ann. Lund Obs. No. 9 (1941) 1.

44 Roberts, M.S.: Int. Astron. Union. Symp. **44**, External galaxies and quasi-stellar objects (Evans, D.S., ed.), Reidel, Dordrecht (1972) 12.
45 Sandage, A.R.: Astrophys. J. **150** (1967) L177; **176** (1972) 21; **178** (1972) 1; **178** (1972) 25; **182** (1973) 711; **202** (1975) 563.
46 Sandage, A.R., Tammann, G.A.: Astrophys. J. **196** (1975) 313.
47 Sandage, A.R., Tammann, G.A.: Astrophys. J. **197** (1975) 265.
48 Sandage, A.R., Visvanathan, N.: Astrophys. J. **223** (1978) 707.
49 Shapley, H.: Harvard Ann. **88**, No. 4 (1934).
50 Shapley, H., Ames, A.: Harvard Ann. **88**, No. 2 (1932), =SA, [18] in 9.1.1.5.
51 Stromberg, C.: Astrophys. J. **61** (1925) 352.
52 Thonnard, N., Rubin, V.C., Ford, W.K., Roberts, M.S.: Astron. J. **83** (1978) 1565.
53 Tully, R.B.: Astron. J. **73** (1968) 5205.
54 Tully, R.B.: Mon. Not. R. Astron. Soc. **159** (1972) 35p.
55 Van den Bergh, S.: Publ. David Dunlap Obs. **2**, No. 6 (1960).
56 Wegner, G.: Astrophys. Space Sci. **60** (1979) 15.
57 Wyatt, S.P., Brown, F.G.: Astron. J. **60** (1955) 415.

## 9.1.3 Qualitative classification of galaxies

### 9.1.3.1 The classical Hubble-Sandage classification

Two broad morphological types characterize most galaxies: elliptical and spiral forms. The remaining small percentage of galaxies that cannot be characterized in this way are generally relegated to the irregular class. The earliest classification schemes that have had a continuing influence on modern systems were independently developed by Lundmark [27] and Hubble [19].

Discussions of this early history of galaxy classifications: [17, 6].

Modern reviews of the existing classification schemes: [51, 69, 13].

Ellipticals

Elliptical galaxies show very little or no internal structure and they are simply characterized by their apparent ellipticity. In general, ellipticals with major and minor axes, $a$ and $b$, respectively, are typed En where $n = 10(a-b)/a$. Thus E0 galaxies are apparently circular, E7 galaxies are the most flattened. When n exceeds 7, other structures appear (such as a cusp in the light distribution or rings of apparent obscuration), and the galaxy is designated as lenticular or of the S0 class.

Spirals

In Hubble's original scheme [20], spirals were subdivided into three types or stages (Sa, Sb, and Sc) characterized either by an increasing openness of the arms or by the decreasing importance of the light contributed by the bulge compared to the disk. Even before this time Curtis [9] had paved the way for independently dividing all spirals into two families according to whether they possessed central bar structures (barred spirals) or not. These two form criteria are the basis of the so-called "tuning fork" diagram for galaxies; see LB, NS, Vol. VI/1, p. 668.

Extension of the basic Hubble scheme started with Holmberg [18], who introduced "late" and "early" subtypes for the Sb and Sc galaxies, see Table 1. De Vaucouleurs [11] also extended Hubble's classification to include explicitly much more ragged spirals, type Sd, and those bordering on the irregular classification – having no nucleus and only a hint of spiral structure – type Sm (= Spirals of Magellanic type).

Lenticulars, S0 galaxies

Hubble [20] introduced S0 galaxies as a morphological transition case between ellipticals and early-type spirals. A detailed description of these lenticular galaxies and a possible sub-classification scheme is given in [50]; see Table 2. Others [4, 70] have suggested that S0's actually run parallel to the ordinary spiral stages but form one extreme of gas-depleted disk systems (S0a, S0b, S0c) coupled to the gas-rich spirals by yet another intermediate sequence of "anaemic" spirals (Aa, Ab, Ac). De Vaucouleurs [11] suggests that lenticulars may be subdivided as well, and his "early"- and "late"-type classification is also outlined in Table 2.

Table 1. Holmberg's extended Hubble classification for late-type spirals [18].

| Type | Definition | Examples |
|---|---|---|
| $Sb^-$ | Extended nuclear region representing a considerable fraction of the total luminosity | NGC 224 |
| | Mean surface brightness is comparatively high | NGC 3031 |
| | Symmetrical and rather closed system of spiral arms with no pronounced contrast between the arms and the main body | |
| | In most cases there is no appreciable resolution | |
| $Sb^+$ | Comparatively small nuclear region | NGC 3952 |
| | Mean surface brightness lower than $Sb^-$ | NGC 3992 |
| | Arm system is open and symmetrical, with good contrast against the main body | |
| $Sc^-$ | Small, sometimes semi-stellar nucleus | NGC 5194 |
| | Mean surface brightness about the same as for $Sb^+$ | NGC 5457 |
| | More or less symmetrical, open and rather pronounced spiral arms | |
| | Resolution well advanced | |
| $Sc^+$ | No prominent nuclear region | NGC 598 |
| | Mean surface brightness less than for $Sc^-$ | |
| | Confused and loosely defined spiral arm system (short arms) | |
| | Highly resolved | |

Table 2. Sub-divisions of S0/lenticular galaxies.

| Type | Definition | Examples |
|---|---|---|
| (a) de Vaucouleurs [11] | | |
| $S0^-$ | Traces of structure can be found in the smooth lens and envelope; a small nucleus may be present | NGC 5273<br>NGC 7166 |
| $S0^0$ | A weak trace of a ring appears at the edge of the lens; a distinct nucleus and outer envelope may be present | NGC 1553<br>NGC 4459 |
| $S0^+$ | A well defined ring is present separating the inner nuclear bulge from outer incipient spiral structure | NGC 2855<br>NGC 7702 |
| (b) Sandage [50] | | |
| $S0_1$ | The existence of an outer envelope flattened to a fundamental plane defines this subtype; the ellipticals of the central section of such galaxies are flatter than E7 | NGC 1201<br>NGC 3245<br>NGC 4762 |
| $S0_2$ | The first appearance of a circular absorption pattern or a true depletion of material in the envelope defines this intermediate subtype | NGC 542<br>NGC 3065<br>NGC 4111 |
| $S0_3$ | A strong internal circular ring defines this extreme of the S0 class | NGC 5866<br>NGC 3032<br>NGC 4459 |

### Irregulars

Holmberg [18] first divided irregular galaxies into two types: Type I irregulars are highly resolved systems similar to the Magellanic Clouds, while Type II irregulars show a smoothed "amorphous" distribution of light, often broken by irregular patches of obscuration. While Type I irregulars have subsequently found a natural home as Im galaxies following the late spirals of class Sd and Sm [11], Sandage and Brucato [52] have suggested that the original Irr II galaxies can be further sub-divided, and a single class should be abandoned. The properties of the Irr II class galaxies (also called I0 in [10]) have been summarized in [26]. Those few objects which show no spiral structure, but have unresolved disks, are now termed amorphous galaxies [52]. Examples are NGC 3077, NGC 5253, and M82.

### 9.1.3.2 Alternate systems and modifications

#### 9.1.3.2.1 The Yerkes classification

Morgan [40···42] has produced a system for classifying the central concentration of light in a galaxy to give population groups which apparently correlate with the stellar content of the inner parts of the galaxy [43] as judged by integrated spectral types. These population groups a, af, f, fg, g, gk, and k imply early and late spectral-type stars, respectively, as contributing most of the light from the nucleus; however, the designation is found by inspecting the central concentration of monochromatic light: a galaxies have little or no central concentration while k galaxies are highly concentrated. In addition, Morgan identifies form families, which are explained in Table 3; these correspond to the most basic classification, see above. A recent application of this classification system to southern galaxies is given in [15].

Table 3. Form families of the Yerkes system [40, 41]. For peculiar types, see also 9.3.1.

| Form family | Description |
|---|---|
| B | Barred spirals |
| D | Galaxies with rotational symmetry but showing neither spiral structure nor ellipticity |
| cD | Supergiant D galaxies, predominantly found in clusters [38] and embedded in an extensive halo |
| db | Dumb-bell systems |
| E | Ellipticals |
| Ep | Peculiar ellipticals containing conspicuous absorption patches |
| I | Irregulars |
| L | Low surface-brightness systems |
| N | High-luminosity nucleus superimposed on a considerably fainter outer envelope, see also [42] |
| Q | Quasi-stellar objects, see also 9.5.1.2 |
| S | Ordinary spirals |

#### 9.1.3.2.2 Spiral varieties

With reference to the inner structure of galaxies, the manner in which the spiral structure terminates is a valid classification parameter [11]. If the spiral arms continuously circle into the nucleus "(s)" is suffixed to the stage and family classification. If the spiral structure ends in a ring structure "(r)" is added. Transition cases "(rs)" are also found. For more details, see also LB, NS, Vol. VI/1, p. 668.

#### 9.1.3.2.3 Dwarf galaxies

As a precursor to the luminosity classification of all galaxies, van den Bergh's [64] division of the intrinsically lowest luminosity galaxies into four groups is outlined in Table 4. The so-called "nucleated dwarfs" described in [45] must be added to this classification; their prototype is IC 3475 (see Table 4). These specific low surface-brightness objects have luminous knots but no global structure other than nuclei in a number of instances. In fact about thirty percent of the dwarfs in the Virgo cluster have discernable nuclei, unlike the Local Group (see 9.3.3.3) in which none of the dwarfs possess nuclei [46].

Table 4. Classification of dwarf galaxies [64].

| Type | Description | Examples |
|---|---|---|
| D Ir | dwarf irregular | NGC 6822, IC 1613 |
| D Sp | dwarf spiral | NGC 3057 |
| D El | dwarf elliptical | NGC 185<br>NGC 205 |
| D Sph | dwarf spheroidal | Draco, Leo B |
| | nucleated dwarf spheroidal | IC 3475 |

#### 9.1.3.2.4 Luminosity classification

See also 9.1.4.2.

In a series of papers van den Bergh [65, 66, 68] suggested that the coherence of the spiral structure could be used to infer the intrinsic luminosity of spiral galaxies. Systems with well-developed global spiral structure are of type I (supergiants) while those with ragged ill-defined spiral arms are of type V (dwarfs), as in the stellar luminosity-class nomenclature. The basic criteria of this DDO-classification (DDO = David Dunlap Observatory) are outlined in Table 5.

Table 5. DDO luminosity classification for Sc galaxies [65, 66, 68].

| Luminosity | Description | Examples |
|---|---|---|
| Sc I | These supergiants galaxies are characterized by long well-developed arms of relatively high-surface-brightness. | M 101<br>M 51 |
| Sc II | The spiral structure of these bright giant galaxies is less well developed as compared to Sc I galaxies. | NGC 3184<br>NGC 3319 |
| Sc III | Short patchy arms extend from a fairly high-surface-brightness main body. | NGC 2403<br>NGC 672 |
| S IV | Outer regions give only a hint of spiral structure emanating from a relatively low-surface-brightness disk. | NGC 247<br>NGC 2500 |
| S V | Dwarf spirals; only a hint of spiral structure is seen in these low-surface-brightness objects. | DDO 122 |

Calibrations of the absolute magnitudes of Sc galaxies as a function of luminosity class (see 9.1.4.2) are controversial; discussion in [67] and [53]. De Vaucouleurs [13] has also discussed this problem, but in terms of his Luminosity Index which is a simple average of his stage and the DDO luminosity class. The absolute magnitudes of Sb galaxies as a function of luminosity class have been recently discussed in [58].

Irregular galaxies may be roughly typed according to intrinsic luminosity [64, 68] through Holmberg's [18] relation between absolute magnitude and apparent surface brightness. Similarly, giant ellipticals have a higher mean surface brightness than dwarf ellipticals [59]. In addition, though, ellipticals show a correlation of their luminosity with colour [5, 12] in the sense that giant ellipticals are redder than their dwarf counterparts.

#### 9.1.3.2.5 Byurakan nuclear types, bright nuclei galaxies, and Seyfert galaxies

See also 9.5.

Kalloglyan and Tovmassian [22] have introduced a classification of galaxies judged solely on the degree of central concentration of galaxy images (Table 6); examples of this system are also to be found in [61···63; 47···49, 44, 21, 39].

Keel and Weedman [25] provided a survey of 448 so-called Bright-Nuclei (BN) galaxies of Type 4 and 5 (Table 6), drawn from the Byurakan master list [8] of nuclear types for 711 galaxies. They draw attention to 10 galaxies which have nuclei morphologically resembling Seyfert galaxies, were originally defined by their optical appearance [55] but now have a strict spectroscopic classification [24] outlined in Table 7. In this respect it is worth noting that the N galaxies [38] now satisfy Seyfert's original morphological criteria as a separate class (see 9.5.1.3).

Table 6. Byurakan nuclear types [8, 22]. (N galaxies)

| Type | Description | Example |
|---|---|---|
| 1 | no central condensation | NGC 4088 |
| 2 | weak central condensation | NGC 5850 |
| 3 | strong central condensation but not stellar | NGC 4442 |
| 4 | stellar nucleus blending into nebulous background | NGC 1300 |
| 5 | strongly stellar nucleus | NGC 3992 |

Table 7. Spectroscopic classification of Seyfert galaxies [24].

| Type | Description | Example |
|---|---|---|
| 1 | Widths of the Balmer emission lines are much broader than the forbidden line widths | NGC 4151 |
| 2 | Balmer lines and forbidden lines have similar widths, typically 500···1000 $km\,s^{-1}$ | NGC 1068 |

#### 9.1.3.2.6 Compact galaxies

According to Zwicky [72] compact is a term that applies to "any galaxy, or any part of a galaxy, whose surface brightness photographically, visually or bolometrically is greater than that which corresponds to 20th magnitude per square second of arc" ($m < 20^{m}/\square''$).

#### 9.1.3.2.7 The numerical code *t* of the morphological types

Heidmann et al. [17a] introduced a numerical code of the morphological types in the revised Hubble system. This code was extended and used in the Second Reference Catalogue of Bright Galaxies, RC2 ([4] in 9.1.1.5). It is defined by:

| Morphological type (Hubble revised) | | Code *t* |
|---|---|---|
| compact (high density) ellipticals | cE | −6 |
| dwarf (low density) ellipticals | dE | −5 |
| normal elliptical systems | E | |
| giant ellipticals with extended optical coronae (in particular the Morgan type cD) | $E^{+}$ | −4 |
| lenticular systems | $L^{-}$ | −3 |
| | L | −2 |
| | $L^{+}$ | −1 |
| irregular systems of type II | I0 | 0 |
| lenticular-spiral systems | S0/a | |
| spiral systems | Sa | 1 |
| | Sab | 2 |
| | Sb | 3 |
| | Sbc | 4 |
| | Sc | 5 |
| | Scd | 6 |
| | Sd | 7 |
| | Sdm | 8 |
| spirals of Magellanic type | Sm | 9 |
| irregular of Magellanic type (=irr of type I) | Im | 10 |
| compact blue irregulars (=isolated extragalactic H II regions) | cI | 11 |

More detailed tables for coding of Hubble types, revised morphological types (de Vaucouleurs [11]), DDO (van den Bergh) types and Yerkes (Morgan) types et al. are given in the RC2.

### 9.1.3.3 Spectroscopic criteria

Seyfert galaxies: see 9.1.3.2.5, Table 7, and 9.5.1.3.

Haro galaxies

Haro [16], using both objective-prism Schmidt plates and three-colour direct plates, discovered 44 decidedly blue or ultraviolet galaxies, one quarter of which had obvious emission.

Markarian galaxies and the Tololo survey

See also 9.3.1.

Using objective-prism techniques Markarian and his associates [31⋯37] have been identifying galaxies with strong ultraviolet continua. They found two main types: (i) the source of the ultraviolet radiation is the nucleus of the galaxy or (ii) sources of emission and ultraviolet continuum radiation are spread throughout the galaxy. The first type further subdivides into Seyfert galaxies and simple bright nucleus (BN) galaxies [71]. The second type includes (a) simple irregular galaxies with active star formation and (b) intrinsically faint systems showing strong high-excitation emission lines. These are the "extragalactic H II regions" in [54] and are also found in the lists of compact galaxies [72] because of their high surface brightness and small physical dimensions.

The Tololo surveys [56] also use objective-prism techniques and line-emission classification criteria. The nature of the objects in the lists of [57, 28···30] is discussed in [7] with the conclusion that, while only about one-third of Markarian-type galaxies have sufficiently strong emission lines to be included in the Tololo listings, the latter contain about 2% class 1 Seyferts, 10% class 2 Seyferts, with the remainder being galaxies with emission lines produced by hot stars.

Fairall [14] has divided the narrow-line emission objects into four groups based on morphological and spectroscopic studies of 47 galaxies from the Tololo surveys. These groups comprise systems where:

(a) the entire galaxy is an emission-line source. These are generally blue ellipticals with $H\beta/\lambda4959$ just less than unity;

(b) the outer regions are emission-line sources. These are generally outlying H II regions in late-type spirals;

(c) only the nucleus is an emission-line source. $H\beta/\lambda4959$ is low and these systems appear to be related to class 2 Seyferts;

(d) multiple systems with high $H\beta/\lambda4959$.

### 9.1.3.4 Peculiar and interacting galaxies

See also 9.3.

A preliminary scheme for grouping peculiar galaxies was given in [1, 2]. Independently Arp and Madore [3] give 24 natural groups listed in Table 8. Karachentsev [23], on the other hand, suggests a much more modest morphological classification for non-equilibrium systems as outlined in Table 9. Because of the very nature of peculiar galaxies none of these systems has found wide application. However, ring galaxies, as a subset, are widely recognized [11] and have been internally classified in [60], as explained in Table 10.

Table 8. Arp-Madore codings and descriptions of peculiar galaxies and associations [3].

| Code | General description |
|---|---|
| 1 | Galaxy with interacting companion(s) |
| 2 | Interacting double |
| 3 | Interacting triple |
| 4 | Interacting quadruple |
| 5 | Interacting quintet |
| 6 | Ring galaxy |
| 7 | Galaxy with a jet |
| 8 | Galaxy with an apparent companion(s) |
| 9 | M51 type |
| 10 | Spiral galaxy with peculiar spiral arm(s) |
| 11 | Three-armed spiral galaxy |
| 12 | Peculiar spiral galaxy |
| 13 | Compact (very high surface-brightness) galaxy |
| 14 | Galaxy with unusual dust absorption |
| 15 | Galaxy with tails or loops |
| 16 | Irregular or disturbed single galaxy |
| 17 | Chain of galaxies (four or more galaxies aligned) |
| 18 | Group of galaxies (four or more galaxies not aligned) |
| 19 | Cluster of galaxies (only very conspicuous clusters) |
| 20 | Dwarf galaxy (low surface-brightness) |
| 21 | Stellar object with associated nebulosity |
| 22 | Miscellaneous |
| 23 | Close pairs (not visibly interacting) |
| 24 | Close triples (not visibly interacting) |

Table 9. Karachentsev's classification of non-equilibrium galaxies [23].

| Type | Explanation |
|---|---|
| LIN | galaxies exhibiting strong interaction in the form of<br>bridges = LIN(br)<br>tails = LIN(ta)<br>bridges and tails = LIN(br+ta) |
| ATM | systems with two or more components in a common halo |
| DIS (n) | systems with signs of distortion in n individual components |

Table 10. Theys' and Spiegel's classification of ring galaxies [60].

| Ring type | Description | Examples |
|---|---|---|
| RE | Crisp, elliptical ring with photographically empty interior | Arp 146, Arp 147, VIIZW466 |
| RN | Elliptical ring with an off-center nucleus | II Hz 4, Lindsay-Shapley Ring |
| RK | Ring with a single, very prominent knot in the ring; large-scale brightness distribution is markedly asymmetrical | I ZW 45, II ZW 28 |

## 9.1.3.5 References for 9.1.3


1 Arp, H.C.: Atlas of Peculiar Galaxies, Calif. Inst. Techn. Pasadena (1966).
2 Arp, H.C.: Astrophys. J. Suppl. **14**, No. 123 (1966) 1.
3 Arp, H.C., Madore, B.F.: Quarterly J. R. Astron. Soc. **18** (1977) 234.
4 Baade, W.: Evolution of Stars and Galaxies (Payne-Gaposchkin, C., ed.), Harvard Univ. Press, Cambridge (1963).
5 Baum, W.: Publ. Astron. Soc. Pacific **71** (1959) 106.
6 Berendzen, R., Hart, R., Seeley, D.: Man Discovers the Galaxies, Science Hist. Publ., New York (1976).
7 Bohuski, T.J., Fairall, A.P., Weedman, D.W.: Astrophys. J. **221** (1978) 776.
8 Byurakan Astrophysical Observatory, Soobshch. Byurakan Obs. **47** (1975) 43.
9 Curtis, H.D.: Lick Obs. Bull. **13** (1918) 12.
10 De Vaucouleurs, G., de Vaucouleurs, A.: Reference Catalogue of Bright Galaxies, Univ. Texas Press, Austin (1964).
11 De Vaucouleurs, G.: Handb. Physik **53** (1959) 295.
12 De Vaucouleurs, G.: Astrophys. J. Suppl. **5** (1961) 233.
13 De Vaucouleurs, G.: in: The Evolution of Galaxies and Stellar Populations (Tinsley, B.M., Larson, R.B., eds.), Yale Univ. Obs., New Haven (1977) p. 43.
14 Fairall, A.P.: Mon. Not. R. Astron. Soc. **191** (1980) 391.
15 Garrison, R.F., Walborn, N.R.: J. R. Astron. Soc. Canada **68** (1974) 117.
16 Haro, G.: Bol. Obs. Tonantzintla y Tacubaya No. 14 (1956) 16.
17 Hart, R., Berendzen, R.: J. Hist. Astron. **2** (1971) 109.
17a Heidmann, J., Heidmann, N., de Vaucouleurs, G.: Mem. R. Astron. Soc. **75** (1972) 85.
18 Holmberg, E.: Medd. Lund Obs. Ser. II, No. 136 (1958).
19 Hubble, E.P.: Astrophys. J. **64** (1926) 321.
20 Hubble, E.P.: The Realm of the Nebulae, Yale Univ. Press, New Haven (1936); Reprinted by Dover Publ. Inc. New York (1958).
21 Iskudaryan, S.G.: Astrofizika **4** (1968) 385.
22 Kalloglyan, A.T., Tovmassian, H.M.: Soobshch. Byurakan Obs. **36** (1964) 31.
23 Karachentsev, I.D.: Soobshch. Spets. Astrofiz. Obs. **1** (1972) 3.
24 Khachikian, E.Y., Weedman, D.W.: Astrophys. J. **192** (1974) 581.
25 Keel, W.C., Weedman, D.W.: Astron. J. **83** (1978) 1.
26 Krienke, O.K., Hodge, P.W.: Astron. J. **79** (1974) 1242.
27 Lundmark, K.: Uppsala Medd. **19b**, No. 8 (1926).
28 MacAlpine, C.M., Lewis, D.W., Smith, S.B.: Astrophys. J. Suppl. **35** (1977) 203.
29 MacAlpine, C.M., Smith, S.B., Lewis, D.W.: Astrophys. J. Suppl. **34** (1977) 95.
30 MacAlpine, C.M., Smith, S.B., Lewis, D.W.: Astrophys. J. Suppl. **35** (1977) 197.
31 Markarian, B.E.: Astrofizika **3** (1967) 55.
32 Markarian, B.E.: Astrofizika **5** (1969) 443.
33 Markarian, B.E.: Astrofizika **5** (1969) 581.
34 Markarian, B.E., Lipovetsky, V.A.: Astrofizika **7** (1971) 511.
35 Markarian, B.E., Lipovetsky, V.A.: Astrofizika **8** (1972) 155.
36 Markarian, B.E., Lipovetsky, V.A.: Astrofozika **9** (1973) 487.
37 Markarian, B.E., Lipovetsky, V.A.: Astrofizika **10** (1974) 307.
38 Matthews, T.A., Morgan, W.W., Schmidt, M.: Astrophys. J. **140** (1964) 35.
39 Mnatsakanyan, M.A.: Astrofizika **9** (1973) 57.
40 Morgan, W.W.: Publ. Astron. Soc. Pacific **70** (1958) 364.
41 Morgan, W.W.: Publ. Astron. Soc. Pacific **71** (1959) 394.
42 Morgan, W.W.: Astron. J. **76** (1971) 1000.
43 Morgan, W.W., Mayall, N.U.: Publ. Astron. Soc. Pacific **69** (1957) 291.
44 Parsamyan, E.S.: Astrofizika **4** (1968) 150.
45 Reaves, G.: Astron. J. **61** (1956) 69.
46 Reaves, G. in: The Evolution of Galaxies and Stellar Populations (Tinsley, B.M., Larson, R.B., eds.), Yale Univ. Obs., New Haven (1977) p. 39.
47 Saakyan, G.S.: Astrofizika **4** (1968) 41.
48 Saakyan, G.S.: Astrofizika **5** (1969) 593.
49 Saakyan, G.S.: Astrofizika **9** (1973) 51.
50 Sandage, A.R.: The Hubble Atlas of Galaxies, Carnegie Institution, Washington, D.C. No. 618 (1961).

51 Sandage, A.R. in: Galaxies and the Universe (Sandage, A.R., Sandage, M., Kristian, J., eds.), Stars and Stellar Systems **9**, Univ. Chicago (1975) p. 1.
52 Sandage, A.R., Brucato, R.: Astron. J. **84** (1979) 472.
53 Sandage, A.R., Tammann, G.A.: Astrophys. J. **194** (1974) 569.
54 Searle, L., Sargent, W.L.E.: Astrophys. J. **173** (1972) 25.
55 Seyfert, C.K.: Astrophys. J. **97** (1943) 28.
56 Smith, M.G.: Astrophys. J. **202** (1975) 591.
57 Smith, M.G., Aguirre, C., Zemelman, M.: Astrophys. J. Suppl. **32** (1976) 217.
58 Stenning, M., Hartwick, F.D.A.: Astron. J. **85** (1980) 101.
59 Strom, K.M., Strom, S.E.: Astron. J. **83** (1978) 73.
60 Theys, J.C., Spiegel, E.A.: Astrophys. J. **208** (1976) 650.
61 Tovmassian, H.M.: Astrofizika **1** (1965) 197.
62 Tovmassian, H.M.: Astrofizika **2** (1966) 317.
63 Tovmassian, H.M.: Astrofizika **3** (1967) 427.
64 Van den Bergh, S.: Publ. David Dunlap Obs. **2**, No. 5 (1959).
65 Van den Bergh, S.: Astrophys. J. **131** (1960) 215.
66 Van den Bergh, S.: Astrophys. J. **131** (1960) 558.
67 Van den Bergh, S.: Publ. David Dunlap Obs. **2**, No. 6 (1960).
68 Van den Bergh, S.: Astron. J. **71** (1966) 922.
69 Van den Bergh, S.: J. R. Astron. Soc. Canada **69** (1975) 57.
70 Van den Bergh, S.: Astrophys. J. **206** (1976) 883.
71 Weedman, D.: Astrophys. J. **183** (1973) 29.
72 Zwicky, F.: Catalogue of Selected Compact Galaxies and of Post-Eruptive Galaxies, L. Speich, Zürich (1971).

## 9.1.4 Properties of galaxies

An extensive review was given by de Vaucouleurs in 1955 [g] with references up to that year.
Modern reviews: [a, e, f].

### 9.1.4.1 Linear dimensions

The linear diameter follows from the angular diameter (see 9.1.2.2) and distance. Both values are somewhat difficult to determine and thus linear diameters are rather uncertain.

The range of linear diameters is about 0.1···50 kpc [18]. Typical range of diameters for different types:

| | |
|---|---|
| elliptical and S0 systems: | 1···50 kpc, |
| spiral systems | 10···30 kpc, |
| irregular systems | 5···20 kpc. |

The range, even for one type, is larger than thought some decades ago, but is still much smaller than the range of stellar diameters (see 4.1.5.2). The diameters for the members of the Local Group are given in 9.3.3.3.

There is a good correlation [18] between the absolute magnitude $M$ (see 9.1.4.2) and the major diameter $A$ in [pc]:

$$M = -6.00 \log A + 7.14 .$$

### 9.1.4.2 Luminosity, absolute magnitude

In order to measure the absolute magnitudes one has to know (1) the apparent magnitude (see 9.1.2.1), (2) the distance (see 9.7.2.2.4 and 9.7.5), (3) the galactic absorption, and (4) the K-correction due to the redshift (see 9.7.2.2.5). A problem in the measurement of the total apparent magnitude is the fact that galaxies have no sharp boundary. Holmberg's photographic photometry [20] is widely used: it gives integrated visual and photographic magnitudes within the 26.5 mag (pg)/□″ isophote. "Holmberg's radius" is the major axis of this isophote (see 9.1.2.2). The surface brightness of the dark sky is about 22.5 mag($B$)/□″. For other systems, e.g. de Vaucouleurs' $B(0)$-systems, see 9.1.2.1.

Table 1 gives the absolute magnitudes for 11 galaxies (5 in the Local Group, 6 in the M 81 group) with good distance moduli derived by Cepheid calibration.

Figure 1 gives the absolute calibration of van den Bergh's luminosity classification, explained in 9.1.3.2.4, valid for spirals and irregular systems.

Table 1. Absolute magnitude for 11 galaxies with good Cepheid distance moduli [e] p. 14.
LC = luminosity class (see 9.1.3.2.4)
$M_{pg}$ = absolute photographic magnitude
$m-M$ = distance modulus

| Galaxy | LC | $M_{pg}$ | $m-M$ |
|---|---|---|---|
| Local Group | | | |
| LMC | III–IV | $-17\overset{M}{.}9$ | $18\overset{m}{.}6$ |
| SMC | IV or IV–V | −16.4 | 19.3 |
| M33 | II–III | −18.3 | 24.6 |
| NGC 6822 | IV–V | −15.8 | 23.9 |
| IC 1613 | V | −14.6 | 24.4 |
| M81 group | | | |
| NGC 2403 | III | −19.0 | 27.6 |
| NGC 2366 | IV–V | −16.3 | 27.6 |
| NGC 4236 | IV | −17.5 | 27.6 |
| IC 2574 | IV–V | −16.7 | 27.6 |
| Ho II | IV–V | −16.5 | 27.6 |
| Ho I | V | −14.4 | 27.6 |

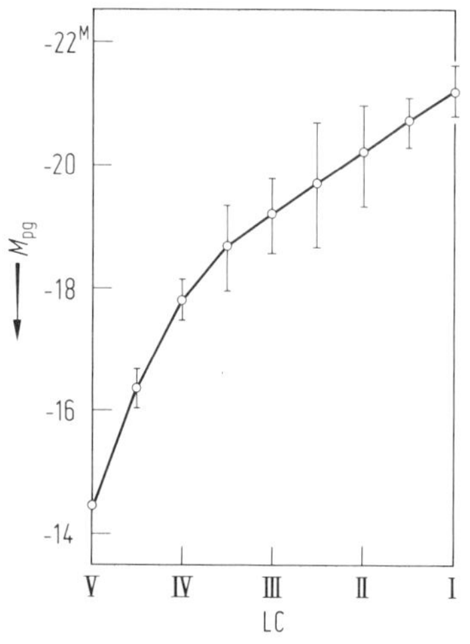


Fig. 1. The Sandage-Tammann calibration of van den Bergh's luminosity classes (LC). Bars show the dispersion about the mean values. $M_{pg}$ is corrected for inclination and absorption [e].

For elliptical galaxies the range in luminosities is much larger than for spirals. On the one hand, there are many faint elliptical dwarf galaxies, for instance in the Local Group, and, on the other, very bright elliptical galaxies in clusters of galaxies. The first-ranked cluster giants have $M_V = -23.3$ (for $H_0 = 50$, see 9.7.5), with only a small scatter for rich clusters ($N > 20$).

The reason for such bright giants within the clusters of galaxies is very probably the effect of "merging of galaxies" (see 9.3.2.2 and 9.3.4).

Bautz and Morgan [3] distinguish three classes of clusters:

BM I: systems with one outstandingly luminous galaxy,
BM II: intermediate type,
BM III: clusters which have no dominant member.

The BM class is not closely related to the richness of the cluster. For BM I clusters the first-ranked galaxy is brighter than the mean of first-ranked galaxies, but the second and third-ranked galaxies are fainter than their means. This suggests that the brightest member in the BM I clusters dominates at the expense of the others. This dominance was probably established at the time of the cluster's formation rather than by later processes.

The disk-to-bulge ratio in S0 spirals has been studied in [6].

The Tully-Fisher relation [38] (see 9.2.5.2 especially Fig. 9) gives a correlation between the absolute magnitude and the 21-cm line profile:

$$M_B = x \log \Delta V_0$$

with $\Delta V_0$ = total width of the 21-cm line $\approx 2\,V_{max}$ ($V_{max}$ = maximum rotational velocity). Different calibrations give the following values for the slope $x$ of this relation:

| | | |
|---|---|---|
| original calibration | [38] | $x = -6.25$ |
| calibration in the IR | [1, 2] | −10 |
| new calibration 1978 | [37] | − 8.6 |
| new calibration 1982 | [30] | − 7.9 |

Also lower and higher values are found in the literature but the most probably range lies between 6.2 and 10.

Overall, the range of absolute magnitudes of galaxies varies from $-8\overset{M}{.}5$ to $-23\overset{M}{.}3$. The range of "typical galaxies" varies from $-15^M$ to $-20^M$. This range is much smaller than the range of absolute magnitudes of the stars (see 4.1.2). But the lower limit of $-8^M$ is certainly due to selection effects. It is not possible, at present, to establish a definite lower limit for the luminosity of stellar systems [18]. Figure 2 shows the luminosity functions from three different authors between 1936 and 1957 and shows the development of our knowledge and the uncertainty of this problem.

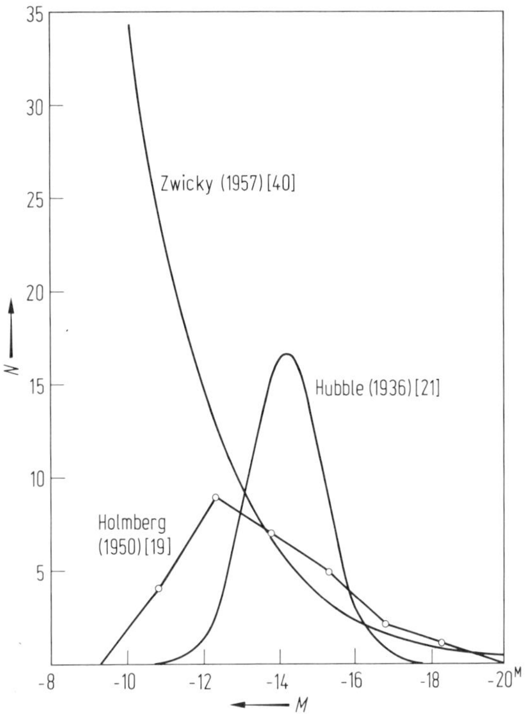


Fig. 2. Luminosity functions from three different authors [18]. $N$ = number of galaxies per interval of $1^m.5$.

For absolute magnitudes of quasars and active nuclei of galaxies, see 9.5.

### 9.1.4.3 Surface brightness

See also 9.1.2.2.

The surface brightness of normal galaxies is only slightly above that of the night sky ($\approx 22^m.5/\square''$ in $B$). This is no accident [11], but rather a selection effect: Except in a fairly narrow range of surface brightness, galaxies of a given absolute magnitude are either too small in apparent size to be distinguished from stars or have too low surface brightness to be seen against the sky, see Fig. 3.

For surface-brightness distribution within the galaxies, see 9.2.3. For ellipticity of galaxies, see 9.2.2.

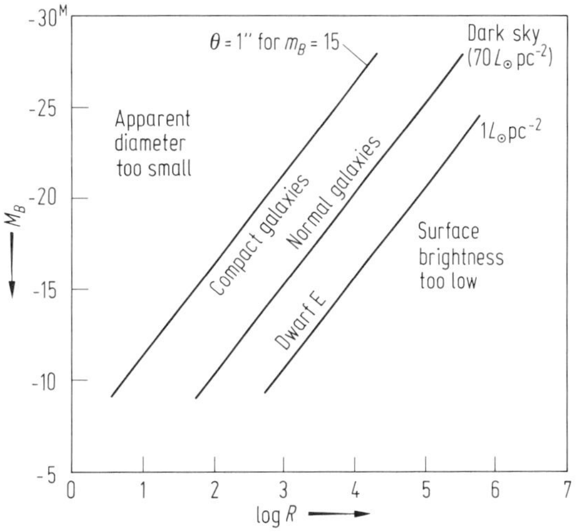


Fig. 3. Location of galaxies in the absolute magnitude – radius diagram [e] adapted from Arp, H.C.: Astrophys. J. **142** (1965) 402. Radius $R$ in [pc], $\theta$ = angular diameter.

### 9.1.4.4 Masses

Reviews: [5, 13].

Methods for estimating galaxy masses from observed kinematical quantities:

(1) rotation curves; the most reliable method, especially for spirals,
(2) line width of the 21-cm H I line,
(3) velocity dispersion of the stars (in elliptical systems),
(4) kinematics of dwarf companions or globular clusters,
(5) kinematics of binary galaxies,
(6) velocity dispersion of galaxies in groups or clusters of galaxies.

(1) Rotation

The motion of ionized gas is measured by its emission lines, mainly in the optical region, and the motion of neutral hydrogen by the 21-cm line. For details of rotation curves, rotation law etc., see 9.2.6.

In order to calculate the masses from the observed velocity curves it is usually necessary to assume that the gas is moving in circular orbits around the centre of the system. The problems of this assumption are discussed in [e] p. 32.

The simpliest way for mass determination is the assumption of Keplerian motion. But the Keplerian decrease of $V$ proportional to $1/\sqrt{R}$ is not observed even in large distances (see 9.2.6).

The velocity $V_{rot}(R)$ is connected with the gravitational potential in the galactic plane $\Phi(R)$ by

$$V_{rot}^3/R = -\partial\Phi/\partial R .$$

Knowing $\partial\Phi/\partial R$, it is possible to determine the mass out to the last observed $V_{rot}(R)$. This can be done either by direct inversion of $V_{rot}(R)$, or by model fitting of the observed $V_{rot}(R)$-curve, or by combination of both these methods. The usual method is "model fitting" (point mass, flattened spheroids, disks with variable density gradients ..., see [5]) with free parameters.

A convenient method for estimating the mass $\mathfrak{M}$ is given by the empirical formula of Bottlinger-Lohmann [26] for gravitational force $F$ at distance $R$ from the center

$$F(R) = \frac{V_{rot}^2}{R} = \frac{aR}{1+bR^3} .$$

The constants $a$ and $b$ can be expressed by the maximum rotational velocity $V_{max}$ and the corresponding distance $R_{max}$:

$$a = 3(V_{max}/R_{max})^2 \quad \text{and} \quad b = 2/R_{max}^3 .$$

By the postulation that at large distance the galaxy acts like a point mass (Keplerian motion) it follows:

$$\mathfrak{M} = \frac{a}{Gb} = \frac{3}{2}\frac{R_{max}V_{max}^2}{G}$$

($G$ = gravitation constant).

A more generalized rotation law is given by Brandt and Belton and is described in more detail in 9.2.6.

Results:

Masses of most of the bright spirals: $10^{10} \cdots 10^{11}\,\mathfrak{M}_\odot$.

Recent observations of CO and H II regions yield a rather flat rotation curve for our own Galaxy up to 16 kpc ([4], see also 9.2.6) and yield a mass of $\approx 3\cdot 10^{11}\,\mathfrak{M}_\odot$ within 16 kpc, a higher value than believed hitherto.

Mass of M 31: $4\cdot 10^{11}\,\mathfrak{M}_\odot$.

Mass of M 33 (Bottlinger-Lohmann law [26] with H I-rotation): $(1.29 \pm 0.23)\cdot 10^{10}\,\mathfrak{M}_\odot$.

(2) Line-width of the 21-cm H I line

Galaxy masses are frequently estimated by the line width of global 21-cm spectra [32], see also 9.2.5.2. This is an important technique as H I profiles are available for a few thousand galaxies. Shostak [37] has suggested that masses determined in this way are a reasonable approximation of the mass within the optical photometric diameter ($\mathfrak{M}_{opt}$) for galaxies which have rotation curves with maxima inside the optical radius and which are flat beyond the maximum (see 9.2.6). Casertano and Shostak [7] found the following line-width-mass relation

$$\mathfrak{M}_{opt}/\mathfrak{M}_{\odot} = 10^{(3.7\pm0.15)} \Delta V_0^2 D(0) r$$

with

$\Delta V_0$ = total profile width (corrected for inclination) of the 21-cm line in $[\mathrm{km\,s^{-1}}] \approx 2V_{max}$,

$D(0)$ = photometric diameter in [arc min] = the isophotal phase-on diameter at which the blue ($B$ band) surface brightness $\mu_B$ has reached 25 mag/□″ (see 9.1.2.2),

$r$ = distance of the galaxy from the sun in [Mpc].

Values for 24 galaxies are given in [7], values for more than 100 galaxies in [37]. The new catalogue of Huchtmeier et al. [22] contains values for more than 4000 galaxies.

(3) Velocity dispersion of the stars

This method is based on the assumption that the virial theorem is valid for a well mixed stellar system:

$$2T + \Omega = 0$$

with $T$ = total kinetic energy, $\Omega$ = potential energy.

The measured velocity dispersions range from 80 km s$^{-1}$ for M 32 to 500 km s$^{-1}$ for M 87, the largest elliptical galaxy in the Virgo cluster. This yields values from 6 to 80 for the mass-to-luminosity ratios (see 9.1.4.5). M 87 has the largest mass known at present: $\approx 3 \cdot 10^{13}\,\mathfrak{M}_{\odot} \approx 300\,\mathfrak{M}_{Galaxy}$.

The velocity dispersion of M 32 (a small companion of M 31) gives a mass for this dwarf elliptical system of $\approx 4 \cdot 10^{9}\,\mathfrak{M}_{\odot}$.

(4) Kinematics of companions

This method has been used for the Galaxy [25] and for M 31 [17], using the kinematics of globular clusters. The results are in good agreement with the masses obtained by the rotation curves.

(5) Binary galaxies

Analogous to the mass determination of binary stars (see 6.1.0.2) but, in addition to the unknown inclination of the relative orbit, the angle between the line connecting the two galaxies and the line of sight is also unknown. Thus much information is lost by projection effects (only the radial component of the velocity can be observed). Therefore this method is only statistically applicable. An investigation yields the following mean values [13]:

for 14 pairs of spiral and irregular systems: $\mathfrak{M} = 2 \cdot 10^{10}\,\mathfrak{M}_{\odot}$,
for 13 pairs of elliptical and S0 systems: $\mathfrak{M} = 6 \cdot 10^{11}\,\mathfrak{M}_{\odot}$.

(6) Velocity dispersion in groups and clusters

Application of the virial theorem (see above) to groups and clusters of galaxies very often yields much higher masses than the other methods, and thus very high mass-to-luminosity-ratios, up to several hundreds. The reason for this behaviour is not yet really understood. Either the virial theorem cannot be applied, which means the groups or clusters are gravitationally not stable. Or there is quite an amount of unobserved mass, especially intergalactic matter, within the clusters. This second explanation of 'missing mass' is the more favoured one today (see also 9.3.4).

Figure 4 shows schematically the relative distribution of galactic masses. Tables 2 and 3 give typical values and ranges for different galaxies and types of galaxies. Masses of the members of the Local Group are given in 9.3.3.3.

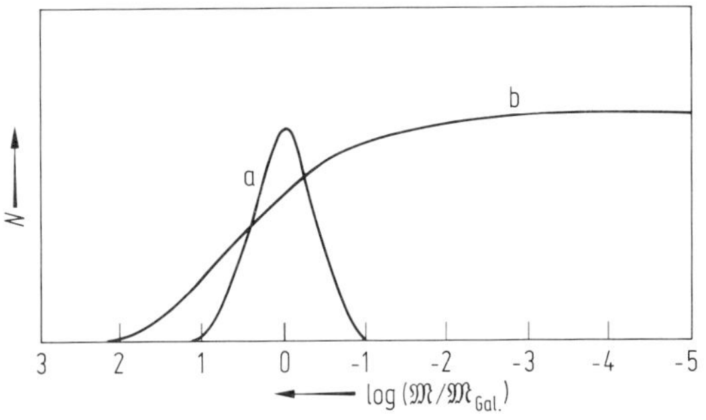


Fig. 4. Schematic distribution of galactic masses [f]. (a) Early determinations, (b) modern data.

Table 2. Approximate masses of some galaxies and types of galaxies [f].

| Type or name | $\mathfrak{M}$ [$10^{10}$ $\mathfrak{M}_\odot$] |
|---|---|
| Massive elliptical | 100 |
| M31 [1]) | 30 |
| Galaxy [1]) | 15···20 |
| Small spiral | 1 |
| Typical Irr I | 1 |
| Dwarf elliptical | 0.0001 |

[1]) Recent observations of extended rotation curves give higher masses, see above.

Table 3. Range of masses and mass-to-luminosity ratios for different types of galaxies [5]. Detailed values for the 58 galaxies are given in [5] Tables 1···3.
$N$ = number of investigated galaxies

| Type | $N$ | $\mathfrak{M}$ [$10^{10}$ $\mathfrak{M}_\odot$] | $\mathfrak{M}/L$ |
|---|---|---|---|
| E, S0 | 9 | 0.36···350 | 10 ···80 |
| Sa | 2 | 1.9 ··· 20 | 3.6··· 7 |
| Sb | 15 | 1.2 ··· 34 | 1.2··· 8.4 |
| Sc | 24 | 0.13··· 27 | 0.4···20 |
| Ir | 8 | 0.07··· 13 | 2 ···11 |

Mean masses (calculated for $H_0 = 75$) [5]

for 13 high-weighted E and S0 systems: $\mathfrak{M} = 79 \cdot 10^{10}\,\mathfrak{M}_\odot$,
for 13 high-weighted S and Ir systems: $\mathfrak{M} = 2.0 \cdot 10^{10}\,\mathfrak{M}_\odot$.

NGC 1961 [type SAB(rs)] is the most massive spiral galaxy now known, with $> 10^{12}\,\mathfrak{M}_\odot$ [34].

Total masses and/or masses of the interstellar hydrogen are also given in [32] (144 galaxies), [37] (169 galaxies) and in [4a, 13].

Masses and $\mathfrak{M}/L$ ratios in 279 double galaxies: [29].

### 9.1.4.5 Mass-to-luminosity ratio

Conventional values are: $\mathfrak{M}/L = 30$ for ellipticals
$= 4$ for spirals
with $\mathfrak{M}$ in [$\mathfrak{M}_\odot$] and $L$ in [$L_\odot$].

The ranges of values for different types of galaxies are given in Table 3.

Mean values for the above-mentioned

13 high-weighted E and S0 systems: $\mathfrak{M}/L = 68$,
13 high-weighted S and Ir systems: $\mathfrak{M}/L = 1.0$.

In particular cases $\mathfrak{M}/L$ can go up to about 100 for elliptical systems. This means that these systems contain a large number of faint objects (white dwarfs, neutron stars, black holes ...), which contribute noticably to the mass but only very little to the luminosity.

A much lower value of $\mathfrak{M}/L = 8.4 \pm 0.44$ for a sample of 32 elliptical galaxies is reported in [36].

Michard [27] found a significant correlation between the $\mathfrak{M}/L$ ratio and the $(U-B)$ colour:

$$\Delta \log(\mathfrak{M}/L) \,/\, \Delta(U-B) = 1.4\,.$$

Single $\mathfrak{M}/L$ values are mostly also given in the references for galaxy masses.

Concerning the apparently extremely high values of $\mathfrak{M}/L$ in groups and clusters: see 9.1.4.4 and 9.3.4.

### 9.1.4.6 Colours

For references to measured colours, see 9.1.2.3.

Integral photoelectric photometry of about 1000 galaxies in two or three colours [9, 10].

Infrared observations of giant elliptical systems, colours $(V-K)$ [16].

The intrinsic colours $(B-V)_0$ of galaxies range from $+1^{\mathrm{m}}$ (giant elliptical systems) to $+0^{\mathrm{m}}.4$ (irregular systems).

Figure 5 shows the colours of the different types of galaxies in the two-colour diagram, in comparison with the main-sequence stars. The range of colours of galaxies is much smaller than the range of colours of stars (see 4.1.2.1). In Fig. 6 normal galaxies, quasars (see 9.5.1.2) and the main sequence are shown in the two-colour diagram. For comparison the curve of the black body is also given. The figures demonstrate how stars, normal galaxies and quasars can be distinguished by their colours.

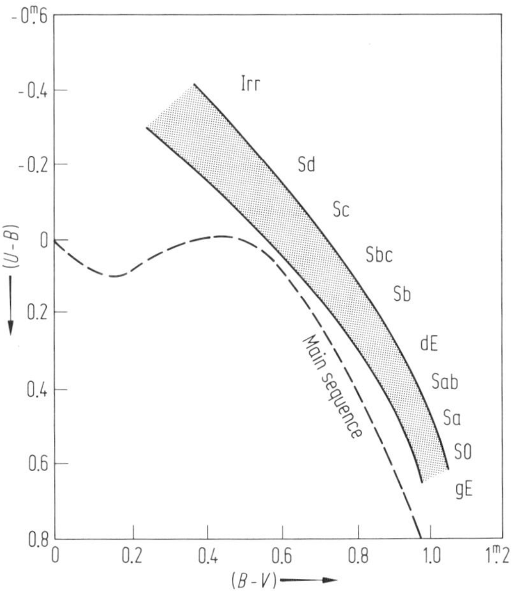


Fig. 5. Correlation of colours and types of galaxies [from Mitton, S.: Exploring the Galaxies, Ch. Scribner's sons, New York (1976)].

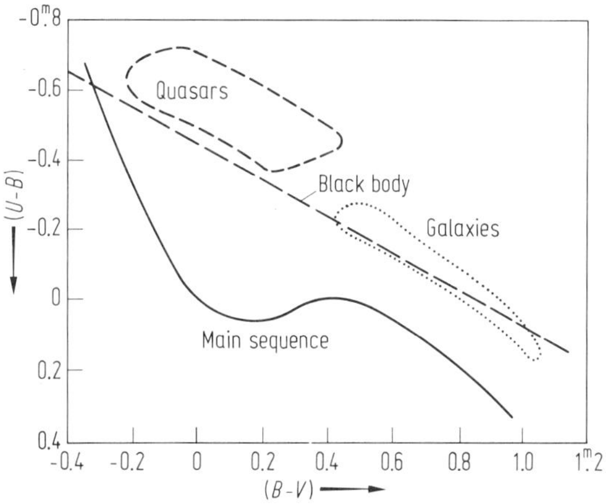


Fig. 6. Normal galaxies, quasars, black body and main sequence in the two-colour diagram (from Mitton, S., see Fig. 5).

Further observations in the infrared: elliptical systems [24]; normal and peculiar systems [31]; far IR observations [12, 33]; several papers in [d].

For X-ray observations: e.g. [15, 28] and ref. therein, and several papers in [c].

## 9.1.4.7 References for 9.1.4

### General references


a Galaxies and the Universe (Sandage, A., Sandage, M., Kristian, J., eds.), = Stars and Stellar Systems Vol. **9**, Univ. of Chicago press (1975).

b Objects of high redshift (Abell, G.O., Peebles, P.J.E., eds.), Int. Astron. Union Symp. **92**, Reidel, Dordrecht (1978).

c X-ray astronomy with the Einstein Satellite (Giacconi, R., ed.), Astrophys. and Space Science Library, Vol. **87**, Reidel, Dordrecht (1981).

d Infrared Astronomy (Wynn-Williams, D.P., Cruikshank, D.P., eds.), Int. Astron. Union Symp. **96**, Reidel, Dordrecht (1981).

e Freeman, K.C.: Observational determination of the overall features. In: Freeman, K., Larson, R.B., Tinsley, B.: Galaxies, Sixth advanced course of the Swiss Society of Astronomy and Astrophysics (Martinet, L., Mayor, M., eds.), Geneva Observatory (1976).

f Taylor, R.J.: Galaxies, Structure and Evolution, Wykeham Publ. Ltd., London (1978).

g De Vaucouleurs, C.: General physical properties of external galaxies. In: Handb. Physik (Flügge, S., ed.) **53** (1959) 311.


### Special references


1 Aaronson, M., Huchra, J., Mould, J.: Astrophys. J. **229** (1979) 1.

2 Aaronson, M., Mould, J., Huchra, J.: Astrophys. J. **237** (1980) 655.

3 Bautz, L.P., Morgan, W.W.: Astrophys. J. **162** (1970) L149.

4 Blitz, L.: Astrophys. J. **231** (1979) L115; erratum in **234** (1979) L172.

4a Brosche, P.: Astron. Astrophys. **23** (1973) 259.

5 Burbidge, E.M., Burbidge, G.: The masses of Galaxies. In: [a] p. 81.

6 Burstein, D.: Astrophys. J. **234** (1979) 435.

7 Casertano, S.P.R., Shostak, G.S.: Astron. Astrophys. **81** (1980) 371.

8 De Vaucouleurs, G.: Mon. Not. R. Astron. Soc. **113** (1953) 134.
9 De Vaucouleurs, G.: Astrophys. J. Suppl. **5**, No. 48 (1961) 233.
10 De Vaucouleurs, G., de Vaucouleurs, A.: Mem. R. Astron. Soc. **77** (1972) 1.
11 De Vaucouleurs, G. in: The Formation and Dynamics of Galaxies (Shakeshoff, J.R., ed.), Int. Astron. Union Symp. **58**, Reidel, Dordrecht (1974) p. 3.
12 Drapatz, S., Haser, L., Hofmann, R., Rothermel, H.: Mitt. Astron. Ges. **50** (1980) 109.
13 Faber, S.M., Gallagher, J.S.: Annu. Rev. Astron. Astrophys. **17** (1979) 135.
14 Freeman, K.C.: Astrophys. J. **160** (1970) 811.
15 Giacconi, R., Tananbaum, H.: Science **209** (1980) 865.
16 Grasdalen, G.L. in: [b] p. 26.
17 Hartwick, F.D.A., Sargent, W.: Astrophys. J. **190** (1974) 283.
18 Holmberg, E.: Magnitudes, Colors, Surface Brightness, Intensity Distributions, Absolute Luminosities, and Diameters of Galaxies. In: [a] p. 123.
19 Holmberg, E.: Medd. Lund Astron. Obs. Ser. II, No. 128 (1950).
20 Holmberg, E.: Medd. Lund Astron. Obs. Ser. II, No. 136 (1958).
21 Hubble, E.: Contr. Mt. Wilson Obs. No. 548 and 549 (1936).
22 Huchtmeier, W.K., Richter, O.G., Bohnenstengel, H.D., Hauschildt, M.: in preparation (1982).
23 King, I.R.: Astron. J. **71** (1966) 64.
24 Lebofsky, M.J. in: [b] p. 251.
25 Lohmann, W.: Sitz. Ber. Österr. Akad. Wiss. II (1961) 169, 171.
26 Lohmann, W.: Astrophys. Space Science **29** (1974) 61.
27 Michard, R.: Astron. Astrophys. **91** (1980) 122.
28 Penston, M.V. in: [b] p. 247.
29 Peterson, S.D.: Astrophys. J. **232** (1979) 20.
30 Richter, O.G., Huchtmeier, W.K.: in preparation (1982).
31 Rieke, G.H., Lebofsky, M.J.: Annu. Rev. Astron. Astrophys. **17** (1979) 477.
32 Roberts, M.S.: Radio Observations of Neutral Hydrogen in Galaxies. In: [a] p. 309.
33 Rounan, D., Viallefond, F., Drapatz, S., Lena, P., Puget, J.L.: Mitt. Astron. Ges. **50** (1980) 13.
34 Rubin, V.C., Ford, W.K. jr., Roberts, M.S.: Astrophys. J. **230** (1979) 35.
35 Rubin, V.C., Burstein, D., Thonnard, N.: Astrophys. J. **242** (1980) L149.
36 Schechter, P.L.: Astron. J. **85** (1980) 801.
37 Shostak, G.S.: Astron. Astrophys. **68** (1978) 321.
38 Tully, R.B., Fisher, J.R.: Astron. Astrophys. **54** (1977) 661.
39 Wilson, C.: Astron. J. **80** (1975) 175.
40 Zwicky, F.: Morphological Astronomy, Springer, Berlin (1957).

## 9.2 Internal structure and dynamics of galaxies

For intrinsic integral properties, see 9.1.4.

| | |
|---|---|
| linear dimensions: | 9.1.4.1 |
| luminosity, absolute magnitude: | 9.1.4.2 |
| surface brightness: | 9.1.4.3 |
| masses: | 9.1.4.4 |
| mass-to-luminosity ratio: | 9.1.4.5 |
| color: | 9.1.4.6 |

### 9.2.1 Stellar and gaseous content of normal galaxies

Reviews are given by Sandage [S 3] and by Spinrad and Peimbert [S 10]. The stellar and gaseous content of galaxies varies systematically along the morphological sequence (see 9.1.3) from E to Sm. Variations along the spirals from Sa to Sd [S 3]

1) increasing absolute luminosity of the brightest stars in regions of spiral arms,
2) increasing percentage of mass in form of gas and dust,
3) increasing size and number of H II regions in the spiral arms,
4) progressively bluer $(B-V)$ and $(U-B)$ colors, indicating progressively earlier stars that contribute most of the light.

Already broad-band photographic or photoelectric colors show that hot stars are important contributors to the blue light in the centers of Irr and Sc systems and in the outer spiral-arm regions of many Sc, Sb and possibly Sa systems. No early-type stars are needed to explain the colors of centers of most Sa, Sb and E systems.

A more quantitative approach to galaxy population studies are 1) photoelectric narrow-band measurements of colors (continuum) and line indices sensitive to stellar temperature, luminosity and abundance differences, and 2) equivalent widths from low- or medium-dispersion slit spectra.

The spectroscopic categories of the low-dispersion approach to the determination of the stellar content are summarized in Table 1.

Model luminosity functions (see 8.2.2) of fainter stars for M 31 and M 32 are shown in Fig. 1.

Table 1. Low-dispersion spectroscopic categories for galaxies [M3], extended by [S10].

| Category | Description | Spectroscopic identification | Typical galaxies |
|---|---|---|---|
| Orion | H II regions, blue stars; often irregularly shaped galaxy | Strong emission lines like the Orion Nebula, a hot-star continuum, He I absorption line, other indicators of types B···F in the blue ($\lambda$ 3820 of He is a good indicator of early B stars) | NGC 4214, NGC 4449, LMC bar, M82 core |
| Intermediate | Nuclear regions of Sc galaxies, main bodies of giant spirals. Yields types f and fg | Blue spectral-type, near F8, very composite spectrum. $\lambda$ 3727 [O II] emission common | NGC 5194, NGC 4321 |
| Amorphous | Centers of big Sb, Sa systems. Main bodies of giant E galaxies | The type K0 in most cases, type closer to M0 in the deep red. Emission lines weak | M31, M81, NGC 4472 |
| Weak-lined | Metal-poor population of dwarf E system | Globular-cluster-like; H lines intermediate, metals weak, no emission lines | Dwarf E's like NGC 205, possibly NGC 5195 |

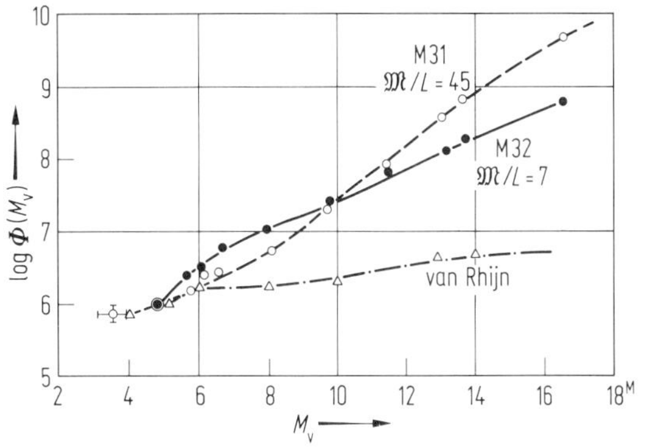


Fig. 1. Model luminosity functions $\Phi$ for M 31 and M 32 compared with van Rhijn's luminosity function for the solar vicinity (see 8.2.2) (faint end: $M_v > 4$) [S 10]. The most left symbol of M 31 with error bars is an extrapolation of the giant branch in M 31 back to the star's original main-sequence position. $\mathfrak{M}/L$ = mass-to-luminosity ratio (see 9.1.4.5).

Infrared studies of stellar content of galaxies: [A 3]. The IR is important as most of the radiation of normal galaxies is emitted in the region $\lambda > 1$ µm.

Review of integrated energy distribution of galaxies: [W 3].

Helium abundance [S 10]: 5 Sc systems $N(\mathrm{He})/N(\mathrm{H}) = 0.120$,
5 Irr systems $= 0.095$.

## 9.2.2 The ellipticity of galaxies

The frequency function of true ellipticities of spheroidal systems

$$e = 1 - c/a = 1 - q_0$$

can be derived from the observed frequency function of apparent ellipticities

$$\varepsilon = 1 - b/a = 1 - q$$

under the assumption of random orientation of the spin axis [D 1], with $a$ = true major axis, $b$ = apparent minor axis (projection effect), $c$ = true minor axis of the flattened spheroid.

An investigation of isophotal diameters ($\mu_B = 25.0\,\mathrm{mag}/\square''$) of more than 2000 galaxies gives the following conclusions:

(1) Elliptical galaxies: The distribution of true ellipticities among E galaxies is definitely not uniform (as was supposed in earlier papers) up to 2/3, i.e. $q_0 > 1/3$. In particular spherical galaxies ($q_0 = 1$) are rare. On the other hand the sharp cutoff of the observed (apparent) ellipticity at type E 7 indicates that no elliptical galaxies with a high degree of flattening (like the spirals) exist. The best-fit Gaussian model gives $\langle e \rangle \approx 0.36$ with a small cosmic dispersion $\sigma = 0.1$.

(2) Lenticular galaxies tend to be more flattened than ellipticals but the analysis suggest that two groups are present:
   a major group ($\approx 90\%$ of the sample) with $\langle e \rangle = 0.65$
   a minor group ($\approx 10\%$) with $\langle e \rangle = 0.35$

(3) Spiral galaxies from S0 to Sm have ellipticity functions similar to the lenticulars with
   a major group ($\approx 70\%$) with $\langle e \rangle = 0.7 \cdots 0.8$
   a minor group ($\approx 30\%$) with $\langle e \rangle = 0.4$
   i.e. spirals having small bulges or large bulges respectively.

$q_0$ decreases smoothly along the Hubble sequence from E (morphological parameter $t = -5$; see 9.1.3.2.7) to Sd ($t = 7$).

Beyond stage Sd the ellipticity decreases rapidly so that Magellanic spirals have typically $q_0 = 0$.

Examples are given in [D 4].

The existence of thick disks [B 15] and extremely thin disks [G 1] has been shown observationally.

Intrinsic flattening of 168 E, 267 S0 and SB0, and 254 ordinary spirals are investigated in [S 2].

The ratio of the two components (spheroidal and flat) of the luminosity distribution (see 9.2.3) varies smoothly as a function of the morphological parameter $t$ along the spiral sequence and it suggests that the Hubble sequence from Sa to Sd is basically an angular momentum sequence. The low velocity of rotation found even in flat systems, as compared with the velocity dispersion, ruled out models attributing the flattening only on rotation.

Flattening is a dynamical property which cannot change significantly in times less than the relaxation time (about $10^{12} \cdots 10^{14}$ years). Therefore the difference in intrinsic flattening between E and S galaxies shows that one type cannot evolve into the other. A basic difference must have existed between these groups already at the time of their formation [S 3].

The isophotes within one elliptical galaxy are not necessarily concentric, the ellipticity can change from the central to the outer parts. Five classes can be distinguished [B 4]: increasing ellipticity, decreasing, with maximum, with minimum and constant. Furthermore, the systems can be twisted, i.e. the direction of the major axis of the isophote can change. Table 2 gives some examples.

Bertola and Galetta [B 3] describe 5 galaxies with dust lanes crossing a luminous elliptical-like body along the minor axis (NGC 1947, 5128, 5363, and the galaxies associated with Cyg A and PKS 1934–63). They suggest a new class of galaxies: prolate stellar structures cut equatorially by gaseous planes. The dynamics of these systems is complicated and not yet understood.

Table 2. Examples for ellipticity trend and twisting [B4].

| Galaxy | Classification | Ellipticity trend | Twisting |
|---|---|---|---|
| NGC 205 | E5 pec | maximum | present |
| NGC 1265 | S0 | increasing | suspected |
| NGC 1270 | E3 | maximum | |
| NGC 1278 | E2 pec | minimum | |
| NGC 1281 | E5 | maximum | |
| CR 32 | E1 | increasing | present |
| IC 312 | E6 | maximum | |
| NGC 4125 | E6 pec | maximum | present |
| NGC 4486 | E0 pec | increasing | present |
| NGC 4494 | E1 | constant | |

## 9.2.3 Luminosity distribution

For total luminosity, see 9.1.4.2.

Review: see Holmberg [H 3].

Elliptical systems

The radial surface-brightness distribution of the brighter elliptical galaxies ($M_V < -15$) and compact dwarf systems follows closely the empirical law [D 2, D 3]

$$\log I(\varrho) = -3.33\,(\varrho^{1/4} - 1) \qquad (1)$$

with $\varrho = R/R_e$

$R_e$ = effective radius = radius (or major axis) of that isophote inside which half of the total light is emitted

$I(\varrho)$ = surface brightness at distance $\varrho$ from the center along the major axis.

The formula fails only in the innermost and in the outermost parts. Another representation is given by King's model [K 2].

The distribution is similar to an isothermal sphere. The similarity of $I(\varrho)$ for all normal ellipticals means that they are all now in a fairly similar dynamical state.

For variation of the ellipticity of the isophotes within individual galaxies, see 9.2.2.

The specific intensity $I_e$ *at* $R_e$ or the corresponding surface brightness $\mu_e$ in [mag per unit solid angle] is related to the average surface brightness $\mu'_e$ *within* $R_e$; for an $\varrho^{1/4}$ distribution

$$\mu'_e = \mu_e - 1.40\,. \qquad (2)$$

In the denser systems the $\varrho^{1/4}$ law apparently applies right up to the center. Here, according to Eq. (1), the surface brightness should be $2.5 \times 3.33 = 8.3$ mag brighter than at $R_e$, thus

$$\mu_0 = \mu_e - 8.3 = \mu'_e - 6.9\,. \qquad (3)$$

The range of $\mu'_e$ in dense systems is from $\approx 19.5$ to $\approx 21.5$ mag$(B)/\square''$ (corresponding to about $1200 \cdots 200\, L_\odot\, \mathrm{pc}^{-2}$).

Supergiant elliptical systems show an elliptical-like core in an extended outer envelope.

Carter [C 3] investigated the optical extent of four giant elliptical and cD galaxies, and in two of them he found no indication of the convergence of the total luminosity of the galaxy.

Low-density dwarf spheroidal systems ($M_V > -15$) do not obey the $\varrho^{1/4}$ law. The light concentration is slight and the central core with radius $R \approx R_e/2$ has a nearly constant surface brightness. But the surface brightness drops off more rapidly in the outer parts than for giant systems. The Fornax system (dE, see 9.3.3.3) has $\mu'_e = 24.8$ mag$/\square''$ ($\approx 9\, L_\odot\, \mathrm{pc}^{-2}$) or about 1% of the corresponding luminosity density in the compact dwarf E 2 system M 32. Preliminary photometries of Sculptor and other nearby dwarfs indicate even lower densities. Large numbers of dwarfs with such densities might remain undetectable by current techniques.

Spiral systems

Edge-on spirals and S0 galaxies clearly show two main structural components, a flat disk and a spherical bulge. The bulge varies in its relative importance from dominant to zero (see Fig. 3). Luminosity distribution of the bulge (characteristic for ellipticals, see above):

$$\log I_1(\varrho) \propto \varrho^{1/4}, \tag{4}$$

luminosity distribution of the flat component (characteristic for disks of late spirals)

$$I_2(R) = I_0\,\mathrm{e}^{-\alpha R} \quad \text{or} \quad \log I_2(R) = \mathrm{const} - \alpha R \tag{5}$$

i.e. an exponential luminosity law. $\alpha$ is the inverse of the scale length $\Lambda$; it is measured by the photometric gradient

$$G(R) = \mathrm{d}(\log I)/\mathrm{d}R, \quad \text{thus} \quad \Lambda = 0.4343/G(R). \tag{6}$$

For a purely exponential law: effective radius $R_e = 1.6785\,\Lambda$. The relations corresponding to Eq. (3) are:

$$\mu_0 = \mu_e - 1.82 = \mu'_e - 1.12. \tag{7}$$

From an investigation of 36 exponential disks Freeman [F 5] arrived at the following conclusion:

a) $\Lambda$ has a range of one order of magnitude from 0.5 to 5 kpc for types earlier than Sc.
b) For types later than Sc the maximum value of $\Lambda$ decreases from 5 kpc at Sc (morphological parameter $t=5$) to 1 kpc at Im ($t=10$), confirming earlier results on the dependence of galaxy diameters on Hubble type.
c) For 28 out of 36 spirals and S0 galaxies the exponential disks have nearly the same intensity scale: $\langle\mu_0\rangle = 21.65$ mag/□″ with a very small standard deviation of $0^m.3$, in spite of a large range of $5^M$ in absolute magnitudes and independent of morphological type from $L^-$ ($t=-3$) to Im ($t=10$), Fig. 2. However, see critical comments in [B 15].

Figure 3 shows the radial luminosity distribution for three galaxies with different ratios of spherical and exponential components, and for comparison a $R^{1/4}$-distribution.

Almost all disk galaxies including the Magellanic irregulars show the exponential disk. Thus in all these systems the outer parts, containing most of the angular momentum, have reached a similar dynamical state.

Some S0 galaxies show up clearly a third, flat, lens-like component.

The luminosity profiles of high surface-brightness disks dip below the projected exponential component, see for instance M 83 in Fig. 3. Freeman [F 4] supposes that those M 83-like galaxies are systems with a lens component: the dip is the result of adding a flat lens component to the bulge and exponential components, as schematically shown in Fig. 4.

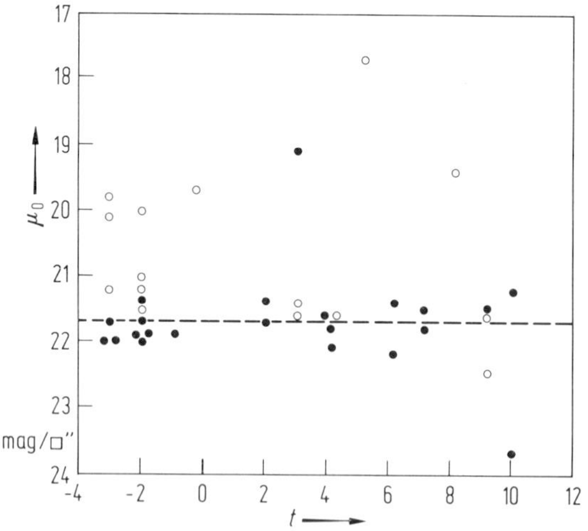


Fig. 2. Corrected face-on central surface brightness $\mu_0$ for exponential disks versus the morphological type $t$ (see 9.1.3.2.7) [F 3]. $\mu_0$ is independent of the type, but some individual galaxies are aberrant. Full circles: type I luminosity profiles; open circles: type II luminosity profiles, as explained in Fig. 3.

Fig. 3: next page.

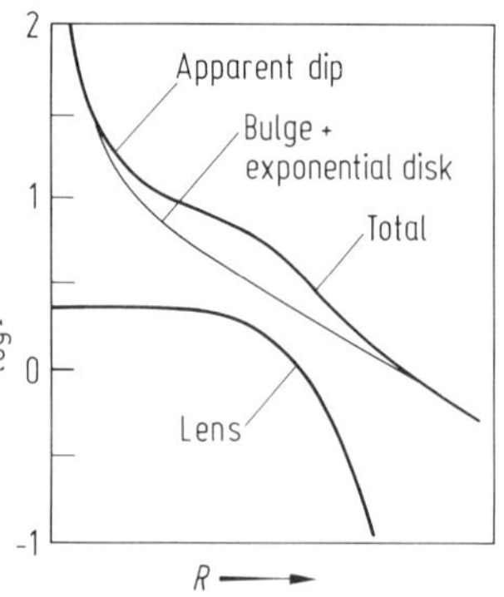


Fig. 4. Possible explanation of the "dip" (M 83 type, see Fig. 3) by superposition of bulge-, exponential- and lens components [F 4].

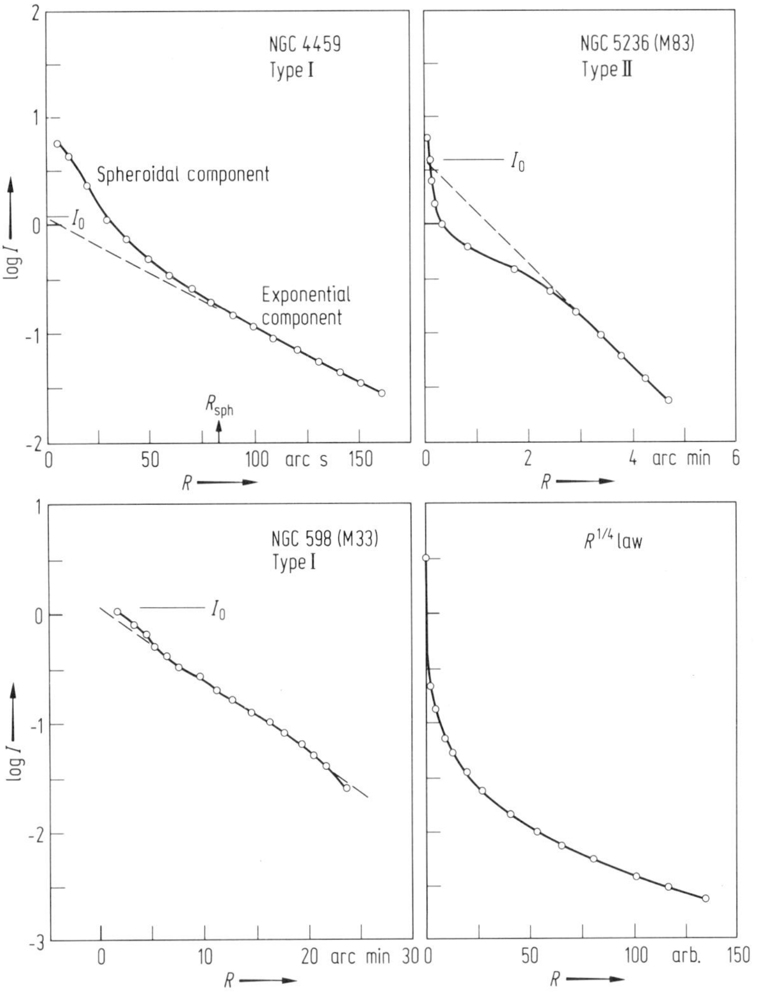


Fig. 3. Radial luminosity distribution for NGC 4459 (type S0), M 83 (Sc), and M 33 (Scd). The $R^{1/4}$ distribution is also shown [F 3].

More recent observations of luminosity distribution in 25 spiral systems are given in [B 7].

The local mass-to-luminosity ratio has been investigated in 6 spiral galaxies [B 8] by combining 21 cm-line studies and optical surface photometry. Inspite of the large uncertainties in the modelling procedures it is concluded that this ratio increases significantly in the outer parts. Local values of $\mathfrak{M}/L$ may be of the order $10^2 \cdots 10^3$.

Barred spirals

Surface photometry of two barred spirals (NGC 7479, 7743) [O 1] shows four components: (1) central bulge, (2) bar, (3) spiral arms, (4) underlying disk. The luminosity distribution of the bar follows the $R^{1/4}$ law.

## 9.2.4 Spiral structure

The density-wave spirals (see 8.4.2.4) are usually well represented by logarithmic spirals with only slowly changing pitch angle [K 1] (see Fig. 5):

$$R = R_0 \exp[\varkappa(R)\Phi] \quad \text{or} \quad \log R = \mathrm{A} + \mathrm{B}\Phi$$

with $\varkappa = \mathrm{tg}\mu(R)$; $\Phi$ = winding angle; pitch angle $\mu$ being nearly constant. Pitch angles in different galaxies vary between a few and about 30°; the change of the pitch angle within any system is at most ±4° [K 1].

Material arms, on the other hand, possess a nearly hyperbolic form [S 6]

$$R\Phi = \mathrm{const} \cdot V_{\mathrm{rot}}(R),$$

dictated by the rotation curve of the galaxy (see 9.2.7).

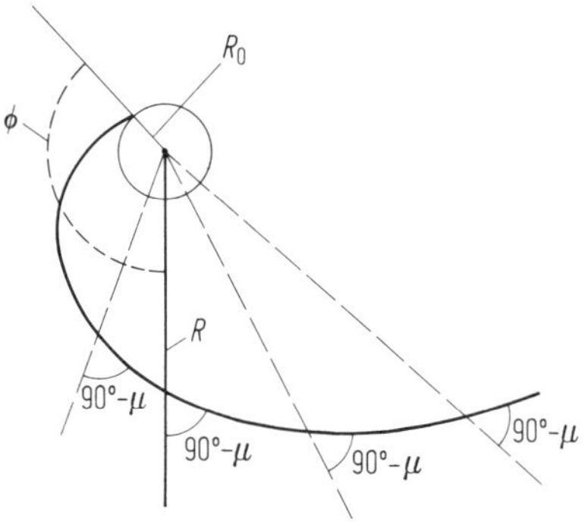


Fig. 5. The logarithmic spiral. $\Phi$ = winding angle; $\mu$ = pitch angle; $90° - \mu$ = characteristic angle.

Kennicutt [K 1] investigated 113 objects and considered two functions: logarithmic spiral with constant pitch (i.e. linear in $\log R$ versus $\Phi$) and hyperbolic spiral (i.e. linear in $R\Phi$ versus $R$) with the result that neither logarithmic nor hyperbolic spirals accurately represent the spiral arms; but, within the limitations set by arm distortions, they may serve as adequate interpolation functions in normal spiral systems. The question of the mathematical form is irrelevant for the majority of real spiral arms.

The pitch angles correlate weakly with the arm structures and with the bulge-to-disk ratio, i.e. the Hubble classification criteria are less closely coupled than thought previously [R 3]. But the maximum rotation velocity $V_{\max}$ (see 9.2.6) is well correlated with the arm pitch, suggesting that the shape of spiral pattern is mainly dictated by kinematic parameters, independent of the physical origin of the arms, Figs. 6 and 7.

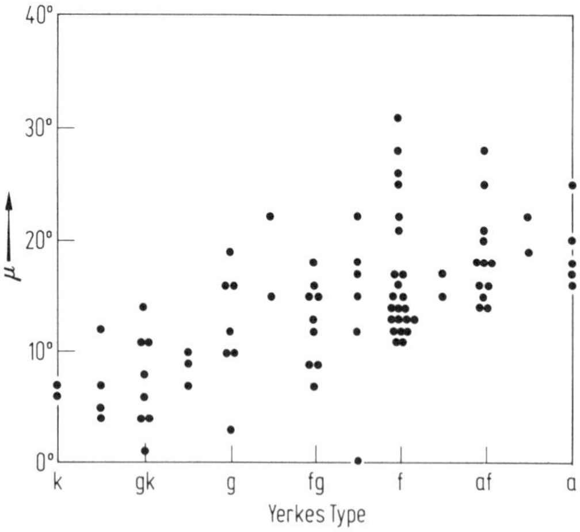


Fig. 6. The measured pitch angles vs. Yerkes-type galaxies [K 1]. For Yerkes type, see 9.1.3.2.1.

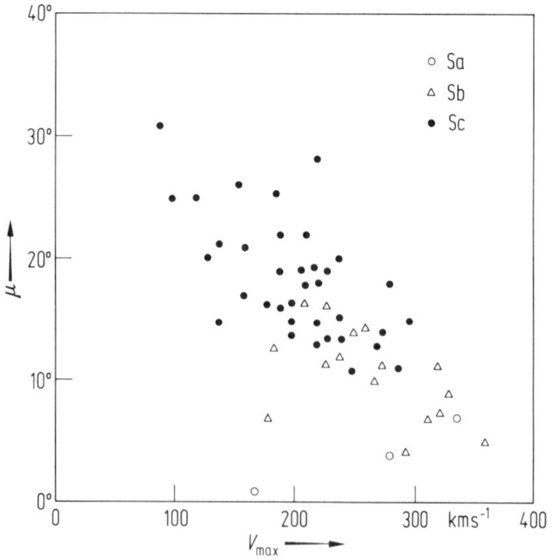


Fig. 7. Correlation between pitch angle $\mu$ and maximum rotational velocity $V_{\max}$ (see 9.2.6) for spiral galaxies [K 1].

In strongly barred galaxies the arms are systematically distorted. Observational constraints of driving mechanism for spiral density waves: [K 4].

## 9.2.5 Radio radiation of normal galaxies

For galaxies with extreme strong radio radiation, see 9.6: Extragalactic radio sources.

### 9.2.5.1 Radio-continuum structure

Reviews

Large-scale radio-continuum structure of E and S0 galaxies: [E 2], and of spiral galaxies: [V 6].

Observations

Only a selection out of the large amount of radio continuum observations at different frequencies can be given here. 17 radio-continuum surveys of M 31 during 1951···1975 and 30 radio-continuum observations of 26 spiral galaxies for which the HWBW of the telescope is less than about 1/5 of the Holmberg radius (see 9.1.2.2 and 9.1.4.2) are collected in [V 4].

Compendium of radio measurements of bright galaxies: [H 1].

Radio continuum survey of isolated galaxies: [A 4].

Radio continuum observations of the nuclei of normal galaxies: [E 3, V 2].

Extended radio emission aligned with compact nuclear sources in normal galaxies: [J 1].

Several contributions dealing with individual galaxies are found in [b].

The radio continuum radiation of spirals in multiple systems is about a factor 2.5 higher in the center than in isolated spirals, but there is no significant difference in the disks [H 11].

Radio-continuum properties of spiral galaxies (based on a sample of 280 galaxies) [H 10]:

(1) In general more than 90% of the total radio power at 1415 MHz is emitted in the disk component.
(2) The median radio power of the disk component as well as of the central sources is directly proportional to the mean optical luminosity.
(3) The radio emissivity in the disk is independent of the morphological *family* (see 9.1.3). Central sources in barred spirals are on the average a factor 2 more powerful than those in ordinary spirals.
(4) The radio emissivity of the disk depends on the morphological *stage*. Compared to Sb/Sbc/Sc galaxies it is a factor 2···3 lower in earlier types and may be as much as a factor 2 lower in the late-type spirals. The radio power of the central sources decreases with the morphological stage. In early-type spirals they are at least a factor 10 stronger than in the late-type spirals.
(5) The disk emissivity, in terms of radio power per unit light, does not depend on color.
(6) The mean ratio of optical to radio size of the disk component is similar for the different morphological types except for early-type spirals which have smaller radio disks.
(7) There is no relation between the radio power of the central source and of the disk component.

On the basis of these results Hummel [H 10] concludes that the main sources of relativistic electrons belong to the old disk population and that in general the density waves and the central sources play an unimportant role in determining the radio-continuum emissivity in the disks of spiral galaxies. The radio power of the central sources depends on the central density and possibly on the occurrence of large non-circular gas motions in the central parts of spiral galaxies.

### 9.2.5.2 Neutral hydrogen (H I) in galaxies

Observations of interstellar neutral hydrogen are based on the 21-cm line (see 7.1.6.1). Intensity observations give information about the amount and the distribution of H I; for rotation curves of galaxies based on Doppler observations, see 9.2.6.

Review (methods, results, correlation with other parameters): [R 1].

Observations of integral spectra for many galaxies are given in Table 3, next page.

H I envelopes of late-type galaxies: [H 6].

About 30% of 100 investigated large galaxies have H I at distances from the center of more than 2 Holmberg radii (see 9.1.2.2 and 9.1.4.2) [H 8].

Table 3. Integral H I observations.

| Objects | Ref. |
|---|---|
| 241 DDO galaxies | Fisher, J.R., Tully, R.B.: Astron. Astrophys. **44** (1975) 151 |
| 169 late type spirals | Shostak, G.S.: Astron. Astrophys. **68** (1978) 321 |
| 40 spiral systems (Sa) | Bottinelli, L., Gouguenheim, L., Paturel, G.: Astron. Astrophys. **88** (1980) 32 |
| 99 blue compact galaxies | Gordon, D., Gottesman, S.T.: Astron. J. **86** (1981) 161 |
| 119 blue compact dwarf galaxies | Thuan, T.X., Martin, G.E.: Astrophys. J. **247** (1981) 823 |
| 461 Nilson dwarf galaxies | Thuan T.X., Seitzer, P.O.: Astrophys. J. **231** (1979) 327, 680 |
| 46 southern galaxies | Whiteoak, J.B., Gardner, F.F.: Australian J. Phys. **30** (1977) 187 |
| 112 bright galaxies | Dickel, J.R., Rood, H.J.: Astrophys. J. **223** (1978) 391 |
| 58 Seyfert galaxies | Heckman, T.M., Balick, B., Sullivan, W.T. III: Astrophys. J. **224** (1978) 745 |
| 1787 nearby galaxies | Fisher, J.R., Tully, R.B.: Astrophys. J. Suppl. **47** (1981) 139 |
| 39 Virgo-cluster galaxies | Huchtmeier, W.K., Tammann, G.A., Wendker, H.J.: Astron. Astrophys. **46** (1976) 381 |
| 83 spiral galaxies (Sa) | Huchtmeier, W.K.: Astron. Astrophys. **110** (1982) 121 |
| 36 S0 galaxies | Biermann, P., Clarke, J.N., Fricke, K.J.: Astron. Astrophys. **75** (1979) 7 |
| 279 binary systems | Peterson, S.D.: Astrophys. J. Suppl. **40** (1979) 527 |

A total of 7400 H I observations of about 4200 different galaxies have been collected from the literature in [H 9].

Detailed information on H I distribution in galaxies is available for about 100 galaxies measured with the aperture synthesis or single-dish telescopes (see also references for rotation curves, 9.2.6).

In general the H I extends to about 1 to 2 times the Holmberg diameter (26.5 mag/□″) [B 8a]. Detailed studies with high resolution yield considerable asymmetries in the surface density as well as in the extend of neutral hydrogen and in the velocity field. Galaxies with bulges generally have a central H I deficiency. $H_2$ fills up the hole in later types (e.g. M 101). A hole in the gas (H I and $H_2$) remains in Sa and Sb galaxies (e.g. M 31, M 81).

H I observations of smooth-arm spiral galaxies obviously being gas-deficient: [W 4].

Observations of dwarf galaxies and irregular galaxies: [H 7].

H I survey of M 33: [B 1, H 5].

Three-dimensional distribution of H I in M 31: [W 1]; a warp in the H I distribution at the extreme NE and SW of M 31 is reported in [E 4].

Warps in the H I distribution have been observed in the case of some edge-on spiral galaxies [S 11].

The distribution and the velocity field of H I in 20 bright northern galaxies combined with optical *UBV* observations is discussed in [V 7].

Many short papers of H I observations are presented in [h].

In 6 nearby spheroidal dwarf systems (members of the Local Group, see 9.3.3.3) no H I was detected. This gives an upper limit of the H I content of dwarf spheroidal galaxies as low as a few parts in $10^4$ by mass [K 3].

The H I mass and the ratio of hydrogen mass to total mass for 140 galaxies are given in [R 1]. The distance-independent ratio $\mathfrak{M}_{HI}/L$ is a function of the morphological type, too, increasing to later type galaxies. The main result is shown in Fig. 8.

Integral properties like mass, luminosity, size, maximal rotation for 169 late-type galaxies determined from H I observations are given in [S 9]. For integral properties, see also 9.1.4.

### Tully-Fisher relation

Tully and Fisher [T 3] found a relation between the absolute magnitude (luminosity, see 9.1.4.2) and the global width of the H I profile for spiral galaxies, i.e. between luminosity and total mass (Fig. 9). This gives a new possibility for distance determination (see 9.7). This relation was studied in the infrared [A 1, A 2]. Observational uncertainties (inclination, luminosity correction for internal extinction) and possible systematic effects (differences between morphological types) [R 4] leave some calibration problems. The slope of the relations lies between −6.25 [T 3] and −10 [A 1, A 2]; see also 9.1.4.2.

A similar correlation (with absolute magnitudes corrected for total internal absorption) was published by Sandage and Tammann ([S 3a] Fig. 1).

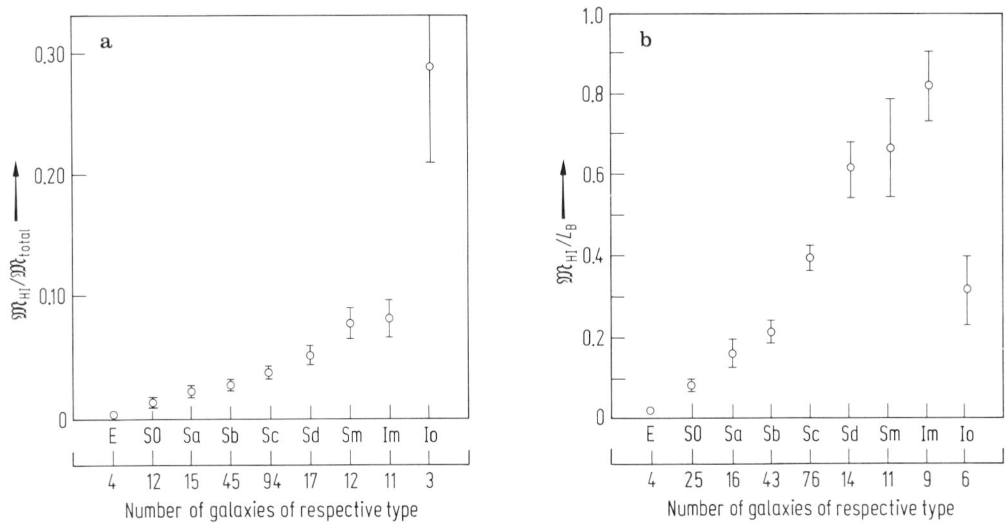


Fig. 8. The average hydrogen content for different morphological types of galaxies [L 1]. (a) Ratio hydrogen mass to total mass, (b) ratio mass to (blue) luminosity. Number of galaxies for each type is indicated. The error bars represent the standard deviation of the sample mean.

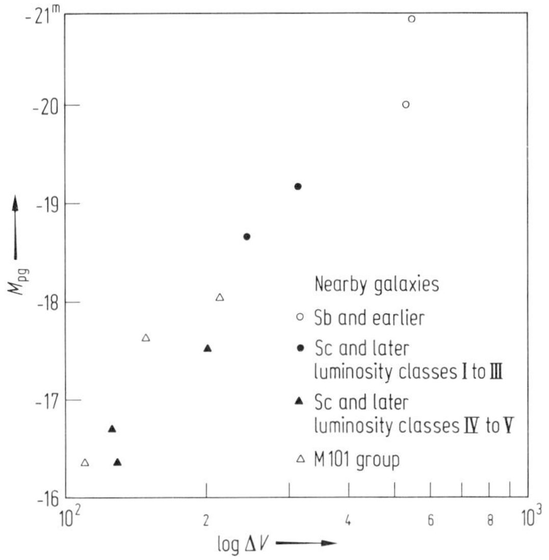


Fig. 9. The Tully-Fisher relation: global width of H I profile $\Delta V$ vs. absolute magnitude for spiral galaxies [T 3]. $\Delta V$ in [km $s^{-1}$].

## 9.2.6 Rotation, kinematics, dynamics

Reviews: Rotation of normal galaxies [B 13].

Kinematics of elliptical and S0 systems [C 1].

Stellar dynamics and structure of galaxies [F 3].

Kinematics of spiral and irregular galaxies [V 5].

See also general references in 9.2.7.

Observations of rotation curves

Rotation curves are derived from radial-velocity measurements of emission lines in the optical region and from the 21-cm line of neutral hydrogen (see also 9.2.5.2). Figure 10 shows typical rotation curves for 25 galaxies including our own Galaxy (see 8.4.1.3).

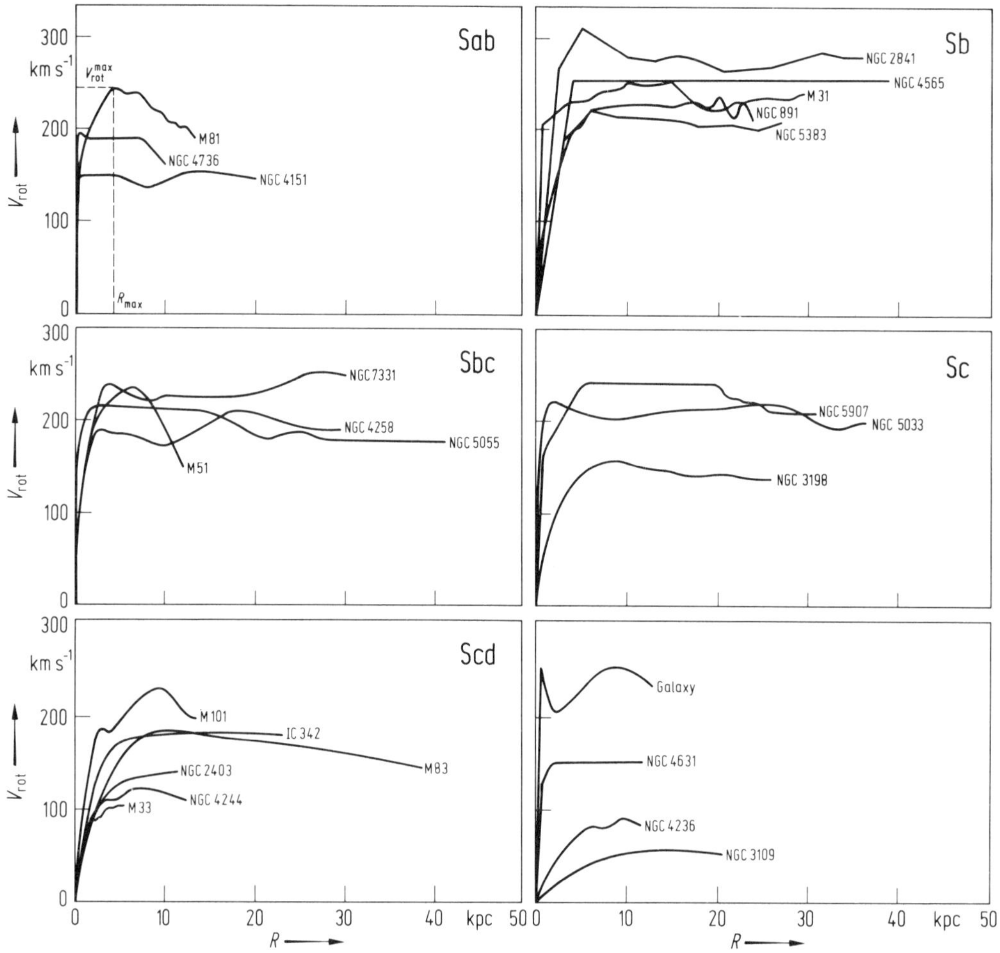


Fig. 10. Rotation curves of 25 galaxies of various types [B 8a]. The parameters $V_{rot}^{max}$ (maximum rotational velocity) $= V_{max}$ in 9.2.4) and $R_{max}$ (the corresponding distance from the center) are shown for M 81 on the first picture left above.

Rotation curves of galaxies [R 2, R 3, R 7, R 8, P 3, P 4, P 6, K 5].

References to rotation curves of 69 barred and normal spiral galaxies are published in [K 4].

List of 48 galaxies for which the velocity field has been measured by sampling many points over a large fraction of their disks by different methods (slit spectra, interferometry, H I 21-cm line) [V 5].

Some observations of the Andromeda Nebula M 31: Rotation of nuclear region [P 2]; velocity dispersion [W 2]; velocity dispersion in the bulge [M 2]; anomalous motion of spiral arms [S 8]; kinematics within M 31 [R 5]. For distribution of H I within M 31, see 9.2.5.2.

Extended observations in the outer parts of galaxies (sensitive H I and optical observations) show, that the rotation curves remain rather flat [R 7]. No significant decrease in rotational velocities is observed in outer parts as would be expected in case of Keplerian rotation. (Exceptions like M 81 are galaxies disturbed gravitationally by close companions). See Fig. 11.

Rotation curves in the outer parts by H I observations [S 1, R 2, H 4].

Extended rotation curve of the Sab galaxy NGC 7217 [P 5].

Extended rotation curves of 10 high-luminosity spirals [R 7].

Rubin et al. [R 8] investigate the rotational properties of 21 Sc galaxies (in some cases up to more than 100 kpc distance from the center). In no case did they find a significant decrease but mostly increasing rotation curves. NGC 801 for instance shows a constant $V_{rot}$ up to 50 kpc. They suggest that Sc spirals of all luminosity classes have significant masses beyond the optical image. Also the rotation curve of our own Galaxy is nowadays believed to be flat perhaps out to 16 kpc [B 6], contrary to Schmidt's model (Keplerian in the outer parts, see 8.4.1.3). See also [G 2, B 6].

Altogether there is some evidence for massive coronae of galaxies [O 2, E 1]. But see also the critical comments in [B 14] and [V 3]. Van den Bergh [V 3] points to the facts (1) that motions in the outer parts might not be circular, (b) that there might happen some "spillover", i.e. emission from the bright central area received a side lobe of the antenna.

The barred spiral NGC 7723 (type SBb) shows a constant angular velocity of 63 km $s^{-1}$ $kpc^{-1}$ (i.e. solid rotation) in the bar region and a constant linear velocity of 210 km $s^{-1}$ in the spiral region [C 4]. Large deviations from circular orbits in the SBbII galaxy NGC 5383 are reported in [P 1].

Rotation and mass distribution in pairs: NGC 935 and IC 1801 (close pair of spirals) [B 5]; NGC 672 and IC 1727 (interacting barred spirals) [C 2]; rotation curves for two other pairs are shown in Fig. 11. These pairs illustrate the lack of the Tully-Fisher relation (see 9.2.5.1).

A correlation between maximum rotational velocity $V_{rot}^{max}$ and morphological type is reported in [B 11], see Fig. 12.

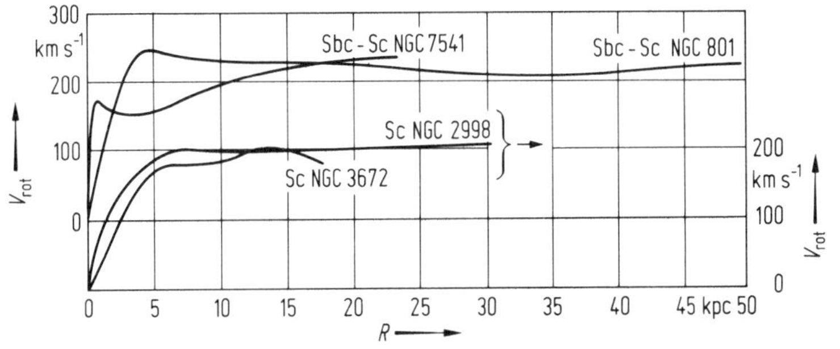


Fig. 11. Extended rotation curves for two pairs of galaxies [R 7].

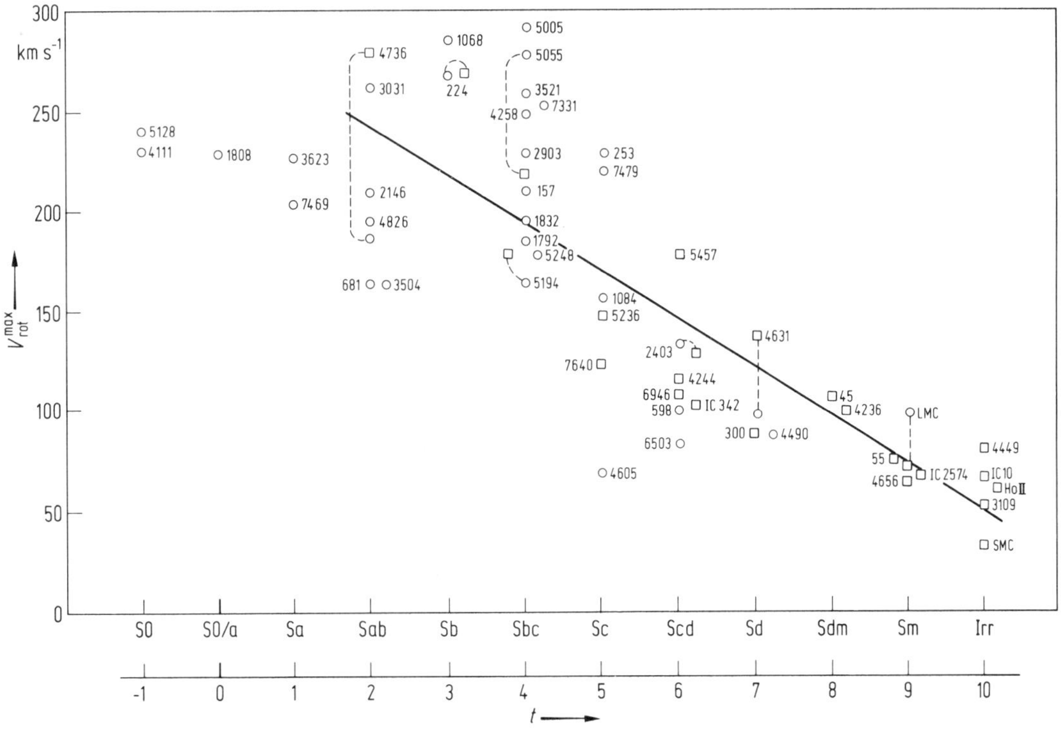


Fig. 12. Maximum rotational velocity vs. morphological type [B 11]. Open circles: optical observations; squares: 21-cm observations. Numbers are NGC numbers.

The velocity law (see also 9.1.4.4)

Brandt and Belton [B 9] have introduced a generalized rotational velocity law to analyse the observed rotational curves

$$V_{rot}(R)=\frac{AR}{(1+B^n R^n)^{3/2n}} \tag{1}$$

with $V_{rot}(R)$ = circular velocity at distance $R$ from the center,

$n$ = shape parameter, numerical index,

$A = 3^{3/2n} V_{rot}^{max}/R_{max}$; $B = 2^{1/n}/R_{max}$,

$V_{rot}^{max}$ = maximum rotational velocity
$R_{max}$ = corresponding distance from the axis of rotation } see Fig. 10.

From (1) follows the gravitational attraction at distance $R$

$$F(R)=\frac{V_{rot}^2(R)}{R}=\frac{A^2 R}{(1+B^n R^n)^{3/n}}. \tag{2}$$

Thus for small $R/R_{max}$: $F(R) \propto R$ (constant angular velocity, solid rotation).
for large $R/R_{max}$: $F(R) \propto 1/R^2$ (Keplerian motion).

With $n=3$, Eq. (2) yields the Bottlinger-Lohmann formula described in 9.1.4.4., p. 250.

Dynamical considerations indicate that the larger the degree of mass concentration toward the center of the galaxy, the greater the $n$ value.

The function $R_{max}/R_{Ho}$ ($R_{Ho}$ = Holmberg radius, see 9.1.2.2 and 9.1.4.2) increases with the morphological type, from ≈0.3 for spirals of type Sab to ≈2 for irregular systems [H 4].

Further dynamical parameters

The above method has been extended and modified by Takase and Kinoshita [T 1] to calculate the projected mass, the angular momentum and the rotational energy:

Projected mass $\mathfrak{M}'(R)$ = mass contained in the cylinder of radius $R$ whose axis coincides with the rotational axis

$$=\int_0^R \mathrm{d}\mathfrak{M}'(R) \tag{3}$$

Angular momentum $$Q(R)=\int_0^R R\, V_{rot}(R)\,\mathrm{d}\mathfrak{M}'(R). \tag{4}$$

Rotational energy $$T(R)=\tfrac{1}{2}\int_0^R V_{rot}^2(R)\,\mathrm{d}\mathfrak{M}'(R). \tag{5}$$

The relationships between angular momentum, rotational energy and total mass are shown in Fig. 13.

A similar investigation of statistical characteristics of the dynamical structure of 21 flat galaxies [M 1] yields the relation

$$\log Q = 1.91 \log \mathfrak{M} + 0.35$$

with angular momentum $Q$ in $[\langle G(10^{11}\,\mathfrak{M}_\odot)^3\,\mathrm{kpc}\rangle^{1/2}]$ ($G$ = gravitational constant) and mass $\mathfrak{M}$ in $[10^{11}\,\mathfrak{M}_\odot]$.

Tables of functions for use in calculating mass distribution, density and mass-surface density are given in [B 10].

The mass-angular momentum relation is also reflected in a relation between optical luminosity and angular momentum density (= angular momentum per unit mass). Vettolani et al. [V 8] found the following relations:

$$\log A = (2.63 \pm 0.02) - (0.29 \pm 0.02)\, M_{20}$$

with a correlation coefficient $r = 0.93$ for a sample of 89 late-type galaxies, studied by [S 9], and

$$\log A = (2.80 \pm 0.02) - (0.28 \pm 0.03)\, M_{20}$$

with $r = 0.88$ for a sample of 38 bright galaxies ($M < -17$) out of low surface-brightness galaxies studied by [F 2].

Here $A$ = angular momentum density in [km kpc $s^{-1}$],

$M_{20} = M + 20$ ($M$ = absolute magnitude).

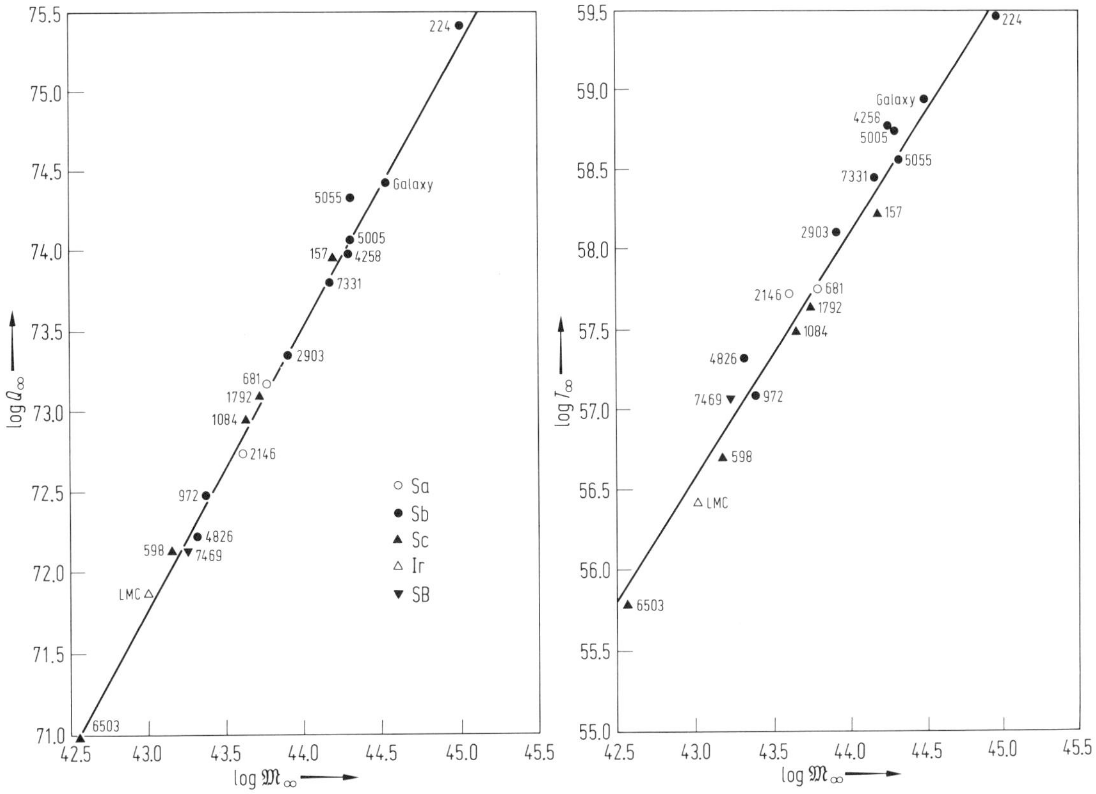


Fig. 13. Statistical relationship of total angular momentum $Q$ in [g $cm^2$ $s^{-1}$] and of the total kinetic energy $T$ in [g $cm^2$ $s^{-2}$] to the total mass $\mathfrak{M}_\infty$ in [g] for galaxies of different type [T 1].

Further investigations:

Angular momentum per unit mass for spiral galaxies [S 4].

Angular momentum of 17 spiral galaxies (methods and correlations) [N 1].

A list of galaxies with dynamical parameters ($V_{rot}$, $L$, $R$, $\mathfrak{M}$, $\mathfrak{M}/L$ etc.) [F 1].

Dynamical parameters of about 40 galaxies with known rotation curves are studied in [B 12]. Figure 14 shows the relation between $D_v/R_{max}$ ($D_v$ = optical diameter) and morphological type (Fig. 14: next page).

In a high-resolution study of gas flow in barred spirals [V 1] the post-shock outflow could be confirmed.

An investigation of the ring-like galaxy NGC 4736 [S 5] suggests that the ring represents the Lindblad resonance (see 8.2.4.2).

The dynamics of the spiral system M 81 has been studied and in particular the high-resolution H I observations are compared with the density theory in [V 9, V 10].

Density-wave theory: [R 6, T 2] (see also 8.4.2.4).

Roche limit in galaxies: [R 0].

Many papers about kinematics of galaxies and dynamical models are found in [g].

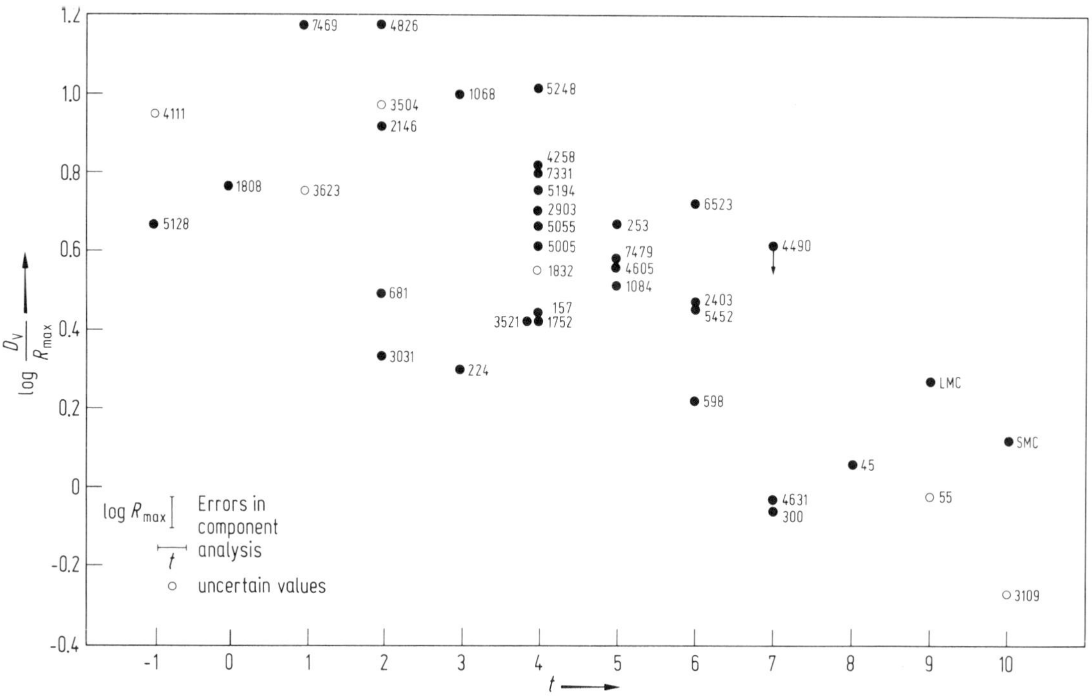


Fig. 14. The ratio of optical diameter $D_v$ to $R_{max}$ vs. morphological type (see 9.1.3.2.7) [B 12].

# 9.2.7 References for 9.2

## General references with review articles


a The Formation and Dynamics of Galaxies (Shakeshaft, J.R., ed.), Int. Astron. Union Symp. **58**, Reidel, Dordrecht (1974).

b Structure and Properties of Nearby Galaxies (Berkhuijsen, E.M., Wielebinski, R., eds.), Int. Astron. Union Symp. **77**, Reidel, Dordrecht (1978).

c Stellar Systems (Flügge, S., ed.), Hdb. Physik **59**, Springer, Berlin (1959).

d Galaxies and the Universe (Sandage, A., Sandage, M., Kristian, J., eds.) = Stars and Stellar Systems Vol. IX, Univ. Chicago Press (1975).

e Freeman, K., Larson, R.B., Tinsley, B.: Galaxies, Sixth advanced course of Swiss Soc. of Astron. and Astrophys. (Martinet, L., Major, M., eds.), Geneva Observatory (1976).

f Taylor, R.J.: Galaxies, Structure and Evolution, Wykeham Publ. Ltd., London (1978).

g Photometry, Kinematics and Dynamics of Galaxies (Evans, D.S., ed.), University of Texas (1979).

h The Structure and Evolution of normal Galaxies (Fall, S.M., Lynden-Bell, D., eds.), Cambridge Univ. Press (1981).


## Special references


A 1 Aaronson, M., Huchra, J., Mould, J.: Astrophys. J. **229** (1979) 1.

A 2 Aaronson, M., Mould, J., Huchra, J.: Astrophys. J. **237** (1980) 655.

A 3 Aaronson, M. in: Infrared Astronomy, Int. Astron. Union Symp. **96** (Wynn-Williams, C.E., Cruitshank, D.P., eds.), Reidel, Dordrecht (1981) p. 297.

A 4 Adams, M.T., Jensen, E.B., Stocke, J.T.: Astron. J. **85** (1980) 1010.

B 1 Baldwin, J.E. in: [b] p. 191.

B 2 Berkhuijsen, E.M. in: [b] p. 149.

B 3 Bertola, F., Galletta, G.: Astrophys. J. **226** (1978) L115.

B 4 Bertola, F., Galletta, G.: Astron. Astrophys. **77** (1979) 363.
B 5 Blackman, C.P.: Mon. Not. R. Astron. Soc. **178** (1977) 15.
B 6 Blitz, L.: Astrophys. J. **231** (1979) L115; erratum in **234** (1979) L172.
B 7 Boroson, T.: Astrophys. J. Suppl. **46** (1980) 177.
B 8 Bosma, A., van der Kruit, D.C.: Astron. Astrophys. **79** (1979) 281.
B 8a Bosma, A.: Ph. Thesis, Univ. Groningen (1978).
B 9 Brandt, J.C., Belton, M.J.S.: Astrophys. J. **136** (1962) 352.
B 10 Brandt, J.C., Scheer, L.S.: Astron. J. **70** (1965) 471.
B 11 Brosche, P.: Astron. Astrophys. **13** (1971) 293.
B 12 Brosche, P.: Astron. Astrophys. **23** (1973) 259.
B 13 Burbidge, E.M., Burbidge, G.R. in: [d] p. 81.
B 14 Burbidge, G.: Astrophys. J. **196** (1975) L7.
B 15 Burstein, D. in: [g] p. 81.
C 1 Capaccioli, M. in: [g] p. 165.
C 2 Carozzi-Meyssonnier, N.: Astron. Astrophys. Suppl. **47** (1982) 237.
C 3 Carter, D.: Mon. Not. R. Astron. Soc. **178** (1977) 137.
C 4 Chevalier, R.A., Flurenlid, I.: Astrophys. J. **225** (1978) 67.
D 1 De Vaucouleurs, G. in: [c] p. 275 ch. II, b.
D 2 De Vaucouleurs, G. in: [c] p. 311.
D 3 De Vaucouleurs, G.: Mon. Not. R. Astron. Soc. **113** (1953) 134.
D 4 De Vaucouleurs, G., Freeman, K.C.: Vistas in Astron. **14** (1972) 163.
E 1 Einasto, J., Kaasik, A., Saar, E.: Nature **250** (1974) 309.
E 2 Ekers, R.D. in: [b] p. 49.
E 3 Ekers, R.D.: Highlights in Astron. **5** (1980) 143.
E 4 Emerson, D.T., Newton, K. in: [b] p. 183.
F 1 Faber, S.M., Gallagher, J.S.: Annu. Rev. Astron. Astrophys. **17** (1979) 135.
F 2 Fisher, J.R., Tully, R.B.: Astron. Astrophys. **44** (1975) 151.
F 3 Freeman, K.C. in: [d] p. 409.
F 4 Freeman, K.C. in: [e] p. 1.
F 5 Freeman, K.C.: Astrophys. J. **160** (1970) 811.
G 1 Goad, J.W., Roberts, M.S.: Astrophys. J. **250** (1981) 79.
G 2 Gunn, J.E., Knapp, G.R., Tremaine, S.D.: Astron. J. **84** (1979) 1181.
H 1 Haynes, R.E., Huchtmeier, W.K.H., Siegman, B.C., Wright, A.E.: A Compendium of Radio Measurements of Bright Galaxies, CSIRO Division of Radiophysics, Melbourne (1975).
H 2 Heidmann, N.: Astrophys. Lett. **3** (1968) 153.
H 3 Holmberg, E. in: [d] p. 123.
H 4 Huchtmeier, W.K.: Astron. Astrophys. **45** (1975) 259.
H 5 Huchtmeier, W.K. in: [b] p. 197.
H 6 Huchtmeier, W.K. in: [b] p. 264 (discussion to [V 3]).
H 7 Huchtmeier, W.K. in: ESO/ESA workshop on Dwarf Galaxies (Tarenghi, M., Kjär, K., eds.), Geneva (1980).
H 8 Huchtmeier, W.K., Richter, O.G.: Astron. Astrophys. **109** (1982) 331.
H 9 Huchtmeier, W.K., Richter, O.G., Bohnenstengel, H.-D., Hauschildt, M.: (1982) in preparation.
H 10 Hummel, E.: Astron. Astrophys. **93** (1981) 93.
H 11 Hummel, E.: Astron. Astrophys. **96** (1981) 111.
J 1 Jones, D.L., Sramek, R.A., Terzian, Y.: Astrophys. J. **247** (1981) L57.
K 1 Kennicutt, R.C. Jr.: Astron. J. **86** (1981) 1847.
K 2 King, I.R.: Astron. J. **71** (1966) 64.
K 3 Knapp, G.R., Kerr, F.J., Bowers, P.F.: Astron. J. **83** (1978) 360.
K 4 Kormendy, J., Norman, C.A.: Astrophys. J. **233** (1979) 539.
K 5 Krumm, N., Salpeter, N.N.: Astron. Astrophys. **56** (1979) 465.
L 1 Li Zong-yun, Liu Ru-liang: Acta Astrophys. Sinica **1** (1981) 28; Translation: Chinese Astron. **5** (1981) 205.
M 1 Miyamoto, M., Satoh, C., Ohashi, M.: Astrophys. Space Science **67** (1980) 147.
M 2 Monnet, G., Pellet, A., Simien, F. in: [b] p. 159.
M 3 Morgan, W.W., Osterbrock, D.E.: Astron. J. **74** (1969) 515.
N 1 Nordsieck, K.H.: Astrophys. J. **184** (1973) 719, 735.

O 1 Okamura, S.: Publ. Astron. Soc. Japan **30** (1978) 91.
O 2 Ostriker, J.P., Peebles, P.J.E., Yahil, A.: Astrophys. J. **193** (1974) L 1.
P 1 Peterson, C.J., Rubin, V.C., Ford, K.W., jr., Thonnard, N.: Astrophys. J. **219** (1978) 31.
P 2 Peterson, C.J.: Astrophys. J. **221** (1978) 80.
P 3 Peterson, C.J.: Astrophys. J. **222** (1978) 84.
P 4 Peterson, C.J.: Astrophys. J. **226** (1978) 75.
P 5 Peterson, C.J., Rubin, V.C., Ford, E.K., jr., Roberts, M.S.: Astrophys. J. **226** (1978) 770.
P 6 Peterson, C.J.: Astron. J. **85** (1980) 226.
R 0 Robe, H.: Astron. Astrophys. **97** (1981) 182.
R 1 Roberts, M.S. in: [d] p. 309.
R 2 Roberts, M.S.: The rotation curves of galaxies, in: Dynamics of stellar systems, Int. Astron. Union Symp. **69** (Hayli, A., ed.), Reidel, Dordrecht (1975) p. 331.
R 3 Roberts, M.S.: Comments Astron. **6** (1976) 105.
R 4 Roberts, M.S.: Astron. J. **83** (1978) 1026.
R 5 Roberts, M.S., Whitehurst, R.N. in: [b] p. 169.
R 6 Roberts, W.W., jr., Roberts, M.S., Shu, F.H.: Astrophys. J. **196** (1975) 381.
R 7 Rubin, V.C., Ford, W.K., jr., Thonnard, N.: Astrophys. J. **225** (1978) L 107.
R 8 Rubin, V.C., Ford, W.K., jr., Thonnard, N.: Astrophys. J. **238** (1980) 471.
R 9 Rogstad, D.H., Wright, M.H.C., Lockhart, I.A.: Astrophys. J. **204** (1976) 703.
S 1 Salpeter, E.E. in: [b] p. 23.
S 2 Sandage, A., Freeman, K.C.: Astrophys. J. **160** (1970) 83.
S 3 Sandage, A. in: [d] p. 1.
S 3a Sandage, A., Tammann, G.A.: Astrophys. J. **210** (1976) 7.
S 4 Savchenko, V.P.: Astron. Zh. **50** (1974) 1177: engl. transl.: Soviet. Astron. **17** (1974) 743.
S 5 Schommer, R.A., Sullivan, W.T. III: Astrophys. Lett. **17** (1976) 191.
S 6 Seiden, P.E., Gerola, H.: Astrophys. J. **233** (1979) 56.
S 7 Shane, W.W., Bystedt, I. in: [b] p. 97.
S 8 Shane, W.W. in: [b] p. 180 (discussion to [W 1]).
S 9 Shostak, G.S.: Astron. Astrophys. **68** (1978) 321.
S 10 Spinrad, H., Peimbert, M. in: [d] p. 37.
S 11 Sancisi, R.: Astron. Astrophys. **53** (1976) 159.
T 1 Takase, B., Kinoshita, H.: Publ. Astron. Soc. Japan **19** (1967) 409.
T 2 Thielheim, K.O.: Astrophys. Space Science **76** (1981) 363.
T 3 Tully, R.B., Fisher, J.R.: Astron. Astrophys. **54** (1977) 661.
V 1 Van Albada, G.D., Roberts, W.W., jr.: Astrophys. J. **246** (1981) 740.
V 2 Van de Hulst, J.M.: Highlights Astron. **5** (1980) 177.
V 3 Van den Bergh, S. in: [b] p. 247.
V 4 Van der Kruit, P.C., Allen, R.J.: Annu. Rev. Astron. Astrophys. **14** (1976) 417.
V 5 Van der Kruit, P.C., Allen, R.J.: Annu. Rev. Astron. Astrophys. **16** (1978) 103.
V 6 Van der Kruit, P.C. in: [b] p. 33.
V 7 Van der Kruit, P.C., Searle, L. in: [g] p. 93.
V 8 Vettolani, G., Marano, B., Zamorani, G., Bergamini, R.: Mon. Not. R. Astron. Soc. **193** (1980) 269.
V 9 Visser, H.C.D. in: [b] p. 105.
V 10 Visser, H.C.D.: Astron. Astrophys. **88** (1980) 149, 159.
W 1 Whitehurst, R.N., Roberts, M.S., Cram, T.R. in: [b] p. 175.
W 2 Whitemore, B.C.: Bull. American Astron. Soc. **12** (1980) 492.
W 3 Whiteford, A.E. in: [d] p. 159.
W 4 Wilkerson, M.S.: Astrophys. J. **240** (1980) L 115.

# 9.3 Galaxies with special peculiarities; pairs, groups and clusters of galaxies

As some species of active galaxies, interacting galaxies and binary galaxies overlap widely, they are considered here in the same chapter.

## 9.3.1 Galaxies with special peculiarities

In this section surveys and other observations of some types of active galaxies and galaxies with other peculiarities are summarized. For catalogues, see also 9.1.1.

The following types of galaxies with strong active nuclei and strong radio radiation are not included here and are discussed in separate sections:

Quasars, Seyfert galaxies, Liners, BL Lac objects in 9.5; Radio galaxies in 9.6

The various lists of peculiar galaxies overlap widely. Useful cross-referencing has been done in RC2 ([4] in 9.1.1.5) and in UCG ([14] in 9.1.1.5). Using the accurate positions for galaxies in [D 12, G 2, G 3] and the cross-referencing of the UCG, the positions of many galaxies mentioned in the lists below can be found.

Arakelian

High surface-brightness galaxies have been listed by Arakelian [A 18]. A general classification of many galaxies in terms of the central surface brightness is given in [A 19]. Spectral observations of galaxies from the original Arakelian list: [A 20, D 9, D 8, O 4]. A comparison between surface brightness and radio emission: [A 21].

Arp

Arp published a catalogue of peculiar galaxies in form of a photographic atlas [A 25]. H I measurements of these galaxies: [P 10]. Radio-continuum survey at three frequencies: [S 37].

Fairall

Compact and bright-nucleus galaxies were listed by Fairall [F 2]. Their colours and spatial distribution are discussed in [F 3].

Haro

A list of galaxies with intense ultraviolet radiation has been compiled by Haro [H 1b]. Spectroscopic and photometric studies: [D 15, D 16]. Neutral hydrogen study: [B 27].

Markarian

Since 1967 Markarian has been publishing a series of lists containing galaxies with strong ultraviolet continuum found with an objective-prism survey (Ref. [11, 12, 12a] in 9.1.1.5). Recently Kazarian has started a similar series [K 6]. Markarian's list contains a large number of Seyfert galaxies and the corresponding studies of Markarian galaxies will be mentioned in 9.5. Accurate positions of the first 700 Markarian galaxies: [P 9, K 9]. Spectral and photometric observations: [W 1a, D 7, A 14, A 15, A 16, U 1b, D 5, K 10, D 6, S 6, N 1a, W 2, B 25, H 18] and others. Infrared observations: [N 1a, A 12]. Radio-continuum observations: [K 9a, T 10, S 32, S 37, K 8, S 1, S 2, B 12, B 18] and others. Neutral hydrogen studies: [B 28, B 29, B 17]. Combination of optical and radio studies: [S 37, B 29]. Surface brightness, emission lines and colour were compared in [A 17]. Morphological studies: [K 1, B 26], many structural peculiarities are noted. The general nature of Markarian galaxies has been discussed in [M 4, M 5, M 6, S 18, B 14, H 20]. Markarian galaxies within pairs: see 9.3.2.1.

Sérsic-Pastoriza

Southern peculiar galaxies have been listed and studied in [P 3, S 19, S 20, S 21, S 22, S 23, A 5].

Zwicky

Compact galaxies have been listed by Zwicky in eight lists circulated privately [Z1] and published in [Z5, Z6]. Further descriptions: [Z2, Z3]. Descriptive spectroscopic survey of many of these galaxies: [S4]. Several of Zwicky's galaxies were studied in great detail: [S5, S17, K11, B8, C7, Z4, O1] and many others.

ESO-Uppsala survey

Many interesting galaxies listed in the ESO-Uppsala survey of southern galaxies (Ref. [10] in 9.1.1.5), have been studied [W 3, W 4, W 5, W 6, B 23, B 24, A 22, B 9].

Tololo survey

At CTIO (Cerro Tololo Inter-American Observatory) an extensive objective-prism survey is being carried out in order to find emission-line galaxies similar to the Markarian galaxies: [S 29, M 1, M 2]. Photometric and spectroscopic observations of these galaxies: [B 22]. Radio studies: [P 8].

Elliptical galaxies with dust lanes

Recently elliptical galaxies crossed by dust lanes have been identified: [B 10, K 12, K 13]. Many of these objects are radio sources.

Amorphous galaxies (M 82-type)

Amorphous galaxies have been defined by Sandage and Brucato [S 3] and are discussed in 9.4.4.2.

## 9.3.2 Pairs of galaxies

### 9.3.2.1 Observations

A general review has been given by Page [P 2].

| Lists of binary galaxies | Further observations and discussions of these binaries |
|---|---|
| 1 Holmberg [H 15, H 16] | [H 17] |
| 2 Page [P 1] | [P 2] |
| 3 Karachentsev [K 2] | [S 33, S 34, K 5] |
| 4 Noerdlinger [N 2] | [A 23, N 3, N 4] |
| 5 Turner [T 14] | [T 15, K 4] |
| 6 Peterson [P 11] | [P 12] |
| 7 Rood/Leir [R 5] | – |

Apart from radial velocity determinations of galaxies (optical or from H I, e.g. [S 38, G 1b, D 8a]), radio-continuum observations have been made and show increasing radio emission with decreasing distance [S 33, S 34, H 21, C 9a]. The sensitive radio-continuum survey at Arecibo [C 9, D 13] covers many binary galaxies from all samples.

Further investigations: orientation and size comparisons: [K 3, A 23, N 4], colour and morphology comparisons: [M 13, T 7, N 3].

The catalogues of de Vaucouleurs (RCBG1), de Vaucouleurs et al. (RCBG2 or RC2), Nilson (UGC), Zwicky et al. (CGCG), and Vorontsov-Velyaminov et al. (MCG) contain many useful notes on binary galaxies; for references, see [3, 4, 14, 23, and 21] in 9.1.1.5.

Interacting galaxies

An atlas of interacting galaxies has been published in two parts by Vorontsov-Velyaminov [V 8, V 15]. These lists have considerable overlap with Arp's atlas of peculiar galaxies [A 25]. A further list of interacting galaxies similar to M 51 is given in [V 12]. Further studies of these galaxies: [V 10, V 11, V 17, V 14, V 16, A 24, A 10, V 18, B 9a, S 30a, D 10] and others. Compact galaxies in the morphological catalogue are described in [V 16].

The Atlas of peculiar galaxies [A 25] contains many pairs of galaxies with long tails.
Many of the binary galaxies in Karachentsev's list [K 2] are interacting galaxies. Radio-continuum studies: [S 33, S 34]. Many of the pairs are contained in the radio-continuum surveys [C 9, D 12]; some of them contain compact radio nuclei [C 10, B 20].

Markarian galaxies within pairs have been described in [V 7, H 7, C 4, C 5].

A number of interacting galaxies have compact radio nuclei, as shown by VLBI measurements [C 10, B 20, P 19, B 29c, B 29a, G 7b, J 1c] or flat radio spectra [C 9, S 34].

X-ray sources, sometimes compact, have been discovered in interacting galaxies [G 8]. The HEAO-A and B Einstein X-ray observatories can be expected to find many more sources of this class (e.g. [M 7, B 11, B 20]).

The proportion of interacting or morphologically peculiar galaxies among the Seyfert galaxies (see 9.5) is higher than normal [A 4], and may indicate that interaction can lead to the Seyfert phenomenon. We note that the Seyfert galaxy is the brighter one in the Karachentsev-pairs with known Seyferts in all cases.

Ring galaxies

A number of known galaxies show ring structure, presumably caused by collision with a second galaxy.

A recent review of ring galaxies has been given by Chatterjee [C 8]. Many ring galaxies are in the lists of peculiar and interacting galaxies of Vorontsov-Velyaminov, Zwicky and Arp (see above). Further lists: [V 9, V 10, V 13, C 2]. Observational studies: [T 3, T 4, L 7, T 5, D 11]. Study of neutral hydrogen in ring galaxies: [S 27]. Search for radio-continuum emission: [G 4].

Double-nuclei galaxies

Galaxies with double nuclei are listed and described in [M 11], where attention is drawn to radio galaxies with extended envelopes (type D of Morgan [M 15]); such galaxies often contain double or multiple nuclei. There are also galaxies with double or multiple nuclei among the Markarian galaxies (see 9.3.1) [P 13, P 14]. In his studies of merging galaxies, Schweizer also mentions galaxies with double nuclei [S 15].

## 9.3.2.2 Theoretical concepts

Non-interacting binary galaxies

Mass of galaxies determined from binary galaxies: [T 15, T 17, P 12] and in earlier work referred to therein. The mass distribution in galaxies: [Y 2, P 12]. A review of the masses and $\mathfrak{M}/L$ ratios of galaxies obtained by this and other methods: [F 1]. See also 9.1.4.5 and 9.1.4.6.

Collisions of spherical galaxies

Encounters of spherical galaxies have been calculated: [A 9, S 9, G 1, L 1a, R 1a, B 13, A 7, S 11, B 16, R 7] and others. Substantial mass loss and change of internal energy occurs in slow encounters only; such encounters, however, can lead to tidal capture (see below).

Tidal interaction

Tidal interaction between disk galaxies can produce long tails observed in e.g. NGC 4038/39 (cf. the lists [1] and [21] in 9.1.1.5). Calculations to demonstrate this were performed by [P 15, P 16, T 2, Y 1, W 13, E 1a]; the most extensive exploration of tidal interaction was done by Toomre and Toomre [T 8] who model NGC 4038/39 successfully in detail. These calculations were criticized in [F 8]. Ring galaxies can be understood in terms of penetrating encounters of a disk galaxy with another galaxy or with an intergalactic cloud [L 7, T 4, F 7, H 13, O 5, H 14, T 9, M 14]. Tidal interaction can lead to capture of two galaxies into a bound binary system [A 9, A 10, S 8, S 10, S 11]. Tidal interaction can increase the angular momentum of proto-galaxies, a process proposed by [P 6, H 17a, P 7] to account for the observed high angular momentum in galaxies; this process was further investigated in [F 4c, D 1b, T 6, B 21]. Tidal interaction also leads to a thickening of initially flat galaxies [M 3].

Dynamical friction

Dynamical friction can slow down the orbital motion of two galaxies around each other and finally lead to merging (see below): [S 31, S 7, H 9, L 2, O 6, T 12, W 10, B 33, A 6, S 36d, O 8].

Merging of galaxies

Galaxies that spiral into each other due to dynamical friction finally merge; slow direct collisions can also lead to mergers. The deep potential wells of the nuclear regions survive longest and may thus explain the double-nuclei galaxies. Calculation and discussion: [W 9, W 10, W 11, F 3b, R 7, C 9b, C 2d, A 1a, M 11a, M 14a, R 8, W 12d, W 12b, W 12c, W 12a, J 1a]. An observational sequence of galaxies corresponding to the evolution from a binary galaxy via a merger to a single galaxy is outlined in [T 8, S 15]. Similarly to interacting galaxies, a number of merging galaxies show active nuclei [S 16].

Merging of a large galaxy with a small one is called cannibalism; in this way a large galaxy can grow at the expense of smaller galaxies in a cluster (see also 9.1.4.2). The significance of merging to the evolution of the dominant galaxies in groups and clusters has been demonstrated by [O 7, H 4, O 8].

Mass exchange

Interaction and penetration of two galaxies lead to an exchange of stars and gas [T 8]. Such gas can feed an active nucleus. Several interacting galaxies are known to show active nuclei (e.g. [S 34, C 9, C 10, B 20]), both in the radio and X-ray range.

Mass exchange between disk galaxies during penetrating encounters can be caused by differential ram pressure [S 13].

Bursts of star formation

Interaction between galaxies inevitably leads to strong shocks in the gas which trigger star formation on a short time scale. The resulting bursts of star formation are considered in 9.4.4.3.

Active nuclei

Mass exchange between interacting galaxies and merging of galaxies can lead to an increase of low-angular-momentum gas; such low-angular-momentum gas can accrete to a nuclear "engine" via an accretion disk [L 5, T 11, L 6]. Thus, the occurrence of active radio, X-ray or Seyfert nuclei in interacting or merging galaxies may be understood. Active nuclei are discussed in 9.5.

## 9.3.3 Groups of galaxies

There is evidence that field galaxies may not exist [H 19, S 30]. A new study of field galaxies has been started [K 7].

The luminosity function of galaxy systems of all sizes – from field galaxies to large clusters – has been calculated in [B 4, H 7a].

The morphology of field galaxies has been included in a general study by Dressler [D 14].

### 9.3.3.1 Definition

A real group fulfills three conditions: a) The crossing time of a galaxy is less than about a third of the Hubble time; this is necessary in order to attain virial equilibrium. b) The number density of the galaxies in the group must be noticeably larger than in the field, in order to exclude statistical fluctuations. c) There should be no correlation between redshift and apparent magnitude of the galaxies; otherwise the group would consist of independent galaxies which fortuitously lie behind each other in space. These conditions are discussed in [T 13]. For verification one can calculate the energy of a group (e.g. [M 8]).

Critical lists of groups: de Vaucouleurs [D 3] and Turner and Gott [T 16]. Further lists of groups and clusters: Table 1.

Data and descriptions of 54 groups nearer than 31 Mpc: [D 2]. Further studies: [R 9, R 10].

Table 1. Lists of groups and clusters of galaxies

| Author | Year | Ref. |
|---|---|---|
| Abell, G.O. | 1958 | Astrophys. J. Suppl. **3** 211 |
| Baier, F.W., Tiersch, H. | 1975 | Astrofizika **11** 221 (Engl. **11** 146) |
| | 1976 | Astrofizika **12** 7, 409 (Engl. **12** 1, 263) |
| | 1978 | Astrofizika **14** 279 (Engl. **14** 157) |
| | 1979 | Astrofizika **15** 33 (Engl. **15** 24) |
| Braid, McGillivray | 1978 | Mon. Not. R. Astron. Soc. **182** 241 |
| Duus, A., Newell, B. | 1977 | Astrophys. J. Suppl. **35** 209 |
| Gunn, J.E., Oke, J.B. | 1975 | Astrophys. J. **195** 255 |
| Humason, M.L., Mayall, N.U., Sandage, A.R. | 1956 | Astron. J. **61** 97 |
| Klemola, A.R. | 1969 | Astron. J. **74** 804 |
| Petrosyan, M.B. | 1974 | Astrofizika **10** 471 (Engl. **10** 291) |
| | 1978 | Astrofizika **14** 631 (Engl. **14** 356) |
| Rood, H.J., Dickel, J.H. | 1978 | Astrophys. J. **224** 724 |
| Rose, J.A. | 1976 | Astron. Astrophys. Suppl. **23** 109 |
| Sandage, A. | 1975 | Astrophys. J. **202** 563 |
| Sandage, A., Tammann, G.A. | 1975 | Astrophys. J. **196** 313 |
| Sandage, A., Kristian, J., Westphal, J.A. | 1976 | Astrophys. J. **205** 688 |
| Sersic, J.L. | 1974 | Astrophys. Space Science **28** 365 |
| Shakhbazyan, R.K. | 1973 | Astrofizika **9** 495 (Engl. **9** 296) |
| Shakhbazyan, R.K., Petrosyan, M.B. | 1974 | Astrofizika **10** 13 (Engl. **10** 6), **10** 327 (Engl. **10** 202) |
| Snow, T.P., Jr. | 1970 | Astron. J. **75** 237 |
| Turner, E.L., Gott, J.R. III | 1976 | Astrophys. J. Suppl. **32** 409 |
| Vaucouleurs, G. de | 1976 | Astrophys. J. **203** 33 |
| Visvanathan, N., Griersmith, D. | 1979 | Astrophys. J. **230** 1 |
| Zwicky, F., et al. | 1961···1968 | Catalogue of galaxies and clusters, 6 Volumes, Ref. see [23] in 9.1.1.5 |

### 9.3.3.2 Structure

In the central region of groups there is often a dominant galaxy (see 9.1.4.2) and the galaxies there are of earlier Hubble type [B 11b, P 17]. There is a general correlation between number density of galaxies and Hubble type [D 14].

So far no intergalactic neutral hydrogen has been found in the groups (e.g. [B 17, L 4, M 9, H 6, G 4a]) but cf. [H 3, H 6a].

Hot intergalactic gas has so far been found in few groups only [B 19, S 14b]; these groups fit well to the $L_X$-velocity dispersion relation for clusters (e.g. [H 12]).

Masses

N-body calculations as well as application of the virial theorem to groups lead to mass-to-luminosity ratios $\mathfrak{M}/L_B$ in the order of 100 [G 5, G 6, R 4, D 1a, M 9a] (see also 9.1.4.5).

### 9.3.3.3 The Local Group

Our own Galaxy is a member of a small cluster or group of galaxies usually called the "Local Group". The Local Group consists of perhaps some dozens of galaxies; three of them are spirals and the rest dwarf ellipticals and irregular systems.

Table 2 gives some data of the members of the Local Group. Upper part: the 17 members known well for long; these are described in detail by van den Bergh [V 3].

Lower part: 7 newly identified members whose membership is not definite in all cases [D 2, F 3a].

In addition, de Vaucouleurs [D 2] considers the following galaxies as members of the Local Group: Sextans B (= DDO 70, Im), Serpentis system (dE) and Capricornus system (dE; perhaps a globular cluster).

Table 2. The Local Group.

Name: DDO = 243 dwarf galaxies, David-Dunlap Observatory [V 2, V 5]
NGC = New General Catalogue (see 9.1.1.5)
M = Messier Catalogue (see 9.1.1.5)
IC = Index Catalogue (see 9.1.1.5)

Type: (a) van den Bergh classification [V 3], see 9.1.3.2.4
(b) de Vaucouleurs classification [D 2]

$B$ = total apparent blue magnitude (corrected for galactic extinction) [D 4]
$M$ = total absolute magnitude, blue [D 2], visual [V 3]
$r$ = distance in [kpc], see [D 2]
$D$ = diameter in [kpc], from [D 2]; values in ( ) from [V 4]
$\mathfrak{M}$ = mass [A 11]

Comparison of this Table with the corresponding Table in LB, NS, Vol. VI/1 (1965) p. 674 shows some remarkable differences due to new observations (especially for $r$ and $D$).

| Name | Type (a) | (b) | $B$ | $M_B$ | $M_V$ | $r$ [kpc] | $D$ [kpc] | log ($\mathfrak{M}/\mathfrak{M}_\odot$) |
|---|---|---|---|---|---|---|---|---|
| Galaxy | Sb/Sc | Sbc? | – | $(-18^{\mathrm{M}}.8)$ | $(-20^{\mathrm{M}})$ | (10) | >20 | 11.3 |
| LMC (Large Magellanic Cloud) | Ir or SBc III–V | SB(s)m | $0^{\mathrm{m}}.1$ | −18.1 | −18.5 | 50 | 6.5 | 10.0 |
| SMC (Small Magellanic Cloud) | Ir IV–V | IB(s)m | 2.2 | −16.0 | −16.8 | 50 | 2.9 | 9.3 |
| UMi system = DDO 199 | spheroidal | dE | | | − 8.8 | 80 | (0.3) | 5 |
| Sculptor system | spheroidal | dE | | −11.2 | −11.7 | 110 | (0.7) | 6.5 |
| Draco system = DDO 208 | spheroidal | dE | | | − 8.6 | 60 | (0.3) | 5 |
| Fornax system | spheroidal | dE | 8.8 | −12.9 | −13.6 | 230 | (1.6) | 7.3 |
| Leo II system = DDO 93 | spheroidal | dE | | − 9.1 | − 9.4 | 230 | (0.3) | 6.0 |
| Leo I system = DDO 74 | spheroidal | E4 | | −10.7 | −11.0 | 230 | (0.6) | 6.6 |
| NGC 6822 = DDO 209 | Ir IV–V | IB(s)m | 8.5 | −14.8 | −15.7 | 500 | 2.1 | 8.5 |
| IC 1613 = DDO 8 | Ir V | Im | 9.8 | −14.2 | −14.8 | 660 | 2.0 | 8.4 |
| Andromeda Nebula = M31 = NGC 224 | Sb I–II | SA(s)b | 3.6 | −20.3 | −21.1 | 690 | 20.0 | 11.5 |
| M32 = NGC 221 | E2 | E2 | 8.8 | −15.6 | −16.4 | 690 | 0.6 | 9.5 |
| NGC 205 | E6p | $E^+5p$ | 8.4 | −15.8 | −16.4 | 690 | 1.5 | 9.9 |
| NGC 185 | dE0 | $E^+3p$ | 9.5 | −14.7 | −15.2 | 690 | 1.0 | 9 |
| NGC 147 = DDO 3 | dE4 | E5p | 9.8 | −14.4 | −14.9 | 690 | 1.0 | 9 |
| M33 = NGC 598 | Sc II–III | SA(s)cd | 5.8 | −18.3 | −18.9 | 720 | 10.5 | 10.1 |
| UMa system | | dE | | | | 120 | | |
| Sextans C | | dE | | | | 140 | | |
| Pegasus system | | dE | | | | 170 | | |
| WLM system [1]) = DDO 221 | | Im | 10.8 | −13.7 | | 870 | 1.3 | |
| Sextans A = DDO 75 | | IBm | 11.6 | −13.6 | | 1000 | 1.5 | |
| Leo A = Leo III = DDO 69 | | Im | 12.7 | −12.3 | | 1100 | | |
| IC 10 | | SB(s)m | 10.0 | −17.3 | | 1260 | 1.3 | |

[1]) Wolf-Lundmark-Melotte system.

Motion of the local standard of rest (LSR; see 8.4.1.2.3) relative to the centroid of the Local Group [Y 3]:

$$V(\mathrm{LSR}) = 300\ \mathrm{km\,s^{-1}}\ \text{towards}\ l = 107°,\ b = -8°.$$

Considering the velocity residuals of other galaxies, the authors [Y 3] conclude that, of the galaxies sometimes mentioned,

IC 342, NGC 6949, NGC 404, Maffei 1, Maffei 2 are certainly not members of the Local Group,
IC 10, WLM system, Leo A, DDO 210, IC 5152, Pegasus dwarf DDO 216 are probably members (the first three are included in Table 2),
Sextans A, Sextans B, DDO 187, GR 8 (=DDO 155), NGC 3109 are possible but unlikely members (the first one is included in Table 2).

Similar results, also based on solar motion solution, are given in [D 4] and, based on a H I survey of DDO dwarfs, in [F 6].

A new Sculptor-type dE galaxy in Carina, found in the ESO southern sky survey, is considered a possible member in [C 1c].

The Local Group itself appears to be an outlying member of a local supercluster which is centered in the Virgo cluster [B 27, D 1d, D 2, D 3, Y 3a, T 1]. Dust in the Local-Group galaxies has been studied in [G 1a].

The author thanks Dr. H.H. Voigt for this section.

## 9.3.4 Clusters of galaxies

Definition

As in the case of the groups, very careful definitions are required to pick out real clusters; the most restrictive conditions have been used by Abell [A 1b]; general data for Abell clusters have been assembled in [S 8, L 3]. The largest list of clusters has been compiled by Zwicky et al. ([23] in 9.1.1.5). Table 1 in 9.3.3.1 contains further lists of groups and clusters.

Types

The major types of clusters have been reviewed in [B 3]; there are three major sequences; Rood and Sastry types [R 2], Bautz and Morgan types [B 7] (see also 9.1.4.2) and Oemler types [O 2]. These three sequences run largely parallel.

Clusters of galaxies have been reviewed in general by [A 3, V 6, B 3, G 10, S 35, S 36, S 25, B 5, W 8].

Structure

The regular clusters have a density law similar to that of the elliptical galaxies [R 7]; the structure in general is reviewed in [B 3, B 5]. Studies of space and velocity distributions: [N 5, C 2c, W 13a, H 13a, H 13b, Q 2, W 5a, B 5a].

Masses

The virial theorem applied to clusters leads to $\mathfrak{M}/L_B$ values between 100 and 200 [T 1a, W 9, B 3, B 4, B 5]. It follows that about 10% of the total mass of a cluster exists in the form of galaxies, 10% in the form of enriched intergalactic gas, and 80% of the mass is of unknown nature (see also 9.1.4.5). The large extended halos of galaxies may consist of the same material as the intergalactic gas; tidal stripping in clusters strips such halos off (e.g. [H 10, B 16, S 35, S 36]).

Total masses of clusters lie in the range of about $10^{15}\,\mathfrak{M}_\odot$, e.g. [P 5], and slightly higher, depending on their true extent. Length scales are 3···10 Mpc, and thus correspond to the extent of the largest known radio galaxies (see 9.6.4).

Morphology

Parallel to the sequence of cluster types, the morphology of the galaxies changes from the dense spherical clusters which are dominated by elliptical and S0 galaxies to the spiral-rich clusters of rather loose structure [O 2, B 5, C 2a, D 14a, H 6b, S 36e, G 4c, C 3a]. A corresponding radial variation of morphology [M 12] can be understood theoretically with ram pressure stripping [H 11, G 4a]. Cluster morphology also shows the general dependence of galaxy morphology on environment density [D 14].

Intergalactic gas

The general nature of intergalactic gas has been reviewed in [F 5]. X-ray observations have revealed large amounts of intergalactic gas, enriched to nearly normal abundance and with equal mass to that of galaxies [M 17, S 28, J 1, P 18, U 1a, P 20, F 6b, M 19, L 1b, P 8c, F 1a, C 1b, H 12b, M 11b, M 14 b, U 1, M 2a]. The typical temperature of the gas corresponds to the motion of the galaxies, and is thus in the order of $10^8$ K.

Corresponding to the sequence of cluster types there is a sequence of spatial structures of the intergalactic gas: cD clusters have a smooth and spherical distribution of the gas, whereas the loose clusters show emission around each of the major galaxies only [J 1b].

The X-ray luminosity $L_X$ of clusters is correlated with the velocity of the galaxies (e.g. [H 12]). With the assumption that an approximately constant portion of the total mass belongs to intergalactic gas, this relation can be understood [S 24]. Including the only detected small group of galaxies, the relation is true for more than 4 powers of ten in $L_X$ [B 19].

The continuum of the gas emission can often be represented well only by a two-temperature model [M 18].

The intergalactic gas also shows X-ray line emission, most often the iron lines at 6.7 and 7.9 keV, but occasionally also lines of Mg, Si, and S [M 18]. It follows from the strength of these lines that the abundance of these elements in the intergalactic gas is between 1/3 of normal and normal. This is usually interpreted to indicate that all this gas was produced in galaxies by normal evolution of stars and then ejected or swept from the galaxies [B 15, Y 5, H 11].

Some clusters also show X-ray emission from the central galaxy, possibly due to inverse Compton effect with the microwave background (e.g. [H 8]).

Intergalactic gas should be observable by a weakening of the microwave background [G 9, P 7b, R 1, B 21a]. By correlating such an observation with X-ray emission, cosmological parameters can be derived [S 26, C 6].

The existence of head-tail radio galaxies also points to the existence of intergalactic gas in many clusters [Y 4, H 3, H 12a, S 16a].

There are no known intergalactic clouds of hydrogen [B 1, H 5, T 1b].

Time scales

Time scales of clusters have been discussed by Bahcall [B 3]. The dynamical friction [W 7] leads to mass segregation [Q 1]. The cooling of the intergalactic gas leads to accretion into the main galaxy [M 10]. Collisions between galaxies lead to mass exchange and increase the effects of relaxation [S 14].

Dominant galaxies (see also 9.1.4.2)

The dominant galaxies in clusters are usually gE or cD galaxies [M 11], see also 9.1.3.2.1. Even poor clusters often have a dominant cD galaxy [M 16, A 8, S 12]. Such very large galaxies [O 3, C 3] provide the deepest potential well in a cluster, and thus the intergalactic gas there is densest, has the shortest cooling time, and may accrete [M 10]. Accretion may lead to the feeding of active nuclei and is quite common in dominant galaxies [M 11, P 17].

Radio emission

The number of radio surveys of clusters of galaxies is very large: [V 1, A 13, B 30, B 31, B 31a, H 1a, W 12e, H 1, H 2, H 2a, G 3b, P 8b, V 1b, K 13a, H 2b, G 3a, P 8a, W 1, V 1a, C 1a, S 27a] and many other papers referenced therein. The observations show that the radio emission consists of two components, identifiable radio galaxies and head-tail radio galaxies on the one hand, and diffuse emission on the other hand. This diffuse emission is currently interpreted as the remains of old radio-galaxy heads.

## 9.3.5 Superclusters

A band of groups of galaxies in the sky indicate the plane of the local supercluster [D 2, D 3, B 2]. In general, clusters seem to tend to form higher order aggregates [A 2, R 6, R 6a, G 7c, F 6a, B 11c, C 8a, S 14a, E 1].

Evolution

N-body calculations can be used to demonstrate how galaxies group and cluster together as a function of time [G 7, S 7a, K 6a, B 11a, W 12f, L 1, F 4b, G 4b, A 1, T 18, D 1, B 17a, G 7a, C 1, S 36a, S 36b, S 36c, F 4d, R 1b, A 13a, C 2b, N 1, J 1, L 3a]. Observations of clusters of galaxies with high redshift can be used to discuss the evolution of the galaxies [K 14, B 32, H 11, K 9b].

General distribution

The general distribution of galaxies in space and (in projection) on the sky can be described by a covariance function, which turns out to be a power law, and thus shows that there are no preferred scales in the clustering of galaxies (e.g. [P 4, W 12, P 7a, F 4]).

Field galaxies

The properties of isolated or field galaxies have been studied and found to reflect their presumably peaceful history [H 9, S 30, K 7, B 4, D 14, B 11b, D 1c, B 29b, A 3a].

# 9.3.6 References for 9.3

## General references


a Galaxies and the Universe (Sandage, A., Sandage, M., Kristian, J., eds.), = Stars and Stellar Systems **9**, Univ. of Chicago Press (1975).
b Int. Astron. Union Symp. **63**, Confrontation of Cosmological Theories with Observational Data (Longair, M.S., ed.), Reidel, Dordrecht (1974).
c Int. Astron. Union Symp. **69**, Dynamics of Stellar Systems (Hayli, A., ed.), Reidel, Dordrecht (1975).
d Int. Astron. Union Symp. **77**, Structure and Properties of Nearby Galaxies (Berkhuijsen, E.M., Wielebinski, R., eds.), Reidel, Dordrecht (1978).
e Int. Astron. Union Symp. **79**, The Large Scale Structure of the Universe (Longair, M.S., Einasto, J., eds.), Reidel, Dordrecht (1978).


Lists of groups and clusters of galaxies are compiled in 9.3.3, Table 1.

## Special references

Remark to the cited Russian references: in ⟨···⟩ the page in the corresponding Englisch translation is given:

Astron. Zh. – Engl. translation: Soviet Astron.
Pis'ma Astron. Zh. – Engl. translation: Soviet Astron. Lett.
Astrofizika – Engl. translation: Astrophysics


A 1 Aarseth, S.J., Gott, J.R.III, Turner, E.L.: Astrophys. J. **228** (1979) 664.
A 1a Aarseth, S.J., Fall, S.M.: Astrophys. J. **236** (1980) 43.
A 1b Abell, G.O.: Astrophys. J. Suppl. **3** (1958) 211.
A 2 Abell, G.O. in: [b] p. 79.
A 3 Abell, G.O.: Clusters of Galaxies, in [a] p. 601.
A 3a Adams, M.T., Jensen, E.B., Stocke, J.T.: Astron. J. **85** (1980) 1010.
A 4 Adams, T.F.: Astrophys. J. Suppl. **33** (1977) 19.
A 5 Agüero, E.L.: Publ. Astron. Soc. Pacific **83** (1971) 310.
A 6 Ahmend, F.: Astrophys. Space Science **60** (1979) 493.
A 7 Albada, T.S. van, van Gorkom, J.H.: Astron. Astrophys. **54** (1977) 121.
A 8 Albert, C.E., White, R.A., Morgan, W.W.: Astrophys. J. **211** (1977) 309.
A 9 Alladin, S.M.: Astrophys. J. **141** (1965) 768.
A 10 Alladin, S.M., Potdar, A., Sastry, K.S. in: [c] p. 167.
A 11 Allen, C.W.: Astrophysical Quantities, 3rd ed., The Athlone Press, London (1972) p. 287.
A 12 Allen, D.A.: Astrophys. J. **207** (1976) 367.
A 13 Andernach, H., Waldthausen, H., Wielebinski, R.: Astron. Astrophys. Suppl. **41** (1980) 339.
A 13a Andernach, H., Schallwich, D., Haslam, C.G.T., Wielebinski, R.: Astron. Astrophys. Suppl. **43** (1981) 155.
A 14 Arakelyan, M.A., Dibai, E.A., Esipov, V.F.: Astrofizika **6** (1970) 39 ⟨14⟩; **8** (1972) 33 ⟨17⟩; **8** (1972) 177 ⟨106⟩; **8** (1972) 329 ⟨197⟩; **9** (1973) 319 ⟨180⟩; **9** (1973) 325 ⟨183⟩.
A 15 Arakelyan, M.A., Dibai, E.A., Esipov, V.F., Markarian, B.E.: Astrofizika **6** (1970) 357 ⟨189⟩; **7** (1971) 177 ⟨102⟩.
A 16 Arakelyan, M.A., Dibai, E.A., Lyuti, B.M.: Astrofizika **8** (1972) 473 ⟨280⟩.
A 17 Arakelyan, M.A.: Astrofizika **10** (1974) 507 ⟨321⟩; **12** (1976) 559 ⟨366⟩.
A 18 Arakelyan, M.A.: Publ. Bjurakan Obs. **47** (1975) 3.
A 19 Arakelyan, M.A.: Publ. Bjurakan Obs. **47** (1975) 43.
A 20 Arakelyan, M.A., Dibai, E.A., Esipov, V.F.: Astrofizika **11** (1975) 15 ⟨8⟩; **11** (1975) 377 ⟨254⟩; **12** (1976) 195 ⟨122⟩; **12** (1976) 683 ⟨456⟩.
A 21 Arakelyan, M.A.: Astrofizika **13** (1977) 245 ⟨130⟩.
A 22 Ardeberg, A., Bergvall, N.: Astron. Astrophys. **61** (1977) 493.
A 23 Arigo, R., Czuhai, K., Hubbard, E., Noerdlinger, P., Wisner, K.: Astrophys. J. **223** (1978) 410.
A 24 Arkhipova, V.P., Esipov, V.F., Savel'eva, M.V.: Astron. Zh. **53** (1976) 53 ⟨**20**, 521⟩.
A 25 Arp, H.C.: Atlas of peculiar galaxies, Calif. Inst. Techn., Pasadena (1966).

B 1 Baan, W.A., Haschick, A.D., Burke, B.F.: Astrophys. J. **225** (1978) 339.
B 2 Bahcall, J.N., Joss, P.C.: Astrophys. J. **203** (1976) 23.
B 3 Bahcall, N.A.: Annu. Rev. Astron. Astrophys. **15** (1977) 505.
B 4 Bahcall, N.A.: Astrophys. J. **232** (1979) 689.
B 5 Bahcall, N.A. in: Joint Discussion on Extragalactic High Energy Physics, Int. Astron. Union, General Assembly, Montreal (1979).
B 5a Bahcall, N.A.: Astrophys. J. **247** (1981) 787.
B 6 Baier, F.W., Petrosyan, M.B., Tiersch, H., Shakhbazyan, R.K.: Astrofizika **10** (1974) 327 ⟨202⟩.
B 7 Bautz, L.P., Morgan, W.W.: Astrophys. J. **162** (1970) L149.
B 8 Bergeron, J.: Astrophys. J. **211** (1977) 62.
B 9 Bergvall, N.A.S., Ekman, A.B.G., Lauberts, A., Westerlund, B.E., Borchkadze, T.M., Breysacher, J., Laustsen, S., Muller, A.B., Schuster, E.-H., Surdej, J., West, R.M.: Astron. Astrophys. Suppl. **33** (1978) 243.
B 9a Bergvall, N.: Astron. Astrophys. **97** (1981) 302.
B 10 Bertola, F., Galletta, G.: Astrophys. J. **226** (1978) L115.
B 11 Bhavsar, S.P., Ulmer, M.P., Cruddace, R.G., Wood, K., Meekins, J., Yentis, D., Evans, W., Smathers, H., Byram, E.T., Chubb, T.A., Friedman, H.: Nature **281** (1979) 462.
B 11a Bhavsar, S.P., Gott, J.R. III, Aarseth, S.J.: Astrophys. J. **236** (1980) 724.
B 11b Bhavsar, S.P.: Astrophys. J. **246** (1981) L5.
B 11c Bhavsar, S.P., Gott, J.R. III, Aarseth, S.J.: Astrophys. J. **246** (1981) 656.
B 12 Bieging, J.H., Biermann, P., Fricke, K.J., Pauliny-Toth, I.I.K., Witzel, A.: Astron. Astrophys. **60** (1977) 353.
B 13 Biermann, P., Silk, J.: Astron. Astrophys. **48** (1976) 287.
B 14 Biermann, P., Fricke, K.J.: Astron. Astrophys. **54** (1977) 461.
B 15 Biermann, P.: Astron. Astrophys. **62** (1978) 255.
B 16 Biermann, P., Wielen, R. in: [e] p. 121.
B 17 Biermann, P., Clarke, J.N., Fricke, K.J.: Astron. Astrophys. **75** (1979) 7; **75** (1979) 19.
B 17a Biermann, P., Shapiro, S.L.: Astrophys. J. **230** (1979) L33.
B 18 Biermann, P., Clarke, J.N., Fricke, K.J., Pauliny-Toth, I.I.K., Schmidt, J., Witzel, A.: Astron. Astrophys. **81** (1980) 235.
B 19 Biermann, P., Kronberg, P.P., Madore, B.F.: Astrophys. J. **256** (1982) L37.
B 20 Biermann, P., Kronberg, P.P., Preuss, E., Schilizzi, R.T., Shaffer, D.B.: Astrophys. J. **250** (1981) L49.
B 21 Binney, J., Silk, J.: Mon. Not. R. Astron. Soc. **188** (1979) 273.
B 21a Birkinshaw, M., Gull, S.F., Moffet, A.T.: Astrophys. J. **251** (1981) L69.
B 22 Bohuski, T.J., Fairall, A.P., Weedman, D.W.: Astrophys. J. **221** (1978) 776.
B 23 Borchkadze, T.M., West, R.M.: Astrofizika **13** (1977) 605 ⟨358⟩.
B 24 Borchkadze, T.M., Breysacher, J., Laustsen, S., Schuster, H.-E., West, R.M.: Astron. Astrophys. Suppl. **30** (1977) 35.
B 25 Börngen, F., Kalloglyan, A.T.: Astrofizika **10** (1974) 159 ⟨97⟩; **11** (1975) 5 ⟨1⟩; **11** (1975) 369 ⟨249⟩.
B 26 Börngen, F., Kalloglyan, A.T.: Astrofizika **11** (1975) 617 ⟨414⟩.
B 27 Bottinelli, L., Chamaraux, P., Gouguenheim, L., Heidmann, J.: Astron. Astrophys. **29** (1973) 217.
B 28 Bottinelli, L., Gouguenheim, L., Heidmann, J.: Astron. Astrophys. **22** (1973) 281.
B 29 Bottinelli, L., Duflot, R., Gouguenheim, L., Heidmann, J.: Astron. Astrophys. **41** (1975) 61.
B 29a Breugel, W.J.M. van, Schilizzi, R.T., Hummel, E., Kapahi, V.K.: Astron. Astrophys. **96** (1981) 310.
B 29b Brosch, N., Shaviv, G.: Astrophys. J. **253** (1982) 526.
B 29c Brown, R.L., Neff, S.G.: Astrophys. J. **241** (1980) 561.
B 30 Burns, J.O., Owen, F.N.: Astron. J. **84** (1979) 1478.
B 31 Burns, J.O., White, R.A., Hanisch, R.J.: Astron. J. **85** (1980) 191.
B 31a Burns, J.O., White, R.A., Hough, D.H.: Astron. J. **86** (1981) 1.
B 32 Butcher, H., Oemler, A., jr.: Astrophys. J. **219** (1978) 18.
B 33 Byrd, G.G.: Astrophys. J. **231** (1979) 32.

C 1 Caldwell, C.N., Oemler, A., Jr.: Astron. J. **86** (1981) 1424.
C 1a Cane, H.V., Erickson, W.C., Hanisch, R.J., Turner, P.J.: Mon. Not. R. Astron. Soc. **196** (1981) 409.
C 1b Canizares, C.R., Clark, G.W., Markert, T.H., Berg, C., Smedira, M., Bardes, D., Schnopper, H., Kalata, K.: Astrophys. J. **234** (1979) L33.
C 1c Cannon, R.D., Hawarden, T.G., Tritton, S.B.: Mon. Not. R. Astron. Soc. **180** (1977) 81p.
C 2 Cannon, R.D., Lloyd, C., Penston, M.V.: Observatory **90** (1970) 153.

C 2a Capelato, H.V., Gerbal, D., Mathez, G., Mazure, A., Salvador-Solé, E., Sol, H.: Astrophys. J. **241** (1980) 521.
C 2b Capelato, H.V., Gerbal, D., Methez, G., Mazure, A., Roland, J., Salvador-Solé, E.: Astron. Astrophys. **96** (1981) 235.
C 2c Capelato, H.V., Gerbal, D., Mathez, G., Mazure, A., Salvador-Solé, E.: Astrophys. J. **252** (1982) 433.
C 2d Carnevali, P., Cavaliere, A., Santangelo, P.: Astrophys. J. **249** (1981) 449.
C 3 Carter, D.: Mon. Not. R. Astron. Soc. **178** (1977) 137.
C 3a Carter, D., Metcalfe, N.: Mon. Not. R. Astron. Soc. **191** (1980) 325.
C 4 Casini, C., Heidmann, J.: Astron. Astrophys. **47** (1976) 371.
C 5 Casini, C., Heidmann, J.: Astron. Astrophys. Suppl. **24** (1976) 473; **34** (1978) 91.
C 6 Cavaliere, A., Danese, L., DeZotti, G.: Astron. Astrophys. **75** (1979) 322.
C 7 Chamaraux, P.: Astron. Astrophys. **60** (1977) 67.
C 8 Chatterjee, T.K.: Bull. Astron. Soc. India **7** (1979) 32.
C 8a Chincarini, G., Rood, H.J., Thompson, L.A.: Astrophys. J. **249** (1981) L47.
C 9 Condon, J.J., Dressel, L.L.: Astrophys. J. **221** (1978) 456.
C 9a Condon, J.J., Condon, M.A., Gisler, A., Puschell, J.J.: Astrophys. J. **252** (1982) 102.
C 9b Copper, G.R., Miller, R.H.: Astrophys. J. **254** (1982) 16.
C 10 Crane, P.C.: Astron. J. **84** (1979) 281.

D 1 DaCosta, L.N., Knobloch, E.: Astrophys. J. **230** (1979) 639.
D 1a Danese, L., DeZotti, G., Giuricin, G., Mardirossian, F., Mezzetti, M., Ramella, M.: Astrophys. J. **244** (1981) 777.
D 1b Danks, A.C., Alcaino, G.: Astron. Astrophys. **98** (1981) 223.
D 1c Davis, M., Huchra, J., Latham, D.W., Tonry, J.: Astrophys. J. **253** (1982) 423.
D 1d De Vaucouleurs, G.: Astron. J. **63** (1958) 253.
D 2 De Vaucouleurs, G.: Nearby groups of galaxies, in [a] p. 557.
D 3 De Vaucouleurs, G.: Astrophys. J. **203** (1976) 33.
D 4 De Vaucouleurs, G., Peters, W.L., Corwin, H.G., Jr.: Astrophys. J. **211** (1977) 319.
D 5 Denisyuk, E.K., Lipovetskii, V.A.: Astrofizika **10** (1974) 315 ⟨195⟩; Pis'ma Astron. Zh. **3** (1977) 7 ⟨3⟩.
D 6 Denisyuk, E.K., Lipovetskii, V.A., Afanas'ev, V.L.: Astrofizika **12** (1976) 665 ⟨442⟩.
D 7 Dibai, E.A.: Astrofizika **6** (1970) 350 ⟨185⟩.
D 8 Dibai, E.A., Doroshenko, V.T., Terebizh, V.Yu.: Astrofizika **12** (1976) 689 ⟨459⟩.
D 8a Dickel, J.R., Rood, H.J.: Astron. J. **85** (1980) 1003.
D 9 Doroshenko, V.T., Terebizh, V.Yu.: Astrofizika **11** (1975) 631 ⟨422⟩.
D 10 Dostal', V.A.: Pis'ma Astron. Zh. **3** (1977) 60 ⟨30⟩.
D 11 Dostal', V.A., Metlov, V.G.: Astron. Zh. **56** (1979) 3 ⟨23, 1⟩.
D 12 Dressel, L.L., Condon, J.J.: Astrophys. J. Suppl. **31** (1976) 187.
D 13 Dressel, L.L., Condon, J.J.: Astrophys. J. Suppl. **36** (1978) 53.
D 14 Dressler, A.: Astrophys. J. **236** (1980) 351.
D 14a Dressler, A.: Astrophys. J. Suppl. **42** (1980) 565.
D 15 DuPuy, D.L.: Publ. Astron. Soc. Pacific **80** (1968) 29.
D 16 DuPuy, D.L.: Astron. J. **75** (1970) 1143.

E 1 Einasto, J., Joeveer, M., Saar, E.: Mon. Not. R. Astron. Soc. **193** (1980) 353.
E 1a Eneev, T.M., Kozlov, N.N., Sunyaev, R.A.: Astron. Astrophys. **22** (1973) 41.

F 1 Faber, S., Gallagher, J.S.: Annu. Rev. Astron. Astrophys. **17** (1979) 135.
F 1a Fabian, A.C., Hu, E.M., Cowie, L.L., Grindlay, J.: Astrophys. J. **248** (1981) 47.
F 2 Fairall, A.P.: Mon. Not. Astron. Soc. South Africa **27** (1968) 67; **29** (1970) 48; Mon. Not. R. Astron. Soc. **180** (1977) 391.
F 3 Fairall, A.P.: Mon. Not. R. Astron. Soc. **153** (1971) 383.
F 3a Fairall, A.P.: Mon. Not. R. Astron. Soc. **196** (1981) 11P.
F 3b Fall, S.M.: Nature **281** (1979) 200.
F 4 Fall, S.M.: Rev. Mod. Phys. **51** (1979) 21.
F 4a Fanti, C., Fanti, R., Feretti, L., Ficarra, A., Gioia, I.M., Giovannini, G., Gregorini, L., Mantovani, F., Marano, B., Padrielli, L., Parna, P., Tomasi, P., Vettolani, G.: Astron. Astrophys. **105** (1982) 200.
F 4b Farouki, R., Shapiro, S.L.: Astrophys. J. **241** (1980) 928.
F 4c Farouki, R., Shapiro, S.L.: Astrophys. J. **243** (1981) 32.
F 4d Farouki, R.T., Salpeter, E.E.: Astrophys. J. **253** (1982) 512.

F 5 Field, G.B.: Mitt. Astron. Ges. **47** (1980) 7.
F 6 Fisher, J.R., Tully, R.B.: Astron. Astrophys. **44** (1977) 151.
F 6a Ford, H.C., Harms, R.J., Ciardullo, R., Bartok, F.: Astrophys. J. **245** (1982) L53.
F 6b Forman, W., Bechtold, J., Blair, W., Giacconi, R., Van Speybroeck, L., Jones, C.: Astrophys. J. **243** (1981) L133.
F 7 Freeman, K.C., de Vaucouleurs, G.: Astrophys. J. **194** (1974) 569.
F 8 Fridman, A.M.: Pis'ma Astron. Zh. **4** (1978) 207 ⟨113⟩.

G 1 Gallagher, J.S. III, Ostriker, J.P.: Astron. J. **77** (1972) 288.
G 1a Gallagher, J.S., Hunter, D.A.: Astron. J. **86** (1981) 1312.
G 1b Gallagher, J.S., Knapp, G.R., Faber, S.M.: Astron. J. **86** (1981) 1781.
G 2 Gallouet, L., Heidmann, N.: Astron. Astrophys. Suppl. **3** (1971) 325.
G 3 Gallouet, L., Heidmann, N., Dampierre, F.: Astron. Astrophys. Suppl. **12** (1973) 89; **19** (1975) 1.
G 3a Gavazzi, G.: Astron. Astrophys. **72** (1979) 1.
G 3b Gavazzi, G., Perola, G.C.: Astron. Astrophys. **84** (1980) 228.
G 4 Ghigo, F.D.: Astron. J. **85** (1980) 215.
G 4a Giovanelli, R., Chincarini, G.L., Haynes, M.P.: Astrophys. J. **247** (1981) 383.
G 4b Gisler, G.R.: Astrophys. J. **228** (1979) 385.
G 4c Gisler, G.R.: Astron. J. **85** (1980) 623.
G 5 Gott, J.R. III: Highlights of Astronomy **4** pt. 1 (1977) 271.
G 6 Gott, J.R. III, Turner, E.L.: Astrophys. J. **213** (1977) 309.
G 7 Gott, J.R. III: Comments **8** (1979) 55.
G 7a Gott, J.R. III, Turner, E.L., Aarseth, S.J.: Astrophys. J. **234** (1979) 13.
G 7b Graham, D.A., Weiler, K.W., Wielebinski, R.: Astron. Astrophys. **97** (1981) 388.
G 7c Gregory, S.A., Thompson, L.A., Tifft, W.G.: Astrophys. J. **243** (1981) 411.
G 8 Griffiths, R.E., Doxsey, R.E., Johnston, M.D., Schwartz, D.A., Schwarz, J., Blades, J.C.: Astrophys. J. **230** (1979) L21.
G 9 Gull, S.F.: Highlights of Astronomy **4** pt. 1 (1977) 341.
G 10 Gursky, H., Schwartz, D.A.: Annu. Rev. Astron. Astrophys. **15** (1977) 541.

H 1 Hanisch, R.J., Erickson, W.C.: Astron. J. **85** (1980) 183.
H 1a Hanisch, R.J., White, R.A.: Astron. J. **86** (1981) 806.
H 1b Haro, G.: Bol. Obs. Tonantzintla y Tacubaya **14** (1956) 8.
H 2 Harris, D.E.: Highlights of Astronomy **4** pt. 1 (1977) 321.
H 2a Harris, D.E., Kapahi, V.K., Ekers, R.D.: Astron. Astrophys. Suppl. **39** (1980) 215.
H 2b Harris, D.E., Lari, C., Vallée, J.P., Wilson, A.S.: Astron. Astrophys. Suppl. **42** (1980) 319.
H 3 Hart, L., Davies, R.D., Johnson, S.C.: Mon. Not. R. Astron. Soc. **191** (1980) 269.
H 4 Hausman, M.A., Ostriker, J.P.: Astrophys. J. **224** (1978) 320.
H 5 Haynes, M.P., Brown, R.L., Roberts, M.S.: Astrophys. J. **221** (1978) 414.
H 6 Haynes, M.P., Roberts, M.S.: Astrophys. J. **227** (1979) 767.
H 6a Haynes, M.P.: Astron. J. **86** (1981) 1126.
H 6b Heckman, T.M.: Astrophys. J. **250** (1981) L59.
H 7 Heidmann, J., Kalloglyan, A.T.: Astrofizika **9** (1973) 71 ⟨37⟩; **11** (1975) 229 ⟨156⟩.
H 7a Heiligman, G.M., Turner, E.L.: Astrophys. J. **236** (1980) 745.
H 8 Helmken, H., Delvaille, J.P., Epstein, A., Geller, M.J., Schnopper, H.W., Jernigan, J.G.: Astrophys. J. **221** (1978) L43.
H 9 Henon, M. in: Dynamical structure and evolution of stellar systems (Martinet, L., Mayor, M., eds.), Geneva Obs., Geneva (1973) p. 183.
H 10 Hickson, P., Richstone, D.O., Turner, E.L.: Astrophys. J. **213** (1977) 323.
H 11 Himmes, A., Biermann, P.: Astron. Astrophys. **86** (1980) 11.
H 12 Hintzen, P., Scott, J.S.: Astrophys. J. **232** (1979) L145.
H 12a Hintzen, P., Scott, J.S.: Astrophys. J. **239** (1980) 765.
H 12b Hintzen, P., Scott, J.S., McKee, J.D.: Astrophys. J. **242** (1980) 857.
H 13 Hockney, R.W., Brownrigg, D.K.K.: Mon. Not. R. Astron. Soc. **167** (1974) 351.
H 13a Hoessel, J.G., Gunn, J.E., Thuan, T.X.: Astrophys. J. **241** (1980) 486.
H 13b Hoessel, J.G.: Astrophys. J. **241** (1980) 493.
H 14 Hohl, F.: Astron. J. **81** (1976) 30.
H 15 Holmberg, E.: Ann. Obs. Lund No. 6 (1937) 1.

H 16 Holmberg, E.: Medd. Lund. Astron. Obs. Ser. II, No. 117 (1946); No. 136 (1958).
H 17 Holmberg, E.: Medd. Lund. Astron. Obs. Ser. I, No. 186 (1954).
H 17a Hoyle, F.: IUTAM-Int. Astron. Union Symp., Problems of Cosmical Aerodynamics (1949) p. 195.
H 18 Huchra, J.P.: Astrophys. J. Suppl. **35** (1977) 171.
H 19 Huchra, J.P., Thuan, T.X.: Astrophys. J. **216** (1977) 694.
H 20 Huchra, J.P.: Astrophys. J. **217** (1977) 928.
H 21 Hummel, E.: Astron. Astrophys. **96** (1981) 111.

J 1 Jones, B.T.: Rev. Mod. Phys. **48** (1976) 107.
J 1a Jones, B.J.T., Efstathiou, G.: Mon. Not. R. Astron. Soc. **189** (1979) 27.
J 1b Jones, C., Mandel, B., Schwarz, J., Forman, W., Murray, S.S., Harnden, F.H., Jr.: Astrophys. J. **231** (1979) L21.
J 1c Jones, D.L., Sramek, R.A., Terzian, Y.: Astrophys. J. **246** (1981) 28.

K 1 Kalloglyan, A.T.: Astrofizika **7** (1971) 521 ⟨311⟩.
K 2 Karachentsev, I.D.: Soobshch. Spec. Astron. Obs. No. 7 (1972).
K 3 Karachentsev, I.D., Fesenko, V.I.: Astron. Zh. **52** (1975) 659 ⟨**19**, 400⟩.
K 4 Karachentsev, I.D., Fesenko, B.I.: Astrofizika **15** (1979) 217 ⟨147⟩.
K 5 Karachentsev, I.D., Sargent, W.L.W., Zimmermann, B.: Astrofizika **15** (1979) 25 ⟨19⟩.
K 6 Kazarian, M.A.: Astrofizika **15** (1979) 5 ⟨1⟩; **15** (1979) 193 ⟨117⟩.
K 6a Kent, S.M.: Astrophys. J. **245** (1981) 805.
K 7 Kirshner, R.P., Oemler, A., Jr., Schechter, P.L.: Astron. J. **83** (1978) 1549 [Erratum: **84** (1979) 1088]; **84** (1979) 951.
K 8 Kojoian, G., Sramek, R.A., Dickinson, D.F., Tovmassian, H.M., Purton, C.R.: Astrophys. J. **203** (1976) 323.
K 9 Kojoian, G., Elliott, R., Tovmassian, H.M.: Astron. J. **83** (1978) 1545.
K 9a Kojoian, G., Tovmanssian, H.M., Dickinson, D.F., St. Clair Dinger, A.: Astron. J. **85** (1980) 1462.
K 9b Koo, D.C.: Astrophys. J. **251** (1981) L75.
K 10 Kopylov, I.M., Lipovetskii, V.A., Pronik, V.I., Chuvaev, K.K.: Astrofizika **10** (1974) 483 ⟨305⟩; **12** (1976) 189 ⟨119⟩.
K 11 Kormendy, J.: Astrophys. J. **214** (1977) 359.
K 12 Kotanyi, C.G.: Astron. Astrophys. **74** (1979) 156.
K 13 Kotanyi, C.G., Ekers, R.D.: Astron. Astrophys. **73** (1979) L1.
K 13a Kotanyi, C.G.: Astron. Astrophys. Suppl. **41** (1980) 421.
K 14 Kron, R.G., Spinrad, H., King, I.R.: Astrophys. J. **217** (1977) 951.

L 1 Larson, R.B., Tinsley, B.M., Caldwell, C.N.: Astrophys. J. **237** (1980) 692.
L 1a Lauberts, A.: Astron. Astrophys. **33** (1974) 231.
L 1b Lea, S.M., Reichert, G., Mushotzky, R., Baity, W.A., Gruber, D.E., Rothschild, R., Primini, F.A.: Astrophys. J. **246** (1981) 369.
L 2 Lecar, M. in: [c] p. 161.
L 3 Leir, A.A., van den Bergh, S.: Astrophys. J. Suppl. **34** (1977) 381.
L 3a Lightman, A.P., Shapiro, S.L.: Rev. Mod. Phys. **50** (1978) 437.
L 4 Lo, K.Y., Sargent, W.L.W.: Astrophys. J. **227** (1979) 756.
L 5 Lüst, R.: Z. Naturforsch. **7a** (1952) 87.
L 6 Lynden-Bell, D., Pringle, J.E.: Mon. Not. R. Astron. Soc. **168** (1974) 603.
L 7 Lynds, R., Toomre, A.: Astrophys. J. **209** (1976) 382.

M 1 MacAlpine, G.M., Smith, S.B., Lewis, D.W.: Astrophys. J. Suppl. **34** (1977) 95; **35** (1977) 197.
M 2 MacAlpine, G.M., Lewis, D.W., Smith, S.B.: Astrophys. J. Suppl. **35** (1977) 203.
M 2a Maccacaro, T., Feigelson, E.D., Fener, M., Giacconi, R., Gioia, I.M., Griffiths, R.E., Murray, S.S., Zamorani, G., Stocke, J., Liebert, J.: Astrophys. J. **253** (1982) 504.
M 3 Marchant, A.B., Shapiro, S.L.: Astrophys. J. **215** (1977) 1.
M 4 Markarian, B.E.: Astrofizika **9** (1973) 5 ⟨1⟩.
M 5 Markarian, B.E.: Astron. Astrophys. **58** (1977) 139.
M 6 Markarian, B.E., Stepanyan, Dzh.A.: Astrofizika **13** (1977) 627 ⟨372⟩.
M 7 Marshall, F.E., Boldt, E.A., Holt, S.S., Mushotzky, R.F., Pravdo, S.H., Rothschild, R.E., Serlemitsos, P.J.: Astrophys. J. Suppl. **40** (1979) 657.
M 8 Materne, J., Tammann, G.A.: Astron. Astrophys. **37** (1974) 383.
M 9 Materne, J., Huchtmeier, W.K., Hulsbosch, A.N.M.: Mon. Not. R. Astron. Soc. **186** (1979) 563.

M 9a Materne, J.: Astron. Astrophys. **86** (1980) 51.
M 10 Mathews, W.G., Bregman, J.H.: Astrophys. J. **224** (1978) 308.
M 11 Matthews, T.A., Morgan, W.W., Schmidt, M.: Astrophys. J. **140** (1964) 35.
M 11a McGlynn, T.A., Ostriker, J.P.: Astrophys. J. **241** (1980) 915.
M 11b McKee, J.D., Mushotzky, R.F., Boldt, E.A., Holt, S.S., Marshall, F.E., Pravdo, S.H., Serlemitsos, P.J.: Astrophys. J. **242** (1980) 843.
M 12 Melnick, J., Sargent, W.L.W.: Astrophys. J. **215** (1977) 401.
M 13 Metik, L.P., Pronik, I.I.: Astron. Zh. **55** (1978) 249 ⟨**22**, 148⟩.
M 14 Metlov, V.G.: Pis'ma Astron. Zh. **4** (1978) 250 ⟨136⟩.
M 14a Miller, R.H., Smith, B.F.: Astrophys. J. **253** (1982) 58.
M 14b Mitchell, R., Mushotzky, R.: Astrophys. J. **236** (1980) 730.
M 15 Morgan, W.W.: Publ. Astron. Soc. Pacific **70** (1958) 364.
M 16 Morgan, W.W., Kayser, S., White, R.A.: Astrophys. J. **199** (1975) 545.
M 17 Mushotzky, R.F., Serlemitsos, P.J., Smith, B.W., Boldt, E.A., Holt, S.S.: Astrophys. J. **225** (1978) 21.
M 18 Mushotzky, R.F., Smith, B.W. in: Joint discussion on extragalactic high energy physics, Int. Astron. Union General Assembly, Montreal (1979).
M 19 Mushotzky, R.F., Holt, S.S., Smith, B.W., Boldt, E.A., Serlemitsos, P.J.: Astrophys. J. **244** (1981) L47.

N 1 Nepveu, M.: Astron. Astrophys. **98** (1981) 65.
N 1a Neugebauer, G., Becklin, E.E., Oke, J.B., Searle, L.: Astrophys. J. **205** (1976) 29.
N 2 Noerdlinger, P.D.: Astrophys. J. **197** (1975) 545.
N 3 Noerdlinger, P.D.: Astrophys. J. **229** (1979) 470.
N 4 Noerdlinger, P.D.: Astrophys. J. **229** (1979) 877.
N 5 Noonan, T.W.: Astrophys. J. Suppl. **45** (1981) 613.

O 1 O'Connell, R.W., Thuan, T.X., Goldstein, S.J.: Astrophys. J. **226** (1978) L11.
O 2 Oemler, A., Jr.: Astrophys. J. **194** (1974) 1.
O 3 Oemler, A., Jr.: Astrophys. J. **209** (1976) 693.
O 4 Osterbrock, D.E., Phillips, M.M.: Publ. Astron. Soc. Pacific **89** (1977) 251.
O 5 Ostriker, J.P., Peebles, P.J.E.: Astrophys. J. **186** (1973) 467.
O 6 Ostriker, J.P., Tremaine, S.D.: Astrophys. J. **202** (1975) L113.
O 7 Ostriker, J.P., Hausman, A.: Astrophys. J. **217** (1977) L125.
O 8 Ostriker, J.P., Turner, E.L.: Astrophys. J. **234** (1979) 785.

P 1 Page, T.: Astrophys. J. **116** (1952) 63; **159** (1970) 791.
P 2 Page, T.: Binary Galaxies, in [a] p. 541.
P 3 Pedreros, M.: Publ. Astron. Soc. Pacific **90** (1978) 14.
P 4 Peebles, P.J.E., Hauser, M.G.: Astrophys. J. Suppl. **28** (1974) 37.
P 5 Peebles, P.J.E. in: Physical Cosmology (Balian, R., Audouze, J., Schramm, D.M. eds.), Les Houches Lectures 1979, North-Holland, Amsterdam (1980) p. 213.
P 6 Peebles, P.J.E.: Astrophys. J. **155** (1969) 393.
P 7 Peebles, P.J.E.: Astron. Astrophys. **11** (1971) 377.
P 7a Peebles, P.J.E.: The Large Structure of the Universe, Princeton Univ. Press (1980).
P 7b Peebles, P.J.E.: Astrophys. J. **243** (1981) L119.
P 8 Penston, M.V., Fosbury, R.A.E., Ward, M.J., Wilson, A.S.: Mon. Not. R. Astron. Soc. **180** (1977) 19.
P 8a Perola, G.C., Valentijn, E.A.: Astron. Astrophys. **73** (1979) 54.
P 8b Perola, G.C., Tarenghi, M., Valentijn, E.A.: Astron. Astrophys. **84** (1980) 245.
P 8c Perrenod, S.C., Henry, J.P.: Astrophys. J. **247** (1981) L1.
P 9 Peterson, S.D.: Astron. J. **78** (1973) 811.
P 10 Peterson, S.D., Shostak, G.S.: Astron. J. **79** (1974) 767.
P 11 Peterson, S.D.: Astrophys. J. Suppl. **40** (1979) 527.
P 12 Peterson, S.D.: Astrophys. J. **232** (1979) 20.
P 13 Petrosyan, A.R., Saakyan, K.A., Khachikyan, E.E.: Astrofizika **14** (1978) 69 ⟨36⟩.
P 14 Petrosyan, A.R., Saakyan, K.A., Khachikyan, E.E.: Astrofizika **15** (1979) 209 ⟨142⟩.
P 15 Pfleiderer, J., Siedentopf, H.: Z. Astrophys. **51** (1961) 201.
P 16 Pfleiderer, J.: Z. Astrophys. **58** (1963) 12.
P 17 Phillips, M.M.: Astrophys. J. **236** (1980) L45.
P 18 Piccinotti, G., Mushotzky, R.F., Boldt, E.A., Holt, S.S., Marshall, F.E., Serlemitsos, P.J., Shafer, R.A.: Astrophys. J. **253** (1982) 485.

P 19 Preuss, E., Pauliny-Toth, I.I.K., Witzel, A., Kellermann, K.I., Shaffer, D.B.: Astron. Astrophys. **54** (1977) 297.

P 20 Primini, F.A., Basinska, E., Howe, S.K., Lang, F., Levine, A.M., Rothschild, R., Baity, W.A., Gruber, D.E., Knight, F.K., Matteson, J.L., Lea, S.M., Reichert, G.A.: Astrophys. J. **243** (1981) L13.

Q 1 Quintana, H.: Astron. J. **84** (1979) 15.

Q 2 Quintana, H., Lawrie, D.G.: Astron. J. **87** (1982) 1.

R 1 Rephaeli, H.: Astrophys. J. **245** (1981) 351.

R 1a Richstone, D.O.: Astrophys. J. **200** (1975) 535.

R 1b Roland, J.: Astron. Astrophys. **93** (1981) 407.

R 2 Rood, H.J., Saastry, G.N.: Publ. Astron. Soc. Pacific **83** (1971) 313.

R 3 Rood, H.J., Page, T.L., Kintner, E.C., King, I.R.: Astrophys. J. **175** (1972) 627.

R 4 Rood, H.J., Dickel, J.R.: Astrophys. J. **224** (1978) 724; **233** (1979) 418.

R 5 Rood, H.J., Leir, A.A.: Astrophys. J. **231** (1979) L3.

R 6 Rood, H.J.: Astrophys. J. **233** (1979) 431.

R 6a Rood, H.J., Struble, M.F.: Astrophys. J. **252** (1982) L7.

R 7 Roos, N., Norman, C.A.: Astron. Astrophys. **76** (1979) 75.

R 8 Roos, N.: Astron. Astrophys. **95** (1981) 349.

R 9 Rose, J.A.: Astrophys. J. **231** (1979) 10.

R 10 Rose, J.A., Graham, J.A.: Astrophys. J. **231** (1979) 320.

S 1 Sanamyan, V.A., Gopal-Krishna: Astrofizika **11** (1975) 637 ⟨425⟩.

S 2 Sanamyan, V.A., Kandalyan, R.A.: Astrofizika **14** (1978) 623 ⟨352⟩.

S 3 Sandage, A.R., Brucato, R.: Astron. J. **84** (1979) 472.

S 4 Sargent, W.L.W.: Astrophys. J. **160** (1970) 405.

S 5 Sargent, W.L.W., Searle, L.: Astrophys. J. **162** (1970) L155.

S 6 Sargent, W.L.W.: Astrophys. J. **173** (1972) 7.

S 7 Saslaw, W.C., de Young, D.S.: Astrophys. J. **170** (1971) 423.

S 7a Saslaw, W.C., Aarseth, S.J.: Astrophys. J. **253** (1982) 470.

S 8 Sastry, G.N., Rood, H.J.: Astrophys. J. Suppl. **23** (1971) 371.

S 9 Sastry, K.S.: Astrophys. Space Science **16** (1972) 284.

S 10 Sastry, K.S., Alladin, S.M.: Astrophys. Space Science **7** (1970) 261.

S 11 Sastry, K.S., Alladin, S.M.: Astrophys. Space Science **46** (1977) 285.

S 12 Schild, R., Davis, H.: Astron. J. **84** (1979) 311.

S 13 Schmutzler, T.: Diplomarbeit Univ. Bonn (1980).

S 14 Schmutzler, T., Biermann, P.: preprint (1982).

S 14a Schuch, N.J.: Mon. Not. R. Astron. Soc. **196** (1981) 695.

S 14b Schwartz, D.A., Schwarz, J., Tucker, W.: Astrophys. J. **238** (1980) L59.

S 15 Schweizer, F. in: [d] p. 279.

S 16 Schweizer, F.: Astrophys. J. **237** (1980) 303.

S 16a Scott, J.S., Holman, G.D., Ionson, J.A., Papadopoulos, K.: Astrophys. J. **239** (1980) 769.

S 17 Searle, L., Sargent, W.L.W.: Astrophys. J. **173** (1972) 25.

S 18 Searle, L., Sargent, W.L.W., Bagnuolo, W.G.: Astrophys. J. **179** (1973) 427.

S 19 Sérsic, J.L., Pastoriza, M.: Publ. Astron. Soc. Pacific **77** (1965) 287; **79** (1967) 152.

S 20 Sérsic, J.L., Pastoriza, M., Carranza, G.J.: Astrophys. Lett. **2** (1968) 45.

S 21 Sérsic, J.L., Agüero, E.L.: Astrophys. Space Science **19** (1972) 387.

S 22 Sérsic, J.L., Carranza, G., Pastoriza, M.: Astrophys. Space Science **19** (1972) 469.

S 23 Sérsic, J.L.: Publ. Astron. Soc. Pacific **85** (1973) 103; Astrophys. Space Science **28** (1974) 365; **39** (1976) 477.

S 24 Silk, J., Tarter, J.: Astrophys. J. **183** (1973) 387.

S 25 Silk, J. in: [e] p. 179.

S 26 Silk, J., White, S.D.M.: Astrophys. J. **226** (1978) L103.

S 27 Silverglate, P.R., Krumm, N.: Astrophys. J. **224** (1978) L99.

S 27a Simon, A.J.B.: Mon. Not. R. Astron. Soc. **188** (1979) 637.

S 28 Smith, B.W., Mushotzky, R.F., Serlemitsos, P.J.: Astrophys. J. **227** (1979) 37.

S 29 Smith, M.G., Aguirre, C., Zemelman, M.: Astrophys. J. Suppl. **32** (1976) 217.

S 30 Soneira, R.M., Peebles, P.J.E.: Astrophys. J. **211** (1977) 1.

S 30a Spitzer, L., Jr., Baade, W.: Astrophys. J. **113** (1951) 413.

S 31 Spitzer, L., Jr.: Astrophys. J. **158** (1969) L139.
S 32 Sramek, R.A., Tovmassian, H.M.: Astrophys. J. **196** (1975) 339; **207** (1976) 725.
S 33 Stocke, J.T.: Astron. J. **83** (1978) 348.
S 34 Stocke, J.T., Tifft, W.G., Kaftan-Kassim, M.A.: Astron. J. **83** (1978) 322.
S 35 Strom, S.E., Strom, K.M.: Astrophys. J. **225** (1978) L93.
S 36 Strom, S.E., Strom, K.M. in: [d] p. 69.
S 36a Struble, M.F.: Astron. J. **84** (1979) 27.
S 36b Struble, M.F., Bludman, S.A.: Astron. J. **84** (1979) 40.
S 36c Struble, M.F.: Astron. J. **84** (1979) 50.
S 36d Struble, M.F., Rood, H.J.: Astrophys. J. **251** (1981) 471.
S 36e Struble, M.F., Rood, H.J.: Astron. J. **87** (1982) 7.
S 37 Sulentic, J.W.: Astron. J. **81** (1976) 582.
S 38 Sulentic, J.W.: Astrophys. J. **252** (1982) 439.

T 1 Tammann, G.A., Yahil, A., Sandage, A.: Astrophys. J. **234** (1979) 775.
T 1a Tarter, J., Silk, J.: Quarterly J. R. Astron. Soc. **15** (1974) 122.
T 1b Tarter, J.C., Wright, M.C.H.: Astron. Astrophys. **76** (1979) 127.
T 2 Tashpulatov, N.: Astron. Zh. **46** (1969) 1236 ⟨**13**, 968⟩; **47** (1970) 277 ⟨**14**, 227⟩.
T 3 Theys, J.C., Spiegel, E.A.: Astrophys. J. **208** (1976) 650.
T 4 Theys, J.C., Spiegel, E.A.: Astrophys. J. **212** (1977) 616.
T 5 Thompson, L.A.: Astrophys. J. **211** (1977) 684.
T 6 Thuan, T.X., Gott, J.R. III: Astrophys. J. **216** (1977) 194.
T 7 Tomov, A.N.: Astron. Zh. **55** (1978) 944 ⟨**22**, 540⟩.
T 8 Toomre, A., Toomre, J.: Astrophys. J. **178** (1972) 623.
T 9 Toomre, A. in: [e] p. 109.
T 10 Tovmassian, H.M., Terzian, Y.: Publ. Astron. Soc. Pacific **86** (1974) 649.
T 11 Trefftz, E.: Z. Naturforsch. **7a** (1952) 99.
T 12 Tremaine, S.D.: Astrophys. J. **203** (1976) 72.
T 13 Turner, E.L., Sargent, W.L.W.: Astrophys. J. **194** (1974) 587.
T 14 Turner, E.L.: Astrophys. J. **208** (1976) 20.
T 15 Turner, E.L.: Astrophys. J. **208** (1976) 304.
T 16 Turner, E.L., Gott, J.R. III: Astrophys. J. Suppl. **32** (1976) 409.
T 17 Turner, E.L., Ostriker, J.P.: Astrophys. J. **217** (1977) 24.
T 18 Turner, E.L., Aarseth, S.J., Gott, J.R. III, Blanchard, N.T., Mathieu, R.D.: Astrophys. J. **228** (1979) 684.

U 1 Ulmer, M.P., Shulman, S., Evans, W.D., Johnson, W.N., McNutt, D., Meekins, J., Share, G.H., Yentis, D., Wood, K., Byram, E.T., Chubb, T.A., Friedman, H.: Astrophys. J. **235** (1980) 351.
U 1a Ulmer, M.P., Kowalski, M.P., Cruddace, R.G., Johnson, M., Meekins, J., Smathers, H., Yentis, D., Wood, K., McNutt, D., Chubb, T., Byram, E.T., Friedman, H.: Astrophys. J. **243** (1981) 681.
U 1b Ulrich, M.-H.: Astrophys. J. **163** (1971) 441.

V 1 Valentijn, E.A. in: Joint discussion on extragalactic high energy astrophysics, Int. Astron. Union, General Assembly, Montreal (1979).
V 1a Valentijn, E.A.: Astron. Astrophys. Suppl. **38** (1979) 319.
V 1b Valentijn, E.A.: Astron. Astrophys. **89** (1980) 324.
V 2 Van den Bergh, S.: Publ. David Dunlap Obs. II, No. 5 (1959).
V 3 Van den Bergh, S.: The galaxies of the Local Group, Comm. David Dunlap Obs. No. 195 (1962) = J. R. Astron. Soc. Canada **62** (1962) No. 4.
V 4 Van den Bergh, S.: Landolt-Börnstein, NS, Vol. VI/1 (1965) p. 674.
V 5 Van den Bergh, S.: Astron. J. **71** (1966) 922.
V 6 Van den Bergh, S.: Vistas in Astronomy **21** (1977) 71.
V 7 Vardanyan, R.A., Melik-Alaverdyan, Yu.K.: Astrofizika **11** (1975) 21 ⟨12⟩.
V 8 Vorontsov-Velyaminov, B.A.: Atlas of interacting galaxies, p. I, Moscow Univ. (1959).
V 9 Vorontsov-Velyaminov, B.A.: Astron. Zh. **37** (1960) 381 ⟨**4**, 365⟩.
V 10 Vorontsov-Velyaminov, B.A.: Astron. Zh. **52** (1975) 491 ⟨**19**, 299⟩.
V 11 Vorontsov-Velyaminov, B.A.: Pis'ma Astron. Zh. **1**, No. 2 (1975) 3 ⟨23⟩; **1** No. 11 (1976) 3 ⟨215⟩.
V 12 Vorontsov-Velyaminov, B.A.: Astron. Zh. **52** (1975) 692 ⟨**19**, 422⟩.
V 13 Vorontsov-Velyaminov, B.A.: Pis'ma Astron. Zh. **2** (1976) 520 ⟨204⟩.
V 14 Vorontsov-Velyaminov, B.A.: Pis'ma Astron. Zh. **2** (1976) 515 ⟨201⟩.

V 15 Vorontsov-Velyaminov, B.A.: Atlas of interacting galaxies, pt. II, Astron. Astrophys. Suppl. **28** (1977) 1.

V 16 Vorontsov-Velyaminov, B.A.: Astron. Zh. **54** (1977) 254 ⟨**21**, 254⟩.

V 17 Vorontsov-Velyaminov, B.A.: Pis'ma Astron. Zh. **3** (1977) 251 ⟨132⟩.

V 18 Vorontsov-Velyaminov, B.A., Arkhipova, V.P.: Astron. Zh. **56** (1979) 225 ⟨**23**, 123⟩.

W 1 Waldthausen, H., Haslam, C.G.T., Wielebinski, R., Kronberg, P.P.: Astron. Astrophys. Suppl. **36** (1979) 237.

W 1a Weedman, D.W., Khachikian, E.E.: Astrofizika **4** (1968) 587 ⟨243⟩; **5** (1969) 113 ⟨51⟩.

W 2 Weedman, D.W.: Astrophys. J. **183** (1973) 29.

W 3 West, R.M.: Astron. Astrophys. **53** (1976) 435.

W 4 West, R.M.: Astron. Astrophys. Suppl. **27** (1977) 73.

W 5 West, R.M., Borchkadze, T.M., Breysacher, J., Laustsen, S., Schuster, H.-E.: Astron. Astrophys. Suppl. **31** (1978) 55.

W 5a West, R.M., Frandsen, S.: Astron. Astrophys. Suppl. **44** (1981) 329.

W 6 Westerlund, B.E., Bergwall, N.A.S., Ekman, A.B.G., Lauberts, A.: Astron. Astrophys. Suppl. **31** (1978) 427.

W 7 White, S.D.M.: Mon. Not. R. Astron. Soc. **174** (1976) 19.

W 8 White, S.D.M.: Highlights of Astronomy **4**, pt. I (1977) 265.

W 9 White, S.D.M.: Comments on Astrophys. **7** (1977) 95.

W 10 White, S.D.M.: Mon. Not. R. Astron. Soc. **184** (1978) 185.

W 11 White, S.D.M.: Astrophys. J. **229** (1979) L9.

W 12 White, S.D.M.: Mon. Not. R. Astron. Soc. **186** (1979) 145.

W 12a White, S.D.M.: Mon. Not. R. Astron. Soc. **189** (1979) 831.

W 12b White, S.D.M., Valdes, F.: Mon. Not. R. Astron. Soc. **190** (1980) 55.

W 12c White, S.D.M.: Mon. Not. R. Astron. Soc. **191** (1980) 1P.

W 12d White, S.D.M.: Mon. Not. R. Astron. Soc. **195** (1981) 1037.

W 12e White, R.A., Burns, J.O.: Astron. J. **85** (1980) 117.

W 12f Wilkerson, M.S.: Astrophys. J. **240** (1980) L115.

W 13 Wright, A.E.: Mon. Not. R. Astron. Soc. **157** (1972) 309.

W 13a Wyckoff, S., Wehinger, P.A., Spinrad, H., Boksenberg, A.: Astrophys. J. **240** (1980) 25.

Y 1 Yabushita, S.: Mon. Not. R. Astron. Soc. **153** (1971) 97.

Y 2 Yahil, A.: Astrophys. J. **217** (1977) 27.

Y 3 Yahil, A., Tammann, G.A., Sandage, S.: Astrophys. J. **217** (1977) 903.

Y 3a Yahil, A., Sandage, A., Tammann, G.A.: Astrophys. J. **242** (1980) 448.

Y 4 Young, D.S.: Annu. Rev. Astron. Astrophys. **14** (1976) 447.

Y 5 Young, D.S.: Astrophys. J. **223** (1978) 47.

Z 1 Zwicky, F.: Eight lists of compact galaxies and compact parts of galaxies; eruptive and post-eruptive galaxies. Privately circulated, published in [Z 5, Z 6].

Z 2 Zwicky, F.: Astrophys. J. **140** (1964) 1467; **143** (1966) 192.

Z 3 Zwicky, F.: Adv. Astron. Astrophys. **5** (1967) 267.

Z 4 Zwicky, F.: Adv. Astron. Astrophys. **7** (1970) 228.

Z 5 Zwicky, F.: Catalogue of selected compact galaxies and of post-eruptive galaxies, L. Speich, Zürich (1971).

Z 6 Zwicky, F., Sargent, W.L.W., Kowal, C.T.: Astron. J. **80** (1975) 545.

# 9.4 Evolution of galaxies

The evolution of galaxies has been the central point of a large number of reviews and conferences [a···m, 13, 75, 76]. Many aspects of galactic evolution have been mentioned also in other chapters.

## 9.4.1 Formation of galaxies

Reviews: [9, 13, 18].

Many properties of galaxies are due to the way they were formed. For example, the radial light distribution in elliptical systems (see 9.2.3) or the disk-to-bulge ratio in spiral systems cannot have changed greatly since their formation as the time scale for stellar dynamical relaxation processes in galaxies is longer than the age of the universe.

Furthermore, elliptical systems are found preferentially in dense clusters of galaxies, while spiral systems are found mostly in the field or in loose groups. This difference between the morphological types must reflect differences in the way in which spirals and ellipticals formed.

But our knowledge of galaxy formation is still very limited and the ideas are uncertain and often speculative.

### Protogalaxies

The origin of galaxies is closely related to the structure and evolution of the early universe. At present, the "Big-Bang" cosmology is almost universally accepted (see 9.7.8) and it is believed that galaxies were formed at some stage of the expansion of the universe by recollapse of dense regions (so-called protogalaxies) of an inhomogeneous pregalactic medium [9, 18].

Gravitational instability and turbulence theories for the origin of galaxies from protogalaxies have been discussed extensively in [16]. A review of recent theories of galaxy formation (inhomogeneities in the early universe; collapse picture; dissipation models for ellipticals; dissipationless collapse for ellipticals; and other models) is given in [13]. Processes of stellar dynamics and gas dynamics relevant for collapsing protogalaxies are distinguished and separately discussed in [18].

Recent investigations of cosmological fluctuations, protogalaxy collapse and related problems: [4, 5, 15, 21···23, 28].

Galaxy formation by the agglomeration of primordial star clusters: [14, 18].

Galaxy formation in an intergalactic medium dominated by explosion of massive stars: [24].

Neutrino dating of galaxy formation: [2].

### Massive neutrinos, the nature of the missing mass

A large number of neutrinos of non-zero mass has a profound influence on the formation of galaxies and their structure [17, 27, 38, 39, 43, 44, 47, 49, 50, 54···58, 61, 64, 68···71, 74, 77, 79]. First models of galaxies with neutrino halos fail to give naturally the flat rotation curves observed [7] (see 9.2.6). A discussion of groups and clusters restricts the possible neutrino mass [52, 71]. Similar arguments can be made using other light particles [63]. The problem cannot be considered solved at the present time.

### Formation of the different morphological types

The clear division of the ellipticals and disks, especially the systematic variation of the stellar content along the sequence of morphological types (9.2.1) indicates differences already in an early stage of formation [9, 30].

The most important parameter in this respect is the angular momentum and the angular-momentum distribution which would control the initial rate of star formation during the protogalaxy collapse toward the disk. Considerations of flattening, stellar content and relaxation times, led Sandage et al. [31] to the following conclusions (summarized in [30], see also [26]):

Stars in the spherical components of all galaxies were formed very rapidly on a time scale comparable to the collapse time of protogalaxies, $\approx 10^9$ years. (The same conclusion was reached by Eggen et al. [8] with other arguments based on the orbital eccentricities of galactic stars.)

Halo stars were formed from matter with low angular momentum per unit mass (spheroidal component) during the collapse time; matter with high angular momentum collapsed to a disk.

The galaxy type is determined by the amount of free gas left over in the disk after collapse. There is no appreciable evolution along the Hubble sequence after the formation of galaxies.

The dominance of the disk in spiral systems shows that their mean angular momentum per unit mass is higher than in ellipticals. Perhaps the higher angular momentum slows the rate of star formation, thereby keeping dust and gas in reserve for new generations of stars in the disk (?).

All stars in the spheroidal components of galaxies should have an age distribution $\Delta T$ small compared to the total age $T$, with $\Delta T/T \lesssim 0.1$. This conclusion agrees with the observational data for globular clusters in our own Galaxy and with the uniformity of light distribution for all ellipticals and for the bulges of spirals (see 9.2.3).

Several scenarios exist to explain the formation of elliptical galaxies: (a) gasdynamical models, (b) stripping of disk systems or irregular systems, (c) merging of disk systems. Detailed stellar dynamic models or gas dynamic models for the formation of elliptical systems and for spiral systems with two or more components are discussed in [18]. The merging hypothesis has recently been explored with some success [37, 40, 46, 48, 51, 53, 62, 67, 73, 78, 80···82].

Origin of S0 galaxies: [6]. They might be spirals which have been swept free of gas during a collision [32].

The importance of some properties of globular clusters for providing information on the formation of galaxies has been pointed out in [12].

The past history of star formation in galaxies is discussed in [34]. For more about star formation in galaxies, see 9.4.4.

If the formation of galaxies is still going on we should also observe young galaxies. This would be interesting for the theories of galaxy formation. Some authors suggest that some active galaxies (e.g. quasars, see 9.5.1.2) may be young galaxies.

An inhomogeneous collapse is discussed in [33] ch. 9.

Evolution of collisionless systems of gravitating masses (formation of clusters of galaxies and of galaxies): [7].

## 9.4.2 Evolution of galaxies

The evolution of galaxies includes not only dynamical properties (like the development of spiral arms, see 9.2.4) but especially the chemical evolution.

A review of the chemical evolution with many references up to 1976 is given by Audouze and Tinsley in [1]. Further reviews of this problem: [e] ch. 7; [18, 25, 26, 33, 35].

To obtain a model for the chemical evolution the following ingredients are needed [26]: the initial conditions; the end product of stellar evolution (which stars eject how much of their mass in the form of various elements after how much time); the initial mass function; the star formation rate as functions of time, gas mass, gas density and probably other parameters; other relevant processes in galactic evolution besides the birth rate and death of stars. (For star formation, see also 9.4.4.)

Main observational facts: colour observations show that the stellar content (see 9.2.1) and population (and thus also the chemical composition) change not only along the morphological sequence but also radially inside the galaxy itself. If our general ideas are correct the initial composition of galaxies was deficient in heavy elements. But at present we do not observe disk stars with this essentially zero heavy-element-content. Stars in globular clusters have much lower content of heavy elements (Z) than most disk stars, but the Z values differ from cluster to cluster (clusters near the center have higher Z than clusters far out), and even in the clusters *no* stars have been observed which are devoid of heavy elements ($Z = 0$). Three factors might be important to reconcile theories and these observations (see [e, 33] and references therein):

(1) Prompt initial enrichment = processes which lead to non-zero Z when the first disk stars are formed. There are three suggestions to realize this: (a) Galaxies are much more massive than generally believed and star formation could have occurred initially in a massive halo (see also 9.2.6). Mass loss from these extreme halo stars could then add heavy elements to the rest of the galaxy before globular clusters and the first disk stars are formed. (b) Globular clusters did not form during the initial collapse of the protogalaxy. Instead galaxies collapsed in gaseous form until massive thermonuclear explosions in their central regions produced heavy elements. (c) The first generation of disk stars consists essentially of massive stars only. This first generation has a short life time and would produce heavy elements. In this model much of the mass of the disks must at

present be in the form of black hole remnants. There is some theoretical support for the idea that star formation in a cloud of pure H and He only leads to a very different initial mass function to that if heavy elements were present. (The possibility of a pregalactic synthesis of some kind seems to be very doubtful [26].)

(2) Metal-enhanced star formation. Because of the kinetic energy cool gas clouds are more likely to collapse to form stars than hotter ones. On the other hand, the presence even of a small amount of heavy elements will enhance the cooling rate of interstellar gas. Thus stars might be formed preferentially in regions of higher than average Z abundance. In this case the average Z of stars of a given age would be higher than the average Z of the interstellar matter out of which they are formed.

(3) Infall = accretion of intergalactic matter (this might be real intergalactic matter or matter from the outer regions of the protogalaxy which takes longer to collapse than the bulk of the mass). Intergalactic matter is supposed to be unprocessed and thus to contain no heavy elements ($Z=0$). If such gas is added to the interstellar matter the amount of heavy elements will be reduced. In this case the average Z of stars will also be higher than the average Z of the interstellar matter. An appropriate combination of initial mass function and accretion rate might be able to explain the numbers of stars with different Z values. For interaction between young galaxies and intergalactic matter, see also [18].

A combination of these three models with arbitrary parameters (within plausible bounds) is capable of providing qualitative (perhaps even quantitative) agreement with the observations.

Larson's dynamical models for the evolution of galaxies [18]:

(a) Elliptical systems: most of the star formation occurs in the first $\approx 10^9$ years. The gas content declines rapidly but all material is retained within the system. After about $10^{10}$ years there is an equilibrium between the two processes, gas collapsing to stars and mass loss from stars, each rate being about $0.1\,\mathfrak{M}_\odot/a$ for a system of $10^{11}\,\mathfrak{M}_\odot$.

(b) Spiral systems: protospirals have a relatively low density and thus a long time scale for star formation.

Some recent papers and reviews:

Nuclear abundance and evolution of interstellar matter: [36].

A code to calculate the evolution of galaxies: [3].

The rate of star formation in galaxies [19] (see also 9.4.4).

Galactic shells: The discovery of shells around galaxies has led to a flurry of interest [41, 59, 60, 72]. It appears possible to explain their properties as a result of merging [45, 65].

The connexion between the variation of the relative hydrogen content (see 9.2.5.2) and the galactic evolution has been pointed out by Roberts [5] ch. 8.

## 9.4.3 References for 9.4.1 and 9.4.2

### General references


a The Formation and Dynamics of Galaxies (Shakeshaft, J.R., ed.), Int. Astron. Union Symp. **58**, Reidel, Dordrecht (1974).

b Galaxies and the Universe (Sandage, A., Sandage, M., Kristian, J., eds.), = Stars and Stellar Systems Vol. IX, Univ. Chicago Press (1975).

c Freeman, K., Larson, R.B., Tinsley, B.M.: Galaxies, Sixth advanced course of Swiss Soc. of Astron. and Astrophys. (Martinet, L., Major, M., eds.), Geneva Observatory (1976).

d Structure and Properties of Nearby Galaxies (Berkhuijsen, E.M., Wielebinski, R., eds.), Int. Astron. Union Symp. **77**, Reidel, Dordrecht (1978).

e Taylor, R.J.: Galaxies, Structure and Evolution, Wykeham Publ. Ltd., London (1978).

f The Structure and Evolution of Normal Galaxies (Fall, S.M., Lynden-Bell, D., eds.), Cambridge Univ. Press (1981).

g Galaktische Struktur und Entwicklung (Buser, R., ed.), Forschungsseminar, Astron. Inst. Univ. Basel (1981).

h Objects of High Redshift (Abell, G.O., Peebles, P.J.E., eds.), Int. Astron. Union Symp. **92**, Reidel, Dordrecht (1980).

i Photometry, Kinematics and Dynamics of Galaxies (Evans, D.S., ed.), Univ. of Texas, Austin (1979).

k Dynamics of Stellar Systems (Hayli, A., ed.), Int. Astron. Union. Symp. **69**, Reidel, Dordrecht (1975).

l The Large Scale Structure of the Universe (Longair, M.S., Einasto, J., eds.), Int. Astron. Union Symp. **79**, Reidel, Dordrecht (1978).
m Structure and Evolution of Galaxies (Setti, G., ed.), NATO Advanced Study Inst. Series C, Vol. 21 (1975).
n The Evolution of Galaxies and Stellar Populations (Tinsley, B.M., Larson, R.B., eds.), Yale Univ. Obs. (1977).


## Special references


1 Audouze, J., Tinsley, B.M.: Annu. Rev. Astron. Astrophys. **44** (1976) 43.
2 Berezinsky, V.S., Ozernoy, L.M.: Astron. Zh. **58** (1981) 505 = Soviet Astron. **25** (1981) No. 3; Astron. Astrophys. **98** (1981) 50.
3 Biermann, P. in: Galactic Evolution (De Loore, C.W.H., ed.), workshop, Brussell (1981) p. 13.
4 Binggeli, B. in: [g] p. 255.
5 Brown, W.K.: Astrophys. Space Science **72** (1980) 15.
6 Burstein, B.: Astrophys. J. **234** (1979) 435.
7 Doroshkevich, A.G., Klypin, A.A.: Astron. Zh. **58** (1981) 225 = Soviet Astron. **25** (1981) No. 2.
8 Eggen, O.J., Lynden-Bell, D., Sandage, A.: Astrophys. J. **136** (1962) 748.
9 Field, G.B. in: [b] p. 359.
10 Freeman, K.C.C. in: [b] p. 409.
11 Freeman, K.C. in: [c] p. 55.
12 Freeman, K.C. in: [f] p. 251.
13 Gott, J.R. III: Annu. Rev. Astron. Astrophys. **15** (1979) 235.
14 Hartquist, T.W. in: Dwarf Galaxies, ESO/ESA workshop (Tarenghi, M., Kjär, K., eds.), Geneva (1980) p. 125.
15 Henriksen, R.N., de Robertis, M.: Astrophys. J. **241** (1980) 241.
16 Jones, B.J.T.: Rev. Mod. Phys. **48** (1976) 107.
17 Klinkhamer, F.R., Norman, C.A.: Astrophys. J. **243** (1981) L1; erratum: **245** (1981) L97.
18 Larson, R.B. in: [c] p. 67.
19 Madore, B.F. in: [f] p. 239.
21 Miller, R.H., Smith, B.F.: Astrophys. J. **244** (1981) 33.
22 Miller, R.H., Smith, B.F.: Astrophys. J. **244** (1981) 467.
23 Ostriker, J.P. in: X-ray Astronomy with the Einstein Satellite (Giacconi, R. ed.), Astrophys. Space Science Library **87**, Reidel, Dordrecht (1981) p. 311.
24 Ostriker, J.P., Cowie, L.L.: Astrophys. J. **243** (1981) L122.
25 Pagel, B.E.J.: Trends in Physics 1978, Fourth Geneva Conference of the European Phys. Soc., Adam Hilger, Bristol (1979) p. 247.
26 Pagel, B.J.E. in: [f] p. 211.
27 Pryor, C., Davis, M., Lecar, M., Witten, E.: Bull. American Astron. Soc. **12** (1980) 861.
28 Rees, M. in: Variability in Stars and Galaxies, Proc. 5$^{th}$ European Regional Meeting, Liège (1980) p. G. 2.1.
29 Roberts, M.S. in: [b] p. 309.
30 Sandage, A. in: [b] p. 1, ch. 11.
31 Sandage, A., Freeman, K.C., Stokes, N.R.: Astrophys. J. **160** (1970) 831.
32 Spitzer, L., Jr., Baade, W.: Astrophys. J. **113** (1951) 413.
33 Tinsley, B.M. in: [c] p. 157.
34 Tinsley, B.M. in: [d] p. 15.
35 Trefzger, C.F. in: [g] p. 201.
36 Wannier, P.G.: Annu. Rev. Astron. Astrophys. **18** (1980) 399.

37 Aarseth, S.J., Fall, S.M.: Astrophys. J. **236** (1980) 43.
38 Bisnovatyi-Kogan, G.S., Novikov, I.D.: Sovj. Astron. A. J. **24** (1980) 516.
39 Bond, J.R., Efstathiou, G., Silk, J.: Phys. Rev. Lett. **45** (1980) 1980.
40 Carnevali, P., Cavaliere, A., Santangelo, P.: Astrophys. J. **249** (1981) 449.
41 Carter, D., Allen, D.A., Malin, D.F.: Nature **295** (1982) 126.
42 Crollolanza, A., Gao Jiang-gong, Ruffini, R.: Nuovo Cimento, Lett. **32** (1981) 411.
43 Doroshkevich, A.G., Zel'dovich, Ya.B., Sunyaev, R.A., Klopov, M.Yu.: Sovj. Astron. Lett. **6** (1980) 252.
44 Doroshkevich, A.G., Kotov, E.V., Novikov, I.D., Polyudov, A.N., Shandarin, S.F., Sigov, Yu.S.: Mon. Not. R. Astron. Soc. **192** (1980) 321.
45 Fabian, A.C., Nulsen, P.E.J., Stewart, G.C.: Nature **287** (1980) 613.
46 Fall, S.M.: Nature **281** (1979) 200.


**Voigt**

47 Fang Li-zhi, Liu Yong-zhen: Chinese Astron. **5** (1981) 381.
48 Farouki, R.T., Shapiro, S.L.: Preprint CRSR No. 782 (1981).
49 Gao Jian-gong, Ruffini, R.: Physics Lett. B **97** (1980) 388; **100** (1980) 47.
50 Gao Jian-gong, Ruffini, R.: Chinese Astron. **5** (1981) 24.
51 Gerhard, O.: Mon. Not. R. Astron. Soc. **197** (1981) 179.
52 Hartwick, F.D.A.: Astrophys. J. Lett. **255** (1982) L91.
53 Jones, B.J.T., Efstathiou, G.: Mon. Not. R. Astron. Soc. **189** (1979) 27.
54 Klinkhamer, F.R.: Astron. Astrophys. **107** (1982) 235.
55 Lewis, B.M.: Proc. Astron. Soc. Australia **4** (1981) 181.
56 Lewis, B.M.: Proc. Astron. Soc. Australia **4** (1981) 182.
57 Lizhi, F., Yongzhen, L.: Nuovo Cimento, Lett. **32** (1981) 129.
58 Lu Tan, Luo Liao-fu, Yang Gou-chen: Chinese Astron. **5** (1981) 377.
59 Malin, D.F.: Nature **277** (1979) 279.
60 Malin, D.F., Carter, D.: Nature **285** (1980) 643.
61 Melott, A.L.: Nature **296** (1982) 721.
62 Ostriker, J.P.: Comments of Astrophys. **8** (1980) 177.
63 Pagels, H., Primack, J.R.: Phys. Rev. Lett. **48** (1982) 223.
64 Qing Cheng-rui, Wu Yong-shi, He Zuo-xiu, Zhang Zhao-xi, Zou Zheng-long: Chinese Astron. **5** (1981) 17.
65 Quinn, P.J.: Astrophys. J. Lett. (1982) in press.
66 Roos, N., Norman, C.A.: Astron. Astrophys. **76** (1979) 75.
67 Roos, N.: Astron. Astrophys. **95** (1981) 349.
68 Ruffini, R.: Nuovo Cimento, Lett. **29** (1980) 161.
69 Sato, H., Takahara, F.: Progress Theor. Phys. **64** (1980) 2029.
70 Schramm, D.N., Steigman, G.: Astrophys. J. **243** (1981) 1.
71 Schramm, D.N., Steigman, G.: Gen. Relativ. Gravitation **13** (1981) 101.
72 Schweizer, F.: Astrophys. J. **237** (1980) 303.
73 Silk, J., Norman, C.: Astrophys. J. **247** (1981) 59.
74 Silk, J.: Nature **297** (1982) 102.
75 Strom, K.M., Strom, S.E.: Science **216** (1982) 571.
76 Toomre, A.: Annu. Rev. Astron. Astrophys. **15** (1977) 437.
77 Tremaine, S., Gunn, J.E.: Phys. Rev. Lett. **42** (1979) 407.
78 Villumsen, J.V.: Mon. Not. R. Astron. Soc. **199** (1982) 493.
79 Wasserman, I.: Astrophys. J. **248** (1981) 1.
80 White, S.D.M.: Mon. Not. R. Astron. Soc. **184** (1978) 185.
81 White, S.D.M.: Mon. Not. R. Astron. Soc. **189** (1979) 831.
82 White, S.D.M.: Mon. Not. R. Astron. Soc. **191** (1980) 1P.

The author wishes to thank Dr. P. Biermann for providing the references [h···n] and [37···82] and some remarks about the massive neutrinos, the galactic shells and the formation of elliptical systems.

## 9.4.4 Star-formation activity in normal galaxies

*Introduction*

We define all those galaxies as normal whose nuclear regions are not known to be dominated by a compact source. Thus we will exclude N galaxies, Seyfert galaxies and radio galaxies, but include galaxies like M 81 and M 82 which have weak compact nuclei.

### 9.4.4.1 Tracers of star formation

Only those tracers are included that can be used near the center of our Galaxy or in neighbouring galaxies.

Molecular clouds, dark clouds and dust lanes

Stars formed by gravitational instability of dense gas, typically found in molecular clouds and dark clouds, e.g. [33···35, 209, 74, 177, 205]. In distant galaxies, dust lanes indicate the presence of large numbers of dark clouds (e.g. in M 51), cf. [126].

IR sources, OH/$H_2O$ maser sources and compact H II regions

Young stars still embedded deeply in dark clouds are visible in the IR, e.g. [207, 73, 197]. Strong OH and $H_2O$ maser lines are emitted by the environment of young massive stars, e.g. [68, 69, 89, 193]. Compact H II regions, visible as thermal radio continuum sources, form around massive stars, e.g. [129, 130, 131, 161, 75]. Compact H II regions can also be observed in the IR, e.g. [208, 97] and in radio recombination lines, e.g. [198, 203, 39].

Bright H II regions

H II regions are the most conspicuous sign of young (massive) stars. They show emission lines of hydrogen and many other elements in the optical range (e.g. Hα, [O II], cf. [143, 81, 83, 85, 46], in the IR (e.g. [Ne II], cf. [2, 12, 117] and in the radio range, e.g. [169, 170, 163, 13, 171]. High resolution radio studies at high frequencies show the thermal radio-continuum emission, e.g. [93, 94]. The dust inside H II regions radiates in the IR, e.g. [183].

Single stars, integrated spectra and colors

In Local-Group galaxies we can observe single stars [6, 7] and thus directly recognize the bright young stars. In more distant galaxies integrated spectra can be used to identify regions of star motion, e.g. [133···135, 57, 139, 202, 194, 147, 201]. At nearly any distance, broad-band colors can be used to discern population characteristics of the underlying stellar population, e.g. [92, 11, 102, 166, 17], see also 9.2.1.

Massive evolved stars

Variable supergiant stars (possibly pre-main sequence) and Mira variables with long periods are young objects on galactic time scales [60]. They have $H_2O$, OH, and SiO maser sources and are also IR sources, e.g. [76, 128, 204].

Supernovae (SN) and supernova remnants (SNR)

Massive stars are believed to explode as supernovae (type II) (see 5.1.3.1) and leave behind supernova remnants, e.g. [186]. Therefore star formation leads to SN and SNR, visible in the radio, optical and X-ray ranges, e.g. [71, 44, 45, 20, 96, 127, 196]. Our lack of understanding of the relation of SN of type I with stellar evolution, and SNR forces the evidence for star formation based on SN and SNR to be critically examined, cf. [141a].

Distributed radio emission

It has been proposed that all nonthermal distributed radio emission in galaxies derives from nonthermal particles originating in SNR, e.g. [120, 25]. Since there is evidence that the distributed radio emission correlates with the old stellar population [112, 113], cf. also [167], a relation of radio emission with recent star formation can only hold if recent star formation dominates strongly [164]. However, if the spectrum of the distributed radio emission is thermal then the presence of H II regions and thus young stars is established, e.g. [8].

Neutral hydrogen (see also 9.2.5.2)

The surface density of neutral hydrogen is well correlated with star formation in other galaxies, e.g. [124, 77, 190]. Thus the presence of H I leads to the expectation that star formation should occur, and obviously, the absence of H I leads to the expectation that star formation cannot occur. Such expectations should be regarded with scepticism.

In conclusion, H II regions are the best indication of star formation in external galaxies. Blue colors of a stellar population of normal metallicity are also useful as indication of recent star formation. Unfortunately, they tell us only about the formation of massive stars.

### 9.4.4.2 Star formation in the nuclear regions of observed galaxies

All references given here should be understood to include all earlier references. All galaxies mentioned by name are meant as examples only, unless otherwise stated.

Elliptical galaxies

Colors (see 9.1.4.6) and spectrophotometry (e.g. [133···136, 57, 64, 173, 200, 202, 139, 178, 147, 54, 1, 175, 176, 179] quite clearly indicate that no stars have been formed in a long time (e.g. [184, 185, 187]). The general absence of observable H I supports this view point (e.g. [88, 105]). All evidence suggests that star formation in elliptical galaxies lasted just for a short time and ceased long ago (see also 9.4.1).

However, there are some noteworthy exceptions. NGC 185 and NGC 205, elliptical companions to M 31, each have dust patches and O stars [80, 83, 84]. Population synthesis models by Williams [202] suggest that in M 32 = NGC 221, NGC 3379, and NGC 4473 star formation continued long after the formation of the galaxy. Among those galaxies classified by Sandage and Visvanathan [157] – excluding peculiar galaxies – eight elliptical galaxies are known to contain H I (NGC 1052, 2974, 3156, 3904, 3962, 4278, 4636, and 5846, cf. the review [70]). NGC 3226, 3773, 4105 have not been classified; they may have to be added to the list. Several of these galaxies show dust patches and/or spectra suggesting the presence of star formation. NGC 1052: [104, 149, 63, 107]; NGC 3226: [105]; NGC 4278: [66]. Dust lanes are also found in other, apparently elliptical galaxies [21, 109, 108]. Whether the presence of gas and/or dust in these galaxies can be taken as indication of star formation, is not clear. As noted above, all the obvious signs of star formation are valid only for massive stars; hence, the exciting possibility has to be considered that in fact star formation occurs but only low-mass stars are generated.

### S0 galaxies

S0 galaxies are an inhomogeneous class of galaxies invented to bridge the gap between ellipticals and spirals; NGC 3115 [178] clearly is a good transition case. The extremely flat disk of NGC 4762 (cf. [156]) makes it difficult to think of it as such a transition case (cf. [19]). The colors and spectrophotometry (same references as for ellipticals) show that their stellar populations are old. Again, the very low upper limits for H I support this viewpoint just as for the ellipticals (e.g. [22, 28, 115]).

However, among those galaxies classified by Sandage and Visvanathan [157] – again excluding the peculiars – 14 contain H I (NGC 1023, 1326, 1533, 2859, 2962, 3032, 3414, 3626, 3941, 4203, 4262, 4958, 5084, and 5102: [65, 206, 9, 103, 22, 24, 28, 115]). Several galaxies not classified in [157] have to be added. Most of these galaxies show no sign of star formation, similar to the ellipticals. Colors bluer than typical suggest star formation in NGC 3032, 3626 [157] and, most noticeably, in NGC 5102 [18, 53, 148]. Several S0 galaxies deviate strongly from a Fisher-Tully relation (see 9.2.5.2) [115, 28, 29], showing line widths too small for their luminosity, suggesting a concentration of H I close to the nucleus [116]. One of these, (NGC 4694), may be a radio source [22, 30]. Several of these galaxies are blue enough to be found in Markarian's lists (see [11] and [12] in 9.1.1.5), suggesting that vigorous star formation occurs in their nuclear regions. Spectrophotometry, H I and radio-continuum mapping is called for.

### Sa galaxies

Sa galaxies show no sign of star formation in their nuclear regions: Red colors, no H II regions, no H I (e.g. [123, 153, 58, 99, 162, 154, 145]). Their nuclear regions are indistinguishable from giant ellipticals [136].

The nuclear colors of, e.g., NGC 2681 and NGC 2798 = VV 50 = Arp 283 = K 195 are evidence for the presence of young stars [98, 99].

### Sb galaxies

Sb galaxies normally have no H I in their center (e.g. [54, 152, 153]), no H II regions (e.g. [82, 85, 123]) and their nuclear colors indicate once again a strong similarity to giant ellipticals (e.g. [136, 147, 99]).

However, there are a large number of exceptions, underlining the transition to Sc galaxies which do have star formation going on in their centers: Keel and Weedman [99] show that the nuclear colors of some Sb's are so red as to suggest reddening, and of others that are blue indicating recent star formation (NGC 4569, 6217, 7714). Lynds [123] notes H II regions in the centers of NGC 4579 and 3351. Pritchet's analysis [147] suggests fairly young stars in the nucleus of M 81 = NGC 3031 and NGC 4736 (cf. also [41, 36]). Several Sb galaxies with nuclear activity are discussed in [155, 114].

### Sc galaxies

Most Sc galaxies show star formation in the center, as demonstrated by H I (e.g. [152, 153]), CO line emission [150, 137], the colors (e.g. [99]), and the spectrum (e.g. [194, 147]). Often, the star formation rate peaks in a ring around the center, as indicated by the H II regions (e.g. [82, 85]).

### Irregular galaxies (LMC-type)

These galaxies show that star formation is strongest in their centers. This is supported by colors and spectra (e.g. [3, 136, 99]), H I (e.g. [54, 152, 153, 49]), and observations of LMC itself (e.g. [83, 4, 5, 56, 95]). The interpretation is often hampered by the lack of a well-defined nucleus. The relation between LMC-type irregulars and dwarf ellipticals is not clear (cf. [27]).

Amorphous galaxies (M 82-type)

Sandage and Brucato [158] define this class corresponding to M 82. It includes NGC 520, 625, 1510, 1531, 1705, 1800, 3034 (=M 82), 3077, 3448 and 5253. M 82 and NGC 520 are both interacting and have compact radio nuclei [174]. They all may be interacting [51]. M 82 and probably all of these galaxies are undergoing intense bursts of star formation (e.g. [15, 168, 110, 191, 72, 50, 55, 140, 10, 78]). NGC 1510 has the light distribution of an elliptical [55].

Our Galaxy

The central region of our Galaxy has been reviewed by Oort [142]. Mezger and Pauls [132] reviewed the evidence of star formation. A detailed discussion is given in [172]. The star formation rate at the center, i.e. inside 800 pc radius, is about 8 % of the entire star formation rate in the Galaxy. From 800 pc to about 4 kpc no detectable star formation is taking place. In this respect our Galaxy resembles the Sc galaxies (among these, it is more similar to Sb's).

Compact galaxies, blue galaxies, narrow-emission line galaxies etc.

These galaxies are discussed in 9.1.3 and 9.5. For most of these galaxies, the data are quite consistent with a strong burst of star formation [16, 32, 37, 38, 43, 59, 106, 121, 122, 125, 138, 141, 144, 159, 160, 180, 181, 192, 199].

## 9.4.4.3 Theoretical interpretation

Along the Hubble sequence

Along the sequence Irr-Sc-Sb-Sa-(S0)-E we find that at first H I and star formation peak in the center (Irr), then a depression in H I develops and star formation peaks in a ring (Sc), then a hole develops in the gas distribution with still some star formation going on right at the center (Sc-Sb, our Galaxy), finally star formation is stopped when the gas vanished (Sb to E) and the hole in the gas distribution gets wider until it encompasses the entire galaxy and no disk is left (E); all this has been discussed by King [102]. This sequence fits very well the models of Talbot and Arnett [182], Larson [118], and Tinsley and Larson [188]. We would like to suggest here that some S0 galaxies with H I should fit in this sequence between E's and Sa's as well and thus should have their H I in a ring around the main body of the galaxy. A correlation between ring features and the presence of H I is indeed suggested by the data [22, 115]. However, those S0 galaxies which have their H I apparently narrowly confined to the central region [116], do not fit the sequence discussed above; we would like to suggest that they are in fact late-type spirals whose outer parts have been stripped of gas (cf. [79]) and which do form stars in their centers.

Bursts of star formation

The active galaxies mentioned in the last paragraph of 9.4.4.2 are understood as extragalactic H II regions (e.g. [165, 62]) or as normal galaxies undergoing intense bursts of star formation (e.g. [166, 23, 26, 87, 119]). This burst often occurs near the center of the galaxy [86].

As an example we can take the well studied galaxy NGC 2146 [14, 61, 42, 111]. Comparison with theoretical models for strong bursts [26] suggests that the burst in NGC 2146 requires more gas than it has possibly available, supposed stars that are made according to the "normal" initial mass function. The gas consumption of the burst is cut down to manageable amounts, if only O stars and possibly early B stars are generated. Rieke and Lebofsky [151] also argue for such a lopsided initial mass function in bursts of star formation on the basis of infrared data, discussing M 82 and NGC 253.

But why should these galaxies experience bursts? The most convenient and most conventional suggestion is that they interact with another galaxy (e.g. [189, 25, 119]). Some galaxies have no visible neighbor, but a lot of circumgalactic gas extending to 100···200 kpc from the galaxy (e.g. NGC 2146, M 101, and IC 10 [61, 90, 91, 47]).

Weak compact nuclei in normal galaxies

Weak compact nuclei were discovered by their radio emission in our Galaxy (e.g. [101]), M 81 and NGC 4594 (e.g. [40, 100]), M 82 [67] and some other galaxies (e.g. [146, 48, 52, 174, 31]). Active nuclei are discussed in 9.5.

*Acknowledgements:* We would like to thank Drs. P.P. Kronberg, H. Kühr, U. Mebold, T. Pauls, P.A. Strittmatter, T.L. Wilson, and A. Winnberg for helpful discussions and comments on this part.

## 9.4.4.4 References for 9.4.4


1 Aaronson, M.: Astrophys. J. **221** (1978) L103.
2 Aitken, D.K., Jones, B., Penman, J.M.: Mon. Not. R. Astron. Soc. **169** (1974) 35P.
3 Alloin, D., Kunth, D.: Astron. Astrophys. **71** (1979) 335.
4 Ardeberg, A.: Astron. Astrophys. **46** (1976) 87.
5 Azzopardi, M., Vigneau, J.: Astron. Astrophys. **56** (1977) 151.
6 Baade, W.: Astrophys. J. **100** (1944) 137.
7 Baade, W.: Astrophys. J. **100** (1944) 147.
8 Baker, J.R., Haslam, C.G.T., Jones, B.B., Wielebinski, R.: Astron. Astrophys. **59** (1977) 261.
9 Balick, B., Faber, S.M., Gallagher, J.S.: Astrophys. J. **209** (1976) 710.
10 Barbieri, C., di Tullio, G.: Astron. Astrophys. **74** (1979) 110.
11 Baum, W.A.: Publ. Astron. Soc. Pacific **71** (1959) 106.
12 Beck, S.C., Lacy, J.H., Baas, F., Townes, C.H.: Astrophys. J. **226** (1978) 545.
13 Bell, M.B., Seaquist, E.R.: Astron. Astrophys. **56** (1977) 461.
14 Benvenuti, P., Capaccioli, M., D'Odorico, S.: Astron. Astrophys. **41** (1975) 91.
15 van den Bergh, S.: Astron. Astrophys. **12** (1971) 474.
16 van den Bergh, S.: Journal R. Astron. Soc. Canada **66** (1972) 237.
17 van den Bergh, S.: Annu. Rev. Astron. Astrophys. **13** (1975) 217.
18 van den Bergh, S.: Astron. J. **81** (1976) 795.
19 van den Bergh, S.: Astrophys. J. **206** (1976) 883.
20 van den Bergh, S.: Astrophys. J. Suppl. **38** (1978) 119.
21 Bertola, F., Galletta, G.: Astrophys. J. **226** (1978) L115.
22 Bieging, J.H., Biermann, P.: Astron. Astrophys. **60** (1977) 361.
23 Bieging, J.H., Biermann, P., Fricke, K.J., Pauliny-Toth, I.I.K., Witzel, A.: Astron. Astrophys. **60** (1977) 353.
24 Bieging, J.H.: Astron. Astrophys. **64** (1978) 23.
25 Biermann, P.: Astron. Astrophys. **53** (1976) 295.
26 Biermann, P., Fricke, K.J.: Astron. Astrophys. **54** (1977) 461.
27 Biermann, P., Shapiro, S.L.: Astrophys. J. **230** (1979) L33.
28 Biermann, P., Clarke, J.N., Fricke, K.J.: Astron. Astrophys. **75** (1979) 7.
29 Biermann, P., Clarke, J.N., Fricke, K.J.: Astron. Astrophys. **75** (1979) 19.
30 Biermann, P., Clarke, J.N., Fricke, K.J., Pauliny-Toth, I.I.K., Schmidt, J., Witzel, A.: Astron. Astrophys. **81** (1980) 235.
31 Biermann, P., Kronberg, P.P., Preuss, E., Schilizzi, R.T., Shaffer, D.B.: Astrophys. J. **250** (1981) L 49.
32 Bohuski, T.J., Fairall, A.P., Weedman, D.W.: Astrophys. J. **221** (1978) 776.
33 Bok, B.J., Reilly, E.F.: Astrophys. J. **105** (1947) 255.
34 Bok, B.J.: Astron. J. **61** (1956) 309.
35 Bok, B.J., Cordwell, C.S., Cromwell, R.H. in: Dark nebulae, globules and protostars (Lynds, B.T., ed.), Univ. of Arizona Press, Tucson (1971) p. 33.
36 Bosma, A., v. d. Hulst, J.M., Sullivan III, W.T.: Astron. Astrophys. **57** (1977) 373.
37 Bottinelli, L., Chamaraux, P., Gouguenheim, L., Heidmann, J.: Astron. Astrophys. **29** (1973) 217.
38 Bottinelli, L., Duflot, R., Gouguenheim, L., Heidmann, J.: Astron. Astrophys. **41** (1975) 61.
39 Brown, R.L., Lockman, F.J., Knapp, G.R.: Annu. Rev. Astron. Astrophys. **16** (1978) 445.
40 de Bruyn, A.G., Crane, P.C., Price, R.M., Carlson, J.B.: Astron. Astrophys. **46** (1976) 243.
41 de Bruyn, A.G.: Astron. Astrophys. **54** (1977) 491.
42 de Bruyn, A.G.: Astron. Astrophys. **58** (1977) 221.
43 Chamaraux, P.: Astron. Astrophys. **60** (1977) 67.
44 Clark, D.H., Caswell, J.L.: Mon. Not. R. Astron. Soc. **174** (1976) 267.
45 Clark, D.H., Stephenson, F.R.: Mon. Not. R. Astron. Soc. **179** (1977) 87P.
46 Cohen, J.G.: Astrophys. J. **203** (1976) 587.
47 Cohen, R.J.: Mon. Not. R. Astron. Soc. **187** (1979) 839.
48 Condon, J.J., Dressel, L.L.: Astrophys. J. **221** (1978) 456.
49 Cottrell, G.A.: Mon. Not. R. Astron. Soc. **177** (1976) 463.
50 Cottrell, G.A.: Mon. Not. R. Astron. Soc. **178** (1977) 577.
51 Cottrell, G.A.: Mon. Not. R. Astron. Soc. **184** (1978) 259.
52 Crane, P.C.: Astron. J. **84** (1979) 281.
53 Danks, A.C., Laustsen, S., v. Woerden, H.: Astron. Astrophys. **73** (1979) 247.

54 Davies, R.D. in: External Galaxies and Quasistellar Objects, Int. Astron. Union Symp. **44** (Evans, D.S., ed.), Reidel, Dordrecht (1972) p. 67.
55 Disney, M.J., Pottasch, S.R.: Astron. Astrophys. **60** (1977) 43.
56 Elliott, K.H., Goudis, C., Meaburn, J., Tebbutt, N.J.: Astron. Astrophys. **55** (1977) 187.
57 Faber, S.M.: Astron. Astrophys. **20** (1972) 361.
58 Faber, S.M., Balick, B., Gallagher, J.S., Knapp, G.R.: Astrophys. J. **214** (1977) 383.
59 Fairall, A.P.: Mon. Not. R. Astron. Soc. **153** (1971) 383.
60 Feast, M.W.: Mon. Not. R. Astron. Soc. **125** (1963) 367.
61 Fisher, J.R., Tully, R.B.: Astron. Astrophys. **53** (1976) 397.
62 Fisher, J.R., Tully, R.B.: Astron. Astrophys. **84** (1979) 62.
63 Fosbury, R.A.E., Mebold, U., Goss, W.M., Dopita, M.A.: Mon. Not. R. Astron. Soc. **183** (1978) 549.
64 Frogel, J.A., Persson, S.E., Aaronson, M., Matthews, K.: Astrophys. J. **220** (1978) 75.
65 Gallagher, J.S., Faber, S.M., Balick, B.: Astrophys. J. **202** (1975) 7.
66 Gallagher, J.S., Knapp, G.R., Faber, S.M., Balick, B.: Astrophys. J. **215** (1977) 463.
67 Geldzahler, B.J., Kellermann, K.I., Shaffer, D.B., Clark, B.G.: Astrophys. J. **215** (1977) L5.
68 Genzel, R., Downes, D.: Astron. Astrophys. Suppl. **30** (1977) 145.
69 Genzel, R., Downes, D.: Astron. Astrophys. **61** (1977) 117.
70 Gouguenheim, L.: Review paper at Austin conference (1979).
71 Gorenstein, P., Tucker, W.H.: Annu. Rev. Astron. Astrophys. **14** (1976) 373.
72 Gottesman, S.T., Weliachew, L.: Astrophys. J. **211** (1977) 47.
73 Grasdalen, G.L., Strom, K.M., Strom, S.E.: Astrophys. J. **184** (1973) L53.
74 Grasdalen, G.L. in: H II regions and related topics (Wilson, T.L., Downes, D., eds.), Springer, Berlin (1975) p. 30.
75 Habing, H.J. in: H II regions and related topics, op. cit. [74] (1975) p. 156.
76 Habing, H.J. in: The Interaction of Variable Stars with their Environment, Int. Astron. Union Coll. No. 42, (Kippenhahn, R., Rahe, J., Strohmeier, W., eds.), Astron. Inst. Univ. Erlangen-Nürnberg, Bd. XI, No. 121 (1977) p. 401.
77 Hamajima, K., Tosa, M.: Publ. Astron. Soc. Japan **27** (1975) 561.
78 Hawarden, T.G., van Woerden, H., Mebold, U., Goss, W.M., Peterson, B.A.: Astron. Astrophys. **76** (1979) 230.
79 Himmes, A., Biermann, P.: Astron. Astrophys. **86** (1980) 11.
80 Hodge, P.W.: Astron. J. **68** (1963) 691.
81 Hodge, P.W.: Astrophys. J. **156** (1969) 847.
82 Hodge, P.W.: Astrophys. J. Suppl. No. 157, **18** (1969) 73.
83 Hodge, P.W.: Astrophys. J. **78** (1973) 807.
84 Hodge, P.W.: Astrophys. J. **182** (1973) 671.
85 Hodge, P.W.: Astrophys. J. Suppl. **27** (1974) 113.
86 Huchra, J.P.: Astrophys. J. Suppl. **35** (1977) 171.
87 Huchra, J.P.: Astrophys. J. **217** (1977) 928.
88 Huchtmeier, W.K., Tammann, G.A., Wendker, H.J.: Astron. Astrophys. **57** (1977) 313.
89 Huchtmeier, W.K., Witzel, A., Kühr, H., Pauliny-Toth, I.I.K., Roland, J.: Astron. Astrophys. **64** (1978) L21.
90 Huchtmeier, W.K., Witzel, A.: Astron. Astrophys. **74** (1979) 138.
91 Huchtmeier, W.K.: Astron. Astrophys. **75** (1979) 170.
92 Humason, M.L., Mayall, N.U., Sandage, A.R.: Astron. J. **61** (1956) 97.
93 Israel, F.P., V. d. Kruit, P.C.: Astron. Astrophys. **32** (1974) 363.
94 Israel, F.P., Goss, W.M., Allen, R.J.: Astron. Astrophys. **40** (1975) 421.
95 Israel, F.P., Koornneef, J.: Astrophys. J. **230** (1979) 390.
96 Iye, M., Kodaira, K.: Publ. Astron. Soc. Japan **27** (1975) 411.
97 Jennings, R.E. in: H II regions and related topics, op. cit. [74] (1975) p. 137.
98 Joly, M., Andrillat, Y.: Astron. Astrophys. **50** (1976) 279.
99 Keel, W.C., Weedman, D.W.: Astron. J. **83** (1978) 1.
100 Kellermann, K.I., Shaffer, D.B., Pauliny-Toth, I.I.K., Preuss, E., Witzel, A.: Astrophys. J. **210** (1976) L121.
101 Kellermann, K.I., Shaffer, D.B., Clark, B.G., Geldzahler, B.J.: Astrophys. J. **214** (1977) L61.
102 King, I.R.: Publ. Astron. Soc. Pacific **83** (1971) 377.
103 Knapp, G.R., Gallagher, J.S., Faber, S.M., Balick, B.: Astron. J. **82** (1977) 106.
104 Knapp, G.R., Gallagher, J.S., Faber, S.M.: Astron. J. **83** (1978) 139.
105 Knapp, G.R., Kerr, F.J., Williams, B.A.: Astrophys. J. **222** (1978) 800.

106 Kormendy, J.: Astrophys. J. **214** (1977) 359.
107 Koski, A.T., Osterbrock, D.E.: Astrophys. J. **203** (1976) L49.
108 Kotanyi, C.G.: Astron. Astrophys. **74** (1979) 156.
109 Kotanyi, C.G., Ekers, R.D.: Astron. Astrophys. **73** (1979) L1.
110 Kronberg, P.P., Wilkinson, P.N.: Astrophys. J. **200** (1975) 430.
111 Kronberg, P.P., Biermann, P.: Astrophys. J. **243** (1981) 89.
112 v. d. Kruit, P.C., Allen, R.J.: Annu. Rev. Astron. Astrophys. **14** (1976) 417.
113 v. d. Kruit, P.C., Allen, R.J., Rots, A.H.: Astron. Astrophys. **55** (1977) 421.
114 v. d. Kruit, P.C.: Astron. Astrophys. **61** (1977) 171.
115 Krumm, N., Salpeter, E.E.: Astrophys. J. **227** (1979) 776.
116 Krumm, N., Salpeter, E.E.: Astrophys. J. **228** (1979) 64.
117 Lacy, J.H., Baas, F., Townes, C.H., Geballe, T.R.: Astrophys. J. **227** (1979) L17.
118 Larson, R.B.: Mon. Not. R. Astron. Soc. **176** (1976) 31.
119 Larson, R.B., Tinsley, B.M.: Astrophys. J. **219** (1978) 46.
120 Lequeux, J.: Astron. Astrophys. **15** (1971) 42.
121 Lüst, R.: Z. Naturforsch. **7a** (1952) 87.
122 Lynden-Bell, D., Pringle, J.E.: Mon. Not. R. Astron. Soc. **168** (1974) 603.
123 Lynds, B.T.: Int. Astron. Union Symp. **44**, op. cit. [54] (1972) p. 56.
124 Madore, B.F., v. d. Bergh, S., Rogstad, D.H.: Astrophys. J. **191** (1974) 317.
125 Markarian, B.E.: Astron. Astrophys. **58** (1977) 139.
126 Mathewson, D.S., v. d. Kruit, P.C., Brown, W.N.: Astron. Astrophys. **17** (1972) 468.
127 Maza, J., van den Bergh, S.: Astrophys. J. **204** (1976) 519.
128 Merrill, K.M. in: Int. Astron. Union Coll. No. **42**, op. cit. [76] (1977) p. 446.
129 Mezger, P.G., Henderson, A.P.: Astrophys. J. **147** (1967) 471.
130 Mezger, P.G., Höglund, B.: Astrophys. J. **147** (1967) 490.
131 Mezger, P.G., Schraml, J., Terzian, Y.: Astrophys. J. **150** (1967) 807.
132 Mezger, P.G., Pauls, T. in: The Large Scale Characteristics of the Galaxy. Int. Astron. Union Symp. **84** (1978) 357.
133 Morgan, W.W., Mayall, N.U.: Publ. Astron. Soc. Pacific **69** (1957) 291.
134 Morgan, W.W.: Publ. Astron. Soc. Pacific **70** (1958) 364.
135 Morgan, W.W.: Publ. Astron. Soc. Pacific **71** (1959) 92.
136 Morgan, W.W., Osterbrock, D.E.: Astron. J. **74** (1969) 515.
137 Morris, M., Lo, K.Y.: Astrophys. J. **223** (1978) 803.
138 Neugebauer, G., Becklin, E.E., Oke, J.B., Searle, L.: Astrophys. J. **205** (1976) 29.
139 O'Connell, R.W.: Astrophys. J. **206** (1976) 370.
140 O'Connell, R.W., Mangano, J.J.: Astrophys. J. **221** (1978) 62.
141 O'Connell, R.W., Thuan, T.X., Goldstein, S.J.: Astrophys. J. **226** (1978) L11.
141a Oemler, A., jr., Tinsley, B.M.: Astron. J. **84** (1979) 985.
142 Oort, J.H.: Annu. Rev. Astron. Astrophys. **15** (1977) 295.
143 Osterbrock, D.E.: Astrophysics of gaseous nebulae, Freeman, San Francisco, Calif. (1974).
144 Penston, M.V., Fosbury, R.A.E., Ward, M.J., Wilson, A.S.: Mon. Not. R. Astron. Soc. **180** (1977) 19.
145 Peterson, C.J., Rubin, V.C., Ford, W.K. Jr., Roberts, M.S.: Astrophys. J. **226** (1978) 770.
146 Preuss, E., Pauliny-Toth, I.I.K., Witzel, A., Kellermann, K.I., Shaffer, D.B.: Astron. Astrophys. **54** (1977) 297.
147 Pritchet, C.: Astrophys. J. Suppl. **35** (1977) 397.
148 Pritchet, C.: Astrophys. J. **231** (1979) 354.
149 Reif, K., Mebold, U., Goss, W.M.: Astron. Astrophys. **67** (1978) L1.
150 Rickard, L.J., Turner, B.E., Palmer, P.: Astrophys. J. **218** (1977) L51.
151 Rieke, G.H., Lebofsky, M.J.: Annu. Rev. Astron. Astrophys. **17** (1979) 477.
152 Roberts, M.S.: Int. Astron. Union Symp. **44**, op. cit. [54] (1972) p. 12.
153 Roberts, M.S. in: Galaxies and the Universe, Vol. 9 of Stars and stellar systems, (Sandage, A.R., Sandage, M., Kristian, J., eds.), Univ. of Chicago Press, Chicago (1975) p. 309.
154 Rubin, V.C., Ford, W.K., jr., Strom, K.M., Strom, S.E., Romanishin, W.: Astrophys. J. **224** (1978) 782.
155 Rubin, V.C., Ford, W.K., jr., Roberts, M.S.: Astrophys. J. **230** (1979) 35.
156 Sandage, A.R.: The Hubble Atlas of Galaxies, Carnegie Institution of Washington, Washington, D.C. (1961).
157 Sandage, A.R., Visvanathan, N.: Astrophys. J. **223** (1978) 707.
158 Sandage, A.R., Brucato, R.: Astron. J. **84** (1979) 472.

159 Sargent, W.L.W.: Astrophys. J. **160** (1970) 405.
160 Sargent, W.L.W.: Astrophys. J. **173** (1972) 7.
161 Schraml, J., Mezger, P.G.: Astrophys. J. **156** (1969) 269.
162 Schweizer, F.: Astrophys. J. **220** (1978) 98.
163 Seaquist, E.R., Bell, M.B.: Astron. Astrophys. **60** (1977) L1.
164 Seaquist, E.R., Davis, L., Bignell, R.C.: Astron. Astrophys. **63** (1978) 199.
165 Searle, L., Sargent, W.L.W.: Astrophys. J. **173** (1972) 25.
166 Searle, L., Sargent, W.L.W., Bagnuolo, W.G.: Astrophys. J. **179** (1973) 427.
167 Segalovitz, A.: Astron. Astrophys. **61** (1977) 59.
168 Sérsic, J.L., Carranza, G., Pastoriza, M.: Astrophys. Space Sci. **19** (1972) 469.
169 Shaver, P.A., Churchwell, E., Rots, A.H.: Astron. Astrophys. **55** (1977) 435.
170 Shaver, P.A., Churchwell, E., Walmsley, C.M.: Astron. Astrophys. **64** (1978) 1.
171 Shaver, P.A.: Astron. Astrophys. **68** (1978) 97.
172 Smith, L.F., Biermann, P., Mezger, P.G.: Astron. Astrophys. **66** (1978) 65.
173 Spinrad, H., Peimbert, M. in: Galaxies and the Universe, op. cit [153] (1975) p. 37.
174 Stocke, J.T., Tifft, W.G., Kaftan-Kassim, M.A.: Astron. J. **83** (1978) 322.
175 Strom, K.M., Strom, S.E.: Astron. J. **83** (1978) 73.
176 Strom, K.M., Strom, S.E.: Astron. J. **83** (1978) 1293.
177 Strom, S.E., Strom, K.M., Grasdalen, G.L.: Annu. Rev. Astron. Astrophys. **13** (1975) 187.
178 Strom, S.E., Strom, K.M., Goad, J.W., Vrba, F.J., Rice, W.: Astrophys. J. **204** (1976) 684.
179 Strom, S.E., Strom, K.M.: Astron. J. **83** (1978) 732.
180 Sulentic, J.W.: Astrophys. J. Suppl. **32** (1976) 171.
181 Sulentic, J.W.: Astron. J. **81** (1976) 582.
182 Talbot, R.J., jr., Arnett, W.D.: Astrophys. J. **197** (1975) 551.
183 Thronson, H.A., jr., Campbell, M.F., Harvey, P.M.: Astron. J. **83** (1978) 1581.
184 Tinsley, B.M., Gunn, J.E.: Astrophys. J. **203** (1976) 52.
185 Tinsley, B.M., Gunn, J.E.: Astrophys. J. **206** (1976) 525.
186 Tinsley, B.M. in: Supernovae (Schramm, D.N., ed.), Reidel, Dordrecht (1977) p. 117.
187 Tinsley, B.M.: Astrophys. J. **222** (1978) 14.
188 Tinsley, B.M., Larson, R.B.: Mon. Not. R. Astron. Soc. **186** (1979) 503.
189 Toomre, A., Toomre, J.: Astrophys. J. **178** (1972) 623.
190 Tosa, M., Hamajima, K.: Publ. Astron. Soc. Japan **27** (1975) 501.
191 Tovmassian, H.M., Sramek, R.A.: Astrofizika **12** (1976) 21.
192 Trefftz, E.: Z. Naturforsch. **7a** (1952) 99.
193 Turner, B.E.: Astron. Astrophys. Suppl. **37** (1979) 1.
194 Turnrose, B.E.: Astrophys. J. **210** (1976) 33.
195 de Vaucouleurs, G., de Vaucouleurs, A., Corwin, H.G., jr.: Second Reference Catalogue of Bright Galaxies, Texas University Press, Austin (1976).
196 Vettolani, G., Zamorani, G.: Mon. Not. R. Astron. Soc. **178** (1977) 693.
197 Vrba, F.J., Strom, K.M., Strom, S.E., Grasdalen, G.L.: Astrophys. J. **197** (1975) 77.
198 Walmsley, C.M. in: H II regions and related topics, op. cit. [74] (1975) p. 17.
199 Weedman, D.W.: Astrophys. J. **183** (1973) 29.
200 Whitford, A.E. in: Galaxies and the Universe, op. cit. [153] (1975) p. 159.
201 Whitford, A.E.: Astrophys. J. **211** (1977) 527.
202 Williams, T.B.: Astrophys. J. **209** (1976) 716.
203 Wilson, T.L. in: H II regions and related topics, op. cit. [74] (1975) p. 39.
204 Winnberg, A. in: Int. Astron. Union Coll. **42**, op. cit. [76] (1977) p. 495.
205 Winnewisser, G., Churchwell, E., Walmsley, C.M.: Modern aspects of microwave spectroscopy, Academic Press, New York (1979).
206 van Woerden, H., Goss, W.M., Mebold, U., Siegman, B., Hawarden, T.G.: Proc. Astron. Soc. Australia **3** (1976) 68.
207 Wynn-Williams, C.G., Becklin, E.E., Neugebauer, G.: Mon. Not. R. Astron. Soc. **160** (1972) 1.
208 Wynn-Williams, C.G., Becklin, E.E.: Publ. Astron. Soc. Pacific **85** (1974) 5.
209 Zuckerman, B., Palmer, P.: Annu. Rev. Astron. Astrophys. **12** (1974) 279.

# 9.5 Quasars and active galactic nuclei

## 9.5.0 Abbreviations

AGN = active galactic nuclei
QSO = quasi-stellar object = quasar
BSO = blue stellar object
NLRG = narrow-line radio galaxy
BLRG = broad-line radio galaxy
Liner = low-ionization nuclear emission-line region

## 9.5.1 Definition and classification

### 9.5.1.1 Active galactic nuclei

Active galactic nuclei (AGNs) is a common name for several types of objects, including quasars (QSOs), Seyfert galaxies, radio galaxies with active nuclei, BL Lac objects and Liners. They all show energy output not usually associated with normal stellar processes and centered in a small nuclear region. We classify an object in one of these groups if *at least* one, and preferably two, of the following criteria are fulfilled:

1. Compact nuclear region, brighter than the corresponding region in galaxies of similar Hubble type
2. Nonthermal nuclear continuum emission
3. Nuclear emission lines indicating non-stellar excitation mechanism
4. Variable continuum and/or emission lines.

Many Seyferts, radio galaxies and quasars show all 4 properties. BL Lac objects show Nos. 1, 2, and 4. Liners show No. 3 and sometimes Nos. 2 or 1.

There is some confusion about the group called N galaxies [101] which have brilliant star-like nuclei containing most of the light of the system, and a faint underlying galaxy. Many of these are radio galaxies (see 9.6) and will be classified as such. Others will be classified here as Seyfert galaxies. X-ray galaxies may be another subgroup of AGNs [60].

Throughout this subsection we assume a Friedmann universe (see 9.7.2.2.3) with a deceleration parameter $q_0=0$ and Hubble constant $H_0=75\ \mathrm{km\,s^{-1}\,Mpc^{-1}}$. The luminosity distance is then:

$$D_L=\frac{cz}{H_0}(1+\tfrac{1}{2}z)$$

and the look-back time:

$$\tau=\frac{z}{1+z}t_0,$$

where $z=\frac{\Delta\lambda}{\lambda}$ is the observed redshift and $t_0$ is the age of the universe. For other cases with $q_0>0$:

$$D_L=\frac{c}{H_0 q_0^2}\{q_0 z+(q_0-1)[(2q_0 z+1)^{1/2}-1]\}.$$

(For more detailed explanations of the symbols, see 9.7.)

The observed flux at frequency $\nu/(1+z)$, $f_{\nu/(1+z)}$ [$\mathrm{erg\,s^{-1}\,cm^{-2}\,Hz^{-1}}$] is related to the flux at the source $F_\nu$ [$\mathrm{erg\,s^{-1}\,Hz^{-1}}$] as follows:

$$F_\nu=\frac{4\pi D_L^2}{1+z}f_{\nu/(1+z)} \quad \text{with} \quad D_L \text{ in [cm]}.$$

Comprehensive reviews of the AGNs can be found in [1, 2, 27, 113, 118, 126, 128, 132], and in many articles in Vol. 17 of Phys. Scr. (1978).

### 9.5.1.2 Quasars

Quasars are the most luminous active nuclei. Their visual luminosity ranges from $10^{45}$ to about $10^{48}\ \mathrm{erg\,s^{-1}}$ ($-31 \lesssim M_V \lesssim -24$). Quasar images are stellar-like on ordinary photographic plates, but deep photographs under good seeing conditions reveal an underlying nebulosity in many small $z$ cases. This is interpreted as an underlying galaxy [48, 54, 138]. The galaxy is much fainter than the quasar and has a typical magnitude of ordinary ellipticals or spirals.

Lowest-redshift quasars have $z \approx 0.1$. The highest-redshift quasar discovered to date is PKS 2000–330 with $z = 3.78$. This corresponds to a look-back time of 0.79 $t_0$ for a $q_0 = 0$ cosmology.

The cosmological interpretation of quasar redshift is confirmed by their association with galaxies of similar redshift [117], by their Hubble diagram [100] and by galaxies under quasars. This strengthens early suggestions [101] that quasars are luminous nuclei of distant galaxies. A considerable number of quasars have been found to lie close to galaxies with very different redshift [46], and various suggestions have been made for a non-cosmological interpretation of their redshift [5, 17]. The significance of the association can only be established by statistical methods. Recently discovered cases of gravitational lenses [125] give more support to the cosmological hypothesis.

Quasars are discovered in several ways:

1. Radio survey (stellar objects with unusual radio properties e.g. [68]),
2. Two-colour method (BSO = blue stellar objects e.g. [16])
3. Proper-motion survey at high galactic latitude [57]
4. X-ray survey (e.g. [42])
5. Variability surveys (blue variable stellar objects, [123])
6. Objective-prism and transmission-grating surveys (e.g. [64, 112]).

More than 1400 quasars are catalogued ($z, m_v$, coordinates, spectral information) in [46]. Spectrophotometric data on over a hundred of these are available [11, 41, 60a, 78, 80, 85, 92, 94, 97, 115, 121] including emission-line fluxes and relative intensities. Table 1 is a representative list.

Table 1. Representative list of quasars with spectrophotometric data.

$$z = \frac{\Delta\lambda}{\lambda} = \text{redshift}$$

$m_v$ = approximate visual magnitude (not taking into account variability)

| Quasar | α (1950) | δ (1950) | z | $m_v$ | Ref. | Comments *) |
|---|---|---|---|---|---|---|
| Q 0002 − 422 | $00^h02^m16^s$ | −42°14′ | 2.758 | $17^m.4$ | 85 | |
| PHL 938 | 00 58 20 | 01 55 | 1.95 | 17.2 | 11 | |
| 4C 25.05 | 01 23 57 | 25 44 | 2.34 | 17.5 | 11 | |
| PHL 1093 | 01 37 23 | 01 17 | 0.262 | 17.1 | 8 | a, var |
| PHL 1194 | 01 48 52 | 09 03 | 0.298 | 17.5 | 8 | a |
| RN 8 | 02 10 49 | 86 05 | 0.184 | 19.0 | 8 | |
| Q 0242 − 410 | 02 42 02 | −41 04 | 2.214 | 18.1 | 85 | |
| Q 0324 − 407 | 03 24 29 | −40 47 | 3.056 | 17.6 | 85 | |
| PKS 0424 − 13 | 04 24 48 | −13 10 | 2.16 | 17.5 | 11 | |
| Q 0453 − 423 | 04 53 48 | −42 21 | 2.661 | 17.3 | 85 | |
| Q 0551 − 366 | 05 51 02 | −36 38 | 2.307 | 17.0 | 85 | |
| OH 471 | 06 42 53 | 44 55 | 3.39 | 18.5 | 11 | |
| PKS 0736 + 01 | 07 36 43 | 01 44 | 0.192 | 16.5 | 8 | |
| 4C 05.34 | 08 05 19 | 04 41 | 2.86 | 18.2 | 11 | var ? |
| 0938 + 119 | 09 38 32 | 11 59 | 3.19 | 19.0 | 11 | |
| 3C 232 | 09 55 25 | 32 38 | 0.533 | 15.8 | 92 | |
| Ton 490 | 10 11 06 | 25 04 | 1.63 | 15.4 | 11 | |
| PKS 1217 + 02 | 12 17 39 | 02 20 | 0.240 | 16.5 | 8 | |
| 3C 273 | 12 26 33 | 02 20 | 0.158 | 12.8 | 8 | a |
| Q 1246 − 057 | 12 46 29 | −05 43 | 2.212 | 17.0 | 85 | b |
| B 340 | 13 04 48 | 34 40 | 0.184 | 17.0 | 8 | var |
| 1331 + 170 | 13 31 10 | 17 04 | 2.08 | 16.0 | 11 | |
| 3C 323.1 | 15 45 31 | 21 01 | 0.264 | 16.7 | 8 | var |
| 4C 29.50 | 17 02 11 | 29 51 | 1.92 | 19.1 | 11 | |
| 3C 351 | 17 04 03 | 60 49 | 0.371 | 15.3 | 92 | |
| Q 2116 − 358 | 21 16 22 | −35 49 | 2.341 | 17.0 | 85 | |
| PKS 2135 − 14 | 21 35 01 | −14 46 | 0.200 | 15.5 | 8 | a |
| 2256 + 017 | 22 56 25 | 01 48 | 2.66 | 18.5 | 11 | |

*) a = underlying galaxy b = broad absorption lines var = variable

The number of quasars brighter than a given limiting magnitude per square degree, $N(<m)$, can be expressed as $N(<m)=C\cdot 10^{am}$ with $0.6\lesssim a\lesssim 0.9$ [40, 83, 85, 104]. For $m=19.5$, $N\approx 7$ with an uncertainty of a factor 2. Evolution in time may also be important for quasars and a $N(z)\propto \exp(10\tau)$, (with $\tau$ = look-back time, see 9.5.1.1) has been suggested [103]. However, there is evidence for a real deficiency of quasars with $z>3$ [84]. The $V/V_{max}$ test ($V$ = volume; explained in more detail in 9.7.3.3) gives a mean value of 0.6···0.7 [85, 131] compared with 0.5 expected for a uniform distribution.

### 9.5.1.3 Seyfert galaxies

Seyfert galaxies are galaxies with bright nuclei that show the spectroscopic characteristics of this group (9.5.3). The nucleus-to-galaxy visual luminosity ratio tends to increase with redshift. For nearby Seyferts it is less than 0.1. For objects with $z>0.1$ it can exceed 0.5.

The first few Seyfert galaxies were described by Seyfert [105]. Many more have been found in objective-prism surveys (e.g. [65]) and among Zwicky's compact galaxies (see 9.3.1). About 1% of all bright spirals are Seyfert galaxies [126]. Their fraction among Markarian blue galaxies (see 9.3.1) is about 10%. Morphological studies of the underlying galaxies show deficiency of ellipticals and some with peculiar shapes [3].

Seyfert galaxies are classified into two major groups, according to their spectroscopic properties [86]: Seyfert 1 galaxies with broad (≈5000 km s$^{-1}$ FWHM) permitted lines, and Seyfert 2 galaxies with permitted and forbidden lines of about equal width (300···1000 km s$^{-1}$ FWHM). Classification into subgroups and spectral analysis is given in 9.5.3. About 90 Seyfert galaxies are listed and described in [126]. Their nuclear visual luminosity ranges from $10^{42}$ to more than $10^{45}$ erg s$^{-1}$. Brightest objects are Seyfert 1 galaxies of highest redshift. Spectrophotometric data and emission-line lists are available for many objects [13, 14, 28, 37, 51, 58, 81, 86, 92, 111, 137]. Table 2 lists several of these.

Table 2. Representative list of Seyfert galaxies with spectrophotometric data.

$z = \frac{\Delta\lambda}{\lambda}$ = redshift

$m_v$ = estimated nuclear visual magnitude not corrected for aperture size or variability
var = variable continuum and/or lines

| Galaxy | α (1950) | δ (1950) | $z$ | $m_v$ | Ref. | Comments |
|---|---|---|---|---|---|---|
| Seyfert 1 galaxies | | | | | | |
| Mrk 335 | $00^h03^m45^s$ | 19°55′ | 0.025 | $14^m.2$ | 86 | var |
| I Zw 1 | 00 51 00 | 12 25 | 0.061 | 14.3 | 86 | |
| Mrk 376 | 07 10 36 | 45 47 | 0.056 | 16.0 | 86 | |
| Mrk 79 | 07 38 47 | 49 56 | 0.020 | 13.4 | 86 | var |
| Mrk 10 | 07 43 07 | 61 03 | 0.029 | 15.0 | 86 | |
| Mrk 110 | 09 21 44 | 52 30 | 0.036 | 16.1 | 86 | |
| NGC 3227 | 10 20 47 | 20 07 | 0.0033 | 13.5 | 86 | |
| NGC 3516 | 11 03 24 | 72 50 | 0.0093 | 13.1 | 86 | var |
| NGC 4151 | 12 08 01 | 39 41 | 0.0033 | 12.0 | 14 | var |
| Mrk 236 | 12 58 18 | 61 55 | 0.052 | 17.0 | 86 | |
| Mrk 279 | 13 51 52 | 69 33 | 0.0307 | 15.4 | 86 | |
| Mrk 290 | 15 34 45 | 58 04 | 0.0308 | 15.6 | 86 | |
| Mrk 486 | 15 35 21 | 54 43 | 0.039 | 15.0 | 86 | |
| Mrk 509 | 20 41 26 | −10 54 | 0.0355 | 13.0 | 86 | |
| NGC 7469 | 23 00 44 | 08 36 | 0.0167 | 13.6 | 86 | var |
| Mrk 541 | 23 53 30 | 07 15 | 0.041 | 15.5 | 86 | var |
| Seyfert 2 galaxies | | | | | | |
| Mrk 1 | $01^h13^m19^s$ | 32°50′ | 0.016 | $16^m.6$ | 51 | |
| NGC 1068 | 02 40 07 | −00 14 | 0.00363 | 10.5 | 51 | |
| Mrk 612 | 03 21 10 | −03 19 | 0.02022 | 16.5 | 111 | |
| III Zw 55 | 03 38 38 | −01 28 | 0.0246 | 14.0 | 51 | |
| Mrk 3 | 06 09 48 | 71 03 | 0.0137 | 13.8 | 51 | |

continued

Table 2 (Seyfert galaxies), continued

| Galaxy | α (1950) | δ (1950) | z | $m_v$ | Ref. | Comments |
|---|---|---|---|---|---|---|
| Mrk 78 | $07^h37^m56^s$ | 65°18′ | 0.0375 | $15^m.6$ | 51 | |
| Mrk 622 | 08 04 21 | 39 09 | 0.02283 | 15.6 | 111 | |
| Mrk 34 | 10 30 52 | 60 17 | 0.051 | 14.8 | 51 | |
| Mrk 176 | 11 29 54 | 53 14 | 0.0269 | 15.5 | 51 | |
| Mrk 270 | 13 39 41 | 67 55 | 0.009 | 15.0 | 51 | |
| Mrk 463E | 13 53 40 | 18 37 | 0.0505 | 16.0 | 111 | |
| Mrk 533 | 23 25 24 | 08 30 | 0.02873 | 16.0 | 111 | |

### 9.5.1.4 Radio galaxies with active nuclei (see also 9.6)

Many radio galaxies have strong emission lines in their spectra and are classified as active nuclei. They are divided into:

1. Narrow-line radio galaxies (NLRGs) with forbidden and permitted lines of the same width ($\approx 500\ \mathrm{km\,s^{-1}}$ FWHM)
2. Broad-line radio galaxies (BLRGs) with very broad permitted lines ($\approx 8000\ \mathrm{km\,s^{-1}}$ FWHM) and narrow forbidden lines.

Many BLRGs are morphologically N galaxies and resemble in many aspects Seyfert 1 galaxies. Nuclear luminosities of active radio galaxies cover a similar range to that of Seyfert galaxies.

A comprehensive list of radio galaxies is given in [18]. Spectrophotometric data are given in [25, 38, 51, 88, 140]. Several of these are listed in Table 3.

Table 3. Representative list of radio galaxies with active nuclei and spectrophotometric data.

$z = \frac{\Delta\lambda}{\lambda}$ = redshift

$m_v$ = approximate nuclear visual magnitude not taking into account variability and different aperture used

Class. = morphological classification (see 9.1.3.2.1, Table 3)

| Galaxy | α (1950) | δ (1950) | z | $m_v$ | Ref. | Class. |
|---|---|---|---|---|---|---|
| BLRGs | | | | | | |
| 3C 109 | $04^h10^m55^s$ | 11°15′ | 0.306 | $18^m.0$ | 140 | N |
| 3C 120 | 04 30 32 | 05 15 | 0.033 | 14.6 | 140 | N−S |
| 3C 227 | 09 45 07 | 07 39 | 0.0855 | 16.3 | 88 | N |
| 3C 234 | 09 58 57 | 29 02 | 0.1846 | 17.1 | 38 | N |
| 3C 287.1 | 13 29 04 | 25 24 | 0.2156 | 18.5 | 38 | N |
| PKS 1417–19 | 14 17 02 | −19 15 | 0.1195 | 17.5 | 38 | N |
| 4C 35.37 | 15 31 45 | 35 52 | 0.1565 | 17.5 | 38 | N |
| 3C 332 | 16 14 44 | 30 09 | 0.1515 | 16.0 | 38 | var |
| 3C 381 | 18 32 28 | 47 24 | 0.1614 | 17.5 | 38 | ND |
| 3C 382 | 18 33 12 | 32 39 | 0.0586 | 15.4 | 88 | D3 |
| 3C 390.3 | 18 45 38.8 | 79 43 | 0.0569 | 15.4 | 88 | var, N |
| 3C 445 | 22 21 15 | −02 21 | 0.0568 | 15.8 | 88 | N |
| NLRGs | | | | | | |
| 3C 33 | $01^h06^m14^s$ | 13°04′ | 0.0595 | $16^m.3$ | 51 | DE4 |
| 3C 98 | 03 56 10 | 10 18 | 0.0306 | 14.8 | 25 | ED3 |
| 3C 178 | 07 22 33 | −09 30 | 0.0079 | 16.1 | 25 | Sc |
| 3C 184.1 | 07 32 20 | 70 20 | 0.1182 | 17 | 51 | D |
| 3C 192 | 08 02 38 | 24 16 | 0.0598 | 16.2 | 25 | DE1 |
| 3C 327 | 15 59 56 | 02 06 | 0.1039 | 16.3 | 25 | DE3–4 |
| 3C 433 | 21 21 30 | 24 52 | 0.1025 | 15.7 | 51 | D4 |
| 3C 452 | 22 43 33 | 39 25 | 0.082 | 16.6 | 51 | ED1 |
| PKS 2322−12 | 23 22 43 | −12 24 | 0.0821 | 15.8 | 25 | D5 |

### 9.5.1.5 Liners

Liners (low-ionization nuclear emission-line regions) are ordinary galaxies showing emission-line spectra that are not generally associated with normal stars [43]. About 1/3 of all bright galaxies belong to this group, and the Liner phenomenon is observable down to the limit of detection (at least two orders of magnitude fainter than the faint Seyfert galaxies).

Liners are characterized by several spectroscopic properties that are discussed in 9.5.3. Some, but not all of them, show evidence of a weak nonthermal visual continuum. Their line excitation mechanism is not yet fully explained.

Lists of emission-line fluxes are given in [25, 44, 55, 111]. Some of the Liners are listed in Table 4.

Table 4. Representative list of Liners with spectrophotometric data.

$$z = \frac{\Delta\lambda}{\lambda} = \text{redshift}$$

$m_v$ = approximate visual magnitude (3″···6″ diameter aperture)
Class. = Hubble types (see 9.1.3.1)

| Galaxy | α (1950) | δ (1950) | z | $m_v$ | Ref. | Class. |
|---|---|---|---|---|---|---|
| Mrk 1158 | $01^h32^m07^s$ | 34°47′ | 0.0151 | 16ᵐ2 | 111 | |
| NGC 1052 | 02 38 37 | −08 28 | 0.0048 | 13.2 | 33 | E4 |
| Ark 160 | 08 17 52 | 19 31 | 0.019 | 17.6 | 111 | |
| NGC 2841 | 09 18 35 | 51 11 | 0.0022 | 13.5 | 44 | Sa |
| NGC 2911 | 09 31 05 | 10 23 | 0.0106 | 15.3 | 44 | E2 |
| NGC 3031 | 09 51 30 | 69 18 | −0.0001 | 12.4 | 44 | Sa |
| NGC 3758 | 11 33 48 | 21 52 | 0.0296 | 16.6 | 111 | |
| NGC 3998 | 11 55 20 | 55 44 | 0.0038 | 13.3 | 44 | E2 |
| NGC 4036 | 11 58 54 | 62 10 | 0.0046 | 14.0 | 44 | E6 |
| NGC 4278 | 12 17 36 | 29 34 | 0.0022 | 13.6 | 44 | E1 |
| NGC 5005 | 13 08 37 | 37 19 | 0.0033 | 14.1 | 111 | Sb |
| NGC 5077 | 13 16 53 | −12 24 | 0.0094 | 14.4 | 44 | E3 |
| NGC 5371 | 13 53 33 | 40 42 | 0.0086 | 15.0 | 44 | Sb |
| Mrk 298 | 16 03 18 | 17 56 | 0.0345 | 16.2 | 51 | Sb |
| Mrk 700 | 17 01 21 | 31 31 | 0.034 | 15 | 51 | Irr |
| NGC 6764 | 19 07 01 | 50 51 | 0.008 | 15.5 | 51 | SBb |

### 9.5.1.6 BL Lac objects

BL Lac objects are named after the first object of this kind, BL Lac (2200+420). They are stellar-like objects with extremely weak emission lines. Common properties are steep power-law continuum (see 9.5.2.1) ($1 \lesssim \alpha \lesssim 6$ with mean value of $\approx 2$), large rapid light variation at all wavelengths, and strongly, variably polarized continuum. The nonthermal source luminosity is typically $-26 \lesssim M_v \lesssim -21$.

An underlying galaxy has been detected in several BL Lac objects. This allows determination of their cosmological redshift by means of stellar spectral features and weak emission lines (e.g. [72]). More than 50 BL Lac objects are listed in [46]. Extensive reviews of their properties can be found in [132]. Table 5 gives several examples.

Table 5. Representative list of BL Lac objects.

$z_{em} = \frac{\Delta\lambda}{\lambda}$ = redshift for weak emission lines, if visible
$z_{gal}$ = redshift for underlying galaxy
$m_v$ = typical visual magnitude; all objects are highly variable
$z_{abs}$ = redshift of different systems of absorption lines

| Object | $\alpha$ (1950) | $\delta$ (1950) | $z_{em}$ | $z_{gal}$ | $m_v$ | $z_{abs}$ |
|---|---|---|---|---|---|---|
| PKS 0215+015 | $02^h15^m13^s$ | 01°31′ | | | $18^m.3$ | 0.41, 0.465, 1.345 |
| AO 0235+164 | 02 35 53 | 16 24 | | | 15.5 | 0.524, 0.852 |
| PKS 0521−365 | 05 21 14 | −36 30 | | 0.55 | 15.0 | |
| PKS 0548−323 | 05 48 50 | −32 17 | | 0.069 | 15.5 | |
| OJ 287 | 08 51 57 | 20 18 | 0.306? | | 14.0 | |
| 4C 22.25 | 09 57 34 | 22 48 | | | 18.0 | |
| Mkn 421 | 11 01 41 | 38 29 | | 0.308 | 13.5 | |
| Mkn 180 | 11 33 30 | 70 25 | | 0.0458 | 15.0 | |
| AP Lib | 15 14 45 | −24 11 | | 0.049 | 15.0 | |
| Mkn 501 | 16 52 12 | 39 50 | | 0.034 | 13.8 | |
| BL Lac | 22 00 40 | 42 02 | | 0.0688 | 14.5 | |

## 9.5.2 Continuum radiation

### 9.5.2.1 Optical and ultraviolet radiation

All active nuclei, except several of the Liners, show evidence of nonthermal nuclear energy radiation. The flux of this continuum can be fitted, over a limited wavelength range, by a power-law of the form $f_\nu = c\nu^{-\alpha}$, where $\alpha$, the spectral index, varies typically between 0.3···2.5. Change of slope and/or additional components are commonly observed when the full 1000···8000 Å wavelength range is studied [78, 121]. There are also some "bumps" that are not yet fully understood, such as a broad feature centered at ≈3000 Å. Ultraviolet observations of some Seyfert 1 galaxies show a double-component UV continuum, with a steep long wavelength ($\lambda > 2000$ Å) and a flatter shorter wavelength ($\lambda < 2000$ Å) part [89]. Steepening of the continuum of high-redshift quasars at $\lambda < 1200$ Å has also been observed [82a, 41]. BL Lac continua are smoother and steeper than those of other AGNs.

### 9.5.2.2 Infrared radiation

Measurements of many Seyfert galaxies and quasars at several infrared wavelengths are available [78, 90, 98, 99, 114]. The total energy output in the 1···30 μm wavelength range is somewhat larger than the optical luminosity for Seyfert galaxies, and about the same for quasars. Energy distribution at infrared wavelengths can be separated into contribution from 1) normal galaxy distribution, 2) thermal emission from dust, and 3) nonthermal power-law continuum. A single power-law for Seyfert 1 galaxies from the ultraviolet to 10 μm has been suggested but does not give a good fit since the infrared continuum is much steeper [98].

Contribution of the dust component is thought to be significant [99, 114] and is confirmed, in several cases, by known dust features. The estimated dust temperature is several hundred degrees and is given by [27]:

$$T_d \approx 1700\, L^{0.2} R^{-0.4} \quad \text{in [K]},$$

where $L$ in $[10^{46}\ \mathrm{erg\,s^{-1}}]$ is the optical/ultraviolet continuum luminosity, and $R$ in [pc] is the distance from the nucleus of the dust particles. $T_d \approx 400 \cdots 800$ K gives $R \approx 1$ pc for Seyferts and $R \approx 10$ pc for quasars, which is similar to the dimension of the broad-line region.

Thermal contribution and the amount of dust is larger in Seyfert 2 galaxies and in Liners. Balmer line ratios in these groups indicate reddening of $A_v \approx 1 \cdots 3$ [44, 98] (see 4.1.3.2.2). According to new ideas, X-ray galaxies may be reddened Seyfert 1 galaxies [60].

Some BL Lac objects have extremely steep and smooth optical/infrared continuum distributions, and thermal contribution there must be small.

### 9.5.2.3 Radio emission

Most quasars are radio quiet, but many of the early discovered ones are strong radio sources. About half of all quasars found in low-frequency surveys (3C, 4C, PKS, B2) have symmetric double-lobe radio structure, and many of these have variable flat-spectrum nuclear components [70]. Most optically selected quasars are radio quiet to the limit of detection (18 out of 247 have been discovered at a level of 10 mJy at 2380 MHz [116]).

Seyfert galaxies are weak compact radio objects, with a steep nonthermal nuclear source [29]. Type 2 are generally more luminous than type 1 at 21 cm. VLA (= Very Large Array, see 1.9.2.2) observations [122] show double or triple nuclear structure on a 1″ scale in several Seyferts, that may be related to the forbidden-line gas.

Liners are probably associated with compact nuclear radio sources [43]. BL Lac objects are variable nonthermal radio sources. Radio galaxies (see 9.6) are similar to radio-loud quasars.

Superluminal expansion has been observed in the centre of several quasars and radio galaxies [22]. Dimensions involved are $\approx 10$ pc. For more details, see 9.6.

### 9.5.2.4 X and γ radiation

Quasars, Seyfert 1 galaxies and BLRGs are strong X-ray emitters. It has been suggested that much and maybe most of the observed soft X-ray background is due to quasars [142]. Soft X-ray measurements [142] of quasars show that their 0.25···4.5 keV flux is somewhat smaller than the optical flux, with a clear correlation between both. Radio-loud quasars tend to be more luminous X-ray sources. A similar relation holds for Seyfert 1 galaxies [53]. Hard X-ray data are available for several Seyferts and BLRGs, but only for very few quasars [66, 71, 134···136]. The 2···50 keV flux of Seyferts can be fitted by $f_\nu \propto \nu^{-\alpha}$ with $\langle\alpha\rangle \approx 0.7$. Extrapolation to 80 keV gives a hard-X-ray flux that exceeds the optical/ultraviolet flux. There seems to be a continuous sequence of decreasing X-ray luminosity from Seyfert 1 to intermediate Seyfert galaxies to narrow-line X-ray galaxies to Seyfert 2 galaxies [60, 71a]. This may indicate that the narrow-line X-ray galaxies should perhaps be included among AGNs. In all of these the $H_\beta$ intensity is proportional to the X-ray luminosity and so is the 3.5 μm flux.

Several X-ray galaxies have large X-ray absorbing columns [47], and it has been suggested that the observed differences in flux are due to obscuration [49, 60]. BL Lac objects are strongly variable X-ray sources with two-component continua [136, 50]. Liners are below the limit of detection for existing hard X-ray detectors, but soft X-rays have been detected in several of them.

Observations of X-ray variability are described in [119a].

Only very few measurements of AGNs at energies larger than several MeV are available [12, 119]. Perhaps a large fraction of the total energy output of AGNs is in this band, although upper limits suggest steepening of the energy spectrum at the low-energy γ-ray region at least for some objects.

### 9.5.2.5 Variability

Many AGNs are variable in most wavelength bands [36, 39, 61, 69, 71, 90, 120, 123, 136]. They can be divided into several groups:

1. Violent variables (VV): Most of them are BL Lac objects and several are radio quasars. They vary by several magnitudes on a time scale of days to months, and tend to show steep nonthermal continua, and compact nuclear radio components with complex structure.

2. Small-amplitude continuum variables: Many quasars, Seyfert 1 galaxies and BLRGs belong to this group. They vary on a time scale of months to years with an amplitude of about $0^m.2 \cdots 1^m$.

3. Emission-line variables: Several Seyfert 1 galaxies and BLRGs show broad emission-line variations on a time scale of several months which may be correlated with continuum variation [23, 15a, 61, 77, 89, 141]. Emission line and continuum variations in the Seyfert galaxy Akn 120 have been reported in [49a, 32a]; for previous references, see [23].

4. Non-variable AGNs: Liners, NLRGs and Seyfert 2 galaxies as well as many broad-line objects belong to this group.

All variables change in an irregular way and no period has been found.

Table 6. Representative list of variable AGNs.
VV = violent variables
$\Delta t$ = typical time scale for variability

| Object | Class | $\Delta t$ | $\Delta m$ | Variability characteristics |
|---|---|---|---|---|
| 3C 273 | quasar | several years | $0^m.2 \cdots 1^m$ | all wavelengths |
| PKS 0736+01 | quasar | weeks | $\approx 0.5$ | IR, optical continuum not lines |
| 3C 345 | quasar, VV | days···weeks | $\gtrsim 1$ | IR, optical continuum, radio |
| 3C 454.3 | quasar, VV | days···weeks | $\approx 2$ | IR, optical continuum, radio |
| BL Lac | BL Lac | days···months | $>1$ | all wavelengths |
| 2A 1219+305 | BL Lac | $\approx 4$ months | $\approx 1$ | optical, X ray |
| Mkn 421 | BL Lac | $\approx 3$ months | 2···3 | optical, X ray |
| NGC 6814 | Seyfert 1 | 100 sec···2 years | $\approx 1$ | lines, X ray |
| 3C 390.3 | BLRG | 1 year | $\approx 0.5$ | lines, optical continuum |
| NGC 3516 | Seyfert 1 | $\approx 2$ months | 0.2···0.5 | lines, optical continuum |
| NGC 4151 | Seyfert 1 | $<1$ month | $\approx 0.5$ | optical, IR, X ray, absorption and emission lines |
| NGC 5548 | Seyfert 1 | $\approx 2$ years | $\approx 0.5$ | optical, IR, X ray, lines |
| NGC 1068 | Seyfert 2 | 10 days | 0.3 | optical, IR, X ray |

The shortest time-scale variation of $\approx 100$ sec is reported for X-ray emission of NGC 6814, but this seems to be unusual among Seyfert 1 galaxies. Infrared variability is typically of smaller magnitude, probably mainly of the nonthermal component [114]. The majority of known variables have low redshift and are radio sources. This may be due to some selection effect except for the VV objects which may indeed be more common among radio-loud AGNs [34, 15a, 122a]. Recently discovered X-ray variability of narrow-emission-line galaxies [71a] suggests that they also can be classified as AGNs.

### 9.5.2.6 Polarization

A polarization study of AGNs is reported in [4]. Most Seyfert 1 galaxies have low linear polarization (1···2%), which can differ between the line and continuum. Quasar continua are even less polarized (average less than 1%), but in some cases, mainly in those of the violent variables, a high degree of polarization is reported ($\approx 10$%). The continua of BL Lac objects are highly polarized (up to 30%) and the degree of polarization varies in time. There is a general correlation between optical polarization, variability and smooth continua. The rapidly variable, highly polarized objects are sometimes called "Blazers" [4].

## 9.5.3 Spectral lines

### 9.5.3.1 Emission lines

Emission lines are the distinctive sign of AGNs. Quasars, Seyfert 1 galaxies and BLRGs show extremely broad permitted lines, of up to about 10000 km s$^{-1}$ FWHM, if interpreted as Doppler motion. Zero-intensity width, of more than 30000 km s$^{-1}$, have been observed. Forbidden narrow lines, of $\approx 500$ km s$^{-1}$ FWHM are seen in most, but not all, of them [86, 126]. In some cases there is a distinct narrow component on top of the permitted lines. This has been the basis for dividing Seyfert galaxies into intermediate subgroups. For example, Seyfert 1.5 shows sharp $H_\beta$ of about 0.1 the intensity of the broad component. In Seyfert 1.8 the narrow component is stronger than the broad one, and in Seyfert 1.9 the broad component is so weak that it is only detected under $H_\alpha$

[87]. This suggests a smooth transition in spectral properties between Seyfert 1 and Seyfert 2 galaxies. Many of the 1.8 and 1.9 classes are strong-X-ray galaxies [109, 124]. The narrow-line component in BLRGs is more distinct than in most Seyferts [38]. There is a noticeable blue asymmetry in the narrow lines of high-ionization species in many Seyferts [45].

Broad-line objects and Seyfert 2 galaxies exhibit a large range of ionization stages, from O I to the O VI and Fe XI. Resonance, forbidden, recombination and intercombination lines are observed. Ultraviolet studies with the IUE satellite [13, 81, 137] show that Seyfert 1 galaxies and BLRGs are similar, although not identical, to quasars in spectral properties.

The composite spectrum of all groups is shown in Table 7. Most of the individual cases have intensity values within a factor of 2 of the mean listed. (No reddening corrections have been applied.)

Figure 1 shows spectra of three typical objects.

Table 7. Composite observed emission-line spectra of AGNs [8, 11, 13, 25, 38, 41, 44, 51, 81, 85, 88, 92, 94, 111, 121, 137, 139].
n = narrow line
b = broad line
$L$ = luminosity
$EW$ = equivalent width relative to the continuum in a 3″···6″ aperture
: = highly uncertain values

| Line | Quasars | Seyferts type 1 | BLRGs | Seyferts type 2 +NLRGs | Liners |
|---|---|---|---|---|---|
| | Relative intensities | | | | |
| Lα (b+n) | 500 | 500 | 600 | 200 ? | |
| N V 1240 (b+n) | 150 | 150 | 150 ? | | |
| O IV]+Si IV 1402 (b+n) | 150 | 150 | 150 ? | | |
| C IV 1549 (b+n) | 250 | 400 | 400 ? | | |
| N IV] 1486 | 30 | – | – | | |
| C III] 1909 | 120 | 60 | 60 ? | | |
| Fe II 2300···2600 | 130 | 130 | – | | |
| Mg II 2798 (b+n) | 130 | 130 | 130 | | |
| [Ne V] 3426 (n) | | 20: | 20: | 5 | |
| [O II] 3727 (n) | 10 | 10 | 20: | 20 | 40 |
| [Ne III] 3869 (n) | 30: | 30 | 30 | 10 | 3 |
| $H_\gamma$ (n) | } 40 | } 40 | } 40 | 3.7 | 3.7 |
| $H_\gamma$ (b) | | | | | |
| Fe II 4400···4600 | 30 | 30 | 10: | | – |
| He II 4686 (n) | } 20 | } 33 | } 15: | 2.6 | 0.6 |
| He II 4686 (b) | | | | | |
| $H_\beta$ (n) | } 100 | 10 | 10 | 10 | 10 |
| $H_\beta$ (b) | | 90 | 90 | | |
| [O III] 5007 (n) | 70 | 80 | 120 | 110 | 15 |
| He I 5876 (n) | } 18 | } 18 | } 8 | 1.6 | 1.5 |
| He I 5876 (b) | | | | | |
| [O I] 6300 | – | 5 | | 8 | 13 |
| $H_\alpha$ (n) | } 400 | } 360 | 70 | 50 | 60 |
| $H_\alpha$ (b) | | | 550 | – | – |
| [N II] 6584 (n) | – | 10: | 60 | 50 | 70 |
| [S II] 6716, 6731 (n) | – | 10: | 40 | 25 | 60 |
| $H_\alpha$: $L$ [erg s$^{-1}$] | $10^{44\cdots46}$ | $10^{42\cdots44}$ | $10^{42\cdots44}$ | $10^{40\cdots42}$ | $10^{38\cdots41}$ |
| $H_\beta$: $EW$ [Å] | 100 | 100 | 100 | 5···30 | 1···10 |

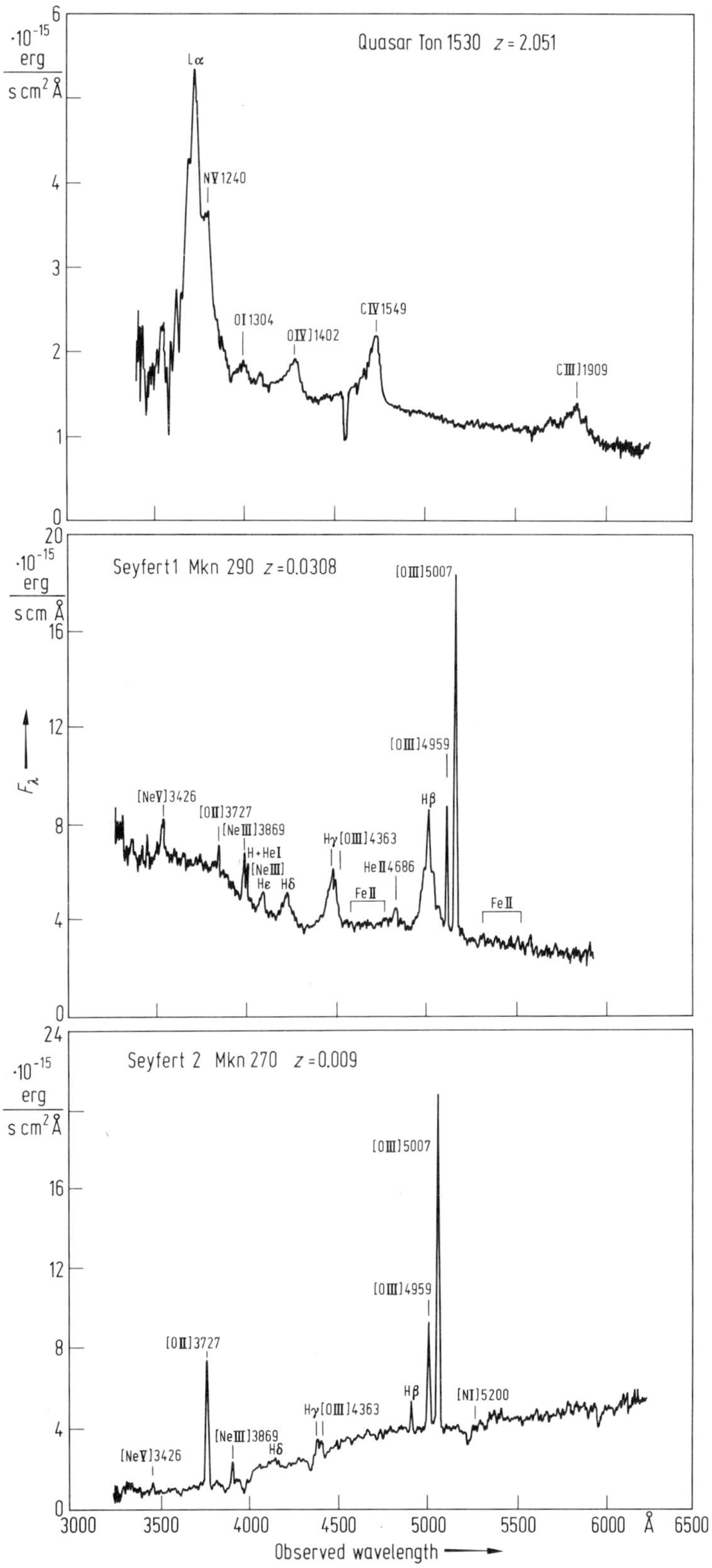


Fig. 1. The observed spectra of three AGNs (Quasar, Seyfert 1, Seyfert 2). $F_{\lambda}$ = flux outside the earth atmosphere.

NLRGs and Seyfert 2 galaxies are very similar. BLRGs and Seyfert 1 galaxies are similar too, but the former have weaker Fe II optical lines, steeper Balmer decrement and broader, less smooth line profiles.

Several correlations have been observed or suggested such as $H_\beta$ intensity with the ionizing flux, Fe II line strength with line width and radio properties, and strong narrow-line components with radio properties [38, 45, 60, 70]. It has been argued [9, 10] that the equivalent width of C IV 1549 and other ultraviolet lines in radio-loud quasars are inversely correlated with the continuum flux. This can be used as a luminosity calibrator. Study of radio-quiet quasars [82] does not confirm this.

### 9.5.3.2 Photoionization models

Photoionization model calculations have been applied to AGNs, with a reasonable degree of success [7, 26, 27 and references therein, 62, 129]. Ionization by a $f_\nu \propto \nu^{-\alpha}$ continuum is assumed. Gas temperature and degree of ionization of the emission-line gas depends critically on the ionization parameter, $U$, defined as:

$$U = \frac{\int_{\nu_{912}} \frac{f_\nu d\nu}{4\pi R^2 h\nu}}{N_e} = \frac{\text{ionizing photon flux}}{\text{electron density}},$$

where $R$ is the distance cloud-continuum source. $10^8 \lesssim U \lesssim 10^9$ cm s$^{-1}$ gives the best value for AGN broad-line regions. There are indications that $U$ (Seyfert 1 galaxies) is larger than $U$ (quasars).

Models accepted today have numerous small ($10^{11} \cdots 10^{13}$ cm) clouds, moving in the gravitation potential of a massive, central body of $\approx 10^9\,\mathfrak{M}_\odot$. The distance from the center is $R \approx 0.1$ pc for Seyfert 1 galaxies and BLRGs, and $R \gtrsim 1$ pc for quasars. $0.3 \cdots 2 \times$ solar abundances, $10^4 \lesssim T_e \lesssim 3 \cdot 10^4$ K, and $10^7 \lesssim N_e \lesssim 10^{10}$ cm$^{-3}$ are best-fit values. Only a small fraction ($\approx 0.1$) of the continuum source is covered with such clouds [21, 27].

Transfer effects, especially in hydrogen lines, are known to be important [32, 55, 72]: $L_\alpha/H_\beta$ is observed to be much smaller than in recombination case B [9, 94, 115, 137] and models involving collisional processes in a low-ionization zone [19, 59, 127], and/or reddening by dust [75, 110] have been suggested to explain this. The excitation of Fe II [24, 74, 92, 93, 130], He I [31, 63, 73] and O I [37, 76] lines, and the "3000 Å bump", are other extensively investigated subjects.

The narrow-line region of Seyfert 1 and 2 galaxies, consists of high-excitation lower-density clouds, of $N_e \approx 10^4$ cm$^{-3}$, at several hundred pc from the center. Liner spectra indicate also low densities but much lower ionization. Figure 2 compares the strength of the *narrow* oxygen and $H_\beta$ lines in the different classes of objects. It suggests that AGNs form a continuous decreasing sequence in $U$, from broad-line objects, over Seyfert 2 galaxies, to Liners. The observed $H_\beta$ equivalent width tends to confirm this [108, 139]. Shock-wave heating is another mechanism suggested for Liners [33, 43, 52].

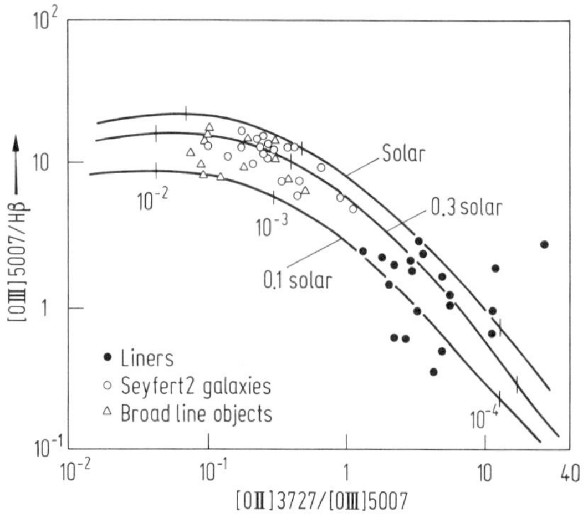


Fig. 2. Observed narrow-line ratios in AGNs. The curves show photoionization model calculations for a $f_\nu \propto \nu^{-1.5}$ ionizing source, and three different metal abundances. The dimensionless ionization parameter $U/c$ varies along the curves and is marked at $10^{-2}$, $10^{-3}$ and $10^{-4}$.

Dust must be an important constituent of AGN emission-line zones, and reddening is evident from the narrow lines of all objects. The case of broad lines is not yet settled, but it is clear that the amount of internal dust, mixed with the broad-line gas, must be small [32, 35].

### 9.5.3.3 Absorption lines

High-redshift quasars show rich absorption-line spectra. They can be classified into redshift systems, according to rules based on line intensities [6]. There are 4 main types of absorption lines [128].

1. Very broad absorption troughs, on the blue side of several emission lines (Table 7). These are probably intrinsic to the source and indicate ejection velocities reaching 0.1 $c$.

2. Sharp-line systems with $(z_{abs}-z_{em})/z_{em} \leqq \pm 0.01$. Material close to the source, perhaps in galaxies in the quasar cluster, has been proposed.

3. Metallic-line systems with large displacement from $z_{em}$. These are sharp ($20 \cdots 100\ km\,s^{-1}$) small column-density ($10^{17} \cdots 10^{21}$ H atoms $cm^{-2}$) absorbing clouds, that range from low to high ionization. Low densities and large distances from the quasars are inferred from the absence of fine-structure lines. Early models, such as [91], preferred the intrinsic hypothesis of gas ejected at relativistic velocities as causing the absorption. Later studies emphasized more the hypothesis of intervening material in halos of galaxies [128]. 21-cm observations [133], and direct observations of the obscuring galaxy [15], confirm the later for at least several well-studied cases.

4. L$\alpha$ systems: Large numbers of sharp hydrogen L$\alpha$ and L$\beta$ absorption lines are observed to the blue of L$\alpha$ (emission) peak. Intervening gas (intergalactic clouds, halos of galaxies) is thought to be the absorbing material [102]. Metal abundances of this gas may be very small, indicating primordial origin.

Absorbing clouds are seen also in several Seyfert 1 galaxies and BLRGs (e.g. [89]), where they are thought to be closely associated with emission-line gas. Absorbing systems of BL Lac are presumably of the same origin as those of quasars. Lack of measurable $z_{em}$ in BL Lac systems, prevents a thorough statistical investigation of this hypothesis.

## 9.5.4 Theoretical models for AGNs

Theoretical models consider the questions of the energy mechanism, spectral distribution, luminosity, masses, and evolution of AGNs. Reviews and references are given in [30, 95, 96] among many others.

Most theories involve a massive compact body at the centre of activity, which can be a dense star cluster, "spinar" or black hole. Black holes are studied more than the others, and may be their final evolution stage. The Schwarzschild radius of a black hole of mass $\mathfrak{M}$ is:

$$R_S \simeq 3 \cdot 10^5\, \mathfrak{M} \quad \text{with} \quad R_S \text{ in [cm]}, \mathfrak{M} \text{ in } [\mathfrak{M}_\odot].$$

If energy output, $L$, is due to mass accretion with efficiency $\varepsilon$, the required accretion rate $\dot{\mathfrak{M}}$ in $[\mathfrak{M}_\odot\, a^{-1}]$ is:

$$\dot{\mathfrak{M}} \simeq 0.1 \varepsilon^{-1} L \quad \text{with} \quad L \text{ in } [10^{46}\, erg\, s^{-1}].$$

With $\varepsilon = 0.1$, quasars must accrete $\approx 10\,\mathfrak{M}_\odot/a$ and Seyfert galaxies $\approx 0.1\,\mathfrak{M}_\odot/a$ that agreement with the observed $L$ is reached. Spherical and non-spherical accretion have been considered. In the latter case an accretion disk is formed around the hole. In this disk matter is transported inward and angular momentum outward, with a rate depending on the viscosity. Original gas supply for accretion can come from ordinary stellar mass-loss processes, or star disruption by tidal forces in a dense stellar core.

Radiation pressure impedes the accretion flow. For spherical accretion onto a central object of mass $\mathfrak{M}$ the critical luminosity is given by the "Eddington limit":

$$L_{Edd} \simeq 1.3 \cdot 10^{38}\, \mathfrak{M} \quad \text{with} \quad L_{Edd} \text{ in } [erg\, s^{-1}] \text{ and } \mathfrak{M} \text{ in } [\mathfrak{M}_\odot].$$

The luminosity can exceed this limit when there is mass loss from the system but $L < L_{Edd}$ if cool gas-absorption processes are more important than electron scattering.

Two-phase models for the atmosphere around the hole have been considered [56]. Cool ($10^4$ K) small gas clouds (high-velocity emission-line gas) are kept in pressure equilibrium in a hot ($\approx 10^8$ K) dilute plasma that confines them. Density distribution of the gas is not known, but it may follow a $N_H \propto R^{-3/2}$ law, expected for infalling gas in a virial balance with the medium.

Photoionization models and line- and continuum-variability studies (9.5.3.2 and 9.5.2.5) give the following consistent set of values:

$\mathfrak{M} \approx 10^8 \cdots 10^{10} \mathfrak{M}_\odot$ (Seyfert galaxies are on the lower side and quasars on the higher),

$R \approx 0.1$ pc for Seyfert galaxies and $\approx 1$ pc for quasars, and

radius of the continuum-radiation source of $\approx 10\, R_S$. (If disks are present, then this is their size.)

Gas of higher electron density ($N_e \gtrsim 10^{11}$ cm$^{-3}$) at smaller $R$ may also be present but does not emit any observable radiation.

Line-profile studies [20, 67, 107] relate such models to observable properties and discuss dynamical properties, such as the role of radiation pressure, cloud stability and their relation to the more distant lower-density gas.

The phase of relativistic particles is responsible for the continuum-emitting processes. Synchrotron radiation and inverse Compton scattering have been suggested. Superluminal expansion and jets are other energetic phenomena related to this phase.

AGNs form a continuous sequence of increasing activity from Liners to quasars. The strength of the nonthermal source, the ionization parameter and/or the appearance of broad lines are several of the variable parameters. It is not yet clear whether all observed properties are continuously varying along this sequence, and what is the inter-relation of the different groups.

## 9.5.5 References for 9.5


1 Active Galactic Nuclei (Hazard, C., Mitton. S. eds.): Nato Advanced Study Institute, Univ. Cambridge (1977).
2 Active Nuclei, Proc. Copenhagen Symp., Phys. Scr. **17** (1978).
3 Adams, T.F.: Astrophys. J. Suppl. **33** (1977) 19.
4 Angel, J.R.P., Stockman, H.S.: Annu. Rev. Astron. Astrophys. **18** (1980) 321.
5 Arp, H.: Coll. Intern. No. 263, CNRS Paris (1977).
6 Bahcall, J.N.: Phys. Scr. **17** (1978) 229.
7 Bahcall, J.N., Kozlovsky, B.Z.: Astrophys. J. **155** (1969) 1077.
8 Baldwin, J.A.: Astrophys. J. **201** (1975) 26.
9 Baldwin, J.A.: Mon. Not. R. Astron. Soc. **178** (1977) 67P.
10 Baldwin, J.A., Burke, W.L., Gaskell, C.M., Wampler, E.J.: Nature (London) **273** (1978) 431.
11 Baldwin, J.A., Netzer, H.: Astrophys. J. **226** (1978) 1.
12 Bignami, G.F., Fichtel, C.F., Hartman, R.C., Thomson, D.J.: Astrophys. J. **232** (1979) 649.
13 Boksenberg, A. et al. (26 authors): Nature (London) **275** (1978) 404.
14 Boksenberg, A., Shortridge, K., Allen, D.A., Fosbury, R.A.E., Penston, M.V., Savage, S.: Mon. Not. R. Astron. Soc. **173** (1975) 381.
15 Boksenberg, A., Sargent, W.L.W.: Astrophys. J. **220** (1978) 42.
15a Bonoli, F., Braccesi, A., Federici, L., Zitelli, V., Formiggini, L.: Astron. Astrophys. Suppl. **35** (1979) 391.
16 Braccesi, A., Formiggini, L., Gandolfi, E.: Astron. Astrophys. **23** (1973) 159.
17 Burbidge, G.R.: Nature **282** (1979) 451.
18 Burbidge, G.R., Crowne, A.H.: Astrophys. J. Suppl. **40** (1979) 583.
19 Canfield, R.C., Puetter, R.C.: Astrophys. J. **243** (1981) 390.
20 Capriotti, E., Foltz, C., Byard, P.: Astrophys. J. **245** (1981) 396.
21 Carswell, R.F., Ferland, G.J.: Mon. Not. R. Astron. Soc. **191** (1980) 55.
22 Cohen, M.H., Pearson, T.J., Readhead, A.C.S., Seielstad, G.A., Simon, R.S., Walker, R.C.: Astrophys. J. **231** (1979) 293.
23 Collin-Souffrin, S. in: Variability in Stars and Galaxies, Fifth European Regional Meeting, Liège (1980) p.C. 1.11.
24 Collin-Souffrin, S., Joly, M., Heidman, N., Dumont, S.: Astron. Astrophys. **72** (1979) 293.
25 Costero, R., Osterbrock, D.E.: Astrophys. J. **211** (1977) 675.
26 Davidson, K.: Astrophys. J. **171** (1972) 213.
27 Davidson, K., Netzer, H.: Rev. Mod. Phys. **51** (1979) 715.
28 De Bruyn, A.G., Sargent, W.L.W.: Astron. J. **83** (1978) 1257.
29 De Bruyn, A.G., Wilson, A.S.: Astron. Astrophys. **64** (1978) 433.
30 Fabian, A.C., Rees, M.J. in: Proc. Int. Astron. Union/Cospar Symp. on X-ray Astronomy (Baity, W.A., Peterson, L.E., eds.), Pergamon Press (1978) p. 381.

31 Feldman, F.R., MacAlpine, G.M.: Astrophys. J. **221** (1978) 486.
32 Ferland, G.J., Netzer, H.: Astrophys. J. **229** (1979) 274.
32a Foltz, C.B., Peterson, B.M., Capriotti, E.R., Byard, P.L., Bertram, R., Lawrie, D.G.: Astrophys. J. **250** (1981) 508.
33 Fosbury, R.A.E., Mebold, U., Goss, W.M., Dopita, M.A.: Mon. Not. R. Astron. Soc. **183** (1972) 549.
34 Gaskell, C.M.: Astrophys. Lett. **21** (1981) 103.
35 Gaskell, C.M., Shields, G.A., Wampler, E.J.: Astrophys. J. **249** (1981) 443.
36 Gilmore, G.: Mon. Not. R. Astron. Soc. **190** (1980) 649.
37 Grandi, S.A.: Astrophys. J. **221** (1978) 501.
38 Grandi, S.A., Osterbrock, D.E.: Astrophys. J. **220** (1978) 783.
39 Grandi, S.A., Tifft, W.G.: Publ. Astron. Soc. Pac. **86** (1974) 873.
40 Green, R.F., Schmidt, M.: Astrophys. J. Lett. **220** (1978) L1.
41 Green, R.F., Pier, J.R., Schmidt, M., Estabrook, F.B., Lane, A.L., Wahlquist, H.D.: Astrophys. J. **239** (1980) 483.
42 Grindlay, J.E., Steiner, J.E., Forman, W.R.: Astrophys. J. Lett. **239** (1980) L43.
43 Heckman, T.M.: Astron. Astrophys. **87** (1980) 152.
44 Heckman, T.M., Balick, B., Crane, C.P.: Astron. Astrophys. Suppl. **40** (1980) 295.
45 Heckman, T.M., Miley, G.K., van Breugel, W.J.M., Butcher, H.R.: Astrophys. J. **247** (1981) 403.
46 Hewitt, A., Burbidge, G.: Astrophys. J. Suppl. **43** (1980) 57.
47 Holt, S.S., Mushotzky, R.F., Becker, R.H., Boldt, E.A., Serlenitsos, P., Szymkowlak, A.E., White, N.E.: Astrophys. J. Lett. **241** (1980) L13.
48 Hutchings, J.B., Crampton, D., Campbell, B., Pritcher, C.: Astrophys. J. **247** (1981) 743.
49 Keel, W.C.: Astron. J. **85** (1980) 189.
49a Kollatschny, W., Fricke, K.J., Schleicher, H., Yorke, H.W.: Astron. Astrophys. **102** (1981) L23.
50 Kondo, Y. et al. (11 authors): Astrophys. J. **243** (1980) 690.
51 Koski, A.T.: Astrophys. J. **223** (1978) 56.
52 Koski, A.T., Osterbrock, J.E.: Astrophys. J. Lett. **203** (1976) L49.
53 Kriss, G.A., Canizares, C.R., Ricker, G.R.: Astrophys. J. **242** (1980) 492.
54 Kristian, J.: Astrophys. J. Lett. **179** (1973) L61.
55 Krolik, J.H., McKee, C.F.: Astrophys. J. Suppl. **37** (1978) 459.
56 Krolik, J.H., McKee, C.F., Tarter, C.B.: Astrophys. J. **249** (1981) 422.
57 Kron, R.G., Chiu, L.T.G.: Publ. Astron. Soc. Pac. **93** (1981) 397.
58 Kunth, D.K., Sargent, W.L.W.: Astron. Astrophys. **76** (1979) 50.
59 Kwan, J., Krolik, J.H.: Astrophys. J. **250** (1981) 478.
60 Lawrence, A., Elvis, M.: Astrophys. J. **256** (1982) 410.
60a Lewis, D.W., MacAlpine, G.M., Weedman, D.W.: Astrophys. J. **233** (1979) 787.
61 Lyutyi, V.M.: Astron. Zh. **54** (1977) 1153; engl. transl.: Soviet Astron. **21** (1977) 655.
62 MacAlpine, G.M.: Astrophys. J. **175** (1972) 11.
63 MacAlpine, G.M.: Astrophys. J. **204** (1976) 694.
64 MacAlpine, G.M., Lewis, D.W.: Astrophys. J. Suppl. **36** (1978) 587.
65 Markarian, B.E.: Astrofizika **5** (1969) 581; engl. transl.: Astrophys. **5** (1969) 286.
66 Marshall, F.E., Mushotzky, R.F., Boldt, E.A., Holt, S.S., Rothschild, R.E., Serlemitsos, P.J.: Nature **275** (1978) 624.
67 Mathews, W.G., Blumenthal, G.R.: Astrophys. J. **214** (1977) 10.
68 Matthews, T.A., Sandage, A.R.: Astrophys. J. **138** (1963) 30.
69 McGimsey, B.Q., Smith, A.G., Scott, R.L., Leacock, R.J., Edwards, P.C., Hackney, R.C., Hackney, K.R.: Astron. J. **80** (1975) 895.
70 Miley, G.K., Miller, J.S.: Astrophys. J. Lett. **228** (1979) L55.
71 Mushotzky, R.F., Marshall, F.E., Boldt, E.A., Holt, S.S., Serlemitsos, P.J.: Astrophys. J. **235** (1980) 377.
71a Mushotzky, R.F.: Astrophys. J. **256** (1982) 92.
72 Netzer, H.: Mon. Not. R. Astron. Soc. **171** (1975) 395.
73 Netzer, H.: Astrophys. J. **219** (1978) 822.
74 Netzer, H.: Astrophys. J. **236** (1980) 406.
75 Netzer, H., Davidson, K.: Mon. Not. R. Astron. Soc. **187** (1979) 871.
76 Netzer, H., Penston, M.V.: Mon. Not. R. Astron. Soc. **174** (1976) 319.
77 Netzer, H., Wills, B.J., Uomoto, A.K., Rybski, P.M., Tull, G.R.: Astrophys. J. Lett. **232** (1979) L155.
78 Neugebauer, G., Oke, J.B., Becklin, E.E., Matthews, K.: Astrophys. J. **230** (1979) 79.
79 Oke, J.B., Gunn, J.E.: Astrophys. J. Lett. **189** (1974) L5.

80 Oke, J.B., Neugebauer, G., Becklin, E.E.: Astrophys. J. **159** (1970) 341.
81 Oke, J.B., Zimmerman, B.: Astrophys. J. Lett. **231** (1979) L13.
82 Osmer, P.S.: Astrophys. J. **214** (1977) 1.
82a Osmer, P.S.: Astrophys. J. **227** (1979) 18.
83 Osmer, P.S.: Astrophys. J. Suppl. **42** (1980) 523.
84 Osmer, P.S.: Astrophys. J. **247** (1981) 762.
85 Osmer, P.S., Smith, M.G.: Astrophys. J. Suppl. **42** (1980) 333.
86 Osterbrock, D.E.: Astrophys. J. **215** (1977) 733.
87 Osterbrock, D.E.: Astrophys. J. **249** (1981) 462.
88 Osterbrock, D.E., Koski, A.T., Phillips, M.M.: Astrophys. J. **206** (1976) 898.
89 Penston, M.V., et al. (15 authors): Mon. Not. R. Astron. Soc. **196** (1981) 857.
90 Penston, M.V., Penston, M.J., Selmes, R.A., Becklin, E.E., Neugebauer, G.: Mon. Not. R. Astron. Soc. **169** (1974) 357.
91 Perry, J.J., Burbidge, E.M., Burbidge, G.R.: Publ. Astron. Soc. Pac. **90** (1978) 337.
92 Phillips, M.M.: Astrophys. J. Suppl. **38** (1978) 187.
93 Phillips, M.M.: Astrophys. J. **226** (1978) 736.
94 Peutter, R.C., Smith, H.E., Willner, S.P., Pipher, J.L.: Astrophys. J. **243** (1981) 345.
95 Rees, M.J.: Quasar Theories, in: 8th Texas Symp. on Relativistic Astrophys., New York, Acad. of Science (1977) p. 613.
96 Rees, M.J. in: X-ray Astronomy (Giacconi, R., Setti, G., eds.), NATO Advanced Study Inst. Ser. C, Vol. 60, Reidel, Dordrecht (1980) p. 339.
97 Richstone, D.O., Schmidt, M.: Astrophys. J. **235** (1980) 361.
98 Rieke, G.H.: Astrophys. J. **226** (1978) 550.
99 Rieke, G.H., Lebofsky, M.J.: Annu. Rev. Astron. Astrophys. **17** (1979) 477.
100 Sandage, A.: Astrophys. J. **178** (1972) 25.
101 Sandage, A.R.: Astrophys. J. **180** (1973) 687.
102 Sargent, W.L.W., Young, P.J., Boksenberg, A., Tytler, D.: Astrophys. J. Suppl. **42** (1980) 41.
103 Schmidt, M.: Phys. Scr. **17** (1978) 329.
104 Setti, G., Woltjer, L.: Ann. New York Acad. Science **224** (1973) 8.
105 Seyfert, C.K.: Astrophys. J. **97** (1943) 28.
106 Shields, G.A.: Astrophys. J. **204** (1976) 230.
107 Shields, G.A. in: Proc. Pittsburgh Conference on BL Lacertae Objects (Wolfe, A., ed.), Univ. Pittsburgh (1978) p. 257.
108 Shuder, J.M.: Astrophys. J. **244** (1981) 12.
109 Shuder, J.M.: Astrophys. J. (1982) in press.
110 Shuder, J.M., MacAlpine, G.M.: Astrophys. J. **230** (1979) 348.
111 Shuder, J.M., Osterbrock, D.E.: Astrophys. J. **250** (1981) 55.
112 Smith, M.G.: Astrophys. J. **202** (1975) 591.
113 Smith, M.G.: Vistas in Astron. **21** (1978) 321.
114 Solfer, B.T., Neugebauer, G.: Int. Astron. Union Symp. **96**, Infrared Astronomy (Wynn-Williams, C.G., Cruikshank, D.P., eds.), Reidel, Dordrecht (1980) 329.
115 Solfer, B.T., Neugebauer, G., Oke, J.B., Metthews, K.: Astrophys. J. **243** (1981) 369.
116 Sramek, R.A., Weedman, D.W.: Astrophys. J. **238** (1980) 435.
117 Stockton, A.N.: Astrophys. J. **223** (1978) 747.
118 Strittmatter, P.A., Williams, R.E.: Annu. Rev. Astron. Astrophys. **14** (1976) 307.
119 Swanenberg, B.N. et al. (12 authors): Nature **275** (1978) 298.
119a Tennant, A.F., Mushotzky, R.F., Boldt, E.A., Swank, J.H.: Astrophys. J. **251** (1981) 15.
120 Tritton, K.P., Selmes, R.A.: Mon. Not. R. Astron. Soc. **153** (1971) 453.
121 Ulrich, M.H. et al. (20 authors): Mon. Not. R. Astron. Soc. **192** (1980) 561.
122 Ulvestad, J.S., Wilson, A.S., Sramek, R.A.: Astrophys. J. **247** (1981) 419.
122a Uomoto, A.K., Wills, B.J., Wills, D.: Astron. J. **81** (1976) 905.
123 Usher, P.D.: Astrophys. J. **222** (1978) 40.
124 Veron, P., Veron, M.P., Zuiderwijk, E.J.: Astron. Astrophys. **102** (1981) 116.
125 Walsh, D., Carswell, R.F., Weymann, R.J.: Nature **279** (1979) 381.
126 Weedman, D.W.: Annu. Rev. Astron. Astrophys. **15** (1977) 69.
127 Weisheit, J.C., Shields, G.A., Tarter, C.B.: Astrophys. J. **245** (1981) 406.
128 Weyman, R.J., Carswell, R.F., Smith, M.G.: Annu. Rev. Astron. Astrophys. **19** (1981) 41.
129 Williams, R.E.: Astrophys. J. **147** (1967) 556.

130 Wills, B.J., Netzer, H., Uomoto, A.K., Wills, D.: Astrophys. J. **237** (1980) 319.
131 Wills, D., Lynds, R.: Astrophys. J. Suppl. **36** (1978) 317.
132 Wolfe, A.M. (ed.): Pittsburgh Conference on BL Lacertae Objects, Univ. Pittsburgh (1978).
133 Wolfe, A.M.: Phys. Scr. **21** (1980) 744.
134 Worrall, D.M., Boldt, E.A., Holt, S.S., Serlemitsos, P.J.: Astrophys. J. **240** (1980) 421.
135 Worrall, D.M., Mushotzky, R.F., Boldt, E.A., Holt, S.S., Serlemitsos, P.J.: Astrophys. J. **232** (1978) 683.
136 Worrall, D.M., Boldt, E.A., Holt, S.S., Mushotzky, R.F., Serlemitsos, P.J.: Astrophys. J. **243** (1980) 53.
137 Wu, C.C., Bogges, A., Gull, T.R.: Astrophys. J. **242** (1980) 14.
138 Wyckoff, S., Wehinger, P.A., Gehren, T.: Astrophys. J. **247** (1981) 750.
139 Yee, H.K.C.: Astrophys. J. **241** (1980) 894.
140 Yee, H.K.C., Oke, J.B.: Astrophys. J. **226** (1978) 753.
141 Yee, H.K.C., Oke, J.B.: Astrophys. J. **248** (1981) 472.
142 Zamorani, G., Henry, J.P., Maccacaro, T., Tananbaum, H., Soltan, A., Avni, Y., Liebert, J., Stocke, J., Strittmatter, P.A., Weyman, R.J., Smith, M.G., Condon, J.J.: Astrophys. J. **245** (1981) 357.

# 9.6 Extragalactic radio sources

General remarks

This section deals with the radio properties of distant extragalactic objects with strong radio radiation. Nearby normal galaxies (see 9.1; 9.2), clusters of galaxies (see 9.3) and the line emission of extragalactic sources (see 9.5) are not discussed here. Furthermore the cosmological results of radioastronomical work are not treated in this section (see 9.7).

It should be noted that the progress in this field is extremely rapid, especially since the introduction of the technique of Very Long Baseline Interferometry in 1967.

The survey of literature for this section was concluded in spring 1982 (only published results are given in the references).

Special units used in this section:

radio flux density measured at the earth:

1 Jy (Jansky) $= 10^{-26}\,\mathrm{W\,m^{-2}\,Hz^{-1}}$
1 mJy (Millijansky) $= 10^{-29}\,\mathrm{W\,m^{-2}\,Hz^{-1}}$

angular diameter:
1 m.a.s. or mas = milliarcsec = $0\overset{''}{.}001$.

## 9.6.1 Observational methods

The angular resolution of radio telescopes [H 1] defines their main application:

1. single-dish telescopes are mainly used for surveys, overall radio spectra, polarization and variability measurements,
2. conventional interferometers [H 1, F 1, R 2] are used for survey work (esp. at low frequencies), the determination of accurate radio positions, and structural information on the scale of $\geqslant 1''$.
3. Very Long Baseline Interferometry (VLBI) [H 1, F 1, R 2] is used for the determination of structures and their variation on the scale of milliarcseconds (m.a.s.).

## 9.6.2 Surveys

Basic aims

There are basically two types of radioastronomical surveys:

1. Sky surveys for radio sources. They differ in frequency, sky coverage, and "depth", i.e. the limiting flux density above which the survey is complete. The basic aim of these surveys is the study of the number-flux density relation (see cosmology, 9.7.6).
2. Surveys for specific classes and properties of sources which can be defined by radioastronomical, optical, X-ray, and other criteria. The basic aim of these surveys is the investigation of physical processes in extragalactic sources and their statistics.

### 9.6.2.1 Radioastronomical sky surveys

Radioastronomical source surveys and catalogues prior to 1975 are tabulated in [K 2]. Common prefixes of these catalogues, e.g. 3C, 4C, NRAO, OA, etc. are also readily found in [K 2]. Nowadays, sources are usually named according to the IAU-convention in the form HHMM ± DD.D indicating the hours and minutes of right ascension and the sign and the degrees of declination up to the first decimal point. More recent radio surveys are found in [A 1, B 6, J 2, K 3, L 2, M 7, P 3, P 4, S 3, S 4, V 3, W 4]. Table 1 lists some representative (strong) source surveys; Table 2 gives representative (deep) source surveys, and Table 3 contains the 50 strongest extragalactic radio sources at 6 cm wavelength ($|b| > 10°$).

Table 1. List of representative strong radio-source surveys.
$\nu$ = frequency
$\delta$-range = declination range
$S_{lim}$ = limiting flux density

| Name | $\nu$ [MHz] | $\delta$-range | $S_{lim}$ [Jy] | Ref. |
|---|---|---|---|---|
| 3C | 159 | −10°⋯+70° | 8 | E1 |
| 3CR | 178 | − 5°⋯+90° | 9 | B7 |
| 4C *) | 178 | − 7°⋯+80° | 2 | P5, G15 |
| Pks | 408 | −90°⋯+20° | 1 | see in E2 |
| B2 | 408 | +24°⋯+40.3° | 0.25 | C2⋯C4 |
| PksF | 2700 | −80°⋯+25° | 0.2⋯0.6 | see in B8 |
| MPIfR/NRAO | 5000 | − 5°⋯+90° | 0.25⋯0.8 | see in K3 |

*) A revision of part of the 4C catalogue has been given in [V4].

For Table 2, see p. 318.

Table 3. The 50 strongest extragalactic sources at 5 GHz at the epoch of the 5 GHz surveys taken from a recent compilation of the radio sources with 5 GHz flux densities in excess of 1 Jy [K 6] in the northern hemisphere at $|b| > 10°$.
(1) source name HHMM ± DD, see text
(2) other name
(3) Idf. = identification
GAL: galaxy
QSO: quasi-stellar object (see 9.5.1.2)
EF: empty field
LAC: BL Lac object (see 9.5.1.6)
(4) Mag. = optical magnitude
(5) redshift $z = \frac{\Delta\lambda}{\lambda}$
(6) spectral index $\alpha$ between 11 and 6 cm (see 9.6.2.3)
(7) $S$ = flux density at 6 cm.

Table 3

| Name (1) | (2) | Idf. (3) | Mag. (4) | z (5) | α (11−6) (6) | S [Jy] (7) |
|---|---|---|---|---|---|---|
| 1228+12 | 3C274.0 | GAL | $9^{m}.6$ | 0.004 | −0.79 | 72.2 |
| 0316+41 | 3C84.0 | GAL | 12.7 | 0.018 | +0.66 | 60.0 |
| 1226+02 | 3C273.0 | QSO | 13.0 | 0.158 | −0.01 | 44.59 |
| 2251+15 | 3C454.3 | QSO | 16.1 | 0.859 | +0.5 | 24.0 |
| 0433+29 | 3C123 | GAL | 21.0 | 0.218 | −0.86 | 16.21 |
| 1648+05 | 3C348.0 | GAL | 18.5 | 0.154 | −1.06 | 13.59 |
| 2134+00 | | QSO | 18.0 | 1.936 | +0.76 | 12.84 |
| 1641+39 | 3C345.0 | QSO | 16.3 | 0.594 | −0.53 | 10.81 |
| 0430+05 | 3C120 | GAL | 15.0 | 0.032 | +0.32 | 10.33 |
| 1216+06 | 3C270.0 | GAL | 11.7 | 0.007 | −0.62 | 9.31 |
| 0923+39 | 4C39.25 | QSO | 17.0 | 0.698 | +1.0 | 8.73 |
| 0538+49 | 3C147.0 | QSO | 17.8 | 0.545 | −0.76 | 8.3 |
| 1328+30 | 3C286 | QSO | 17.0 | 0.846 | −0.53 | 7.48 |
| 1828+48 | 3C380.0 | QSO | 17.0 | 0.691 | −0.53 | 7.45 |
| 0106+13 | 3C33 | GAL | 15.7 | 0.059 | −0.59 | 7.34 |
| 1409+52 | 3C295.0 | GAL | 20.5 | 0.461 | −0.95 | 6.5 |
| 0831+55 | DA251 | GAL | 17.5 | | −0.46 | 5.65 |
| 0134+32 | 3C48 | QSO | 16.2 | 0.367 | −0.85 | 5.37 |
| 2145+06 | | QSO | 17.5 | 0.99 | +0.21 | 5.0 |
| 0356+10 | 3C98 | GAL | 15.3 | 0.03 | −0.71 | 4.97 |
| 2200+42 | BLLAC | LAC | 14.0 | 0.07 | −0.15 | 4.77 |
| 1845+79 | 3C390.3 | GAL | 15.0 | 0.057 | −0.69 | 4.45 |
| 0809+48 | 3C196.0 | QSO | 17.6 | 0.871 | −0.88 | 4.42 |
| 0518+16 | 3C138 | QSO | 18.8 | 0.759 | −0.69 | 4.16 |
| 0040+51 | 3C20.0 | GAL | | | −0.72 | 4.15 |
| 1633+38 | 4C38.41 | QSO | 18.0 | 1.814 | +0.77 | 4.02 |
| 0528+13 | | QSO | 20.0 | | +0.47 | 3.98 |
| 0951+69 | 3C231.0 | GAL | 9.2 | 0.001 | −0.66 | 3.91 |
| 0220+42 | 3C66.0 | GAL | 14.5 | 0.021 | −0.86 | 3.81 |
| 2230+11 | CTA102 | QSO | 17.3 | | −0.5 | 3.78 |
| 1055+01 | | QSO | 18.0 | 0.888 | +0.34 | 3.77 |
| 2121+24 | 3C433 | GAL | 17.0 | 0.102 | −0.99 | 3.74 |
| 1458+71 | 3C309.1 | QSO | 16.8 | 0.905 | −0.66 | 3.73 |
| 0742+10 | | EF | | | −0.08 | 3.68 |
| 0315+41 | 3C83.1 | GAL | 14.0 | 0.025 | −0.54 | 3.57 |
| 0305+03 | 3C78 | GAL | 14.8 | 0.028 | −0.62 | 3.44 |
| 1559+02 | 3C327.0 | GAL | 15.0 | 0.104 | −0.74 | 3.42 |
| 1928+73 | | QSO | 15.5 | | −0.01 | 3.34 |
| 2243+39 | 3C452.0 | GAL | 16.0 | 0.082 | −0.96 | 3.32 |
| 0814+42 | | QSO | 18.5 | | +0.68 | 3.29 |
| 1328+25 | 3C287 | QSO | 17.7 | 1.055 | −0.59 | 3.26 |
| 0133+47 | | LAC | 19.0 | | +0.62 | 3.22 |
| 1222+13 | 3C272.1 | GAL | 9.3 | 0.003 | −0.66 | 3.12 |
| 0851+20 | OJ287 | LAC | 14.5 | | −0.06 | 3.06 |
| 0316+16 | CTA21 | EF | | | −0.95 | 3.02 |
| 1345+12 | | GAL | 17.0 | 0.122 | −0.41 | 3.01 |
| 1404+28 | OQ208 | GAL | 14.0 | 0.077 | +0.75 | 2.97 |
| 0106+01 | | QSO | 18.4 | 2.107 | +0.26 | 2.93 |
| 0403+76 | | EF | | | −0.58 | 2.82 |
| 2335+26 | 3C465 | GAL | 13.3 | 0.03 | −0.68 | 2.8 |

Table 2. List of some representative (deep) source surveys.
$\nu$ = frequency
Instr. = instrument used
Position = central position or region
$S_{lim}$ = limiting flux density

| Name | $\nu$ MHz | Instr. | Area ster | Position α | Position δ | $S_{lim}$ mJy | Ref. | Comments |
|---|---|---|---|---|---|---|---|---|
| 5C1 | 408 | one-mile telescope | $3.83 \cdot 10^{-3}$ | $09^h40^m$ | 50° | 25 | K4 | only some have been measured at 1407 MHz |
| 5C2 | 408 | one-mile telescope | $3.83 \cdot 10^{-3}$ | 11 00 | 49°40′ | 11.5 | P6 | |
| 5C3 | 408 | one-mile telescope | $3.83 \cdot 10^{-3}$ | 00 40 | 41 00 | 12 | P7 | |
| 5C4 | 408 | one-mile telescope | $3.83 \cdot 10^{-3}$ | $12^h57^m20^s$ | 28 15 | 2.2 | W5 | Coma cluster |
| | 1407 | one-mile telescope | $0.24 \cdot 10^{-3}$ | | | 1.3 | | |
| 5C5 | 408 | one-mile telescope | $3.83 \cdot 10^{-3}$ | 09 40 | 47 00 | 8.7 | P8 | |
| | 1407 | one-mile telescope | $0.24 \cdot 10^{-3}$ | | | 1.8 | | |
| 5C6 | 408 | one-mile telescope | $3.83 \cdot 10^{-3}$ | 02 14 | 32 | 10 | P9 | |
| | 1407 | one-mile telescope | $0.24 \cdot 10^{-3}$ | | | 1.5 | | |
| 5C7 | 408 | one-mile telescope | $3.83 \cdot 10^{-3}$ | 08 17 | 27 | 10 | | |
| | 1407 | one-mile telescope | $0.24 \cdot 10^{-3}$ | | | 1.5 | | |
| 5C9 | 408 | one-mile telescope | $3.83 \cdot 10^{-3}$ | 05 00 | 88 30 | 20 | W6 | |
| | 1407 | one-mile telescope | $0.24 \cdot 10^{-3}$ | | | 1.7 | | |
| 5C10 | 408 | one-mile telescope | $3.83 \cdot 10^{-3}$ | 10 39 | 54.5 | 9.8 | S6 | |
| | 1407 | one-mile telescope | $0.24 \cdot 10^{-3}$ | | | 1.7 | | |
| MC | 408 | Molonglo Cross | 0.025 | var. | −20 | 90 | R3 | |
| deep | | Molonglo Cross | 0.005 | var. | −62 | | R10 | |
| | 610 | Westerbork | 0.03 | | | 10 | H5, V5, K5, H2 | different fields (see ref.) |
| | 1400 | half-mile telescope (Cambridge) | 0.007 | | | 30 | H3 | 7 fields at the north celestial pole |
| | 1400 | Westerbork | 0.033 | | | 10 | V6 W7 | different fields*) |
| | 4755 | NRAO 91 m | $9.56 \cdot 10^{-3}$ | $7^h05 \cdots 18^h$ | 35° | 15 | L2 | |
| | 4850 | MPI 100 m | $4.6 \cdot 10^{-3}$ | 5C9 region | | 14 | P10 | |
| | 4850 | MPI 100 m | $7.14 \cdot 10^{-3}$ | var. and $14^h15^m$ | ~30°20′ 14°58′ | 10···18 | P3 | |
| | 4850 | MPI 100 m | $6.6 \cdot 10^{-3}$ | 5C6 region | | 20 | M7 | |
| | 4995 | Westerbork | $0.71 \cdot 10^{-3}$ | 89 different fields | | down to 4.5 for some fields | W4 | |

*) around strong galaxies and quasars

### 9.6.2.2 Surveys for specific classes and properties of sources

Optically selected quasars (see 9.5.1.2)

Surveys for the radio emission of optically selected quasars [A 2, C 5···C 8, S 7···S 11] have shown that $\lesssim 10\%$ of those objects are detectable at wavelengths $\gtrsim 2.8$ cm. At 1 mm wavelength some "radio quiet" quasars have been detected [S 10]. Statistical investigations are not compatible with relativistic beam models as the explanation of "radio quiet" quasars, e.g. [S 11].

Surveys for variability

Flux-density variations both at high and low frequencies have been monitored, e.g. [C 9, E 3, E 4, F 2···F 5, G 3, H 4, P 12, S 12, S 13, T 1, W 9]. As an example, time variations of 3C 120 are given in Fig. 1.

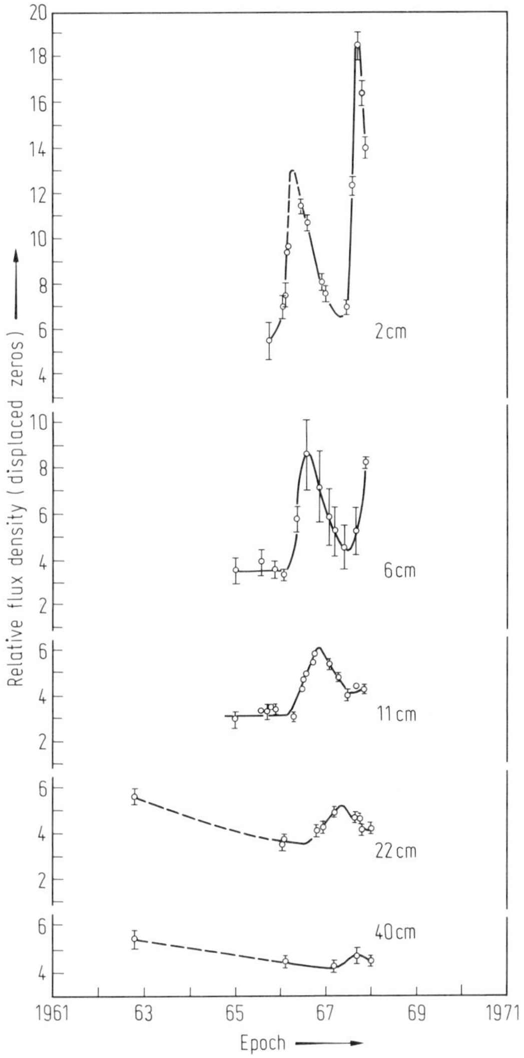


Fig. 1. Radio flux density variations of the galaxy 3C 120 (=0430+05) from 1962 to 1968 [K 14].

Surveys for polarization

Recent work on the polarization characteristics of extragalactic radio sources is found in [A 3, G 4, L 4, P 13, S 14, S 15]. In general the overall linear polarization at radio wavelengths is $\lesssim 5\%$. Only in some cases (e.g. BL Lac and 3C 286) does it exceed 10%. Variations in linear polarization can occur both in percentage and position angle. The circular polarization of radio sources is $\leqq 1\%$ [R 4].

Surveys of the structure of radio sources

For many extragalactic radio sources structural information on the arc second to arc minute scale is available. Table 4 lists investigations of the structure of sources found in well-known surveys. Surveys for the m.a.s. structure of samples of sources have been carried out [G 6]. Work on the m.a.s. structure of the sources from the 1-Jy catalogue [K 6] is in progress. Surveys for 'hot spots' in extended lobes of radio sources are found in [K 9].

Table 4. Structure investigations of sources in wellknown catalogues.
Cat. = catalogue (see Table 1)
$N$ = number of investigated sources
$\nu$ = frequency
$\theta$ = half-power beamwidth

| Cat. | $N$ | $\nu$ [GHz] | $\theta$ | Instrument | Ref. |
|---|---|---|---|---|---|
| 3CR | 82 | 0.408 | 80″ | 1-mile telescope | M9 |
| | | 1.407 | 23″ | | |
| | 60 | 0.408 | 80″ | 1-mile telescope | M10 |
| | | 1.407 | 23″ | | |
| | 78 | 0.408 | 80″ | 1-mile telescope | E5 |
| | | 1.407 | 23″ | | |
| | 166 | 0.178 | ≈ 1′ | 5-km interferometer | B7 |
| | 57 | 0.96 | ≈ 2″.7 | Jodrell Bank interferometer | B9 |
| | 40 | 2.7 | 4″···7″ | 5-km interferometer | L5 |
| | | 15.4 | 1″···2″ | | |
| 4C | 81 | 1.415 | 24″× 24″ cosec $\delta$ | Westerbork | T2 |
| | | (4.995) | 6″× 6″ cosec $\delta$ | | |
| | 84 | 2.695 | 9″ | NRAO interferometer | R5 |
| | | 8.085 | 3″ | | |
| | 57 | 2.7 | 9″ | NRAO interferometer | P14 |
| | | 8.1 | 3″ | | |
| B2 | 66 | 5 | 6″× 6″ cosec $\delta$ | Westerbork | G5 |
| | 31 | 1.415 | 24″× 24″ cosec $\delta$ | Westerbork | F6 |
| S4 | 180 | 5 | 6″× 6″ cosec $\delta$ | Westerbork | G6, K7···K9 |

## Optical identifications

Optical identifications of radio sources are usually carried out on the basis of positional coincidence. This is especially easy for compact sources, i.e. sources with flat or inverted radio spectra. The necessary positional accuracy of about 1″ is readily reached by conventional interferometers.

Figure 2 shows the relative identification content per flux density interval for sources of the MPIfR/NRAO 6-cm surveys and the 1-Jy catalogue.

The classes are:
GAL: galaxy
EF: empty field
QSO: quasi-stellar object (some BL Lac objects are included).

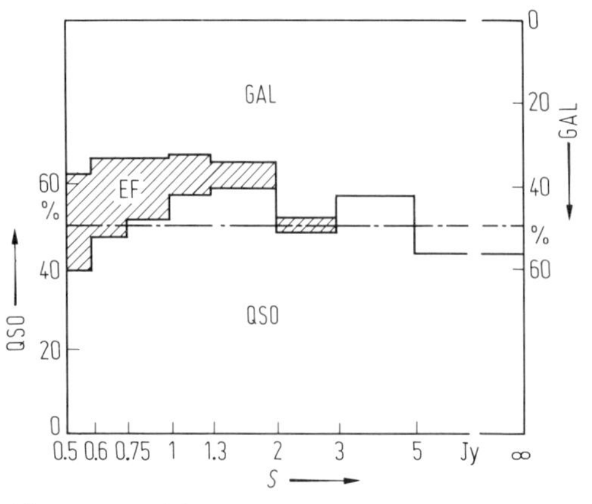


Fig. 2. Relative content of the identification classes per flux density interval for the MPIfR/NRAO 6-cm survey [W 11]. GAL: galaxy; EF: empty field; QSO: quasistellar object.

About 60% of the quasars are spectroscopically verified. A list of optical identifications of radio sources is given in [V 7]. (This catalogue is updated yearly and available on request from the authors.)

### 9.6.2.3 Radio spectra

Multifrequency observations of radio sources establish their radio spectra, see e.g. [S 16]. The flux-density scale commonly adopted is given by [B 10]. Two types of spectral information are generally used:

1. Two-point spectral indices; most useful for statistical investigation.
   For definition:
   The distribution of the radio power with frequency is often approximated by the power law
   $P \propto \nu^{\alpha}$
   with $\alpha$ = spectral index.
   For more details, see also 9.6.3.1.2.
2. Many-point spectra of sources over a wide frequency range (preferably single epoch spectra); useful for detailed studies of individual sources and statistical investigations.

Spectral index distributions

Figure 3 shows the distribution of the two-point spectral index $\alpha$ (11 cm-6 cm) for the extragalactic sources of the 3CR survey and the MPIfR/NRAO 6-cm surveys.

Two types of sources are readily distinguished:

1. Sources with 'steep spectra' ($\alpha \approx -0.8$); these appear in both, the high- and the low-frequency surveys. Identified with extended objects (mainly galaxies) with diameter $\theta > 1''$.
2. Sources with 'flat spectra' ($\alpha \approx 0$); found in large proportions only in the high-frequency surveys. These sources contain compact components ($\theta \ll 1''$) and are mainly identified with QSO's and BL Lac's.

Table 5 gives the median spectral indices for different classes of sources from the MPIfR/NRAO 6-cm surveys with declinations $> 35°$ and the percentages of sources with flat spectra ($\alpha > -0.5$). Different spectral behaviour for morphologically different types and a flux-density dependent behaviour are established.

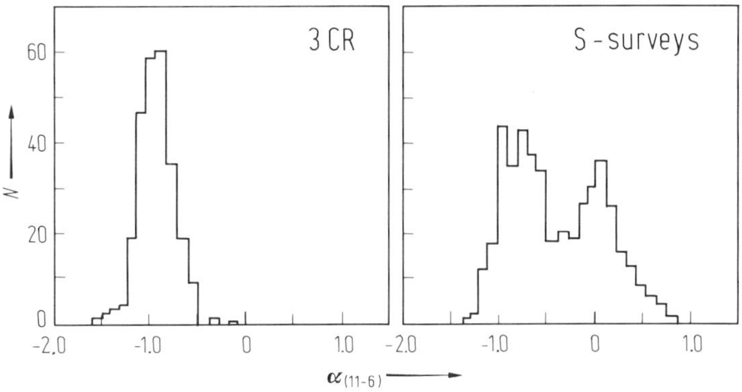


Fig. 3. Distribution of the two-point spectral index (11 cm, 6 cm) for extragalactic sources [W 11]. Left: 3CR surveys; right MPIfR/NRAO 6-cm surveys.

Table 5. Spectral indices for different classes of sources from the MPIfR/NRAO 6-cm survey; declination $> 35°$. For symbols of selection, see Table 3.
$S$ = flux density
$\bar{\alpha}(11-6)$ = median spectral index
$\alpha > -0.5$ = percentage of sources with flat spectra ($\alpha > -0.5$)

| Selection | $S$ [Jy] | $N$ | $\bar{\alpha}(11-6)$ | $\alpha(11-6) > -0.5$ [%] |
|---|---|---|---|---|
| All | <0.6 | 713 | $-0.74 \pm 0.04$ | 33 |
| All | ≧0.6 | 257 | $-0.38 \pm 0.08$ | 56 |
| QSO | <0.6 | 71 | $-0.27 \pm 0.12$ | 65 |
| QSO | ≧0.6 | 127 | $-0.06 \pm 0.04$ | 77 |
| GAL | <0.6 | 68 | $-0.83 \pm 0.07$ | 27 |
| GAL | ≧0.6 | 90 | $-0.77 \pm 0.05$ | 25 |
| EF | <0.6 | 56 | $-0.90 \pm 0.08$ | 19 |
| EF | ≧0.6 | 37 | $-0.48 \pm 0.10$ | 51 |

Individual radio spectra

Multifrequency radio observations in the range from ≈10 MHz to ≈90 GHz are available for many sources, especially for the stronger sources of 3CR, MPIfR/NRAO and Pks surveys.

The commonly used classification scheme (e.g. [K 10]) is based on a phenomenological description of the radio spectra and a division into four classes (see Table 6). Characteristic examples of such radio spectra are given in Fig. 4. For interpretation of radio spectra of compact sources, see 9.6.6.3.4.

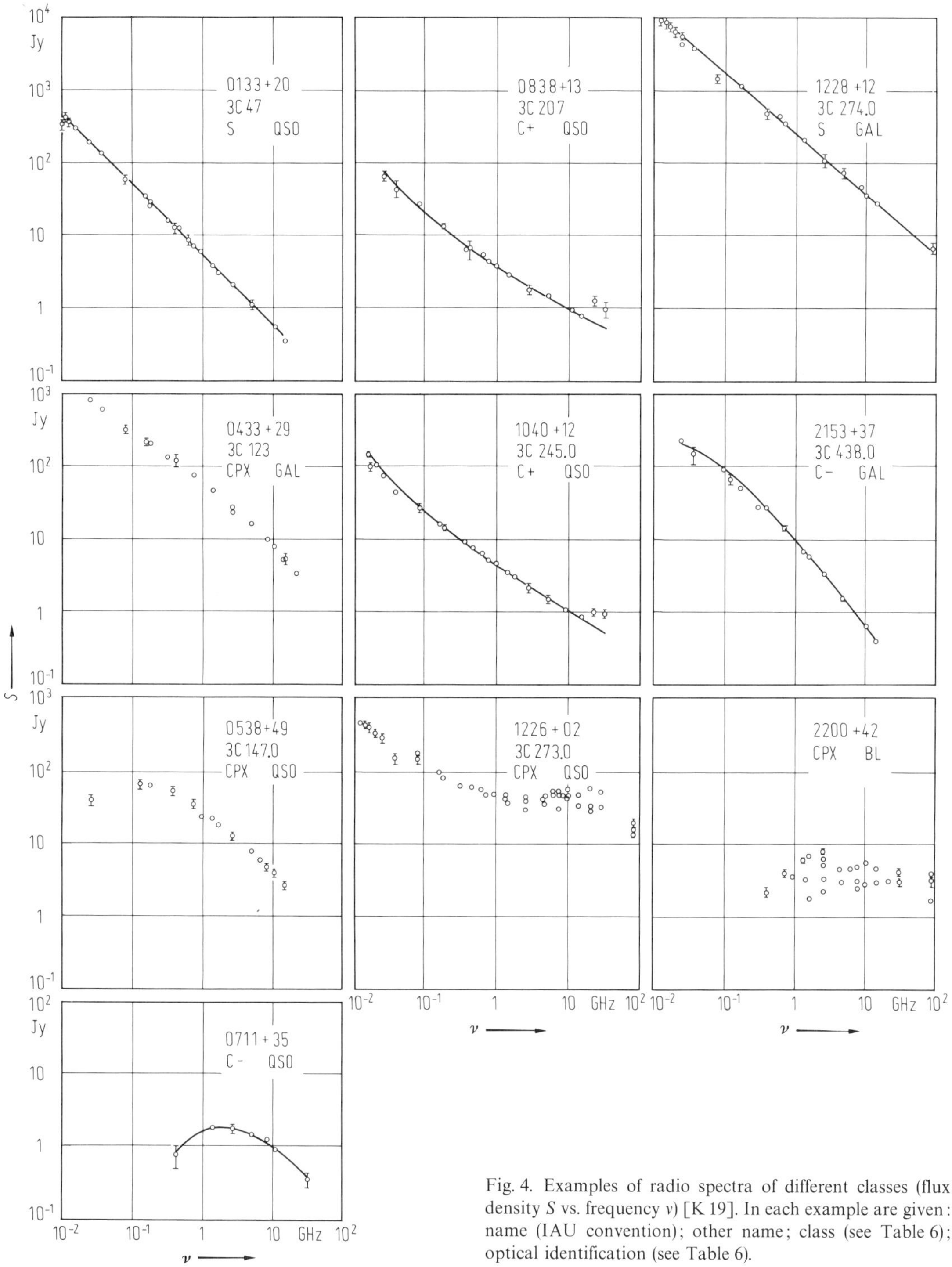


Fig. 4. Examples of radio spectra of different classes (flux density $S$ vs. frequency $\nu$) [K 19]. In each example are given: name (IAU convention); other name; class (see Table 6); optical identification (see Table 6).

Table 6. Scheme of radio spectral classes [K10].
Opt. idf. = optical identification (see Table 3)
Var. = radio variability

| Spectr. class | Shape | Slope | Opt. idf. | Size of source | Explanation | Var. | Examples |
|---|---|---|---|---|---|---|---|
| S | straight | steep $\alpha \approx -0.8$ | GAL | extended $\gg 1''$ | optically thin synchrotron sources | no | Virgo A, 3C47, (3C274) |
| $C^-$ | convex | a) steepening | GAL | extended | 'Aging' | no | 3C123, 3C438 |
| | | b) low-frequency cut offs | QSO | compact | synchrotron self-absorption | some | 3C147, 0711+35 |
| $C^+$ | concave | steep and straight at lower frequencies, flattening or rising at high frequencies | QSO, active GAL | extended + compact components | extended lobes + compact self-absorbed nucleus | yes | 3C245, 3C207 |
| Cpx | complex | varying | QSO, LAC | compact | several compact *) self-absorbed components, sometimes extended lobes present | yes | 3C273, BL Lac |

*) The existence of compact components in many sources has been shown by VLBI observations. Different explanations for Cpx spectra have been given by [C12, D4, M11].

## 9.6.3 Basic relations

The physics of nonthermal radio-continuum radiation in relation to compact and extended extragalactic sources is presented in great detail in [B 12, B 13, D 1, J 4, K 10, K 11, L 6···L 8, M 1, M 12, M 13, O 4, P 1, P 21···P 23, R 9, S 19, S 22].

### 9.6.3.0 List of symbols

| | |
|---|---|
| $B$ | magnetic field [G] or [μG] |
| | $B_\perp$, $B_\parallel$ components perpendicular or parallel to the path of the electron |
| $B_{eq}$ | equivalent magnetic field with energy density $u_{ph,o}$ |
| $c$ | velocity of light; except in Eqs. (34) and (35): constant of proportionality |
| $c_1 \cdots c_5$ | constants, see Table 7 |
| $e$ | charge of electron |
| $E$ | kinetic energy of electron [eV] or [erg] |
| $F_X$ | flux density in the X-ray range [erg s$^{-1}$ cm$^{-2}$ Hz$^{-1}$] |
| $F_R$ | flux density in the radio range [Jy$=10^{-26}$ W m$^{-2}$ Hz$^{-1}$] or [mJy] |
| $H_0$ | Hubble constant (see 9.7.5) |
| $I_\nu$ | intensity [erg cm$^{-2}$ s$^{-1}$ Hz$^{-1}$ sterad$^{-1}$] |
| $K$ | constant of proportionality, Eq. (6) |
| $k$ | constant, defined in 9.6.3.4 or Boltzmann's constant |
| $L$ | luminosity [erg s$^{-1}$] |
| $L_C$ | Compton luminosity |
| $L_s$ | synchrotron luminosity |
| $l$ | thickness, geometrical path length [pc] |
| $m_e$ | rest mass of electron |
| $m_p$ | rest mass of proton |

continued

List of symbols, continued

| | |
|---|---|
| $N(E)$ | number density of electron with energy $E$ [$cm^{-3} erg^{-1}$] |
| $n(E, \theta)$ | number density of electrons with energy $E$ and pitch angle $\theta$ [$cm^{-3} erg^{-1} sterad^{-1}$] |
| $n_e$ | number density of electrons [$cm^{-3}$] |
| $n_i$ | number density of ions [$cm^{-3}$] |
| $P_\nu(E)$ | spectral distribution of radiation of single electron |
| $p$ | electron spectral index (see Table 9); functions $a(p)$, $b(p)$, $c(p)$, and $g(p)$: see Table 11 |
| $Q$ | injection rate of relativistic electrons |
| $q$ | constant of proportionality (see 9.6.3.3) |
| $q_0$ | deceleration constant (see 9.7.6) |
| $R$ | radius of the source |
| $S_\nu$ | flux density [Jy] or [mJy] |
| $S_m$ | maximum flux density [Jy] or [mJy] |
| $T$ | temperature [K] |
| $T_B$ | brightness temperature [K] |
| $t$ | time |
| $u_e$ | electron energy density [$erg\,cm^{-3}$] |
| $u_m$ | magnetic energy density [$erg\,cm^{-3}$] |
| $u_{ph}$ | energy density of photons [$erg\,cm^{-3}$] |
| $u_{ph,0}$ | energy density of the 2.7 K microwave background radiation [$erg\,cm^{-3}$] |
| $U_B$ | total magnetic energy |
| $U_e$ | total energy of electrons |
| $U_p$ | total energy of protons |
| $U_{tot}$ | total energy (particles and magnetic field) |
| $V$ | source volume |
| $Z$ | charge number of atoms (ions) ($Z \cdot e$ = charge of ions) |
| $z = \frac{\Delta\lambda}{\lambda}$ | redshift |
| $\alpha$ | (radio) spectral index (see Table 9) |
| $\gamma$ | $\frac{E}{m_e c^2}$ |
| $\varepsilon_\nu$ | emission coefficient |
| $\theta$ | pitch angle |
| $\kappa_\nu$ | absorption coefficient [$cm^{-1}$] |
| $\lambda$ | wavelength [mm, cm...] |
| $\nu$ | frequency [MHz] or [GHz] |
| $\nu_c$ | characteristic or critical frequency of the radiation of an electron |
| $\nu_m$ | turnover or cutoff frequency |
| $\nu_{th}$ | cutoff frequency of thermal absorption |
| $\nu_{TS}$ | Tsytovitch frequency |
| $\Pi(\alpha)$ | linear polarization fraction |
| $\tau$ | containment time scale |
| $\tau_\nu$ | optical depth at frequency $\nu$ |
| $\tau_l$ | optical depth of the geometrical depth $l$ |
| $\phi$ | angular diameter (radius) of the source [mas = milli arc sec] or [arc sec], polarization angle in Eq. (21) |
| $\Delta\Omega$ | solid angle of the source |

## 9.6.3.1 Synchrotron radiation

### 9.6.3.1.1 Emission by a single electron

The characteristic or critical frequency $\nu_c$ of the radiation of an electron of energy $E$, mass $m_e$, and charge $e$ which moves in a magnetic field $B$ is

$$\nu_c = c_1 B_\perp E^2 . \qquad (1)$$

Here $B_\perp = B \sin\theta$ is the component of the magnetic field perpendicular to the path of the electron, $\theta$ is called the pitch angle. For a homogeneous field and a random pitch-angle distribution $\langle B_\perp^2 \rangle = \frac{2}{3} B^2$. Numerical values for $c_1$ and other constants appearing in this section are compiled in Table 7. $B_\perp$ is measured in [Gauss] [1]). Convenient units for radio sources are MHz, μG, and GeV. With these units for $\nu_e$, $B_\perp$, and $E$:

$$\nu_c = 16.08\, B E^2 . \qquad (2)$$

The spectral distribution of this radiation is [G 7, J 4, M 13, P 21]

$$P_\nu(E) = c_2 B_\perp F\left(\frac{\nu}{\nu_c}\right) \qquad (3)$$

where

$$F\left(\frac{\nu}{\nu_c}\right) \approx \begin{cases} 2.13(\nu/\nu_c)^{1/3} & \text{for} \quad \nu < 0.01\nu_c \\ (\frac{\pi}{2}\nu/\nu_c)^{1/2} e^{-\nu/\nu_c} & \text{for} \quad \nu > 20\nu_c . \end{cases}$$

The function $F(X)$ is tabulated in [P 21, W 10]. The total radiated power is

$$-\dot{E} = \int_0^\infty P_\nu(E)\,d\nu = c_3 B_\perp^2 E^2 . \qquad (4)$$

The synchrotron radiation by protons (mass $m_p$) of the same energy has therefore less power by a factor $(m_p/m_e)^4$. The radiation is elliptically polarized. The ellipse is oriented perpendicular to the projection of $B$ onto the plane normal to the line of sight. The polarization is linear only in the direction of the electron velocity vector and becomes circular at large angles to it. The electric vector as seen by the observer rotates clockwise for the wave vector lying inside the pitch angle and counterclockwise otherwise.

As a result the radiation in the case of an isotropic distribution of the magnetic pitch angles is nearly linearly polarized perpendicular to the projected field.

Table 7. Some constants.

$e$ = electron charge = $4.802 \cdot 10^{-18}\ \text{cm}^{3/2}\ \text{g}^{1/2}\ \text{s}^{-1}$
$m_e$ = electron mass = $9.107 \cdot 10^{-28}$ g
$c$ = velocity of light = $2.998 \cdot 10^{10}\ \text{cm s}^{-1}$.

| Constant | Definition | Value [CGS] |
|---|---|---|
| $c_1$ | $\frac{3e}{4\pi m_e^3 c^5}$ | $6.27 \cdot 10^{18}$ |
| $c_2$ | $\frac{\sqrt{3}e^3}{m_e c^2}$ | $2.34 \cdot 10^{-23}$ |
| $c_3$ | $\frac{2e^4}{3m_e^4 c^7}$ | $2.37 \cdot 10^{-3}$ |
| $c_4$ | $\frac{\pi}{2c^2} c_1^{-1/2}$ | $6.97 \cdot 10^{-31}$ |
| $c_5$ | $\frac{c_1^{1/2}}{c_3}$ | $1.06 \cdot 10^{12}$ |

[1]) If not otherwise indicated CGS units are used throughout this section.

### 9.6.3.1.2 Radiation from an ensemble of electrons

Energy distribution and density

If $n(E,\theta)$ is the number density of electrons with energies around $E$ and pitch angles around $\theta$ then for an isotropic distribution of pitch angles the energy distribution is

$$N(E)=\int n(E,\theta)\,\mathrm{d}\Omega=4\pi n(E) \tag{5}$$

which is generally assumed to be a power law

$$N(E)=KE^{-p},\ E_1 \leqq E \leqq E_2\,. \tag{6}$$

The electron density and electron energy densities are

$$n_e=\int_{E_1}^{E_2} N(E)\,\mathrm{d}E \quad \text{and} \quad u_e=\int_{E_1}^{E_2} E\cdot N(E)\,\mathrm{d}E,\ \text{respectively}. \tag{7}$$

Emission coefficient $\varepsilon$

$$\varepsilon_\nu(\theta)=\int_{E_1}^{E_2} P_\nu(E)\,n(E,\theta)\,\mathrm{d}E \tag{8}$$

or

$$\varepsilon_\nu=\int_{E_1}^{E_2} P_\nu(E)\,N(E)\,\mathrm{d}E \ \text{in the case of isotropy}. \tag{9}$$

Numerically,

$$\varepsilon_\nu=a(p)\,(2c_1)^{\frac{p-1}{2}}\,KB_\perp^{\frac{p+1}{2}}\,\nu^{-\frac{p-1}{2}}, \tag{10}$$

$a(p)$ is a slowly varying function of $p$ and is tabulated in Table 11, p. 333; $\alpha=-(p-1)/2$ is the slope of the power-law spectrum and is called the spectral index.

Polarization

For an isotropic distribution of pitch angles of the electrons with respect to a large-scale magnetic field the radiation of the whole ensemble of electrons is approximately linearly polarized perpendicular to the projected field. Then a power-law distribution produces a frequency-independent linear-polarization fraction

$$\Pi(\alpha)=\frac{p+1}{p+7/3}=\frac{-\alpha+1}{-\alpha+5/3} \tag{11}$$

which is $\approx 70\%$ for $p=2$.

## 9.6.3.2 Absorption mechanisms and plasma effects

a) Radiative transfer

Along the line of sight the specific intensity $I_\nu$ changes according to

$$\mathrm{d}I_\nu=(\varepsilon_\nu/\kappa_\nu-I_\nu)\,\mathrm{d}\tau_\nu \tag{12}$$

where $\kappa_\nu$ is the absorption coefficient and $\tau_\nu=\int_0^l \kappa_\nu \mathrm{d}l'$ is the optical depth for a slab of thickness $l$. For details, see [M 15, T 4]. The flux density is defined as $S_\nu=I_\nu\Delta\Omega$ where $\Delta\Omega=\frac{\pi}{4}\phi^2$ is the solid angle of the source and $\phi$ the angular diameter of the source. Dimensions:

$$I_\nu \text{ in } [\mathrm{erg\,cm^{-2}\,s^{-1}\,Hz^{-1}\,sterad^{-1}}]$$

$$S_\nu \text{ in } [\mathrm{Jy}=10^{-26}\,\mathrm{W\,m^{-2}\,Hz^{-1}}=10^{-23}\,\mathrm{erg\,cm^{-2}\,Hz^{-1}\,s^{-1}}]$$

$$\kappa_\nu \text{ in } [\mathrm{cm^{-1}}].$$

The solution of Eq. (12) is

$$S_\nu = S_{\nu 0} e^{-\tau_\nu} + \int_0^{\tau_\nu} \frac{\varepsilon_\nu}{\kappa_\nu} e^{-t_\nu} dt_\nu \tag{13}$$

$$= S_{\nu 0} e^{-\tau_\nu} + \Delta\Omega \frac{\varepsilon_\nu}{\kappa_\nu}(1 - e^{-\tau_\nu}) \quad \text{if} \quad \frac{\varepsilon_\nu}{\kappa_\nu} \approx \text{const}.$$

For an optically thin source:

$$S_\nu \approx \varepsilon_\nu l \Delta\Omega \propto \nu^\alpha, \tag{14}$$

for an optically thick source:

$$S_\nu \approx \frac{\varepsilon_\nu}{\kappa_\nu} \Delta\Omega. \tag{15}$$

b) Thermal absorption

The free-free absorption coefficient due to thermal electrons in the Coulomb field of ions in the synchrotron source is given by [G 8, O 5, S 20]

$$\kappa_\nu = \frac{16\pi Z^2 e^6 n_e n_i}{3(2\pi m_e k)^{3/2} c} T^{-3/2} \nu^{-2} \cdot \ln\left[0.2362 \left(\frac{2kT}{m_e}\right)^{3/2} \cdot \frac{m_e}{\pi Z e^2 \nu}\right]$$
$$\approx 0.9786 n_e^2 T^{-3/2} \nu^{-2.1} \quad \text{for} \quad T \approx 10^4\,\text{K}, Z = 1, \tag{16}$$

where $n_i$ is the number density of the ions with mean charge $Z \cdot e$ and $n_e \approx n_i$ is the number density of the thermal electrons. The optical thickness is then

$$\tau_\nu = \int \kappa_\nu dl = T^{-3/2} \nu^{-2.1} \int n_e^2 dl, \tag{17}$$

where $E.M. = \int n_e^2 dl$ is the emission measure in CGS units. Frequently $E.M.$ is given in the unit [pc cm$^{-6}$] with 1 pc $\approx 3 \cdot 10^{18}$ cm. A low-frequency cutoff is produced if $\tau_\nu \approx \kappa_\nu l \approx 1$; i.e. at a frequency

$$\nu_{th} \approx (l n_e^2 T^{-3/2})^{1/2.1} \tag{18}$$

or

$$\nu_{th} \approx \tfrac{1}{2} n_e l^{1/2} \quad \text{with} \quad \nu_{th} \text{ in [MHz]}, n_e \text{ in [cm}^{-3}], l \text{ in [pc]}.$$

The optically thick part of the spectrum has the flux density

$$S_\nu = \frac{\varepsilon_\nu}{\kappa_\nu} \Delta\Omega \propto \nu^{\alpha + 2.1} = \nu^{-p/2 + 2.6}. \tag{19}$$

c) Razin-Tsytovitch effect

For a refractive index of the plasma less than 1, relativistic electrons emit Cerenkov radiation. The intensity of the synchrotron radiation will then be drastically reduced for $\nu < \nu_{Ts}$ with the Tsytovitch frequency

$$\nu_{Ts} = \frac{4c \cdot n_e \cdot e}{3B_\perp} \approx 20 \frac{n_e}{B_\perp} \tag{20}$$

where $n_e$ is the number density of the thermal electrons. Only for small sources ($l \approx$ some pc) with weak magnetic fields ($B \approx$ some μG) having $B_\perp < 40 l^{-1/2}$, when $\nu_{Ts} > \nu_{th}$, is the Tsytovitch effect important, cf. [R 6, T 3, T 4].

d) Faraday rotation

When a wave propagates through a plasma having a thermal electron density $n_e$ and a magnetic field $B_{||}$ parallel to the wave vector the plane of polarization rotates by

$$\phi = R.M.\lambda^2 \quad \text{with} \quad \phi \text{ in [rad]}, \lambda \text{ in [cm]}. \tag{21}$$

$R.M.$ is the rotation measure in [rad/cm$^2$],

$$R.M. \approx 81 \int n_e B_{||} dl \tag{22}$$

with $l$ the geometrical path length in [pc]. The rotation is clockwise if $B_{||}$ is positive. Faraday rotation generally causes depolarization of the radiation with respect to the intrinsic polarization degree. For details, see [L 6, P 21, P 22].

e) Synchrotron self-absorption

Here synchrotron photons are reabsorbed by the radiating electrons within the source [B 13, G 10, P 21, P 19]. The absorption coefficient is

$$\kappa_\nu(\theta) = -\frac{c^2}{2\nu^2}\int_0^\infty E^2\left(\frac{d}{dE}\left[\frac{n(E,\theta)}{E^2}\right]\right)P_\nu(E)\cdot dE. \tag{23}$$

Numerically, the self-absorption coefficient for an isotropic power-law distribution of electrons is [P 21]

$$\kappa_\nu = b(p)(2c_1)^{\frac{p+4}{2}} K B_\perp^{\frac{p+2}{2}} \nu^{-\frac{p+4}{2}}. \tag{24}$$

The slowly varying function $b(p)$ is contained in Table 11, p. 333. In the optically thick case

$$S_\nu = c(p) B_\perp^{-1/2} \nu^{5/2} \Delta\Omega \tag{25}$$

where $c(p)$ is also tabulated in Table 11, p. 333.

Approximately,

$$S_\nu = c_4 B_\perp^{-1/2} \nu^{5/2} \phi^2 \tag{26}$$

since the emitted spectrum is of Rayleigh-Jeans type

$$S_\nu = \frac{2k\cdot T_b\cdot \nu^2}{c^2}\Delta\Omega \tag{27}$$

with brightness temperature $T_b$, and the kinetic energy of the electrons radiating at frequency $\nu$ is approximately

$$E = \left(\frac{\nu}{c_1 B_\perp}\right)^{1/2} \approx kT_b. \tag{28}$$

For $c_1$ and $c_4$, see Table 7.

The spectrum turns over at low frequencies due to selfabsorption. In the absence of thermal absorption the spectrum at low frequencies cannot get steeper than $\alpha = 2.5$. The turnover or cutoff frequency $\nu_m$ and maximum flux density $S_m$ are reached when $\tau_\nu \approx \kappa_\nu \cdot l \approx 1$ implying

$$\nu_m \approx 2c_1 (l b(p))^{\frac{2}{p+4}} K^{\frac{2}{p+4}} B_\perp^{\frac{p+2}{p+4}} \tag{29}$$

where $l$ is the geometrical thickness of the source in [cm]. $S_m$ follows from Eq. (26) for $\nu = \nu_m$.

In the optically thin part the spectrum is approximately $S_\nu = S_m(\nu/\nu_m)^\alpha$.

The angular diameter $\phi$ of a self-absorbed source is connected with $\nu_m$, $S_m$ through

$$\phi = \left(\frac{4}{\pi}\cdot\frac{1}{c(p)}\right)^{1/2} \cdot S_m^{1/2} \cdot B_\perp^{1/4} \cdot \nu_m^{-5/4} \tag{30}$$

or

$$\phi \approx B^{1/4} S_m^{1/2} \nu_m^{-5/4} (1+z)^{1/4}, \tag{31}$$

where $\phi$ in [milliarcsec], $S_m$ in [Jy], $\nu_m$ in [GHz] are measured quantities and $z = \Delta\lambda/\lambda$ is the redshift; $B$ is given in $[10^{-4}$ Gauss].

A source which is optically thick at 1 GHz must therefore exhibit a milliarcsec (mas) structure. The maximum observed in the spectra of many very compact sources is most likely due to synchrotron self-absorption. Figure 6 in [M 13] displays the relation between $\nu_m$, $S_m$, and $\phi$ for various spectral indices $\alpha$.

## 9.6.3.3 Energy losses and evolution of source spectra

### 9.6.3.3.1 Energy loss rates and time scales

Energy loss mechanisms for relativistic electrons are:

(i) ionization losses with nearly energy-independent rates,

(ii) bremsstrahlung, adiabatic expansion of the source, and leakage of particles from the source, all processes with rates nearly $\propto E$,

(iii) synchrotron and inverse Compton losses with rates $\propto E^2$.

The loss rates $-\dot{E}$ and time scales $E/|\dot{E}|$ for the individual processes are listed in Table 8. Generally $-\dot{E}=a+bE+cE^2$ and the change of the electron distribution is governed by

$$\frac{\partial N}{\partial t}+\frac{\partial}{\partial E}(\dot{E}N)=Q-\frac{1}{\tau}N, \qquad (32)$$

where $Q=q\cdot E^{-p}$ is the injection rate of relativistic electrons and $\tau$ the containment time scale [G 9, K 10, K 15, M 14, P 21, S 16].

Table 8. Energy loss rates and time scales [M13, P1, P21, T4].
For explanation of symbols, see 9.6.3.0.
$B$ in [μG], $\nu$ in [GHz].
$\gamma = E/m_e c^2$

| Mechanism | Loss rate [erg/s] | Approximate time scale [a] | Remarks | Ref. |
|---|---|---|---|---|
| Ionization | $8.43\cdot 10^{-20} n_e (\log\gamma+2.72)$ | $\frac{3.08\cdot 10^5\gamma}{n_e(\log\gamma+2.72)}$ | atomic hydrogen | L7, S22 |
| | $2.81\cdot 10^{-20} n_e \left(\log\frac{\gamma}{n_e}+31.87\right)$ | $\frac{2.9\cdot 10^4\gamma}{n_e}$ | fully ionized plasma | L8, G11, G12 |
| Bremsstrahlung | $3.16\cdot 10^{-16} n_p E(\log\gamma+0.16)$ | $\frac{10^8}{n_p \log\gamma}$ | fully ionized plasma | B13, G13 |
| Synchrotron radiation | $2.37\cdot 10^{-3} B_\perp^2 E^2$<br>$1.58\cdot 10^{-3} B^2 E^2$ | $\frac{2.45\cdot 10^{13}}{B\gamma}$ | uniform magnetic field $B$ and isotropic pitch angle distribution | L8, K10 |
| | $4.0\cdot 10^{-2} u_m E^2$ | $\frac{1.44\cdot 10^9}{B^{3/2}\nu^{1/2}(1+z)^{1/2}}$ | $z$ = source redshift<br>$\nu$ = observed frequency | |
| | $u_m = B^2/8\pi$ magnetic energy density in source | $\frac{1}{u_m\gamma}$ | | |
| Inverse Compton collision | $4.0\cdot 10^{-2} u_{ph} E^2$<br>$1.58\cdot 10^{-3} B_{eq}^2 E^2$ | $\frac{1}{u_{ph}\gamma}$ | $\gamma h\nu_{ph} \ll m_e c^2$<br>Thomson cross section | L8, G14, F7 |
| With microwave background photons | $B_{eq}=(8\pi u_{ph})^{1/2}$<br>$=3.2(1+z)^2$ | $\frac{(1+z)^4}{u_{ph,0}\gamma}$ | $u_{ph,0}=4\cdot 10^{-13}$ erg cm$^{-3}$ | R7, L3 |
| | | $\frac{2.39\cdot 10^{12}}{(1+z)^4\gamma}$ | $\nu$ = observed frequency | R7, L3 |
| | | $\frac{1.42\cdot 10^8 B^{1/2}}{\nu^{1/2}(1+z)^{9/2}}$ | | |
| + synchrotron | $4.0\cdot 10^{-2}(u_m+u_{ph})E^2$<br>$1.58\cdot 10^{-3}(B^2+B_{eq}^2)E^2$ | $\frac{1.44\cdot 10^9 B^{1/2}}{(B^2+10.24(1+z)^4)\nu^{1/2}(1+z)^{1/2}}$ | | |
| Adiabatic expansion losses | $E\dot{R}/R$ | $R/\dot{R}$ | | S23 |
| Leakage | $E/\tau_l(E)$ | $\tau_l(E)$ | | L7 |

### 9.6.3.3.2 Equilibrium spectra

In the steady state $\left(\frac{\partial N}{\partial t}=0\right)$

$$N(E)=\tau q E^{-p} \quad \text{if} \quad E \ll \tau \cdot |\dot{E}| \text{ (losses unimportant)} \tag{33}$$

$$N(E)=\frac{q}{p-1}\frac{E^{-p+1}}{a+bE+cE^2} \quad \text{if} \quad E \gg \tau \cdot |\dot{E}| \text{ (losses important)}. \tag{34}$$

A bend in the spectrum occurs around $E_b=\tau|\dot{E}|$. Ionization losses flatten, synchrotron and inverse Compton losses steepen the radio spectrum while bremsstrahlung and expansion losses leave the spectrum unchanged. The changes in $p$ and $\alpha$ are listed in Table 9 and are shown in Fig. 5.

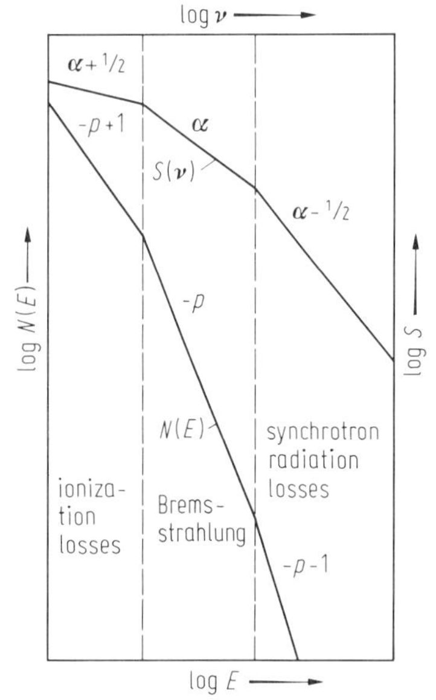


Fig. 5. Energy loss rates. Upper curve (top and right): flux density $S(\nu)$; Lower curve (left and below): electron density $N(E)$; $\alpha$=radio spectral index; $p$=electron spectral index. For more details and explanation, see text and Table 9.

Table 9. Equilibrium spectra for steady losses and injection.

| Loss mechanism | Electron spectral index | Radio spectral index |
|---|---|---|
| Losses unimportant ($E \ll E_b$) | $-p$ | $\alpha=-(p-1)/2$ |
| Ionization | $-p+1$ | $\alpha+\frac{1}{2}$ |
| Bremsstrahlung/ adiabatic expansion | $-p$ | $\alpha$ |
| Synchrotron/ inverse Compton | $-p-1$ | $\alpha-\frac{1}{2}$ |

### 9.6.3.3.3 Synchrotron and Compton losses

An initial energy distribution $N(E,0)=KE^{-p}$, $E_1<E<E_2$ subject to synchrotron losses $cE^2$ changes as

$$N(E,t)=KE^{-p}\mathrm{e}^{-t/\tau}(1-cEt)^{p-2} \tag{35}$$

in the energy range $E_1(1+cE_1t)^{-1}<E<E_2(1+cE_2t)^{-1}$ with a cutoff energy $E_c \leqq (ct)^{-1}=12.5\cdot 10^9 t\, B^{-2}$ GeV or a cutoff frequency $\nu_c(t)=2.06\, B^{-3}\,(10^9 t)^{-2}$ GHz ($B$ in [μG], $t$ in [a]) [K 10] where $B$ stands for $B+B_{eq}$ if also inverse Compton losses are substantial (see 9.6.3.3.5).

The spectrum is given in Table 10.

Table 10. Spectrum evolution with synchrotron and Compton losses (random pitch-angle distribution).

| Frequency range | Electron spectral index | Radio spectral index |
|---|---|---|
| $\nu<\nu_c(t)$ | $-p$ | $\alpha=-(p-1)/2$ |
| $\nu>\nu_c(t)$ | $-\frac{1}{3}(4p+5)$ | $\frac{4}{3}\alpha-1$ |

### 9.6.3.3.4 Inverse Compton limit

The frequency $\nu$ changes in a single Compton collision to

$$\nu' = \tfrac{4}{3}\gamma^2 \quad \text{where} \quad \gamma = \frac{E}{m_e c^2}. \tag{36}$$

[L 7] provided $\frac{h\nu}{m_e c^2} < 1 \ll \gamma$ when the Thomson cross section may be used [L 8]. For general $\gamma$, see [B 13]. Thus, scattering of a 300 GHz photon ($\lambda = 1$ mm) and an 1 GeV electron ($1.6 \cdot 10^{-3}$ erg, $\gamma = 2 \cdot 10^3$) produces soft X-ray photons of 6 keV ($1.5 \cdot 10^{18}$ Hz). Scattered (comptonized) photons may occur in the infrared, optical, UV, X-ray, and $\gamma$-ray spectral regions [B 14, J 5, K 16···K 18, S 21]. If inverse Compton losses in the field of the synchrotron radiation (self-Compton) become comparable to the synchrotron losses the radiation may be quenched by catastrophic losses due to higher-order scattering [K 16]. For first-order scattering (optically thin source) the ratio of Compton to synchrotron (radio) luminosity is (cf. Table 8)

$$\frac{L_C}{L_s} = \frac{u_{ph}}{B^2/8\pi} \tag{37}$$

where $u_{ph} = L_s/4\pi R^2 c$ is the synchrotron radiation density within the source of radius $R$. With Eqs. (26) and (28)

$$\frac{L_s}{4\pi R^2 c}\,\frac{8\pi}{B^2} \approx \frac{S_m \nu_m}{\phi^2 c}\,\frac{8\pi}{B^2} \approx \nu \left(\frac{T_b}{10^{12.6}}\right)^5 \tag{38}$$

where $\phi$ is the angular radius of the source and $\nu_m$, $S_m$, $T_b$ are frequency, flux density and brightness temperature at the cutoff of its spectrum, respectively. Inverse Compton losses become catastrophic when e.g. $\nu = 300$ GHz ($\lambda = 1$ mm) and $T_b \approx 10^{12}$ K [O 4, K 18]. The self-Compton effect may determine the overall spectra of radio galaxies and QSO's from the mm to the X-ray range with overall slope $\approx -1$ [J 5, P 11, Z 1].

### 9.6.3.3.5 Inverse Compton effect and X-ray observations

For first-order scattering of relativistic electrons with the microwave background photons

$$\frac{L_C}{L_s} = \frac{F_X}{F_R} = \frac{u_{ph}}{B^2/8\pi} = \frac{u_{ph,0}}{B^2/8\pi}(1+z)^4 = \left(\frac{B}{B_{eq}}\right)^{-2}(1+z)^4 \tag{39}$$

where $z$ is the redshift, $u_{ph,0} = 4 \cdot 10^{-13}$ erg cm$^{-3}$ the energy density of the 2.7 K microwave background, and $B_{eq} = 3.2$ µG the equivalent magnetic field with energy density $u_{ph,0}$. $F_X$ and $F_R$ are the fluxes in the X-ray and radio range, respectively. The monochromatic approximation [F 7] and the steady-state solution [Eqs. (33), (34)] yield

$$\frac{F_X}{F_R} = \left(\frac{B}{B_{eq}}\right)^{-2}(1+z)^{5/2}\left(\frac{\nu_X p_R B}{\nu_R \cdot p_X}\right)^{3/2} \cdot \frac{N(\sqrt{\nu_X/p_X}, z)}{N(\sqrt{(1+z)\nu_R/p_R B}, z)} \tag{40}$$

where the functions $N$ are to be chosen according to $E \lessgtr E_b = (\tau c(z))^{-1}$. Here $c(z) = 1.6 \cdot 10^{-14}(1+z)^4$ and $\nu_R = p_R \cdot B \cdot E^2$, $\nu_X = p_X(1+z)E^2$ are the frequencies observed in the radio and X-ray ranges, respectively [B 15]. By comparison of the X-ray and radio fluxes in extended sources the magnetic field distribution may be determined from

$$B = [6.6 \cdot 10^{-40}(4800)^\alpha (1+z)^{3-\alpha} F_R F_X^{-1} \nu_R^{-\alpha} E_X^\alpha]^{\frac{1}{1-\alpha}} \text{ Gauss} \tag{41}$$

where dimensions $F_R$[Jy], $F_X$[erg cm$^{-2}$ Hz$^{-1}$], $\nu_R$[GHz], $E_X$[keV] are adopted here [M 1].

#### 9.6.3.3.6 Adiabatic expansion

During an adiabatic isotropic expansion $R=R(t)$, magnetic flux conservation for a large-scale tangled field and for each individual particle orbit [M 13] implies $B \propto R^{-2}$ and $E \propto B^{1/2} \propto R^{-1}$, and therefore $K \propto R^{-p-2}$ in Eq. (10). Then from Eqs. (11) and (24)

$$\varepsilon_\nu \propto R^{-2p-3} \cdot \nu^{-\frac{(p-1)}{2}}, \tag{42}$$

$$\kappa_\nu \propto R^{-2p-4} \cdot \nu^{-\frac{(p+4)}{2}}, \tag{43}$$

and

$$S_\nu \propto R^3 \varepsilon_\nu \propto R^{-2p} \cdot \nu^{-\frac{p-1}{2}} \quad \text{for} \quad \nu > \nu_m \tag{44}$$

$$S_\nu \propto R^2 \varepsilon_\nu / \kappa_\nu \propto R^3 \nu^{5/2} \quad \text{for} \quad \nu < \nu_m \tag{45}$$

for self-absorption, where

$$\nu_m \propto R^{-\frac{4p+6}{p+4}}, \tag{46}$$

$$S_m \propto R^{-\frac{7p+3}{p+4}} \propto \nu_m^{\frac{7p+3}{4p+6}}, \tag{47}$$

are the frequency and flux density, respectively, at the turnover of the spectrum [when $\tau_\nu \approx \kappa_\nu R(t) \approx 1$] which occurs at progressively lower frequencies and flux levels as the source expands [K 10, M 13, S 23].

### 9.6.3.4 Energy considerations

a) Minimum energy of particles and fields

Assuming a source volume $V$ to be occupied by relativistic particles and the magnetic field $B$, the total energy of the electrons is

$$\begin{aligned} U_e = V u_e &= \frac{VK}{2-p}(E_2^{2-p} - E_1^{2-p}) \quad \text{for} \quad p \neq 2 \\ &= VK \ln \frac{E_2}{E_1} \quad \text{for} \quad p = 2 \end{aligned} \tag{48}$$

and the synchrotron luminosity

$$\begin{aligned} L &= \frac{c_3 VK B^2}{3-p}(E_2^{3-p} - E_1^{3-p}) \quad \text{for} \quad p \neq 3 \\ &= c_3 VK B^2 \ln \frac{E_2}{E_1} \quad \text{for} \quad p = 3 \end{aligned} \tag{49}$$

where Eq. (4) for $\dot{E}$ has been adopted with $B$ instead of $B_\perp$. (For $c_3$, see Table 7.) Replacing $E$ by the emitted critical frequency from Eq. (1) and eliminating $K$ from Eqs. (48) and (49) gives

$$U_e = g(p) L B^{-3/2} \tag{50}$$

where

$$g(p) = c_5 \frac{2\alpha+2}{2\alpha+1} \frac{\nu_2^{\alpha+1/2} - \nu_1^{\alpha+1/2}}{\nu_2^{\alpha+1} - \nu_1^{\alpha+1}} \tag{51}$$

and $\alpha = -\frac{1}{2}(p-1) \neq -\frac{1}{2}$ or $-1$. $g(p)$ is tabulated in Table 11, p. 333 for $\nu_1 = 10^{11}$ Hz, $\nu_2 = 10^7$ Hz and in [P 21]; $c_5$ is given in Table 7.

The energy of the protons is accounted for by putting $U_p = (1+k) U_e$; $k$ is of order 50 to 100 [B 16, R 8]. Then the total energy of particles and the magnetic field is

$$U_{tot} = (1+k) g(p) L B^{-3/2} + \tfrac{1}{6} B^2 R^3 \tag{52}$$

where $R$ is the characteristic radius of the source.

The minimum energy with respect to $B$ is

$$U_{\text{tot}}^{\text{min}} = \tfrac{7}{4}(1+k)\,g(p)\,L B_{\text{min}}^{-3/2} \tag{53}$$

where

$$B_{\text{min}} = (\tfrac{9}{2}(1+k)\,g(p)\,L)^{2/7} R^{-6/7} \tag{54}$$

and therefore

$$U_{\text{tot}}^{\text{min}} = \tfrac{7}{18}(\tfrac{9}{2}g(p)\cdot(1+k)L)^{4/7} R^{9/7}, \tag{55}$$

$$U_{\text{B}}^{\text{min}} = \tfrac{3}{4}(1+k)\,U_{\text{e}} = \tfrac{3}{7}U_{\text{tot}}^{\text{min}}. \tag{56}$$

There is nearly equipartition between particle and field energies under minimum conditions.

The minimum energy condition seems to be established in extended sources, while in compact sources the magnetic field might become much larger than the minimum field [O 4].

b) Source energies and observed quantities

In terms of the observed angular diameter $\phi$ and the turnover frequency $\nu_{\text{m}}$ and flux density $S_{\text{m}}$, the luminosity and radius of the source are $L \approx 4\pi d_L^2 S_{\text{m}} \nu_{\text{m}}$ and $R = \frac{1}{2}\phi d_{\text{A}}$ where $d_L = \frac{c}{H_0}\left(z + (1-q_0)\frac{z^2}{2}\right)$ and $d_{\text{A}} = (1+z)^{-2} d_L$ are the luminosity and angular distances, respectively, provided the redshift $z \ll 1$ or $q_0 = 0$ [W 8], see also 9.7.2.2.4. For numerical evaluations a Hubble constant $H_0 = 50$ (km s$^{-1}$)/Mpc is assumed here. With $\nu_{\text{m}} = \nu_{\text{obs}}(1+z)$ and $g(p) \approx c_5 \nu_{\text{m}}^{-1/2}(1+z)^{-1/2}$ [For $c_5$, see Table 7, p. 325.]

$$U_{\text{tot}}^{\text{min}} \approx 2\cdot 10^{58}(1+k)^{4/7}(S_{\text{m}}\nu_{\text{m}}^{1/2})^{4/7}\phi^{9/7}\cdot\frac{\left(z+\frac{z^2}{2}(1-q_0)\right)^{17/7}}{(1+z)^{20/7}}\ \text{erg} \tag{57}$$

$$B_{\text{min}} \approx 2.5\cdot 10^{-5}(1+k)^{2/7}(S_{\text{m}}\nu_{\text{m}}^{1/2})^{2/7}\cdot\phi^{-6/7}\cdot\frac{(1+z)^{11/7}}{\left(z+(1-q_0)\frac{z^2}{2}\right)^{2/7}}\ \text{Gauss} \tag{58}$$

where GHz, Jy, arc sec are used as units for $\nu_{\text{m}}$, $S_{\text{m}}$, $\phi$, respectively.

Without the assumptions of equipartition

$$B \approx c_4^2 S_{\text{m}}^{-2} \nu_{\text{m}}^5 \phi^4 (1+z)^{-1} \tag{59}$$

the ratio of magnetic to electron energy in terms of observed spectral data may be written

$$\frac{U_B}{U_{\text{e}}} \approx \frac{1}{24\pi}\frac{c_4^7}{c_5}\left(\frac{c}{H_0}\right)\frac{\left(z+(1-q_0)\frac{z^2}{2}\right)}{(1+z)^9}\frac{\nu_{\text{m}}^{17}\phi^{17}}{S_{\text{m}}^8} \tag{60}$$

i.e. depends extremely sensitively on angular size, cutoff frequency, flux density, and redshift [O 4, W 3]. For $c_4$ and $c_5$, see Table 7, p. 325.

Table 11. Some functions of $p$ (CGS units).

Definitions:

$$a(p) = \frac{c_2}{16\pi}\Gamma\left(\frac{3p-1}{12}\right)\Gamma\left(\frac{3p+7}{12}\right)\left(\frac{p+7/3}{p+1}\right)$$

$$b(p) = \frac{c_2}{128\pi}\left(\frac{c}{c_1}\right)^2\left(p+\frac{10}{3}\right)\Gamma\left(\frac{3p+2}{12}\right)\Gamma\left(\frac{3p+10}{12}\right)$$

$$c(p) = (2c_1)^{-5/2} a(p)/b(p) \approx 2c^{-2} c_1^{-1/2}$$

$$g(p) = c_5\frac{3-p}{2-p}\,\frac{\nu_2^{\frac{2-p}{2}} - \nu_1^{\frac{2-p}{2}}}{\nu_2^{\frac{3-p}{2}} - \nu_1^{\frac{3-p}{2}}} \approx c_5 \nu_{\text{m}}^{-1/2}.$$

The constants $c_1$, $c_2$, $c_5$ are given in Table 7, p. 325. $\Gamma$ is Euler's Gamma function [A 6].

| $p$ | $a(p)$ | $b(p)$ | $c(p)$ | $g(p)$ |
|---|---|---|---|---|
| 0.5 | $2.66\cdot10^{-22}$ | $1.62\cdot10^{-40}$ | $2.95\cdot10^{-30}$ | $5.57\cdot10^{6}$ |
| 1.0 | $4.88\cdot10^{-23}$ | $1.18\cdot10^{-40}$ | $7.43\cdot10^{-31}$ | $6.62\cdot10^{6}$ |
| 1.5 | $2.26\cdot10^{-23}$ | $9.69\cdot10^{-41}$ | $4.19\cdot10^{-31}$ | $9.03\cdot10^{6}$ |
| 2.0 | $1.37\cdot10^{-23}$ | $8.61\cdot10^{-41}$ | $2.86\cdot10^{-31}$ | $1.60\cdot10^{7}$ |
| 2.5 | $9.68\cdot10^{-24}$ | $8.10\cdot10^{-41}$ | $2.15\cdot10^{-31}$ | $3.34\cdot10^{7}$ |
| 3.0 | $7.52\cdot10^{-24}$ | $7.97\cdot10^{-41}$ | $1.69\cdot10^{-31}$ | $7.10\cdot10^{7}$ |
| 3.5 | $6.29\cdot10^{-24}$ | $8.16\cdot10^{-41}$ | $1.38\cdot10^{-31}$ | $1.24\cdot10^{8}$ |
| 4.0 | $5.56\cdot10^{-24}$ | $8.55\cdot10^{-41}$ | $1.17\cdot10^{-31}$ | $1.69\cdot10^{8}$ |
| 4.5 | $5.16\cdot10^{-24}$ | $9.24\cdot10^{-41}$ | $1.00\cdot10^{-31}$ | $2.01\cdot10^{8}$ |
| 5.0 | $4.98\cdot10^{-24}$ | $1.03\cdot10^{-40}$ | $8.68\cdot10^{-32}$ | $2.23\cdot10^{8}$ |
| 5.5 | $4.97\cdot10^{-24}$ | $1.16\cdot10^{-40}$ | $7.69\cdot10^{-32}$ | $2.39\cdot10^{8}$ |
| 6.0 | $5.11\cdot10^{-24}$ | $1.34\cdot10^{-40}$ | $6.85\cdot10^{-32}$ | $2.50\cdot10^{8}$ |

## 9.6.4 Extended sources

With the exception of a few nearby galaxies which are intrinsically weak radio emitters (e.g. M 31, M 33) and a few "giant radio galaxies" (e.g. 3C 236, DA 240) the angular sizes of extended radio sources are smaller than $\approx 10'$. Thus structural information can usually be obtained only by interferometers. For interpretation of extended sources, see 9.6.6.1.

### 9.6.4.1 Structure of extended sources

Most of the extended sources exhibit radiation coming from two components or "lobes" aligned through the optically visible galaxy or quasar. At high frequencies compact components (self-absorbed at lower frequencies) at the central position often become visible. Also "jets" pointing towards the outer lobes become detectable at the high spatial resolutions in use nowadays.

Several classification schemes for the morphology of extended sources have been proposed. Table 12 gives a detailed scheme. Figure 6 gives typical examples for such sources.

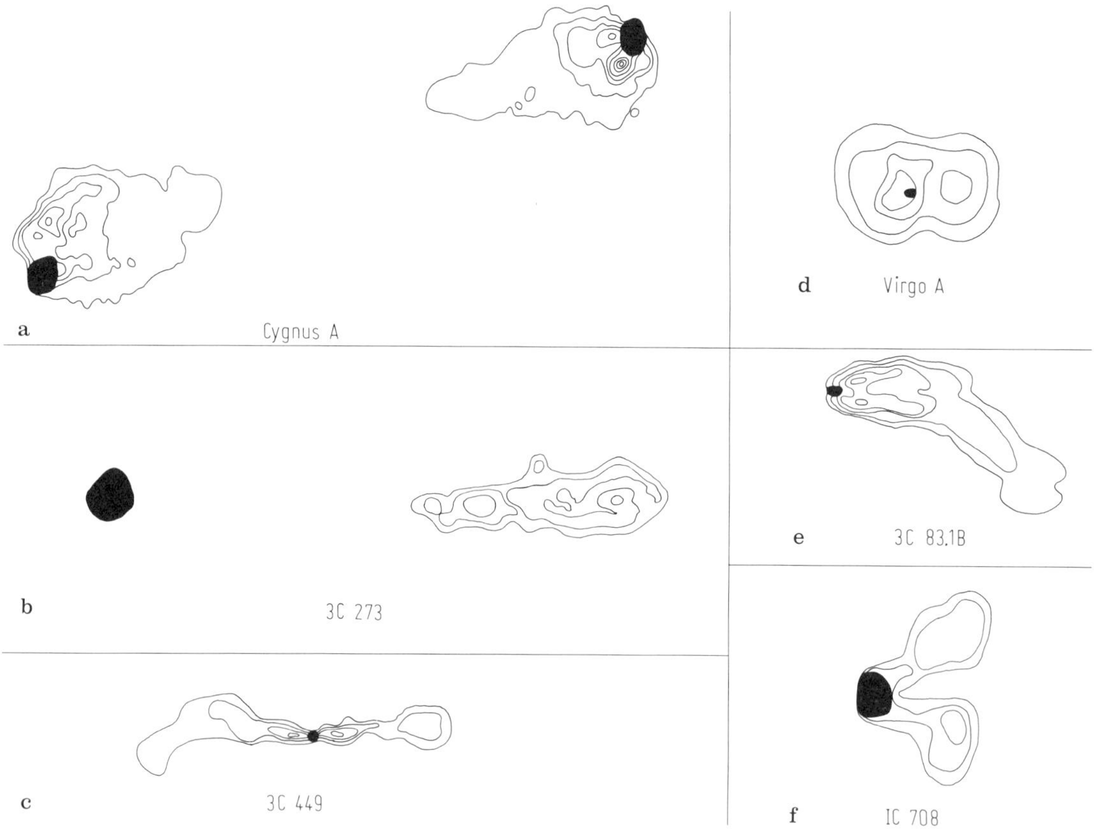


Fig. 6. Typical examples (not to scale) for extended radio sources of different morphological type, see Table 12 [M 1].

Table 12. Scheme of the morphology of extended radio sources [M1]
Examples in boldface are shown in Fig. 6.

| Designation | Other designation | Examples | Occurrence | Characteristic Properties |
|---|---|---|---|---|
| Ia | | | | |
| narrow edge-brightened double sources | classic doubles, symmetric doubles, Cygnus A-type doubles, Type I doubles, triplets, | 3C452<br>**CygA** (Fig. 6a) | ≈70% of strong-source surveys at low frequencies (3C, 4C, Parkes *) | length-to-width ratio $\gtrsim 4$; symmetric, edgebrightened lobes (due to hot spots), flat-spectrum central components associated with the nucleus of a galaxy or QSO, collinearity of lobes and central component, alignment of the orientation of the core and the overall structure |
| Ib | | | | |
| asymmetric doubles | D2 doubles, C-sources | **3C273** (Fig. 6b), | ≈10% of all extended QSO's | as under Ia but flux ratio between opposite lobes usually exceeding 10:1, galaxies as "core-components" are rarer |
| II | | | | |
| narrow edge-darkened double sources | (3C31-types) | CenA (inner part), **3C449** (Fig. 6c) 3C31 | | decreasing brightness towards the edges, radio cores with collimated radio jets towards the outer structure (3C31-types) |
| III | | | | |
| wide double sources | | Fornax A, **VirA** (Fig. 6d) Cen A (outer part), 3C310 | | fatter lobes, length-to-width ratio of 2 or 3 |
| IV | | | | |
| narrow-tailed sources | head-tail sources, NAT (narrow angle tail) | **3C83.1B** (Fig. 6e) (NGC1265), IC310, 3C129, 4C39.49 | | distinction between "twin-tailed" and "single-tailed" is resolution dependent; radio cores, often resolved into narrow jets |
| V | | | | |
| wide-tailed sources | WAT (wide angle tail) | 3C465, **IC708** (Fig. 6f), NGC6034, 4C4821 | | between I and IV |

*) Flux density dependent; at $S_{178} < 10$ Jy the percentage decreases.

### 9.6.4.2 Structure of lobes

In addition to the diffuse emission in numerous lobes, "hot spots" have been detected. These are small regions of high brightness temperature near the outer edges of the lobes. Typical sizes for hot spots are a few kpc [K 1, R 1].

For large sample of 3C sources a correlation between the emitted power $P$ at 1.4 GHz and the relative intensity of the hot spots has been shown [J 1]. For sources with $10^{25} < P < 10^{26}$ WHz$^{-1}$ about 10 to 20% of the lobe-intensity is contained in hot spots, for sources with $P \approx 10^{28}$ WHz$^{-1}$ virtually all flux of the extended components is contained in hot spots.

No structure as compact as in the nuclei has yet been found in the extended regions.

### 9.6.4.3 "Jets"

In a number of extended radio sources narrow "jets" of radio emission connecting the nucleus and the outer lobes have been detected, e.g.

| | |
|---|---|
| NGC 315 | [B 2, B 3] |
| 3C 31 | [B 4] |
| NGC 6251 | [W 2] |
| 3C 449 | [P 2] |
| 0844+31 | [V 1] |
| 3C 83.1B/NGC 1265 | [M 4, O 1] |
| 3C 129 | [V 2] |

Typical length scales of jets range from about 3 kpc in Virgo A (=NGC 4486) to about 340 kpc in NGC 315. Gaps of up to a few kpc between the nuclear core and the detectable jet are common, e.g. 3C 449, 3C 83.1B/NGC 1265. Large-scale bending of jets is discussed in 9.6.4.5. Wiggles on scales of some 10 kpc are detected in some jets, e.g. 3C 449, 3C 83. Asymmetric jets, e.g. one-sided jets, are frequently observed. In general the flux density ratios of opposite jets are considerably larger than those of opposite lobes. For interpretation, see 9.6.6.4.

### 9.6.4.4 Properties of extended radio sources

1. Physical sizes

a) Narrow edge-brightened doubles (collimated sources)

The physical size $l$ of a collimated double is taken as the projected separation between the brightness peaks at the outer lobes. For sources with 1.4 GHz-powers between $10^{23}$ and $10^{25}$ W Hz$^{-1}$: $l \approx 80$ kpc (10%), for sources with $10^{25} \leqq P_{1.4\,\mathrm{GHz}} < 10^{26.2}$ W Hz$^{-1}$: $l \approx 460$ kpc [P 1, G 1]. Sources with $l > 600$ kpc are rare (5%). The largest source known to date is 3C 236 with $l \approx 5.6$ Mpc (e.g. [W 1]).

b) Narrow edge-darkened sources and wide doubles (relaxed sources)

The physical size $l$ of a relaxed double is poorly defined, since there is a continuous decrease of brightness towards the edges. There is a correlation between total power (typically $< 10^{25}$ W Hz$^{-1}$ at 1.4 GHz) and the distance from the nucleus to the peak of the brightness distribution in the lobes, which ranges from a few to $\approx 160$ kpc [B 1]. The largest sources in this class comprise Cen A and DA 240 with linear sizes of the order of 2 and 3 Mpc.

c) Tailed sources

The lengths of tails vary from $\approx 20$ kpc to several hundred kpc (600 kpc for 3C 129). The widths of the sources lie in the range 10 to 60 kpc.

2. Symmetry

Asymmetries both in the flux density of the two components of double sources and in the distance from the central core of the outer lobes are observed.

75% of a sample of 3CR sources show a flux density ratio of their outer components of less than 2. Only 10% show a ratio $> 5$ [M 2].

Generally the distances from the core component to both outer components are roughly equal. They rarely differ by more than 50%.

The D2-sources can be interpreted as highly asymmetric doubles with a flux ratio between opposite lobes exceeding 10:1. Only one outer component and the core exhibit detectable radio emission. 10% of all extended quasars belong into this class [S 1, M 3]; D2-galaxies are much rarer.

3. Polarization

At short wavelengths the percentage polarization is not strongly influenced by Faraday depolarization (at $\lambda \leqq 6$ cm, rotation $\lesssim 10°$). Observations with sufficient resolution exhibit locally high-percentage polarizations up to $\approx 30\%$ of the theoretical value of synchrotron emission (in a few cases even higher).

The integrated polarization of the 5 GHz emission of Cyg A is $\approx 1\%$, whereas locally up to 25% has been measured. In sources like 3C 452 the distribution of polarized emission is more regular and the integrated polarization can be $\approx 10\%$ over the whole source. Multifrequency investigations of polarized emission from extended sources (incl. Faraday rotation) have shown that the electron densities are of the order of $5 \cdot 10^{-3}$ to $10^{-5}$ cm$^{-3}$ assuming equipartition. This leads to masses of the thermal plasma of the order of $10^3$ to $10^{10}$ $\mathfrak{M}_\odot$ in extreme cases.

4. Spectral indices (see 9.6.2.3)

The radio spectra of the diffuse lobes are well represented by a power law of index $-0.7 > \alpha > -1.3$.

In some cases the spectral index steepens with increasing distance from the nucleus [M 5, V 3].

The spectra of hot spots resemble the overall spectra of the surrounding lobe but are sometimes slightly flatter [G 2, B 5].

Spectral indices of jets range from $-0.5$ to $-0.8$.

5. Magnetic field and energy

Magnetic field strengths calculated under the assumption of equipartition are typically

$10^{-6} \cdots 2 \cdot 10^{-4}$ Gauss for the diffuse emission

$10^{-5} \cdots 5 \cdot 10^{-4}$ Gauss for hot spots

Example: Cyg A:

$B_{eq} \approx 1.5 \cdot 10^{-4}$ Gauss in lobes

$B_{eq} \approx 4.5 \cdot 10^{-4}$ Gauss in hot spots.

Minimum total energies are in the range

$10^{49} \cdots 10^{53}$ Ws for lobes

$10^{48} \cdots 10^{50}$ Ws for hot spots

Example: Cyg A:

$U_{min} \approx 4.6 \cdot 10^{51}$ Ws for lobes

$U_{min} \approx 5.3 \cdot 10^{49}$ Ws for hot spots.

6. Lifetimes

"Kinematical" ages derived from the average separation of nuclear and extended components and translational velocities from symmetry arguments [L 1] lead to characteristic ages greater than $2 \cdot 10^6$ years for the extended components of strong double sources.

Derived synchrotron lifetimes are of the order of $\geqq 10^5$ years, e.g. [P 1].

### 9.6.4.5 Alignments on large scale

In sources like Cyg A [K 13], 3C 111 [P 18], 3C 236 [S 2], and NGC 6251 [C 1] structures from the innermost core region (see 9.5.1.4) out to the extended lobes are aligned within a few degrees. This phenomenon occurs in both collimated (narrow edge-brightened) and relaxed doubles.

In e.g. NGC 6251 the preferential direction of the nuclear (2 pc) component is within 5° of the main axis defined by the 100 kpc jet, corresponding to an alignment on the scale $10^5:1$.

In e.g. 3C 236 the preferential direction of the nuclear (1 kpc) component is well aligned with the axis of the extended double, indicating a constant position angle of the collimation over some $10^7$ years.

In contrast severe misalignments are observed in e.g. Vir A and Cen A.

The preferential directions defined by the nuclear and the outermost components differ by 70° (Vir A) and 45° (Cen A), possibly indicating a precession of the collimation during the sources' lifetime [M 6].

A sequence of shapes of extended radio sources indicating rotational processes or shear in the intergalactic medium [M 1] is given in Fig. 7.

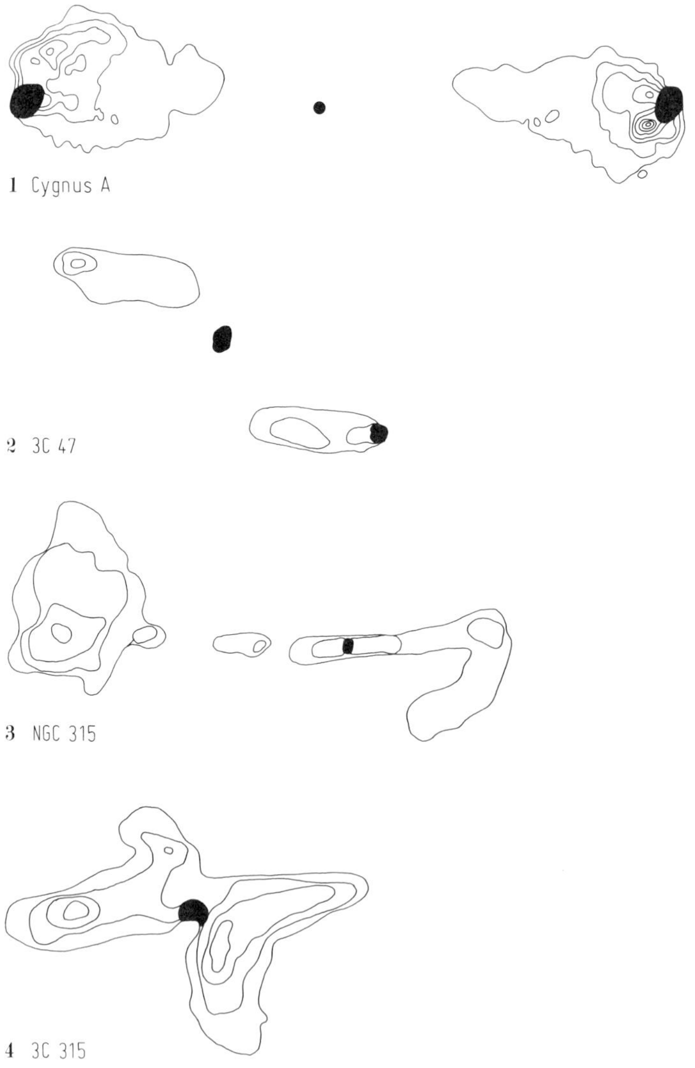


Fig. 7. A sequence of extended radio sources (not to scale) indicating rotational processes or shear in the intergalactic medium [M 11].

## 9.6.5 Compact sources

Compact radio sources have sizes of the order of 0.1···100 milliarcsec corresponding to cutoff frequencies of $\gtrsim$100 MHz. Linear sizes are of the order of 0.01 to several hundred parsec. These radio sources are coincident with nuclei of active galaxies, QSO's, and BL Lac objects (see 9.5) (for some, no optical counterpart has yet been found). For interpretation of compact radio nuclei see 9.6.6.3.

### 9.6.5.1 Structure of compact sources

Structural information of the radio emission on the m.a.s. scale can be obtained only by VLBI observations. Thus, only compact components with brightness temperatures $>10^9$ K have been investigated. Due to present observational limitations the dynamic range is limited to roughly 1:10.

Morphology

The structure of compact extragalactic objects can be classified as follows:

1. a single unresolved component with size $\leqq$0.1 m.a.s.;
   example: nucleus of M 81 (linear size $\leqq$1300 A.U.) [K 12]
2. slightly resolved single component; size: some m.a.s.;
   example: OJ 287 [P 15]
3. core-halo or core-jet structures: compact core of size $\lesssim$0.5 m.a.s. and associated extended emission;
   example: 3C 147, NGC 315 [S 17]
4. double sources; both symmetric, like CTD 93, 4C 39.25, and asymmetric, like 3C 395 [P 15, P 16].
   Component separations range from a few to $\approx$50 m.a.s. or even larger.
5. more complex structures, like 3C 84 [P 17].

In general the structures are wavelength-dependent because of the different radio spectra of the individual components.

Properties

The fraction of the total flux density contained in the compact nucleus varies from a few percent (Cyg A) [K 13] to essentially 100 percent (some BL Lac objects like 0454+84) [B 11].

The compact structures are in general wavelength dependent.

The peak brightness temperatures are typically $10^{11\pm1}$ K.
The magnetic fields are of the order of $10^{-3\pm1}$ G.

The energies in the relativistic particles are $E_p \approx 10^{50} \cdots 10^{56}$ erg.

The energies in the magnetic field are $E_m \approx 10^{50} \cdots 10^{54}$ erg.

Radio luminosities range from $10^{37}$ to $10^{46}$ erg s$^{-1}$ corresponding to a volume emissivity of 1 mW km$^{-3}$.

In the case of extended double sources with a compact central component the direction defined by the extended and compact structures are in good agreement ("direction memory"). Examples: 3C 111, Cyg A [K 13, P 18].

In the case of sources with an asymmetric outer structure (D2) a significant bending is observed between the structures on the m.a.s. scale and the largescale structure. Example: 3C 345, 3C 273 [P 15].

### 9.6.5.2 Variability of compact sources

Intensity variations

Intensity variations have been detected over the wavelength range of about 3 mm to about 1 m (for references, see [K 11]). Two main types of flux-density variability are observed, which can be explained by the canonical model of variable sources (see [K 14]):

1. Intensity outbursts of similar amplitude occur simultaneously over a wide wavelength range. This behaviour can be explained by prolonged injection (or acceleration) of relativistic particles into an optically thin synchrotron source (intensity increase) and subsequent adiabatic expansion or radiation losses (intensity decrease). Example: BL Lac.
2. Intensity outbursts occur first at short wavelengths and propagate with decreasing amplitude towards longer wavelengths. This pattern can be explained by adiabatic expansion of an initially opaque synchrotron source. Examples: 3C 120, 3C 454.3.

Some sources exhibit low-frequency flux density variations without correspondingly strong high-frequency variations (see e.g. [D 2, J 3]).

Explanations for this pattern using

a) extrinsic phenomena (free-free absorption, scintillation),

b) intrinsic phenomena (nonrelativistic evolution, coherent synchrotron processes, relativistic effects etc.) are discussed in [D 2, J 3].

Time scales for intensity variations vary from weeks (possibly shorter) to years.

Polarization variability

Time scales for significant changes in radio polarization (degree and position angle) appear to be shorter than for intensity variations. No clear correlation between polarization and intensity variability has been shown [A 3, A 4, A 5].

Structural variability

The data discussed here come exclusively from VLBI observations. Thus only variability in compact components with brightness temperatures $>10^9$ K has been studied. Most compact radio sources show intensity variations, and, since their brightness distribution is usually complex, their structure is also time variable. An example for such variations is given in Fig. 8 which shows the well-studied core of 3C 84 (NGC 1275) at different epochs [P 17]. The relative motions of the individual components of 3C 84 did not exceed 0.5 $c$ during the period 1972 to 1980 [P 17].

Table 13 lists the sources for which apparent superluminal expansion with $5<v/c<45$ has been quoted ($H_0=50$ km s$^{-1}$ Mpc$^{-1}$, $q_0=0$; see 9.7.5 and 9.7.6).

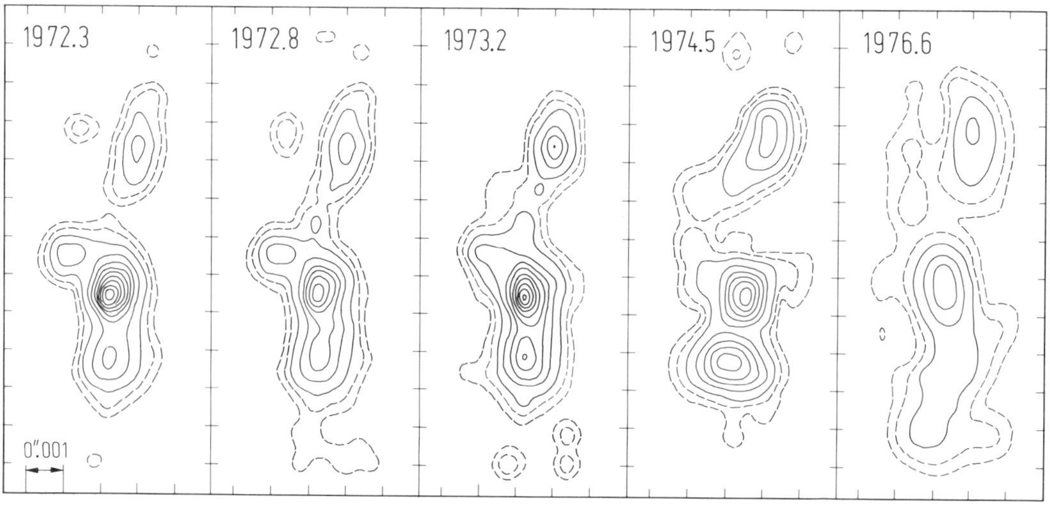


Fig. 8. Structural changes in the nucleus of 3C 84 (=NGC 1275) at $\lambda=2.8$ cm [P 17].

Table 13. Sources with apparent superluminal expansion.
Type: see Table 3
$z=\dfrac{\Delta\lambda}{\lambda}$
Inf. since: structural information available since

| Source | Type | $z$ | Superluminal expansion | Inf. since |
|---|---|---|---|---|
| 3C 120 | GAL | 0.033 | yes | 1972 |
| 3C 179 | QSO | 0.843 | yes | 1979 |
| 3C 273 | QSO | 0.158 | yes | 1970 |
| 3C 279 | QSO | 0.538 | yes | 1970 |
| 3C 345 | QSO | 0.595 | yes | 1970 |
| BL Lac | LAC | 0.07 | marginal | 1973 |

A review of observational data prior to 1977 is found in [C 11]. Evidence for superluminal expansion for BL Lac is found in [P 24] and for 3C 179 in [P 20]. A review of the characteristics and explanations of the phenomenon of superluminal expansion is given in [K 11] and references therein. See also 9.6.6.4.2.

### 9.6.5.3 Correlations of radiospectra with other properties

a) Correlation of emitted radio power (in rest frame of source) with spectral index [V 8], [S 18]. Radio galaxies with high monochromatic intrinsic radio power have steeper spectral indices.
b) Correlation of radio spectrum and the existence of compact components. Virtually all sources with flat spectra ($\alpha > -0.5$) contain compact components [K 11].
c) Correlation of radio spectrum and intensity variability [A 4]. Virtually all sources with flat radio spectra exhibit intensity variations at cm-wavelengths; at dm- and low m-wavelengths $\approx 40\%$ of these sources exhibit variations [D 3].
d) Usually Optically Violent Variable (OVV) sources show large radio variations (at high and/or low frequencies) [D 3, K 11].
e) The mm intensities of flat-spectrum radio sources are correlated with the intensities in the X-ray domain [O 3].
f) The correlation between the angular size of radio sources and their redshift is discussed in [D 1, G 16].

## 9.6.6 Interpretation of extended and compact radio sources (see p. 396)

## 9.6.7 References for 9.6.1···9.6.5

A 1 Amirkhanyan, V.R., Gorshkov, A.G., Kapustkin, A.A., Konnikova, V.K., Lazutkin, A.N., Larinov, M.G., Nikanorov, A.S., Sidonenkov, V.N., Ugol'kova, L.S., Khromov, O.I.: Astron. Tsirk. No. 1099 (1980) 2.
A 2 Arp, H.: Astrophys. J. **239** (1980) 463.
A 3 Aller, H.D., Aller, M.F., Hodge, P.E.: Astron. J. **86** (1981) 325.
A 4 Altschuler, D.R., Wardle, J.F.C.: Mem. R. Astron. Soc. **82** (1976) 1.
A 5 Andrew, B.H., McLeod, J.M., Harvey, G.A., Medd, W.J.: Astron. J. **83** (1978) 863.
A 6 Abramowitz, M., Stegun, I.A.: Handbook of Mathematical Functions, Dover, New York (1968).

B 1 Birkinshaw, M.R., Laing, R., Scheuer, P., Simon, A.: Mon. Not. R. Astron. Soc. **185** (1978) 39P.
B 2 Bridle, A.H., Fomalont, E.B.: Astron. Astrophys. **52** (1976) 107.
B 3 Bridle, A.H., Davis, M.M., Fomalont, E.B., Willis, A.G., Strom, R.G.: Astrophys. J. **228** (1979) L9.
B 4 Burch, S.F.: Mon. Not. R. Astron. Soc. **187** (1979) 187.
B 5 Burch, S.F.: Mon. Not. R. Astron. Soc. **186** (1979) 519.
B 6 Braude, S.Ya., Megn, A.V., Sokolov, K.P., Tkachenko, A.P., Sharykin, N.K.: Astrophys. Space Sci. **64** (1979) 73.
B 7 Bennett, A.S.: Mem. R. Astron. Soc. **68** (1961) 163.
B 8 Bolton, J.G., Wright, A.E., Savage, A.: Australian J. Phys. Suppl. **46** (1979) 1.
B 9 Bedford, N.H., Kerr, A.J., Mathur, S.H., Morrison, I., Spencer, R.E., Stannard, D.: Mon. Not. R. Astron. Soc. **195** (1981) 245.
B 10 Baars, J.W.M., Genzel, R., Pauliny-Toth, I.I.K., Witzel, A.: Astron. Astrophys. **61** (1977) 99.
B 11 Biermann, P., Duerbeck, H., Eckart, A., Fricke, K., Johnston, K.J., Kühr, H., Liebert, J., Pauliny-Toth, I.I.K., Schleicher, H., Stockman, H., Strittmatter, P.A., Witzel, A.: Astrophys. J. **247** (1981) L53.
B 12 Burbidge, G.R., Burbidge, E.M.: Quasistellar Objects, Freeman, San Francisco (1967).
B 13 Blumenthal, G., Gould, R.: Rev. Mod. Phys. **42** (1970) 237.
B 14 Burbidge, G.R., Jones, T.W., O'Dell, S.L.: Astrophys. J. **193** (1974) 43.
B 15 Biermann, P., Schlickeiser, R.: Int. Cosmic Ray Conference, Paris, Conf. paper, OG-Section, Vol. **2** (1981) 252.
B 16 Burbidge, G.R.: Astrophys. J. **129** (1959) 849.

C 1 Cohen, M.H., Readhead, A.C.S.: Astrophys. J. **233** (1979) L101.
C 2 Colla, G., Fanti, C., Fanti, R., Ficarra, A., Formiggini, L., Gondolfi, E., Grueff, G., Lari, C., Padrielli, L., Roffi, G., Tomasi, P., Vigotti, M.: Astron. Astrophys. Suppl. **1** (1970) 281.

C 3 Colla, G., Fanti, C., Fanti, R., Ficarra, A., Formiggini, L., Gandolfi, E., Lari, C., Marano, B., Padrielli, L., Tomasi, P.: Astron. Astrophys. Suppl. **7** (1972) 1.
C 4 Colla, G., Fanti, C., Fanti, R., Ficarra, A., Formiggini, L., Gandolfi, E., Gioia, I., Lari, C., Marano, B., Padrielli, L., Tomasi, P.: Astron. Astrophys. Suppl. **11** (1973) 291.
C 5 Condon, J.J., Condon, M.A., Mitchell, K.J., Usher, P.D.: Astrophys. J. **242** (1980) 486.
C 6 Condon, J.J., O'Dell, S.L., Puschell, J.J., Stein, W.A.: Nature **283** (1980) 357.
C 7 Condon, J.J., O'Dell, S.L., Puschell, J.J., Stein, W.A.: Astrophys. J. **246** (1981) 624.
C 8 Condon, J.J., Condon, M.A., Jauncey, D.L., Smith, M.G., Turtle, A.J., Wright, A.E.: Astrophys. J. **244** (1981) 5.
C 9 Condon, J.J., Ledden, J.E., O'Dell, S.L., Dennison, B.: Astron. J. **89** (1979) 1.
C 10 Condon, J.J. in: Pittsburgh Conference on BL Lac Objects, Univ. Pittsburgh Press (1978) 21.
C 11 Cohen, M.H., Kellermann, K.I., Shaffer, D.B., Linfield, R.P., Moffet, A.T., Romney, J.D., Seielstad, G.A., Pauliny-Toth, I.I.K., Preuss, E., Witzel, A., Schilizzi, R.T., Geldzahler, B.J.: Nature **268** (1977) 405.
C 12 Condon, J.J., Dressel, L.L.: Astrophys. Lett. **15** (1973) 203.

D 1 De Young, D.S.: Annu. Rev. Astron. Astrophys. **14** (1976) 447.
D 2 Dennison, B., Broderick, J.J., Ledden, J.E., O'Dell, S.L., Condon, J.J.: Astron. Astrophys. **86** (1981) 1604.
D 4 De Bruyn, A.G.: Astron. Astrophys. **52** (1976) 439.

E 1 Edge, D.D., Shakeshaft, J.R., McAdam, W.B., Baldwin, J.E., Archer, S.: Mem. R. Astron. Soc. **68** (1959) 37.
E 2 Ekers, J.A. (ed.): Australian J. Phys. Suppl. **7** (1969).
E 3 Efanov, V.A., Moiseev, I.G., Nesterov, N.S., Tiuri, M., Urpo, S.: Pisma Astron. Zh. **6** (1980) 340 (Engl. Transl. in Soviet Astron. Lett. **6**).
E 4 Epstein, E.E., Landau, R., Rather, J.D.G.: Astron. J. **85** (1980) 1427.
E 5 Elsmore, B., Mackay, C.D.: Mon. Not. R. Astron. Soc. **146** (1969) 361.

F 1 Fomalont, E.B., Wright, E.B.: in: Galactic and Extragalactic Radio Astronomy (Kellermann, K.I., Verschuur, G.L., eds.), Springer, Berlin (1974).
F 2 Fanti, R., Ficarra, A., Mantovani, F., Padrielli, L., Weiler, K.: Astron. Astrophys. Suppl. **36** (1979) 359.
F 3 Fisher, J.R., Erickson, W.C.: Astrophys. J. **242** (1980) 884.
F 4 Fanti, C., Ficarra, A., Gregorini, L., Matovani, F., Olori, M.C.: Astron. Astrophys. **97** (1981) 251.
F 5 Flett, A.M., Henderson, C.: Mon. Not. R. Astron. Soc. **194** (1981) 961.
F 6 Fanti, R., Feretti, L., Giovannini, G., Padrielli, L.: Astron. Astrophys. Suppl. **35** (1979) 169.
F 7 Felten, J.E., Morrison, P.: Astrophys. J. **146** (1966) 686.

G 1 Gavazzi, G., Perola, G.C.: Astron. Astrophys. **66** (1978) 407.
G 2 Gopal-Krishna: Mon. Not. R. Astron. Soc. **181** (1977) 247.
G 3 Gregory, P.C., Taylor, A.R.: J. R. Astron. Soc. Canada **73** (1979) 304.
G 4 Gioia, L.M., Gregorini, L.: Astron. Astrophys. Suppl. **36** (1979) 347.
G 5 Grueff, G., Kotanyi, C., Schiavo-Campo, P., Tanzella, Nitti, G., Vigotti, M.: Astron. Astrophys. Suppl. **44** (1981) 241.
G 6 Gopal-Krishna, Preuss, E., Schilizzi, R.T.: Nature **288** (1980) 344.
G 7 Ginzburg, V.L., Syrovatskii, S.I.: Annu. Rev. Astron. Astrophys. **3** (1965) 297.
G 8 Ginzburg, V.L.: The Propagation of Electromagnetic Waves in Plasmas, Pergamon Press (1964).
G 9 Ginzburg, V.L., Syrovatskii, S.I.: The Origin of Cosmic Rays, Pergamon Press, London (1964).
G 10 Ginzburg, V.L., Sazonov, V.N., Syrovatskii, S.I.: Soviet Phys.-Usp. **11** (1968) 34.
G 11 Gould, R.: Physica **60** (1972) 145.
G 12 Gould, R.: Physica **62** (1972) 555.
G 13 Gould, R.: American J. Phys. **38** (1970) 189.
G 14 Ginzburg, V.L., Syrovatskii, S.I.: Soviet Phys.-JETP **19** (1964) 1255.
G 15 Gower, J.F.R., Scott, P.F., Willis, P.: Mem. R. Astron. Soc. **71** (1967) 49.
G 16 Gunn, J.E., Longair, M.S., Rees, M.J.: Observational Cosmology, 8$^{th}$ Adv. Course Swiss Soc. Astron. Astrophys. Saas-Fee, Geneva Obs. (1978).

H 1 Hachenberg, O., in: 1.9 = Landolt-Börnstein NS, Vol. VI/2a (1981) p. 50.
H 2 Harris, D.E., Miley, G.K.: Astron. Astrophys. Suppl. **34** (1978) 117.
H 3 Halliday, J.: Mon. Not. R. Astron. Soc. **179** (1977) 111.
H 4 Hine, R.G., Scheuer, P.A.G.: Mon. Not. R. Astron. Soc. **193** (1980) 285.
H 5 Harris, D.E., Bahcall, A., Strom, R.G.: Astron. Astrophys. **60** (1977) 27.

I 1 Int. Astron. Union Symposium **74** (Jauncey, D.L., ed.): Radio Astronomy and Cosmology, Reidel, Dordrecht (1976): references given on page 398.
I 2 Int. Astron. Union Symposium **47** (Longair, J.E., Rees, M.S., eds.) Observational Cosmology (1978).

J 1 Jenkins, C.J., McEllin, M.: Mon. Not. R. Astron. Soc. **180** (1977) 219.
J 2 Joshi, M.N., Singal, A.K.: Mem. Astron. Soc. India **1** (1980) 49.
J 3 Jones, T.W., Burbidge, G.R.: Astrophys. J. **186** (1973) 291.
J 4 Jackson, J.: Classical Electrodynamics, Wiley, New York (1962).
J 5 Jones, T.W., O'Dell, S.L., Stein, W.A.: Astrophys. J. **188** (1974) 353.

K 1 Kapahi, V.K.: Astron. Astrophys. **67** (1978) 157.
K 2 Kesteven, M.J.L., Bridle, A.H.: J. R. Astron. Soc. Canada **71** (1977) 21.
K 3 Kühr, H., Pauliny-Toth, I.I.K., Witzel, A., Schmidt, J.: Astron. J. **86** (1981) 854.
K 4 Kenderdine, S., Ryle, M., Pooley, G.G.: Mon. Not. R. Astron. Soc. **134** (1966) 189.
K 5 Katgert, J.K.: Astron. Astrophys. Suppl. **31** (1978) 409.
K 6 Kühr, H., Witzel, A., Pauliny-Toth, I.I.K., Nauber, U.: Astron. Astrophys. Suppl. **45** (1981) 367.
K 7 Kapahi, V.K.: Astron. Astrophys. **74** (1979) L11.
K 8 Kapahi, V.K.: Astron. Astrophys. Suppl. **43** (1981) 381.
K 9 Kapahi, V.K., Schilizzi, R.T.: Nature **277** (1979) 610.
K 10 Kellermann, K.I. in: Galactic and Extragalactic Radio Astronomy (Kellermann, K.I., Verschuur, G.L., eds.), Springer, Berlin (1974).
K 11 Kellermann, K.I., Pauliny-Toth, I.I.K.: Annu. Rev. Astron. Astrophys. **19** (1981) 373.
K 12 Kellermann, K.I., Shaffer, D.B., Pauliny-Toth, I.I.K., Preuss, E., Witzel, A.: Astrophys. J. **210** (1976) L121.
K 13 Kellermann, K.I., Downes, A.J.B., Pauliny-Toth, I.I.K., Preuss, E., Shaffer, D.B., Witzel, A.: Astron. Astrophys. **97** (1981) L1.
K 14 Kellermann, K.I., Pauliny-Toth, I.I.K.: Annu. Rev. Astron. Astrophys. **6** (1968) 417.
K 15 Kardashev, N.: Sov. Astron. **6** (1962) 317.
K 16 Kompaneets, A.S.: Soviet Phys.-JETP **4** (1957) 730.
K 17 Katz, J.I.: Astrophys. J. **206** (1976) 916.
K 18 Kellermann, K.I., Pauliny-Toth, I.I.K.: Astrophys. J. Lett. **155** (1969) L71.
K 19 Kühr, H., Nauber, U., Pauliny-Toth, I.I.K., Witzel, A.: A Catalogue of Radio Sources, 2nd edition, MPIfR Preprint No. 55 (1981).

L 1 Longair, M.S., Ryle, J.M.: Mon. Not. R. Astron. Soc. **188** (1979) 625.
L 2 Ledden, J.E., Broderick, J.J., Condon, J.J., Brown, R.L.: Astron. J. **85** (1980) 780.
L 3 Laan, H., van der, Perola, G.C.: Astron. Astrophys. **3** (1969) 468.
L 4 Laing, R.A.: Mon. Not. R. Astron. Soc. **195** (1981) 245.
L 5 Laing, R.A.: Mon. Not. R. Astron. Soc. **195** (1981) 261.
L 6 Lang, K.R.: Astrophysical Formulae, Springer, Berlin (1980).
L 7 Longair, M.S.: High Energy Astrophysics, Cambridge (1981).
L 8 Landau, L., Lifshitz, I.: The Classical Theory of Fields, Addison Wesley (1962).

M 1 Miley, G.: Annu. Rev. Astron. Astrophys. **18** (1980) 165.
M 2 Mackay, C.D.: Mon. Not. R. Astron. Soc. **154** (1971) 209.
M 3 Miley, G.K., Hartsuijker, A.P.: Astron. Astrophys. Suppl. **34** (1978) 129.
M 4 Miley, G.K., Perola, G.C.: Astron. Astrophys. **45** (1975) 223.
M 5 Miley, G.K., Laan, H., van der, Wellington, K.J. in: The Formation and Dynamics of Galaxies (Shakeshaft, J.R., ed.), Int. Astron. Union Symp. **58**, Reidel, Dordrecht (1974) 109.
M 6 Miley, G.K. in: Proc. NATO Summer School: Physics of Nonthermal Radio Sources (1976) 1.
M 7 Maslowski, J., Pauliny-Toth, I.I.K., Witzel, A., Kühr, H.: Astron. Astrophys. **95** (1981) 285.
M 8 Mihalas, D.: Stellar Atmospheres, Freeman, San Francisco, 2nd ed. (1978).
M 9 MacDonald, G.H., Kenderdine, S., Neville, A.C.: Mon. Not. R. Astron. Soc. **138** (1968) 259.
M 10 Mackay, C.D.: Mon. Not. R. Astron. Soc. **145** (1969) 31.
M 11 Marsher, A.P.: Astrophys. J. **216** (1976) 244.
M 12 Moffet, A.T.: Annu. Rev. Astron. Astrophys. **4** (1966) 145.
M 13 Moffet, A.T. in: Stars and Stellar Systems, Vol. 9, Galaxies and the Universe (Sandage, A., Sandage, M., Kristian, J. eds.), Gordon and Breach (1975) p. 211.
M 14 Melrose, D.: Astrophys. Space Sci. **5** (1969) 131.

O 1 Owen, F.N., Burns, J.O., Rudnick, L.: Astrophys. J. **226** (1978) L119.
O 2 O'Dell, S.L. in: Pittsburgh Conference on BL Lac Objects, Univ. Pittsburgh Press (1978) 312.
O 3 Owen, F.N.,Helfand, P.J., Spangler, S.R.: Astrophys. J. **250** (1981) L55.
O 4 O'Dell, S.L. in: Active Galactic Nuclei (Hazard, C., Mitton, S., eds.), Cambridge Univ. Press (1979) p. 95.
O 5 Oster, L.: Rev. Mod. Phys. **33** (1961) 525.

P 1 Perola, G.C.: Fund. Cosmic Phys. **7** (1981) 59.
P 2 Perley, R.A., Johnston, K.J.: Astron. J. **84** (1979) 1247.
P 3 Pauliny-Toth, I.I.K., Steppe, H., Witzel, A.: Astron. Astrophys. **85** (1980) 329.
P 4 Porcas, R.W., Urry, C.M., Browne, I.W.A., Cohen, A.M., Daintree, E.J., Walsh, D.: Mon. Not. R. Astron. Soc. **191** (1980) 607.
P 5 Pilkington, J.D.H., Scott, P.F.: Mem. R. Astron. Soc. **69** (1965) 183.
P 6 Pooley, G.G., Kenderdine, S.: Mon. Not. R. Astron. Soc. **139** (1968) 529.
P 7 Pooley, G.G.: Mon. Not. R. Astron. Soc. **144** (1969) 101.
P 8 Pearson, T.J.: Mon. Not. R. Astron. Soc. **171** (1975) 475.
P 9 Pearson, T.J., Kus, A.J.: Mon. Not. R. Astron. Soc. **182** (1973) 273.
P 10 Pauliny-Toth, I.I.K., Witzel, A., Preuss, E., Baldwin, J.E., Hills, R.E.: Astron. Astrophys. Suppl. **34** (1978) 253.
P 11 Pacini, F., Salvati, M.: Astrophys. J. Lett. **225** (1978) L99.
P 12 Pollach, J.T., Pica, A.J., Smith, A.G., Leacock, R.J., Edwards, P.L., Scott, R.L.: Astron. J. **84** (1979) 1658.
P 13 Parma, P., Weiler, K.W.: Astron. Astrophys. **96** (1981) 412.
P 14 Pottasch, R.I., Wardle, J.F.C.: Astron. J. **84** (1979) 707.
P 15 Pauliny-Toth, I.I.K., Preuss, E., Witzel, A., Graham, D., Kellermann, K.I., Rönnäng, B.: Astron. J. **86** (1981) 371.
P 16 Phillips, R.B., Mutel, R.L.: Astrophys. J. **236** (1980) 89.
P 17 Preuss, E., Kellermann, K.I., Pauliny-Toth, I.I.K., Witzel, A., Shaffer, D.B.: Astron. Astrophys. **79** (1979) 268.
P 18 Pauliny-Toth, I.I.K., Preuss, E., Witzel, A., Kellermann, K.I., Shaffer, D.B.: Astron. Astrophys. **52** (1976) 471.
P 19 Pacholczyk, A.G., Swihart, T.L.: Astrophys. J. **150** (1967) 647.
P 20 Porcas, R.W.P.: Nature **294** (1981) 47.
P 21 Pacholczyk, A.G.: Radio Astrophysics, Freeman, San Francisco (1970).
P 22 Pacholczyk, A.G.: Radio Galaxies, Pergamon Press, New York (1976).
P 23 Pacini, F., Ryter, C., Strittmatter, P.A.: Extragalactic High Energy Astrophysics, Saas Fee (1979).
P 24 Phillips, R.B., Mutel, R.L.: Astrophys. J. **257** (1982) L19.

R 1 Readhead, A.C.S., Hewish, A.: Mon. Not. R. Astron. Soc. **176** (1976) 571.
R 2 Rogers, A.E.E., Pooley, G., Moran, J.M. in: Methods of Experimental Physics. Vol. **12**C, Academic Press, New York (1976).
R 3 Robertson, J.G.: Australian J. Phys. **30** (1977) 209.
R 4 Ryle, M., O'Dell, D.M., Waggett, P.C.: Mon. Not. R. Astron. Soc. **173** (1975) 9.
R 5 Rudnick, L., Adams, M.T.: Astron. J. **84** (1979) 437.
R 6 Razin, V.: Radiofisika **3** (1960) 584.
R 7 Rees, M.J., Setti, G.: Nature **219** (1968) 127.
R 8 Ramaty, R., Lingenfelter, R.E.: J. Geophys. Res. **71** (1966) 3687.
R 9 Rybicki, G.B., Lightman, A.P.: Radiative Processes in Astrophysics, Wiley, New York (1979).
R 10 Robertson, J.G.: Australian J. Physics **30** (1977) 231.

S 1 Stannard, D., Niell, D.S.: Mon. Not. R. Astron. Soc. **179** (1977) 719.
S 2 Schilizzi, R.T., Miley, G.K., van Ardenne, A., Baud, B., Bååth, L., Rönnäng, B.O., Pauliny-Toth, I.I.K.: Astron. Astrophys. **77** (1979) 1.
S 3 Subrahmanya, C.R., Gopal-Krishna: Mem. Astron. Soc. India **1** (1979) 2.
S 4 Singhal, A.K., Gopal-Krishna, Venugopal, V.R.: Mem. Astron. Soc. India **1** (1979) 14.
S 6 Schuch, N.J.: Mon. Not. R. Astron. Soc. **196** (1981) 695.
S 7 Savage, A., Bolton, J.G.: Mon. Not. R. Astron. Soc. **188** (1979) 599.
S 8 Sramek, R.A., Weedman, D.W.: Astrophys. J. **238** (1980) 435.
S 9 Smith, M.G., Wright, A.E.: Mon. Not. R. Astron. Soc. **191** (1980) 871.
S 10 Sherwood, W.A., Kreysa, E., Schultz, G.V.: Mitt. Astron. Ges. **52** (1981) 138.
S 11 Strittmatter, P.A., Hill, P., Pauliny-Toth, I.I.K., Steppe, A., Witzel, A.: Astron. Astrophys. **88** (1980) L12.

S 12 Spangler, S.R., Cotton, W.D.: Astron. J. **86** (1981) 730.
S 13 Schaefer, B.E.: Publ. Astron. Soc. Pacific **92** (1980) 255.
S 14 Simard-Normandin, N., Kronberg, P., Neidhöfer, J.: Astron. Astrophys. Suppl. **40** (1980) 319.
S 15 Simard-Normandin, N., Kronberg, P., Neidhöfer, J.: Astron. Astrophys. Suppl. **43** (1981) 19.
S 16 Scheuer, P.A.G., Williams, P.J.S.: Annu. Rev. Astron. Astrophys. **6** (1968) 321.
S 17 Simon, R.S., Readhead, A.C.S., Moffet, A.T., Wilkinson, P.N., Anderson, B.: Astrophys. J. **236** (1980) 707.
S 18 Saikia, D.J.: Mon. Not. R. Astron. Soc. **197** (1981) 1097.

S 19 Scheuer, P.A.G.: Plasma Astrophysics (Sturrock, I.A., ed.), Varenna Summer School, Course 38 (1967) 262.
S 20 Scheuer, P.A.G.: Mon. Not. R. Astron. Soc. **120** (1960) 231.
S 21 Syunyaev, R.A.: Soviet Astron.-A. J. **15** (1971) 190.
S 22 Spitzer, L.: Physical Processes in the Interstellar Medium, J. Wiley, New York (1978).
S 23 Shklovsky, I.S.: Soviet Astron.-A. J. **4** (1960) 355.

T 1 Taylor, A.R., Gregory, P.C.: J. R. Astron. Soc. Canada **73** (1979) 305.
T 2 Tielens, A.G.G.M., Miley, G.K., Willis, A.G.: Astron. Astrophys. Suppl. **35** (1979) 153.
T 3 Tsytovitch, V.N.: Annu. Rev. Astron. Astrophys. **11** (1973) 363.
T 4 Tucker, W.H.: Radiation Processes in Astrophysics, MIT Press (1975).

V 1 Van Breugel, W.J.M.: Astron. Astrophys. **81** (1980) 275.
V 2 Van Breugel, W.J.M., Miley, G.K.: Nature **265** (1977) 315.
V 3 Venkatakrishna, K.L., Swarup, G.: Mem. Astron. Soc. India **1** (1979) 25.
V 4 Véron, M.P., Véron, P.: Astron. Astrophys. Suppl. **40** (1980) 191.
V 5 Valentijn, E.A., Perola, G.C., Jaffe, W.J.: Astron. Astrophys. Suppl. **28** (1977) 333.
V 6 Van Vliet, W., Harten, R., Miley, G.K., Albers, H.: Astron. Astrophys. **47** (1976) 345.
V 7 Véron, M.P., Véron, P.: Astron. Astrophys. Suppl. **18** (1974) 309.
V 8 Véron, M.P., Véron, P., Witzel, A.: Astron. Astrophys. **18** (1972) 82.
V 9 Valentijn, E.A., Perola, G.C.: Astron. Astrophys. **63** (1978) 29.

W 1 Willis, A.G., Strom, R.G., Wilson, A.S.: Nature **250** (1974) 625.
W 2 Waggett, P.C., Warner, P.J., Baldwin, J.E.: Mon. Not. R. Astron. Soc. **181** (1977) 465.
W 3 Woltjer, L. in: Quasars and Active Nuclei of Galaxies (Ulfbeck, J., ed.), Physics Scriota **17** (1978) 275.
W 4 Willis, A.G., Miley, G.K.: Astron. Astrophys. Suppl. **37** (1979) 397.
W 5 Wilson, M.A.G.: Mon. Not. R. Astron. Soc. **155** (1972) 385.
W 6 Waggett, P.C.: Mon. Not. R. Astron. Soc. **181** (1977) 547.
W 7 Willis, A.G., Oosterbaan, C.E., de Ruiter, H.R.: Astron. Astrophys. Suppl. **25** (1977) 453.
W 8 Weinberg, S.: Gravitation and Cosmology, J. Wiley, New York (1972).
W 9 Webber, J.C., Yang, K.S., Swenson, G.W.: Astron. J. **85** (1980) 1434.
W 10 Westfold, K.C.: Astrophys. J. **130** (1959) 241.
W 11 Witzel, A. in: Proc. 5th Göttingen-Jerusalem Symp. (Fricke, K.J., Shaham, J., eds.), Abhandl. Akad. Wiss. Göttingen (1981) p. 99.

Z 1 Zamorani, G., Henry, J.P., Maccacaro, T., Tananbaum, H., Soltan, A., Avni, T., Liebert, J., Stocke, J., Strittmatter, P.A., Weymann, R.J., Smith, M.G., Condon, J.J.: Astrophys. J. **245** (1981) 357.


*Acknowledgement.* We thank A. Eckart and J. A. Zensus for help in the preparation of this article.

# 9.7 Cosmology

## 9.7.1 List of symbols

| | |
|---|---|
| $a$ | Stefan-Boltzmann constant = radiation density constant |
| $(B-V)$ | colour index: blue minus visual magnitude |
| $c$ | velocity of light |
| $e$ | electron charge |
| $G$ | constant of gravitation |
| $H$ | Hubble parameter |
| $H_0$ | Hubble constant; present value of $H$ |
| $h$ | (in distance-dependent quantities): the ratio of the true Hubble constant $H_0$ and the value $50\,\mathrm{km\,s^{-1}\,Mpc^{-1}}$, which is used in this section. (Note: some authors define $h=H_0/100$.) |
| $L$ | luminosity in $[\mathrm{erg\,s^{-1}}]$ |
| $\mathfrak{L}$ | luminosity density in $[L_\odot\,\mathrm{Mpc^{-3}}]$ |
| $\lambda$ | wavelength |
| $\Lambda$ | cosmological constant |
| $\mathfrak{M}$ | mass |
| $\mathfrak{M}_\odot$ | solar mass |
| $m$ | apparent magnitude |
| $m_e$ | mass of electron |
| $m_R$ | apparent red magnitude |
| $N$ | number of objects per square degree |
| $n$ | number density |
| $\nu$ | frequency |
| $\Omega$ | density parameter |
| $\Omega_0$ | present density parameter |
| $q$ | deceleration parameter |
| $q_0$ | deceleration constant; present value of $q$ |
| $R, R(t)$ | cosmic scale factor (radius of curvature) |
| $R_0, R(t_0)$ | present value of $R(t)$ |
| $\varrho$ | density of matter |
| $\varrho_0$ | present density |
| $S$ | radio flux density, normally in $[10^{-26}\,\mathrm{W\,m^{-2}\,Hz^{-1}}]=1$ Jy (Jansky) |
| $T$ | temperature |
| $T_0$ | present temperature of microwave background |
| $t$ | cosmic time |
| $t_0$ | expansion age of universe (Friedmann time) |
| $v$ | observed recession velocity (radial velocity) |
| $v_0$ | corrected recession velocity |
| $Y$ | fractional mass abundance of helium |
| $z=\dfrac{\Delta\lambda}{\lambda}$ | redshift |

## 9.7.2 Friedmann cosmologies [42]

The Friedmann models play a particular rôle because of their simplicity and because they conform to all present observations.

### 9.7.2.1 Basic assumptions

(1) The known laws of physics including general relativity (cf. 9.7.3.1) apply everywhere and at all times.
(2) The universe began in a (hot) Big Bang (cf. 9.7.3.4).
(3) The universe is approximately isotropic and homogeneous when large volumes of space (cosmological principle) are considered. If the Copernican principle is adopted, there must be a family of Fundamental Observers (i.e. observers freed from local effects) having the same view of the universe as we; in that case homogeneity follows from isotropy [13] (cf. 9.7.3.5).

## 9.7.2.2 Basic equations [5]

### 9.7.2.2.1 Line element

The Robertson-Walker metric [43, 44] gives the most general line elements in Riemannian geometry if the universe is isotropic and homogeneous:

$$ds^2 = c^2 dt^2 - R^2(t) du^2,$$

$$du^2 = \frac{dr^2}{1-kr^2} + r^2(d\theta^2 + \sin^2\theta \, d\varphi^2), \qquad (1)$$

where $r$, $\theta$, and $\varphi$ are dimensionless, comoving, time-independent space coordinates. $R(t)$ is the (time-dependent) cosmic scale factor, and $k = -1, 0$, or $+1$ is the space curvature.

### 9.7.2.2.2 Einstein field equations

$$\frac{\dot{R}^2}{R^2} + \frac{2\ddot{R}}{R} + \frac{8\pi G p}{c^2} = -\frac{kc^2}{R^2} + \Lambda c^2 \qquad (2)$$

$$\frac{\dot{R}^2}{R^2} - \frac{8\pi G \varrho}{3} = -\frac{kc^2}{R^2} + \frac{\Lambda c^2}{3} \qquad (3)$$

with: $p$ isotropic, hydrodynamic pressure of matter and radiation
$\varrho$ density of matter and energy
$\Lambda$ cosmological constant [1, 45].

### 9.7.2.2.3 The general Friedmann equation

One defines [46, 47]:

Hubble parameter: $$H \equiv \frac{\dot{R}}{R} \qquad (4)$$

deceleration parameter: $$q \equiv -\frac{\ddot{R}}{RH^2}. \qquad (5)$$

The present, observable values of $H$ and $q$ are denoted $H_0$ (Hubble constant) and $q_0$ (deceleration constant). The subscript zero refers to present values throughout.

For $\Lambda = 0$ and $p = 0$ one obtains from Eqs. (1)···(5)

$$\varrho_0 = \frac{3H_0^2 q_0}{4\pi G} \qquad (6)$$

$$\frac{kc^2}{R_0^2} = H_0^2(2q_0 - 1), \qquad (7)$$

or from Eqs. (3) and (4) the general Friedmann equation:

$$\frac{kc^2}{R_0^2} = \frac{8\pi G \varrho_0}{3} - H_0^2. \qquad (8)$$

Present observations do not require models with $\Lambda \neq 0$. The approximation $p = 0$ is justified for epochs after the decoupling of matter and radiation. The present pressure,

$p_0$ = radiation pressure $(a \cdot T_0^4/3)$ + pressure due to random motion of galaxies $(\varrho_0 \sigma^2(v_{rd}))$

with $\sigma(v_{rd})$ = velocity dispersion, is negligible [5].

For radiation-filled epochs, see [5].

One of the fundamental problems of cosmology is to determine the function $R(t)$. This can be done by a series expansion in terms of $R_0(t_0)$, $\dot{R}_0$, and $\ddot{R}_0$ [3, 7, 46]. But Eq. (8) can be integrated directly [48] as it contains only two fundamental constants, the present values $H_0$ and $q_0$. $H_0$ and $q_0$ determine the full history of the large-scale mass distribution in a Friedmann universe after the epoch of decoupling of matter and radiation.

The different Friedmann models

| Curvature | Space | $q_0$ | $\Omega_0$ [3]) | Density $\varrho_0$ | Expansion |
|---|---|---|---|---|---|
| $k=+1$ | closed | $>1/2$ | $>1$ | $>\frac{3H_0^2}{8\pi G}$ | turns eventually into contraction |
| $k=0$ [1]) | flat (Euclidean) | 1/2 | 1 | $\varrho_{0,c}=\frac{3H_0^2}{8\pi G}$ [4]) | stops in infinite future |
| $k=-1$ [2]) | open | $0\leqq q_0<1/2$ | $0\leqq\Omega_0<1$ | $<\frac{3H_0^2}{8\pi G}$ | forever |

[1]) Considered by Einstein and de Sitter [49].
[2]) Considered by Milne [50].
[3]) The "density parameter" $\Omega\equiv\varrho/\varrho_c$, where $\varrho_c$ is the critical closure density (i.e. for the case $q_0=1/2$).
[4]) With $H_0=50\,\mathrm{km\,s^{-1}\,Mpc^{-1}}$, the present critical density becomes $\varrho_{0,c}=4.7\cdot10^{-30}\,\mathrm{g\,cm^{-3}}$.

### 9.7.2.2.4 Distances

The various distances are discussed in [5, 13]. For definition of redshift, $z$, see (9.7.4)

(1) Luminosity distance [48] (for bolometric luminosities):

$$d_\mathrm{L}=\frac{c}{H_0q_0^2}\{q_0z_0+(q_0-1)[\sqrt{1+2q_0z}-1]\},\quad \text{for}\quad q_0>0 \tag{9}$$

$$=\frac{cz}{H_0}\left(1+\frac{z}{2}\right)\quad \text{for}\quad q_0=0. \tag{10}$$

(2) Angular-diameter distance (for metric diameters):

$$d_\mathrm{A}=(1+z)^{-2}d_\mathrm{L}. \tag{11}$$

(3) Proper-motion distance:

$$d_\mathrm{M}=(1+z)^{-1}d_\mathrm{L}. \tag{12}$$

(4) Parallax distance [51]

$$d_\mathrm{p}=\frac{c[zq_0+(q_0-1)(-1+\sqrt{2q_0z+1})]}{H_0[q_0^4(1+z)^2-(2q_0-1)\{zq_0+(q_0-1)(-1+\sqrt{2q_0z+1})\}^2]^{1/2}}\quad q_0>0. \tag{13}$$

For other distances, see [52].

### 9.7.2.2.5 The magnitude–redshift relation [48, 5]

The apparent bolometric magnitude $m_\mathrm{bol}$ of a source with absolute magnitude $M_\mathrm{bol}$ at redshift $z$ is given by:

$$m_\mathrm{bol}=5\log\frac{c}{q_0^2}\{q_0z+(q_0-1)[\sqrt{1+2q_0z}-1]\}+C_\mathrm{bol},\quad \text{for}\quad q_0>0 \tag{14}$$

$$m_\mathrm{bol}=5\log cz(1+\tfrac{1}{2}z)+C_\mathrm{bol},\quad \text{for}\quad q_0=0, \tag{15}$$

where $C_\mathrm{bol}=-5\log H_0+M_\mathrm{bol}+25\,[\mathrm{mag}]$ and $M_\mathrm{bol}$ is the absolute bolometric magnitude, which is related to the total luminosity of the source by $L\approx10^{-0.4M_\mathrm{bol}}\cdot3.02\cdot10^{35}\,\mathrm{erg\,s^{-1}}$, $c$ in $[\mathrm{km\,s^{-1}}]$ and $H_0$ in $[\mathrm{km\,s^{-1}\,Mpc^{-1}}]$.

For small redshift $z$:

$$m_{\mathrm{bol}} = 5 \log cz + C_{\mathrm{bol}}. \tag{14', 15'}$$

If heterochromatic magnitudes $m_{\mathrm{het}}$, $M_{\mathrm{het}}$ are used, e.g. photographic ($m_{\mathrm{pg}}$, $M_{\mathrm{pg}}$), blue ($m_B$, $M_B$) or red ($m_R$, $M_R$) instead of $m_{\mathrm{bol}}$ and $M_{\mathrm{bol}}$, $m_{\mathrm{bol}}$ is to be substituted by $m_{\mathrm{het}} - K_{\mathrm{het}}$ and $C_{\mathrm{bol}}$ by $C_{\mathrm{het}} = -5 \log H_0 + M_{\mathrm{het}} + 25$.

The $K_{\mathrm{het}}$-correction allows for the photometric effect of stretching the energy distribution curve of the emitter and shifting it through the respective filter band:

$$K = 2.5 \log(1+z) + 2.5 \log \frac{\int_0^\infty I(\lambda) S(\lambda) \mathrm{d}\lambda}{\int_0^\infty I\left(\frac{\lambda}{1+z}\right) S(\lambda) \mathrm{d}\lambda} \text{ [mag]}. \tag{16}$$

$I(\lambda)$ is the incident energy flux per unit wavelength and $S(\lambda)$ is the photometer response function.

$K$-corrections for blue ($B$), visual ($V$), and red ($R$) magnitudes are determined for bright elliptical galaxies in [53···56].

#### 9.7.2.2.6 The count–magnitude relation [57, 5]

(1) Optical objects: The integrated number $N(m)$ of objects, uniformly distributed in space, per square degree, which are brighter than apparent magnitude $m$ is given by:

$$N(m) = \frac{2\pi n_0}{Q H_0^3} \begin{cases} (1-2q_0)^{-3/2} [P\sqrt{1+P^2} - \operatorname{arc\,sinh} P], & q_0 < 1/2 \\ (2q_0-1)^{-3/2} [\arcsin P \mp P\sqrt{1-P^2}], & q_0 > 1/2 \end{cases} \tag{17}$$

where

$$P = \frac{A\sqrt{k(2q_0-1)}}{q_0(1+A) - (q_0-1)\sqrt{1+2A}} \tag{18}$$

and

$$A = 10^{0.2(m-K-C)}, \tag{19}$$

with

$$C = M + 25 + 5 \log(c/H_0). \tag{20}$$

$n_0$ is the number of objects per local unit volume [$\mathrm{Mpc}^3$], and $Q = 41252.96$ is the number of square degrees of the whole sky. $M$ is a properly defined mean absolute magnitude of the objects counted and must be independent of redshift and cosmic time $t$.

For the Euclidean case ($q_0 = 1/2$):

$$N(m) = \frac{4\pi n_0 A^3}{3 Q H_0^3} \{1/2[1 + A + \sqrt{1+2A}]\}^{-3}. \tag{21}$$

Equations (17) and (21) can be directly used for bolometric magnitudes $m_{\mathrm{bol}}$ [inserting $M_{\mathrm{bol}}$ and $K=0$ in Eq. (19)]. For heterochromatic magnitudes it must be noted that the $K$-correction is $z$-dependent, or that one can calculate the (unobservable) quantity $N(m_{\mathrm{het}} - K_{\mathrm{het}})$ as function of $m_{\mathrm{het}} - K_{\mathrm{het}}$ [5].

For bright magnitudes ($z \ll 1$), Eqs. (17) and (21) become

$$\log N(m_\lambda) = 0.6 m_\lambda + \mathrm{const}_\lambda. \tag{22}$$

(2) Radio sources [58···61]: The integrated number $N(S)$ of radio sources per square degree with apparent radio flux density brighter than $S[\mathrm{W\,m^{-2}\,Hz^{-1}}]$ and properly defined mean absolute radio luminosity $L_\nu[\mathrm{W\,Hz^{-1}}]$ is approximately given by:

$$N(S_\nu)=\frac{4\pi}{3Q}\varrho_0\left(\frac{L_\nu}{4\pi S}\right)^{3/2}\left\{1-\frac{3}{2}(1+\alpha)\underbrace{\frac{H_0}{c}\left(\frac{L_\nu}{4\pi S}\right)^{1/2}}_{\approx z}\pm\cdots\right\}, \tag{23}$$

where $\varrho$ is the number of sources per unit volume and $\alpha$ is the spectral index defined by $L_\nu \propto \nu^{-\alpha}$ (spectral curvature neglected). (Notice that in 9.6.2.3 the spectral index $\alpha$ is defined with another sign.) In the presence of a (typically) broad luminosity function $\varphi(L)$ one must integrate over $\varphi(L)$. For the exact form of Eq. (23), see [51].

The slope of a $\log N$ versus $\log S$ diagram is given by [52]:

$$\beta=\frac{\mathrm{d}\log N}{\mathrm{d}\log S}\approx-\frac{3}{2}+\frac{3}{4}(1+\alpha)z\pm\ldots; \tag{24}$$

because $\alpha>0$ (typically $\approx 0.75$) it follows that $|\beta|$ is always smaller than 1.5, unless $L$ increases with $z$.

Frequently, differential counts are used. In a static Euclidean universe one has

$$\left(\frac{\mathrm{d}N}{\mathrm{d}S}\right)_0=-\frac{2\pi}{Q}\varrho_0\left(\frac{L_\nu}{4\pi}\right)^{3/2}S^{-5/2}. \tag{25}$$

Hence [61]

$$\frac{\mathrm{d}N}{\mathrm{d}N_0}=\frac{\mathrm{d}N/\mathrm{d}S}{(\mathrm{d}N/\mathrm{d}S)_0}=\frac{2c(1+z)^{-(3/2)(1+\alpha)}}{H_0(2q_0z+1)^{1/2}\left[D(1+\alpha)+2(1+z)\dfrac{\mathrm{d}D}{\mathrm{d}z}\right]}, \tag{26}$$

where $D=d_L(1+z)^{-1}$ and $d_L$ is given by Eqs. (9) and (10).

#### 9.7.2.2.7 Angular diameter–redshift relation [5]

(1) Diameters of rigid rods (metric diameter). A rigid rod of length $l$ [Mpc] will have the apparent angular size $\theta$ given by

$$\theta=\frac{l(1+z)^2}{d_L}, \tag{27}$$

where $d_L$ is given by Eqs. (9) and (10).

(2) Isophotal diameters. The change of bolometric surface brightness $B$ is given, independent of $q_0$, by

$$B=\frac{L}{4\pi y(1+z)^4}, \tag{28}$$

where $y$ is the proper area perpendicular to the line of sight and $L/4\pi y$ is the surface brightness at the source. If $L_{\nu,1+z}$ is the monochromatic luminosity at frequency $\nu(1+z)^{-1}$, the monochromatic surface brightness is given by

$$B_\nu=\frac{L_{\nu,1+z}}{4\pi y(1+z)^3}. \tag{28'}$$

The $z$-dependence of $B$ requires that the isophotes of a galaxy move toward the nucleus as $z$ increases. The isophotal diameter $\theta_\mathrm{i}$ therefore depends on the surface brightness distribution within the galaxy. For the outer parts of elliptical galaxies

$$\theta_\mathrm{i}=\theta(\text{metric})(1+z)^{-2}f(K), \tag{29}$$

approximately [5, 116], where $f(K)$ is a $K$-like correction to the measured diameters to reduce them to a specific proper wavelength.

## 9.7.3 Observations supporting basic assumptions

### 9.7.3.1 Support for general relativity (GR) [62, 63]

Observational tests of GR are expressed in terms of the "space curvature parameter", $\gamma$, of the post-Newtonian formalism (cf. [15] p. 1072):

| Effect | $\gamma$ | Method | Ref. |
|---|---|---|---|
| Gravitational redshifts | $1.01 \pm 0.06$ | solar potassium line | 64 |
| Deflection of radiation by the sun | $1.007 \pm 0.009$ | interferometry of radio sources | 65 |
| Advance of the perihelion of Mercury | $1.003 \pm 0.005$ | radar reflections | 66 |
| Time delay of electromagnetic waves | $1.0011 \pm 0.0007$ | Viking spacecraft, transmitters on Mars | 67, 67a |
| Equality of inertial and passive gravitational mass | within $\pm 1.5\%$ | lunar laser ranging | 68, 69 |

In addition, the model of the binary pulsar PSR 1913+16 (see 5.6.2.7) is entirely consistent with 6 or 7 GR effects (transverse Doppler shift, gravitational redshift, gravitational propagation delay, advance of periastron, nonellipticity of the orbit, decrease of orbital period due to gravitational radiation, and possibly spin-orbit coupling) [70, 71].

### 9.7.3.2 Limits of the annual variation of some physical constants

(1) Gravitational constant $G$:

$dG/G \lesssim 3\cdot 10^{-11}\,a^{-1}$ from lunar laser ranging [72], if the rate of atomic clocks is constant (involving $h$, $c$, $e$, and $m_e$).

$dG/G = (-0.5 \pm 2)\cdot 10^{-11}\,a^{-1}$ from fossil data [73] (if masses of solar system bodies constant).

$dG/G \leqq 8\cdot 10^{-12}\,a^{-1}$ from palaeomagnetism of planets and of the moon [74] (if masses of solar system bodies constant).

(2) Velocity of light $c$:

$dc/c = (+1.2 \pm 1.5)\cdot 10^{-12}\,a^{-1}$ from aberration constant of galaxies at $z \approx 0.2$ [75⋯78b].

$dc/c \lesssim 10^{-11}\,a^{-1}$ from lunar laser ranging [72, 79], if $G$=const. and depending on atomic clocks.

(3) Planck's constant $h$ and velocity of light $c$:

$\frac{d(hc)}{hc} = (0.2 \pm 1.6)\cdot 10^{-12}\,a^{-1}$ from spectrum of microwave background radiation [80].

$\frac{d(hc)}{hc} < (-3 \pm 4)\cdot 10^{-13}\,a^{-1}$ from energy measurements of old and young photons [78, 81]

$\frac{d(hc)}{hc} < 2\cdot 10^{-14}\,a^{-1}$ from zero-redshift of telluric night sky lines in spectra of redshifted objects [78].

(4) A combination of physical constants:
If $\alpha = e^2/\hbar c$ is the Sommerfeld fine-structure constant with $\hbar = h/2\pi$, $g_p$ the gyromagnetic ratio of the proton, $m_e/m_p$ the ratio of electron-to-proton mass, and $X = \alpha^2 g_p m_e/m_p$, then

$dX/X \lesssim 3\cdot 10^{-14}\,a^{-1}$ from optical and 21-cm redshifts of quasars [82].

### 9.7.3.3 Evidence for cosmic evolution

See also [36] p. 289⋯406.

Epoch of galaxy formation

The small range in $(B-V)$ colours of bright elliptical (E) galaxies requires that (most) E galaxies were formed at one definite time in the past [83].

Colour and spectral evolution of galaxies

(1) Field galaxies: It has been suggested that bright galaxies with $z \geqq 0.3$ may be exceptionally blue [84].

(2) Cluster galaxies: The percentage of blue galaxies increases with increasing redshift (for $z \gtrsim 0.2$) [85···87]. Spectra of some galaxies with $z \approx 0.6$ are exceptionally blue [88].

(3) Radio galaxies: Appear very blue for $0.4 \lesssim z \lesssim 0.7$ [89, 90].

Counts of galaxies and radio sources

(1) Optical counts: Counts of galaxies at faint levels give marginal evidence for luminosity evolution (brighter galaxies in the past) or density evolution (more galaxies per proper unit volume in the past) [91···93].

(2) Radio sources: The steep slope [$|\beta| > 1.5$, see Eq. (24)] of counts of sources with high flux densities ($S \gtrsim 2$ Jy) [94, 59, 95, cf. also 34] strongly suggests source evolution [61]. The differential counts at lower flux densities can be modelled assuming source evolution [96, 61].

Quasars

(1) Counts of radio-quiet quasars (QSOs): The number increases much faster than $\log N(m) \propto 0.6m$ [see Eq. (22)], which requires an increase of the comoving volume density roughly with $r^{3/2}$ ($r$ = distance) [97, 98] or with $(1+z)^n$, where n = 7.5···8.8 [99].

(2) The $V/V_{max}$ test [100] of radio quasars (QSRs); the test is based on a comparison of the volume V, within which an object is observed, and the maximum volume $V_{max}$, throughout which the object would still have entered the catalogue under investigation. The test indicates density evolution of QSRs [101...103]. The increase of density $\varrho$ can be approximated by

$$\varrho = \varrho_0^{k(t-t_0)/t_0},$$

where $t$ is the cosmic epoch, $t_0$ the age of the universe, $k = 10 \cdots 12$ for steep-spectrum QSRs, and $k = 4$ for flat-spectrum QSRs [98].

X-ray sources

The increase of sources at faint levels requires an increase of the comoving space density by a factor of $(1+z)^n$, where $n \geqq 3$ [104].

Diameters of galaxy clusters

Cluster diameters at different redshifts $z$ may reveal diameter evolution [105]; the distance-independent definition of cluster sizes is, however, dubious.

### 9.7.3.4 Observational evidence for a Big Bang

In general relativity an isotropic and homogeneous universe with constant mass density and constant space curvature is not possible unless $\Lambda \neq 0$ [106, 1]. Theoretically, if only basic assumptions are made, a singularity becomes inevitable [107].

The magnitude-redshift relation ("Hubble diagram") of first-ranked cluster galaxies (standard candles)

The magnitudes ($m_R$) of first-ranked cluster galaxies obey Eq. (14) within a narrow magnitude scatter (after correction for galactic absorption and $K$-correction $\sigma_m \approx 0\overset{m}{.}4$; after further correction for cluster properties, i.e. richness and Bautz-Morgan class (cf. 9.1.4.2), $\sigma_m \approx 0\overset{m}{.}3$); for history, see [108; 109···111]. The same applies for the mean magnitudes of the ten brightest cluster galaxies [112] and, with somewhat larger scatter, for radio galaxies [113, 114, 111]. (Because of the width of the luminosity function of quasars the interpretation of their Hubble diagram is complex, cf. [115].)

The diameter–redshift relation

(1) First-ranked cluster galaxies: The isophotal angular diameters of first-ranked cluster galaxies at different redshifts $z$ obey Eq. (29) [116].

(2) Radio sources: The angular separations of components of double radio sources follow Eq. (27) [117, 118].

(3) Clusters of galaxies: The characteristic angular metric sizes of clusters of galaxies follow Eq. (27) [119, 105].

The abundance of the lightest elements

Hot Big Bang nucleosynthesis of the isotopes $^2$D (deuterium), $^4$He, and possibly $^7$Li [120···126] must be inferred, because nucleosynthesis in stars or spallation by cosmic rays can hardly account for their observed abundances [127]. The origin of $^3$He is ambiguous [128, 129].

The mass fraction of $^4$He is (with a few explicable exceptions) at least $Y \approx 0.24$ everywhere, in stars and in the interstellar gas (including old population objects and the gas in metal-poor galaxies) [130···132]. This large amount of He cannot have been produced by stellar nucleosynthesis without straining observational limits set by the surface brightness of galaxies [133, 134]; for additional arguments for the primordial origin of $^4$He, see [132].

The deuterium-hydrogen ratio in the solar system and the interstellar gas is $D/H = 1.5 \cdot 10^{-5}$ [135], making D one of the 12 most frequent isotopes in the Galaxy [136]. Because D is effectively destroyed in stars, it is likely to be of primordial origin [137···140].

The microwave background radiation (MBR)

The predicted MBR [141...143b], the relic radiation of a primordial fireball in an expanding universe, exhibits the required properties, i.e. black-body spectrum (of $T \approx 2.7 \cdots 3.0$ K) [144···150] and isotropy (cf. 9.7.3.5), within the observational accuracy. (For the dipole moment of the MBR, see 9.7.4.3.2.) The very high energy density of the MBR is evidence for a primordial fireball. See also [150a, 150b].

### 9.7.3.5 Isotropy

Upper limits to any anisotropy of the universe are set by the following observations:

Counts of galaxies

(1) Nearby counts: For $m_{pg} \lesssim 13\overset{m}{.}5$ a northern-hemisphere excess of galaxies exists [151], caused by the Virgo cluster and its halo ("Supergalaxy", "Metagalaxy", "Virgo complex") [152···157].

(2) Counts at intermediate magnitudes: Counts to $m \lesssim 20^m$ [158···163] are affected by galaxy clustering to scales of $\lesssim 100\,h$ Mpc, but are compatible with large-scale isotropy and homogeneity [164···166]. The integral counts $N(m)$ follow $\log N(m) \propto 0.6\,m$ [cf. Eq. (22)], after correction for the $K$-correction and cosmological effects, proving in favour of a constant mean space density of galaxies, and contradicting hierarchical world models [157].

(3) Counts at faint levels ($m \gtrsim 20^m$): Because of considerable practical problems (Galactic absorption, magnitude standards) the evidence given by different authors diverges. For a review, see [91]; at median redshifts $z \approx 0.5$ no definite deviations from isotropy are detected.

Counts of radio sources

Counts at different frequencies show, except for the strongest sources, no significant deviations from isotropy [167···170, 61]. Density fluctuations are less than 5···10% over linear scales of 1 Gpc ($H_0 = 50$ km s$^{-1}$ Mpc$^{-1}$) [171].

The X-ray background

A substantial fraction of the X-ray background comes from point sources (mainly quasars with redshifts $z \approx 1 \cdots 3$) [172···176]. Fluctuations of the extragalactic background within $10° \times 10°$ bins are between 1.8 and 2.8% [104]. (For the dipole moment of the X-ray background, see 9.7.4.3.2.)

The microwave background radiation

At microwave wavelengths ($\lambda = 0.1 \cdots 6$ cm) the temperature fluctuations are less than $\Delta T // T =$ a few times $10^{-3} \cdots$ few times $10^{-5}$ K over angular scales $2' \cdots 2°$ [177, 178]. These small fluctuations would require an excessive number of individual sources (with unprecedented properties). The origin of the microwave background is therefore generally ascribed to the relic radiation of the primordial fireball. Then the fluctuations refer to redshifts $z \approx 1100$ (epoch of the recombination of hydrogen), unless the universe has been re-ionized in later epochs [179···181]. No quadrupole anisotropy has been found larger than $\Delta T/T \lesssim 10^{-3}$ K [148, 182]. These observations set the most stringent limits to any deviations from large-scale isotropy and homogeneity at the moment. (For the dipole moment, see 9.7.4.3.2.)

## 9.7.4 Redshifts

See also [108].

### 9.7.4.1 Definition of Doppler shift

When line radiation of frequency $\nu_e$ is emitted from an object travelling at a velocity $V$ with respect to an observer at rest (i.e. comoving with the expanding universe), the observed frequency, $\nu_{obs}$ is [183···185, 51]

$$\begin{aligned} \nu_{obs} &= \nu_e (1-(V/c)\cos\theta)^{-1}(1-V^2/c^2)^{1/2} \\ &\approx \nu_e(1-V/c) \quad \text{for} \quad v \ll c, \quad \theta \ll \pi/2, \end{aligned} \tag{30}$$

where $\theta$ is the angle between the velocity vector and the line of sight; the term $v = V\cos\theta$ is the radial velocity. The redshift is defined as [186]

$$z \equiv \frac{\lambda_{obs}-\lambda_e}{\lambda_e} = \frac{\nu_e - \nu_{obs}}{\nu_{obs}} = \frac{R(t_{obs})}{R(t_1)} - 1, \tag{31}$$

where

$\lambda_{obs}$ = observed wavelength
$\lambda_e$ = emitted (laboratory) wavelength
$R(t)$ = radius of curvature at time $t$
$t_{obs}$ = time of observation
$t_1$ = time at which the light was emitted.

Because in an expanding universe $R(t_{obs}) > R(t_1)$ the light of (field) galaxies is redshifted. Some authors [26b] distinguish between redshift due to expansion, and Doppler shift.

Radio astronomers have frequently called the term $(\nu_e - \nu_{obs})\nu_e^{-1}$ "redshift"; this usage is not recommended [187].

The observed redshift $z$ is frequently transformed into a radial velocity by means of

$$v \approx cz \quad \text{for} \quad z \ll 1, \tag{32}$$

or of the special relativistic Doppler formula

$$v = c\frac{(z+1)^2-1}{(z+1)^2+1}. \tag{33}$$

This practice is not strictly correct [186].

In 1980 about $10^5$ extragalactic redshifts were available with typical mean errors (expressed as velocities) of $\sigma_v = 10 \cdots 300\ \mathrm{km\,s^{-1}}$ (for optical observations) or $\sigma_v = 5 \cdots 50\ \mathrm{km\,s^{-1}}$ (for 21-cm observations). Compilations of redshifts are contained, e.g., in [188···190].

### 9.7.4.2 Corrections

The observed redshifts are affected by the peculiar motion of the observer. Because the velocities involved are small, the observed velocity $v$ [from Eq. (32) or (33)] can be corrected to a good approximation by simply adding the local peculiar velocity vector $\Delta v$.

(1) Motion of the earth around the sun:

$$\Delta v_{hel} = 30\cos B\ [\mathrm{km\,s^{-1}}] \tag{34}$$

where $B$ is the angle between the terrestrial velocity vector and the line of sight. All published extragalactic redshifts (velocities) are expected to represent heliocentric values.

(2) Motion of the sun around the Galactic centre: Occasionally published extragalactic velocities are reduced to the Galactic centre. Assuming a solar velocity vector of 250 $\mathrm{km\,s^{-1}}$, one obtains

$$\Delta v_{Gal} = 250\cos A\ [\mathrm{km\,s^{-1}}], \tag{35}$$

where $A$ is the angle between the apex of the solar rotation (at Galactic coordinates $l = 90°$, $b = 0°$) and the line of sight.

(3) Motion of the sun with respect to the centroid of the Local Group (see 9.3.3.3): Several determinations of the solar motion with respect to Local Group members have been made [e.g. 191···193]; a typical determination gives $(308 \pm 23)\,\mathrm{km\,s^{-1}}$ toward $l = 105° \pm 5°$, $b = -7° \pm 4°$ [194]; this corresponds to a correction of

$$\Delta v = -79 \cos l \cos b + 296 \sin l \cos b - 36 \sin b \,[\mathrm{km\,s^{-1}}] \tag{36}$$

(cf. also [189]).

The official correction $\Delta v$ [187] is a hybrid of Eq. (35) and (36) and assumes a solar velocity of $300\,\mathrm{km\,s^{-1}}$ toward $l \equiv 90°$ and $b \equiv 0°$. With $A$ as in Eq. (35) the correction is

$$\Delta v = 300 \cos A \,[\mathrm{km\,s^{-1}}]. \tag{37}$$

This correction is most frequently used.

The sign of the corrections $\Delta v$ is defined such that the corrected velocity $v_0$ is given by

$$v_0 = v + \Delta v. \tag{38}$$

### 9.7.4.3 Peculiar motions of galaxies

Observed peculiar velocities $v_{\mathrm{pec}}$ of galaxies, which are not due to the expansion of the universe, depend on the matter density of the surrounding medium, thereby suggesting that they are mainly induced by density fluctuations and are not of primordial origin [195]. (Primordial peculiar velocities are damped out with the expansion of the universe.)

#### 9.7.4.3.1 Random peculiar motions

(1) Member galaxies of the Local Group (see 9.3.3.3): the standard deviation, $\langle v_R \rangle$, of the radial components of the peculiar velocities is $\langle v_R \rangle \lesssim 35 \cdots 80\,\mathrm{km\,s^{-1}}$ [196].

(2) Motion of Local Group relative to nearest field galaxies: $v_{\mathrm{pec}} \lesssim 20$ ($< 100!$) $\mathrm{km\,s^{-1}}$ [196].

(3) Nearby field galaxies: $\langle v_R \rangle \lesssim 25 \cdots 50\,\mathrm{km\,s^{-1}}$ [196].

(4) Member galaxies within nearby, well-defined groups of galaxies: $\langle v_R \rangle \lesssim 25 \cdots 85\,\mathrm{km\,s^{-1}}$ [196, 196a]. In more distant, possibly larger and less reliably defined groups: $\langle v_R \rangle = 30 \cdots 550\,\mathrm{km\,s^{-1}}$ [197].

(5) Member galaxies of clusters: $\langle v_R \rangle \approx 400 \cdots 2500\,\mathrm{km\,s^{-1}}$ [198]. Note that for distant clusters the conventional way of computing $\langle v_R \rangle$ is inadequate [199].

(6) Centres of clusters of galaxies: $\langle v_R \rangle \lesssim 100\,\mathrm{km\,s^{-1}}$ [200].

#### 9.7.4.3.2 Streaming motions

(1) Virgocentric velocity: Streaming motions are expected in the neighbourhood of density enhancements. The only such motion detected so far is the velocity component of $\approx (220 \pm 50)\,\mathrm{km\,s^{-1}}$ of the Local Group (see 9.3.3.3) centroid toward the Virgo cluster (see Table 1; cf. also [201, 202]). A large Virgocentric velocity of $\approx 480\,\mathrm{km\,s^{-1}}$ of the Local Group centroid suggested by 21-cm line width data of cluster galaxies [203], has not been confirmed by corresponding data of field galaxies [204].

Several other indications for local peculiar streaming motions (or non-Doppler redshifts) are rejected by [205]. A proposed motion of $\approx 400\,\mathrm{km\,s^{-1}}$ of the centroid of the Local Group roughly in the direction of the Galactic plane [206] (Rubin-Ford effect) is hardly significant [194, 202] and, according to [207], is caused by an anisotropy of the apparent magnitudes of the original sample. Complex streaming motions, as suggested in [208], depend on the absence of selection bias (Malmquist effect), and are contradicted by the approximate linearity of the local expansion field [209].

(2) Motions with respect to the cosmic background: Measurements of the dipole moment of the microwave background radiation are interpreted as a local peculiar motion (Table 1), which is significantly larger than the Virgocentric velocity vector and $\approx 40°$ away from the Virgo cluster. A dipole moment of the X-ray background is consistent with this observation [210]. A large volume including the Virgo cluster and the Local Group, must therefore partake of the motion relative to the background. The origin of this peculiar motion can only be inferred to be a distant small-amplitude density enhancement of large scale length ($> 50$ Mpc) [cf. 210a]. Such long-wave density fluctuations are known to exist [210b, 307].

Table 1. Peculiar velocity $v_{LG}$ of the centroid of the Local Group with respect to the Virgo cluster and the microwave background.

| $v_{LG}$ [km s$^{-1}$] | Apex | Method | Ref. |
|---|---|---|---|
| 174 ± 74 | Virgo [1]) | comparison Virgo – Coma cluster | 196 |
| 280 ± 80 | Virgo | relative to Shapley-Ames galaxies | 211 |
| 190 ± 130 | Virgo | elliptical galaxies | 212 |
| 522 [2]) | $l = 274°$ [2]) $b = +36°$ | microwave background | 213 |
| 653 [3]) | $l = 277°$ [3]) $b = +32°$ | microwave background | 214 |

[1]) The Virgo cluster is at Galactic longitude $l \approx 284°$, Galactic latitude $b \approx 74°$.
[2]) The mean of the northern and southern hemisphere data for the solar peculiar motion is $(300 \pm 40)$ km s$^{-1}$ toward $l = 250° \pm 10°$, $b = 63° \pm 6°$ [213]. This velocity is reduced to the centroid of the Local Group with Eq. (36).
[3]) The solar peculiar motion is $(401 \pm 19)$ km s$^{-1}$ toward $l = 268° \pm 4°$, $b = 51° \pm 3°$ [214]. This is reduced to the Local Group with Eq. (36).

### 9.7.4.4 The nature of extragalactic redshifts

Except for the peculiar motions listed in 9.7.4.3, observed extragalactic redshifts are generally interpreted as the Doppler effect of an expanding universe [cf. Eq. (31)]. For discussions of other views, see e.g. [215⋯219].

There is observational evidence that even the largest observed redshifts, i.e. those of quasars, are cosmological:

(1) The distinction between quasars and N-galaxies (see 9.1.3.2.1) depends on observational parameters; N-galaxies have no detectable noncosmological redshift component [220].

(2) Some quasars are members of galaxy clusters with more than 20 member galaxies, whose mean redshift agrees closely with the quasar redshift [221, 222].

(3) The steep slope of the $\log N(m)$ versus $m$ relation of quasars [Eq. (22)] would remain unexplained if quasars were local [98]. With cosmological redshifts the slope is expected in the presence of luminosity evolution (cf. 9.7.3.3).

(4) The visible quasars ($\lesssim 20\cdots21^{m}$) account for a substantial fraction of the observed X-ray background (cf. 9.7.3.5). If they were local the integrated flux of all quasars in the universe would be excessive [172, 173].

Future tests on the nature of redshifts may involve the different $(1+z)$-dependence of the metric and isophotal surface brightness for an expanding and a tired-light model [cf. Eqs. (27) and (28′)] [22, 223], and the time dilatation [a factor $(1+z)$] of supernova light curves [224].

## 9.7.5 The determination of the Hubble constant $H_0$

Because of non-Hubble motions of galaxies (cf. 9.7.4.3), distances must be obtained for galaxies beyond recession velocities $v_0 \approx 1000$ km s$^{-1}$ to determine $H_0$. Few, if any, distance determinations can reach such distances in one step; therefore absolute as well as relative distance determinations are important. The multiplicity of necessary steps is largely responsible for the discrepancy of published values of $H_0$.

In Table 2 various relative distance determinations are compiled for four particularly significant stepping stones within the extragalactic distance scale: the M 81 group of galaxies, the M 101 group (for members of these groups, see e.g. [225]), the Virgo cluster (comprising the galaxies with $v_0 < 2700$ km s$^{-1}$ within 6° from the cluster centre [226])], and the Coma cluster of galaxies [227]. The relative distances are given as differences of the distance moduli $\Delta(m-M)^0$ (the symbol $^0$ indicates that the distance moduli are corrected for galactic absorption) between (1) the M 81 group and the mean modulus for the Large [LMC] and Small [SMC] Magellanic Clouds (M 81-Magell.), (2) the M 101 group and the M 81 group (M 101 – M 81), (3) the Virgo cluster and the M 101 group (Virgo – M 101), (4) the Coma cluster and the Virgo cluster (Coma – Virgo). Where necessary the modulus differences are reduced to a modern model of Galactic absorption [83, 228⋯230].

For several Local Group galaxies (see 9.3.3.3) good distances are known from Cepheid variables (also for NGC 2403 in the M 81 group), e.g. [245]. These galaxies are the primary calibrators of the extragalactic distance scale. Particularly good distance determinations are available for the Magellanic Clouds. From Cepheids, RR

Lyrae stars, novae, and brightest stars one obtains $(m-M)^0 = 18^m\!.7$ ($\pm 0^m\!.2$) for LMC and $(m-M)^0 = 19^m\!.1$ ($\pm 0^m\!.2$) for SMC [245···247, 230]. Combining the mean distance modulus of $(m-M)^0 = 18^m\!.9$ for LMC and SMC with the relative moduli in Table 1 yields the absolute distance moduli of Table 3.

Table 2. Relative distance moduli of groups and clusters of galaxies. For explanation, see text.

| | $\Delta(m-M)^0$ | | | | |
|---|---|---|---|---|---|
| | M81 −Magell. | M101 −M81 | Virgo −M101 | Coma −Virgo | Ref. |
| | $8^m\!.66$ | $1^m\!.64$ | $2^m\!.30$ | $3^m\!.88$ $\pm 0^m\!.12$ | 231, 230 |
| | 8.98 | 1.30 | 2.20 | | 232 |
| | 9.34 | 0.80 | | | 233 |
| | 8.91 | 0.70 | 2.18 | | 234 |
| | >8.01 | 1.90 | | | 235 |
| | | 1.18 | | | 236 |
| | 8.66 | 1.64 | (1.60) [1]) | | 237, 238 |
| | | 1.65 | 1.94 | | 239 |
| | | | 1.83 | | 240 |
| | | | >2.2 [2]) | | 230 |
| | | | | 4.01 | 241 |
| mean | 8.91 ±0.13 | 1.35 ±0.15 | 2.11 ±0.08 | 3.90 | |

Footnotes to Table 2:
[1]) The value is based on the 21-cm line width of spiral galaxies. There is evidence [242···244, 230] that the method is unreliable for cluster galaxies.
[2]) Based on the requirement that the brightest Sc spirals in the Virgo cluster should be at least as bright as the nearby Sc galaxy, M101.

Footnotes to Table 3:
[1]) Cumulative mean errors from Table 2, assuming for the mean modulus of the Magellanic Clouds an error of $\pm 0^m\!.2$.
[2]) The observed mean velocity of the Virgo cluster is $\langle v_0 \rangle = (967 \pm 53)$ km s$^{-1}$ (from 160 members within 6° of the cluster centre) [248]. To obtain the cosmic value of $H_0$ this value is corrected by the local Virgocentric infall motion of $(220 \pm 50)$ km s$^{-1}$ (cf. 9.7.4.3.2).
[3]) The value is unreliable because it could be influenced by the peculiar motion of the group.
[4]) The value is not corrected for the radial component of the expected Virgocentric motion of the M101 group.

Table 3. Distance moduli of groups and clusters of galaxies.

| | M81 group | M101 group | Virgo | Coma |
|---|---|---|---|---|
| $(m-M)^0$ | $27^m\!.81 \pm 0.24$ [1]) | $29.16 \pm 0.28$ [1]) | $31.27 \pm 0.29$ [1]) | $35.17 \pm 0.32$ [1]) |
| Distance [Mpc] | 3.6 | 6.8 | $17.9 \pm 2.5$ | $108 \pm 16$ |
| $v_0$ [km s$^{-1}$] | $240 \pm 22$ [196a] | $368 \pm 23$ [196a] | $1187 \pm 73$ [2]) | $6960 \pm 63$ [198] |
| $H_0$ [km s$^{-1}$ Mpc$^{-1}$] | (67) [3]) | (54) [4]) | $66 \pm 10$ | $64 \pm 11$ |

The unweighted means in Table 2 and 3 indicate that the value of $H_0$ should lie between 55 and 80 km s$^{-1}$ Mpc$^{-1}$. Physical diameter determinations of supernovae (Baade-Wesselink method and variations thereof) [249···251] and the evidence from historical supernovae consistently yield $H_0 = (52 \pm 5)$ km s$^{-1}$ Mpc$^{-1}$ [230], cf. [252]. Other methods yield $H_0 = 44$ [253], 64 [254], 80 [255, 256], and 93 km s$^{-1}$ Mpc$^{-1}$ [257]. The prerequisite that the Galaxy and M 31 be not excessively large compared to field galaxies requires $H_0 < 75$ km s$^{-1}$ Mpc$^{-1}$ [258]. Globular clusters as extragalactic distance indicators are discussed in [259]. Distances from X-ray clusters and their microwave background decrement (Sunyaev-Zel'dovich effect) are still inconclusive [260, 261]. The value of $H_0$ derived from superluminous radio sources [262] is model-dependent and has been refuted [263].

Selection bias of catalogued galaxies, in the sense that their luminosity increases with distance (Malmquist effect), necessarily leads to an apparent increase of $H_0$ with distance [264, 265, 209], unless special precautions are taken. Suggestions that $H_0$ increases up to ≈100 km s$^{-1}$ Mpc$^{-1}$ between $v_0 = 1000$ and ≈4000 km s$^{-1}$ [266, 267] are therefore unreliable [230].

The best weighted mean is probably $H_0 = 55 \cdots 60$ with an uncertainty of ≈10 km s$^{-1}$ Mpc$^{-1}$. This value depends on the adopted distances in the Local Group, particularly on those of LMC and SMC. These distances refer to the Galactic distance scale and to a Hyades modulus of $(m-M)^0 = 3^m\!.03$ [268]. Because the Hyades modulus probably has to be increased to $(m-M)^0 = 3^m\!.30$ [269], the distances to the local calibrating galaxies have to be increased by ≈13%, and hence the value of $H_0$ has to be decreased by the same percentage. As a consequence the rounded value of

$$H_0 = 50 \text{ km s}^{-1} \text{ Mpc}^{-1}$$

is recommended as realistic (see also [269a]).

The distances of the Virgo and Coma clusters lead directly to the absolute magnitudes of the respective first-ranked and of the ten brightest cluster galaxies. The absolute magnitudes can be inserted into the Hubble diagram of the first-ranked and of the ten brightest cluster galaxies respectively [Eqs. (14) and (15); cf. 9.7.3.4], to show that the cosmic value of $H_0$ remains constant to within $\approx 15\%$ even for large redshifts ($v_0 \approx 25\,000\ \mathrm{km\,s^{-1}}$) [270, 271, 230].

In this section all distance-dependent quantities are reduced to $H_0 = 50\ \mathrm{km\,s^{-1}\,Mpc^{-1}}$.

## 9.7.6 The determination of the deceleration constant $q_0$

The determination of $q_0$ via the relations discussed in 9.7.2.2.5···9.7.2.2.7 are known as "cosmological tests", cf. [5, 272].

### 9.7.6.1 The classical tests for $q_0$

#### 9.7.6.1.1 The magnitude-redshift relation

The Hubble diagram of first-ranked cluster galaxies taken as standard candles (cf. 9.7.3.4) yields a formal value of $q_0 = 1.6 \pm 0.4$ [113, 111]. A different interpretation of possible selection effects leads to a substantially smaller value [110, 272]. The formal value must be corrected for

(1) Luminosity evolution: Because giant stars contribute substantially to the optical luminosity of E galaxies [273] they must have been brighter in earlier epochs, and thus necessitate a correction $\Delta q_0$ of the formal value of $q_0$ [274, 272]. Present estimates give $\Delta q_0 \lesssim -1.4$ [275].

(2) Dynamical evolution: Dynamical friction in clusters leads to galaxy mergers; the luminosity of the first-ranked galaxy therefore increases with time [276···278, 272]. Involved surface brightness measurements suggest $1 < \Delta q_0 < 1.7$ [279].

(3) Intergalactic absorption: Upper limits to this effect can be set by the visibility of quasars up to redshifts $z \gtrsim 3$.

(4) Imperfect imaging of a universe with density fluctuations [280···285]: The effect tends to average out for unbiased samples of galaxies [286].

Allowing for the above uncertainties it can be concluded only that the true value of $q_0$ is $\lesssim 2$.

Other prospective standard candles suggested are a characteristic luminosity $L^*$ in the luminosity function of cluster galaxies [287], originally proposed in [288], and flat-radio-spectrum quasars, using the strength of certain resonance lines as a luminosity calibrator [289]. The latter method suggests $q_0 \gtrsim 0.5$ with no evolution corrections applied [290]. Possibly the maximum luminosity of Type I supernovae is relatively free of luminosity evolution [224].

#### 9.7.6.1.2 The count–magnitude relation

Counts of extragalactic objects of low magnitudes or low flux densities are possibly dominated by evolutionary effects (cf. 9.7.3.3) [291, 91, 93]. Allowing for standard evolutionary effects, counts in two colors, suggest a small $q_0$ ($q_0 \approx 0$) [292].

#### 9.7.6.1.3 The angular diameter–redshift relation

The definition of a metric galaxy diameter of sufficient size poses severe difficulties. A method proposed by [293] relies on a specific shape of the luminosity profile of galaxies. If measurements of apparent angular diameters of first-ranked cluster E-type galaxies were to be carried out for $z > 0.4$ they might contribute to the determination of $q_0$, if the evolutionary effects are considered [116]. The evidence from the angular size of radio sources is still inconclusive (for a discussion, see [34], p. 109···147). Characteristic sizes of clusters of galaxies (or their core radii [294]) can be used to derive a formal value of $q_0 = 0.25 \pm 0.5$, which does not allow for evolutionary effects [119] (cf. 9.7.3.3).

#### 9.7.6.1.4 Other methods

Other methods proposed for the determination of $q_0$ include:

(1) The count–angular size relation [295].

(2) The redshift distribution by apparent magnitude [295].

(3) Distance determinations of supernovae at $z \gtrsim 0.2$ via the Baade-Wesselink method [250, 296, 297].

(4) The night-sky brightness: The first correct derivation of the significance of the low night-sky brightness for the large-scale structure of the universe by J.-P. Loys de Cheseaux in 1744, repeated by W. Olbers in 1823 (the "Loys de Cheseaux-Olbers Paradox") [1], is considered to be the first observation relevant to modern cosmology, e.g. [298]. Measurements of the extragalactic night-sky component contain information on $q_0$ [299], but the expected differences in optical light are small [300]. Actual measurements are difficult [301···302a], and their interpretation is dominated by the cosmic turn-on time of galaxies and their luminosity evolution [303].

### 9.7.6.2 The determination of the mean matter density

The mean matter density $\varrho$ is related to $q$ in a pressure-free universe by Eq. (6)

$$q_0 = \frac{4\pi G \varrho_0}{3H_0^2}. \tag{39}$$

The determination of $\varrho_0$ is equivalent to the determination of $q_0$, if $\Lambda = 0$. [Note that $q_0$ from Eq. (39) is independent of the value of $H_0$, because $\varrho_0 \propto H_0^2$.] Values of $q_0$ derived from Eq. (39), are generally expressed in terms of the density parameter $\Omega_0$ (cf. 9.7.2.2.3), where $\Omega_0 = 2q_0$. See also [272].

If $\Lambda \neq 0$, the relation between $\Omega_0$ and $q_0$ is given by

$$\Omega_0 = 2q_0 + \frac{2\Lambda}{3H_0^2}. \tag{40}$$

An independent determination of $\Omega_0$ (including also hidden mass) and $q_0$ therefore leads to the value of $\Lambda$.

#### 9.7.6.2.1 $\Omega_0$ from the mean luminosity density

The value of $\Omega_0$ follows from the mean large-scale luminosity density, $\mathfrak{L}$, residing in galaxies, and an appropriate mass-to-luminosity ratio of galaxies, $\mathfrak{M}/L$.

(1) The mean luminosity density, $\mathfrak{L}_B$: the large-scale luminosity density is given in [157] for:

| | |
|---|---|
| E and S0 galaxies: | $\mathfrak{L}_B = 2.2 \cdot 10^7\, L_\odot\, \text{Mpc}^{-3}$ |
| Spiral galaxies: | $\mathfrak{L}_B = 5.6 \cdot 10^7\, L_\odot\, \text{Mpc}^{-3}$ |
| Same after correction for internal absorption: | $= 8.8 \cdot 10^7\, L_\odot\, \text{Mpc}^{-3}$. |

The values of $\mathfrak{L}_B$ refer to $B$-magnitudes, after correction for galactic absorption. $\mathfrak{L}$ scales with $H_0$ like $h$. For independent determinations of $\mathfrak{L}$, see [304···307].

(2) The mass-to-luminosity ratio of galaxies (see also 9.1.4.5): The mass-to-luminosity ratio $\mathfrak{M}/L_B$ in units of solar mass $\mathfrak{M}_\odot$ and solar blue luminosity $L_{B\odot}$, after full correction for internal absorption in the case of spirals, are best represented (the values scale with $H_0$ like $h$):

| | | |
|---|---|---|
| Rotation curves of spirals: | $\mathfrak{M}/L_B \gtrsim 3 \cdots 5$ | [308] |
| Pairs and groups of galaxies: (mainly spirals) | $\approx 25$ | [308···310] |
| Large clusters of galaxies: (mainly ellipticals). | $\approx 325$ | [308] |

(3) The mean mass density, $\varrho_0$: Taking as a first approximation, the value $\mathfrak{M}/L$ in pairs and groups of galaxies as being representative for spirals and the value $\mathfrak{M}/L$ in large clusters for elliptical galaxies, and by combining these with the appropriate values $\mathfrak{L}$ under (1) one obtains (the values scale with $H_0$ like $h^2$):

$\varrho_0$ (spirals) $= 2.2 \cdot 10^9\, \mathfrak{M}_\odot\, \text{Mpc}^{-3}$

$\varrho_0$ (ellipticals) $= 7.2 \cdot 10^9\, \mathfrak{M}_\odot\, \text{Mpc}^{-3}$

---

sum: $\varrho_0 = 9.4 \cdot 10^9\, \mathfrak{M}_\odot\, \text{Mpc}^{-3}$

$= 6.4 \cdot 10^{-31}\, \text{g cm}^{-3}$.

From this it follows (independent of $H_0$):

$$\Omega_0 = \varrho_0/\varrho_{0,c} = 0.14 .$$

(For $\varrho_{0,c}$, cf. 9.7.2.2.3). The uncertainty of $\Omega$ is at least a factor of 2.

### 9.7.6.2.2 $\Omega_0$ from the Virgocentric flow model

Deviations of a local expansion field, characterized by $H_0$ (local), from the global expansion field, characterized by $H_0$ (cosmic), provide a method for the mass determination of density fluctuations, the amplitude of which can be determined by galaxy counts [151]. Taking the local streaming velocity of $(220 \pm 50)\, \text{km s}^{-1}$ toward the Virgo cluster (cf. 9.7.4.3.2) as the gravitational effect of the excess mass within a sphere centered on the Virgo cluster and the Local Group at its periphery, and combining this with the density contrast $\Delta\varrho_0/\varrho_0 = 3 \pm 0.2$ within that sphere, determined from galaxy counts [157, 311] yields [211]

$$\Omega \approx 0.08 \pm 0.04 .$$

This method measures all visible and invisible mass which exhibit the same density contrast $\Delta\varrho_0/\varrho_0$ as galaxies within the Virgo complex.

The value $\Omega_0 = 0.08$ implies a mean mass-to-luminosity ratio of $\mathfrak{M}/L_B \approx 50$ for galaxies in the Virgo complex. The high value of $\mathfrak{M}/L_B \approx 325$ found in rich clusters therefore cannot apply to all galaxies

### 9.7.6.2.3 $\Omega_0$ from galaxy correlations

Deviations of the random angular distribution of galaxies can be expressed by the angular pair correlation function $\xi(r)$ which can be transformed into a spatial pair correlation function [312].

$\xi(r)$ and information on the random motions of galaxies can be used to impose a "cosmic energy condition" (sometimes called "cosmic virial theorem", although also applied to matter not in virial equilibrium) on an ensemble of galaxies [313, 314]; this leads to estimates of $\Omega_0$ varying between [315]

$$0.07 \lesssim \Omega_0 \lesssim 0.3 .$$

A variation of this method applies the triplet correlation function and assumes dynamical equilibrium for the galaxies under consideration (therefore correctly called "cosmic virial theorem"). This results in similar estimates of $\Omega_0$ [316, 315]. For more details, see e.g. [317⋯320]; for a review, see [315].

### 9.7.6.2.4 $\Omega_0$ from the primordial nucleosynthesis

The yield of light elements, particularly $^2$D and $^4$He, during the primordial fireball nucleosynthesis (at a temperature $T \approx 10^9$ K) depends on the matter density $\varrho$ in that epoch ($t \approx 220$ s) [120⋯126]. That density which gives the proper yield to agree with the observed abundance can be scaled down to the present cosmic background radiation temperature (cf. 9.7.3.4), thus giving a determination of the density $\varrho_0$ and hence of $\Omega_0$.

The yield of $^4$He is relatively insensitive to density. With an adopted primordial He abundance of Y = 0.24⋯0.27 (cf. 9.7.3.4) the permissible range of $\Omega_0$ (scaling like $h^2$) is [121]

$$0.01 \leqq \Omega_0 \leqq 0.35 .$$

The yield of $^2D$ is quite sensitive to density. The observed hydrogen-deuterium ratio ($D/H=1.5\cdot10^{-5}$ by number, $X_D=2.1\cdot10^{-5}$ by mass [135]) is probably affected by destruction in stars [321]; thus the primordial value is $X_D \gtrsim 3\cdot10^{-5}$. This value requires [322]

$$0.04 \leqq \Omega_0 \leqq 0.08 .$$

The determination of $\Omega_0$ can be invalidated by assuming non-zero lepton numbers [323], but then the $^4$He abundance of $Y \approx 0.24$ becomes a result of chance, whereas in canonical Big Bang models it is a necessity [324]. The yield of light elements is also sensitive to primordial shearing velocity fields [324]. See also [140].

The combined observational evidence favours $\Omega_0 \approx 0.08$. A value of $\Omega_0=0.1$ is recommended for practical purposes. In any case the data strongly suggest $\Omega_0 \ll 1$, and that the universe is open. Final proof is still lacking: sufficient amounts of dark matter to close the universe would escape detection if they were uniformly distributed or clumped independently of the distribution of galaxies. This, however, would render difficult the formation of galaxy clusters [311].

The possible contribution of relic neutrinos to the total mass is limited if neutrinos with rest masses $m_\nu c^2 \gtrsim 3$ eV participate in gravitational clustering [325]; in that case their mass contribution is allowed for in the mass-to-luminosity ratio of rich clusters (9.7.6.2.1) and in the value of $\Omega_0$ from the Virgocentric flow model (9.7.6.2.2). A mass determination of the electron neutrino of $m_\gamma c^2 \approx 30$ eV was reported [326, 327]. If, however, the electron neutrino has $m_\nu c^2 \lesssim 3$ eV and does not cluster, its contribution to the total density is $\Omega_{0,\nu} \lesssim 0.15$ (or three times this value if muonic and tau neutrinos have similar masses) [328]. Thus neutrinos may dominate the universe, but can hardly close it. See also [329, 330].

## 9.7.7 Constituents of the universe

Present values are given under (1)···(4). Values with * are derived by theoretical considerations; all others are observed.

(1) Observable matter:

Matter density, $\varrho_0$:

$$\varrho_0 = \Omega_0 \varrho_{0,c}$$
$$= 4.7\cdot10^{-31}\, h^2\, \mathrm{g\,cm^{-3}} \qquad (\Omega_0=0.1).$$

Number density of baryons, $n_b$:

$$n_b \approx 3\cdot10^{-7}\,\mathrm{cm^{-3}} \qquad (\Omega_0=0.1).$$

The total mass of the universe, $\mathfrak{M}_T$:

for $q_0 \leqq 0.5$: $\mathfrak{M}_T$ infinite

Mass, $\mathfrak{M}_T$, within present particle horizon, $r_h$, [13] ($r_h$ is of order of $cH_0^{-1}$):

$$\mathfrak{M}_T^* \approx \tfrac{1}{2}\Omega_0 c^3 H_0^{-1} G^{-1}$$
$$\approx 10^{22}\,\Omega_0 h\, \mathfrak{M}_\odot$$

for $q_0 > 0.5$:

$$\mathfrak{M}_T^* = \frac{3\pi c^3}{8H_0 G}\frac{\Omega_0}{(\Omega_0-1)^{3/2}}$$
$$= 1.5\cdot10^{23}\frac{\Omega_0}{(\Omega_0-1)^{3/2}}\, h\, \mathfrak{M}_\odot$$
$$\approx 10^{80}\ \text{baryons}.$$

(2) Unobservable matter: limits to the mass of the intergalactic gas, outside the clusterings of galaxies, are discussed e.g. in [180, 331].

(3) Number density of photons, $n_\gamma$:

$$n_\gamma = 6\cdot 10^7 \Omega_0^{-1} n_b = 6\cdot 10^8 n_b \quad (\Omega_0 = 0.1)$$

$$n_\gamma \gtrsim 2.4\cdot 10^9 n_b \quad [322].$$

Table 4. Approximate estimates of the energy and number density of photons of the isotropic background radiation [61].

| Wavelength range | Energy density of radiation eV cm$^{-3}$ | Number density of photons cm$^{-3}$ |
|---|---|---|
| Radio | | |
| a) metre wavelengths | $3\cdot 10^{-8}$ | 0.3 |
| b) microwave background radiation | 0.25 | 400 |
| Infrared | $\approx 10^{-2}\cdots 10^{-3}$ | $\approx 0.1\cdots 1$ |
| Optical | $\approx 3\cdot 10^{-3}$ | $\approx 10^{-3}$ |
| X-ray | | |
| a) soft X-rays (energies $\varepsilon < 1$ keV) | $\approx 10^{-4}\cdots 10^{-5}$ | $\approx 3\cdot 10^{-7}\cdots 3\cdot 10^{-8}$ |
| b) hard X-rays ($1 < \varepsilon < 200$ keV) | $4\cdot 10^{-4}$ | $10^{-8}$ |
| $\gamma$-rays | | |
| a) soft $\gamma$-rays ($1 < \varepsilon < 10$ MeV) | $10^{-4}\cdots 10^{-5}$ | $3\cdot 10^{-11}\cdots 3\cdot 10^{-12}$ |
| b) hard $\gamma$-rays ($\varepsilon < 30$ MeV) | $10^{-5}$ | $10^{-12}$ |

(4) Number density of neutrinos, $n_\nu$:

$$n_\nu^* = \tfrac{6}{22} n_\gamma \text{ (for electron neutrinos)}$$

For more details, see [322].

## 9.7.8 The time scale of the universe

### 9.7.8.1 The Friedmann time

The expansion time since the Big Bang, the "Friedmann time" $t_0$, is smaller than the "Hubble time", $H_0^{-1}$, in all realistic Friedmann models. The ratio of these two times, $t_0 H_0$, is calculated and tabulated as a function of $q_0$ (also for a radiation-filled universe) in [5]. Some typical values are shown in Table 5.

### 9.7.8.2 The look-back time

Objects of different redshifts $z$ sample the universe at different epochs. The corresponding look-back times, $\tau$, are calculated and tabulated in [322]. Some typical values are given in Table 5.

Table 5. The Friedmann time, $t_0$, for different values of $q_0$, and the look-back time $\tau$ for different values of $q_0$ and $z$. To obtain $t_0$ the tabulated values $t_0 H_0$ must be multiplied by $H_0 = 6.2\cdot 10^{17}$ s $= 1.96\cdot 10^{10}$ a ($H_0 = 50$ km s$^{-1}$ Mpc$^{-1}$).

| $q_0$ | $t_0 H_0$ | $\tau/t_0$ | | | | |
|---|---|---|---|---|---|---|
| | | $z=0.05$ | $z=0.1$ | $z=0.5$ | $z=1$ | $z=4$ |
| 0 | 1 | 0.05 | 0.09 | 0.33 | 0.50 | 0.80 |
| 0.1 | 0.85 | 0.06 | 0.11 | 0.39 | 0.57 | 0.87 |
| 0.5 | 2/3 | 0.07 | 0.13 | 0.46 | 0.65 | 0.91 |
| 1.0 | 0.57 | 0.08 | 0.15 | 0.50 | 0.68 | 0.92 |

### 9.7.8.3 The age of the universe

The oldest objects observed in the Galaxy set a lower limit to the age of the universe [333]. In particular

(1) Globular clusters: Fitting the colour-magnitude diagrams of the best-observed, extremely metal-poor halo globular clusters M 15 and M 92 [334] to a set of theoretical isochrones gives an age (nearly independent of the adopted helium abundance, Y) of [335]

$$14 \cdots 16 \cdot 10^9 \text{ years}.$$

Similar fitting, but allowing for differences of the respective RR Lyrae variables, yields ages of the globular clusters M 15 and M 3 of [335a]

$$(17.3 \pm 3.5) \cdot 10^9 \text{ years}.$$

(2) Radioactive elements: The observed abundances of $^{232}Th$, $^{235}U$, and $^{238}U$ in the solar system indicate an age of $7 \cdots 8 \cdot 10^9$ years, if they were formed by some sudden nucleosynthesis in the past [333, 336]. Assuming a steady-state production rate until the solar system was isolated increases the age to $\approx 12 \cdot 10^9$ years [333]. Models of the variable production rate (as function of the Galactic star formation and supernova rate) which include the additional evidence from $^{129}I$ and $^{244}Pu$, lead to an age of the first radioactive elements of $(12 \pm 3) \cdot 10^9$ years [337, 338].

The isotopes $^{187}Re$ and $^{187}Os$ provide an independent cosmochronometer [339]. Allowing for a wide range of Galactic chemical evolution models yields an age of $7 \cdots 20 \cdot 10^9$ years [340] and $13 \cdots 22 \cdot 10^9$ years [341]. The accuracy of the method is questioned by [342].

The combined evidence favours an age of the Galaxy of $(15 \pm 3) \cdot 10^9$ years. The value is to be compared with a Friedmann time $t_0 = 16.7 \cdot 10^9$ years ($H_0 = 50 \text{ km s}^{-1} \text{ Mpc}^{-1}$, $\Omega_0 = 0.1$).

Note: even an uncomfortably low age of the Galaxy of $10^{10}$ years would require $H_0 < 100 \text{ km s}^{-1} \text{ Mpc}^{-1}$ for all Friedmann models.

Observational evidence for a turn-on time of quasars or galaxies – from the apparent redshift cutoff at $z \approx 3.5$ for quasars [343], and from the high extragalactic night-sky brightness [301; compare, however 302, 302a] – is not stringent.

### 9.7.8.4 The history of the Canonical Big Bang

The history of the Friedmann universe can be traced back with considerable reliability through the lepton era (cosmic time $t > 10^{-4}$ s), where the effects of photons and leptons (electrons, neutrinos, muons etc. and their corresponding antiparticles) dominate the universe [13, 384]. The preceding hadron era ($t \gtrsim 10^{-7}$ s) is less well understood, as the situation is much more complicated because of strong interactions between hadrons (protons, neutrons, quarks, mesons etc. and their corresponding antiparticles) [22, 386, 26]. The very early epochs ($t \gtrsim 10^{-42}$ s) pose formidable problems and investigation is only beginning [386, 387]. The epoch $t \lesssim 10^{-42}$ s is still speculative [388, 386]. In Table 6 the major events are listed. (Cf. also [126, 181, 150b] etc.).

Table 6. The history of the Friedmann universe. (For explanations, see p. 365.)

| Era | $t$ s | $z$ | $T$ K | Density [g cm$^{-3}$] radiation | Density [g cm$^{-3}$] matter | Event |
|---|---|---|---|---|---|---|
| | 0 | ∞ | ∞ | ∞ | | Singularity. Quantum gravitational effects. |
| | ($10^{-43}$) | | $10^{32}$ | $10^{93}$ | | Planck time. Gravitational waves decouple. Strong interactions and weak and electromagnetic interactions separate out from a grand unified theory of interactions. Era of baryon production. Freezing of slight asymmetry between baryons and antibaryons (1 additional part in $\approx 6\cdot 10^5$) |
| | $\approx 10^{-35}$ | | $10^{26}$ | | | |
| | | | $10^{16}$ | | | Weak and electromagnetic interactions separate out from a gauge theory of interactions. Era of intermediate vector boson production. |
| Hadron era | $\approx 10^{-7}$ | | $6\cdot 10^{14}$ | $5\cdot 10^{24}$ | | Photons, leptons, antileptons, quarks, antiquarks, and gluons constantly form out of the radiationfield and annihilate. They form an almost ideal gas described by quantum chromodynamics. At lower temperatures equation of state unknown because of strong interactions; probable phase transition. Quarks combine to make hadrons most frequent particles. As temperature cools below threshold temperature of (anti-)hadrons they annihilate. |
| Lepton era | $10^{-4}$ | $8\cdot 10^{11}$ | $1.6\cdot 10^{12}$ | $3\cdot 10^{14}$ | $3\cdot 10^{5}$ | Lightest hadron (pi meson) annihilates. Mainly leptons and photons survive, coexisting in nearly perfect *thermal equilibrium* (this determines the later evolution, independent of the earlier history). |
| | 1 | $5\cdot 10^{9}$ | $10^{10}$ | $5\cdot 10^{5}$ | $7\cdot 10^{-2}$ | Neutrinos decouple (from now on they only contribute to the gravitational field). |
| | 3 | $3\cdot 10^{9}$ | $6\cdot 10^{9}$ | $6\cdot 10^{4}$ | $2\cdot 10^{-2}$ | The temperature becomes less than threshold temperature of electrons and positrons; they rapidly annihilate, leaving only enough $e^-$ to balance charge of protons. During lepton era the proton-neutron ratio of 1:1 shifts through neutron decay to $\approx 6:1$. Deuterium still unstable, blocking nucleosynthesis through two-particle reactions. |
| Radiation era | $1.8\cdot 10^{2}$ | $4\cdot 10^{8}$ | $10^{9}$ | 12 | $3\cdot 10^{-5}$ | |
| | $2.2\cdot 10^{2}$ | $3.6\cdot 10^{8}$ | $9\cdot 10^{8}$ | 8 | $2\cdot 10^{-5}$ | $^2$D becomes stable, build-up of $^3$He, $^4$He, $^7$Li. The $^4$He yield depends on number of neutrons left. |
| | $2\cdot 10^{13}$ | 1100 | 3000 | $1\cdot 10^{-22}$ | $6\cdot 10^{-22}$ | Free electrons are bound by hydrogen "re"-combination. The universe becomes transparent (presumably for all later epochs). Matter and radiation decouple. |
| Matter era | $5\cdot 10^{13}$ | 740 | 2000 | $1.9\cdot 10^{-22}$ | $1.9\cdot 10^{-22}$ | Matter and radiation densities are equal. The universe becomes matter-dominated. |
| | $1\cdot 10^{15}$ | 100 | 300 | $1\cdot 10^{-25}$ | $5\cdot 10^{-25}$ | Galaxies and clusters of galaxies begin to form (possibly in reversed order). |
| | $6\cdot 10^{16}$ | 5 | 11 | $2\cdot 10^{-31}$ | $6\cdot 10^{-29}$ | |
| | $5.3\cdot 10^{17}$ | 0 | 2.7 | $6.5\cdot 10^{-34}$ | $4.7\cdot 10^{-31}$ | Present |

To Table 6: The numerical values are scaled to $H_0=50$ [km s$^{-1}$ Mpc$^{-1}$] and $\Omega_0=0.1$, i.e. for an expansion age (Friedmann time) of $t_0=16.7\cdot10^9$ years $=5.3\cdot10^{17}$ s. The expansion factor $R/R_0$ is given in terms of redshift, $z=R/R_0-1$, Eq. (31). The temperature $T$ is the photon temperature. A present cosmic microwave background temperature of $T_0=2.7$ K is adopted. The radiation density includes the neutrino contribution (assuming zero rest mass for neutrinos).

## 9.7.9 Other cosmologies

The following list of non-Friedmann cosmologies is incomplete; it gives a selection of those theories which are of particular interest to the astrophysicist.

Newtonian cosmology was shown to provide [347, 348] a good approximation to general-relativistic models, except for the curvature of space and hence for the propagation of light [1, 4, 13, 18].

### 9.7.9.1 Other pressure-free, uniform world models of general relativity

| Model | Characterization | Original source |
|---|---|---|
| de Sitter | $R(t)=R_0 \exp tH_0\sqrt{\lambda_0}$ *) | 344 |
| Einstein – de Sitter | $R(t)=R_0\,(\frac{3}{2}\,tH_0)^{2/3}$ | 49 |
| Milne | $R(t)=ct$ | 50 |
| static (Einstein) | $R\ =c^2/\Lambda,\ \Lambda=4\pi G\varrho_0$ | 345 |

*) $\lambda_0=\dfrac{\Lambda}{3H_0^2}$.

The Lemaître model [349] with positive curvature, $k=1$, and cosmological constant $\Lambda\neq0$, expands from a Big Bang, slows down for a period, and then either expands to infinity ($\Lambda>\Lambda_{crit}$), or contracts back into a singularity ($\Lambda<\Lambda_{crit}$). The model with a logarithmically infinite standstill is known as the Lemaître-Eddington solution [350]. The phase of slow expansion has been thought of as the epoch of galaxy formation. Present observations do not call for a $\Lambda$-term, but do not exclude it. For $\Lambda$, see also [351].

### 9.7.9.2 Models with variable mass and/or variable physical constants

Eight fundamental "constants" are known which seem to define the physical properties of the universe [352], i.e. Planck's constant $h$, the velocity of light $c$, the electron charge $e$, the electron mass $m$, Fermi's constant of weak interactions $g$, the constant of gravitation $G$, the Hubble constant $H$, and the mean density $\varrho$ of matter (not considering $\Lambda$ here), – the choice of the eight "constants" can still be slightly modified. It is possible to combine these eight to form five dimensionless numbers. To allow for an evolving universe at least two of the eight cannot really be constant with time. In all standard models the two variable ones are assumed to be $H$ and $\varrho$. But secular changes of any two or more fundamental "constants" can be assumed to construct evolving world models. Cosmologies of the latter type are for instance:

#### 9.7.9.2.1 Steady-state model

Introduced by [353] and [354] the steady-state theory postulates the "Perfect Cosmological Principle", which requires the universe not only to be homogeneous and isotropic, but to look the same to a fundamental observer at any given time [4, 24]. In an expanding universe this implies continuous mass creation. The observational evidence for cosmic evolution (9.7.3.3) is in contradiction to the Perfect Cosmological Principle.

#### 9.7.9.2.2 Dirac-Jordan cosmology

In Dirac's [355] Big Bang cosmology the "Large Numbers" [356, 180, 357] are taken as a significant feature of the universe. This requires $G$ to decrease with cosmic time $t$: $G \propto t^{-1}$. In a later revision, the additive mass creation model [358], it was assumed in addition that the mass in the universe increases with $t^2$. Both versions of the theory have severe difficulties to account properly for the evolution of stars [359].

Similarly Jordan [360] allowed $G$ to be a function of a scalar field and abandoned the law of conservation of mass. Even a later, improved version of his theory [361] found little acclaim [362, 13].

Dirac's third version of his theory is the multiplicative creation [358]: $G$ decreases with time and the mass increases, but new matter is created where some already exists and the new matter has the same chemical composition and physical properties as the already existing matter. These assumptions have developed into a scale-covariant theory of gravitation [363···367a]. The observed low upper limits of a variation of $G$ (9.7.3.2) may not invalidate the theory, because the interpretation of the observations becomes complex. The theory is elusive to observational tests.

#### 9.7.9.2.3 Scalar-tensor theory

Starting with Mach's principle, Brans and Dicke [368] developed a Big Bang cosmology with non-constant $G$. The variable $G$ appears in a scalar field which is added to the tensor field of general relativity [13, 15]. The theory makes a number of testable predictions which usually are not far from those of general relativity. The test centered for several years on the excess motion of Mercury's perihelion as caused by a controversial solar quadrupole moment [369···371]. The more recent tests of general relativity (9.7.3.1) essentially disqualify the solar-tensor theory.

#### 9.7.9.2.4 Hoyle-Narlikar theory

Like the Brans-Dicke cosmology the theory of Hoyle and Narlikar [372] suggests a Machian theory of gravitation. According to its authors the expanding universe is indistinguishable from a shrinking universe, in which the mass of all elementary particles increases with time while their charge remains constant. Redshifts are merely the effect of lower masses of the atoms at earlier epochs. There has been no Big Bang singularity. The gravitational constant, $G$, decreases with cosmic time: $G \propto t^{-1/2}$; this [24, 373] and the line width of quasars [374] may eventually enable an observational check of the theory; see also [374a].

### 9.7.9.3 Matter-antimatter cosmologies

Omnès [375, 376] has proposed that preparation of matter and antimatter took place after the hadron era. The present universe consists of large regions of matter and antimatter. For observational problems of this theory, see [377]. Even larger observational difficulties are encountered by a matter-antimatter cosmology with no Big Bang, as originally proposed by Alfvén and Klein [378···380a].

### 9.7.9.4 Tired-light models

Many suggestions have been put forward to explain extragalactic redshifts by non-Doppler effects. Few theories have been offered. The best studied theory is the FIB model of Barnothy and Forro-Barnothy [381···383]. It presents a static solution of the Friedmann equation, and conforms with Mach's principle and with the perfect cosmological principle (cf. 9.7.2.1). Signal carriers (photons, neutrinos, gravitons) change their energy and wavelength relative to the stable elementary particles with time. The experimentum crucis [382] has not confirmed the theory [78, 81].

## 9.7.10 References for 9.7

### 9.7.10.1 General references

**a) Historical**

1 North, J.D.: The Measure of the Universe, Clarendon Press, Oxford (1965).

### b) Modern cosmology


1a Tolman, R.C.: Relativity, Thermodynamics, and Cosmology, Clarendon Press, Oxford (1934).

2 Hubble, E.: The Realm of the Nebulae, Yale University Press, New Haven, and Oxford University Press, London (1936). Reprinted by Dover Publications, Inc., New York (1958).

3 Heckmann, O.: Theorien der Kosmologie, Springer, Berlin (1942), reprinted (1968).

3a Couderc, P.: L'expansion de l'univers, Presses Universitaires de France, Paris (1950).

4 Bondi, H.: Cosmology, Cambridge University Press, London (1952), 2nd ed. (1960).

5 Sandage, A.: The Ability of the 200-inch Telescope to discriminate between Selected World Models, in: Astrophys. J. **133** (1961) 355.

6 Heckmann, O., Schücking, E.: Relativistic Cosmology, in: Gravitation, An Introduction to Current Research (Witten, L., ed.), Wiley, New York (1962) p. 438.

7 McVittie, G.C.: General Relativity and Cosmology, Chapman and Hall, London, 2nd ed. (1965).

8 Zeld'ovich, Ya.B.: Survey of Modern Cosmology, in: Advances in Astronom and Astrophysics, Vol. 3 (Kopal, Z., ed.), Academic Press, New York (1965).

9 Schücking, E.L.: Cosmology, in: Relativistic Theory and Astrophysics; 1. Relativity and Cosmology (Ehlers, J., ed.), American Mathematical Society, Providence (1967).

10 Robertson, H.P., Noonan, T.W.: Relativity and Cosmology, W. B. Saunders, Philadelphia (1968).

11 Andrillat, H.: Introduction à l'étude des cosmologies, Armand Colin, Paris (1970).

12 Peebles, P.J.E.: Physical Cosmology, Princeton University Press, Princeton (1971).

13 Weinberg, S.: Gravitation and Cosmology, Wiley, New York (1972).

14 Hawking, S.W., Ellis, G.F.R.: The Large Scale Structure of Space-Time, University Press, Cambridge (1973).

15 Misner, C., Thorne, K., Wheeler, J.: Gravitation, Freeman, San Francisco (1973).

16 Ryan, M.P., Shepley, L.C.: Homogeneous Relativistic Cosmologies, Princeton University Press, Princeton (1975). (Includes a chronological list of theoretical papers on cosmology).

17 Sexl, R.U., Urbantke, H.K.: Gravitation und Kosmologie, Bibliographisches Institut, Mannheim (1975).

18 Sciama, D.: Modern Cosmology, Cambridge University Press, Cambridge (1975).

19 Zel'dovich, Ya.B., Novikov, I.D.: Structure and Evolution of the Universe (in Russian), Nauka, Moskva (1975).

20 Berry, M.: Principles of Cosmology and Gravitation, Cambridge University Press, Cambridge (1976).

21 Rowan-Robinson, M.: Cosmology, Clarendon Press, Oxford (1977).

22 Weinberg, S.: The First Three Minutes, A. Deutsch, London (1977).
German translation: Die ersten drei Minuten, Piper, München, 2nd ed. (1978).

23 Landsberg, P.T., Evans, D.A.: Mathematical Cosmology, Clarendon Press, Oxford (1977).

24 Narlikar, J.V.: Lectures on General Relativity and Cosmology, Macmillan, London (1979).

25 Raychaudhuri, A.K.: Theoretical Cosmology, Clarendon Press, Oxford (1979).

26 Silk, J.: The Big Bang, Freeman, San Francisco (1980).

26a Peebles, P.J.E.: The Large-Scale Structure of the Universe, Princeton University Press, Princeton (1980).

26b Harrison, E.R.: Cosmology, Cambridge University Press, Cambridge (1981).

26c Raine, D.J.: The Isotropic Universe, Adam Hilger, Bristol (1981).


### c) Symposia, Lecture Notes, and Compendia


27 Int. Astron. Union Symp. **15**: Problems of Extra-Galactic Research (McVittie, G.C., ed.), Macmillan, London (1962).

28 General Relativity and Cosmology, 47th Course of the International School Enrico Fermi (Sachs, B.K., ed.), Academic Press, New York (1971).

29 Int. Astron. Union Symp. **44**: External Galaxies and Quasi-Stellar Objects (Evans, D.S., ed.), Reidel, Dordrecht (1972).

30 Relativity, Astrophysics and Cosmology, 1972 Summer School at the Banff Centre (Israel, W., ed.), Reidel, Dordrecht (1973) = Astrophys. and Space Sci. Library **38** (1973).

31 Cargèse Lectures in Physics, Vol. 16 (A Series of lectures on General Relativistic Cosmology given in 1971) (Schatzman, E., ed.), Gordon and Breach, London (1973).

32 Int. Astron. Union Symp. **63**: Confrontation of Cosmological Theories with Observational Data (Longair, M.S., ed.), Reidel, Dordrecht (1974).

33 Galaxies and the Universe (Sandage, A., Sandage, M., Kristian, J., ed..), University of Chicago Press, Chicago (1975) = Stars and Stellar Systems, Vol. IX.

34 Int. Astron. Union Symp. **74**: Radio Astronomy and Cosmology (Jauncey, D.L., ed.), Reidel, Dordrecht (1977).
35 Int. Astron. Union Coll. **37**: Décalages vers le rouge et expension de l'univers (Balkowski, C., Westerlund, B.E., eds.), CNRS, Paris (1977).
36 Int. Astron. Union Symp. **79**: The Large Scale Structure of the Universe (Longair, M.S., Einasto, J., eds.), Reidel, Dordrecht (1978).
36a Théories cosmologiques, 20e cours des Chercheurs en Physique (Hauck, B., Jalanti, T., Bouvier, P., eds.), Lausanne (1978).
37 Observational Cosmology, 8th Saas-Fee Course 1978 (Maeder, A., Martinet, L., Tammann, G., eds.), Geneva Observatory, Geneva (1978).
38 General Relativity (Hawking, S.W., Israel, W., eds.), Cambridge University Press, Cambridge (1979).
39 The Universe at Large Redshifts, Proceedings of the Copenhagen Symposium 1979; in: Physica Scripta **21** (1980) No. 5.
40 Int. Astron. Union Symp. **92**: Objects of High Redshift (Abell, G.O., Peebles, P.J.E., eds.), Reidel, Dordrecht (1980).
41 Physical Cosmology, 1979 Les Houches Summer School (Balian, R., Audouze, J., Schramm, D.N., eds.), North-Holland, Amsterdam (1980).
41a La galaxie-l'univers extragalactique, Bureau de Longitudes, Editions Gauthier-Villars, Paris (1980).
41b Astrophysics and Elementary Particles, Common Problems, International Meeting, Academia Nazionale dei Lincei, Rome (1980).
41c Cosmology and Particles, Sixteenth Rencontre de Moriond, Editions Frontières, Dreux (1981).
41d Astrophysical Cosmology, Vatican Study Week (Brück, H.A., Coyne, G.V., Longair, M.S., eds.), (1982).

## 9.7.10.2 Special references

42 Friedmann, A.: Z. Physik **10** (1922) 377.
43 Robertson, H.P.: Proc. Nat. Acad. Sci. **15** (1929) 822.
44 Walker, A.G.: Proc. London Math. Soc. **42** (1936) 90.
45 Zel'dovich, Ya.B., Novikov, I.D.: Relativistic Astrophysics, Vol. 1, Stars and Relativity, University of Chicago Press, Chicago (1971) p. 28ff.
46 Robertson, H.P.: Publ. Astron. Soc. Pacific **67** (1955) 82.
47 Hoyle, F., Sandage, A.: Publ. Astron. Soc. Pacific **68** (1956) 301.
48 Mattig, W.: Astron. Nachr. **284** (1958) 109.
49 Einstein, A., de Sitter, W.: Proc. Nat. Acad. Sci. **18** (1932) 213.
50 Milne, E.A.: Relativity, Gravitation and World Structure, Clarendon Press, Oxford (1935).
51 Lang, K.R.: Astrophysical Formulae, Springer, Berlin (1974).
52 Hoerner, S. v.: Cosmology, in: Galactic and Extra-Galactic Radio Astronomy (Verschuur, G.L., Kellermann, K.I., eds.), Springer, New York (1974) p. 353.
53 Oke, J.B., Sandage, A.: Astrophys. J. **154** (1968) 21.
53a Oke, J.B., Schild, R.: Astrophys. J. **161** (1970) 1015.
54 Whitford, A.E. in: [33] p. 159.
54a Pence, W.: Astrophys. J. **203** (1976) 39.
55 Kristian, J., Sandage, A., Westphal, J.A.: Astrophys. J. **221** (1978) 383.
56 Code, A.D., Welch, G.A.: Astrophys. J. **228** (1979) 95.
57 Mattig, W.: Astron. Nachr. **285** (1959) 1.
58 Ryle, M.: Annu. Rev. Astron. Astrophys. **7** (1969) 527.
59 Jauncey, D.L.: Annu. Rev. Astron. Astrophys. **13** (1975) 23.
60 Scheuer, P.A.G. in: [33] p. 725.
61 Longair, M.S. in: [37] p. 127.
62 Shapiro, I.I.: American Inst. Phys. Conference Proceedings 1971, No. 2 (1972) p. 286.
63 Shakeshaft, J.R.: Observatory **99** (1979) 122.
64 Snider, J.L.: Phys. Rev. Lett. **28** (1972) 853.
65 Fomalont, E.B., Sramek, R.A.: Phys. Rev. Lett. **36** (1976) 1475.
66 Shapiro, I.I., Pettengill, G.H., Ash, M.E., Ingalls, R.P., Campbell, D.B., Dyce, R.B.: Phys. Rev. Lett. **28** (1972) 1594.
67 Shapiro, I.I. et al.: J. Geophys. Res. **82** (1977) 4329.

67a Cain, D.L., Anderson, J.D., Keesey, M.S.W., Komarek, T., Laing, P.A., Lau, E.L.: Bull. American Astron. Soc. **10** (1978) 396.
68 Williams, J.G. et al.: Phys. Rev. Lett. **36** (1976) 551.
69 Shapiro, I.I., Counselman, C.C., King, R.W.: Phys. Rev. Lett. **36** (1976) 555 and 1068.
70 Taylor, J.H., McCulloch, P.M.: Ann. New York Acad. Sci. **336** (1980) 442.
71 Taylor, J.H., Fowler, L.A., McCulloch, P.M.: Nature **277** (1979) 437.
72 Williams, J.G., Sinclair, W.S., Yoder, C.F.: Geophys. Research Letters **5** (1978) 943.
73 Blake, G.M.: Mon. Not. R. Astron. Soc. **178** (1977) 41P.
74 McElhinny, M.W., Taylor, S.R., Stevenson, D.J.: Nature **271** (1978) 316.
75 Biesbroeck, G. van: Astrophys. J. **75** (1932) 64.
76 Heckmann, O.: Astron. Nachr. **272** (1942) 215.
77 Dieckvoss, W.: Astron. Nachr. **282** (1955) 206.
78 Baum, W.A., Florentin-Nielsen, R.: Astrophys. J. **209** (1976) 319.
78a Heckmann, O.: Observatory **79** (1959) 105; Mitt. Hamburger Sternwarte **10** (1960) No. 110.
78b Dickens, R.J., Malin, S.R.C.: Observatory **85** (1965) 260.
79 Lerch, F.J., Laubacher, R.E., Klosko, S.M., Smith, D.E., Kolenkiewicz, R., Putney, B.H., Marsh, J.G., Brownd, J.E.: Geophys. Res. Lett. **5** (1978) 1031.
80 Blake, G.M.: Mon. Not. R. Astron. Soc. **181** (1977) 47P.
81 Solheim, J.-E., Barnes, T.G., Smith, H.J.: Astrophys. J. **209** (1976) 330.
82 Tubbs, A.D., Wolfe, A.M.: Astrophys. J. Lett. **236** (1980) L105.
83 Sandage, A.: Astrophys. J. **183** (1973) 711.
84 Turner, E.L. in: [40] p. 71.
85 Butcher, H., Oemler, A.: Astrophys. J. **219** (1978) 18; **226** (1978) 559.
86 Butcher, H., Oemler, A., Wells, D. in: [40] p. 49.
87 Spinrad, H. in: [40] p. 39.
88 Oke, J.B.: Scientific Research with the Space Telescope (Longair, M.S., Warner, J.W., eds.), NASA, Washington (1980) p. 309 = Int. Astron. Union Coll. **54**.
89 Katgert, P., de Ruiter, H.R., van der Laan, H.: Nature **280** (1979) 20.
90 van der Laan, H., Katjert, P., de Ruiter, H.R. in: [39] p. 669.
91 Kron, R.G. in: [39] p. 652, and references therein.
92 Kron, R.G. in: [40] p. 9.
93 Tinsley, B.M. in: [41]. p. 161.
94 Véron, P.: Astron. Astrophys. Suppl. **30** (1977) 131.
95 Kellerman, K.I. in: [39] p. 664.
96 Wall, J.V., Pearson, T.J., Longair, M.S. in: [34] p. 269.
97 Green, R.F., Schmidt, M.: Astrophys. J. Lett. **220** (1978) L1.
98 Schmidt, M., Green, R.F. in: [40], p. 73.
99 Braccesi, A., Zitelli, V., Bònoli, F., Formiggini, L.: Astron. Astrophys. **85** (1980) 80.
100 Schmidt, M.: Astrophys. J. **151** (1968) 393.
101 Schmidt, M.: Astrophys. J. **162** (1970) 371.
102 Schmidt, M. in: [36] p. 281.
103 Wills, D., Lynds, C.R.: Astrophys. J. Suppl. **36** (1978) 317.
104 Schwartz, D.A. in: [39] p. 644.
105 Hickson, P., Adams, P.J.: Astrophys. J. Lett. **234** (1979) L91.
106 Einstein, A.: Sitzber. Preuss. Akad. Wiss. (Phys.-Math. Kl.) (1917) 142.
107 Hawking, S.W., Penrose, R.: Proc. Roy. Soc. **314**A (1970) 529.
108 Sandage, A. in: [33] p. 761.
109 Sandage, A., Hardy, E.: Astrophys. J. **183** (1973) 743.
110 Gunn, J.E., Oke, J.B.: Astrophys. J. **195** (1975) 255.
111 Kristian, J., Sandage, A., Westphal, J.A.: Astrophys. J. **221** (1978) 383.
112 Weedman, D.W.: Astrophys. J. **203** (1976) 6.
113 Sandage, A.: Astrophys. J. **178** (1972) 1.
114 Smith, H.E. in: [34] p. 279.
115 Turner, E.L.: Astrophys. J. **230** (1979) 291.
116 Sandage, A.: Astrophys. J. **173** (1972) 485.
117 Wardle, J.F.C., Miley, G.K.: Astron. Astrophys. **30** (1974) 305.
118 Ekers, R.D., Miley, G.K. in: [34] p. 109.
119 Bruzual, G., Spinrad, H.: Astrophys. J. **220** (1978) 1; **222** (1978) 1119.

**Tammann**

120 Peebles, P.J.E.: Astrophys. J. **146** (1966) 542.
121 Wagoner, R.V., Fowler, W.A., Hoyle, F.: Astrophys. J. **148** (1967) 3.
122 Doroshkevich, A.G., Novikov, I.D., Sunyaev, R.A., Zeldovich, Ya.B.: Highlights of Astronomy **2** (1971) 318.
123 Peebles, P.J.E. in: [12] p. 240.
124 Weinberg, S. in: [13] p. 545.
125 Wagoner, R.V.: Astrophys. J. **179** (1973) 343.
126 Rees, M.J. in: [37] p. 270.
127 Reeves, H.: Annu. Rev. Astron. Astrophys. **12** (1974) 437.
128 Rood, R.T., Wilson, T.L., Steigman, G.: Astrophys. J. Lett. **227** (1979) L 97.
129 Rood, R.T., Steigman, G., Tinsley, B.M.: Astrophys. J. Lett. **207** (1976) L 57.
130 Danziger, I.J.: Annu. Rev. Astron. Astrophys. **8** (1970) 161.
131 Peimbert, M., Torres-Peimbert, S.: Astrophys. J. **203** (1976) 581.
132 Greenstein, J.L.: Physica Scripta **21** (1980) 759.
133 Burbidge, G.: Publ. Astron. Soc. Pacific **70** (1958) 83.
134 Hoyle, F., Tayler, R.J.: Nature **203** (1964) 1108.
135 Laurent, C., Vidal-Madjar, A., York, D.G.: Astrophys. J. **229** (1979) 923.
136 Cameron, A.G.W.: Explosure Nucleosynthesis (Schramm, D.N., Arnett, W.D., eds.), University of Texas Press, Austin (1973) p. 1.
137 Geiss, J., Reeves, H.: Astron. Astrophys. **18** (1972) 126.
138 Penzias, A.A.: American Scientist **66** (1978) 291.
139 Eichler, D.: Astrophys. J. **229** (1979) 39.
140 Reeves, H. in: [41] p. 443.
141 Alpher, R.A., Herman, R.: Phys. Rev. **75** (1949) 1089.
142 Gamov, G.: Danske Videnskab Selskab Mat. Fys. Medd. **27** (1953) No. 10.
143 Penzias, A.A., Wislon, R.W.: Astrophys. J. **142** (1965) 419; **142** (1965) 1149.
143a Penzias, A.A.: Science **205** (1979) 549.
143b Wilson, R.W. in: [39] p. 599.
144 Thaddeus, P.: Annu. Rev. Astron. Astrophys. **10** (1972) 305.
145 Muehlner, D.: Infrared and Submillimeter Astronomy (Fazio, G.G., ed.), Reidel, Dordrecht (1977) p. 143.
146 Clegg, P.E.: Infrared Astronomy (Setti, G., Fazio, G.G., eds.), Reidel, Dordrecht (1978) p. 181.
147 De Zotti, G.: Physics of the Expanding Universe (Demiański, M., ed.), Springer, Berlin (1979) p. 165.
148 Longair, M.S. in: [37] p. 129.
149 Wilkinson, D.T. in: [39] p. 606.
150 Richards, P.L. in: [39] p. 610.
150a Weiss, R.: Annu. Rev. Astron. Astrophys. **18** (1980) 489.
150b Sunyaev, R.A., Zel'dovich, Ya.B.: Annu. Rev. Astron. Astrophys. **18** (1980) 537.
151 Sandage, A., Tammann, G.A., Hardy, E.: Astrophys. J. **172** (1972) 253.
152 Shapley, H., Ames, A.: Harvard Ann. **88** (1932) 2.
153 Reiz, A.: Lund Obs. Ann. No. 9 (1941).
154 de Vaucouleurs, G.: Vistas in Astronomy 2 (1955) 1584.
155 Jones, B.J.T.: Mon. Not. R. Astron. Soc. **174** (1976) 429.
156 Bahcall, J.N., Joss, P.C.: Astrophys. J. **203** (1976) 23.
157 Yahil, A., Sandage, A., Tammann, G.A.: Astrophys. J. **242** (1980) 448.
158 Hubble, E.: Astrophys. J. **64** (1926) 321; **79** (1934) 8; **84** (1936) 517.
159 Mayall, N.U.: Lick Obs. Bull. No. 458 (1934).
160 Shane, C.D., Wirtanen, C.A.: Publ. Lick Obs. **22** (1967) Part 1.
161 Rudnicki, K., Dworak, T.Z., Flin, P., Raranowski, B., Sendrakowski, A.: Acta Cosmologica **1** (1973) 7.
162 Rainey, G.W.: Ph. D. dissertation, University of California at Los Angeles (1977).
163 Brown, G.S.: Astron. J. **84** (1979) 1647.
164 Shane, C.D., Wirtanen, C.A.: Astron. J. **59** (1954) 285.
165 Groth, E.J., Peebles, P.J.E.: Astrophys. J. **217** (1977) 385.
166 Peebles, P.J.E. in: [36] p. 217.
167 Willis, A.G., Oosterbaan, C.E., Le Poole, R.S., de Ruiter, H.R., Strom, R.G., Valentijn, E.A., Katgert, P., Katgert-Merkelijn, J.K. in: [34] p. 39.
168 Wall, J.V. in: [34] p. 55.
169 Pauliny-Toth, I.I.K. in: [34] p. 63.
170 Webster, A. in: [34] p. 75.

171 Fanti, C. in: [40] p. 145.
172 Setti, G., Woltjer, L.: Astron. Astrophys. **76** (1979) L 1.
173 Tananbaum, H. et al.: Astrophys. J. Lett. **234** (1979) L 5.
174 Murray, S.S. in: [39] p. 684.
175 Bookbinder, J., Cowie, L.L., Krolik, J.H., Ostriker, J.P., Rees, M.: Astrophys. J. **237** (1980) 647.
176 Cavaliere, A., Danese, L., De Zotti, G., Franceschini, A.: Astron. Astrophys. **85** (1980) L 9.
177 Boynton, P.E. in: [36] p. 317.
178 Partridge, R.B. in: [39] p. 624.
179 Weinberg, S. in: [13] p. 506.
180 Rees, M., Ruffini, R., Wheeler, J.A.: Black Holes, Gravitational Waves and Cosmology: An Introduction to Current Research, Gordon and Breach, New York (1974), p. 195.
181 Rees, M. in: [39] p. 614.
182 Smoot, G.F. in: [39] p. 619.
183 Doppler, C.: Abh. königl. Böhm. Ges. Wiss. **2** (1842) 467.
184 Einstein, A.: Ann. Phys. **17** (1905) 891; **18** (1905) 639.
185 Gill, T.P.: The Doppler Effect, Logos Press, London (1965).
186 Weinberg, S. in: [13] p. 415.
187 Transactions of the Int. Astron. Union XVI B (1976) p. 213.
188 de Vaucouleurs, G., de Vaucouleurs, A., Corwin, H.G.: Second Reference Catalogue of Bright Galaxies, University of Texas Press, Austin (1976).
189 Sandage, A., Tammann, G.A.: The Revised Shapley-Ames Catalogue, Carnegie Institution of Washington, Washington (1981).
190 Tanzella-Nitti, G., Vettolani, G., Palumbo, G.G.C.: preprint (1980).
191 Humason, M.L., Wahlquist, H.D.: Astron. J. **60** (1955) 254.
192 de Vaucouleurs, G., Peters, W.L., Corwin, H.G.: Astrophys. J. **211** (1977) 319.
193 Lynden-Bell, L., Lin, D.N.C.: Mon. Not. R. Astron. Soc. **181** (1977) 37.
194 Yahil, A., Tammann, G.A., Sandage, A.: Astrophys. J. **217** (1977) 903.
195 Peebles, P.J.E.: Fundamental Interactions in Physics and Astrophysics (Iverson, G., Perlmutter, A., Mintz, S., eds.), Plenum Press, New York (1973) p. 318.
196 Tammann, G.A., Sandage, A., Yahil, A. in: [39] p. 630.
196a Tammann, G.A., Kraan, R. in: [36] p. 71.
197 Gott, J.R., Turner, E.L.: Astrophys. J. **213** (1977) 309.
198 Danese, L., De Zotti, G., di Tullio, G.: Astron. Astrophys. **82** (1980) 322.
199 Harrison, E.R.: Astrophys. J. Lett. **191** (1974) L 51.
200 Sandage, A.: Astrophys. J. **178** (1972) 1.
201 Peebles, P.J.E.: Astrophys. J. **205** (1976) 318.
202 Schechter, P.L.: Astron. J. **82** (1977) 569.
203 Aaronson, M., Mould, J., Huchra, J., Sullivan, W.T., Schommer, R.A., Bothun, G.D.: Steward Obs. Preprint No. 255 (1980).
204 Schechter, P.L.: private communication (1980).
205 Rubin, V.C. in: [35] p. 119.
206 Rubin, V.C., Thonnard, N., Ford, W.K., Roberts, M.S.: Astron. J. **81** (1976) 719.
207 Yahil, A.: Talk presented at the Scientific Meeting of Commission 28, Int. Astron. Union General Assembly, Montreal (1979).
208 de Vaucouleurs, G., Bollinger, G.: Astrophys. J. **233** (1979) 433.
209 Sandage, A., Tammann, G.A., Yahil, A.: Astrophys. J. **232** (1979) 352.
210 Fabian, A.C., Warwick, R.S., Pye, J.P. in: [39] p. 650.
210a White, S.D.M. in: [39] p. 640.
210b Tarenghi, M., Chincarini, G., Rood, H.J., Thompson, L.A.: Astrophys. J. **235** (1980) 724.
211 Yahil, A., Sandage, A., Tammann, G.A. in: [41] p. 127.
212 Schlechter, P.L.: Astron. J. **85** (1980) 801.
213 Smoot, G.F. in: [40] p. 321.
214 Cheng, E.S., Boughn, S., Wilkinson, D.T.: Bull. American Astron. Soc. **12** (1980) 489.
215 The Redshift Controversy (Field, G.B., Arp, H., Bahcall, J.N., eds.), W.A. Benjamin Inc., Reading (1973).
216 Rees, M.J. in: [35] p. 563.
217 Burbidge, G. in: [40] p. 99.
218 Arp, H.: Astrophys. J. **236** (1980) 63, and references therein.
219 Tifft, W.G.: Astrophys. J. **236** (1980) 70, and references therein.

220 Sandage, A.: Astrophys. J. **180** (1973) 687.
221 Phillips, M.M.: Astrophys. J. Lett. **236** (1980) L 45.
222 Petrosian, V.: Astrophys. J. Lett. **209** (1976) L 1.
223 Sandage, A.: Large Space Telescope, a New Tool for Science, American Inst. Aeronautics Astronautics, 12th Aerospace Science Meeting, Washington (1974) p. 19.
224 Tammann, G.A.: Astronomical Uses of the Space Telescope (Macchetto, F., Pacini, F., Tarenghi, M., eds.), ESA/ESO, Geneva (1979) p. 329.
225 Kraan-Korteweg, R.C., Tammann, G.A.: Astron. Nachr. **300** (1979) 181.
226 Sandage, A., Tammann, G.A.: Astrophys. J. Lett. **207** (1976) L 1.
227 Rood, H.J., Page, T.L., Kintner, E.C., King, I.R.: Astrophys. J. **175** (1972) 627.
228 Sandage, A., Visvanathan, N.: Astrophys. J. **223** (1978) 707.
229 Burstein, D., Heiles, C.: Astrophys. J. **225** (1978) 40.
230 Tammann, G.A., Sandage, A., Yahil, A. in: [41] p. 53.
231 Sandage, A., Tammann, G.A.: Astrophys. J. **210** (1976) 7.
232 de Vaucouleurs, G.: Astrophys. J. **227** (1979) 729.
233 de Vaucouleurs, G.: Astron. Astrophys. **79** (1979) 274.
234 Wray, J.D., de Vaucouleurs, G.: Astron. J. **85** (1980) 1.
235 Humphreys, R.M.: preprint (1980).
236 Melnick, J.: Astrophys. J. **213** (1977), 15; Astron. Astrophys. **70** (1978) 157.
237 Tully, R.B., Fisher, J.R.: Astron. Astrophys. **54** (1977) 661.
238 Fisher, J.R., Tully, R.B.: Comm. Astrophys. **7** (1977) 85.
239 Kennicutt, R.C.: Astrophys. J. **228** (1979) 696; **228** (1979) 704.
240 Mould, J., Aaronson, M., Huchra, J.: Astrophys. J. **238** (1980) 458.
241 Michard, R.: Astron. Astrophys. **74** (1979) 206.
242 Roberts, M.S.: Astron. J. **83** (1978) 1026.
243 Freeman, K.C.: Scientific Research with the Space Telescope (Longair, M.S., Warner, J.W., eds.), NASA, Washington (1980) p. 292 = Int. Astron. Union Coll. **54**.
244 Rubin, V.C., Ford, W.K., Thonnard, N.: Astrophys. J. **238** (1980) 471.
245 Sandage, A., Tammann, G.A.: Astrophys. J. **190** (1974) 525.
246 Martin, W.L., Warren, P.R., Feast, M.W.: Mon. Not. R. Astron. Soc. **188** (1979) 139.
247 Crampton, D.: Astrophys. J. **230** (1979) 717.
248 Kraan-Korteweg, R.C.: private communication (1980).
249 Kirshner, R.P., Kwan, J.: Astrophys. J. **193** (1974) 27.
250 Schurmann, S.R., Arnett, W.D., Falk, S.W.: Astrophys. J. **230** (1979) 11.
251 Branch, D.: Mon. Not. R. Astron. Soc. **186** (1979) 609.
252 Branch, D., McCall, M., Rybski, P., Uomoto, A., Wills, B., Wills, D.: Bull. American. Astron. Soc. **11** (1979) 694.
253 Paturel, G.: Astron. Astrophys. **71** (1979) 19.
254 Visvanathan, N.: Proc. Astron. Soc. Australia **3** (1979), Contribution 309.
255 Bottinelli, L., Gouguenheim, L.: Astron. Astrophys. **51** (1976) 275.
256 Shostak, G.S.: Astron. Astrophys. **68** (1978) 321.
257 van den Bergh, S.: Structure and Evolution of Galaxies (Setti, G., ed.), Reidel, Dordrecht (1975) p. 247.
258 Tammann, G.A.: Roy. Greenwich Obs. Bull. No. 182 (1976) 135.
259 Harris, W.E., Racine, R.: Annu. Rev. Astron. Astrophys. **17** (1979) 241.
260 Cavaliere, A., Danese, L., De Zotti, G.: Astron. Astrophys. **75** (1979) 322.
261 Birkinshaw, M.: Mon. Not. R. Astron. Soc. **187** (1979) 847.
262 Lynden-Bell, D.: Nature **270** (1977) 396.
263 Lynden-Bell, D.: private communication (1980).
264 Oleak, H., Schmidt, K.-H.: Astron. Nachr. **298** (1977) 33.
265 Teerikorpi, P.: Astron. Astrophys. **45** (1975) 117; **50** (1976) 455.
266 de Vaucouleurs, G., Bollinger, G.: Astrophys. J. **233** (1979) 433.
267 Aaronson, M., Mould, J., Huchra, J., Sullivan, W.T., Schommer, R.A., Bothun, G.D.: Astrophys. J. **239** (1980) 12.
268 van Bueren, H.G.: Bull. Astron. Inst. Netherlands **11** (1952) 385.
269 Hanson, R.B.: Star Clusters (Hesser, J.E., ed.), Reidel, Dordrecht (1980) p. 71 = Int. Astron. Union Symp. **85**.
269a Sandage, A., Tammann, G.A. in: [41b] p. 23.
270 Sandage, A.: Astrophys. J. Lett. **152** (1968) L 149.

271 Tammann, G.A. in: [35] p. 43.
272 Gunn, J.E. in: [37] p. 1.
273 Baldwin, J.R., Danziger, I.J., Frogel, J.A., Persson, S.E.: Astrophys. Lett. **14** (1973) 1.
274 Tinsley, B.M.: Astrophys. J. Lett. **173** (1972) L 93.
275 Tinsley, B.M., Gunn, J.E.: Astrophys. J. **203** (1976) 52.
276 Ostriker, J.P., Tremaine, S.D.: Astrophys. J. Lett. **202** (1975) L 113.
277 Hausman, M.A., Ostriker, J.P.: Astrophys. J. **224** (1978) 320.
278 Gunn, J.E., Tinsley, B.M.: Astrophys. J. **210** (1976) 1.
279 Hoessel, J.G.: Ph. D. Thesis, California Institute of Technology (1980).
280 Zel'dovich, Ya.B.: Astron. Zh. **41** (1964) 19.
281 Bertotti, B.: Proc. Roy. Soc. London A **294** (1966) 195.
282 Kantowski, R.: Astrophys. J. **155** (1969) 89.
283 Refsdal, S.: Astrophys. J. **159** (1970) 357.
284 Dyer, C.C., Roeder, R.C.: Astrophys. J. Lett. **174** (1972) L 115.
285 Roeder, R.C.: Astrophys. J. **196** (1975) 671.
286 Weinberg, S.: Astrophys. J. Lett. **208** (1976) L 1.
287 Schechter, P.: Astrophys. J. **203** (1976) 297.
288 Abell, G.O.: Annu. Rev. Astron. Astrophys. **3** (1965) 1.
289 Baldwin, J.A., Burke, W.L., Gaskell, C.M., Wampler, E.J.: Nature **273** (1978) 431.
290 Wampler, E.J. in: [40] p. 119.
291 Brown, G.S., Tinsley, B.M.: Astrophys. J. **194** (1974) 555.
292 Bruzual, G., Kron, R.G.: Astrophys. J. **241** (1980) 25.
293 Baum, W.A. in: [29] p. 393.
294 Bahcall, N.A.: Scientific Research with the Space Telescope (Longair, M.S., Warner, J.W., eds.), NASA, Washington (1980) p. 307 = Int. Astron. Union Coll. **54**.
295 Peebles, P.J.E. in: Int. Astron. Union Coll. **54** (1980) 295; see [294].
296 Wagoner, R.V.: Astrophys. J. Lett. **214** (1977) L 5.
297 Wagoner, R.V. in: [41] p. 179.
298 Harrison, E.R.: Phys. Today **27** (1974) No. 2, 30.
299 Baum, W.A.: Publ. Astron. Soc. Pacific **68** (1956) 118.
300 Sandage, A., Tammann, G.A.: Annu. Rep. Dir. Mt. Wilson and Palomar Obs. 1964–65, p. 35.
301 Mattila, K.: Astron. Astrophys. **47** (1976) 77.
302 Dube, R.R., Wickes, W.C., Wilkinson, D.T.: Astrophys. J. **232** (1979) 333.
302a Schnur, G.F.O.: Two Dimensional Photometry (Crane, P., Kjär, K., eds.), ESO, Geneva (1980) p. 365.
303 Tinsley, B.M.: Astrophys. J. **220** (1978) 816.
304 Oort, J.H.: La structure et l'évolution de l'univers (Stoops, R., ed.), Institut International de Physique, Solvay, Bruxelles (1958) p. 163.
305 Gott, J.R., Turner, E.L.: Astrophys. J. **209** (1976) 1.
306 Felten, J.E.: Astron. J. **82** (1977) 861.
307 Kirshner, R.P., Oemler, A., Schechter, P.L.: Astron. J. **84** (1979) 951.
308 Faber, S.M., Gallagher, J.S.: Annu. Rev. Astron. Astrophys. **17** (1979) 135.
309 Karachentsev, I.D., Fesenko, B.I.: Astrofisika **15** (1979) 217.
310 Materne, J.: Astron. Astrophys. **86** (1980) 91.
311 Yahil, A., Sandage, A., Tammann, G.A. in: [39] p. 635.
312 Limber, D.N.: Astrophys. J. **117** (1953) 134.
313 Fall, S.M.: Mon. Not. R. Astron. Soc. **172** (1975) 23 P.
314 Peebles, P.J.E.: Astrophys. Space Sci. **45** (1976) 3.
315 Fall, S.M.: Rev. Mod. Phys. **51** (1979) 21.
316 Peebles, P.J.E.: Astrophys. J. Lett. **205** (1976) L 109.
317 Sargent, W.L.W., Turner, E.L.: Astrophys. J. Lett. **212** (1977) L 3.
318 Geller, M.J., Davis, M.: Astrophys. J. **225** (1978) 1.
319 Seldner, M., Peebles, P.J.E.: Astrophys. J. Lett. **214** (1977) L 1.
319a Peebles, P.J.E.: Astron. J. **84** (1979) 730.
320 Peebles, P.J.E. in: [41] p. 213.
321 Davis, M., Groth, E., Peebles, P.J.E.: Astrophys. J. Lett. **212** (1977) L 107.
322 Yang, J., Schramm, D.N., Steigman, G., Rood, R.T.: Astrophys. J. **227** (1979) 697.
323 Yahil, A., Beaudet, G.: Astrophys. J. **206** (1976) 26.
324 Olson, D.W., Silk, J.: Astrophys. J. **226** (1978) 50.

325 Tremaine, S., Gunn, J.E.: Phys. Rev. Lett. **42** (1979) 407.
326 Lyubimov, V.A., Novikov, E.G., Nozik, V.Z., Tretyakov, E.F., Kozik, V.S.: Phys. Lett. **94B** (1980) 266.
327 Kozik, V.S., Lyubimov, V.A., Novikov, E.G., Nozik, V.Z., Tretyakov, E.F.: Sovj. J. Nuclear Phys. **32** (1980), 152.
328 Schramm, D.N., Steigman, G.: Gen. Relativ. Gravitation **13** (1981) 101.
329 Gunn, J.E., Lee, B.W., Lerche, I., Schramm, D.N., Steigman, G.: Astrophys. J. **223** (1978) 1015.
330 Zel'dovich, Ya.B., Syunyaev, R.A.: Astron. Zh. Lett. **6** (1980) 451.
331 Field, G.B.: Mitt. Astron. Ges. **47** (1980) 7.
332 Sandage, A.: Astrophys. J. **134** (1961) 916.
333 Sandage, A.: Galaxies and the Universe, Vetlesen Symp. 1966 (Woltjer, L., ed.), Columbia University Press, New York (1968) p. 75.
334 Sandage, A.: Astrophys. J. **162** (1970) 841.
335 Demarque, P., McCleve, R.D.: The Evolution of Galaxies and Stellar Populations (Tinsley, B.M., ed.), Yale University Observatory, New Haven (1977) p. 199.
335a Sandage, A., Katem, B., Sandage, M.: Astrophys. J. Suppl. **46** (1981) 41.
336 Unsöld, A.: Naturwissenschaften **63** (1976) 443.
337 Fowler, W.A. in: Cosmology, Fusion, and Other Matters (Reines, F., ed.), Colorado Associated Universities Press, Boulder (1972) p. 67.
338 Fowler, W.A.: Cosmochemistry (Milligan, W.O., ed.), Robert A. Welch Foundation, Houston (1978).
339 Clayton, D.D.: Astrophys. J. **139** (1964) 637.
340 Hainebach, K.L., Schramm, D.N.: Astrophys. J. **212** (1977) 347.
341 Luck, J.-M., Birck, J.-L., Allegre, C.-J.: Nature **283** (1980) 256.
342 Woosley, S.E., Fowler, W.A.: Astrophys. J. **233** (1979) 411.
343 Sandage, A.: Astrophys. J. **178** (1972) 25.
344 De Sitter, W.: Mon. Not. R. Astron. Soc. **78** (1917) 3 (and references therein).
345 Einstein, A.: Sitzber. Preuss. Akad. Wiss. (Phys.-Math. Kl.) **142** (1917).
346 Stabell, R., Refsdal, R.: Mon. Not. R. Astron. Soc. **132** (1966) 379.
347 Milne, E.A.: Quart. J. Math. (Oxford Ser.) **5** (1934) 64.
348 McCrea, W.H., Milne, E.A.: Quart. J. Math. (Oxford Ser.) **5** (1934) 73.
349 Lemaître, G.: Ann. Soc. Sci. Bruxelles **47**A (1927) 49.
350 Eddington, A.S.: Mon. Not. R. Astron. Soc. **90** (1930) 668.
351 McCrea, W.: Quart. J. R. Astron. Soc. **12** (1971) 140.
352 Dyson, F.J.: Aspects of Quantum Theory (Salam, A., Wigner, E.F., eds.), Cambridge University Press, Cambridge (1972) p. 213.
353 Bondi, H., Gold, T.: Mon. Not. R. Astron. Soc. **108** (1948) 252.
354 Hoyle, F.: Mon. Not. R. Astron. Soc. **108** (1948) 372; **109** (1949) 365.
355 Dirac, P.: Proc. R. Soc. London A**165** (1938) 199.
356 Harison, E.R.: Phys. Today **25** (1972) No. 12, 30.
357 Carter, B. in: [32] p. 291.
358 Dirac, P.A.M.: Proc. R. Soc. London A**338** (1974) 439.
359 Maeder, A.: Astron. Astrophys. **56** (1977) 359; **57** (1977) 125.
360 Jordan, P.: Die Herkunft der Sterne, Wissenschaftliche Verlagsgesellschaft, Stuttgart (1947).
361 Jordan, P.: Schwerkraft und Weltall, 2nd edition, Friedrich Vieweg & Sohn, Braunschweig (1955) = Die Wissenschaft, Vol. 107.
362 Heckmann, O.: Z. f. Astrophys. **40** (1956) 278.
363 Dirac, P.A.M.: Proc. R. Soc. London A**333** (1973) 403.
364 Canuto, V., Adams, P.J., Hsieh, S.H., Tsiang, E.: Phys. Rev. D**16** (1977) 1643.
365 Canuto, V.M., Owen, J.R.: Astrophys. J. Suppl. **41** (1979) 301 and references therein.
366 Maeder, A. in: [36a] p. 257.
367 Bouvier, P., Maeder, A.: Astron. Astrophys. **79** (1979) 158 and references therein.
367a Maeder, A. in: [41] p. 533.
368 Brans, C., Dicke, R.H.: Phys. Rev. **124** (1961) 925.
369 Dicke, R.H., Goldenberg, H.M.: Astrophys. J. Suppl. **27** (1974) 131.
370 Hill, H.A., Stebbins, R.T.: Astrophys. J. **200** (1975) 471.
371 Chapman, G.A., McGuire, T.E.: Astrophys. J. **217** (1977) 657.
372 Hoyle, F., Narlikar, J.V.: Action at a Distance in Physics and Cosmology, Freeman, San Francisco (1974).
373 Newton, R.R.: The Moon's Acceleration and its Physical Origins, Johns Hopkins University Press, Baltimore, Vol. 1 (1979).

374 Narlikar, J.V. in: [35] p. 497.
374a Banothy, J.M., Tinsley, B.M.: Astrophys. J. **182** (1973) 343.
375 Omnès, R.: Phys. Rev. Lett. **23** (1969) 38.
376 Omnès, R. in: [36a] p. 191.
377 Steigman, G. in: [32] p. 347; Annu. Rev. Astron. Astrophys. **14** (1976) 339.
378 Alfvén, H., Klein, O.: Arkiv Fisik **23** (1962) 187.
379 Alfvén, H.: Problems of Physics and Evolution of the Universe (Mirzoyan, L.V., ed.), Publishing House Armenian Academy of Sciences, Erevan (1978) p. 9, 38.
380 Alfvén, H.: Rev. Medd. Phys. **37** (1965) 652.
380a Alfvén, H.: Astrophys. Space Sci. **66** (1979) 23.
381 Barnothy, J.M., (Barnothy-) Forro, M.: Csillagaszati Lapok **7** (1944) 65.
382 Barnothy, J.M., Barnothy, M.F. in: [29] p. 478
383 Barnothy, M.F., Barnothy, J.M.: Bull. American Astron. Soc. **10** (1978) 690.
384 Kundt, W. in: Springer Tracts in Modern Physics (Höhler, G., ed.), Springer, Berlin **60** (1971) 1.
386 Weinberg, S. in: [39] p. 773.
387 Barrow, J.D., Silk, J.: Scientific American **242** (1980) No. 4, p. 98.
388 Sciama, D.W. in: [39] p. 769.

# Appendix

## 3.3.3.10 Comet Halley

The comets in general are treated in Vol. VI/2a in section 3.3.3. After publication of this volume more details about the planned comet Halley missions in 1985/86 were known. Because of the importance of and the interest in these missions, we add an appendix about comet Halley and the missions and about other future comets suitable for similar missions.

This contribution may also be taken as a more detailed example for a periodic comet.

### 3.3.3.10.1 Orbital calculations, historical apparitions

First correct prediction of return of a comet by E. Halley (1656···1742) [9], who noted the similarities in the orbits of the comets in 1531, 1607, and 1682, and predicted its return for "about the year 1758". Halley also suspected that the comet of 1456 was an earlier apparition of the same comet.

The discoverers of comet Halley before 1682 are not known. In 1682, the comet was detected with the naked eye on August 26 by Picard and La Hire in Paris, and independently on August 30 by Flamsteed, Greenwich. On its next apparition, it was first sighted on December 25, 1758, by Georg Palitsch near Dresden with an 8-foot telescope, and independently by Charles Messier in Paris. In 1835, the comet was first noticed on August 5 by Etienne Dumouchel in Rome, and independently on August 20 by W. Struve in Dorpat. Recovery of the last return in 1910: see 3.3.3.10.2.

Recently [22] the long-term motion of comet Halley was numerically integrated back to 1404 B.C. by incorporating planetary perturbations and non-gravitational effects and including ancient Chinese observations. In 1404 B.C., the (computed) comet passed the earth within 0.04 AU; without additional observations prior to 1404 B.C., the orbit could not be rectified again. Non-gravitational forces are responsible [10] for a deceleration of the comet's mean motion of about 4 days/(period)$^2$.

Comet Halley observations prior to 240 B.C. have not yet been identified in ancient Chinese records. Apparitions after 240 B.C. were generally more favorable than those between 240 B.C. and 1404 B.C. [22]. For the 29 apparitions from 240 B.C. to 1910, 14 had an earth-comet-distance of less than 0.25 AU while the comet was in the dark sky, for the 16 apparitions from 315 B.C. to 1404 B.C., only two such (1266 B.C. and 1404 B.C.). The favorable apparition in 164 B.C., with a minimum distance from earth of 0.1 AU, is apparently not recorded.

The osculating orbital elements for comet Halley from 1910 back to 240 B.C. are given in Table 1. For more details of orbital elements, see 3.3.3.1 in Vol. VI/2a.

Table 1. Osculating orbital elements for comet Halley [22].

$T$ = time of perihelion passage in ephemeris time [E.T.] (see 2.2.3)
$q$ = perihelion distance
$e$ = eccentricity
$P$ = period
$\omega$ = argument of perihelion measured in the plane of the comet's orbit
$i$ = inclination of the orbit to the ecliptic
$\Omega$ = longitude of ascending node

| Year | $T$ [E.T.] | $q$ [AU] | $e$ | $P$ [a] | $\omega$ | $\Omega$ | $i$ | Epoch [E.T.] |
|---|---|---|---|---|---|---|---|---|
| 1910 II | 1910 Apr. 20.17771 | 0.5871888 | 0.9672968 | 76.08 | 111°.71703 | 57°.84670 | 162°.21507 | 1910 May 9.0 |
| 1835 III | 1835 Nov. 16.43871 | 0.5865423 | 0.9673860 | 76.27 | 110.68555 | 56.80251 | 162.25518 | 1835 Nov. 18.0 |
| 1759 I | 1759 Mar. 13.06075 | 0.5844466 | 0.9676792 | 76.89 | 110.68990 | 56.52871 | 162.36893 | 1759 Mar. 21.0 |
| 1682 | 1682 Sep. 15.28069 | 0.5826084 | 0.9679230 | 77.41 | 109.20541 | 54.85222 | 162.26569 | 1682 Aug. 31.0 |
| 1607 | 1607 Oct. 27.54063 | 0.5836150 | 0.9674895 | 76.06 | 107.53140 | 53.05354 | 162.90204 | 1607 Oct. 24.0 |
| 1531 | 1531 Aug. 26.23846 | 0.5811975 | 0.9677499 | 76.50 | 106.95724 | 52.34044 | 162.91385 | 1531 Aug. 14.0 |
| 1456 | 1456 Jun. 9.63257 | 0.5797014 | 0.9679974 | 77.10 | 105.81647 | 51.15021 | 162.88607 | 1456 Jun. 28.0 |
| 1378 | 1378 Nov. 10.68724 | 0.5762013 | 0.9683723 | 77.76 | 105.27668 | 50.30348 | 163.10897 | 1378 Nov. 5.0 |
| 1301 | 1301 Oct. 25.58194 | 0.5727097 | 0.9689307 | 79.14 | 104.48199 | 49.43575 | 163.07179 | 1301 Nov. 9.0 |
| 1222 | 1222 Sep. 28.82294 | 0.5742108 | 0.9688444 | 79.12 | 103.83087 | 48.58845 | 163.18782 | 1222 Oct. 15.0 |
| 1145 | 1145 Apr. 18.56090 | 0.5747921 | 0.9687853 | 79.02 | 103.68573 | 48.33830 | 163.22004 | 1145 Apr. 2.0 |
| 1066 | 1066 Mar. 20.93405 | 0.5744956 | 0.9688655 | 79.26 | 102.45543 | 46.90873 | 163.10814 | 1066 Mar. 8.0 |
| 989 | 989 Sep. 5.68757 | 0.5819144 | 0.9678887 | 77.14 | 101.46581 | 45.84533 | 163.39474 | 989 Aug. 19.0 |
| 912 | 912 Jul. 18.67429 | 0.5801559 | 0.9680692 | 77.45 | 100.75913 | 44.93122 | 163.30679 | 912 Jul. 14.0 |
| 837 | 837 Feb. 28.27000 | 0.5823182 | 0.9678055 | 76.90 | 100.08403 | 44.21516 | 163.44258 | 837 Mar. 10.0 |
| 760 | 760 May 20.67126 | 0.5818368 | 0.9678541 | 77.00 | 99.98016 | 43.97218 | 163.43860 | 760 Jun. 2.0 |
| 684 | 684 Oct. 2.76682 | 0.5795841 | 0.9681495 | 77.62 | 99.13197 | 43.08465 | 163.41338 | 684 Sep. 29.0 |
| 607 | 607 Mar. 15.47581 | 0.5808315 | 0.9680396 | 77.47 | 98.78209 | 42.54593 | 163.47190 | 607 Mar. 18.0 |
| 530 | 530 Sep. 27.12998 | 0.5755915 | 0.9687113 | 78.90 | 97.56504 | 41.26006 | 163.38977 | 530 Oct. 8.0 |
| 451 | 451 Jun. 28.24911 | 0.5737438 | 0.9689123 | 79.29 | 97.01122 | 40.49602 | 163.47468 | 451 Jun. 25.0 |
| 374 | 374 Feb. 16.34230 | 0.5771940 | 0.9685857 | 78.76 | 96.49409 | 39.86451 | 163.53760 | 374 Mar. 1.0 |
| 295 | 295 Apr. 20.39842 | 0.5759148 | 0.9687528 | 79.13 | 95.22565 | 38.39767 | 163.36268 | 295 Apr. 25.0 |
| 218 | 218 May 17.72347 | 0.5814660 | 0.9679755 | 77.37 | 94.13158 | 37.19436 | 163.56891 | 218 Apr. 29.0 |
| 141 | 141 Mar. 22.43405 | 0.5831377 | 0.9678439 | 77.23 | 93.67835 | 36.50620 | 163.43259 | 141 Mar. 24.0 |
| 66 | 66 Jan. 25.96014 | 0.5851046 | 0.9675458 | 76.55 | 92.63672 | 35.41600 | 163.57158 | 66 Feb. 6.0 |
| 12 B.C. | − 11 Oct. 10.84852 | 0.5871999 | 0.9673664 | 76.33 | 92.54399 | 35.19064 | 163.58392 | − 11 Oct. 8.0 |
| 87 B.C. | − 86 Aug. 6.46171 | 0.5856047 | 0.9676769 | 77.12 | 90.76383 | 33.30553 | 163.33505 | − 86 Aug. 23.0 |
| 164 B.C.*) | −163 Nov. 12.56604 | 0.5845470 | 0.9676686 | 76.88 | 89.09882 | 31.35152 | 163.69946 | −163 Nov. 15.0 |
| 240 B.C. | −239 May 25.11796 | 0.5853647 | 0.9675871 | 76.75 | 88.09919 | 30.09811 | 163.46207 | −239 Jun. 7.0 |

*) Apparition not recorded.

Orbital calculations indicate (a) that the comet's spin axis has been fixed in space without noticeable precessional motion back to at least 87 B.C. [22], and (b) that its gas production rate seems to have remained essentially constant [5].

The tail lengths of comet Halley for the apparitions 1759, 1835 and 1910 are given in Fig. 1.

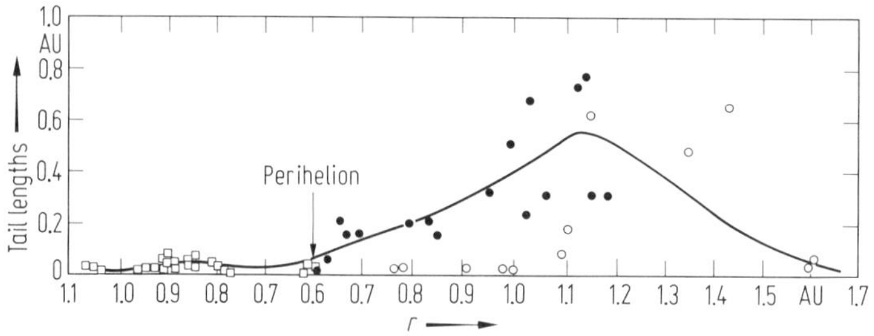


Fig. 1. Comet Halley's tail length [21]. ○ 1759 data; □ 1835 data; ● 1910 data. The greatest heliocentric distance $r$ at which the tail has been photographed was 2.3 AU (preperihelion) and 1.6 AU (post-perihelion).

Summary of historical data [21]:

| | |
|---|---|
| Earliest recorded apparition | 240 B.C. |
| Last apparition observed | August 1909···June 1911 |
| Number of recorded apparitions (From 240 B.C. to 1910 A.D.; only the 164 B.C. apparition was not recorded) | 28 |
| Total number of apparitions (considering the probability of its changing orbit by close encounters with planets) | at least 100 |
| Shortest period between returns to perihelion | 74.42 years (1835···1910) |
| Longest period between returns to perihelion | 79.25 years (451···530) |
| Closest approach to earth | 0.04 AU (April 11, 837) |
| Longest angular tail length recorded | 93° (mid-April 837) |
| Brightest apparent magnitude recorded (approximate) | $-3^{\mathrm{m}}.5$ (April 11, 837) |

Summary of orbital characteristics [21]:

| | |
|---|---|
| Location of orbit pole (ecliptical coordinates) | $\lambda = 328^\circ.15$ <br> $\beta = -72^\circ.24$ |
| Location of perihelion | $\lambda = 305^\circ.32$ <br> $\beta = 16^\circ.45$ |
| Heliocentric distance of orbit nodes | $r(\Omega) = 1.81$ AU <br> $r(\mho) = 0.85$ AU |
| Distance of perihelion and aphelion above or below orbit plane | $Z(q) = 0.17$ AU <br> $Z(Q) = 9.99$ AU |
| Orbital velocity | $V = 29.8\left(\frac{2}{r}-\frac{1}{a}\right)^{1/2}$ km s$^{-1}$ |
| at perihelion ($r=q$) | $V_q = 54.55$ km s$^{-1}$ |
| at aphelion ($r=Q$) | $V_Q = 0.91$ km s$^{-1}$ |

### 3.3.3.10.2 The 1910 apparition

Recovery: By Max Wolf, Heidelberg, on September 11, 1909. Subsequently, two pre-discovery images were found on plates taken August 24, 1909, in Helwan, Egypt, and on September 9, 1909, in Greenwich.

At this time, heliocentric distance $r=3.4$ AU, geocentric distance $\Delta=3.6$ AU, visual magnitude $m_v=15^{\mathrm{m}}$, with coma of several arc seconds.

Last photograph: May 30, 1911 at Lowell Observatory, at $r=5.28$ AU, with photographic magnitude $m_{pg}=18^{\mathrm{m}}.0$. During the preceding five days, the comet faded by more than one magnitude.

General appearance: discussed in [2, 15, 13].

Atlas of comet Halley photographs and spectra: [14].

Tail length: see Fig. 1.

Spectroscopy: First spectrogram of comet Halley on October 22, 1909.

Emission identified (: uncertain)
$C_2$(2–0), (1–0), (0–0), (0–1), (0–2):, (0–3): bands
CN(0–0), (0–1), (0–2): bands

CH emission
$C_3$ 4050 Å group (then called C+H or Raffety bands)
Na D lines
$CO^+$(1–0), (2–0), (3–0) bands

$N_2^+$ emission
Continuum-to-emission intensity appears to be "normal" [6].

Ratio of dust to gas production rates by mass, $\mu=0.5$.

Light curve: Discussion of pre- and post-perihelion total brightness $m_{tot}$ as function of heliocentric distance $r$ is given by [11]:

$$m_{tot}=m_0+5\log\Delta+2.5n\log r$$

with heliocentric distance $r$ and geocentric distance $\Delta$ in [AU], $n$=brightness parameter, $m_0$ – brightness for $r=\Delta=1$ AU.

The values of $m_0$ and $n$ changed 1909···1911 from pre- to post-perihelion:

pre-perihelion: $m_0=5^{\rm m}.49\pm0^{\rm m}.07$ $n=5.15\pm0.10$
($\Delta r=2.76$ to 0.59 AU)

post-perihelion: $m_0=5^{\rm m}.44\pm0^{\rm m}.05$ $n=2.70\pm0.17$
($\Delta r=0.59$ to 1.61 AU)

($\Delta r$ gives the range of $r$ of the observations used).

Values for the 1910 visual magnitudes are shown in Fig. 2.

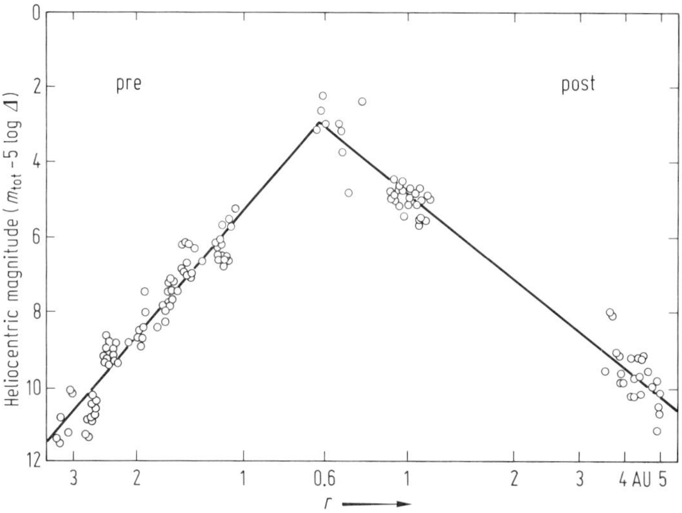


Fig. 2. Light curve of comet Halley 1910 II [11]. Lines represent power-law solutions. Gaps in data represent periods when the comet was not observable.

The predicted values of $m_{tot}$ for the years 1982 to 1987 (determined empirically from the 1910/11 estimated magnitudes) are given in Table 2.

Size of nucleus: Values usually quoted in literature refer to the "photometric nucleus", i.e., are much larger than the "true nucleus". Upper limits of 5(+11, −4) km diameter were derived using comet Bennett as prototype [12]. Unsuccessful recovery attempts lead to a diameter of less than 13 km, assuming a geometric albedo of 0.1 [13].

Mass: $\mathfrak{M}=6.5\cdot10^{16}$ g, assuming a bulk density of 1 g cm$^{-3}$, and a nominal diameter of 5 km.

Rotation: $P_{rot}=10^{\rm h}19^{\rm m}$ [19]; rotation is direct; rotational axis nearly perpendicular to orbital plane.

Further physical characteristics [21]:

| | |
|---|---|
| Photometric behavior | Brighter post-perihelion |
| | Fountain effect from nucleus sunward |
| | Spherical halos expanding from nucleus (0.1 to several km s$^{-1}$) |
| | Jets and streamers showing evidence of direct ejection |
| | Sudden outbursts |
| | Maximum visual coma diameter: about $4\cdot10^5$ km |
| Spectroscopic data | CN, $C_2$, $C_3$, CH, $CO^+$, $N_2^+$, Na D, strong continuum |
| | $^{12}C/^{13}C$ isotopic bands in the Swan system of $C_2$ |
| Tail structures | Plasma and dust tail present |
| | Plasma tail begins to form about 1.5 AU before perihelion |
| | Dust tail begins to form near perihelion |
| | Motion of fine streamers and disconnection events in plasma tail |
| | Numerous envelopes showing "closing umbrella" phenomenon |
| | Maximum visual tail length reached 5···6 weeks post perihelion |
| Associated meteor streams | η-Aquarid (early May) and Orionid (late October) |

## 3.3.3.10.3 The 1985/86 apparition

### 3.3.3.10.3.1 Ephemerides and orbital data

A schematic drawing of Halley's orbit from 1910 to 1986 is shown in Fig. 3. In Fig. 4 the inner part of the orbit (1985/86) is given in more detail.

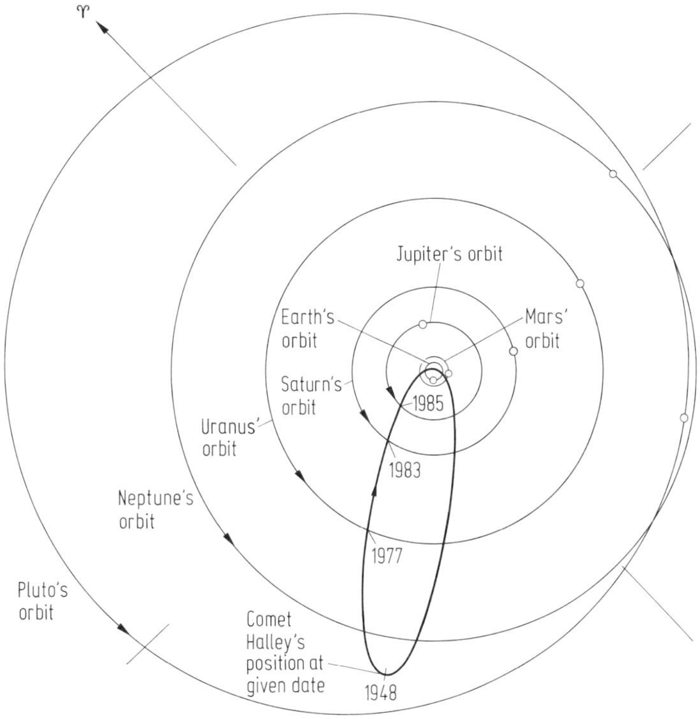


Fig. 3. Schematic drawing of Halley's orbit 1910···1986.

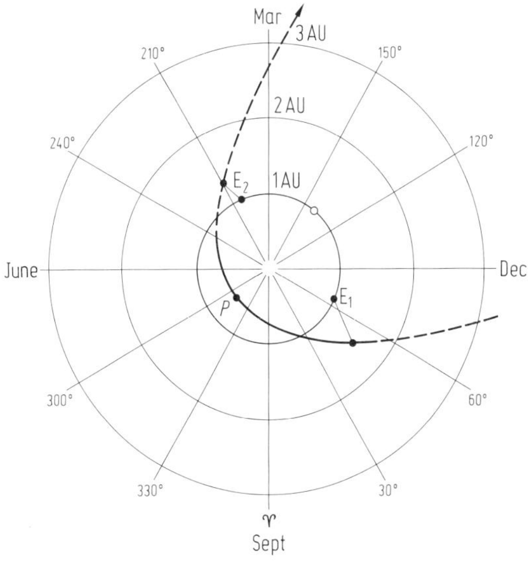


Fig. 4. Schematic drawing of earth's and comet Halley's orbit in 1985/86. $P$ = perihelion of comet Halley (passage: Febr. 9, 1986; heliocentric distance $r = 0.59$ AU; geocentric distance $\Delta = 1.54$ AU). ○ = position of earth at perihelion passage of comet Halley. $E_1$ = position of earth at pre-perihelion closest approach (Nov. 27, 1985; $r = 1.5$ AU; $\Delta = 0.62$ AU). $E_2$ = position of earth at post-perihelion closest approach (April 11, 1986; $r = 1.3$ AU; $\Delta = 0.42$ AU). The comet will cross the ecliptic on Nov. 9, 1985 ($r = 1.8$ AU; $\Delta = 0.9$ AU) and on March 11, 1986 ($r = 0.85$ AU; $\Delta = 1.0$ AU).

The ephemerides and some orbital and physical data for the time from January 1982 through March 1987 are given in Table 2, next page.

Table 2. Ephemerides and some orbital and (expected) physical data for 1982...1987 *) [21].

Date: Y = year, M = month, D = day
J.D. = Julian date (ephemeris time E.T.)
$\alpha, \delta$ (1950) = geocentric right ascension and declination referred to the mean equator and equinox of 1950.0. Light time corrections have been applied.
$\alpha, \delta$ (apparent) = apparent geocentric right ascension and declination. Light time, annual aberration, and nutation corrections have been applied to the ephemeris date.
$\Delta$ = geocentric distance of comet in [AU]
$V(\Delta)$ = geocentric radial velocity of comet in [km s$^{-1}$]
$r$ = heliocentric distance of comet in [AU]
$V(r)$ = heliocentric radial velocity of comet in [km s$^{-1}$]
$m_{tot}$ = total magnitude = $5.0+5.0\log\Delta+13.1\log r$ pre-perihelion. Post-perihelion, $m_{tot}$ is determined empirically from the 1910/11 magnitude estimates (see Fig. 2)
$m_N$ = nuclear magnitude = $7.5+5.0\log\Delta+10.0\log r$
$\theta$ = elongation = angle sun-earth-comet; $\varphi$ = phase angle = angle sun-comet-earth — referred to the ecliptic and equinox 1950.0

| Date Y | M | D | J.D. | $\alpha$ (1950) | $\delta$ (1950) | $\alpha$ (apparent) | $\delta$ (apparent) | $\Delta$ AU | $V(\Delta)$ km s$^{-1}$ | $r$ AU | $V(r)$ km s$^{-1}$ | $m_{tot}$ | $m_N$ | $\theta$ | $\varphi$ |
|---|---|---|---|---|---|---|---|---|---|---|---|---|---|---|---|
| 1982 | 1 | 1 | 2444970.5 | 6ʰ59ᵐ.927 | + 8°26′.87 | 7ʰ1ᵐ.675 | + 8°24′.09 | 11.70 | −12.01 | 12.66 | − 9.20 | — | 23ᵐ.9 | 164°.8 | 1°.2 |
| | 4 | 1 | 2445060.5 | 6 37.230 | +10 7.76 | 6 38.989 | +10 5.88 | 12.15 | 19.81 | 12.17 | − 9.47 | — | 23.8 | 88.9 | 4.7 |
| | 7 | 1 | 2445151.5 | 6 52.392 | +11 1.51 | 6 54.152 | +10 59.02 | 12.65 | − 7.35 | 11.66 | − 9.77 | — | 23.7 | 12.7 | 1.1 |
| | 10 | 1 | 2445243.5 | 7 11.305 | + 9 49.08 | 7 13.080 | + 9 45.80 | 11.28 | −38.89 | 11.14 | −10.10 | — | 23.2 | 79.3 | 5.1 |
| 1983 | 1 | 1 | 2445335.5 | 6 48.156 | + 9 10.08 | 6 49.986 | + 9 7.72 | 9.63 | −11.85 | 10.59 | −10.45 | — | 22.7 | 166.0 | 1.3 |
| | 4 | 1 | 2445425.5 | 6 20.965 | +11 5.33 | 6 22.791 | +11 4.22 | 10.07 | 18.57 | 10.04 | −10.84 | — | 22.5 | 85.1 | 5.7 |
| | 7 | 1 | 2445516.5 | 6 39.731 | +12 11.61 | 6 41.564 | +12 9.69 | 10.45 | −10.33 | 9.46 | −11.27 | — | 22.4 | 11.0 | 1.2 |
| | 10 | 1 | 2445608.5 | 7 1.765 | +10 55.99 | 7 3.611 | +10 53.08 | 8.94 | −40.95 | 8.85 | −11.76 | — | 21.7 | 81.5 | 6.4 |
| 1984 | 1 | 1 | 2445700.5 | 6 29.101 | +10 12.81 | 6 30.984 | +10 11.35 | 7.25 | −11.31 | 8.21 | −12.32 | — | 20.9 | 166.8 | 1.6 |
| | 4 | 1 | 2445791.5 | 5 53.837 | +12 31.69 | 5 55.737 | +12 31.95 | 7.69 | 16.37 | 7.54 | −12.96 | — | 20.7 | 77.7 | 7.4 |
| | 7 | 1 | 2445882.5 | 6 19.585 | +14 3.75 | 6 21.504 | +14 2.80 | 7.84 | −15.75 | 6.84 | −13.72 | — | 20.3 | 10.1 | 1.5 |
| | 10 | 1 | 2445974.5 | 6 46.414 | +12 52.29 | 6 48.349 | +12 50.07 | 6.07 | −44.54 | 6.09 | −14.64 | — | 19.3 | 86.1 | 9.4 |
| 1985 | 1 | 1 | 2446066.5 | 5 45.486 | +12 2.31 | 5 47.453 | +12 3.07 | 4.34 | − 8.55 | 5.28 | −15.81 | — | 17.9 | 162.3 | 3.2 |
| | 2 | 1 | 2446097.5 | 5 13.735 | +12 39.48 | 5 15.705 | +12 41.82 | 4.35 | 8.48 | 4.99 | −16.27 | — | 17.7 | 126.2 | 9.2 |
| | 3 | 1 | 2446125.5 | 4 56.114 | +13 35.33 | 4 58.086 | +13 38.53 | 4.55 | 14.54 | 4.73 | −16.74 | — | 17.5 | 94.4 | 12.1 |
| | 4 | 1 | 2446156.5 | 4 51.342 | +14 51.96 | 4 53.322 | +14 55.40 | 4.79 | 10.90 | 4.42 | −17.30 | — | 17.4 | 62.8 | 11.6 |
| | 5 | 1 | 2446186.5 | 4 58.386 | +16 8.75 | 5 0.380 | +16 11.86 | 4.90 | 0.56 | 4.12 | −17.92 | — | 17.1 | 35.6 | 8.2 |
| | 6 | 1 | 2446217.5 | 5 13.472 | +17 19.72 | 5 15.489 | +17 22.10 | 4.78 | −13.45 | 3.79 | −18.64 | — | 16.7 | 10.5 | 2.8 |
| | 7 | 1 | 2446247.5 | 5 32.133 | +18 13.66 | 5 34.173 | +18 15.13 | 4.43 | −27.65 | 3.46 | −19.43 | — | 16.1 | 16.1 | 4.7 |
| | 8 | 1 | 2446278.5 | 5 52.227 | +18 52.54 | 5 54.290 | +18 53.01 | 3.81 | −40.94 | 3.10 | −20.39 | 14ᵐ.3 | 15.3 | 40.3 | 12.2 |
| | 9 | 1 | 2446309.5 | 6 8.608 | +19 20.73 | 6 10.692 | +19 20.38 | 2.98 | −51.07 | 2.73 | −21.52 | 13.1 | 14.2 | 66.0 | 19.7 |
| | 10 | 1 | 2446339.5 | 6 11.520 | +19 59.98 | 6 13.629 | +19 59.47 | 2.04 | −55.78 | 2.35 | −22.82 | 11.4 | 12.8 | 94.5 | 25.2 |
| | 11 | 1 | 2446370.5 | 5 22.481 | +21 46.70 | 5 24.635 | +21 48.72 | 1.08 | −48.73 | 1.92 | −24.41 | 8.9 | 10.5 | 136.7 | 20.7 |
| | 12 | 1 | 2446400.5 | 1 6.800 | +13 43.44 | 1 8.697 | +13 54.99 | 0.63 | 10.10 | 1.49 | −26.09 | 6.3 | 8.2 | 132.1 | 29.5 |

continued

Table 2, continued

| Date Y | M | D | J.D. | α (1950) | δ (1950) | α (apparent) | δ (apparent) | Δ AU | V(Δ) km s⁻¹ | r AU | V(r) km s⁻¹ | $m_{tot}$ | $m_N$ | θ | φ |
|---|---|---|---|---|---|---|---|---|---|---|---|---|---|---|---|
| 1986 | 1 | 1 | 2446431.5 | 22h15m.913 | − 2°29′.01 | 22h17m.749 | − 2°18′.33 | 1.16 | 33.68 | 1.01 | −26.51 | 5m.4 | 7m.9 | 55°.5 | 53°.4 |
| | 1 | 10 | 2446440.5 | 21 56.697 | − 4 23.97 | 21 58.548 | − 4 13.79 | 1.32 | 29.18 | 0.87 | −25.29 | 4.8 | 7.5 | 41.4 | 48.0 |
| | 1 | 20 | 2446450.5 | 21 38.695 | − 6 12.52 | 21 40.563 | − 6 2.87 | 1.47 | 20.64 | 0.74 | −21.73 | 4.1 | 7.0 | 26.7 | 36.9 |
| | 2 | 1 | 2446462.5 | 21 17.484 | − 8 30.66 | 21 19.379 | − 8 21.70 | 1.56 | 4.65 | 0.62 | −11.64 | 3.2 | 6.4 | 10.4 | 16.7 |
| | 2 | 5 | 2446466.5 | 21 10.166 | − 9 23.38 | 21 12.074 | − 9 14.67 | 1.56 | − 2.04 | 0.60 | − 6.57 | 3.0 | 6.2 | 6.8 | 11.4 |
| | 2 | 10 | 2446471.5 | 21 0.942 | −10 35.20 | 21 2.867 | −10 26.83 | 1.54 | −10.79 | 0.59 | 0.49 | 2.9 | 6.1 | 8.5 | 14.4 |
| | 2 | 15 | 2446476.5 | 20 51.816 | −11 53.65 | 20 53.759 | −11 45.62 | 1.50 | −19.20 | 0.60 | 7.47 | 3.1 | 6.2 | 14.5 | 24.4 |
| | 2 | 20 | 2446481.5 | 20 42.941 | −13 18.75 | 20 44.907 | −13 11.06 | 1.43 | −26.58 | 0.63 | 13.53 | 3.6 | 6.3 | 21.5 | 35.1 |
| | 2 | 25 | 2446486.5 | 20 34.311 | −14 51.37 | 20 36.301 | −14 44.03 | 1.35 | −32.57 | 0.68 | 18.22 | 4.3 | 6.4 | 28.7 | 44.6 |
| | 3 | 1 | 2446490.5 | 20 27.437 | −16 12.43 | 20 29.448 | −16 5.37 | 1.27 | −36.38 | 0.72 | 20.99 | 4.7 | 6.6 | 34.5 | 51.1 |
| | 3 | 10 | 2446499.5 | 20 10.638 | −19 50.87 | 20 12.712 | −19 44.50 | 1.06 | −42.13 | 0.84 | 24.73 | 5.0 | 6.9 | 48.1 | 61.5 |
| | 3 | 20 | 2446509.5 | 19 43.551 | −25 46.62 | 19 45.735 | −25 41.44 | 0.81 | −43.97 | 0.99 | 26.40 | 4.6 | 7.0 | 65.4 | 66.2 |
| | 4 | 1 | 2446521.5 | 18 23.586 | −38 35.50 | 18 26.074 | −38 34.25 | 0.53 | −34.33 | 1.17 | 26.80 | 4.1 | 6.8 | 95.5 | 57.9 |
| | 4 | 10 | 2446530.5 | 15 25.059 | −47 29.12 | 15 27.616 | −47 36.69 | 0.42 | − 4.64 | 1.31 | 26.60 | 4.0 | 6.8 | 131.0 | 35.1 |
| | 4 | 20 | 2446540.5 | 12 4.691 | −32 58.06 | 12 6.570 | −33 10.31 | 0.51 | 34.93 | 1.47 | 26.15 | 4.8 | 7.7 | 147.8 | 21.4 |
| | 5 | 1 | 2446551.5 | 10 54.816 | −18 21.86 | 10 56.602 | −18 33.61 | 0.80 | 51.23 | 1.63 | 25.56 | 6.1 | 9.1 | 128.6 | 28.9 |
| | 6 | 1 | 2446582.5 | 10 23.923 | − 6 33.81 | 10 25.739 | − 6 44.92 | 1.80 | 56.64 | 2.07 | 23.84 | 8.1 | 11.9 | 90.5 | 29.3 |
| | 7 | 1 | 2446612.5 | 10 33.198 | − 4 59.43 | 10 35.023 | − 5 10.68 | 2.74 | 51.03 | 2.47 | 22.37 | 9.6 | 13.6 | 64.2 | 21.7 |
| | 8 | 1 | 2446643.5 | 10 50.539 | − 5 51.99 | 10 52.363 | − 6 3.53 | 3.56 | 40.56 | 2.86 | 21.11 | 11.2 | 14.8 | 39.9 | 13.2 |
| | 9 | 1 | 2446674.5 | 11 9.214 | − 7 48.30 | 11 11.040 | − 8 0.07 | 4.17 | 27.02 | 3.23 | 20.04 | 12.6 | 15.7 | 18.2 | 5.6 |
| | 10 | 1 | 2446704.5 | 11 25.423 | −10 9.15 | 11 27.255 | −10 21.08 | 4.52 | 12.61 | 3.57 | 19.17 | 13.6 | 16.3 | 16.7 | 4.6 |
| | 11 | 1 | 2446735.5 | 11 37.170 | −12 42.20 | 11 39.014 | −12 54.23 | 4.61 | − 1.41 | 3.90 | 18.38 | 14.0 | 16.7 | 40.0 | 9.4 |
| | 12 | 1 | 2446765.5 | 11 39.958 | −14 54.76 | 11 41.815 | −15 6.86 | 4.50 | −11.13 | 4.22 | 17.71 | 14.1 | 17.0 | 67.4 | 12.5 |
| 1987 | 1 | 1 | 2446796.5 | 11 29.439 | −16 18.94 | 11 31.303 | −16 31.08 | 4.26 | −13.17 | 4.53 | 17.10 | 14.1 | 17.2 | 99.3 | 12.4 |
| | 2 | 1 | 2446827.5 | 11 3.340 | −15 55.63 | 11 5.196 | −16 7.63 | 4.09 | − 3.78 | 4.83 | 16.56 | 14.4 | 17.4 | 133.9 | 8.5 |
| | 3 | 1 | 2446855.5 | 10 32.199 | −13 37.14 | 10 34.044 | −13 48.72 | 4.16 | 12.78 | 5.09 | 16.11 | — | 17.7 | 158.1 | 4.2 |

*) A more detailed, and after the recovery continuously up-dated, table with smaller time intervals can be delivered by the author on request. The ephemerides change by perturbations by the large planets and due to non-gravitational effects. They will be recalculated after the recovery of the comet.

The following osculating orbital elements are consistent with the ephemerides in Table 2 (for explanation, see Table 1):

| | |
|---|---|
| Epoch | J.D. 2446480.50000 ≙ 1986, Febr. 19.00000 (E.T.) |
| Perihelion passage | J.D. 2446471.16128 ≙ 1986, Febr. 9.66128 (E.T.) |
| Perihelion distance | 0.5870959 AU |
| Eccentricity | 0.9672671 |
| Argument of perihelion | 111°.85336 |
| Longitude of ascending node | 58°.15313 |
| Inclination | 162°.23779 |

### 3.3.3.10.3.2 Ground-based data

The expected viewing conditions in 1985/86 are discussed in [21]. In the following some selected Figures are given.

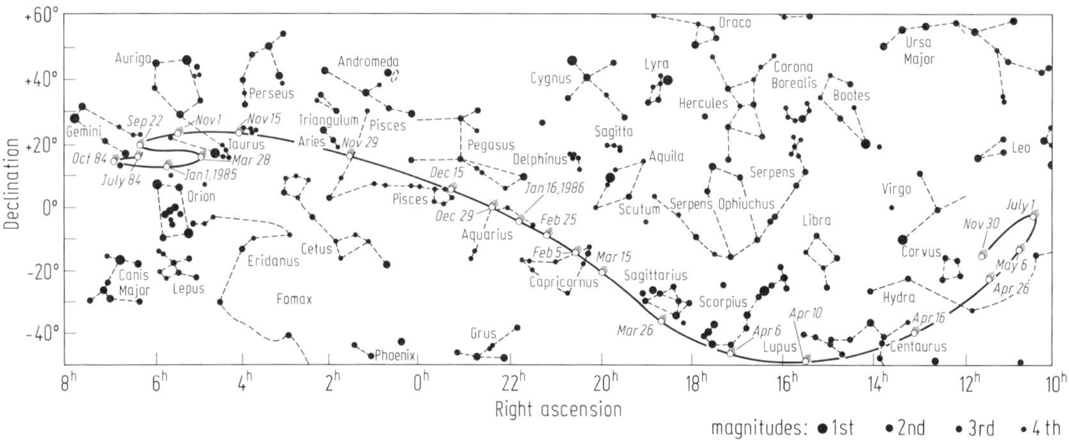


Fig. 5. Path of Halley's comet on the celestial sphere during 1984 July to 1986 November. Note the comet's retrograde loops and its transition from a primarily northern hemisphere object in late 1985 to a southern hemisphere one in early 1986 [18].

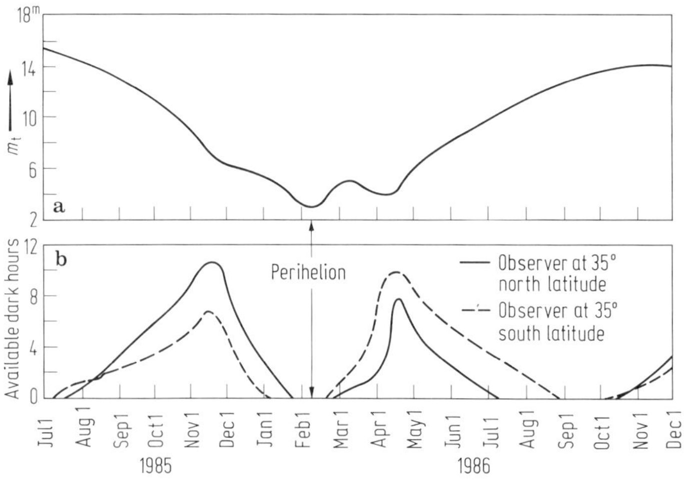


Fig. 6. Comet Halley 1985/86 ground-based observing conditions [21]. a) predicted apparent total magnitude $m_{tot}$. b) number of hours when comet is above, and sun is more than 18° below the local horizon.

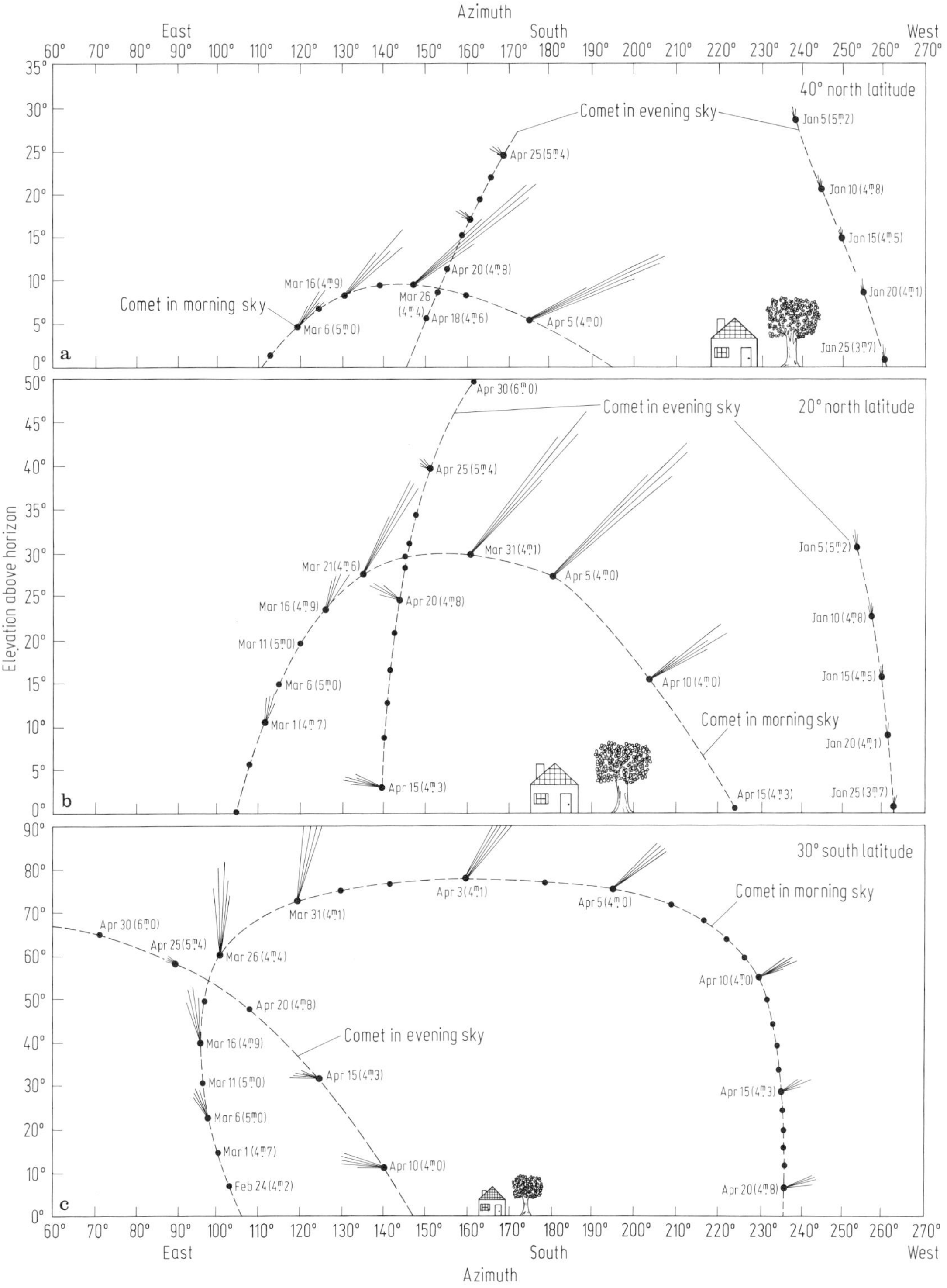


Fig. 7. Comet Halley observing conditions for ground-based observers in 1986. Comet positions are given for the beginning of morning astronomical twilight or end of evening astronomical twilight. Approximate total visual magnitudes are given in parantheses following dates. Viewing with binoculars and ideal observing conditions are assumed [21]. Fig. 7a: for observers at 40° north latitude, Fig. 7b: for observers at 20° north latitude, Fig. 7c: for observers at 30° south latitude.

Ground-based activities:

Recovery attempts: Starting in November 1977; unsuccessful as of middle of Sept. 1982.

International Halley Watch (IHW) [18]: The purpose of the IHW is to promote worldwide cooperation, communication, and standardization (where useful) in studies of Halley's comet. Coordination and communication extend from the present planning and preparation phase through the observation period to include archival publication of all scientific Halley observations. Lead centers of the IHW are at the Jet Propulsion Laboratory in Pasadena, California, and the Remeis-Observatory in Bamberg, West Germany.

### 3.3.3.10.3.3 Expected physical data; models

See also summary of physical characteristics 3.3.3.10.2 and Table 2.

Nominal models: Described and referenced in [13]. Current models assume radius of nucleus of $R_N = 2.5$ km; mean density of 1 g cm$^{-3}$; composition of material leaving nucleus surface to be 50% volatiles by mass with 5/6 $H_2O$ by number and 1/6 other material with mass 44 amu; surface temperature assumed to be 185 K. Resulting total gas production rate is given in Table 3. Table 4 gives the general nature of the dust environment for $r = 1.53$ AU (pre-perihelion) and $r = 0.90$ AU (post-perihelion).

Table 3. Gas production predictions for comet Halley in 1985···86 [13].
$\Delta$ = geocentric distance
$r$ = heliocentric distance

| Date | | $\Delta$ AU | $r$ AU | Nominal gas production particles/s |
|---|---|---|---|---|
| 1985 | Oct. 26 | 1.23 | 2.0 | $1.19 \cdot 10^{28}$ |
| | Nov. 23 | 0.63 | 1.6 | $2.98 \cdot 10^{28}$ |
| | Nov. 28 | 0.62 | 1.53 | $3.55 \cdot 10^{28}$ |
| | Dec. 20 | 0.92 | 1.2 | $8.70 \cdot 10^{28}$ |
| 1986 | Jan. 2 | 1.18 | 1.0 | $1.68 \cdot 10^{29}$ |
| | Feb. 4 | 1.56 | 0.6 | $7.44 \cdot 10^{29}$ |
| | Feb. 9.7 | 1.54 | 0.587 | perihelion |
| | Feb. 15 | 1.50 | 0.6 | $6.12 \cdot 10^{29}$ |
| | Mar. 7 | 1.14 | 0.8 | $1.98 \cdot 10^{29}$ |
| | Mar. 21 | 0.79 | 1.0 | $1.56 \cdot 10^{29}$ |
| | Apr. 29 | 0.74 | 1.6 | $5.56 \cdot 10^{28}$ |
| | May 27 | 1.63 | 2.0 | $2.73 \cdot 10^{28}$ |
| | July 3 | 2.80 | 2.5 | $1.16 \cdot 10^{28}$ |
| | Aug. 12 | 3.82 | 3.0 | $5.59 \cdot 10^{27}$ |
| | Sept. 25 | 4.47 | 3.5 | $2.87 \cdot 10^{27}$ |

Table 4. Nominal near nucleus dust parameters [13].

$d_{part}$ = particle diameter
$\dot{Q}$ = production rate
$n_{part}$ = particle density at $10^3$ km
$F_{part}$ = particle flux at $10^3$ km
fluence = total fluence on $10^3$ km closest approach
$V_{appr}$ = approximate velocity

| $d_{part}$ cm | $\dot{Q}$ part. s$^{-1}$ | $n_{part}$ part. m$^{-3}$ | $F_{part}$ m$^{-2}$ s$^{-1}$ | Fluence *) m$^{-2}$ | $V_{appr}$ m s$^{-1}$ |
|---|---|---|---|---|---|
| | | 1.53 AU pre-perihelion | | | |
| $0.9 \cdots 5.0 \cdot 10^{-4}$ | $6.4 \cdot 10^{16}$ | $2.9 \cdot 10^{1}$ | $1.0 \cdot 10^{4}$ | $9.1 \cdot 10^{7}$ | 350 |
| $0.5 \cdots 10.0 \cdot 10^{-3}$ | $7.7 \cdot 10^{14}$ | $3.5 \cdot 10^{-1}$ | $1.2 \cdot 10^{2}$ | $2.6 \cdot 10^{6}$ | 150 |
| $0.1 \cdots 10.0 \cdot 10^{-1}$ | $2.7 \cdot 10^{10}$ | $1.2 \cdot 10^{-5}$ | $4.3 \cdot 10^{-3}$ | $3.4 \cdot 10^{2}$ | 40 |
| | | 0.90 AU post-perihelion | | | |
| $0.9 \cdots 5.0 \cdot 10^{-4}$ | $4.1 \cdot 10^{17}$ | $1.6 \cdot 10^{2}$ | $6.5 \cdot 10^{4}$ | $4.9 \cdot 10^{8}$ | 420 |
| $0.5 \cdots 10.0 \cdot 10^{-3}$ | $3.9 \cdot 10^{15}$ | $2.2 \cdot 10^{0}$ | $6.2 \cdot 10^{2}$ | $7.0 \cdot 10^{6}$ | 280 |
| $0.1 \cdots 10.0 \cdot 10^{-1}$ | $1.4 \cdot 10^{11}$ | $1.9 \cdot 10^{-4}$ | $2.2 \cdot 10^{-2}$ | $5.8 \cdot 10^{2}$ | 120 |

*) The fluence is the total accumulation of material on a surface passing through the coma at a velocity that is large relative to the intrinsic velocity of the dust.

The encounter model and the model of the cometary coma with the trajectory of "Giotto" (see 3.3.3.10.4) are shown in Figs. 8 and 9.

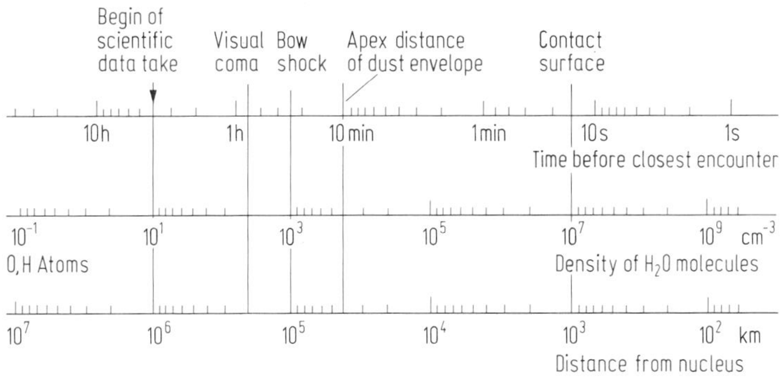


Fig. 8. Comet Halley encounter model (cf. also Fig. 9).

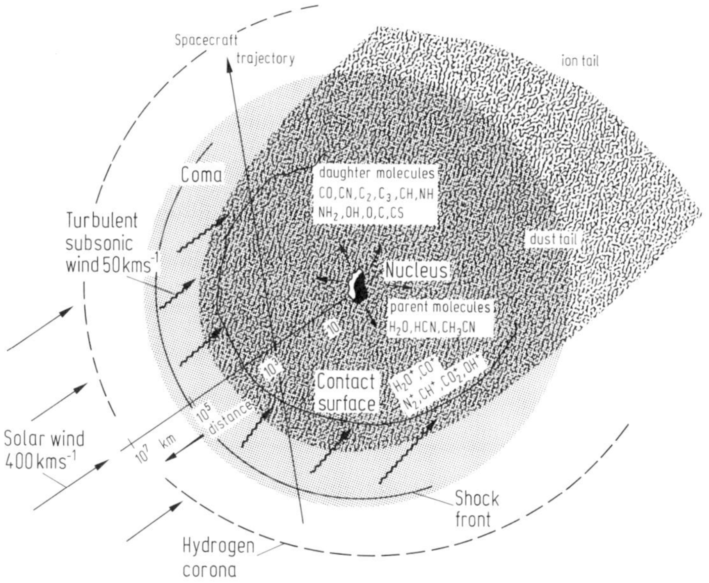


Fig. 9. Model of cometary coma and spacecraft trajectory (schematic). The spacecraft will be targetted to pass the nucleus on the sunward side.

Meteor showers associated with comet Halley: The orbital elements of the annual May η Aquarid and the October Orionid meteor showers are very similar to those of comet Halley and suggest comet Halley as their "parent" body [8], (see also LB, NS, Vol. VI/1, p. 197). The average rate is 10···20 per hour. Yeomans [20] predicts shower maximum dates of 1985 October 24.2 (Orionids) and 1986 May 8.5 (η Aquarids).

### 3.3.3.10.4 Comet Halley missions

The 1985/86 apparition of Halley's comet provides the first opportunity to send a special comet-spacecraft to this comet and observe it "in situ" on this "once in a lifetime" occasion. Besides the mystique and fascination that for thousands of years have been associated with this comet, there are numerous scientific reasons why Halley's comet was chosen to be the first candidate for a comprehensive comet mission.

It is the "freshest" comet with a period of less than 200 years. It is the only short-period comet to display the full range of almost all known cometary phenomena. It is the only really active comet with a well-established orbit; its gas production rate is about 100 times greater than that of any other periodic comet. And it provides the unique opportunity to compare the properties of a bright, active comet during two successive appearances.

The ESA Science Program Committee decided at its meeting on July 8/9, 1980, to fly a mission called "Giotto" to Halley's comet. The spacecraft's design is based on the existing Geos spacecraft. It will be launched from Kourou (French Guiana) in July 1985 and fly by the comet in March 1986, about a month after the comet's perihelion passage (February 9, 1986). The transfer trajectory lies in the ecliptic plane; it is inside the earth's orbit and has a closest distance from the sun of 0.7 AU. The trajectory of Giotto is shown in Fig. 10. The encounter duration (Fig. 9) during which measurements can be performed, lasts four hours.

The Japanese Institute of Space and Astronautical Science in Tokyo plans to send its first interplanetary probe named "Planet-A", into a heliocentric orbit to take large-scale images of the comet in the vacuum ultraviolet region, and to measure the solar wind plasma near the comet. Launch date: middle of August 1985. Encounter date: early March 1986.

Two Soviet spacecrafts, called "Vega", will carry Venus entry probes to the planet's vicinity and will then, as a result of a Venus swing-by, be directed into a trajectory to Halley's comet; see Fig. 12. They will encounter the comet near the descending node, about 270 days after the Venus swing-by. Launch date: end of December 1984. Encounter date: early March 1986.

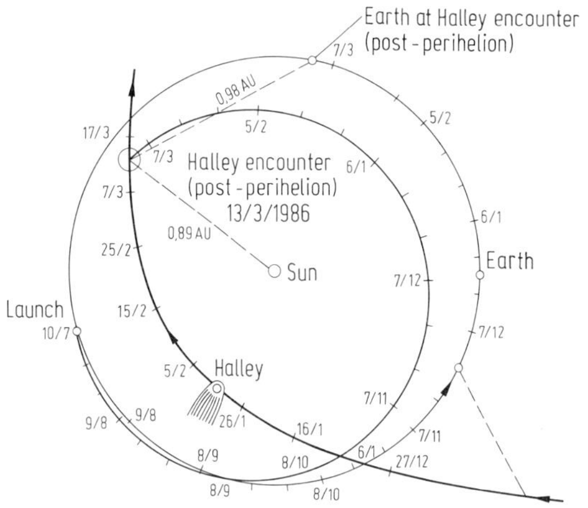


Fig. 10. Reference trajectory for Giotto from launch on July 10, 1985 (10/7) to post-perihelion Halley encounter on March 13, 1986 (13/3).

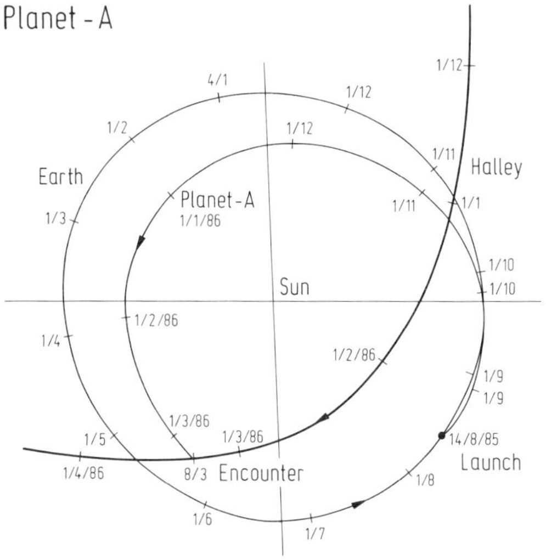


Fig. 11. The Japanese Planet-A spacecraft heliocentric trajectory.

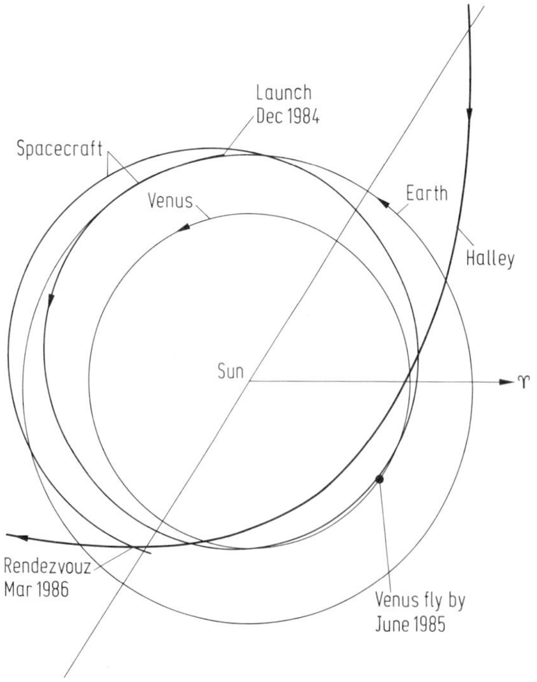


Fig. 12. Trajectory of the VEGA-mission (USSR).

In Table 5 important data of the four spacecraft encounters with comet Halley in March 1986 are summarized.

Table 5. Spacecraft encounters with Halley in March 1986.

| Spacecraft parameter | Vega 1 | Planet-A | Giotto | Vega 2 |
|---|---|---|---|---|
| Flyby date | 8 March | 8 March | 13 March | ~16 March |
| Heliocentric distance, AU | 0.83 | 0.83 | 0.89 | ~0.93 |
| Flyby velocity, km/s | 77.7 | ~70 | 68.7 | ~77 |
| Phase angle to Sun at which coma traversed | 110° | 105···110° | 107.2° | ~110° |
| Minimum distance in the orbital plane from cometary nucleus, km (+ sunward, − antisunward side) | ~$10^4$ | $10^4$···$10^5$ | +500 | perhaps $3\times10^3$ km |
| Distance above/below comet orbit plane at closest approach, km | ~$10^4$ | $10^4$···$10^5$ | *) | perhaps $3\times10^3$ km |

*) To be determined.

Giotto mission scientific objectives

To provide the elemental and isotopic composition of the volatile components in the cometary coma, in particular to identify the parent molecules,
to characterize the physical processes and chemical reactions that occur in the cometary atmosphere and ionosphere,
to determine the elemental and isotopic composition of the cometary dust particles,
to measure the total gas production rate and the dust flux and size/mass distribution and to derive the dust to gas ratio,
to investigate the macroscopic system of plasma flows resulting from the interaction between the cometary and the solar wind plasma,
to provide numerous images of the comet nucleus with a resolution down to 50 m. From these the nucleus size and rotation may be deduced and its mass may be estimated.

Giotto scientific payload

To accomplish these objectives the Giotto spacecraft will carry:
a camera for imaging the inner coma and the nucleus,
neutral, ion and dust mass spectrometers for composition measurements,
a dust impact detector for studies of the dust environment,
electron and ion plasma analyzers and a magnetometer for plasma studies,
an optical probe for in situ measurements of the cometary atmosphere.

## 3.3.3.10.5 Other future comets

Table 6 contains a chronological list of the predicted perihelia of short-period comets from 1982 through 2002. The purpose is to have at hand a quick reference from which the geometry of any perihelion in the near future of any of the short-period comets is readily available.

Table 6. Chronological list of perihelia of periodic comets 1982···2002 [1].
$r$ = heliocentric distance of perihelion $\lambda, \beta$ = ecliptical coordinates (1950) of perihelion
$\Delta$ = geocentric distance of perihelion $\theta$ = elongation = angle sun-earth-comet at perihelion passage

| Name | Date | $r$ [AU] | $\Delta$ [AU] | $\lambda$ | $\beta$ | $\theta$ |
|---|---|---|---|---|---|---|
| 1982 | | | | | | |
| Grigg-Skjellerup | 1982 May 15.0 | 0.989 | 0.372 | −147°99 | − 0°24 | 76°02 |
| Perrine-Mrkos | 1982 May 18.6 | 1.303 | 2.305 | 47.04 | 4.09 | 5.98 |
| Vaisala 1 | 1982 Jul. 30.5 | 1.800 | 2.516 | −178.14 | 8.59 | 36.24 |
| D'Arrest | 1982 Sep. 14.1 | 1.291 | 0.735 | − 44.00 | 1.01 | 94.47 |
| Tempel-Swift | 1982 Oct. 22.0 | 1.605 | 0.712 | 44.15 | 3.74 | 139.50 |
| Churyumov-Gerasim | 1982 Nov. 11.1 | 1.306 | 0.417 | 61.60 | 1.39 | 131.94 |
| Gunn | 1982 Nov. 26.9 | 2.459 | 3.399 | − 95.39 | − 3.02 | 15.03 |
| Neujmin 3 | 1982 Dec. 6.1 | 2.060 | 2.870 | − 65.08 | 2.25 | 28.57 |
| 1983 | | | | | | |
| Pons-Winnecke | 1983 Apr. 7.5 | 1.254 | 1.295 | − 94.36 | 2.91 | 64.75 |
| Arend | 1983 May 22.5 | 1.857 | 2.811 | 40.79 | 14.42 | 15.70 |
| Tempel 2 | 1983 Jun. 1.5 | 1.381 | 1.237 | − 50.17 | − 2.34 | 74.99 |
| Dutoit-Neuj-Delpo | 1983 Jun. 6.5 | 1.713 | 1.298 | − 56.27 | 2.56 | 94.76 |
| Oterma | 1983 Jun. 18.2 | 5.471 | 6.059 | 27.12 | 1.61 | 50.64 |
| Tempel 1 | 1983 Jul. 9.8 | 1.491 | 0.952 | −112.61 | 0.18 | 98.42 |
| Kopff | 1983 Aug. 10.3 | 1.576 | 0.921 | − 76.82 | 1.39 | 109.01 |
| Taylor | 1983 Nov. 25.5 | 1.951 | 1.383 | 104.06 | − 1.55 | 109.63 |
| Harrington-Abell | 1983 Dec. 2.0 | 1.785 | 1.332 | 115.74 | 6.70 | 99.67 |
| Johnson | 1983 Dec. 3.2 | 2.302 | 2.738 | − 35.80 | − 6.41 | 54.03 |
| 1984 | | | | | | |
| Smirnova-Chernykh | 1984 Feb. 19.7 | 3.558 | 2.645 | 167.84 | 6.64 | 153.66 |
| Crommelin | 1984 Feb. 20.2 | 0.735 | 0.967 | 84.12 | − 7.63 | 44.11 |
| Tritton | 1984 Mar. 2.0 | 1.438 | 1.493 | 87.92 | 3.75 | 67.29 |
| Encke | 1984 Mar. 27.7 | 0.341 | 0.710 | 160.05 | − 1.24 | 12.45 |
| Clark | 1984 May 29.1 | 1.551 | 0.703 | − 92.26 | − 4.59 | 128.40 |
| Wolf 1 | 1984 May 31.8 | 2.415 | 3.017 | 7.60 | 8.13 | 45.36 |
| Faye | 1984 Jul. 9.9 | 1.593 | 2.224 | 42.54 | − 3.66 | 40.44 |
| Tuttle-Giaco-Kres | 1984 Jul. 28.4 | 1.123 | 1.670 | −157.66 | 7.54 | 41.03 |
| Wild 2 | 1984 Aug. 15.5 | 1.491 | 2.401 | 175.95 | 2.10 | 20.13 |
| Wolf-Harrington | 1984 Sep. 22.7 | 1.616 | 1.769 | 80.72 | − 2.17 | 64.57 |
| Haneda-Camp | 1984 Sep. 26.6 | 1.101 | 0.210 | 11.89 | − 5.17 | 113.03 |
| Neujmin 1 | 1984 Oct. 8.2 | 1.553 | 1.037 | − 26.42 | − 3.20 | 99.44 |
| Schwa-Wach 3 | 1984 Nov. 30.2 | 0.913 | 1.871 | − 92.84 | − 3.32 | 9.47 |
| Arend-Rigaux | 1984 Dec. 1.4 | 1.446 | 0.683 | 91.69 | − 9.11 | 118.95 |
| Schaumasse | 1984 Dec. 7.3 | 1.213 | 1.161 | 137.25 | 9.95 | 68.23 |
| 1985 | | | | | | |
| Tsuchinshan 1 | 1985 Jan. 6.9 | 1.508 | 0.595 | 118.62 | 4.04 | 144.63 |
| Honda-Mrkos-Pajdu | 1985 May 23.9 | 0.542 | 1.551 | 54.46 | − 2.38 | 2.82 |
| Schuster | 1985 Jun. 26.9 | 1.628 | 2.414 | 45.13 | − 2.12 | 30.88 |
| Russell 1 | 1985 Jul. 2.8 | 1.613 | 1.235 | −129.61 | 0.09 | 91.04 |
| Gehrels 3 | 1985 Jul. 7.6 | 3.395 | 4.409 | 110.08 | − 0.77 | 4.04 |
| Tsuchinshan 2 | 1985 Jul. 18.4 | 1.794 | 2.786 | 130.60 | − 2.63 | 9.97 |
| Kowal 2 | 1985 Jul. 18.8 | 1.521 | 2.394 | 76.19 | − 2.54 | 23.79 |
| Giclas | 1985 Jul. 27.7 | 1.732 | 1.922 | 28.56 | − 7.87 | 63.76 |
| Daniel | 1985 Aug. 3.8 | 1.651 | 2.409 | 78.64 | 3.71 | 32.88 |
| Giacobini-Zinner | 1985 Sep. 5.7 | 1.028 | 0.459 | 8.32 | 3.96 | 79.44 |

continued

Table 6, continued

| Name | Date | r [AU] | Δ [AU] | λ | β | θ |
|---|---|---|---|---|---|---|
| 1986 | | | | | | |
| Boethin | 1986 Jan. 23.2 | 1.114 | 1.421 | 37°.43 | 1°.16 | 51°.37 |
| Ashbrook-Jackson | 1986 Jan. 24.3 | 2.307 | 3.059 | − 8.98 | − 2.41 | 33.82 |
| Halley | 1986 Feb. 9.7 | 0.587 | 1.547 | − 54.68 | 16.45 | 8.17 |
| Holmes | 1986 Mar. 14.1 | 2.168 | 3.156 | − 10.48 | 7.48 | 5.58 |
| Wirtanen | 1986 Mar. 19.8 | 1.084 | 1.601 | 77.81 | − 0.79 | 41.74 |
| Kojima | 1986 Apr. 5.2 | 2.414 | 1.964 | 142.59 | − 0.18 | 104.29 |
| Shajn-Schaldach | 1986 May 27.4 | 2.331 | 3.150 | 22.59 | − 3.62 | 30.21 |
| Whipple | 1986 Jun. 25.0 | 3.077 | 3.568 | 23.55 | − 3.72 | 53.73 |
| Wild 1 | 1986 Sep. 29.7 | 1.977 | 2.940 | 166.61 | 4.11 | 12.98 |
| 1987 | | | | | | |
| Forbes | 1987 Jan. 2.5 | 1.474 | 2.454 | − 74.39 | − 4.63 | 3.91 |
| Neujmin 2 | 1987 Apr. 2.4 | 1.271 | 0.640 | 161.97 | − 3.07 | 99.36 |
| Jackson-Neujmin | 1987 May 22.3 | 1.438 | 2.122 | − 0.75 | − 3.98 | 36.35 |
| Grigg-Skjellerup | 1987 Jun. 20.1 | 0.993 | 0.940 | −148.01 | − 0.25 | 60.89 |
| Russell 2 | 1987 Jul. 4.1 | 2.161 | 1.200 | − 70.63 | −11.38 | 154.06 |
| Encke | 1987 Jul. 17.4 | 0.332 | 1.268 | 160.16 | − 1.29 | 10.93 |
| Klemola | 1987 Jul. 22.7 | 1.773 | 1.061 | − 29.27 | 4.69 | 117.15 |
| West-Kohou-Ikemur | 1987 Jul. 26.9 | 1.572 | 2.442 | 83.39 | − 0.08 | 24.16 |
| Gehrels 1 | 1987 Aug. 13.6 | 2.988 | 3.005 | 41.03 | 4.56 | 79.31 |
| Comas Sola | 1987 Aug. 18.8 | 1.830 | 2.682 | 105.16 | 9.20 | 26.33 |
| Wild 3 | 1987 Aug. 26.5 | 2.288 | 2.349 | −108.64 | 0.19 | 74.06 |
| Schwa-Wach 2 | 1987 Aug. 30.6 | 2.071 | 2.972 | 123.55 | − 0.14 | 22.04 |
| Brooks 2 | 1987 Oct. 16.8 | 1.845 | 0.870 | 14.31 | − 1.72 | 162.40 |
| Reinmuth 2 | 1987 Oct. 25.7 | 1.936 | 1.506 | − 18.74 | 4.97 | 99.42 |
| Harrington | 1987 Oct. 30.9 | 1.596 | 1.139 | − 8.33 | − 6.91 | 96.69 |
| Kohoutek | 1987 Nov. 2.5 | 1.766 | 1.286 | 84.71 | 0.46 | 100.83 |
| Denning-Fujikawa | 1987 Nov. 3.9 | 0.779 | 0.439 | 15.41 | − 3.74 | 49.21 |
| Devico-Swift | 1987 Dec. 7.1 | 2.182 | 2.124 | 0.45 | 0.20 | 80.09 |
| Borrelly | 1987 Dec. 18.4 | 1.357 | 0.502 | 68.98 | − 3.36 | 128.96 |
| 1988 | | | | | | |
| Reinmuth 1 | 1988 May 10.0 | 1.869 | 2.230 | 132.04 | 1.83 | 56.33 |
| Finlay | 1988 Jun. 5.9 | 1.094 | 1.719 | 4.01 | − 2.23 | 36.97 |
| Tempel 2 | 1988 Sep. 16.7 | 1.383 | 0.955 | − 50.10 | − 2.36 | 89.73 |
| Longmore | 1988 Oct. 10.0 | 2.409 | 3.385 | −150.66 | − 6.41 | 10.24 |
| 1989 | | | | | | |
| Tempel 1 | 1989 Jan. 4.4 | 1.497 | 2.363 | −112.68 | 0.19 | 21.97 |
| D'Arrest | 1989 Feb. 4.0 | 1.292 | 2.278 | − 43.97 | 0.98 | 0.96 |
| Perrine-Mrkos | 1989 Mar. 1.4 | 1.298 | 1.922 | 46.57 | 4.08 | 38.23 |
| Dutoit 1 | 1989 Mar. 7.8 | 1.294 | 1.730 | − 81.34 | −18.22 | 47.89 |
| Tempel-Swift | 1989 Apr. 12.1 | 1.588 | 2.542 | 44.14 | 3.73 | 14.02 |
| Churyumov-Gerasim | 1989 Jun. 16.5 | 1.300 | 2.269 | 61.65 | 1.40 | 12.99 |
| Pons-Winnecke | 1989 Aug. 19.9 | 1.261 | 1.168 | − 94.36 | 2.90 | 70.27 |
| Gunn | 1989 Sep. 25.6 | 2.472 | 2.784 | − 95.45 | − 3.01 | 61.69 |
| Brorsen-Metcalf | 1989 Sep. 28.9 | 0.478 | 1.012 | 82.21 | 14.75 | 27.46 |
| Dubiago | 1989 Oct. 7.8 | 0.479 | 0.940 | 82.14 | 14.77 | 28.37 |
| Schwa-Wach 1 | 1989 Oct. 18.9 | 5.772 | 4.883 | 1.38 | 7.13 | 150.64 |
| Dutoit-Neuj-Delpo | 1989 Oct. 25.5 | 1.720 | 1.952 | − 56.22 | 2.56 | 61.71 |
| Gehrels 2 | 1989 Nov. 3.7 | 2.348 | 1.357 | 39.05 | − 0.41 | 177.06 |
| Clark | 1989 Nov. 28.6 | 1.556 | 2.496 | − 92.34 | − 4.58 | 13.71 |

continued

Table 6, continued

| Name | Date | r [AU] | Δ [AU] | λ | β | θ |
|---|---|---|---|---|---|---|
| 1990 | | | | | | |
| Kopff | 1990 Jan. 20.4 | 1.585 | 2.544 | − 76°.83 | 1°.39 | 10°.13 |
| Tuttle-Giaco-Kres | 1990 Feb. 8.0 | 1.068 | 1.091 | −157.87 | 8.11 | 61.64 |
| Sanguin | 1990 Mar. 20.8 | 1.811 | 2.784 | − 14.77 | 5.64 | 9.79 |
| Schwa-Wach 3 | 1990 Mar. 27.6 | 0.914 | 1.243 | − 92.84 | − 3.32 | 46.53 |
| Tritton | 1990 Jul. 5.4 | 1.438 | 2.435 | 87.92 | 3.75 | 8.84 |
| Honda-Mrkos-Pajdu | 1990 Sep. 12.7 | 0.541 | 0.923 | 54.47 | − 2.38 | 32.21 |
| Haneda-Camp | 1990 Sep. 14.7 | 1.101 | 0.403 | 11.89 | − 5.17 | 92.78 |
| Wild 2 | 1990 Oct. 16.1 | 1.491 | 2.426 | 175.95 | 2.10 | 15.66 |
| Encke | 1990 Oct. 28.5 | 0.331 | 1.217 | 160.15 | − 1.29 | 12.72 |
| Johnson | 1990 Nov. 19.0 | 2.313 | 2.539 | − 35.73 | − 6.43 | 65.60 |
| Kearns-Kwee | 1990 Nov. 24.2 | 2.215 | 1.408 | 87.23 | 6.70 | 134.61 |
| Taylor | 1990 Dec. 19.4 | 1.951 | 1.054 | 104.06 | − 1.55 | 146.31 |
| 1991 | | | | | | |
| Swift-Gehrels | 1991 Feb. 28.0 | 1.361 | 2.039 | 38.49 | 9.20 | 35.55 |
| Wolf-Harrington | 1991 Apr. 4.9 | 1.608 | 2.204 | 80.80 | − 2.21 | 42.09 |
| Van Biesbroeck | 1991 Apr. 25.0 | 2.401 | 2.248 | − 77.22 | 4.74 | 86.18 |
| Arend | 1991 May 25.2 | 1.850 | 2.794 | 40.80 | 14.45 | 17.03 |
| Harrington-Abell | 1991 May 31.9 | 1.761 | 2.560 | 116.23 | 6.68 | 30.31 |
| Russell 1 | 1991 Aug. 9.5 | 1.613 | 1.837 | −129.61 | 0.09 | 61.14 |
| Russell 1 | 1991 Aug. 9.5 | 1.613 | 2.564 | −129.61 | 0.09 | 13.56 |
| Tsuchinshan 1 | 1991 Sep. 5.3 | 1.497 | 2.336 | 118.61 | 4.04 | 26.02 |
| Wirtanen | 1991 Sep. 21.1 | 1.083 | 1.354 | 77.87 | − 0.77 | 52.13 |
| Arend-Rigaux | 1991 Oct. 2.4 | 1.438 | 1.660 | 91.75 | − 9.09 | 59.46 |
| Faye | 1991 Nov. 15.6 | 1.593 | 0.644 | 42.57 | − 3.68 | 154.03 |
| 1992 | | | | | | |
| Kowal 2 | 1992 Jan. 21.8 | 1.521 | 1.066 | 76.19 | − 2.54 | 95.67 |
| Giclas | 1992 Apr. 2.2 | 1.732 | 2.699 | 28.56 | − 7.87 | 11.56 |
| Kowal | 1992 Apr. 4.2 | 4.663 | 3.693 | −153.51 | 0.15 | 164.25 |
| Giacobini-Zinner | 1992 Apr. 15.2 | 1.034 | 2.015 | 8.32 | 3.94 | 8.64 |
| Tsuchinshan 2 | 1992 May 16.4 | 1.782 | 2.259 | 130.59 | − 2.63 | 49.76 |
| Grigg-Skjellerup | 1992 Jul. 24.9 | 0.995 | 1.417 | −148.04 | − 0.26 | 44.58 |
| Smirnova-Chernykh | 1992 Aug. 2.9 | 3.573 | 4.434 | 165.66 | 6.63 | 28.29 |
| Neujmin 2 | 1992 Aug. 20.3 | 1.265 | 2.257 | 161.95 | − 3.07 | 8.50 |
| Wolf 1 | 1992 Aug. 28.4 | 2.427 | 1.684 | 7.63 | 8.07 | 126.75 |
| Daniel | 1992 Aug. 31.6 | 1.650 | 2.089 | 78.72 | 3.77 | 50.86 |
| Schuster | 1992 Dec. 14.8 | 1.628 | 1.035 | 45.13 | − 2.12 | 107.37 |
| Gale | 1992 Dec. 17.0 | 1.214 | 2.187 | − 85.77 | − 6.20 | 6.27 |
| 1993 | | | | | | |
| Schaumasse | 1993 Mar. 5.0 | 1.202 | 0.572 | 137.28 | 9.96 | 96.79 |
| Forbes | 1993 Mar. 15.8 | 1.450 | 2.016 | − 75.55 | − 5.43 | 42.57 |
| Holmes | 1993 Apr. 10.6 | 2.177 | 3.076 | − 10.59 | 7.44 | 21.69 |
| Vaisala 1 | 1993 Apr. 29.2 | 1.783 | 1.167 | −178.83 | 8.51 | 110.01 |
| Slaughter-Burnham | 1993 Jun. 19.9 | 2.543 | 3.192 | 29.54 | 5.66 | 42.87 |
| Ashbrook-Jackson | 1993 Jul. 13.9 | 2.316 | 2.012 | − 9.09 | − 2.43 | 93.97 |
| Gehrels 3 | 1993 Nov. 12.1 | 3.383 | 3.038 | 105.56 | − 0.78 | 101.89 |
| Neujmin 3 | 1993 Nov. 13.3 | 2.001 | 2.561 | − 63.18 | 2.17 | 45.81 |
| Shajn-Schaldach | 1993 Nov. 16.0 | 2.344 | 1.577 | 22.57 | − 3.62 | 130.69 |
| West-Kohou-Ikemur | 1993 Dec. 25.4 | 1.578 | 0.629 | 83.46 | − 0.01 | 155.50 |

continued

Table 6, continued

| Name | Date | r [AU] | Δ [AU] | λ | β | θ |
|---|---|---|---|---|---|---|
| 1994 | | | | | | |
| Chernykh | 1994 Jan. 19.7 | 2.568 | 2.553 | 40°80 | – 5°72 | 79°80 |
| Schwa-Wach 2 | 1994 Jan. 24.2 | 2.070 | 1.086 | 123.80 | – 0.12 | 179.11 |
| Encke | 1994 Feb. 9.4 | 0.331 | 0.686 | 160.16 | – 1.29 | 9.68 |
| Tempel 2 | 1994 Mar. 16.8 | 1.484 | 2.310 | – 47.87 | – 3.05 | 26.14 |
| Kojima | 1994 Jun. 8.3 | 2.423 | 2.973 | 143.51 | – 0.19 | 48.51 |
| Tuttle | 1994 Jun. 27.0 | 0.998 | 1.969 | 106.06 | –21.52 | 12.04 |
| Reinmuth 2 | 1994 Jun. 29.7 | 1.893 | 1.714 | – 18.86 | 5.01 | 83.59 |
| Kohoutek | 1994 Jun. 30.3 | 1.775 | 2.775 | 84.79 | 0.45 | 8.20 |
| Tempel 1 | 1994 Jul. 3.0 | 1.494 | 0.848 | –112.78 | 0.20 | 106.16 |
| Wild 3 | 1994 Jul. 16.8 | 2.288 | 1.678 | –108.64 | 0.19 | 114.00 |
| Russell 2 | 1994 Aug. 17.6 | 2.161 | 1.469 | – 70.63 | –11.38 | 119.98 |
| Harrington | 1994 Aug. 22.1 | 1.572 | 0.775 | – 8.28 | – 6.95 | 122.71 |
| Brooks 2 | 1994 Sep. 1.5 | 1.843 | 1.185 | 14.12 | – 1.70 | 114.04 |
| Borrelly | 1994 Oct. 31.8 | 1.365 | 0.734 | 68.91 | – 3.37 | 103.42 |
| Whipple | 1994 Dec. 21.1 | 3.092 | 2.823 | 23.35 | – 3.68 | 96.48 |
| 1995 | | | | | | |
| Van Houten | 1995 Jan. 6.5 | 4.215 | 3.857 | 42.89 | 4.28 | 104.70 |
| Devico-Swift | 1995 Apr. 4.8 | 2.145 | 3.126 | 0.19 | 0.20 | 9.35 |
| Finlay | 1995 May 4.6 | 1.036 | 1.932 | 4.95 | – 2.18 | 19.34 |
| Clark | 1995 May 31.5 | 1.552 | 0.681 | – 92.46 | – 4.56 | 131.73 |
| Tuttle-Giaco-Kres | 1995 Jul. 20.0 | 1.062 | 1.520 | –157.78 | 8.12 | 44.21 |
| Schwa-Wach 3 | 1995 Jul. 22.4 | 0.910 | 0.533 | – 92.87 | – 3.32 | 63.24 |
| D'Arrest | 1995 Jul. 26.2 | 1.347 | 0.441 | – 43.56 | 0.67 | 130.97 |
| Tempel-Swift | 1995 Aug. 30.5 | 1.584 | 1.527 | 44.02 | 3.76 | 74.14 |
| Reinmuth 1 | 1995 Sep. 3.4 | 1.874 | 2.807 | 132.20 | 1.86 | 18.02 |
| Jackson-Neujmin | 1995 Oct. 5.9 | 1.381 | 0.460 | 0.13 | – 4.62 | 139.33 |
| Longmore | 1995 Oct. 6.0 | 2.399 | 3.361 | 150.61 | – 6.45 | 13.31 |
| Perrine-Mrkos | 1995 Dec. 7.2 | 1.293 | 0.613 | 47.10 | 4.08 | 105.59 |
| Honda-Mrkos-Pajdu | 1995 Dec. 30.1 | 0.533 | 0.694 | 54.53 | – 2.36 | 31.42 |
| 1996 | | | | | | |
| Pons-Winnecke | 1996 Jan. 3.4 | 1.256 | 2.217 | – 94.38 | 2.91 | 9.14 |
| Churyumov-Gerasim | 1996 Jan. 20.6 | 1.300 | 1.135 | 61.61 | 1.39 | 75.32 |
| Dutoit-Neuj-Delpo | 1996 Mar. 15.1 | 1.724 | 2.478 | – 56.38 | 2.57 | 32.51 |
| Comas Sola | 1996 Jun. 11.0 | 1.846 | 2.789 | 105.24 | 9.15 | 17.50 |
| Kopff | 1966 Jul. 2.3 | 1.579 | 0.567 | – 76.91 | 1.40 | 170.61 |
| Gunn | 1996 Jul. 25.2 | 2.462 | 1.769 | – 95.63 | – 2.98 | 121.92 |
| Haneda-Camp | 1996 Sep. 1.9 | 1.101 | 0.610 | 11.89 | – 5.17 | 81.73 |
| Tritton | 1996 Nov. 6.8 | 1.438 | 1.001 | 87.92 | 3.75 | 92.43 |
| Denning-Fujikawa | 1996 Dec. 5.7 | 0.779 | 0.872 | 15.41 | – 3.74 | 49.16 |
| Wild 2 | 1996 Dec. 15.7 | 1.491 | 1.823 | 175.95 | 2.10 | 54.81 |
| 1997 | | | | | | |
| Wirtanen | 1997 Mar. 14.5 | 1.063 | 1.521 | 77.96 | 0.74 | 44.11 |
| Boethin | 1997 May 1.1 | 1.158 | 2.164 | 36.02 | 1.71 | 2.37 |
| Encke | 1997 May 23.5 | 0.331 | 1.018 | 160.16 | – 1.29 | 18.78 |
| Gehrels 2 | 1997 Aug. 7.6 | 1.998 | 2.212 | 42.66 | – 1.40 | 64.53 |
| Grigg-Skjellerup | 1997 Sep. 3.8 | 0.997 | 1.807 | –148.01 | – 0.24 | 25.51 |
| Russell 1 | 1997 Sep. 15.3 | 1.613 | 2.304 | –129.61 | 0.09 | 36.69 |
| Wolf-Harrington | 1997 Sep. 29.4 | 1.582 | 1.642 | 80.84 | – 2.27 | 68.65 |
| Johnson | 1997 Oct. 31.9 | 2.308 | 2.246 | – 36.02 | – 6.36 | 80.95 |
| Taylor | 1997 Dec. 10.4 | 1.951 | 1.156 | 104.06 | – 1.55 | 131.25 |

continued

Table 6, continued

| Name | Date | $r$ [AU] | $\Delta$ [AU] | $\lambda$ | $\beta$ | $\theta$ |
|---|---|---|---|---|---|---|
| 1998 | | | | | | |
| Neujmin 2 | 1998 Jan. 7.3 | 1.270 | 1.086 | 161°93 | − 3°07 | 75°55 |
| Temple-Tuttle | 1998 Feb. 28.0 | 0.977 | 1.471 | 61.72 | 2.24 | 41.23 |
| Tsuchinshan 1 | 1998 Apr. 26.1 | 1.496 | 1.894 | 118.56 | 4.04 | 51.71 |
| Klemola | 1998 Jun. 14.5 | 1.757 | 1.660 | − 29.99 | 4.70 | 78.03 |
| Arend-Rigaux | 1998 Jul. 12.2 | 1.369 | 2.356 | 92.84 | − 8.88 | 10.55 |
| Kowal 2 | 1998 Jul. 26.8 | 1.521 | 2.337 | 76.19 | − 2.54 | 28.31 |
| Giacobini-Zinner | 1998 Nov. 24.7 | 1.033 | 0.906 | 8.41 | 3.90 | 65.98 |
| Harrington-Abell | 1998 Dec. 7.1 | 1.749 | 1.223 | 116.32 | 6.68 | 104.27 |
| Giclas | 1998 Dec. 7.6 | 1.732 | 1.275 | 28.56 | − 7.87 | 99.24 |
| 1999 | | | | | | |
| Tsuchinshan 2 | 1999 Mar. 3.9 | 1.771 | 1.065 | 130.53 | − 2.64 | 118.90 |
| Faye | 1999 May 6.5 | 1.657 | 2.664 | 43.34 | − 3.81 | 2.53 |
| Forbes | 1999 May 8.5 | 1.449 | 1.250 | − 75.47 | − 5.41 | 79.00 |
| Arend | 1999 Aug. 3.0 | 1.917 | 2.206 | 42.06 | 14.35 | 60.23 |
| Tempel 2 | 1999 Sep. 7.7 | 1.482 | 0.821 | − 47.77 | − 3.08 | 107.77 |
| Kearns-Kwee | 1999 Sep. 18.7 | 2.336 | 2.472 | 80.28 | 7.39 | 70.40 |
| Wild 1 | 1999 Dec. 25.1 | 1.961 | 1.942 | 166.52 | 4.07 | 76.45 |
| 2000 | | | | | | |
| Tempel 1 | 2000 Jan. 2.1 | 1.500 | 2.385 | −112.78 | 0.20 | 20.07 |
| Holmes | 2000 May 11.6 | 2.165 | 2.795 | − 10.51 | 7.48 | 42.79 |
| West-Kohou-Ikemur | 2000 Jun. 1.6 | 1.597 | 2.596 | 83.50 | 0.04 | 7.83 |
| Swift-Gehrels | 2000 Jun. 2.0 | 1.361 | 2.275 | 38.49 | 9.20 | 19.39 |
| Schuster | 2000 Jun. 3.8 | 1.628 | 2.569 | 45.13 | − 2.12 | 17.14 |
| Daniel | 2000 Jun. 26.9 | 2.170 | 3.168 | 83.62 | 7.20 | 9.07 |
| Schwa-Wach 3 | 2000 Nov. 16.7 | 0.915 | 1.824 | − 92.88 | − 3.32 | 16.02 |
| Wolf 1 | 2000 Nov. 21.3 | 2.412 | 1.955 | 7.68 | 8.05 | 105.35 |
| Clark | 2000 Dec. 3.2 | 1.559 | 2.517 | − 92.49 | − 4.56 | 10.74 |
| Tuttle-Giaco-Kres | 2000 Dec. 26.3 | 1.050 | 1.646 | −157.67 | 8.16 | 37.35 |
| 2001 | | | | | | |
| Ashbrook-Jackson | 2001 Jan. 6.5 | 2.305 | 2.856 | − 8.95 | − 2.39 | 47.27 |
| Smirnova-Chernykh | 2001 Jan. 11.7 | 3.547 | 3.049 | 163.00 | 6.64 | 112.73 |
| Reinmuth 2 | 2001 Feb. 19.9 | 1.890 | 2.865 | − 18.74 | 5.03 | 7.73 |
| Kohoutek | 2001 Feb. 24.2 | 1.777 | 1.718 | 84.69 | 0.46 | 76.88 |
| Honda-Mrkos-Pajdu | 2001 Apr. 6.1 | 0.529 | 1.451 | 54.54 | − 2.36 | 13.27 |
| Schaumasse | 2001 May 3.6 | 1.205 | 1.518 | 136.46 | 9.93 | 52.38 |
| Shajn-Schaldach | 2001 May 9.1 | 2.330 | 3.270 | 22.64 | − 3.63 | 17.85 |
| Harrington | 2001 Jun. 4.6 | 1.568 | 1.988 | − 8.20 | − 6.96 | 51.27 |
| Wild 3 | 2001 Jun. 6.2 | 2.288 | 1.277 | −108.64 | 0.19 | 173.76 |
| Brooks 2 | 2001 Jul. 20.5 | 1.835 | 1.889 | 14.22 | − 1.74 | 71.27 |
| Borrelly | 2001 Sep. 14.0 | 1.358 | 1.520 | 69.00 | − 3.32 | 61.10 |
| Russell 2 | 2001 Oct. 1.1 | 2.161 | 2.187 | − 70.63 | −11.38 | 75.21 |
| 2002 | | | | | | |
| Tempel-Swift | 2002 Jan. 16.9 | 1.576 | 1.576 | 44.08 | 3.74 | 71.82 |
| Schwa-Wach 2 | 2002 Jan. 18.1 | 3.406 | 2.468 | 131.84 | 1.44 | 159.33 |
| D'Arrest | 2002 Feb. 2.4 | 1.354 | 2.338 | − 43.54 | 0.64 | 2.24 |
| Finlay | 2002 Feb. 6.8 | 1.034 | 1.847 | 4.96 | − 2.18 | 24.54 |
| Gehrels 3 | 2002 Feb. 25.6 | 3.437 | 2.878 | 107.68 | − 0.75 | 116.62 |
| Pons-Winnecke | 2002 May 16.9 | 1.258 | 0.645 | − 94.38 | 2.92 | 96.34 |

continued

Table 6, continued

| Name | Date | $r$ [AU] | $\Delta$ [AU] | $\lambda$ | $\beta$ | $\theta$ |
|---|---|---|---|---|---|---|
| 2002, continued | | | | | | |
| Gehrels 1 | 2002 Jul. 14.0 | 2.963 | 3.447 | 40°.86 | 4°.52 | 53°.81 |
| Devico-Swift | 2002 Aug. 2.1 | 2.145 | 1.708 | 0.31 | 0.22 | 100.92 |
| Dutoit-Neuj-Delpo | 2002 Aug. 11.3 | 1.734 | 0.793 | − 56.41 | 2.56 | 147.18 |
| Churyumov-Gerasim | 2002 Aug. 17.8 | 1.292 | 1.745 | 61.63 | 1.40 | 47.21 |
| Wirtanen | 2002 Aug. 28.7 | 1.058 | 1.624 | 77.99 | − 0.72 | 39.28 |
| Longmore | 2002 Aug. 31.8 | 2.311 | 3.037 | −150.69 | − 6.70 | 36.90 |
| Perrine-Mrkos | 2002 Sep. 11.6 | 1.287 | 1.160 | 47.13 | 4.06 | 72.49 |
| Sanguin | 2002 Sep. 21.0 | 1.811 | 0.863 | − 14.77 | 5.64 | 151.58 |
| Grigg-Skjellerup | 2002 Dec. 3.1 | 1.114 | 1.989 | −147.45 | 0.60 | 19.91 |
| Kopff | 2002 Dec. 12.4 | 1.584 | 2.517 | − 76.96 | 1.40 | 14.61 |
| Reinmuth 1 | 2002 Dec. 24.4 | 1.878 | 1.299 | 132.22 | 1.86 | 109.91 |
| Neujmin 1 | 2002 Dec. 28.0 | 1.552 | 2.231 | − 26.35 | − 3.18 | 36.31 |
| Oterma | 2002 Dec. 30.1 | 5.470 | 5.219 | 27.30 | 1.61 | 99.63 |

Technical criteria for selection of target comet for a mission [3]:

- reliable orbit, in order to predict ephemerides with sufficient accuracy;
- recovery of comet sufficiently long before encounter (at least 100 days) in order to allow adjustment of calculated ephemerides with new observations;
- visibility of comet from earth at encounter in order to complement spacecraft observations with ground-based observations;
- encounter should occur when comet is already active, i.e., not too far from the sun (less than about 1.5 AU);
- low departure hyperbolic velocity in order to allow for large payload mass for spacecraft;
- relative encounter velocity of comet and spacecraft should be as low as possible in order to provide sufficient observing time close to the comet.

Based on these criteria, Boissard et al. [3] list 10 comets with 19 favorable perihelion passages between 1984 and 2000 as candidates for a mission. The orbital characteristics of the 10 comets are listed in Table 7 on page 394.

A comparison of Halley with other comets (Table 8) shows: Comet Halley's total magnitude, $m_0$ (for unit distance $r=\Delta=1$, see 3.3.3.10.2), and gas production rate $\dot{Q}_{H_2O}$ are similar to those of long-period "new" comets. Comet Halley is by two orders of magnitude more active than other comets with a predictable orbit.

Table 8. Comparison of Halley with other comets [7].
$m_0$ = total magnitude (for $r = \Delta = 1$)
$\dot{Q}_{H_2O}$ = production rate ($H_2O$ molecules per sec)

| Group | | Comet | $m_0$ | $\dot{Q}_{H_2O}$ at 1 AU |
|---|---|---|---|---|
| Mission can be planned | short periodic comets | Encke (1990) | 12···13 | $3 \cdot 10^{27}$ |
| | | Tempel 2 (1988) | ≈13 | $1 \cdot 10^{27}$ |
| | | Tuttle-Giacobini-Kresak (1990) | 12 | $\approx 1 \cdot 10^{27}$ |
| | | Honda-Mrkos-Pajdusakova (1991) | 13···14 | $\approx 1 \cdot 10^{27}$ |
| | | Faye (1991) | 11···12 | $\approx 1 \cdot 10^{27}$ |
| | intermediate periodic | Halley (1986) | 5 | $1 \cdot 10^{29}$ |
| Mission cannot be planned | long periodic or "new" comets | Tago-Sato-Kosaka (1969) | 6.4 | $2 \cdot 10^{29}$ |
| | | Bennett (1970) | 3.5 | $2.5 \cdot 10^{29}$ |
| | | Kohoutek (1973) | 6.0 | $1.5 \cdot 10^{29}$ |
| | | West (1976) | 5.0 | $2.3 \cdot 10^{29}$ |

Table 7. Orbital characteristics of candidate comets for future missions (1984···2000) [3].

$T_0$ = date of last perihelion passage
$P$ = period
$q$ = perihelion distance from the sun
$i$ = inclination of the orbit with respect to the ecliptic
$T_1$ = favorable perihelion passage
$V_h$ = departure hyperbolic velocity
$\delta_A$ = launch asymptote declination
(launch opportunities with unfavorable departure conditions are marked with "u")

| Comet | $T_0$ | $P$ a | $q$ AU | $i$ | $T_1$ | $V_h$ km s$^{-1}$ | $\delta_A$ |
|---|---|---|---|---|---|---|---|
| Encke | 1977.63 | 3.31 | 0.341 | 11°.9 | 1984 Mar. 27.0 | u | |
| | | | | | 1987 Jul. 17.0 | u | |
| | | | | | 1990 Nov. 3.6 | 10 | −5 *) |
| | | | | | 1994 Feb. 21.3 | u | |
| | | | | | 1997 Jun. 11.0 | u | |
| | | | | | 2000 Sep. 28.7 | 10 | 0 |
| Tempel 2 | 1978.14 | 5.27 | 1.369 | 12.5 | 1988 Sep. 16.7 **) | 4 | −5 |
| | | | | | 1999 Sep. 6.6 | 3 | −5 |
| Honda-Mrkos-Pajdusakova | 1974.99 | 5.28 | 0.579 | 13.1 | 1990 Sep. 20.0 | 4 | −5 |
| | | | | | 1996 Jan. 17.3 | 10 | −5 |
| Tuttle-Giacobini-Kresak | 1978.98 | 5.58 | 1.124 | 9.9 | 1990 Feb. 6.6 **) | 8 | −5 |
| D'Arrest | 1976.62 | 6.23 | 1.164 | 16.7 | 1995 Jul. 7.0 | 6.5 | −5 |
| Giacobini-Zinner | 1979.12 | 6.52 | 0.996 | 31.7 | 1985 Sep. 4.0 | 3 | −5 |
| | | | | | 1998 Nov. 9.7 | 4 | −5 |
| Borelly | 1974.36 | 6.76 | 1.316 | 30.2 | 1987 Dec. 18.2 | 5 | −5 |
| | | | | | 1994 Oct. 28.1 | 9 | −5 |
| Arend-Rigaux | 1978.09 | 6.83 | 1.442 | 17.9 | 1984 Dec. 1.4 | 6 | −5 *) |
| Crommelin | 1956.82 | 27.89 | 0.743 | 28.9 | 1984 Sep. 1.0 | 5 | −5 |
| Halley | 1910.30 | 76.09 | 0.587 | 162.2 | 1986 Feb. 9.3 | 3 | −5 |

*) Flight time > 500 d.
**) Close approach to Jupiter before perihelion passage.

### 3.3.3.10.6 References for 3.3.3.10


1 Bender, D.: Private Communication (1981).
2 Bobrovnikoff, N.T.: Pub. Lick Observ. **17** (1931) 309
3 Boissard, J.B., Flury, S., Janin, G., Baghi, S.: Review of Ariane capabilities for interplanetary missions, ESA Journal **5** (1981) 1.
4 Brandt, J.C., Donn, B., Greenberg, M., Rahe, J. (eds.): Modern Observational Techniques for Comets, JPL-Publ. (1981) 81.
5 Broughton, R.P.: J. Roy. Astron. Soc. Canada **73** (1979) 24.
6 Donn, B.: A Comparison of the Composition of New and Evolved Comets. In "Comets Asteroids Meteorites: interrelations, evolution and origin" (Delsemme, A.H., ed.), Toledo (1977) p. 15.
7 Giotto-Comet Halley Flyby, Report on the Phase A Study, European Space Agency, ESA SCI (80) **4** (1980).
8 Hajduk, A.: The Core of the Meteor Stream Associated with Comet Halley. In "Solid Particles in the Solar System" (Halliday, I., McIntosh, B.A., eds.), Reidel, Dordrecht (1981) p. 149.
9 Halley, E.: Astronomiae Cometicae Synopsis, Oxford (1705).
10 Michielsen, H.F.: J. Spacecraft and Rockets **5** (1968) 328.

11 Morris, C.S., Green, D.W.E.: Astron. J. **87** (1982) 918.
12 Newburn, R.L., jr.: Physical Models of Comet Halley Based Upon Qualitative Data from the 1910 Apparition. In "The Comet Halley Micrometeroid Hazard" (Longdon, N., ed.), ESA, Paris (1979) p. 147.
13 Newburn, R.L., jr., Yeomans, D.K.: Annu. Rev. Astron. Astrophys. (1982) in press.
14 Rahe, J., Brandt, J.C., Donn, B.: Atlas of Comet Halley Photographs and Spectra, NASA-Goddard Space Flight Center (1982).
15 Rahe, J., Brandt, J.C., Friedman, L.D., Newburn, R.L., jr.: Halley's Comet and Plans for its Observation During its Return in 1985/86. In Proc. 6th European Regional Meeting on Astronomy (Fricke, W., Teleki, H., eds.). Reidel, Dordrecht (1982) p. 307.
16 Rahe, J., Donn, B., Wurm, K.: Atlas of Cometary Forms-Structures Near the Nucleus, NASA-SP-198, Washington (1969).
17 Reinhard, R.: The ESA Mission To Comet Halley. In "Modern Observational Techniques for Comets" (Brandt, J.C., Donn, B., Greenberg, J.M., Rahe, J., eds.), JPL Publ. (1981) p. 81.
18 The International Halley Watch: Report of the Science Working Group NASA-TM 82181, Washington (1980).
19 Whipple, F.L.: Periodic Comet Halley, IAU-Circ. No. 3459 (1980).
20 Yeomans, D.K.: Astron. J. **82** (1977) 435.
21 Yeomans, D.K.: The Comet Halley Handbook: An Observer's Guide, NASA Doc. JPL 400-91, Jet Propulsion Laboratory (1981).
22 Yeomans, D.K., Kiang, T.: Mon. Not. R. Astron. Soc., in press.

# 9.6.6 Interpretation of extended and compact radio sources

## Introduction

In this section a survey of the recent literature on theoretical concepts which have been developed in the area of extragalactic radio sources is presented. Models for the various components of radio sources: extended lobes (see 9.6.4), beams and jets (see 9.6.4.3 and 9.6.4.5), compact cores (see 9.6.5), and central energy sources, are considered. The objects predominantly addressed here are strong radio galaxies, quasars, and BL Lac objects (see also 9.3, 9.4, 9.5) with their radio and overall spectral properties. The wealth of observational phenomena in this field is reflected in the diversity of theoretical interpretations, theories and speculations. These cannot be reviewed here in any detail, and also the literature references must remain far from being comprehensive.

## 9.6.6.1 Extended sources

Models for the extended radio components are reviewed in [B 32, B 33, D 3, F 15, F 24, F 26, L 1, L 18, M 20, M 21, P 13, R 12···R 15, R 22, S 15, S 18, S 59, W 14].

Theories of extended radio sources address problems like: origin and lifetime of hot spots and diffuse lobes, transport of energy from the nucleus to the sources without loss, directive ejection of energy, collimation, confinement of relativistic particles and fields, in situ particle acceleration, internal dynamics and appearance (surface brightness, spectrum, polarization) etc.

Three basic mechanisms have been proposed to explain the origin and persistence of collimated double radio sources (9.6.4.4) over lifetimes of at least $10^6$ years:

(i) Ejection of massive objects from the parent galaxy by the gravitational slingshot mechanism [R 25, S 9, S 14],

(ii) single or multiple ejection of "plasmoids" (thermal plasma, relativistic particles and magnetic fields) from the nucleus [C 8, C 9, D 5],

(iii) continuous flow of energy and particles from the nucleus to the lobes [B 18, B 31, M 29, R 11, S 16].

These models have also been applied to head-tail sources in clusters of galaxies (9.3.4) and to more complex sources (9.6.4.4).

### 9.6.6.1.1 Ejection of massive objects

Coherent objects with masses $\gtrsim 10^6 \mathfrak{M}_\odot$ (supermassive stars [F 16, F 22, F 23, H 13, H 14], spinars [F 14, M 28, O 14, O 15, W 21], black holes with and without accretion disks [L 15, P 24, P 25, R 25] etc.) are supposed to be ejected by the gravitational slingshot mechanism (disruption of an unstable system of three massive objects) [S 10, S 14] or by fission of a single rotating massive object [B 5] from the parent galaxy. They are located at the hot spots [F 23, H 2, M 21, R 5], supply mass, magnetic fields and energy to the lobes, accelerate particles at their surfaces [D 4, O 15, R 25], and gravitationally confine the source [B 45, C 1, F 14]. Collimation and double character are achieved automatically. No direction memory (9.6.4.5) can be maintained. Ejection occurs in the plane of the parent galaxy which is largely in conflict with the observations [B 40, M 21, S 43], cf. however [G 1, M 1]. Compact objects of the kind proposed have hitherto not been detected in the lobes [K 13, S 11, S 13, T 11]. Calculations for collimated sources may be found in [B 51, S 14, V 2, V 3], for head-tail sources in: [V 4, V 5, V 7]. Further pertinent observations are discussed in: [D 6, K 1, O 3, S 11, S 42, S 44, V 6].

### 9.6.6.1.2 Ejection of plasmoids

During one [D 3, D 5] or more [C 10, C 11, C 13, S 38] outbursts of unspecified nature in the nucleus, or by buoyancy [C 31, G 6, R 15], plasma clouds are ejected in opposite directions along the rotation axis of the galaxy, thus accounting for directionality and alignment (9.6.4.5). Basic problems: adiabatic expansion losses during ejection and confinement of the developed sources, particularly the hot spots [L 18, L 21]. Proposed confinement mechanisms [L 21, M 21, P 13]:

(i) external gas pressure of the galactic [G 6, G 8, L 21, M 30, N 5] and intergalactic environment [F 1a, F 12, M 30], particularly in clusters of galaxies where $n_e \lesssim 10^{-3}$ cm$^{-3}$ (9.3.4) [G 8]. The hot spots and tails of Cyg A – which is in a cluster – cannot be confined this way [F 3],

(ii) inertia of thermal gas mixed up with the relativistic particles. The required densities seem too high from the evidence of Faraday rotation and depolarization measurements [D 3, H 3, S 42, V 1] (see 9.6.4.4) and from X-ray constraints [G 8, M 30].

(iii) ram pressure $\varrho_e \cdot V^2$ of a cloud of mainly thermal plasma moving supersonically with speed $V$ ($\approx 0.1c$) through the external medium of density $\varrho_e$($\approx 10^{-5}\,\mathrm{cm}^{-3}$) [D 3, K 4, M 1, M 21, P 13]. Again the amount of plasma required thermal appears to be excessively high.

The ram-pressure model [D 5] has been elaborated and modified extensively: [C 8, C 10, D 2, D 3, M 25, N 2, N 3]. Detailed investigations concern: The role of Rayleigh-Taylor and Kelvin-Helmholtz instabilities: [C 11···C 13, N 2···N 4, B 18]; explosion and collimation calculations: [C 13, M 27, S 5]; applications to head-tail sources like 3C 129 (leading to cloud masses $\approx 10^7\,\mathfrak{M}_\odot$ and velocities $\approx 2000\,\mathrm{km\,s^{-1}}$): [B 33, C 31, J 1, M 23, P 1].

In situ acceleration of relativistic particles is required both by the fine structure of the hot spots (e.g. of 3C 236), where the light travel time exceeds the synchrotron lifetime [B 33, L 18, L 21], and in the tails or trails (e.g. of IC 711, 3C 129 + 3C 129.1) [D 3, P 1, W 16]. Acceleration mechanisms have been considered by [E 3, L 2, N 3], and [B 27, P 1] (see also 9.6.6.2).

#### 9.6.6.1.3 Continuous flow or beam models

In this model, energy, particles, mass, momentum, and magnetic flux are continuously supplied to the radio components over the lifetime of the source [R 11]. Related to this model are the bubble [G 6] and multiple plasmoid [C 9, S 38] models describing weak or intermittent activity, respectively. In the Rees-model (see below) a self-focussed beam along the rotation axis of the compact nucleus of the galaxy funnels fast-moving hot plasma and/or relativistic particles and probably also magnetic fields and electromagnetic waves [B 31, B 33, L 21, R 11, R 23, S 16] into the outer lobes. The beam impinges on the external medium and forms the hot spots in ram pressure equilibrium. There the bulk kinetic energy of the beam is converted into relativistic electrons with a power law spectrum and into magnetic field energy (see 9.6.6.2). The relativistic electrons are not confined in the hot spots but escape and form the extended tails of the source. There the particles cool by synchrotron radiation and inverse Compton scattering of the universal background radiation [C 28, F 3, G 8, M 21, P 14] and the radio spectrum steepens. The lobes thus accumulate matter and energy for $10^6 \cdots 10^8$ years, the total energy content amounting to $10^{60} \cdots 10^{62}$ erg.

The beam model appears to be completely vindicated by the recent high-angular-resolution observations, particularly with the VLA (= Very Large Array), which made the beams visible in many sources [B 39, M 21] and revealed them as a very common feature of extragalactic radio sources. Prominent examples are NGC 6251 [B 21, W 1], 3C 449 [B 21, P 11, P 12], NGC 315 [B 41, B 42]. In NGC 6251 the jet can be followed from 1 pc to several hundred kpc [R 4].

If the galaxy moves through an external medium the beams are bent backwards and radio trails develop as in 3C 129 or in 3C 83.1B (= NGC 1265 in the Perseus cluster) [M 22, O 5].

The beam model proved very successful in explaining all kinds of extragalactic sources. Many theoretical investigations exist on the problems of understanding the generation and collimation of jets, their bending, precession and stability, their substance, magnetic field structure, particle acceleration properties, and spectra (see 9.6.6.2).

### 9.6.6.2 Large-scale jets and beams

Systematics and interpretations of large-size jets (beams) are described in: [B 21, B 32, B 33, B 39, D 3, F 15, F 24, L 18, M 20, M 21, P 13, R 12···R 15, R 21, R 23, S 18, S 24, W 15].

About 80 jets have hitherto been resolved by the VLA in extended sources [B 39, M 21, R 23, R 26, W 14, W 15]. In many cases they are the extension of compact jets observed on VLBI scales. Sometimes their optical and X-ray counterparts have been observed locally, most prominent examples being 3C 273, M 87, Cen A, [A 11, A 12, S 24]; thus overall spectra from the radio up to the X-ray spectral ranges could be constructed [S 24]. There is now little doubt that the "jets" are visible manifestations of the beams postulated originally by Rees [R 11] along which the power is supplied from the parent galactic nucleus to the intermediate and/or outer lobes [B 10, B 32, B 33, L 21, R 13, R 14, R 15, R 26]. Examples: 3C 449, NGC 6251, NGC 1265.

Composition

The "working" substance is in general a mixture of thermal ($n_e \lesssim 10^{-2}\,cm^{-3}$) and relativistic electrons, as well as of magnetic fields and low-frequency electromagnetic waves [B 31, B 33, B 39, L 21, R 11, R 22, R 23, S 16].

Velocity

The jet velocity is uncertain, from $V \approx 500\,km\,s^{-1}$ [B 22] up to the velocity of light and may differ among jets [R 23]. Relativistic outflow is almost certain in "superluminal" sources (9.6.6.4.2), where it causes superluminal delay and/or screen illumination effects [B 26] and in one-sided jets where it may cause exaggeration of the near-side jet [K 26, R 18]. Bending [N 1, S 45] indicates non-relativistic flows [P 19, V 8]. From separation and flux asymmetries $V < 0.2c$ has been inferred for individual sources [L 20, M 1, M 2]. Note that the "mini"-jet of the galactic source SS 433 has a measured velocity of $V = 80000\,km\,s^{-1} = 0.27c$ [M 8].

Radiation

Emission in the radio, optical, and X-ray spectral ranges is most likely due to synchrotron and/or inverse Compton radiation from relativistic electrons accelerated in situ by strong shocks in the jet or in the walls of the beam [B 33] (cf. 9.6.6.4). For spectra of optical jets, see [S 24].

Shock acceleration in jets and hot spots

In-situ production of relativistic particles is unavoidable for synchrotron emission from small-scale structures at sufficiently high frequencies, when the crossing time exceeds the loss time (cf. 9.6.6.3.1), conditions which are now believed to prevail in jets and hot spots [B 33, L 18, L 21, R 23, S 24].

The literature on acceleration mechanisms prior to 1976 is reviewed e.g. by Parker [P 5]. Since then the theory of first-order Fermi acceleration of cosmic rays in the presence of strong shocks has been developed and successfully applied also to radio sources [A 15, A 16, B 7, B 8, B 20, B 27, D 8, K 19, L 19, T 8]. For acceleration in tails, see: [E 3, L 2, N 3, P 1, S 45].

Confinement

Free jets or pressure confinement raise difficulties [B 39]; magnetic self-confinement with toroidal field perpendicular to the jet seems to work [B 10, C 6].

One-sided or invisible jets, gaps

One-sidedness may be caused by Doppler beaming in relativistic flows ("Doppler favouritism") [R 35, B 11, B 12, R 23] or by instabilities in relativistic or subrelativistic flows [B 11, B 12, B 30, F 7, S 46]. An intrinsic mechanism may be flip-flop behaviour by instabilities in or oscillations of the central source [B 5, B 11, B 12, R 23, S 38].

The jet may also be invisible due to the absence of dissipation [B 4, B 5, B 11, B 12, R 23] along the beam up to the hot spots (e.g. in Cyg A)[H 2, S 1] or in a gap on the way (e.g. in 3C 449, NGC 315, 3C 219) [B 11, B 12, M 21, P 11].

Collimation

a) In the "twin exhaust" model a cavity is inflated by the hot plasma or radiation pressure driven outflow. A nozzle forms parallel to the rotation axis of the galaxy (pressure confinement) [B 31, B 33, R 15, W 10] or of the accretion disk (ram pressure confinement by accreting matter). Constraints to the model are imposed by X-ray emission from the confining gas and by the required stability of the configuration over $10^6 \cdots 10^7$ years [B 33, C 5, L 5, L 6].
b) Self-focussing of supersonic hydromagnetic flows may occur in the inner regions of accretion disks: [B 29, B 31, B 34a, K 24, L 24, S 39, Z 4].

Complex sources

Mirror symmetric sources like 3C 449 may reflect orbital motion of the nucleus around a companion galaxy [P 11] while inversion symmetric sources like 4C 26.03 (=NGC 326) may record the precessional motion of the spin axis in a double or multiple nucleus galaxy formed by merging [R 30, R 31] or cannibalism [B 21, O 4, R 19, W 17] (see also 9.3.2.2). Note that the jet in SS 433 is inversion symmetric and caused by precession [H 9, M 15]. Other morphologies can be explained by interaction of the beams and lobes with the external gas [B 33, M 21].

Instabilities

Rayleigh-Taylor and Kelvin-Helmholtz instabilities of beams have been investigated with linear stability analyses [B 11, B 12, B 30, B 36, F 7···F 11, S 46, T 10].

Numerical experiments

Realistic simulations of Blandford-Rees beams are prohibitively difficult at present and hampered by lack of knowledge of microscopic properties of the plasma under cosmic conditions. Numerical gas dynamical experiments are nonetheless important for studying the generation, collimation, stability, dynamics and appearance of jets. Employing the most powerful computers to date very promising results have been obtained recently: [N 7, N 8, R 1, S 47, W 11, W 12]. Alternatively, laboratory experiments might provide a tool for studying the physics of extended radio sources [B 21, R 20, R 23].

## 9.6.6.3 Compact radio nuclei

The physics of compact cores is reviewed in: [A 7, B 33, B 34, B 39, C 22, F 1, F 2, K 8, K 9, K 13, M 12, O 1, O 2, P 2, R 7, R 20···R 22, R 29a, S 19, S 31, S 56, S 57, T 3, T 4, W 8, W 20].

### 9.6.6.3.1 Optical objects

The flux of compact radio sources at $\gtrsim 1$ GHz is dominated by a single bright component with angular size $<0.1$ arcsec or linear size $\lesssim 1$ kpc. Nearly all BL Lac objects [A 6, A 7, S 56], most quasars [S 57] (although only 10% are radio-loud [A 7, C 26, S 54, S 57, W 2]), and many radio galaxies with extended structure contain compact radio nuclei [C 26]. They are often also found in nuclei of active galaxies, normal spirals and ellipticals (see 9.2···9.5) [P 21, P 22]. Properties of compact radio sources do not seem to be correlated with their optical indentification. They usually exhibit flat radio spectra and radio variability [B 37, H 8].

"Blazars" (i.e. optically violent variable quasars and BL Lac objects, cf. [A 7, C 26] and 9.5.2.5) show the extremes of inverted spectra, high polarization, and variability [A 7, C 22].

### 9.6.6.3.2 The canonical model, sizes and energetics

The primary radio emission can for the majority of the compact sources – as with the extended sources – be attributed to incoherent synchrotron radiation by relativistic electrons in a partially ordered magnetic field [B 48].

The energy content of a compact source is strongly dependent on its angular size $\theta$ (see 9.6.3) [B 48, O 2]. Direct or indirect size determinations [O 13] involve:

(i) absence of interstellar scintillation: $\theta \gtrsim 1$ microarcsec,

(ii) VLBI measurements (milliarcsec scale),

(iii) flux density variations: $\theta < c\tau(1+z)/d_L$ with $\tau$ = variability time scale, $z=\frac{\Delta\lambda}{\lambda}$ = redshift, $d_L$ = luminosity distance (see 9.7.2.2.4), $c$ = velocity of light. Variability is reviewed in e.g. [F 2, H 8, K 13], (see also 9.6.5.2),

(iv) the interpretation of a low-frequency turnover due to synchrotron self-absorption: $\theta \approx B^{1/4} S_{\mathrm{m}}^{1/2} \nu_{\mathrm{m}}^{-5/4} (1+z)^{1/4}$ where $\theta$ in [milliarcsec], the turnover flux $S_{\mathrm{m}}$ in [Jy], the turnover frequency $\nu_{\mathrm{m}}$ in [GHz], the magnetic field $B$ in [$10^{-4}$ Gauss] [J 7, O 2],

(v) the rapid inverse Compton losses if $\theta \gtrsim S_{\mathrm{m}}^{1/2} \nu_{\mathrm{m}}^{-1} (1+z)^{1/2}$ is violated (see 9.6.3.3.4) [B 33, K 11]. The intrinsic quantity characterizing a compact radio source is its brightness temperature in its proper frame $T_{\mathrm{b}} \approx 5 \cdot 10^{11} S_{\mathrm{m}} \nu_{\mathrm{m}}^{-2} \theta^{-2} (1+z)$. Condition (v) implies $T_{\mathrm{b}} \lesssim 10^{12}$ K (see 9.6.3.3.4) which is consistent with most VLBI observations.

Generally, VLBI, variability, self-absorption and Compton limit yield comparable angular sizes. The energy content is typically $\approx 10^{56}$ erg which is much higher than the energy released in a single supernova ($\approx 10^{51}$ erg) and much less than the content of extended sources ($\approx 10^{60}$ erg) which are probably fuelled by nuclear activity over timescales of $\gtrsim 10^{6}$ years.

### 9.6.6.3.3 Coherent models

Some rapidly variable sources if assumed at cosmological distances seem to exceed the inverse Compton limit. From variability of the BL Lac object AO0235+164 angular dimensions have been derived [L 4, M 3] which imply $T_b \approx 10^{15}$ K. Brightness temperatures up to $10^{16}$ K and energy contents in excess of $10^{60}$ erg result, if rapid variability occurs in decimeter wavelengths [C 24, C 29, F 2]. In view of such cases coherent radiation processes and comptonization models [B 9, C 14, C 15, C 20, C 21] or scaled-up pulsar type emission mechanisms [A 9, C 4, P 2a, P 3, W 21] have been invoked. Coherent models have been criticized on theoretical and observational grounds [A 6, A 7, B 33, B 34, K 13, R 17, S 17].

The Compton radiation and the energy problems encountered in some sources as well as the apparent superlight velocities (see 9.6.6.4.2) have been taken by some authors as evidence against the cosmological interpretation of the redshifts [B 46, B 47, B 50, H 12, J 4].

### 9.6.6.3.4 Interpretation of radio spectra of compact sources

The compact radio sources show flat (spectral index $\alpha \approx 0$), undulating or inverted spectra sometimes up to 300 GHz ($\lambda = 1$ mm) [C 27, C 30, J 9, K 11···K 13, K 18, O 7, O 8, O 9] (see Fig. 1). Locally the value $\alpha = +2.5$ of a single completely opaque synchrotron source is never reached. At low frequencies they either show a steep

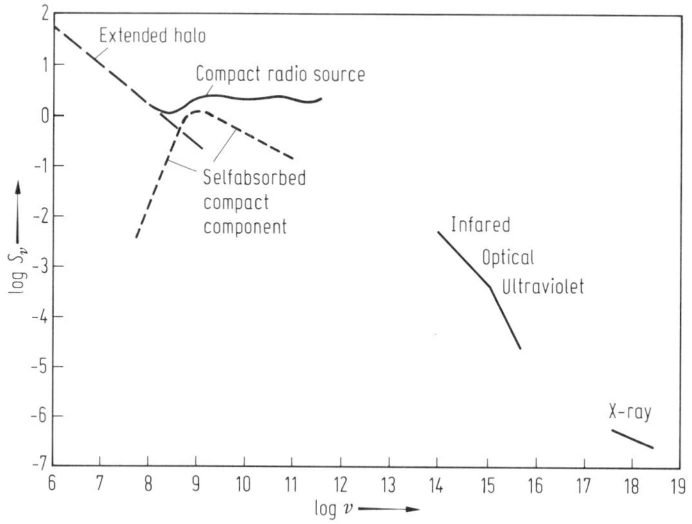


Fig. 1. Schematic representation of the overall continuum spectrum of a compact radio source (quasar or BL Lac object). The flat part of the radio spectrum does not necessarily result from the superposition of several self-absorbed compact components. The often observed equipartition of radiating power over the spectral bands from mm to X-ray wavelengths is assumed in this hypothetical example. $\nu$ in [Hz], $S_\nu$ in [Jy].

spectrum due to an extended halo ($\alpha \approx -0.8$) or a low-frequency cutoff between $\approx 0.4$ and 1.4 GHz indicating the absence of a low-frequency halo [G 5]. Some proposed explanations for the flat spectra within the framework of incoherent synchrotron theory [J 7, O 2] are indicated here:

(i) The standard model assumes several uniform components which become synchrotron self-absorbed at different frequencies and "conspire" to add up to a flattish total spectrum [B 43, B 48, C 30, H 11, J 6, J 7, K 7, K 10, K 11] (cf. Fig. 1).

(ii) Nonthermal nuclei with radial structure of electron density and magnetic field resembling core-halo sources with no preferred direction [B 44, C 23, M 9, S 50]. Local spectral indices $\alpha < +2.5$ are explained this way.

(iii) Optically thin synchrotron emission from relativistic electrons with energy distribution $N(E) \propto E^{-1}$ [M 10].

(iv) Maxwellian energy distribution of relativistic electrons [B 24a, J 5, L 8, S 50, W 7].

(v) The Blandford-Königl compact jet model [B 23, K 16]. In the spirit of the Scheuer-Readhead hypothesis (see 9.6.6.4.3) radio emission of compact sources is assumed to originate from collimated relativistic jets.

The cutoff of the local electron distribution within the jet is determined by equating synchrotron loss time and expansion time which requires continuous acceleration and implies position dependence of the cutoff frequencies within the jet. The integrated spectrum is flat between the cutoff frequencies corresponding to the outer and inner edges of the jet. The innermost parts of the jet produce a steep optical spectrum with the index $\alpha \approx -1$. For extensions of the model, see [R 27, R 28]. Further modifications will have to take the shock acceleration process (cf. 9.6.6.4.5) consistently into account.

Model (v) seems at present to provide the most promising solution to the problems of flat radio spectra of compact sources and also for their optical spectra and beyond (see 9.6.6.3.6).

### 9.6.6.3.5 Correlations between radio, optical and X-ray fluxes

Observations with the Einstein satellite clearly established quasars and BL Lac objects as being strong X-ray emitters at a level of $10^{43} \cdots 10^{47}$ erg s$^{-1}$ [T 3, T 4]. Average spectral indices $\alpha_{oX}$ and luminosity ratios $R_{oX} = l_X/l_o \approx (\nu_X f_X)/(\nu_o f_o)$ (where $\nu$ and $f$ denote frequency and flux density, respectively) between the X-ray and optical bands and, analogously, also $\alpha_{ro}$ and $R_{ro}$ between the radio and optical bands have been discussed in terms of various emission mechanisms [B 14, K 20, S 30, T 2, T 3, Z 1]. For blazars $\alpha_{oX} \approx -1.25$ and for radio-quiet QSO's $\alpha_{oX} \approx -1.5$ [B 23, K 16, P 10, W 13]. For compact millimeter radio sources a very strong correlation between the millimeter and X-ray flux densities is found [K 21, O 6] and reflected in an average spectral index $\alpha_{mX} = -1.02 \pm 0.05$ [O 6] (cf. Fig. 1). Total spectra of the resolved jets in Cen A, M 87, and 3C 273 are similar for the individual knots and show a break $\Delta\alpha \approx 0.5$ between $10^{13}$ and $10^{15}$ Hz [F 4, S 24, S 25] implying synchrotron lifetimes ($\approx 10^2$ years) short compared to the jet travel time ($\approx 10^3$ years) and thereby demand continuous acceleration. Often changes in spectral slope around $10^{15}$ Hz are seen in IUE data of BL Lac objects [B 38, C 7, F 25, G 3, M 5]. X-ray emission mechanisms in galactic nuclei are reviewed in [F 1, G 8, R 20].

### 9.6.6.3.6 Models for optical, UV, and X-ray emission from compact sources

The following emission and radiation transport mechanisms have been considered.

(i) Synchrotron radiation [G 4, R 33]: This process has been found most useful in the interpretation of the optical and X-ray jets [S 24]. Selfconsistent incorporation of shock acceleration may produce the observed break of the spectrum in the optical domain. In the framework of the compact jet model by Blandford and Königl [B 23, K 16] and its extensions [R 27, R 28] synchrotron radiation and inverse Compton effect produce a total integrated emission with a flat radio spectrum ($\alpha \approx 0$) and a power law ($\alpha \approx -1$) up to the X-ray band.

(ii) The Synchrotron-Self-Compton (SSC) effect [J 7, O 2, R 29]: Emission occurs at least in the synchrotron frequency band $\nu_s = 10^6 B \gamma^2$ Hz and simultaneously in the first-order inverse Compton band $\nu_{iC} = \gamma^2 \nu_s$ with typical value of the Lorentz factor $\gamma = 10^2 \cdots 10^3$ and of the magnetic field strength $B \approx 10^2$ G. Generally, $\alpha_{oX} \lesssim -1$ as observed in quasars. Special SSC models like, e.g., the spinar model of Pacini and Salvati [P 2a, P 3], predict $\alpha_{mX} = -1$ as observed in the very compact millimeter sources [B 14, K 21, O 6].

(iii) Thermal comptonization [B 17, B 35, F 1, F 6, I 1, I 3, J 3, K 5, K 17, P 8, P 20, S 35, S 60 ··· S 62, T 1]: Low-energy photons from accretion disks, massive objects or of unspecified origin injected into a hot thermal plasma are Compton scattered into the UV, X, and $\gamma$ bands of the spectrum. At large electron scattering optical depths $\tau_{es} \gg 1$ a Planck spectrum, for $\tau_{es} \lesssim 1$ a power law spectrum results [B 28, B 28a, E 1, E 2, I 2, L 8, L 10 ··· L 13, M 4, M 6, M 7].

(iv) Plasmon-photon scattering and comptonization [B 35, C 20, C 21, F 1, P 15]:
Photons produced from plasma oscillations with $\nu \approx 2\nu_p$ ($\nu_p$ = plasma frequency) are upscattered by the oscillations and comptonized later on as in (iii).

(v) Thermal bremsstrahlung [R 32, R 33]:
Many attempts have been made to build thermal models for the X-ray emission of active galaxies, quasars and jets. The energy requirements are often excessively high, cf. e.g. [E 4, F 4].

The various incoherent synchrotron and inverse Compton models appear at present most attractive while comptonization models and coherent processes cannot be excluded [F 1, R 17, R 20]. For an overview of the various processes, see Fig. 2.

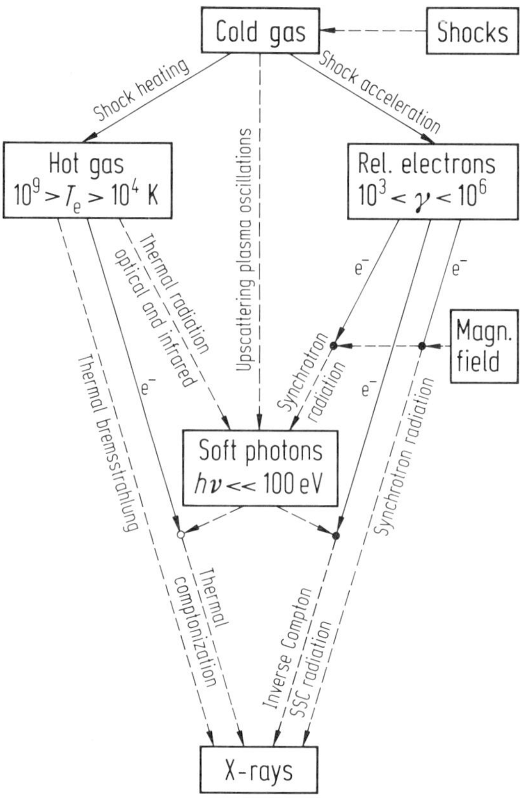


Fig. 2. Overview of continuum radiation processes in galactic nuclei from the radio to the X-ray spectral ranges. Full lines indicate electrons ($e^-$), broken lines photons (or plasma oscillations), cf. also [F 1].

## 9.6.6.4 Compact radio jets

Properties and models of compact jets are reviewed in: [B 21, B 26, B 34, C 16, K 8, K 13, K 14, L 18, P 22, R 3, R 7, R 15, R 20, R 21, R 23, S 19, S 24, S 31, S 38, T 2].

### 9.6.6.4.1 Angular structure

Most strong and variable compact sources (e.g. 3C 273, 3C 345, 3C 147, 3C 380) are of the core-jet type (9.6.5.1) with a flat spectrum core and a one-sided inhomogeneous steep spectrum jet (1···10 milliarcsec) [P 7, P 9, R 2, R 6]. Also in all mapped nuclei of double radio sources VLBI jets are found. They are aligned and often connected by large-scale jets with the outer lobes presumably feeding them with matter and energy from a central "powerhouse" [B 33, R 17], the best examples being NGC 6251, 3C 219, NGC 315, 3C 66B.

Optical counterparts of jets on the kpc scale and their extensions cannot generally be resolved from the ground. Exceptions are the detections in 3C 273 and M 87. More of them may be revealed by the Space Telescope. X-ray counterparts are found by Einstein-HRI observations for 3C 273, M 87, and Cen A [F 4, S 24].

### 9.6.6.4.2 Superluminal motions

Monitoring of compact sources since 1970 with transcontinental VLBI indicated for nearly half of the strong sources [B 23, C 17] and established firmly for some of them (see 9.6.5.2, Table 13) apparent superluminal separation velocities of individual components if cosmological distances were adopted: [C 16···C 18, K 14, P 7, P 16, P 18, S 28, W 18].

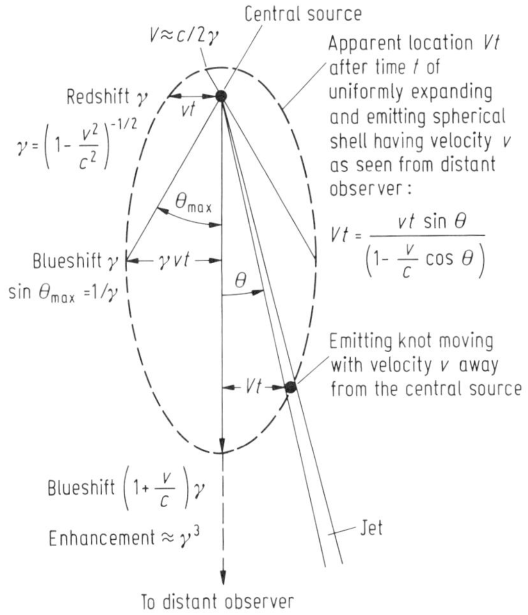


Fig. 3. Kinematics of spherical uniform relativistic expansion of an emitting surface and of relativistic uniform motion of jet inhomogeneities (knots) away from a central source.

Understanding of the phenomenon is incomplete; proposed explanations are numerous and often mutually exclusive. The simplest types of viable models employ Doppler beaming and boosting [B 21, O 12, R 9, R 10, S 37] of the radiation coming from a linearly and relativistically moving source [J 10, K 13, L 1, M 14, T 5, V 9]. For a Lorentz factor $\gamma = (1 - v^2/c^2)^{-1/2} \gg 1$ apparent transverse motions of inhomogeneities up to $\approx \gamma c$ at an angle arcsin $\gamma^{-1}$ and an intensity enhancement by a factor $\approx \gamma^3$ will be observed (see Fig. 3). Blandford et al. [B 26] proposed a general class of models in which a relativistic signal (e.g. a shock) is scattered by an appropriately shaped screen causing the kinematic illusion of superluminal motion to a distant observer having arbitrary orientation to the source. Other models invoke light echoes [L 26, L 28], variations in synchrotron opacity [A 16, E 5], synchrotron curvature radiation along the polar field lines of a magnetic dipole [B 1, M 24, S 4, S 6], etc.

The observational phenomena are reviewed in [K 8, K 9, K 13, P 6, S 31], a classification of models is found in [B 26], kinematics and dynamics are considered in [B 33, M 14, R 15].

### 9.6.6.4.3 The Scheuer-Readhead hypothesis [S 19]

This model provides a unifying interpretation of extragalactic radio sources on the basis that the majority of bright compact radio sources are in fact relativistic or supersonic jets which are observed at small angles to their axes. The jets emerge in opposite directions but often only the near-side jet is visible due to "Doppler favouritism" [R 23]. In this picture, the sequence: radio-quiet quasars, radio-loud quasars, blazars correspond to similarly strong sources, associated with relativistic jets viewed at progressively smaller angles to their axes. The predicted fraction of radio-loud quasars is $\approx 1/\gamma^2$. In this interpretation the extended sources are viewed at large angles to their jets and their differences in power are related to the jet speed: In the strong source Cyg A the jets are relativistic, in NGC 6251 mildly relativistic, and in the weak source 3C 449 subrelativistic, so that both jets are visible in this case. The compact jets produce and end in the faint halo of extended emission seen to surround compact sources [B 23, K 16, P 10, W 13]. For an illustration of the Scheuer-Readhead idea, see Fig. 4.

In blazars the highly variable and strongly polarized beamed emission components outshine the steady and unpolarized isotropic components responsible for photoionizing the emission-line gas near the nucleus. Relativistic jets are also expected to produce the optical and X-ray emission via synchrotron or inverse Compton radiation [K 16, R 28]. Another advantage is that the energy requirements for strong sources (cf. 9.6.6.3.3) are drastically reduced due to the Doppler boosting effect. The constraints on the model provided by optical spectroscopic and polarization measurements are discussed in [A 6, A 7, B 34].

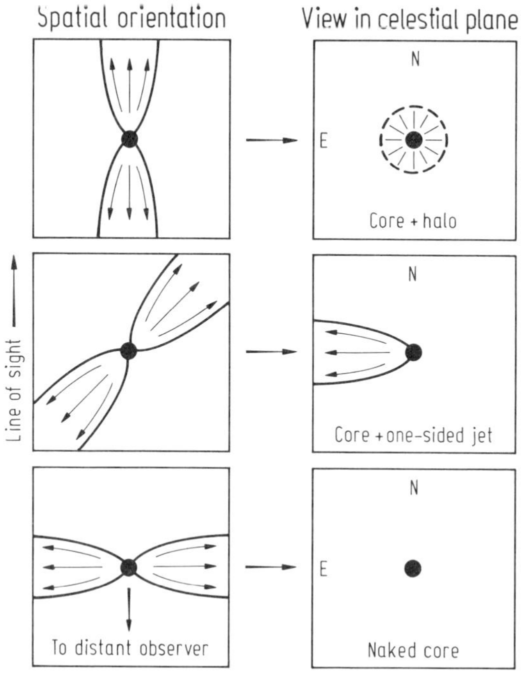


Fig. 4. Schematic illustration of the appearance of differently oriented jet sources as seen from a distant observer in the plane of the sky according to the Scheuer-Readhead hypothesis (cf. [B 21]).

The model owes its great attraction to its potential unifying power for many diverse observations of compact radio sources and quasars. However, there are many objections to it both regarding properties of individual sources [B 13, K 13, S 38, S 48] as well as of statistical nature, e.g. radio observations of optically selected quasars [C 25, K 13, K 23, S 48, S 54, S 58] suggesting that the model cannot be valid in its simple version.

### 9.6.6.4.4 Relativistic bulk motions

Most direct evidence for relativistic bulk motions in the emitting regions of radio jets [R 9, R 10, S 37] is provided by low frequency variability [C 24, C 29, D 0, F 2, J 4] and by apparent superluminal motions [B 26, B 33, M 14, O 1]. Circumstantial evidence for relativistic motions comes from radio polarization observations indicating the absence of significant amounts of thermal plasma [J 8, W 7, W 8], from exaggerated bending of jets by large angles at progressively higher resolution [R 6, R 8], from the one-sidedness of jets if explained by relativistic Doppler beaming [S 19], and from the inferred efficient feeding of extended sources from the nucleus via jets and beams [B 31, S 16] (cf. 9.6.6.2), etc.

The concept of relativistic motions has sometimes been questioned [J 4, P 19, B 22].

### 9.6.6.4.5 In-situ acceleration

Relativistic electrons responsible for the observed synchrotron emission must be accelerated locally within the jet because of synchrotron, expansion, and inverse Compton losses [B 23, B 24, B 27, M 13, R 18]. The action of continuous acceleration is most clearly demonstrated by the presence of short-lived relativistic electrons producing the inhomogeneous optical, X-ray, and VLBI jets [B 38, C 7, F 5, G 3, L 18, R 23, S 24]. The acceleration occurs most probably in strong internal shocks [B 25, J 10, M 11], which may manifest themselves in the patches or knots seen along the radio and optical jets, most impressively in M 87 [A 11, A 12]. The important role of the first-order Fermi acceleration mechanism [L 19] in the presence of strong shocks [L 3, Z 3] has been recognized only recently [A 16, B 7, B 8, B 27, K 19]. Reviews of earlier literature are in [A 10, B 25, L 19, P 5]. Recent developments may be found in [A 2, A 3, A 13···A 15, B 20, B 28, B 28a, D 8, T 8]. For criticism, see [K 24, K 26, M 17].

#### 9.6.6.4.6 Origin of compact jets

Given a fuelling central source (9.6.6.5) and an asymmetric confining environment (shaped e.g. by angular momentum), collimation by the "twin-exhaust mechanism" [B 31] may work on any scale of the emitted cloud, at $\approx 10^{-3}$ pc near the central black hole or massive object, at $\approx 0.01 \cdots 0.1$ pc in the central dense star cluster, at some kpc on galactic sizes. "Scaling laws" down to SS 433 jets [M 8] and jets occurring in regions of star formation have been contemplated by Rees [R 20, R 23]. In galactic nuclei the jets may be confined by ion [A 4, K 15, R 23, R 24, P 16a] or by radiation pressure [A 1, J 2] supported rotating gas tori (doughnuts) [R 21]. Winds or jets from accretion disks are considered in: [B 19, B 25, B 29, B 34, B 34a, L 7, L 25, L 27···L 30, M 18, N 9, P 17, P 25, R 23, S 32, S 39, T 9, Z 4] etc.

The ultimate collimation operates probably in sub-parsec dimensions, presently not accessible to observation, and close to the central collapsed object where high energy particles are produced with great efficiency. Near a rotating (Kerr) black hole, Lense-Thirring precession [B 2, M 26, T 6] imposes symmetry on the accreted material parallel to the hole's rotation axis and thereby guarantees the long-term stability of the jet axis required by the observations [K 13, M 21, W 14]. The VLBI jets would then be secondary phenomena [B 33]. "Cannibalism" or merging [O 4, R 19, R 30, W 17] (see 9.3.2.2) of galaxies may result in double or multiple nuclei orbiting around each other with precessing spin axes and jets tied to them [B 5, B 21] (see 9.6.6.2).

### 9.6.6.5 The central engine

Theoretical models for the ultimate energy source in the nucleus of the parent galaxy of radio sources and quasars are reviewed in: [B 34, L 25, L 27, M 28, O 10, P 2, P 3, R 16, R 17, R 20, R 21, R 23, S 9, T 6, W 21, W 22].

#### 9.6.6.5.1 Gravitation as ultimate source of energy

Cosmological distances imply powers of compact radio sources and quasars as high as $L \approx 10^{46}$ erg s$^{-1}$ corresponding to Eddington luminosities $L_{\rm Edd} = 1.3 \cdot 10^{38} (\mathfrak{M}/\mathfrak{M}_\odot)$ erg s$^{-1}$ of $10^7 \cdots 10^8\, \mathfrak{M}_\odot$ objects [A 7, R 29a, T 3, T 4]; see, however [D 7, S 7, Y 5]. The rapid variability of some of the sources (e.g. AO 0235+164 and OJ 287) indicate [A 7, B 34] dimensions $R$ smaller than those accessible by VLBI, possibly $R \approx 10^{15} \cdots 10^{16}$ cm. The energy typically stored in strong external radio sources of quasars indicates a yield of the central source by $\approx 10^{61} \cdots 10^{62}$ erg, corresponding to a source lifetime of $10^7 \cdots 10^8$ years and a rest mass of at least $10^7 \cdots 10^8\, \mathfrak{M}_\odot$. The binding energy of such an object exceeds already by two orders of magnitude its nuclear energy reservoir. Considerations of this kind [L 25, L 27] demonstrate the irrelevance of nuclear energy sources to quasar theories and suggest gravitational energy release from a deep potential well as the ultimate energy source for these objects – provided exotic energy sources like matter-antimatter annihilation (cf. [S 55]) or "new physics" [A 5, B 47, H 15] are discarded. The massive objects at the center could be (i) a dense star cluster or dense gas-star system of some kind, (ii) a coherent supermassive star, disk or magnetoid, or (iii) a high-mass Schwarzschild or Kerr black hole. The objects under (i) and (ii) may probably be regarded as transition phases [R 16, R 17] in the evolution of an active nucleus towards the formation of a massive central black hole (cf. Fig. 5).

The accumulation of the large-mass at the center of a galactic nucleus constitutes one of the major problems of the theory [C 3, G 7, L 23, M 16, R 17, S 51, Z 3].

Observationally, a high central mass concentration of $\gtrsim 10^9\, \mathfrak{M}_\odot$ reveals itself as a central "cusp" in the surface brightness distribution, e.g. in M 87 and NGC 6251 [D 7, S 7, Y 5]. However, a distinction between a dense cluster or a coherent large-mass object cannot be made at present [D 7, D 9].

#### 9.6.6.5.2 Dense star clusters [S 8, S 9]

Many scenarios have been investigated to account for the observed nonthermal radiation:

Multiple supernova [B 16, C 19, L 23], a dense cluster of pulsars [A 10, B 16], stellar collisions [K 6, S 3, S 12, S 52, S 53], evolution of dense gas-star systems [B 6, B 16, H 1, H 4, H 5, N 6, S 52, S 53, U 1]. Such systems either tend to blow up quickly or eventually develop one or more coherent massive objects at their centers by coalescence [C 19, D 1, S 3, S 27] or by various black-hole feeding processes [F 17, F 20, Y 2···Y 5].

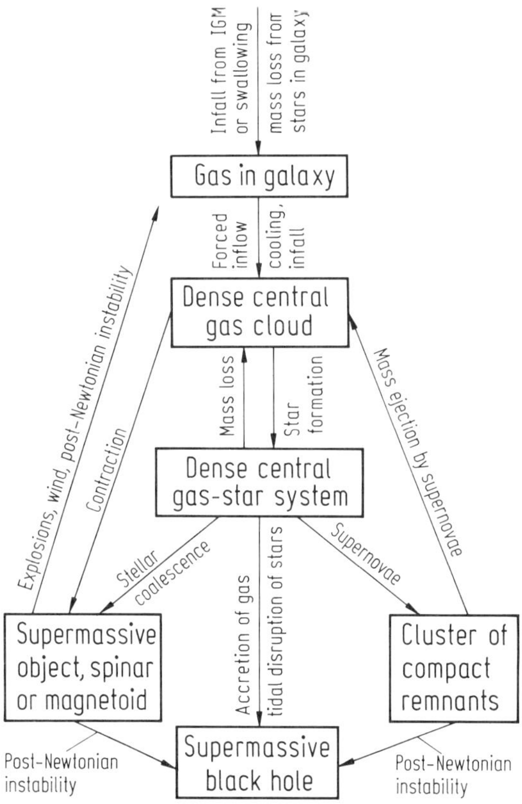


Fig. 5. Possible pathways of the process by which a massive black hole could form in a galactic nucleus which is fed from the outside with mass (gas or stars). Before the final state is reached various stops seem possible. These stages may well be important if not essential periods in the active life of a nucleus (cf. [R 17]).

### 9.6.6.5.3 **Supermassive stars** [H 13, H 14], **rotators** [W 21], **spinars** [M 28, M 29] **or magnetoids** [O 14, O 15]

The amount of binding energy which can be extracted from these objects is limited by the post-Newtonian instability [A 8, F 16, H 10]. In hot (radiation pressure supported) configurations this happens typically already at $\beta^{-1/2} \approx 1000$ Schwarzschild radii, where $\beta$ is the ratio of gas pressure to total pressure [F 16]. They ultimately (after $\approx 10^6$ years) will explode or form massive black holes [F 22, F 23, S 36]. The efficiency of converting rest mass ($\mathfrak{M}c^2$) via gravitational ($E_g$) and rotational ($E_{rot}$) into nonthermal ($E_{nth}$) energy is $\eta = \frac{E_{nth}}{E_{rot}} \frac{E_{rot}}{E_g} \frac{E_g}{\mathfrak{M}c^2}$. Most advantageous are therefore magnetized disks or "pancakes" close to their "Schwarzschild" radii radiating with a pulsar-like mechanism. In this case $\eta \approx 10\%$. Detailed models are described for supermassive stars and rotators in: [F 14, F 16, H 10, H 13, H 14, S 49, W 1b, W 21, W 22]; for spinars and magnetoids in: [B 15, C 2, C 4, F 14, G 2, K 24, M 28, M 29, O 14···O 16, P 2a, P 3, S 49].

### 9.6.6.5.4 **Massive black holes** [L 25, L 27, S 2]

Rotating (Kerr) black holes [M 26] provide the most stable and effective central engines due to their strong gravitational fields and their ability to create high energy particles in their vicinity with high efficiency [B 19, B 25, B 29, B 34, B 34a, L 7, L 25, L 27···L 30, M 18, N 9, P 17, P 25, R 23, S 32, S 35, S 39, T 9, Y 1].

Among the processes discussed in detail for massive black-hole growth and energy extraction from black-holes in galactic nuclei are:
Infall of gas [G 7, L 22, S 51], disk accretion [B 20a, C 2, C 3, F 13, G 9, L 7, L 9, M 18, N 9, O 11, P 17, P 23···P 25, S 33, T 6], spherical accretion [B 3, F 21, M 19, S 20, S 34, T 7, Y 1, Z 2], stellar disruption [F 19, G 9, G 10, H 6, H 7, K 6], loss-cone diffusion [F 18···F 20], nonaxisymmetric flows [N 6, R 30, R 31, S 26], etc.

The "natural" ability of a massive Kerr black hole to form jets [B 2, B 19, B 29, B 34a, R 20, R 23, S 33, S 40, S 41, T 6] in its galactic environment and to maintain their direction for millions of years supports strongly the massive black-hole interpretation of quasars and radio sources. The model has been criticized, e.g. by [K 24, K 25, P 2, P 3].

## 9.6.6.6 Large-scale distribution of radio sources and cosmological evolution

This topic is partly the subject of the chapter on cosmology (see 9.7). Recent results are described in: [J 2a, K 3, K 22, L 16, L 17, S 21···S 23, S 29, S 30, W 2a···W 5, W 19].

Acknowledgement: I thank P. Biermann and A. Witzel for critically reading the manuscript.

## 9.6.6.7 References for 9.6.6

### General references


a Pacholczyk, A.G.: Radio Astrophysics, Freeman, San Francisco (1970).
b Nuclei of Galaxies (O'Connell, D.J.K., ed.), Pont. Acad. Sci. Scripta Varia **35**, North-Holland, Amsterdam, London (1971).
c Galaxies and Relativistic Astrophysics (Barbanis, B., Hadjidemetriou, J.D., eds.), Proceedings of the First European Astronomical Meeting, Vol. 3, Springer, Berlin (1974).
d The Physics of Nonthermal Radio Sources (Setti, G., ed.), Proceedings of the NATO Advanced Study Institute held at Urbino, Italy June 29–July 13, 1975, Reidel, Dordrecht (1976).
e Pacholczyk, A.G.: Radio Galaxies, Pergamon Press, Oxford (1977).
f Quasars and Active Nuclei of Galaxies (Ulfbeck, O., ed.), Proceedings of the Copenhagen Symposium on "Active Nuclei", June 27–July 2, 1977, Physica Scripta **17**, No. 3 (1978).
g Pittsburgh Conference on BL Lac Objects, held at the University of Pittsburgh, April 24–26, 1978 (Wolfe, A., ed.), University of Pittsburgh (1978).
h Active Galactic Nuclei (Hazard, C., Mitton, S., eds.), Cambridge University Press, Cambridge (1979).
i X-Ray Astronomy (Giacconi, R., Setti, G., eds.), Proceedings of the NATO Advanced Study Institute, held at Erice, Sicily, July 1–14 (1979).
k Pacini, F., Ryter, C., Strittmatter, P.A.: Extragalactic High Energy Astrophysics (Blécha, A., Maeder, A., eds.) = 9th Advanced Course, Swiss Society of Astronomy and Astrophysics, Saas Fee (1979).
l Extragalactic High Energy Astrophysics (Laan, H., van der, ed.) = Highlights of Astronomy, Vol. **5**, Int. Astron. Union, Joint Discussion, Reidel, Dordrecht (1980).
m Optical Jets in Galaxies, Proceedings of the 2nd ESO/ESA Workshop on the use of the Space Telescope and co-ordinated ground based research, ESA-SP-162 (1981).
n Longair, M.S.: High Energy Astrophysics, Cambridge University Press, Cambridge (1981).
o Int. Astron. Union Symp. **97**, Extragalactic Radio Sources (Heeschen, D.S., Wade, C.M., eds.), Reidel, Dordrecht (1982).


### Special references


A 1 Abramovicz, M.A., Calvani, M., Nobile, L.: Astrophys. J. **242** (1980) 772.
A 2 Achterberg, A.: Astron. Astrophys. **98** (1981) 161.
A 3 Achterberg, A.: Astron. Astrophys. **97** (1981) 259.
A 4 Allan, P. in: [o] p. 235.
A 5 Ambartsumian, V.A. in: The Structure and Evolution of Galaxies, Proc. 13th Conf. Phys. at Univ. of Brussels; Wiley Interscience, New York (1965) p. 1.

A 6 Angel, J.R.P. et al. in: [g] p. 117.
A 7 Angel, J.R.P., Stockman, H.S.: Annu. Rev. Astron. Astrophys. **18** (1980) 321.
A 8 Appenzeller, I., Fricke, K.J.: Astron. Astrophys. **21** (1972) 285.
A 9 Arons, J., Kulsrud, R.N., Ostriker, J.P.: Astrophys. J. **198** (1975) 683.
A 10 Arons, J., Max, C., McKee, C.F. (eds.): Particle Acceleration Mechanisms in Astrophysics, American Institute of Physics Conf. Proc. **56** (1979).
A 11 Arp, H., Lorre, J.: Astrophys. J. **210** (1976) 58.
A 12 Arp, H. in: [m] p. 53.
A 13 Axford, W.I. in: Origin of Cosmic Rays (Setti, G., Spada, G., Wolfendale, W.A., eds.), Int. Astron. Union Symp. **94** (jointly with Int. Union Pure Appl. Phys.), Reidel, Dordrecht (1981) p. 339.
A 14 Axford, W.I.: Proc. Tenth Texas Symp. on Relativistic Astrophysics, Baltimore, Ann. New York Acad. Sci. **375** (1981) 297.
A 15 Axford, W.I.: Proc. Intern. School Varenna (1981) p. 335, ESA SP-161, p. 425.
A 16 Axford, W.I., Leer, E., Skadron, G.: Proc. 15th Intern. Cosmic Ray Conf. (Plovdiv) **11** (1977) 132.

B 1 Bahcall, J.N., Milgrom, M.: Astrophys. J. **236** (1980) 24.
B 2 Bardeen, J.M., Petterson, H.S.: Annu. Rev. Astron. Astrophys. **18** (1980) 321.
B 3 Begelman, M.C.: Mon. Not. R. Astron. Soc. **187** (1979) 237.
B 4 Begelman, M.C. in: [o] p. 223.
B 5 Begelman, M.C., Blandford, R.D., Rees, M.J.: Nature **287** (1980) 307.
B 6 Begelman, M.C., Rees, M.J.: Mon. Not. R. Astron. Soc. **185** (1978) 847.
B 7 Bell, A.R.: Mon. Not. R. Astron. Soc. **182** (1978) 147.
B 8 Bell, A.R.: Mon. Not. R. Astron. Soc. **182** (1978) 443.
B 9 Benford, G.: Mon. Not. R. Astron. Soc. **179** (1977) 595.
B 10 Benford, G.: Mon. Not. R. Astron. Soc. **183** (1978) 29.
B 11 Benford, G. in: [o] p. 231.
B 12 Benford, G.: Astrophys. J. **247** (1981) 458.
B 13 Biermann, P., Fricke, K.J., Johnston, K.J., Kühr, H., Pauliny-Toth, I.I.K., Strittmatter, P.A., Urbanik, M., Witzel, A.: Astrophys. J. **252** (1982) L1.
B 14 Biermann, P., Duerbeck, H., Eckart, A., Fricke, K.J., Johnston, K.J., Kühr, H., Liebert, J., Pauliny-Toth, I.I.K., Schleicher, H., Stockman, H., Strittmatter, P.A., Witzel, A.: Astrophys. J. **247** (1981) L53.
B 15 Bisnovatyi-Kogan, G.S., Blinnikov, S.I.: Astron. Astrophys. **59** (1977) 111.
B 16 Bisnovatyi-Kogan, G.S., Syunyaev, R.A.: Sov. Astr.-AJ **16** (1972) 201.
B 17 Bisnovatyi-Kogan, G.S., Zeldovich, Ya.B., Syunyaev, B.A.: Sov. Astr.-AJ **15** (1971) 17.
B 18 Blake, G.M.: Mon. Not. R. Astron. Soc. **156** (1972) 67.
B 19 Blandford, R.D.: Mon. Not. R. Astron. Soc. **176** (1976) 465.
B 20 Blandford, R.D. in: Workshop on Particle Acceleration Mechanisms in Astrophysics (Arons, J., Max, C., McKee, C.F., eds.), American Institute of Physics Conf. Proc. **56** (1979) p. 333.
B 20a Blandford, R.D. in: [h] p. 241.
B 21 Blandford, R.D., Begelman, M.C., Rees, M.J.: Sci. American **246**, No. 5 (1982) 84.
B 22 Blandford, R.D., Icke, V.: Mon. Not. R. Astron. Soc. **185** (1978) 527.
B 23 Blandford, R.D., Königl, A.: Astrophys. J. **232** (1979) 34.
B 24 Blandford, R.D., Königl, A.: Astrophys. Lett. **20** (1979) 15.
B 24a Blandford, R.D., McKee, C.F.: Phys. Fluids **19** (1976) 1130.
B 25 Blandford, R.D., McKee, C.F.: Mon. Not. R. Astron. Soc. **180** (1977) 343.
B 26 Blandford, R.D., McKee, C.F., Rees, M.J.: Nature **267** (1977) 211.
B 27 Blandford, R.D., Ostriker, J.P.: Astrophys. J. **221** (1978) L29.
B 28 Blandford, R.D., Payne, D.G.: Mon. Not. R. Astron. Soc. **194** (1981) 1033.
B 28a Blandford, R.D., Payne, D.G.: Mon. Not. R. Astron. Soc. **194** (1981) 1041.
B 29 Blandford, R.D., Payne, D.G.: Mon. Not. R. Astron. Soc. **199** (1982) 883.
B 30 Blandford, R.D., Pringle, J.E.: Mon. Not. R. Astron. Soc. **176** (1976) 433.
B 31 Blandford, R.D., Rees, M.J.: Mon. Not. R. Astron. Soc. **169** (1974) 395.
B 32 Blandford, R.D., Rees, M.J.: Contemp. Phys. **16** (1975) 1.
B 33 Blandford, R.D., Rees, M.J. in: [f] p. 265.
B 34 Blandford, R.D., Rees, M.J. in: [g] p. 328.
B 34a Blandford, R.D., Znajek, R.L.: Mon. Not. R. Astron. Soc. **179** (1977) 433.
B 35 Blumenthal, G.R., Gould, R.J.: Rev. Mod. Phys. **42** (1970) 237.
B 36 Bodo, G., Ferrari, A., Massaglia, S.: Mon. Not. R. Astron. Soc. **196** (1981) 481.

B 37 Brandie, G.W., Bridle, A.H., Kesteven, M.J.L.: Nature **252** (1974) 212.
B 38 Bregman, J.N., Glassgold, A.E., Huggins, P.J.: Proc. Third European IUE Conf., Madrid, ESA SP-176 (1982) 589.
B 39 Bridle, A. in: [o] p. 121.
B 40 Bridle, A.H., Brandie, G.W.: Astrophys. Lett. **15** (1973) 21.
B 41 Bridle, A.H., Davis, M.M., Fomalont, E.B., Willis, A.G., Strom, R.G.: Astrophys. J. **288** (1979) L9.
B 42 Bridle, A.H., Fomalont, E.B.: Astron. Astrophys. **52** (1976) 107.
B 43 Broderick, J.J., Condon, J.J.: Astrophys. J. **202** (1975) 596.
B 44 Bruyn, de, A.G.: Astron. Astrophys. **52** (1976) 439.
B 45 Burbidge, G.R.: Nature **216** (1967) 1287.
B 46 Burbidge, G.R. in: [d] p. 121.
B 47 Burbidge, G.R. in: [f] p. 281.
B 48 Burbidge, G.R., Jones, T.W., O'Dell, S.L.: Astrophys. J. **193** (1974) 43.
B 50 Burbidge, G.R., Stein, W.A.: Comments Astrophys. Space Phys. **6** (1975) 87.
B 51 Byrd, G.G., Valtonen, M.J.: Astrophys. J. **221** (1978) 481.

C 1 Callahan, P.S.: Mon. Not. R. Astron. Soc. **174** (1976) 587.
C 2 Callahan, P.S.: Astron. Astrophys. **59** (1977) 127.
C 3 Carter, B. in: [h] p. 185.
C 4 Cavaliere, A., Pacini, F., Setti, G.: Astrophys. Lett. **4** (1969) 103.
C 5 Cavallo, G., Rees, M.J.: Mon. Not. R. Astron. Soc. **183** (1978) 359.
C 6 Chan, K.L., Henriksen, R.N.: Astrophys. J. **241** (1980) 534.
C 7 Chiapetti, L., Maraschi, L., Tanzi, E.G., Treves, A.: Proc. Third European IUE Conf., Madrid, ESA SP-176 (1982) 581.
C 8 Christiansen, W.A.: Mon. Not. R. Astron. Soc. **145** (1969) 327.
C 9 Christiansen, W.A.: Astrophys. J. **167** (1971) 541.
C 10 Christiansen, W.A.: Mon. Not. R. Astron. Soc. **164** (1973) 211.
C 11 Christiansen, W.A., Frater, N., Watkinson, R.H., O'Sullivan, J.D., Lockhart, I.A., Goss, W.M.: Mon. Not. R. Astron. Soc. **181** (1977) 183.
C 12 Christiansen, W.A., Pacholczyk, A.G., Scott, J.S.: Nature **266** (1977) 593.
C 13 Christiansen, W.A., Scott, J.S.: Astrophys. J. **216** (1977) L1.
C 14 Cocke, W.J., Pacholczyk, A.G.: Nature **256** (1975) 608.
C 15 Cocke, W.J., Pacholczyk, A.G., Hopf, F.A.: Astrophys. J. **226** (1978) 26.
C 16 Cohen, M.H., Unwin, S. in: [o] p. 345.
C 17 Cohen, M.H. et al.: Nature **268** (1977) 405.
C 18 Cohen, M.H. et al.: Astrophys. J. **231** (1979) 293.
C 19 Colgate, S.A.: Astrophys. J. **150** (1967) 163.
C 20 Colgate, S.A., Colvin, J.D., Petschek, A.G.: Astrophys. J. **197** (1975) L105.
C 21 Colgate, S.A., Petschek, A.G. in: [g] p. 349.
C 22 Condon, J.J. in: [g] p.21.
C 23 Condon, J.J., Dressel, L.L.: Astrophys. Lett. **15** (1973) 203.
C 24 Condon, J.J., Ledden, J.D., O'Dell, S.L., Dennison, B.: Astron. J. **84** (1979) 1.
C 25 Condon, J.J., O'Dell, S.L., Puschell, J.J., Stein, W.A.: Astrophys. J. **246** (1981) 624.
C 26 Condon, J.J., O'Dell, S.L., Puschell, J.J., Stein, W.A.: Nature **283** (1980) 357.
C 27 Cook, D.B., Spangler, S.R.: Astrophys. J. **240** (1980) 751.
C 28 Cooke, B.A., Lawrence, A., Perola, G.C.: Mon. Not. R. Astron. Soc. **182** (1978) 661.
C 29 Cotton, W.D., Spangler, S.R.: Astrophys. J. **228** (1979) L63.
C 30 Cotton, W.D., Wittels, J.J., Shapiro, I.I., Marcaide, J., Owen, F.N., Spangler, S.R., Rius, A., Angulo, C., Clark, T.A., Knight, C.A.: Astrophys. J. **238** (1980) L123.
C 31 Cowie, L.L., McKee, C.F.: Astron. Astrophys. **43** (1975) 337.
C 32 Craine, E.R.: A Handbook of Quasistellar and BL Lacertae Objects, Pachart, Tucson (1977).

D 0 Dennison, B., Broderick, J.J., Ledden, J.E., O'Dell, S.L., Condon, J.J.: Astron. J. **86** (1981) 1604.
D 1 De Young, D.S.: Astrophys. J. **153** (1968) 633.
D 2 De Young, D.S.: Astrophys. J. **167** (1971) 541.
D 3 De Young, D.S.: Annu. Rev. Astron. Astrophys. **14** (1976) 447.
D 4 De Young, D.S.: Astrophys. J. **211** (1977) 329.
D 5 De Young, D.S., Axford, W.I.: Nature **216** (1967) 129.
D 6 Downes, A.: Mon. Not. R. Astron. Soc. **190** (1980) 261.

D 7 Dressler, A.: Astrophys. J. **240** (1980) L11.
D 8 Drury, L. O'C.: Phys. Reports (1982) (in press).
D 9 Duncan, M.J., Wheeler, J.C.: Astrophys. J. **237** (1980) L27.

E 1 Eardley, D.M., Lightman, A.P., Payne, D.G., Shapiro, S.L.: Astrophys. J. **224** (1978) 53.
E 2 Eardley, D.M., Lightman, A.P., Shapiro, S.L.: Astrophys. J. **199** (1975) L153.
E 3 Eilek, J.A.: Astrophys. J. **231** (1979) 373.
E 4 Elvis, M., Maccacaro, T., Wilson, A.S., Ward, M.J., Penston, M.V., Fosbury, R.A.E., Perola, G.C.: Mon. Not. R. Astron. Soc. **183** (1978) 129.
E 5 Epstein, R.I., Geller, M.J.: Nature **265** (1977) 219.

F 1 Fabian, A.C., Rees, M.J.: Proc. IAU/COSPAR Symp. on X-Ray Astronomy (Baity, W., Peterson, L.E., eds.), Pergamon Press, Oxford (1978) p. 381.
F 1a Fabian, A.C., Kembhavi, A.K. in: [o] p. 453.
F 2 Fanti, R., Padrielli, L., Salvati, M. in: [o] p. 317.
F 3 Fabbiano, G., Doxsey, R.E., Johnston, M.A., Schwartz, D.A., Schwarz, J.: Astrophys. J. **230** (1979) L67.
F 4 Feigelson, E.D., Schreier, E.J., Delvaille, J.P., Giacconi, R., Grindlay, J.E., Lightman, A.P.: Astrophys. J. **251** (1981) 31.
F 5 Felten, J.E., Morrison, P.: Astrophys. J. **146** (1966) 686.
F 6 Felten, J.E., Rees, M.J.: Astron. Astrophys. **17** (1972) 226.
F 7 Ferrari, A., Massaglia, S., Trussoni, E., Zaninetti, L. in: [o] p. 229.
F 8 Ferrari, A., Trussoni, E., Zaninetti, L.: Astron. Astrophys. **79** (1979) 190.
F 9 Ferrari, A., Trussoni, E., Zaninetti, L.: Mon. Not. R. Astron. Soc. **193** (1980) 469.
F 10 Ferrari, A., Trussoni, E., Zaninetti, L.: Mon. Not. R. Astron. Soc. **196** (1981) 1051.
F 11 Ferrari, A., Trussoni, E., Zaninetti, L. in: [m] p. 82.
F 12 Field, G.B.: Annu. Rev. Astron. Astroph. **10** (1972) 227.
F 13 Flammang, R.A.: Mon. Not. R. Astron. Soc. **199** (1982) 833.
F 14 Flasar, F.M., Morrison, P.: Astrophys. J. **204** (1976) 352.
F 15 Fomalont, E.B. in: Origin of Cosmic Rays (Setti, G., Spada, G., Wolfendale, W.A., eds.), Int. Astron. Union Symp. **94** (jointly with Int. Union Pure Appl. Phys.), Reidel, Dordrecht (1981) p. 111.
F 16 Fowler, W.A. in: High Energy Astrophysics (Gratton, L., ed.), Academic Press, New York (1966) p. 313.
F 17 Frank, J.: Observatory **96** (1976) 198.
F 18 Frank, J.: Mon. Not. R. Astron. Soc. **184** (1978) 87.
F 19 Frank, J.: Mon. Not. R. Astron. Soc. **187** (1979) 883.
F 20 Frank, J., Rees, M.J.: Mon. Not. R. Astron. Soc. **176** (1976) 633.
F 21 Freihoffer, D.: Astron. Astrophys. **100** (1981) 178.
F 22 Fricke, K.J.: Astrophys. J. **183** (1973) 941.
F 23 Fricke, K.J.: Astrophys. J. **189** (1974) 535.
F 24 Fricke, K.J.: Mitt. Astron. Ges. **42** (1977) 59.
F 25 Fricke, K.J., Kollatschny, W.: Astron. Astrophys. (1982) in press.
F 26 Fricke, K.J., Reinhardt, M.: Naturwiss. **62** (1975) 309.

G 1 Gibson, D.M.: Astron. Astrophys. **39** (1975) 377.
G 2 Ginzburg, V.L., Ozernoy, L.M.: Astrophys. Space Sci. **50** (1977) 23.
G 3 Glassgold, A.E., Bregman, J.N., Huggins, P.J.: Proc. Third European IUE Conf., Madrid, ESA SP-176 (1982) 577.
G 4 Gould, R.J.: Astron. Astrophys. **76** (1979) 306.
G 5 Gregorini, L., Mantovani, F., Biermann, P., Witzel, A.: Astron. Astrophys. (1982) in press.
G 6 Gull, S.F., Northover, K.J.E.: Nature **244** (1973) 80.
G 7 Gunn, J.E. in: [h] p. 213.
G 8 Gursky, H., Schwartz, D.A.: Annu. Rev. Astron. Astrophys. **15** (1977) 54.
G 9 Gurzadyan, V.G., Ozernoy, L.M.: Astron. Astrophys. **95** (1981) 39.
G 10 Gurzadyan, V.G., Ozernoy, L.M.: Astron. Astrophys. **86** (1980) 315.

H 1 Hara, T.: Prog. Theoretical Physics Japan **60** (1978) 711.
H 2 Hargrave, P.J., Ryle, M.: Mon. Not. R. Astron. Soc. **166** (1974) 305.
H 3 Haves, P.: Mon. Not. R. Astron. Soc. **173** (1975) 553.
H 4 Heggie, D.C.: Mon. Not. R. Astron. Soc. **186** (1979) 155.
H 5 Heggie, D.C.: Mon. Not. R. Astron. Soc. **188** (1979) 525.
H 6 Hills, J.G.: Nature **254** (1975) 295.

H 7 Hills, J.G.: Mon. Not. R. Astron. Soc. **182** (1978) 517.
H 8 Hine, R.G., Scheuer, P.A.G.: Mon. Not. R. Astron. Soc. **193** (1980) 285.
H 9 Hjellming, R.M., Johnston, K.J. in: [o] p. 197.
H 10 Hoerner, S. von, Saslaw, W.C.: Astrophys. J. **206** (1976) 917.
H 11 Hoyle, F., Burbidge, G.R.: Astrophys. J. **144** (1966) 534.
H 12 Hoyle, F., Burbidge, G.R., Sargent, W.L.W.: Nature **209** (1966) 751.
H 13 Hoyle, F., Fowler, W.A.: Nature **197** (1963) 533.
H 14 Hoyle, F., Fowler, W.A.: Mon. Not. R. Astron. Soc. **125** (1963) 169.
H 15 Hoyle, F., Narlikar, J.V.: Proc. Roy. Soc. London (A) **290** (1966) 177.

I 1 Illarionov, A.F., Kallman, T., McCray, R., Ross, R.: Astrophys. J. **228** (1979) 279.
I 2 Illarionov, A.F., Syunyaev, R.A.: Astrophys. Space Sci. **19** (1972) 61.
I 3 Illarionov, A.F., Sunyayev, R.A.: Sov. Astr.-AJ **16** (1972) 45.

J 1 Jaffe, W.J., Perola, G.C.: Astron. Astrophys. **26** (1973) 423.
J 2 Jaroszyński, M., Abramowicz, M.A., Paczyński, B.: Acta Astron. **30** (1980) 1.
J 2a Jauncey, D.L. (ed.): Radio Astronomy and Cosmology, Int. Astron. Union Symp. **74**, Reidel, Dordrecht (1977).
J 3 Jones, F.C.: Phys. Rev. **167** (1968) 1159.
J 4 Jones, T.W., Burbidge, G.R.: Astrophys. J. **186** (1973) 791.
J 5 Jones, T.W., Hardee, P.E.: Astrophys. J. **228** (1979) 268.
J 6 Jones, T.W., O'Dell, S.L., Stein, W.A.: Astrophys. J. **188** (1974) 353.
J 7 Jones, T.W., O'Dell, S.L., Stein, W.A.: Astrophys. J. **192** (1974) 261.
J 8 Jones, T.W., O'Dell, S.L.: Astron. Astrophys. **61** (1977) 291.
J 9 Jones, T.W., Rudnick, L., Owen, F.N., Puschell, J.J., Ennis, D.J., Werner, M.: Astrophys. J. **243** (1981) 97.
J 10 Jones, T.W., Tobin, W.: Astrophys. J. **215** (1977) 474.

K 1 Kapahi, V.K., Schilizzi, R.T.: Nature **277** (1979) 610.
K 3 Katgert, P., de Ruiter, H.R., van der Laan, H.: Nature **280** (1979) 20.
K 4 Katgert-Merkelijn, J., Lari, C., Padrielli, L.: Astron. Astrophys. Suppl. **40** (1980) 91.
K 5 Katz, J.I.: Astrophys. J. **206** (1976) 910.
K 6 Keenan, D.W.: Mon. Not. R. Astron. Soc. **185** (1978) 389.
K 7 Kellermann, K.I. in: Galactic and Extragalactic Radio Astronomy (Kellermann, K.I., Verschuur, G.L., eds.), Springer, Berlin (1974).
K 8 Kellermann, K.I. in: [d] p. 27.
K 9 Kellermann, K.I. in: [f] p. 257.
K 10 Kellermann, K.I., Pauliny-Toth, I.I.K.: Annu. Rev. Astron. Astrophys. **6** (1968) 417.
K 11 Kellermann, K.I., Pauliny-Toth, I.I.K.: Astrophys. J. **155** (1969) L71.
K 12 Kellermann, K.I., Pauliny-Toth, I.I.K.: Astrophys. Lett. **8** (1971) 153.
K 13 Kellermann, K.I., Pauliny-Toth, I.I.K.: Annu. Rev. Astron. Astrophys. **19** (1981) 373.
K 14 Kellermann, K.I., Shaffer, D.B. in: L'évolution des galaxies et ses implications cosmologiques (Balkowski, C., Westerlund, B.E., eds.), Int. Astron. Union Coll. No. 37 = Centre Nat. Rech. Sci. Coll. No. 263, CNRS, Paris (1977) p. 347.
K 15 Königl, A.: Astrophys. J. **225** (1978) 732.
K 16 Königl, A.: Astrophys. J. **243** (1981) 700.
K 17 Kompaneets, A.S.: Sov. Phys. JETP **4** (1957) 730.
K 18 Kreysa, E., Pauliny-Toth, I.I.K., Schultz, G.V., Sherwood, W.A., Witzel, A.: Astrophys. J. **240** (1980) L17.
K 19 Krymski, G.F.: Dokl. Akad. Nauk. SSR **234** (1977) 1306-8, Engl. transl. in: Sov. Phys. Dokl. **23** 327-8.
K 20 Ku, W.H.M. in: [l] p. 677.
K 21 Ku, W.H.M., Helfand, D.J., Lucy, L.B.: Nature **288** (1980) 323.
K 22 Kühr, H.: Thesis, Univ. Bonn (1980).
K 23 Kühr, H., Strittmatter, P.A., Witzel, A.: Astron. Astrophys. (1982) in press
K 24 Kundt, W.: Astrophys. Space Sci. **62** (1979) 335.
K 25 Kundt, W. in: [o] p. 265.
K 26 Kundt, W., Gopal-Krishna: Astrophys. Space Sci. **75** (1981) 257.

L 1 Laan, H., van der, in: Eighth Texas Symp. on Relativistic Astrophysics (Papagiannis, M.D., ed.), Ann. New York Acad. Sci. **302** (1977) p. 637.
L 2 Lacombe, C.: Astron. Astrophys. **54** (1977) 1.
L 3 Landau, L.D., Lifshitz, E.M.: Fluid Mechanics, Pergamon Press, Oxford (1959).

L 4 Ledden, J.E., Aller, H.D., Dent, W.A.: Nature **260** (1976) 752.
L 5 Lerche, I., Wiita, P.J.: Astrophys. Space Sci. **68** (1980) 207.
L 6 Lerche, I., Wiita, P.J.: Astrophys. Space Sci. **68** (1980) 475.
L 7 Liang, E.P.T., Thompson, K.A.: Mon. Not. R. Astron. Soc. **189** (1979) 421.
L 8 Lightman, A.P., Band, D.L.: Astrophys. J. **251** (1981) 713.
L 9 Lightman, A.P., Eardley, D.M.: Astrophys. J. **187** (1974) L1.
L 10 Lightman, A.P., Giacconi, R., Tananbaum, H.: Astrophys. J. **224** (1978) 375.
L 11 Lightman, A.P., Lamb, D.Q., Rybicki, G.B.: Astrophys. J. **248** (1981) 738.
L 12 Lightman, A.P., Rybicki, G.B.: Astrophys. J. **232** (1979) 882.
L 13 Lightman, A.P., Rybicki, G.B.: Astrophys. J. **236** (1980) 928.
L 14 Lightman, A.P., Shapiro, S.L.: Astrophys. J. **211** (1977) 244.
L 15 Lin, D.C.N., Saslaw, W.C.: Astrophys. J. **217** (1977) 958.
L 16 Longair, M.S. in: Objects of High Redshift (Abell, G.O., Peebles, P.J.E., eds.), Int. Astron. Union Symp. **92**, Reidel, Dordrecht (1980) p. 135.
L 17 Longair, M.S. in: Observational Cosmology (Maeder, A., Martinet, L., Tammann, G., eds.), Geneva Observatory (1978) p. 125.
L 18 Longair, M.S. in: [m] p. 133.
L 19 Longair, M.S.: High Energy Astrophysics, Cambridge University Press, Cambridge (1981).
L 20 Longair, M.S., Riley, J.M.: Mon. Not. R. Astron. Soc. **188** (1979) 625.
L 21 Longair, M.S., Ryle, M., Scheuer, P.A.G.: Mon. Not. R. Astron. Soc. **164** (1973) 243.
L 22 Loose, H.H., Fricke, K.J.: Astrophys. Lett. **21** (1980) 65.
L 23 Loose, H.H., Fricke, K.J. in: The Most Massive Stars, Proc. ESO Workshop (D'Odorico, S., Baade, D., Kjär, H., eds.), München (1982) p. 269.
L 24 Lovelace, R.V.E.: Nature **262** (1976) 649.
L 25 Lynden-Bell, D.: Nature **223** (1969) 690.
L 26 Lynden-Bell, D.: Nature **270** (1977) 396.
L 27 Lynden-Bell, D. in: [f] p. 185.
L 28 Lynden-Bell, D., Liller, W.: Mon. Not. R. Astron. Soc. **185** (1978) 539.
L 29 Lynden-Bell, D., Pineault, S.: Mon. Not. R. Astron. Soc. **185** (1978) 679.
L 30 Lynden-Bell, D., Pineault, S.: Mon. Not. R. Astron. Soc. **185** (1978) 695.

M 1 Mackay, C.D.: Mon. Not. R. Astron. Soc. **151** (1971) 421.
M 2 Mackay, C.D.: Mon. Not. R. Astron. Soc. **162** (1973) 1.
M 3 MacLeod, J.M., Andrew, B.H., Harvey, G.H.: Nature **260** (1976) 751.
M 4 Maraschi, L., Perola, G.C., Reina, C., Treves, A.: Astrophys. J. **230** (1979) 243.
M 5 Maraschi, L., Tanzi, E.G., Tarenghi, M., Treves, A.: Nature **285** (1980) 555.
M 6 Maraschi, L., Perola, G.C., Treves, A.: Astrophys. J. **241** (1980) 910.
M 7 Maraschi, L., Roasio, R., Treves, A.: Astrophys. J. **253** (1982) 312.
M 8 Margon, B., Grandi, S.A., Downes, D.A.: Astrophys. J. **241** (1980) 306.
M 9 Marscher, A.P.: Astrophys. J. **216** (1977) 244.
M 10 Marscher, A.P.: Astron. J. **82** (1977) 781.
M 11 Marscher, A.P.: Astrophys. J. **219** (1978) 392.
M 12 Marscher, A.P. in: [g] p. 365.
M 13 Marscher, A.P.: Astrophys. J. **239** (1980) 296.
M 14 Marscher, A.P., Scott, J.S.: Publ. Astron. Soc. Pacific **92** (1980) 127.
M 15 Martin, P.G., Rees, M.J.: Mon. Not. R. Astron. Soc. **189** (1979) 19P.
M 16 McCray, R. in: [h] p. 227.
M 17 McKee, C.F., Hollenbach, D.J.: Annu. Rev. Astron. Astrophys. **18** (1980) 219.
M 18 Meier, D.L. in: [o] p. 263.
M 19 Mészáros, P.: Astron. Astrophys. **44** (1975) 59.
M 20 Miley, G.K. in: [d] p. 1.
M 21 Miley, G.K.: Annu. Rev. Astron. Astrophys. **18** (1980) 165.
M 22 Miley, G.K., Perola, G.C.: Astron. Astrophys. **45** (1975) 223.
M 23 Miley, G.K., Perola, G.C., van der Kruit, P.C., van der Laan, H.: Nature **237** (1972) 269.
M 24 Milgrom, M., Bahcall, J.N.: Nature **274** (1978) 349.
M 25 Mills, D.M., Sturrock, P.A.: Astrophys. Lett. **5** (1970) 105.
M 26 Misner, C.W., Thorne, K.S., Wheeler, J.A.: Gravitation, Freeman, San Francisco (1973).
M 27 Möllenhoff, C.: Astron. Astrophys. **50** (1976) 105.

M 28 Morrison, P., Cavaliere, A. in: [b] p. 485.
M 29 Morrison, P.: Astrophys. J. **157** (1969) L73.
M 30 Mushotzky, R.F., Serlemitsos, P.J., Smith, B.W., Boldt, E.A., Holt, S.S.: Astrophys. J. **225** (1978) 21.

N 1 Neff, S.G. in: [o] p. 137.
N 2 Nepveu, M.: Astron. Astrophys. **75** (1979) 149.
N 3 Nepveu, M.: Astron. Astrophys. **79** (1979) 40.
N 4 Nepveu, M.: Astron. Astrophys. **84** (1980) 142.
N 5 Norman, C.A., Silk, J.: Astrophys. J. **233** (1979) L1.
N 6 Norman, C.A., Silk, J. in: CECAM Report of Workshop on the Formation, Structure and Evolution of Galaxies, Paris (1981) p. 50.
N 7 Norman, M.L., Smarr, L., Wilson, J.R., Smith, M.D.: Astrophys. J. **247** (1981) 52.
N 8 Norman, M.L., Smarr, L., Winkler, K.H., Smith, M.D.: Astron. Astrophys. (1982) in press.
N 9 Novikov, I., Thorne, K.S. in: Black Holes (DeWitt, C., DeWitt, B.S., eds.), Gordon and Breach, New York (1973) p. 343.

O 1 O'Dell, S.L. in: [g] p. 312.
O 2 O'Dell, S.L. in: [h] p. 95.
O 3 Okoye, S.E., Obinabo, D. in: [o] p. 71.
O 4 Ostriker, J.P., Hausman, A.: Astrophys. J. **217** (1977) L125.
O 5 Owen, F.N., Burns, J.J., Rudnick, L.: Astrophys. J. **226** (1978) L119.
O 6 Owen, F.N., Helfand, D.J., Spangler, S.R.: Astrophys. J. **250** (1981) L55.
O 7 Owen, F.N., Mufson, S.L.: Astron. J. **82** (1977) 776.
O 8 Owen, F.N., Porcas, R.W., Mufson, S.L., Moffat, T.J.: Astron. J. **83** (1978) 685.
O 9 Owen, F.N., Spangler, S.R., Cotton, W.D.: Astron. J. **85** (1980) 357.
O 10 Ozernoy, L.M. in: [c] p. 65.
O 11 Ozernoy, L.M., Reinhardt, M.: Astrophys. Space Sci. **59** (1978) 17.
O 12 Ozernoy, L.M., Sazonov, V.N.: Astrophys. Space Sci. **3** (1969) 395.
O 13 Ozernoy, L.M., Shishov, V.I. in: [o] p. 325.
O 14 Ozernoy, L.M., Usov, V.V.: Astrophys. Space Sci. **13** (1971) 3.
O 15 Ozernoy, L.M., Usov, V.V.: Astrophys. Space Sci. **25** (1973) 149.
O 16 Ozernoy, L.M., Usov, V.V.: Astron. Astrophys. **56** (1977) 163.

P 1 Pacholczyk, A.G., Scott, J.S.: Astrophys. J. **203** (1976) 313.
P 2 Pacini, F. in: [k] p. 1.
P 2a Pacini, F., Salvati, M.: Astrophys. J. **225** (1978) L99.
P 3 Pacini, F., Salvati, M. in: [o] p. 247.
P 4 Paczyński, B., Wiita, P.J.: Astron. Astrophys. **88** (1980) 23.
P 5 Parker, E.N. in: [d] p. 137.
P 6 Pauliny-Toth, I.I.K. in: Origin of Cosmic Rays (Setti, G., Spada, G., Wolfendale, W.A., eds.), Int. Astron. Union Symp. **94**, Reidel, Dordrecht (1981) p. 176.
P 7 Pauliny-Toth, I.I.K., Preuss, E., Witzel, A., Graham, D., Kellermann, K.I., Rönnäng, B.O.: Astron. J. **86** (1981) 371.
P 8 Payne, D.G.: Astrophys. J. **237** (1980) 951.
P 9 Pearson, T.J., Readhead, A.C.S.: Astrophys. J. **248** (1981) 61.
P 10 Perley, R.A. in: [o] p. 175.
P 11 Perley, R.A., Johnston, K.J.: Astron. J. **84** (1979) 1247.
P 12 Perley, R.A., Willis, A.G., Scott, J.S.: Nature **281** (1978) 437.
P 13 Perola, G.C.: Fundam. Cosmic. Phys. **7** (1981) 59.
P 14 Perola, G.C., Reinhardt, M.: Astron. Astrophys. **17** (1972) 432.
P 15 Petschek, A.G., Colgate, S.A., Colvin, J.D.: Astrophys. J. **209** (1976) 356.
P 16 Phillips, R.B., Mutel, R.L.: Astrophys. J. **257** (1982) 19.
P 16a Phinney, E.S. in: Proc. Varenna Workshop on Plasma Astrophysics (Guyenne, T.D., Lévy, G., eds.), ESA SP-161 (1981) p. 337.
P 17 Piran, T.: Mon. Not. R. Astron. Soc. **180** (1977) 45.
P 18 Porcas, R.W.: Nature **294** (1981) 47.
P 19 Potash, R.I., Wardle, J.F.C.: Astron. J. **84** (1979) 707.
P 20 Pozduyakov, L.A., Sobol, I.M., Syunyaev, R.A.: Sov. Astr.-AJ **21** (1977) 708.

P 21 Preuss, E., Pauliny-Toth, I.I.K., Witzel, A., Kellermann, K.I., Shaffer, D.B.: Astron. Astrophys. **54** (1977) 297.
P 22 Preuss, E. in: [m] p. 97.
P 23 Pringle, J.E.: Annu. Rev. Astron. Astrophys. **19** (1981) 137.
P 24 Pringle, J.E., Rees, M.J.: Astron. Astrophys. **21** (1973) 1.
P 25 Pringle, J.E., Rees, M.J., Pacholczyk, A.C.: Astron. Astrophys. **29** (1973) 179.
R 1 Rayburn, D.R.: Mon. Not. R. Astron. Soc. **179** (1977) 603.
R 2 Readhead, A.C.S. in: Objects of High Redshift (Abell, G.O., Peebles, P.J.E., eds.), Int. Astron. Union Symp. **92**, Reidel, Dordrecht (1980) p. 165.
R 3 Readhead, A.C.S.: Sci. American **246**, No. 6 (1982) 38.
R 4 Readhead, A.C.S., Cohen, M.H., Blandford, R.D.: Nature **272** (1978) 131.
R 5 Readhead, A.C.S., Hewish, A.: Mon. Not. R. Astron. Soc. **176** (1976) 571.
R 6 Readhead, A.C.S., Pearson, T.J., Cohen, M.H., Ewing, M.S., Moffat, A.T.: Astrophys. J. **231** (1979) 299.
R 7 Readhead, A.C.S., Pearson, T. in: [o] p. 279.
R 8 Readhead, A.C.S. et al.: Astrophys. J. **231** (1979) 299.
R 9 Rees, M.J.: Nature **211** (1966) 468.
R 10 Rees, M.J.: Mon. Not. R. Astron. Soc. **135** (1967) 345.
R 11 Rees, M.J.: Nature **229** (1971) 312.
R 12 Rees, M.J. in: Proc. Seventh Texas Symp. Relativistic Astrophys., Ann. New York Acad. Sci. **262** (1975) p. 449.
R 13 Rees, M.J. in: Structure and Evolution of Galaxies (Setti, G., ed.), Reidel, Dordrecht (1975) p. 285.
R 14 Rees, M.J.: Comments Astrophys. **6** (1976) 113.
R 15 Rees, M.J. in: [d] p. 107.
R 16 Rees, M.J. in: Proc. Eights Texas Symposium on Relativistic Astrophysics, Ann. New York Acad. Sci. **302** (1977) p. 613.
R 17 Rees, M.J. in: [f] p. 193.
R 18 Rees, M.J.: Mon. Not. R. Astron. Soc. **184** (1978) 61P.
R 19 Rees, M.J.: Nature **275** (1978) 516.
R 20 Rees, M.J. in: [i] p. 339.
R 21 Rees, M.J.: Space Sci. Rev. **30** (1981) 87.
R 22 Rees, M.J. in: Origin of Cosmic Rays (Setti, G., Spada, G., Wolfendale, W.A., eds.), Int. Astron. Union Symp. **94**, Reidel, Dordrecht (1981) p. 139.
R 23 Rees, M.J. in: [o] p. 211.
R 24 Rees, M.J., Begelman, M.C., Blandford, R.D., Phinney, E.S.: Nature **295** (1982) 17.
R 25 Rees, M.J., Saslaw, W.C.: Mon. Not. R. Astron. Soc. **171** (1975) 53.
R 26 Reich, W., Stutz, U., Reif, K., Kalberla, P.M.W., Kronberg, P.P.: Astrophys. J. **236** (1980) L61.
R 27 Reynolds, S.P.: Astrophys. J. **256** (1982) 13.
R 28 Reynolds, S.P.: Astrophys. J. **256** (1982) 38.
R 29 Riegler, G.R., Agrawal, P.C., Mushotzky, R.F.: Astrophys. J. **223** (1979) L47.
R 29a Rieke, G.H., Lebofsky, M.J.: Annu. Rev. Astron. Astrophys. **17** (1979) 477.
R 30 Roos, N.: Astron. Astrophys. **104** (1981) 218.
R 31 Roos, N.: Thesis, Univ. Leiden (1981).
R 32 Rybicki, G.B., Lightman, A.P., Karzas, W.J., Latter, R.: Astrophys. J. Suppl. **6** (1961) 167.
R 33 Rybicki, G.B., Lightman, A.P.: Radiative Processes in Astrophysics, Interscience, New York (1979).
S 1 Saikia, D.J., Wiita, P.J.: Mon. Not. R. Astron. Soc. **200** (1982) 83.
S 2 Salpeter, E.E.: Astrophys. J. **140** (1964) 796.
S 3 Sanders, R.H.: Astrophys. J. **162** (1970) 791.
S 4 Sanders, R.H.: Nature **248** (1974) 390.
S 5 Sanders, R.H.: Astrophys. J. **205** (1976) 335.
S 6 Sanders, R.H., DaCosta, L.N.: Astron. Astrophys. **70** (1978) 477.
S 7 Sargent, W.L.W., Kristian, J., Westphal, J.A.: Astrophys. J. **234** (1979) 76.
S 8 Saslaw, W.C.: Publ. Astron. Soc. Pacific **85** (1973) 5.
S 9 Saslaw, W.C. in: The Formation and Dynamics of Galaxies (Shakeshaft, J.R., ed.), Int. Astron. Union Symp. **58**, Reidel, Dordrecht (1974) p. 305.
S 10 Saslaw, W.C.: Astrophys. J. **195** (1975) 773.
S 11 Saslaw, W.C., Crane, P., Tyson, J.A.: Astrophys. J. **246** (1981) 647.
S 12 Saslaw, W.C., De Young, D.S.: Astrophys. Lett. **11** (1972) 87.

S 13 Saslaw, W.C., Tyson, J.A., Crane, P.: Astrophys. J. **222** (1978) 435.
S 14 Saslaw, W.C., Valtonen, M.J., Aarseth, S.J.: Astrophys. J. **190** (1974) 253.
S 15 Scheuer, P.A.G. in: Plasma Astrophysics (Sturrock, A., ed.), Academic Press, New York (1967) p. 262.
S 16 Scheuer, P.A.G.: Mon. Not. R. Astron. Soc. **166** (1974) 513.
S 17 Scheuer, P.A.G.: Mon. Not. R. Astron. Soc. **177** (1976) 1P.
S 18 Scheuer, P.A.G. in: [d] p. 93.
S 19 Scheuer, P.A.G., Readhead, A.C.S.: Nature **277** (1979) 182.
S 20 Schmid-Burgk, J.: Astrophys. Space Sci. **56** (1979) 191.
S 21 Schmidt, M.: Astrophys. J. **151** (1968) 393.
S 22 Schmidt, M.: Astrophys. J. **209** (1976) L55.
S 23 Schmidt, M., Green, F. in: [o] p. 437.
S 24 Schreier, E.J. in: [m] p. 109.
S 25 Schreier, E.J., Feigelson, E.D., Delvaille, J.P., Giacconi, R., Grindlay, J., Schwartz, D.A., Fabian, A.D.: Astrophys. J. **234** (1979) L39.
S 26 Schwarz, M.P.: Astrophys. J. **247** (1981) 77.
S 27 Seidle, F.G.P., Cameron, A.G.W.: Astrophys. Space Sci. **15** (1972) 44.
S 28 Seielstad, G.A. et al.: Astrophys. J. **229** (1979) 53.
S 29 Setti, G. in: [g] p. 385.
S 30 Setti, G., Woltjer, L.: Astron. Astrophys. **76** (1979) L1.
S 31 Shaffer, D.B. in: [g] p. 68.
S 32 Shakura, N.I., Syunyaev, R.A.: Astron. Astrophys. **24** (1973) 337.
S 33 Shakura, N.I., Syunyaev, R.A.: Mon. Not. R. Astron. Soc. **175** (1976) 613.
S 34 Shapiro, S.L.: Astrophys. J. **180** (1973) 531.
S 35 Shapiro, S.L., Lightman, A.P., Eardley, D.M.: Astrophys. J. **204** (1976) 187.
S 36 Shapiro, S.L., Teukolsky, S.A.: Astrophys. J. **234** (1979) L177.
S 37 Shklovsky, I.S.: Sov. Astron.-AJ **9** (1965) 22.
S 38 Shklovsky, I.S. in: [o] p. 475.
S 39 Shields, G.A., Wheeler, J.C.: Astrophys. J. **17** (1976) L69.
S 40 Sikora, M.: Mon. Not. R. Astron. Soc. **196** (1981) 257.
S 41 Sikora, M., Wilson, D.B.: Mon. Not. R. Astron. Soc. **197** (1981) 529.
S 42 Simkin, S.M.: Astrophys. J. **222** (1978) L55.
S 43 Simkin, S.M.: Astrophys. J. **234** (1979) 56.
S 44 Simkin, S.M., Ekers, R.D.: Astron. J. **84** (1979) 56.
S 45 Smith, M.D., Norman, C.A.: Mon. Not. R. Astron. Soc. **194** (1981) 771.
S 46 Smith, M.D., Norman, C.A.: Mon. Not. R. Astron. Soc. **194** (1981) 785.
S 47 Smith, M.D., Smarr, L., Norman, M.L., Wilson, J.R.: Nature **293** (1981) 277.
S 48 Smith, M.G., Wright, A.E.: Mon. Not. R. Astron. Soc. **191** (1980) 871.
S 49 Sorrell, W.H.: Astrophys. Space Sci. **85** (1982) 3.
S 50 Spangler, S.R.: Astrophys. Lett. **20** (1980) 123.
S 51 Spitzer, L. in: [b] p. 443.
S 52 Spitzer, L., Saslaw, W.C.: Astrophys. J. **143** (1966) 400.
S 53 Spitzer, L., Stone, M.E.: Astrophys. J. **147** (1967) 519.
S 54 Sramek, R.A., Weedman, D.W.: Astrophys. J. **221** (1978) 468.
S 55 Steigman, G., Strittmatter, P.A.: Astron. Astrophys. **11** (1971) 279.
S 56 Stein, W.A. in: [g] p. 1.
S 57 Stein, W.A., O'Dell, S.L., Strittmatter, P.A.: Annu. Rev. Astron. Astrophys. **14** (1976) 173.
S 58 Strittmatter, P.A., Hill, P., Pauliny-Toth, I.I.K., Steppe, H., Witzel, A.: Astron. Astrophys. **88** (1980) L12.
S 59 Strom, R.G., Miley, G.K., Oort, J.H.: Sci. American **233** No. 2 (1975) 26.
S 60 Syunyaev, R.A.: Sov. Astr.-AJ **15** (1970) 190.
S 61 Syunyaev, R.A.: Sov. Phys. – JETP **36** (1973) 663.
S 62 Syunyaev, R.A., Titarchuk, L.G.: Astron. Astrophys. **86** (1980) 121.

T 1 Takahara, F.: Prog. Theor. Phys. **63** (1980) 1551.
T 2 Tananbaum, H. in: [i] p. 291.
T 3 Tananbaum, H., Avni, Y., Branduardi, G., Elvis, M., Fabbiano, G., Feigelson, E., Giacconi, R., Henry, J.P., Pye, J.P., Soltan, A., Zamorani, G.: Astrophys. J. **234** (1979) L9.
T 4 Tananbaum, H., Marshall, H.L. in: [o] p. 269.
T 5 Terrell, J.: Astrophys. J. **213** (1977) L93.

T 6 Thorne, K.S., Blandford, R.D. in: [o] p. 255.
T 7 Thorne, K.S., Flammang, R.A., Żytkow, A.N.: Mon. Not. R. Astron. Soc. **194** (1981) 475.
T 8 Toptygin, I.N.: Space Sci. Rev. **26** (1980) 157.
T 9 Tsuruta, S.: Astron. Astrophys. **61** (1977) 647.
T 10 Turland, B.D., Scheuer, P.A.G.: Mon. Not. R. Astron. Soc. **176** (1976) 421.
T 11 Tyson, J.A., Crane, P., Saslaw, W.C.: Astron. Astrophys. **59** (1977) L15.

U 1 Unno, W.: Publ. Astron. Soc. Japan **23** (1971) 123.

V 1 Vallée, J.P., Kronberg, P.P.: Astron. Astrophys. **43** (1975) 233.
V 2 Valtonen, M.J.: Astron. Astrophys. **46** (1976) 429.
V 3 Valtonen, M.J.: Astron. Astrophys. **46** (1976) 435.
V 4 Valtonen, M.J.: Astrophys. J. **209** (1976) 35.
V 5 Valtonen, M.J.: Astrophys. J. **213** (1977) 356, 648.
V 6 Valtonen, M.J.: Astrophys. J. **225** (1978) 738.
V 7 Valtonen, M.J.: Astrophys. J. **227** (1979) L79.
V 8 Van Groningen, E., Miley, G.K., Norman, C.A.: Astron. Astrophys. **90** (1980) L7.
V 9 Vitello, P., Pacini, F.: Astrophys. J. **220** (1978) 756.

W 1 Waggett, P.C., Warner, P.J., Baldwin, J.E.: Mon. Not. R. Astron. Soc. **181** (1977) 465.
W 1a Wagoner, R.V.: Annu. Rev. Astron. Astrophys. **7** (1969) 553.
W 2 Wall, J.V.: Observatory **95** (1975) 196.
W 2a Wall, J.V. in: Radio Astronomy and Cosmology (Jauncey, D.L., ed.), Int. Astron. Union Symp. **74**, Reidel, Dordrecht (1977) p. 55.
W 3 Wall, J.V., Benn, C. in: [o] p. 441.
W 4 Wall, J.V., Pearson, T.J., Longair, M.S.: Mon. Not. R. Astron. Soc. **193** (1980) 683.
W 5 Wall, J.V., Pearson, T.J., Longair, M.S.: Mon. Not. R. Astron. Soc. **196** (1981) 597.
W 6 Wall, J.V., Scheuer, P.A.G., Pauliny-Toth, I.I.K., Witzel, A.: Mon. Not. R. Astron. Soc. **198** (1982) 221.
W 7 Wardle, J.F.C.: Nature **269** (1977) 563.
W 8 Wardle, J.F.C. in: [g] p. 39.
W 10 Wiita, P.J.: Astrophys. J. **221** (1978) 41.
W 11 Wiita, P.J.: Astrophys. J. **221** (1978) 436.
W 12 Wiita, P.J., Siah, M.J.: Astrophys. J. **243** (1981) 710.
W 13 Wilkinson, P.N. in: [o] p. 149.
W 14 Willis, A.G. in: [f] p. 243.
W 15 Willis, A.G. in: [m] p. 71.
W 16 Wilson, A.S., Vallée, J.P.: Astron. Astrophys. **58** (1977) 79.
W 17 Wirth, A., Smarr, L., Gallagher, T.S.: IAP 82-6 (1982) preprint.
W 18 Wittels, J.J. et al.: Astron. J. **81** (1976) 933.
W 19 Witzel, A. in: Proc. Fifth Göttingen-Jerusalem-Symp. on Astrophysics, Vandenhoek & Ruprecht, Göttingen (1981) p. 99.
W 20 Wolfe, A.M. in: Proc. Ninth Texas Symp; Relativistic Astrophys., Ann. New York Acad. Sci. **336** (1980) p. 12.
W 21 Woltjer, L. in: [b] p. 477.
W 22 Woltjer, L. in: [d] p. 131.

Y 1 Yahel, R.Z. in: Proc. Fifth Göttingen-Jerusalem Symp. on Astrophysics, Vandenhoek & Ruprecht, Göttingen (1981) p. 141.
Y 2 Young, P.L.: Astrophys. J. **212** (1977) 227.
Y 3 Young, P.L.: Astrophys. J. **215** (1977) 36.
Y 4 Young, P.L., Shields, G.A., Wheeler, J.C.: Astrophys. J. **212** (1977) 367.
Y 5 Young, P.L., Westphal, J.A., Kristian, J., Wilson, C.P., Landauer, F.P.: Astrophys. J. **221** (1978) 721.
Y 6 Young, P.L., Sargent, W.L.W., Kristian, J., Westphal, J.A.: Astrophys. J. **234** (1979) 76.

Z 1 Zamorani, G., Henry, J.P., Maccacaro, T., Tananbaum, H., Soltan, A., Avni, T., Liebert, J., Stocke, J., Strittmatter, P.A., Weymann, R.J., Smith, M.G., Condon, J.J.: Astrophys. J. **245** (1981) 357.
Z 2 Zeldovich, Ya.B., Novikov, I.D.: Relativistic Astrophysics, Chicago University Press, Chicago (1971).
Z 3 Zeldovich, Ya.B., Raizer, Yu.P.: Physics of Shock Waves and High Temperature Phenomena, Academic Press, New York (1966).
Z 4 Znajek, R.: Thesis, Cambridge University Press, Cambridge (1977).

# Comprehensive index

This cumulative subject index incorporates key words from all papers in the three subvolumes of Volume 2 of Group VI of the Landolt-Börnstein New Series. An effort has been made to choose key words or free descriptors which are quite similar to those used in the astronomical bibliography Astronomy and Astrophysics Abstracts (AAA). The formation, arrangement, and versatility of the terms are therefore already known to the majority of the astronomical community which is considered as being familiar with the AAA scheme. Nevertheless, there are some important differences between the two services arising from the natural differences between a bibliography and an encyclopaedic review work.

Whenever possible, the key words are formed in such a manner that there are two different and also supplementary items, e.g. the combination of terms

*minor planets* and *colors* .

These pairs of terms were preferably chosen in such a way that they can be inverted in order to increase the variety and the usefulness of this index. In the given example there are the two entries

*minor planets : colors* and *colors : minor planets* .

Exceptions to the rule of inversion of key word combinations are given in all cases where the second term is either a very specific (e.g. *ESP events*) or a general one (e.g. *structure*). The use of substantives is preferred. Due to the encyclopaedic character of the papers many one-term key words describing special features (e.g. *RATAN-600* or *URSIgrams*) have been added. Furthermore, there are some hints for a proper use in the printout. In some cases synonymous entries for (nearly) the same facts were used which is indicated by a *see also* phrase (e.g. *chemical composition see also element abundances* and vice versa). On the other hand there are also excluding remarks like *asteroids see minor planets.*

Abbreviations which are not contained in the leading pages of the three subvolumes but which are considered to be widely known among astronomers (e.g. VLA or PZT) were used as they stand without further explanations. In any case the user is requested to look for synonymous or more specialized entries, as further references to a specific topic might exist elsewhere in this index under another current astronomical term.

The layout arrangement closely follows that of the recent AAA volumes. The sorting routine sorts blanks before hyphens and letters. All programmes are developed by staff members of the AAA department at the Astronomisches Rechen-Institut, Heidelberg. The computations and the printing were carried out on an IBM 360/44 computer.